HANDBOOK ON ENERGY AND CLIMATE CHANGE

Handbook on Energy and Climate Change

Edited by

Roger Fouquet

Principal Research Fellow, Grantham Research Institute on Climate Change and the Environment, London School of Economics, UK

Edward Elgar
Cheltenham, UK • Northampton, MA, USA

Published by
Edward Elgar Publishing Limited
The Lypiatts
15 Lansdown Road
Cheltenham
Glos GL50 2JA
UK

Edward Elgar Publishing, Inc.
William Pratt House
9 Dewey Court
Northampton
Massachusetts 01060
USA

A catalogue record for this book
is available from the British Library

Library of Congress Control Number: 2012943745

This book is available electronically in the ElgarOnline.com
Economics Subject Collection, E-ISBN 978 0 85793 369 0

ISBN 978 0 85793 368 3 (cased)

Typeset by Servis Filmsetting Ltd, Stockport, Cheshire
Printed by MPG PRINTGROUP, UK

Contents

PART IV   CLIMATE AGREEMENTS

PART V   CARBON MITIGATION POLICIES

PART VI   LOW-CARBON BEHAVIOUR AND GOVERNANCE

Contributors

Joseph E. Aldy, Harvard Kennedy School, Harvard University, Cambridge, Massachusetts, USA, Resources for the Future and National Bureau of Economic Research

Edward B. Barbier, Department of Economics and Finance, University of Wyoming, Laramie, Wyoming, USA

Alex Bowen, Grantham Research Institute on Climate Change and the Environment, London School of Economics, UK

Julien Chevallier, Université Paris 8 (LED), Saint-Denis, France

Joanne Evans, Surrey Energy Economics Centre (SEEC), School of Economics, University of Surrey, UK

Nick Eyre, Environmental Change Institute (ECI) and Oriel College, University of Oxford, UK

Massimo Filippini, Centre for Energy Policy and Economics (CEPE), ETH Zurich and Department of Economics, University of Lugano, Switzerland

Roger Fouquet, Grantham Research Institute on Climate Change and the Environment, London School of Economics, UK

Steven A. Gabriel, University of Maryland, College Park, Maryland, USA

Alberto Gago, Research Group in Energy, Innovation and Environment (Rede), University of Vigo and Economics for Energy, Spain

Caterina Gennaioli, Grantham Research Institute on Climate Change and the Environment, London School of Economics, UK

John M. Gowdy, Department of Economics, Rensselaer Polytechnic Institute, Troy, New York, USA

Clemens Haftendorn, German Institute for Economic Research (DIW Berlin), Germany

James D. Hamilton, Department of Economics, University of California, San Diego, USA

Michael Hanemann, Arizona State University and University of California, Berkeley, USA

Ivan Haščič, OECD Environment Directorate

David F. Hendry, Institute for New Economic Thinking at the Oxford Martin School and Economics Department, University of Oxford, UK

Cameron Hepburn, Grantham Research Institute on Climate Change and the Environment, London School of Economics and New College, Oxford, UK

Bjart Holtsmark, Statistics Norway, Oslo, Norway

Franziska Holz, German Institute for Economic Research (DIW Berlin), Germany

Chris Hope, Judge Business School, University of Cambridge, UK

Lester C. Hunt, Surrey Energy Economics Centre (SEEC) and Research Group on Lifestyles Values and Environment (RESOLVE), School of Economics, University of Surrey, UK

Henry D. Jacoby, MIT Sloan School of Management, Massachusetts Institute of Technology, Cambridge, MA, USA

Michael Jefferson, Centre for International Business and Sustainability, London Metropolitan Business School and Department of Economics and International Studies, University of Buckingham, UK

Nick Johnstone, OECD Environment Directorate

John G. Kassakian, Department of Electrical Engineering and Computer Science, Massachusetts Institute of Technology, Cambridge, MA, USA

Claudia Kemfert, German Institute for Economic Research (DIW Berlin), Germany

Snorre Kverndokk, Ragnar Frisch Centre for Economic Research, Oslo, Norway

Xavier Labandeira, Research Group in Energy, Innovation and Environment (Rede), University of Vigo and Economics for Energy, Spain

Henry Lee, Harvard Kennedy School of Government, Harvard University, Cambridge, MA, USA

Humberto Llavador, Universitat Pompeu Fabra, Barcelona, Spain

Grant Lovellette, Belfer Center for Science and International Affairs, Harvard Kennedy School, Harvard University, Cambridge, MA, USA

Ralf Martin, Imperial College Business School, Imperial College, London and Centre for Economic Performance, London School of Economics, UK

Ross McKitrick, Department of Economics, University of Guelph, Ontario, Canada

Arild Moe, Fridtjof Nansen Institute, Lysaker, Norway

Mirabelle Muûls, Grantham Research Institute on Climate Change and the Environment, Imperial College Business School, Imperial College London and Centre for Economic Performance, London School of Economics, UK

Tanya O'Garra, Faculty of Business and Economics, Universidad del País Vasco, Bizkaia, Spain and Grantham Research Institute for Climate Change and the Environment, London School of Economics, UK

Ian Parry, Fiscal Affairs Department, International Monetary Fund

Christian de Perthuis, Climate Economics Chair, Université Paris-Dauphine, Paris, France

Michael G. Pollitt, Electricity Policy Research Group, University of Cambridge, UK

Felix Pretis, Institute for New Economic Thinking at the Oxford Martin School and Economics Department, University of Oxford, UK

Ana Ramos, Research Group in Energy, Innovation and Environment (Rede), University of Vigo and Economics for Energy, Spain

Colin Robinson, University of Surrey, UK

John E. Roemer, Yale University, USA

Knut Einar Rosendahl, UMB School of Economics and Business, Ås and Statistics Norway, Oslo, Norway

Richard Schmalensee, MIT Sloan School of Management, Massachusetts Institute of Technology, Cambridge, MA, USA

Irina Shaorshadze, Electricity Policy Research Group, University of Cambridge, UK

Joaquim Silvestre, University of California, Davis, USA

Paul Stevens, Chatham House, University of Dundee, UK and UCL Australia

Richard S.J. Tol, Department of Economics, University of Sussex, UK and Institute of Environmental Studies and Department of Spatial Economics, Vrije Universiteit, Amsterdam, the Netherlands

Raphaël Trotignon, Climate Economics Chair, Université Paris-Dauphine, Paris, France

Marina Tsygankova, Thomson Reuters Point Carbon, Oslo and Statistics Norway, Oslo, Norway

G. Cornelis van Kooten, Department of Economics, University of Victoria, British Columbia, Canada

Christian von Hirschhausen, German Institute for Economic Research (DIW Berlin), Germany

Introduction[1]
Roger Fouquet

1. THE ROLE OF ECONOMICS IN ADDRESSING CLIMATE CHANGE

Political efforts to address climate change have developed greatly in the last 20 years. In 1992, at the Rio Summit, the UN Framework Convention on Climate Change (UNFCCC) was established. Then, in 1997, despite its flaws, the Kyoto Protocol set targets to curb greenhouse gas emissions on a number of industrialized countries emissions between 2008 and 2012. These developments have been driven by advances in the natural and social sciences over the last 30 years, coupled with more recent media attention and NGO campaigns. They have also coincided with a rising public awareness of the existence of climate change and its physical and economic threats (Whitmarsh, 2011), creating a broader demand for climate stability.

While natural scientists identified the relationship between greenhouse gas concentrations and climate change, and highlighted many of the threats, social scientists and particularly economists have played a crucial role in developing strategies for mitigating climate change (Nordhaus, 1977, 1991; Cline, 1992; IPCC, 2007). Economists have been influential in arguing that the costs of mitigation may not be as great as many industrialists claimed (Porter, 1991; Fischer and Newell, 2008) and that there may be substantial benefits (Stern, 2006; Sterner and Persson, 2008). They have also proposed mechanisms for trading responsibilities and credits related to greenhouse gas emission reductions, which have been a central tool in agreements on targets related to the Kyoto Protocol and certain national climate policies (Dales, 1968; Atkinson and Tietenberg, 1991; Stavins, 1995). At a national level, many governments have introduced taxes to discourage the consumption of high-carbon energy sources (Pearce, 1991; Newbery, 1992; Oates, 1995; Parry and Small, 2005; Nordhaus, 2007; Sterner, 2007). In other words, economists have become highly influential in the global efforts to achieve climate stability.

Yet, to me, this apparent success hides a potential problem. One of the impressions I have formed from talks at conferences, working papers and journal articles over the last decade is that the shift has been associated with a perceived decline in the number of new ideas being presented – intellectual 'blockbusters' that 'challenge or influence the boundaries of knowledge and . . . change the way we think about problems' (Brouthers et al., 2012, p. 960). The hypotheses proposed here, which will be considered more fully in this chapter, are that (i) during the 1990s, there was a growth in research originality in the economics of energy and climate change, (ii) during the 2000s, there was a rapid growth in research production in this field, and (iii) in the last five to seven years, either the originality of research has declined or the originality relative to the quantity has declined.

This may be an inevitable process (Hargadon and Sutton, 1997). After a period of great ideas, which created several new 'research fronts' at the intersection of energy and environmental economics (Upham and Small, 2010), economists are in a phase of refining and applying them. This is a crucial aspect of developing research and converting economic ideas into useable tools for policy makers, and can be responsible for a large research output. However, eventually, declining marginal returns (from using and developing these ideas) tend to set in. In time, new 'research fronts' need to be developed for the discipline to generate new knowledge and be of long-term value (Hall et al., 2005).

Indeed, despite the successes of economic analysis in this field, it is clear that many energy and climate change problems remain that economists (and other social and natural scientists) are not managing to fully resolve. Thus it is proposed that economists investigating energy and climate change issues need to develop new 'research fronts'.

So, with this in mind, this handbook has been constructed around the objectives of displaying some of the best of current thinking in the economics of energy and climate change, and encouraging the formulation of new questions and the development of new ideas. Before outlining the chapters in the handbook, this introduction will briefly highlight some of the most influential ideas in the literature on the economics of energy and climate change – that is, the 'blockbusters' that have created new 'research fronts' and changed the way we think about these issues. This introduction will use bibliometric evidence to examine the trends in related research over the last 40 years, identify and analyse the explosion in energy and climate change research in the last ten years, and consider the validity of this hypothesized rise in original ideas in the literature and then decline (or relative decline) since the explosion in research output.

2.   THE ORIGINS OF THE ECONOMICS OF ENERGY AND CLIMATE CHANGE

Economists' current approach to energy and environmental problems has its roots in the seminal ideas produced between the 1960s and the 1990s that were, no doubt, driven by a broader consciousness about environmental and resource issues (Pearce, 2002). While this is not the place to offer a review of the economics of energy and climate change, it is worth commenting on a few salient 'research fronts' that are so important today (see, e.g., Kula, 1998; Stevens, 2000; Pearce, 2002 for broad reviews of the literature).

In the 1960s, energy economics was driven by an empirical approach to questions, exemplified by the Resources for the Future (RFF) studies of long-run energy consumption and prices (Schurr and Netschert, 1960; Potter and Christy, 1962; Barnett and Morse, 1963; Adelman, 1972). Long before them, Jevons (1865) had considered the economics of coal and Hotelling (1931) had brought some theoretical grounding to possible non-renewable resource price trends. Nevertheless, it was this growing awareness of the economic importance of energy during and after the Second World War, and particularly the oil shock in 1973 (Fisher and Ward, 2000), that led more economists to analyse the role of energy in the economy (Nordhaus, 1980; Hamilton, 1983; Bohi, 1989), the demand for energy (Christensen et al., 1973; Hudson and Jorgensen, 1974; Berndt and Wood, 1975; Griffin and Gregory, 1976; Pindyck, 1979; Dubin and McFadden, 1984; Bhatia, 1987; Griffin, 1993), resource production and its costs (Gordon, 1967; Pindyck, 1978; Dasgupta and Heal, 1979; Slade, 1982; Arrow and Chang, 1982; Krautkraemer, 1998), and energy markets (Penrose, 1957; Nordhaus, 1973; Gately, 1984; Adelman, 1986).

From the 1980s, a number of new issues came to the forefront: energy policy and particularly the liberalization–privatization debate (Joskow and Schmalensee, 1983; Joskow, 1987; Helm et al., 1988; Green and Newbery, 1992; Newbery and Pollitt, 1997; see also Pollitt, 2012), developing economy energy markets (Bhatia, 1987; Pearce and Webb, 1987; Pearson and Stevens, 1987; Asafu-Adjaye, 1999; Soytas and Sari, 2003); the role of technology and efficiency improvements in energy consumption (Khazzoom, 1980; Goldemberg et al., 1985; Jaffe and Stavins, 1994; Nordhaus, 1996; Unruh, 2002; Sorrell and Dimitropoulos, 2008; Fouquet and Pearson, 2012); the potential development of markets for renewable energy technologies and sources, incentivized by feed-in tariffs (Menanteau et al., 2003; Meyer, 2003), competitive tender (Mitchell, 2000), renewable portfolio standards (Wiser et al., 1998; Fischer and Newell, 2008) or a demand for low-polluting energy sources (Fouquet, 1998; Roe et al., 2001; Scarpa and Willis, 2010); and, of course, concerns about the environmental damage associated with energy consumption (Nordhaus, 1977; Shafik and Bandyopadhyay, 1992; Newbery, 1994).

This last issue required the fusion of two disparate literatures and approaches to economic analysis. No doubt because of the lack of data, environmental economics had begun from a more theoretical perspective. The starting points for much of the energy-related environmental economics literature were Pigou (1920), explaining the need to internalize external costs in order to improve market efficiency, and Coase (1960), arguing for the need to ensure well-defined property rights to allow exchanges between polluters and victims.

In the 1960s and 1970s, the analysis of and arguments for market-based instruments to regulate environmental pollution, such as taxes and tradable permits, were developed (Dales, 1968; Baumol and Oates, 1971; Montgomery, 1972; Sandmo, 1975). Weitzman (1974) raised the question about which instrument to use when faced with uncertainty. This was a crucial start to a large literature on the comparative advantages of different instruments – frequently supporting the theoretical virtues of taxation over other instruments (Goulder et al., 1999; Pizer, 1999, 2002; Aldy et al., 2010).

The details of optimal and second-best environmental taxation (which sought to discourage environmentally damaging behaviour rather than simply using taxes as a source of raising revenue (Pearce, 1991; Newbery, 1992; Oates, 1995)) began to be explored in the 1990s, such as the distortionary effects of environmental taxation (Bovenberg and de Mooij, 1995) and possible double dividends (Bovenberg and van der Ploeg, 1994; Parry, 1995; Bovenberg and Goulder, 1996; Allcott et al., 2011).

Meanwhile, the argument for introducing tradable permits was persuading politicians to introduce flexible mechanisms to reduce pollution (Hahn, 1984; Stavins, 1995; Rubin, 1996). In the 1990s, the introduction of the US sulphur dioxide tradable permit scheme was a crucial natural experiment in environmental markets (Stavins, 1998; Schmalensee et al., 1998; Ellerman et al., 2000). This experience has been crucial for the creation of the EU Emissions Trading Scheme, both in convincing politicians of its benefits and in its development (Ellerman et al., 2010; Chevallier, 2012). This understanding is also important for other forms of carbon credit trading, such as the Clean Development Mechanism.

The traditional view of environmental regulation (whether market-based instruments or other policies) had been that requiring firms to reduce pollution externalities would lower their options and profits. Porter (1991) and Porter and van de Linde

(1995) controversially proposed that environmental regulation could actually enhance competitiveness. While the debate continues 20 years later, Jaffe et al. (1995) provided important evidence that environmental regulation did not necessarily harm industrial competitiveness.

Simultaneously, the debate surrounding the environmental Kuznets curve (EKC), which proposed that environmental degradation first increased with economic development and then declined beyond a threshold, was important because it brought a great deal of additional empirical evidence to environmental economics and linked environmental damage with economic growth (Selden and Song, 1992; Stern et al., 1996; Stern, 2004).

Another important strand of empirical environmental economics was the valuation literature, which developed creative methods for getting around the lack of data to estimate and understand the demand for environmental quality (Hanemann et al., 1991; Cropper and Oates, 1992). The development of valuation techniques, as well as refinements in cost–benefit analysis (Barbier et al., 1990; Palmer et al., 1995), have been crucial for measuring the impacts of climate change (Mendelsohn et al., 1994; Manne et al., 1995) and estimating the social cost of carbon (Hope et al., 1993; Tol, 1994, 2005). When linked to the mitigation costs, these can offer long-term strategies related to climate policy (Nordhaus, 1991; Nordhaus and Yang, 1996; Stern, 2006; Weitzman, 2007; Nordhaus, 2008). These articles have also tackled the philosophical, moral and empirical question about discounting future generations (Schelling, 1995; Lind, 1995; Azar and Sterner, 1996; Weitzman, 1998; Groom et al., 2005). This debate has run in parallel with the more game-theoretical literature on environmental agreements (Maler, 1989; Barrett, 1990, 1994; Hoel, 1991; Carraro and Siniscalco, 1993; Nordhaus and Yang, 1996).

Following Tol and Weyant (2006), and other bibliometric studies, Table I.1 provides an insight into the most cited energy and energy-related environmental economics articles in mainstream economic journals (which were found in an extensive search of the ISI/Web of Knowledge (1 March 2012)). Although far from a complete list, it acts as a crude indicator of the most influential articles in this field. Other relevant articles with more than 200 citations in mainstream economic journals include Bovenberg and de Mooij (1995), Palmer et al. (1995), Stern et al. (1996), Carraro and Siniscalco (1993), and Nordhaus and Yang (1996). Naturally, this ignores many highly influential chapters in books, such as Watkins (1992) or Nordhaus (1996), or books, such as Barnett and Morse (1963), Dales (1968), Darmstadter et al. (1971), Baumol and Oates (1975), Pearce et al. (1989), Cline (1992), Nordhaus (1994, 2008), Stern (2006) and IPCC (2007), or books by economists that might influence public attitudes, such as Kahn (2010) and Wagner (2011).

As an example of further key articles in the economics of energy and climate change, Table I.2 presents the most cited relevant articles (again based on ISI/Web of Knowledge) in energy and energy-related environmental economics journals (*Energy Economics* (*EE*), *The Energy Journal* (*EnJ*), *Resource and Energy Economics* (*REE*), *Energy Policy* (*EnPol*), and relevant articles in the *Journal of Environmental Economics and Management* (*JEEM*), *Environmental and Resource Economics* (*ERE*), and the *Review of Environmental Economics and Policy* (*REEP*)). It should be noted that, although no articles from the journals *EE*, *EnJ* and *REE* appear in this short list, they were close behind, with articles featuring in the top 20 or 30 most cited, and *REEP* already has some articles with impressive numbers of citations given that it has only been in existence for the last five years.

Table I.1   Most cited energy and energy-relevant environmental economics articles in mainstream economic journals

Article	Journal	General topic	Citations
Weitzman (1974)	*RES*	Market-based instruments	554
Jaffe et al. (1995)	*JEL*	Environmental policy	419
Green and Newbery (1992)	*JPE*	Electricity liberalization	385
Hamilton (1983)	*JPE*	Oil prices effects	368
Nordhaus (1991)	*EcJ*	Climate policy	358
Montgomery (1972)	*JET*	Tradable permits	296
Joskow (1987)	*AER*	Coal markets	279
Mendelsohn et al. (1994)	*AER*	Carbon impacts	272
Stern (2004)	*WDev*	EKC	247
Baumol and Oates (1971)	*SwJE*	Environmental policy	229

Source: ISI/Web of Knowledge (1 March 2012).

Table I.2   Most cited articles in energy and energy-related environmental economics journals

Article	Journal	General topic	Citations
Selden and Song (1992)	*JEEM*	EKC	432
Unruh (2000)	*EnPol*	Carbon lock-in	202
Stavins (1995)	*JEEM*	Emissions trading	198
McDonald and Schrattenholzer (1999)	*EnPol*	Energy technology	192
Manne et al. (1995)	*EnPol*	CO_2 emissions	181
Weitzman (1998)	*JEEM*	Discounting	163
Greening et al. (2000)	*EnPol*	Rebound effect	159
Tol (2005)	*EnPol*	Carbon impacts	152
Dincer (2002)	*EnPol*	Energy policy	152
Tol (2002)	*ERE*	Carbon impacts	143

Source: ISI/Web of Knowledge (1 March 2012).

Naturally, apart from a few cases in Table I.2, it is too early to identify the seminal pieces in the twenty-first century. This is partly because of this crude bibliometric approach. However, this approach will become even more helpful for identifying seminal articles in the future, as the research output on the subject grows.

3.   RECENT TRENDS IN ENERGY AND CLIMATE CHANGE RESEARCH

This recent phase that seemed to me to be a period of refinement and application of original ideas, rather than appearing to produce a relatively large number of new ideas, was, if one looks at the statistics, a golden age for the economics of energy

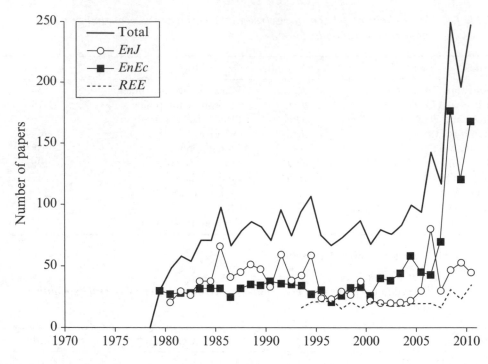

Figure I.1   Trend in energy economics articles (1979–2010)

and climate change. The first decade of the twenty-first century saw a dramatic increase in the production of research related to energy and climate change issues. The number of energy economics academic articles took off in 2005 (see Figure I.1). Before that year, a little under 100 articles were published in the main energy economics journals (*Energy Economics* (*EnEc*), *The Energy Journal* (*EnJ*) and *Resource and Energy Economics* (*REE*)). By the end of the decade, the average was closer to 250 – much of the increase due to the expansion of the journal *Energy Economics*. There are naturally many more energy economics articles, since some get published in mainstream economics journals (for now, it is unclear whether the number of energy economics articles in mainstream journals increased or decreased), and in broader environmental and resource journals (such as *Journal of Environmental Economics and Management, Environmental and Resource Economics, Ecological Economics* and more recently *Review of Environmental Economics and Policy*, and *Economics of Energy and Environmental Policy*).

Table I.3 shows the total number of articles in these journals, plus those published in *Energy Policy*, which was founded by energy economists and provides an important forum for the more policy-related analysis – as shown in Table I.2. These journals published three times more articles in the period 2006–10 than in 2001–05, and 13 times more than in 1976–80.

Figure I.2 shows the *Energy Economics* articles (*EnEcon*, the total from Figure I.1), those published in the journal *Energy Policy* (*EnPol*), and working papers that appeared

Table I.3 *Annual average number of articles in energy economics and policy journals (1973–2010)*

Years	Total	Years	Total	Years	Total	Years	Total
1973–75	25	1981–85	110	1991–95	185	2001–05	228
1976–80	54	1986–90	128	1996–2000	165	2006–10	722

Source: Journal websites.

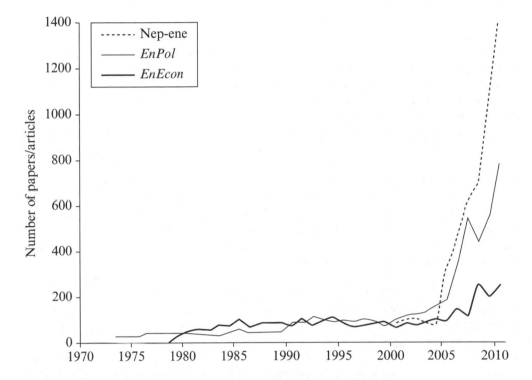

Figure I.2 *Trend in energy-related papers and articles (1973–2010)*

in the weekly nep-ene reports, the list of New Economic Papers dedicated broadly to energy economics. While there was a 3.5-fold increase in articles in specifically energy economics journals in the first decade of the twenty-first century, a ten-fold increase was measured in the broader literature on energy and climate change issues (see Figure I.2). In this period, the number of articles published in the journal *Energy Policy* leapt – rising eight-fold between 2000 and 2010.

Also shown in Figure I.2 are working papers listed in the energy economics report of the New Economic Papers, nep-ene (http://ideas.repec.org/n/nep-ene/), which are new additions to RePEc, the world's largest virtual repository of economic working papers. These working papers went from being one-third of the total articles and papers in 2000–04 to 50 per cent–58 per cent in 2006–10. So, in part, this simply reflects that more working papers are being included in this virtual repository. This is certainly true: more institutions are linked to the RePEc (Research Papers in Economics); thus economics working papers in all sub-disciplines have increased during this decade. However, before 2007, the number of working papers appearing in the energy economics report (nep-ene) was less than 3 per cent of the total working papers (nep-all). This increased to 3.2 per cent in 2007, 3.8 per cent in 2008, 4.7 per cent in 2009 and 6.1 per cent in 2010, without any change in the selection criteria for including working papers in nep-ene. Thus economists are increasingly attracted to producing research related to energy and climate change.

It is important to stress that whereas the first two categories (*EnEcon* and *EnPol*) are peer-reviewed, nep-ene is not – as the reports include mostly working papers by academic and research institutions. Also, a number of the articles in the journals appear as working papers in nep-ene first. Thus there is some double-counting. At the same time, it provides an indicator of the crude amount of research on the economic, social and policy aspects of energy and climate change. So, while many articles or papers are left out, and there is some double-counting, nevertheless, Figure I.2 reflects the spectacular increase that has occurred from 2005.

It is also interesting to identify whether certain issues are becoming more or less topical. Table I.4 shows the breakdown of working papers in nep-ene according the main subject of the paper between 2000 and 2010. For longer-run studies of trends in subjects for environmental, resource and ecological economics, it is worth consulting Fisher and Ward (2000) and Silva and Teixeira (2011). It seems that oil (averaging 18 per cent between 2006 and 2010) and natural gas (averaging 4 per cent) became more important in the second half of the decade, while electricity was less covered (falling from more than 24 per cent between 2002 and 2005, averaging 13 per cent in the second half of the decade). Renewable electricity and biofuels became more important, averaging 9 per cent – biofuels on its own became a hot topic in 2008, reaching 10 per cent of the total working papers. It is interesting to note that, given the potential roles renewable electricity and nuclear power might play in a low-carbon economy, the former (excluding the biofuel papers) has never reached 5 per cent and the latter never more than 1 per cent of the total papers in any year.

Conversely, the environmental Kuznets curve (EKC), and the relationship with growth, which was a hot topic at the beginning of the decade, has become less important. General environmental policies have also been covered less. This had been replaced by a greater focus on climate change policies and agreements. This subsection includes issues not directly related to energy, such as carbon sequestration, and climate change impacts and adaptation – these topics are included as they are seen as some of the external costs of energy use. Environmental and carbon taxes were popular in 2002–03, declining substantially since then. Emissions trading, which also peaked in those years, has maintained an important share of the total. Interestingly, this corresponds with the beginning of the EU Emissions Trading Scheme. There was then a new peak in 2008–10 reflecting analysis of this scheme.

Overall, though, the share of energy working papers (rather than predominantly about the environment or climate change) in nep-ene has increased from under 50 per cent in the first half of the decade to an average of 57 per cent in the second half of the decade. Thus the dramatic overall increase in articles and papers from 2005 signals a rising

Table 1.4 Average shares (%) of different topics in nep-ene reports (2000–10)

	Energy (modelling, forecasts, policies)	Coal	Oil	Gas	Electricity	Renewables & biofuels	Techn. & efficiency	Growth & EKC	Env. policies	Env. & carbon taxes	Emissions trading	Climate policies (incl. sequestration, adaptation)
2000	1.3	0.0	11.3	0.0	17.5	3.8	2.5	7.5	27.5	6.3	3.8	18.8
2001	3.1	0.0	18.6	2.1	10.3	2.1	4.1	7.2	29.9	4.1	3.1	14.4
2002	5.8	0.0	10.6	1.9	35.6	1.9	3.8	2.9	16.3	8.7	3.8	8.7
2003	2.3	0.0	8.1	2.3	23.3	2.3	8.1	2.3	16.3	12.8	10.5	11.6
2004	1.3	1.3	8.9	0.0	25.3	3.8	7.6	2.5	16.5	1.3	12.7	19.0
2005	5.2	0.6	18.5	1.5	24.0	1.2	8.0	6.5	15.4	3.4	5.2	9.8
2006	5.3	0.2	20.8	4.2	18.6	4.7	5.3	4.0	13.1	2.7	7.1	12.9
2007	5.7	1.6	19.9	5.8	13.8	7.1	5.2	3.1	6.2	3.2	6.5	21.7
2008	7.1	1.1	16.9	4.0	13.7	13.8	4.3	2.8	4.4	2.7	9.2	19.8
2009	5.2	0.9	18.4	2.1	9.7	12.4	7.0	2.6	4.3	2.3	8.8	25.7
2010	8.2	0.5	14.8	3.0	10.3	8.9	6.4	4.8	6.2	3.1	9.3	24.0

Source: nep-ene archives (http://ideas.repec.org/n/nep-ene/).

9

interest in oil issues (especially related to resource scarcity and price hikes) and renewables, as well as in climate change.

4.   THE CAUSES OF THE EXPLOSION IN ENERGY ECONOMICS RESEARCH

It is worth dwelling on the causes of the explosion in energy and climate change research. There are general academic trends. Total economic working papers, including those in nep-all, increased from 2700 in 2000 to 23 500 in 2010.

Obviously, as mentioned before, there has been a growth in the use of this system as a virtual repository of research output. While, in 2000, only a fraction of economic institutions were linked to RePEc (which provides the list of new working papers for nep-all), today most working papers written in English (and other European languages) from economic departments and research centres around the world are now included in nep-all, and accessible through New Economic Papers and RePEc.

Also, the knowledge economy has been benefiting from greater investment in research and development (R&D), including economic, social and political knowledge (Jaffe and Trajtenberg, 2005). On the supply side, thanks to computers and the Internet, researchers find it easier to produce a working paper today than in the twentieth century. There is also more pressure for academics to publish than 20 years ago (De Rond and Miller, 2005), although there may be signs that the use of bibliometric indices is reversing the trend towards researchers publishing fewer, more cited articles (Weightman, 2011).

However, the specific growth (indicated by the increasing share of energy and climate change issues in economics working papers) reflects a demand for a better understanding of energy markets and policies, and the many dimensions of climate change. Research funding agencies and private organizations are using their constrained budgets to answer related questions. Academics independent of funding may also be drawn to these topics. So energy economists are in demand for many of the old issues and equally for the environmental angle.

Probably the growth reflects in part the boom years up to 2008. One could say that the income elasticity of demand for energy and climate change research was positive. It seems that it was also greater than one. Certainly the 'research intensity' (i.e. the research output relative to GDP) has risen in the twenty-first century (see Table I.5).

Yet this research intensity indicator associated with energy economics has not always risen and, at times, has fallen. Using a very crude indicator, in 1975, four articles were produced on energy issues per trillion (2010) dollars of global GDP (see Table I.5). This peaked at more than 12 articles per trillion dollars in 1985, following the oil shocks. It then fell back in the 1990s, before reaching 13 articles (and papers) in 2005. This increased rapidly in the second half of the 2000s to almost 40 articles per trillion dollars of GDP.

The variation in this crude research intensity indicator shows that research output (and, no doubt, funding) reflects critical periods in the history of energy rather than more general economic output. The growth from 1975 to 1985 was clearly driven by funding agencies' and researchers' reaction to the oil shocks. This interest slumped with the oil

Table I.5   Energy economics' research intensity (1975–2010)

Year	Research intensity	Year	Research intensity	Year	Research intensity
1975	4.7	1990	7.1	2005	13.2
1980	8.3	1995	5.8	2010	38.9
1985	12.2	2000	7.6		

Sources: Figure I.2, and World Bank (2011).

price in the early 1990s. It took off again with the climate change debate and then more recently the new hike in oil and other energy prices.

At present, crucial for many reading this handbook is whether, when the economy starts growing again, this research output and intensity will begin to rise once more. Indeed, it is very possible that the research output related to energy and climate change issues will plateau at 2010 levels or even start to drop. The decline in the 1990s hints that 2005–10 may have been a golden period for energy economists.

If, on the other hand, energy prices rise further and climate-change-related damage intensifies, the business of energy economics is likely to be rosy. If so, research output will continue to grow in absolute terms and even relative to GDP, and this should generate a great deal of new understanding of energy and climate change issues.

5.   A COMMENT ON RESEARCH ORIGINALITY

The discussion has focused on the quantity of research output measured by the number of articles published, and working papers available on a specific website. This was crude, but the trend seems to be clear: around 2005, research output focusing on energy and energy-related environmental articles and working papers exploded.

However, this tells us nothing about the quality[2] of the research output. At the beginning of the chapter, I proposed that there was a period of growth in the originality of the research related to the economics of energy and climate change during the 1990s, and then, following an explosion in research output, perhaps a decline in the number of original ideas being presented and new 'research fronts' developed (Upham and Small, 2010). Of course, it would be absurd to suggest that there have not been important and even seminal ideas produced since this explosion, around 2005. No doubt there have been, in absolute terms, more new ideas in the last decade than before. However, because of the explosion in papers on the subject, they might be a smaller percentage of the overall total. Thus the researcher attending conferences or flicking (or clicking, now) through journals or working paper lists in search of inspiring new thoughts may be looking in the proverbial haystack.

It would be interesting to consider: (i) did the economics of energy and climate change experience a period of great originality, when many new 'research fronts' were developed and the associated publications were especially 'influential' during the 1990s? (ii) Around 2005, was there a decline in relative 'originality'? A more general set of questions is: (iii) Is there any correlation between resource quantity and originality? (iv) Are

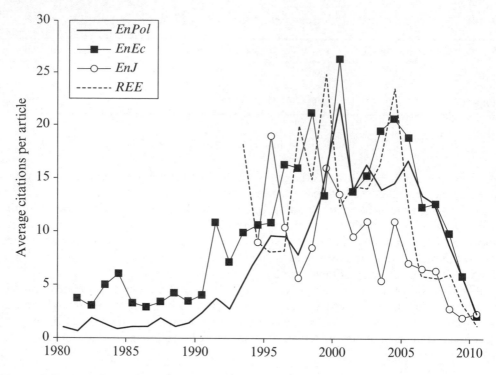

Figure I.3   Trend in annual average citations per article in energy economics and policy journals (1980–2010)

they positively or negatively correlated? (v) Does a rise in originality precede a rise in quantity?

Clearly, these are hard questions to answer. Again, despite the limitations of bibliometric analysis, they might offer some information on the subject. Figure I.3 presents the trends in average citations per article in each year for specific energy-related journals. The journals presented are the same as above (*Energy Economics (EnEc)*, *The Energy Journal (EnJ)*, *Resource and Energy Economics (REE)* and *Energy Policy (EnPol)*). The trends suggest a peak around 2000–2001 with another lower peak in 2005. However, one should be very careful before concluding that hypotheses (i) and (ii) are supported by the evidence.

First, this is a very crude metric of research originality and even influence. There are a number of different possible indicators, such as the impact factor (citations for the journal in the last two years) or five-year impact factor, eigenfactor, h-index for journals, all quite flawed.[3] Also, in each journal, the number of 'blockbuster' articles and their distribution in relation to standard ones varies greatly and, therefore, affects the average citations in each journal (Brouthers et al., 2012).

Second, the older the article, the more time has passed for it to be cited. So, all other things being equal, the oldest articles might be expected to be the most cited. This is not the case. Other factors must also be at work.

Third, present articles build on past ones, and if a growing number of articles is published, the ratio between present and past increases, meaning that past articles will be cited more often. When the growth is especially rapid, such as between 2005 and 2010, the difference between past and present articles is especially pronounced, implying many citations for past articles. So the total number of citations should be expected to increase. However, with each year that passes the total number of past articles to be cited increases. So, all other things being equal, in periods when the growth rate in research output increases, average citations in the preceding period probably increase, and, when the growth rate falls, average citations in preceding periods probably fall.

Fourth, the advent of computers and access to journal articles on the Internet has meant that today's researchers are probably citing more than in the past. So the Internet is helping to increase average citations.

Fifth, except for classic or 'blockbuster' articles, researchers are likely to prefer the use of recent publications with the most up-to-date information and methods of analysis. So, given all this growth (in research output and in the use of the Internet), we might expect the peak to be in 2010.

Sixth, however, articles do not get cited immediately. They take a while to be read, digested, discussed and considered useful. In fact, the peak number of citations for articles in any year seems to be between seven and ten years after publication. The articles published in 2000 for the *Energy Journal* and *Resource and Energy Economics* received the maximum number of citations in 2007, for *Energy Policy* in 2009 and for *Energy Economics* in 2010. Beyond this peak, the number of citations in each year declines, but still adds to the total and average number of citations for any year of publication. So, given that 2011 is the last year for which data are available, a peak around 2000 might be expected (looking at the data from a 2012 perspective).

So, in light of the many caveats, all that can be said is that there has been a growth in the number of citations per article from the early 1990s, which peaked between 2001 and 2005. Let each reader judge whether there was indeed an important rise in research originality from 1995.

Future research might be able to answer this more thoroughly, offering greater insight into research influence, originality and maybe even quality. Over the next ten years it might be interesting to follow the trend in average citations per article in any year. If average citations between 2006 and 2010 do not catch up with those between 2000 and 2005, then one might suspect (although, very tentatively, since there are so many other factors involved and it is a very crude indicator) that the growth in research output coincided with a decline in originality, and that there was a rise in originality preceding the growth in research output.

If this is confirmed, then the discussion and these hypotheses hint at a possible causality between the two variables, and a narrative: driven by a change in general concern for resource and environmental issues, between the 1960s and the early 1990s, mainstream economists applied their expertise to energy and environmental issues (see Table I.1). In response to the oil shocks, energy economics became a specific sub-discipline of economics, with dedicated journals. Inspired by these and more basic ideas from mainstream economists, the growing interest in issues related to developing economies, energy policy, efficiency and environmental concerns led, during the 1990s, to a great deal of original research among more specialized energy and environmental economists. In the language of students of science, new research fronts opened up, with a relatively large number of 'blockbuster' articles produced (Upham and Small, 2010; Brouthers et al., 2012). Although still to be confirmed, the level of 'originality' among energy and environmental economists may have peaked between 1995 and 2005. After this period, the focus was probably on using and applying these ideas, and it generated a great deal of funding and research output.

6.   NEW DIRECTIONS IN THE ECONOMICS OF ENERGY AND CLIMATE CHANGE

Two issues stand out. First, since 2005, we may have entered a growth phase (of research output) that coincides with an emphasis on refinement rather than original thinking. This is a concern because energy security, supply and prices, and of course climate change mitigation and adaptation, are still serious problems that require original ideas to solve them.

Second, for the energy economists seeking new ideas, one challenge is how to find seminal working papers and articles (Brouthers et al., 2012). It is a challenge that is likely to continue and, depending on oil prices and climate change, and the ensuing supply of energy economics, may become even greater. There will be a need to find a way to approach this challenge, such as by bringing these ideas together. Thus, in this handbook, I have taken the responsibility of inviting (what are in my view) creative and controversial thinkers to contribute.

That is certainly not to claim that all the chapters offer or address radically new ideas. Instead they might be providing important overviews or reviews of topics. This handbook is simply an invitation to researchers to present some stimulating thought and debate. Nevertheless, it is hoped that the reader will find here a high concentration of stimulating ideas and, on reading individual chapters, will discover new windows onto possible solutions to energy and climate change problems to have opened up.

The handbook is divided into seven parts: fossil fuel markets, electricity markets, energy policy, climate agreements, carbon mitigation policies, low-carbon behaviour and low-carbon growth.

The chapters in Part I address aspects of the various fossil fuel markets. Hamilton, in the first chapter, provides a long-run perspective on global oil production and prices, and their relationship with economic growth. He indicates that, historically, global oil production has kept up with the growth in demand by finding new geographical areas. In the periods before production had adjusted, prices inevitably rose, with substantial impacts on the global economy. There is evidence that these impacts have become weaker through time, but that there may be less potential for energy production to grow by finding new geographical areas.

In Chapter 2, Stevens presents the history of natural gas markets and considers the role of gas in the global primary energy mix. He discusses the reasons for the rapid growth in natural gas in the last two decades, emphasizing the important role that declining LNG (liquefied natural gas) transport costs have played. He then reviews the markets in different regions, and the important role Russia has and will continue to play in a number of the regional markets. He also discusses the potential impact of advances in shale gas production on future global markets. In Chapter 3, the potential implications of an OPEC-style natural gas cartel are considered by Gabriel et al. They find that, in the foreseeable future, the problems of coordination are too high and the gains from production cuts among GECF (Gas Exporting Countries Forum) members too low (especially for Russia) to develop a genuine natural gas cartel. They do suggest, however, that a more subtle form of supply infrastructure coordination may develop as a result of Russia's strategic interests.

In Chapter 4, Haftendorn et al. warn us of the continued expansion of world demand for coal, and its potential for large carbon dioxide emissions. They suggest that there will probably be a move towards India and Asia. They compare the impacts of different climate policies on the coal trade, and find that global climate policies will be crucial for minimizing carbon dioxide emissions related to coal consumption.

Part II covers different issues related to the market for electricity. Jacoby et al. in Chapter 5 investigate how the future electricity grid in the USA and more generally is likely to evolve, presenting the institutional issues and incentives facing the industry. They identify the key factors that will determine the evolution of the electricity transmission and distribution systems. They present some of the technological developments under way, such as smart meters, and some of the associated threats, including cyber privacy.

In Chapter 6, Johnstone and Haščič focus on the drivers of past R&D into energy storage and grid management. Both for the future of electricity markets and especially given the intermittency associated with some renewable energy sources, energy storage and grid management are seen as important enablers of renewable electricity sources. Also, there is a major role for the public support of such R&D projects because of the difficulty of the innovators capturing the benefits from these advances; the problem of picking winners among the generating technologies is also avoided. In Chapter 7, Lee and Lovellette consider the factors, in the USA and internationally, that will promote the use of electric vehicles. They highlight the huge uncertainties associated with the potential uptake of electric cars. The key variables are trends in oil prices, the extent of biofuels expansion, electricity infrastructure, technological developments, such as battery performance, and consumer expectations about car attributes. They also indicate that these factors vary greatly across countries, and that, in some countries, the factors might all combine to encourage their uptake.

Part III considers the involvement of government in these fossil fuel and electricity markets. Despite their potential for reducing carbon dioxide emissions, barriers to energy efficiency, and the higher costs of renewable energy and nuclear power compared with fossil-fuel-generated electricity, these are central parts of today's energy and climate policy. Evans et al. in Chapter 8 present an original method for identifying macro-level improvements in energy efficiency. They provide new evidence on energy efficiency improvements for a range of different economies, and the potential of efficiency in meeting carbon dioxide emission targets. The chapter also acts as a valuable reminder that trends in energy intensity are frequently not a good indicator of energy efficiency improvements in an economy.

In Chapter 9, van Kooten reviews the current policies in place to promote the development of renewable energy technologies and sources, with a detailed discussion of feed-in tariff (FIT) programmes. He discusses these tariffs in theory and in practice, indicating their widespread adoption across the world and the level of tariffs in different countries for wind power and solar photovoltaic. He argues that feed-in tariffs as an energy and climate policy are inefficient compared to other instruments, and suggests that the higher costs of the tariffs and associated rising electricity prices are starting to reduce their popularity among a number of populations and electorates. In Chapter 10, Jefferson questions the potential contribution that renewable energy sources can make to future (medium-term) energy requirements. He suggests that the rising concern about climate change is giving energy supply industries, especially the nuclear and renewable industries, an opportunity to develop a strong rhetoric (which might involve stretching the truth, and using expert advice and scientific reports to their advantage), which is affecting the public's and government's perception of their potential expansion. He proposes that their potential expansion in the medium term is far more limited than is generally perceived, and places doubt on the ability of the electricity supply industries to achieve renewable energy growth targets.

Robinson, in Chapter 11, reviews the evolution of energy policies and identifies where they might go in the future. In particular, the role of government in energy markets appears to have increased in the mid-twentieth century and declined at the end of the century. He argues that the increased concern about climate change is an important factor reviving government involvement in energy markets.

Part IV is about climate agreements. Hendry and Pretis in Chapter 12 offer new evidence on the first and potentially most important agreement, the general consensus among scientists and even politicians that climate change is affected by anthropogenic activities. They combine long-term and seasonal data to show that, while natural factors are important for seasonal fluctuations, fossil-fuel-related carbon dioxide emissions are crucial for explaining air temperature trends.

In Chapter 13, Holtsmark discusses some of the main reasons why it is difficult to gain broad international support for an effective climate agreement. He also proposes that a set of regional agreements, which has been proposed as a solution to the problem, might lead to inefficiently small emission reductions. He also argues that unilateral pledges by individual or groups of countries to make deep cuts in emissions may in fact make it more difficult, not easier, to achieve an effective international climate agreement. In turn, Tol in Chapter 14 proposes that the Kyoto Protocol may be a satisfactory framework for achieving major reductions in global carbon dioxide emissions. He argues that all the requirements necessary for international climate policy are in place (electoral demand for action, international monitoring of emissions data, a forum to pledge domestic action and review progress, and international flexibility in emissions reduction) and that future Conferences of the Parties (COP) meetings should focus on refining the existing agreements.

Aldy, in Chapter 15, discusses some of the key conditions that will be necessary to achieve a successful post-2012 international climate regime. Three issues he raises are the need to enhance the legitimacy of multilateral climate policy efforts, promote best policy practices, and channel financing for investments in climate change risk mitigation activities in developing countries. He emphasizes the importance of transparency and surveillance to enhance the credibility of international agreements. For instance, this surveillance and individual country legitimacy may also be used to justify financing in that country, ultimately increasing support for carbon finance more generally. Aldy proposes a Bretton Woods-style institution to act as the central architecture for addressing these issues and coordinating international climate change agreements and developments.

Part V discusses different policies for encouraging reductions in carbon dioxide emissions. Given the agreement to raise $100 billion per year by 2020 for financing climate mitigation and adaptation projects in developing countries, Parry in Chapter 16 discusses how the revenue to fund such projects might be raised. He compares the features of different approaches to raising revenue: the effectiveness, revenue raised, cost-effectiveness, cross-country distribution of revenues, and impacts on energy prices. He offers the rationale for carbon pricing, ways of dealing with the opposition to carbon pricing (such as pulling back environmentally less effective taxes) and alternatives to carbon taxes, such as 'feebates' that combine fees and rebates to incentivize less polluting consumption.

In Chapter 17, Hope seeks to identify how high a tax related to climate change might reach. He outlines the role of integrated assessment models in estimating the social costs of carbon. He shows that estimates of the social costs of carbon tend to be based on a subsample of the results, which excludes extreme results. However, given the great uncertainty about climate change, when extreme results are incorporated, estimates of the social costs of carbon increase substantially. Based on a fuller subsample of results, he proposes that a tax related to climate change should probably be well over $100 per tonne. In Chapter 18, McKitrick proposes a state-contingent (i.e. temperature) pricing mechanism for carbon dioxide emissions. One of the purposes of the idea is to find a compromise for two opposing sides – those that fear climate change and, therefore, believe that associated taxes are under-predicted, and those that believe they are over-predicted. It is also an interesting proposal because it creates incentives for future climate change to be predicted with greater accuracy.

In Chapter 19, Gago et al. focus on energy consumption and related emissions in buildings. They argue that conventional policy instruments may not achieve the desired outcomes unless they are packaged together to tackle the existing problems of information, split incentives among agents, uncertainty and access to capital. They propose a policy package based around energy certification of buildings, the use of flexible building codes, smart metering and a new tax on energy inefficiency. In Chapter 20, Gennaioli et al. offer a methodological introduction to the econometric analysis of climate policies and a review of the evidence to date with the objective of providing more robust evaluation of instruments. They suggest that a race between 'hard' incentives (such as forms of carbon pricing) and more 'soft' interventions (such as energy reports, free energy advice or voluntary agreement) is under way. Evidence indicates that either type of policies can achieve reductions in carbon dioxide emissions. Nevertheless, they question whether soft interventions can be a substitute for the typically less popular hard interventions.

In Chapter 21, Chevallier reviews the main characteristics of tradable permits markets, with specific emphasis on the two largest cap-and-trade systems to date: the EU Emissions Trading Scheme (ETS) and the Kyoto Protocol's Clean Development Mechanism. He considers the distortionary effects of allocative mechanisms, identifies the factors determining permit prices and links market fluctuations with broader macro-economic forces. He also considers the potential unification of regional carbon market initiatives into a 'world' carbon market. In Chapter 22, Kverndokk examines some of the barriers to the introduction of tradable permits. He presents evidence on the public's ethical concern about such schemes. This barrier implies that concerned individuals are likely to impede the progress of environmental legislation and may be responsible for greater emissions than otherwise.

In Chapter 23, de Perthuis and Trotignon consider the transition to the next phase (2013–20) of the EU carbon dioxide emissions trading scheme. They analyse proposed reforms to the scheme, including setting a reserve auction price, allowing for set-asides and extending the phase until 2030. A key message of the analysis is that policy clarity will be crucial to the effectiveness of the long-run development of the EU tradable permit market. Finally, they discuss the formation of a carbon regulating authority, similar to central banks, independent of any short-term political pressures, to manage the permit market to achieve emission reductions while minimizing price volatility and promoting market efficiency.

Part VI explores some of the behavioural limitations to a low-carbon economy. Pollitt and Shaorshadze in Chapter 24 review the literature on behavioural economics in the context of energy and low-carbon policies. They focus on consumption and habits, investment in energy efficiency, and the provision of public goods and support for pro-environmental behaviour. The large-scale deployment of smart meters and appliances will be particularly relevant for the application of behavioural economics to energy issues. They propose that a key role this approach can offer is in influencing public perceptions about the affordability of climate policy and in facilitating the creation of a more responsive energy demand.

In Chapter 25, Gowdy argues that the unique social characteristics of the human species can be used to address climate change. In particular, he reviews the fascinating anthropological and neurobiological literature on the sociality of human decision-making and links this economic behaviour related to energy and climate change. In Chapter 26, O'Garra outlines the moral dimensions of polluting behaviour related to climate change. She identifies features that make framing climate change as a moral issue currently problematic, and discusses how they might be re-cast within the structure of the archetypal moral problem, so that individuals may start to perceive climate change as a moral issue. She argues that, as long as the public fails to perceive climate change as a moral issue, the demand for climate stabilization will be weak and little progress will be made towards a low-carbon economy.

In Chapter 27, Eyre explores the need for a decentralized governance structure to achieve certain aspects of a low-carbon transition. He emphasizes that many of the carbon mitigation strategies will involve the involvement of an increasing number of decentralized players, whether households, social collectives or small businesses, investing in and running projects related to energy efficiency improvements or the provision of small-scale and intermittent renewable energy. He proposes that the nature of energy and climate policy may have to adapt to meet the changing dynamics between players, including a greater involvement of local government. Barbier, in Chapter 28, draws on institutional and historical perspectives to explain the continuing policy stalemate related to climate change, proposing that there is a policy failure on a massive scale. With this in mind, he asks whether a global crisis will be required to break the existing institutional structure and introduce new ones that are more suited for preventing climate change.

Part VII considers the longer-term perspective associated with low-carbon economic growth. Hepburn and Bowen, in Chapter 29, provide a conceptual and synthetic analysis of the relationship between economic growth and environmental limits, including those imposed by climate change. They propose that continued economic growth is feasible and desirable, although not without significant changes in its characteristics. These changes need to involve ultimately the reduction of the rate of material output, with continued growth in value being generated by expansion in the 'intellectual economy'. In Chapter 30, Llavador et al. refocus the debate about optimal climate strategies by asking which characteristics, such as consumption, knowledge or education, are desirable and should be promoted when seeking economic growth. Their message (as well as that of the authors of the preceding chapter) is that by broadening the set of values, policy makers might help to lead the economy towards activities that are less energy- and carbon-intensive.

In Chapter 31, Chevallier explores whether a low-carbon transition is possible in China. Given that China is responsible for about one-fifth of global carbon dioxide emissions at present, future climate change will be affected by the economy's ability to stabilize and reduce emissions. Similarly, because it is becoming a global economic leader, its path will have a very large impact on the behaviour of other economies. Furthermore, as a centrally planned economy, its approach may offer new lessons for achieving a transition to a low-carbon economy. Chevallier concludes that, given its industry structure, current energy mix and potential for technological development in renewables, a transition towards a low-carbon economy in China is necessary, and feasible with a possibility of developing domestic emissions trading and joining a future global carbon market

Given the concerns about climate change, on the one hand, and those about the economic implications of a transition to a low-carbon economy on the other, Fouquet, in Chapter 32, offers a speculative discussion about the problems and possibilities created by a climate-changed and low-carbon economy. Taking account of future population growth and economic development, he presents three extreme scenarios of the world during the twenty-first century that focus on, first, climate adaptation, second, an all-nuclear economy and, finally, an all-renewable economy. He highlights that each outcome implies specific constraints and major implications (some positive, some negative) for the future energy system and for the wider economy and even society. He warns against a blind objective of achieving a low-carbon transition at any cost, and instead encourages care about the outcome sought and, thus, the path chosen.

This handbook is far from an exhaustive survey of the issues related to energy and climate change. For one, it has focused on the economic aspects rather than approaching the issues from more technical or other social angles. Naturally, it tries to focus on the interaction between energy and climate change, and accompanies similar endeavours that focus on either energy or climate change (Evans and Hunt, 2009; Dinar and Mendelsohn, 2011; Dryzek et al., 2011). Finally, certain topics have not received the attention of a whole chapter dedicated to them, including: the specifics of energy production, distribution and use in developing economies (see, e.g., Wolfram et al., 2012), nuclear power (see, e.g., Davis, 2012 or Joskow and Parsons, 2012), the Clean Development Mechanism (see, e.g., Popp, 2012) and carbon capture and sequestration (see, e.g., von Hirschhausen et al., 2012). This handbook tries only to offer some insights into where the literature has reached – after the concerns in the 1960s about the long-run relationship between economic growth and resource scarcity, to the many new 'research fronts' created from the 1970s and to the 1990s, to the explosion in research output since 2005 – and where it might be going.

NOTES

1. I would like to thank Peter Pearson for his valuable comments and Claire Fitzgerald for collecting some of the data related to this chapter.
2. Research quality takes on several dimensions – as well as the development of new knowledge and influence on other research, it might include the theoretical and empirical analytical rigour of the research (Tol and Weyant, 2006; Frey and Rost, 2010; Brouthers et al., 2012).
3. For instance, the standard impact factor uses only the citations in the two years after publication, although, below, it will be shown that the peak for average citations occurs with a lag of seven to ten years after publication. Meanwhile, the eigenfactor identifies all the citations in a journal, weighting them according to the 'ranking' of the journal in which the citation occurred, but naturally depends on the ranking methodology. Finally, the h-index favours journals with a large number of issues (Seglen, 1997; Bornmann and Daniel, 2005; Chang and McAleer, 2011).

REFERENCES

Adelman, M.A. (1972), *The World Petroleum Market*, Baltimore, MD: Johns Hopkins University Press.
Adelman, M.A. (1986), 'Scarcity and world oil prices', *Review of Economics and Statistics*, **68**(3), 387–97.
Aldy, J.E., Krupnick, A.J., Newell, R.G., Parry, I.W.H. and Pizer, W.A. (2010), 'Designing climate mitigation policy', *Journal of Economic Literature*, **48**(4), 903–34.
Allcott, H., Mullainathan, S. and Taubinsky, D. (2011), 'Externalizing the internality', Working Paper, Economics Department, New York University.
Arrow, K.J. and Chang, S. (1982), 'Optimal pricing, use and exploration of uncertain natural resource stocks', *Journal of Environmental Economics and Management*, **9**(1), 1–10.
Asafu-Adjaye, J. (1999), 'The relationship between energy consumption, energy prices and economic growth: time series evidence from Asian developing countries', *Energy Economics*, **22**(6), 615–25.
Atkinson, S. and Tietenberg, T. (1991), 'Market failure in incentive-based regulation: the case of emission trading', *Journal of Environmental Economics and Management*, **21**(1), 17–31.
Azar, C. and Sterner, T. (1996), 'Discounting and distributional considerations in the context of the global warming', *Ecological Economics*, **19**(2), 169–84.
Barbier, E.B., Markandya, A. and Pearce, D.W. (1990), 'Environmental sustainability and cost–benefit analysis', *Environment and Planning A*, **22**(9), 1259–66.
Barnett, H.J. and Morse, C. (1963), *Scarcity and Growth: The Economics of Natural Resource Scarcity*, Washington, DC: Resources for the Future.
Barrett, S. (1990), 'The problem of global environmental protection', *Oxford Review of Economic Policy*, **6**(1), 68–79.
Barrett, S. (1994), 'Self-enforcing international environmental agreements', *Oxford Economic Papers*, **46**, 878–94.
Baumol, W.J. and Oates, W.E. (1971), 'Use of standards and prices for protection of the environment', *Swedish Journal of Economics*, **73**(1), 42–54.
Baumol, W.J. and Oates, W.E. (1975), *The Theory of Environmental Policy*, Englewood Cliffs, NJ: Prentice-Hall.
Berndt, E.R. and Wood, D.O. (1975), 'Technology, prices and the derived demand for energy', *Review of Economics and Statistics*, **57**(3), 259–68.
Bhatia, R. (1987), 'Energy demand analysis in developing countries: A review', *The Energy Journal*, **8**, 1–33.
Bohi, D.R. (1989), *Energy Price Shocks and Macroeconomic Performance*, Washington, DC: Resources for the Future.
Bornmann, L. and Daniel, H.D. (2005), 'Does the h-index for ranking of scientists really work?', *Scientometrics*, **65**(3), 391–2.

Bovenberg, A.L. and Goulder L.H. (1996), 'Optimal environmental taxation in the presence of other taxes: general-equilibrium analyses', *American Economic Review*, **86**(4), 985–1000.

Bovenberg, A.L. and de Mooij, R.A. (1995), 'Environmental levies and distortionary taxation', *American Economic Review*, **84**(4), 1085–9.

Bovenberg, A.L. and van der Ploeg, F. (1994), 'Environmental policy, public finance and the labour market in a second-best world', *Journal of Public Economics*, **55**(3), 349–90.

Brouthers, K.D., Mudambi, R. and Reeb, D.M. (2012), 'The blockbuster hypothesis: influencing the boundaries of knowledge', *Scientometrics*, **70**(3), 959–82.

Carraro, C. and Siniscalco, D. (1993), 'Strategies for the international protection of the environment', *Journal of Public Economics*, **52**(3), 309–28.

Chang, C.L. and McAleer, M. (2011), 'How should journal quality be ranked? An application to agricultural, energy, environmental and resource economics', Working Papers in Economics 11/43, University of Canterbury, Department of Economics and Finance.

Chevallier, J. (2012), *Econometric Analysis of Carbon Markets: The European Union Emissions Trading Scheme and the Clean Development Mechanism*, Berlin: Springer.

Christensen, L.R., Jorgensen, D.W. and Lau, L.J. (1973), 'Transcendental logarithmic production frontiers', *Review of Economics and Statistics*, **55**, 28–45.

Cline, W.R. (1992), *The Economics of Global Warming*, Washington, DC: Institute for International Economics.

Coase, R.H. (1960), 'The problem of social cost', *Journal of Law and Economics*, **3**(1), 1–44.

Cropper, M.L. and Oates, W.E. (1992), 'Environmental economics – a survey', *Journal of Economic Literature*, **30**(2), 675–740.

Dales, J. (1968), *Pollution, Property and Prices*, Toronto, Canada: Toronto University Press.

Darmstadter, J., Teitelbaum, P.D. and Polach, J.G. (1971), *Energy in the World Economy: A Statistical Review of Trends in Output, Trade, and Consumption Since 1925*, Baltimore, MD: The Johns Hopkins University Press.

Dasgupta, P. and Heal, G. (1979), *Economic Theory and Exhaustible Resources*, Cambridge: Cambridge University Press.

Davis, L. (2012), 'Prospects for nuclear power', *Journal of Economic Perspectives*, **26**(1), 49–66.

De Rond, M. and Miller, A.N. (2005), 'Publish or perish: bane or boon of academic life?', *Journal of Management Inquiry*, **14**, 321–9.

Dinar, A. and Mendelsohn, R. (2011), *Handbook on Climate Change and Agriculture*, Cheltenham, UK and Northampton, MA, USA: Edward Elgar.

Dincer, I. (2002), 'The role of exergy in energy policy making', *Energy Policy*, **30**(2), 137–49.

Dryzek, J.S., Norgaard, R.B. and Schlosberg, D. (2011), *The Oxford Handbook of Climate Change and Society*, Oxford: Oxford University Press.

Dubin, J.A. and McFadden, D.L. (1984), 'An econometric analysis of residential electric appliance holdings and consumption', *Econometrica*, **52**(2), 345–62.

Ellerman, A.D., Convery, F. and De Perthuis, C. (2010), *Pricing Carbon: The European Emissions Trading Scheme*, Cambridge: Cambridge University Press.

Ellerman, AD., Joskow, P.L., Schmalensee, R., Montero, J.P. and Bailey, E. (2000), *Markets for Clean Air: The US Acid Rain Program*, New York: Cambridge University Press.

Evans, J. and Hunt, L.C. (2009), *International Handbook on the Economics of Energy*, Cheltenham, UK and Northampton, MA, USA: Edward Elgar.

Fischer, C. and Newell, R.G. (2008), 'Environmental and technology policies for climate mitigation', *Journal of Environmental Economics and Management*, **55**(2), 142–62.

Fisher, A.C. and Ward, M. (2000), 'Trends in natural resource economics in JEEM 1974–1997: breakpoint and nonparametric analysis', *Journal of Environmental Economics and Management*, **39**, 264–81.

Fouquet, R. (1998), 'The United Kingdom demand for renewable electricity in a liberalised market', *Energy Policy*, **26**(4), 281–94.

Fouquet, R. and Pearson, Peter J.G. (2012), 'The long run demand for lighting: elasticities and rebound effects in different phases of economic development', *Economics of Energy and Environmental Policy*, **1**(1), 83–100.

Frey, B.S. and Rost, K. (2010), 'Do rankings reflect research quality?', *Journal of Applied Economics*, **13**(1), 1–38.

Gately, D. (1984), 'A ten-year retrospective: OPEC and the world oil market', *Journal of Economic Literature*, **22**(3), 1100–14.

Goldemberg, J., Johansson, T.B. Reddy, A.K.N. and Williams, R.H. (1985), 'An end use oriented energy strategy', *Annual Review of Energy and the Environment*, **10**, 613–88.

Gordon, R.L. (1967), 'A reinterpretation of the pure theory of exhaustion', *Journal of Political Economy*, **75**, 274–86.

Goulder L.H., Parry I.W.H. and Williams, R.C. (1999), 'The cost-effectiveness of alternative instruments for environmental protection in a second-best setting', *Journal of Public Economics*, **72**(3), 329–60.

Green, R.J. and Newbery, D.M. (1992), 'Competition in the British electricity spot market', *Journal of Political Economy*, **100**(5), 929–53.

Greening, L.A., Greene, D.L. and Difiglio, C. (2000), 'Energy efficiency and consumption – the rebound effect – a survey', *Energy Policy*, **28**(6–7), 389–401.

Griffin, J.M. (1993), 'Methodological advances in energy modelling: 1970–1990', *The Energy Journal*, **14**(1), 111–24.

Griffin, J.M. and Gregory, P.R. (1976), 'An inter-country translog model of energy substitution responses', *American Economic Review*, **66**(12), 845–57.

Groom, B., Hepburn, C. and Koundouri, P. (2005), 'Declining discount rates: the long and the short of it', *Environmental & Resource Economics*, **32**(4), 445–93.

Hahn, R.W. (1984), 'Market power and transferable property rights', *Quarterly Journal of Economics*, **99**, 753–65.

Hall, B., Jaffe, A. and Trajtenberg, M. (2005), 'Market value and patent citations', *RAND Journal of Economics*, **36**, 16–38.

Hamilton, James D. (1983), 'Oil and the macroeconomy since World War II', *Journal of Political Economy*, **91**, 228–48.

Hanemann, M.W., Loomis, J. and Kanninen, B. (1991), 'Statistical efficiency of double-bounded dichotomous choice contingent valuation', *American Journal of Agricultural Economics*, **73**(4), 1255–63.

Hargadon, A. and Sutton, R. (1997), 'Technology brokering and innovation in a product development firm', *Administrative Science Quarterly*, **42**, 716–49.

Helm, D., Kay, J. and Thompson, D. (1988), 'Energy policy and the role of the state in the market for energy', *Fiscal Studies*, **9**(1), 41–61.

Hoel, M. (1991), 'Global environmental problems: the effects of unilateral actions taken by one country', *Journal of Environmental Economics and Management*, **20**, 55–70.

Hope, C., Anderson, J. and Wenman, P. (1993), 'Policy analysis of the greenhouse-effect – an application of the PAGE model', *Energy Policy*, **21**(3), 327–37.

Hotelling, H. (1931), 'The economics of exhaustible resources', *Journal of Political Economy*, **39**, 137–75.

Hudson, E. and Jorgensen, D.W. (1974), 'U.S. energy policy and economic growth, 1975–2000', *Bell Journal of Economics*, **5**, 461–514.

IPCC (2007), *Climate Change 2007 – Mitigation of Climate Change: Working Group III Contribution to the Fourth Assessment Report of the IPCC*, Cambridge: Cambridge University Press.

Jaffe, A.B., Peterson, S.R. and Portney, P.R. (1995), 'Environmental regulation and the competitiveness of U.S. manufacturing: what does the evidence tell us?', *Journal of Economic Literature*, **33**(1), 132–63.

Jaffe, A. and Stavins, R. (1994), 'The energy efficiency paradox. What does it mean?', *Energy Policy*, **22**(10), 804–10.

Jaffe, A.B. and Trajtenberg, M. (2005), *Patents, Citations, and Innovations: A Window on the Knowledge Economy*, Cambridge, MA: MIT Press.

Jevons, W.S. (1865), *The Coal Question: An Inquiry Concerning the Progress of the Nation, and the Probable Exhaustion of Our Coal-mines*, London: Macmillan.

Joskow, P.L. (1987), 'Contract duration and relationship-specific investments – empirical evidence from coal markets', *American Economic Review*, **77**(1), 168–85.

Joskow, P.L. and Parsons, J.E. (2012), 'The future of nuclear power after Fukushima', *Economics of Energy and Environmental Policy*, **1**(2), 99–113.

Joskow, P.L. and Schmalensee, R. (1983), *Markets for Power: An Analysis of Electric Utility Deregulation*, Cambridge, MA: MIT Press.

Kahn, M.E. (2010), *Climatopolis: How Our Cities Will Thrive in the Hotter Future*, New York: Basic Books.

Khazzoom, J.D. (1980), 'Economic implications of mandated efficiency in standards for household appliances', *Energy Journal*, **1**(4), 21–40.

Krautkraemer, J.A. (1998), 'Nonrenewable resource scarcity', *Journal of Economic Literature*, **36**, 2065–107.

Kula, E. (1998), *History of Environmental Economic Thought*, London: Routledge.

Lind, R.C. (1995), 'Intergenerational equity, discounting, and the role of cost–benefit analysis in evaluating global climate policy', *Energy Policy*, **23**(4/5), 379–89.

Maler, K.G. (1989), 'International environmental problems', *Oxford Review of Economic Policy*, **6**(1), 80–108.

Manne, A., Mendelsohn, R. and Richels, R. (1995), 'A model for evaluating regional and global effects of GHG reduction policies', *Energy Policy*, **23**(1), 17–34.

McDonald, A. and Schrattenholzer L. (1999), 'Learning rates for energy technologies', *Energy Policy*, **29**(4), 255–61.

Menanteau, P., Finon, D. and Lamy, M.L. (2003), 'Prices versus quantities: choosing policies for promoting the development of renewable energy', *Energy Policy*, **31**(8), 799–812.

Mendelsohn, R., Nordhaus, W.D. and Shaw, D. (1994), 'The impact of global warming on agriculture – a Ricardian analysis', *American Economic Review*, **84**(4), 753–71.

Meyer, N.I. (2003), 'European schemes for promoting renewables in liberalised markets', *Energy Policy*, **31**(7), 665–76.

Mitchell, C. (2000), 'The non-fossil fuel obligation and its future', *Annual Review of Energy and the Environment*, **25**, 285–312.

Montgomery, W.D. (1972), 'Markets in licenses and efficient pollution control programs', *Journal of Economic Theory*, **5**, 395–418.

Newbery, D.M. (1992), 'Should carbon taxes be additional to other transport fuel taxes?', *The Energy Journal*, **13**(2), 47–60.

Newbery, D.M. (1994), 'The impact of sulfur limits on fuel demand and electricity prices in Britain', *The Energy Journal*, **15**(3), 19–42.

Newbery, D.M.G. and Pollitt, M.G. (1997), 'Restructuring and privatisation of the CEGB – was it worth it?', *Journal of Industrial Economics*, **45**(3), 269–304.

Nordhaus, W.D. (1973), 'The allocation of energy resources', *Brookings Papers on Economic Activity*, **4**(3), 529–76.

Nordhaus, W.D. (1977), 'Economic growth and the climate: the carbon dioxide problem', *American Economic Review*, **67**(1), 341–6.

Nordhaus, W.D. (1980), 'The energy crisis and macroeconomic policy', *The Energy Journal*, **1**(1), 12–24.

Nordhaus, W.D. (1991), 'To slow or not to slow: the economics of the greenhouse effect', *The Economic Journal*, **101**(4), 920–37.

Nordhaus, W.D. (1994), *Managing the Global Commons: The Economics of Climate Change*, Cambridge, MA: MIT Press.

Nordhaus, W.D. (1996), 'Do real-output and real-wage measures capture reality? The history of lighting suggests not', in T.F. Bresnahan and R. Gordon (eds), *The Economics of New Goods*, Chicago, IL: Chicago University Press, pp. 27–70.

Nordhaus, W.D. (2007), 'To tax or tot to tax: the case for a carbon tax', *Review of Environmental Economics and Policy*, **1**(1), 26–44.

Nordhaus, W.D. (2008), *A Question of Balance: Weighing the Options on Global Warming Policies*, New Haven, CT: Yale University Press.

Nordhaus, W.D. and Yang, Z.L. (1996), 'A regional dynamic general-equilibrium model of alternative climate-change strategies', *American Economic Review*, **86**(4), 741–65.

Oates, WE. (1995), 'Green taxes: can we protect the environment and improve the tax system at the same time?', *Southern Economic Journal*, **61**, 915–22.

Palmer, K., Oates, W.E. and Portney, P.R. (1995), 'Tightening environmental standards – the benefit–cost or no-cost paradigm', *Journal of Economic Perspectives*, **9**(4), 119–32.

Parry, I.W.H. (1995), 'Pollution taxes and revenue recycling', *Journal of Environment Economics and Management*, **29**(3S), S64–S77.

Parry, I.W.H. and Small, K.A. (2005), 'Does Britain or the United States have the right gasoline tax?', *American Economic Review*, **95**(4), 1276–89.

Pearce D.W. (1991), 'The role of carbon taxes in adjusting to global warming', *Economic Journal*, **101**, 938–48.

Pearce, D.W. (2002), 'An intellectual history of environmental economics', *Annual Review of Energy and the Environment*, **27**, 57–81.

Pearce, D.W., Markandya, A. and Barbier, E.B. (1989), *Blueprint for a Green Economy*, London: Earthscan.

Pearce, D.W. and Webb, M. (1987), 'Rural electrification in developing countries: a reappraisal', *Energy Policy*, **15**(4), 329–38.

Pearson, P.J.G. and Stevens, P.J. (1987), 'Energy transitions in the third world', *Energy Policy*, **15**(2) 170–72.

Penrose, E.T. (1957), 'Profit sharing between producing countries and oil companies in the Middle East', *Economic Journal*, **69**(3), 238–54.

Pigou, A.C. (1920), *The Economics of Welfare*, London: Macmillan.

Pindyck, R.S. (1978), 'The optimal exploration and production of nonrenewable resources', *Journal of Political Economy*, **86**, 841–61.

Pindyck, R.S. (1979), 'Interfuel substitution and the industrial demand for energy: an international comparison', *Review of Economics and Statistics*, **61**(2), 169–79.

Pizer, W.A. (1999), 'The optimal choice of climate change policy in the presence of uncertainty', *Resource and Energy Economics*, **21**(3–4), 255–87.

Pizer, W.A. (2002), 'Combining price and quantity controls to mitigate global climate change', *Journal of Public Economics*, **85**(3), 409–34.

Pollitt, M.G. (2012), 'The role of policy in energy transitions: lessons from the energy liberalisation era', *Energy Policy Special Issue on Past and Prospective Energy Transitions*, **50**, 120–29.

Popp, D. (2012), 'International technology transfer, climate change and the Clean Development Mechanism', *Review of Environmental Economics and Policy*, **5**(1), 131–52.

Porter, M.E. (1991), 'America's green strategy', *Scientific American*, **264**(4), 168.

Porter, M.E. and van der Linde, C. (1995), 'Toward a new conception of the environment–competitiveness relationship', *Journal of Economic Perspectives*, **9**(4), 97–118.

Potter, N. and Christy, F.T. (1962), *Trends in Natural Resource Commodities*, Baltimore, MD: Johns Hopkins University Press for Resources for the Future.

Roe, B., Teisl, M.F., Levy, A. et al. (2001), 'US consumers' willingness to pay for green electricity', *Energy Policy*, **29**(11), 917–25.

Rubin, J. (1996), 'A model of intertemporal emission trading, banking, and borrowing', *Journal of Environmental Economics and Management*, **31**, 269–86.

Sandmo, A. (1975), 'Optimal taxation in presence of externalities', *Swedish Journal of Economics*, **77**(1), 86–98.

Scarpa, R. and Willis, K. (2010), 'Willingness-to-pay for renewable energy: primary and discretionary choice of British households for micro-generation technologies', *Energy Economics*, **32**(1), 129–36.

Schelling, T.C. (1995), 'Intergenerational discounting', *Energy Policy*, **23**(4/5), 395–401.

Schmalensee, R., Joskow, P.L., Ellerman, A.D., Montero, J.P. and Bailey, A.M. (1998), 'An interim evaluation of sulfur dioxide emissions trading', *The Journal of Economic Perspectives*, **12**, 53–68.

Schurr, S. and Netschert, B. (1960), *Energy in the American Economy*, Baltimore, MD: Johns Hopkins University Press.

Seglen, P.O. (1997), 'Why the impact factor of journals should not be used for evaluating research', *British Medical Journal*, **314**(7079), 498–502.

Selden, T.M. and Song, D.Q. (1992), 'Environmental-quality and development: is there a Kuznets curve for air-pollution emissions?', *Journal of Environmental Economics and Management*, **27**(2), 147–62.

Shafik, N. and Bandyopadhyay, S. (1992), 'Economic growth and environmental quality: time-series and cross-country evidence', World Bank Policy Research Working Paper, No. 904. World Bank, Washington, DC.

Silva, E. and Teixeira, A.A.C. (2011), 'A bibliometric account of the evolution of EE in the last two decades: is ecological economics (becoming) a post-normal science?', *Ecological Economics*, **70**(5), 849–62.

Slade, M.E. (1982), 'Trends in natural resource commodity prices: an analysis of the time domain', *Journal of Environmental and Economic Management*, **9**, 122–37.

Sorrell, S. and Dimitropoulos, J. (2008), 'The rebound effect: microeconomic definitions, limitations and extensions', *Ecological Economics*, **65**(3), 636–49.

Soytas, U. and Sari, R. (2003), 'Energy consumption and GDP: causality relationship in G-7 countries and emerging markets', *Energy Economics*, **25**(1), 33–7.

Stavins, R.N. (1995), 'Transaction costs and tradeable permits', *Journal of Environmental Economics and Management*, **29**(2), 133–48.

Stavins, R.N. (1998), 'What can we learn from the grand policy experiment? Lessons from SO_2 allowance trading', *Journal of Economic Perspectives*, **12**(3), 69–88.

Stern, D.I. (2004), 'The rise and fall of the environmental Kuznets curve', *World Development*, **32**(8), 1419–39.

Stern, D.I., Common, M.S. and Barbier, E.B. (1996), 'Economic growth and environmental degradation: the environmental Kuznets curve and sustainable development', *World Development*, **24**(7), 1151–60.

Stern, N.H. (2006), *The Stern Review: The Economics of Climate Change*, Cambridge: Cambridge University Press.

Sterner, T. (2007), 'Fuel taxes: an important instrument for climate policy', *Energy Policy*, **35**(6), 3194–202.

Sterner, T. and Persson, U.M. (2008), 'An even Sterner review: introducing relative prices into the discounting debate', *Review of Environmental Economics and Policy*, **2**(1), 61–76.

Stevens, P.J. (2000), 'Introduction', in P.J. Stevens (ed.), *The Economics of Energy*, Cheltenham, UK and Northampton, MA, US: Edward Elgar.

Tol, R.S.J. (1994), 'The damage costs of climate-change – a note on tangibles and intangibles, applied to DICE', *Energy Policy*, **22**(5), 436–8.

Tol, R.S.J. (2002), 'Estimates of the damage costs of climate change. Part 1: benchmark estimates', *Environmental & Resource Economics*, **21**(1), 47–73.

Tol, R.S.J. (2005), 'The marginal damage costs of carbon dioxide emissions: an assessment of the uncertainties', *Energy Policy*, **33**(16), 2064–74.

Tol, R.S.J. and Weyant, J.P. (2006), 'Energy economics' most influential papers', *Energy Economics*, **28**, 405–9.

Unruh, G.C. (2002), 'Understanding carbon lock-in', *Energy Policy*, **28**(12), 817–30.

Upham, S.P. and Small, H. (2010), 'Emerging research fronts in science and technology: patterns of new knowledge development', *Scientometrics*, **83**(1), 15–38.

Von Hirschhausen, C., Herold, J. and Oei, P.Y. (2012), 'How a "low-carbon" innovation can fail – tales from a "lost decade" for carbon capture, transport, and sequestration (CCTS)', *Economics of Energy and Environmental Policy*, **1**(2), 115–24.

Wagner, G. (2011), *But Will the Planet Notice?*, New York: Hill & Wang/Farrar, Strauss & Giroux.

Watkins, G.C. (1992), 'The economic analysis of energy demand: perspectives of a practitioner', in D. Hawdon (ed.), *Energy Demand: Evidence and Expectations*, London: Surrey University Press, pp. 29–96.

Weightman, A.L. (2011), 'Using bibliometrics to define the quality of primary care research', *British Medical Journal*, **342**, 560–61.

Weitzman, M.L. (1974), 'Prices vs quantities', *The Review of Economic Studies*, **41**(5), 477–91.

Weitzman, M.L. (1998), 'Why the far-distant future should be discounted at its lowest possible rate', *Journal of Environmental Economics and Management*, **36**(3), 201–8.

Weitzman, M.L. (2007), 'A review of the Stern Review on the economics of climate change', *Journal of Economic Literature*, **45**(3), 703–24.

Whitmarsh, L. (2011), 'Scepticism and uncertainty about climate change: dimensions, determinants and change over time', *Global Environmental Change*, **21**, 690–700.

Wiser, R., Pickle, S. and Goldman, C. (1998), 'Renewable energy policy and electricity restructuring: a California case study', *Energy Policy*, **26**(6), 465–75.

Wolfram, C., Shelef, O. and Gertler, P. (2012), 'How will energy demand develop in the developing world?', *Journal of Economic Perspectives*, **26**(1), 49–66.

World Bank (2011), *World Development Indicators*, Washington, DC: World Bank.

PART I

FOSSIL FUEL MARKETS

1 Oil prices, exhaustible resources and economic growth*
James D. Hamilton

1 OIL PRICES AND THE ECONOMICS OF RESOURCE EXHAUSTION

One of the most elegant theories in economics is Hotelling's (1931) characterization of the price of an exhaustible natural resource. From the perspective of overall social welfare, production today needs to be balanced against the consideration that, once consumed, the resource will be unavailable to future generations. One option for society would be to produce more of the commodity today, invest the current marginal benefits net of extraction costs in some other form of productive capital, and thereby accumulate benefits over time at the rate of interest earned on productive capital. An alternative is to save the resource so that it can be used in the future. Optimal use of the resource over time calls for equating these two returns. This socially optimal plan can be implemented in a competitive equilibrium if the price of the resource net of marginal production cost rises at the rate of interest. For such a price path, the owner of the mine is just indifferent between extracting a little bit more of the resource today and leaving it in the ground to be exploited at higher profit in the future.

This theory is compelling and elegant, but very hard to reconcile with the observed behavior of prices over the first century and a half of the oil industry. Figure 1.1 plots the real price of crude petroleum since 1860. Oil has never been as costly as it was at the birth of the industry. Prior to Edwin Drake's first oil well in Pennsylvania in 1859, people were getting illuminants using very expensive methods.[1] The term kerosene, which we still use today to refer to a refined petroleum product, was actually a brand name used in the 1850s for a liquid manufactured from asphalt or coal, a process which was then, as it still is now, quite expensive.[2] *Derrick's Handbook* (1898) reported that Drake had no trouble selling all the oil his well could produce in 1859 at a price of $20 per barrel. Given the 24-fold increase in estimates of consumer prices since 1859, that would correspond to a price in 2010 dollars a little below $500 per barrel. As drillers producing the new-found 'rock oil' from other wells brought more of the product to the market, the price quickly fell, averaging $9.31 per barrel for 1860 (the first year shown in Figure 1.1). In 2010 prices, that corresponds to $232 per barrel, still far above anything seen subsequently. Even ignoring the initial half-century of the industry, the price of oil in real terms continued to drop from 1900 to 1970. And despite episodes of higher prices in the 1970s and 2000s, throughout the period from 1992 to 1999, the price of oil in real terms remained below the level reached in 1920.

There are two traditional explanations for why Hotelling's theory appears to be at odds with the long-run behavior of crude oil prices. The first is that although oil is in principle an exhaustible resource, in practice the supply has always been perceived to be

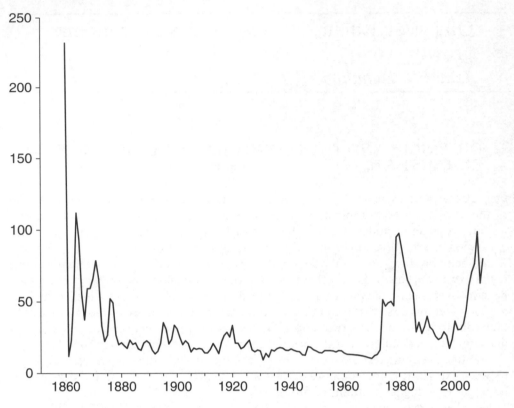

Data source: 1861–2010 from BP, *Statistical Review of World Energy 2010*; 1860 from Jenkins (1985, Table 18) (which appears to be the original source for the early values of the BP series) and *Historical Statistics of the United States*, Table E 135-166, Consumer Prices Indexes (BLS), all items, 1800 to 1970.

Figure 1.1   Price of oil in 2010 dollars per barrel, 1860–2010

so vast, and the date at which it will finally be exhausted has been thought to be so far into the future, that finiteness of the resource had essentially no relevance for the current price. This interpretation can be reconciled with the Hotelling solution if one hypothesizes a tiny rent accruing to owners of the resource that indeed does grow at the rate of interest, but in practice has always been sufficiently small that the observed price is practically the same as the marginal extraction cost.

A second effort to save Hotelling's theory appeals to the role of technological progress, which can lower marginal extraction costs (e.g. Slade, 1982), lead to discovery of new fields (Dasgupta and Heal, 1979; Arrow and Chang, 1982), or allow the exploitation of resources previously thought not to be economically accessible (Pindyck, 1978). In generalizations of the Hotelling formulation, these can give rise to episodes or long periods in which the real price of oil is observed to fall, although eventually the price will begin to rise according to these models. Krautkraemer (1998) has a nice survey of theories of this type and examination of their empirical success at fitting the observed data.

Although it can sometimes be helpful to think about technological progress in broad, abstract terms, there is also much insight to be had from looking in some detail at the specific factors that allowed global oil production to increase almost without interruption over the last 150 years. For this purpose, I begin by examining some of the long-run trends in US oil production.

1.1   Oil Production in the USA, 1859–2010

Certainly the technology for extracting oil from beneath the earth's surface has evolved profoundly over time. Although Drake's original well was steam-powered, some of the early drills were driven through rock by foot power, such as the spring-pole method. The workers would kick a heavy bit at the end of the rope down into the rock, and spring action from the compressed pole would lift the bit back up. After some time at this, the drill would be lifted out and a bucket lowered to bail out the debris. Of course subsequent years produced rapid advances over these first primitive efforts – better sources of power, improved casing technology, and vastly superior knowledge of where oil might be found. Other key innovations included the adoption of rotary drilling at the turn of the century, in which circulating fluid lifted debris out of the hole, and secondary recovery methods first developed in the 1920s, in which water, air or gas is injected into oil wells to repressurize the reservoir and allow more of the oil to be lifted to the surface.

Figure 1.2 plots the annual oil production levels for Pennsylvania and New York, where the industry began, from 1862 to 2010. Production increased by a factor of 10 between 1862 and 1891. However, it is a mistake to view this as the result of application of better technology to the initially exploited fields. Production from the original Oil Creek District in fact peaked in 1874 (Williamson and Daum, 1959, p. 378). The production gains instead came primarily from the development of new fields, most importantly the Bradford field near the Pennsylvania–New York border, but also from Butler, Clarion and Armstrong Counties. Nevertheless, it is unquestionably the case that better drilling techniques than used in Oil Creek were necessary in order to reach the greater depths of the Bradford formation.

One also sees quite clearly in Figure 1.2 the benefits of the secondary recovery methods applied in the 1920s, which succeeded in producing much additional oil from the Bradford formation and elsewhere in the state. However, it is worth noting that these methods never lifted production in Pennsylvania back to where it had been in 1891. In 2010 – with the truly awesome technological advances of the century and a half since the industry began, and with the price of oil five times as high (in real terms) as it had been in 1891 – Pennsylvania and New York produced under 4 million barrels of crude oil. That is only 12 percent of what had been produced in 1891 – 120 years before – and about the level that the sturdy farmers with their spring-poles were getting out of the ground back in 1868.

Although Pennsylvania was the most important source of US oil production in the nineteenth century, the nation's oil production continued to increase even after Pennsylvanian production peaked in 1891. The reason is that, later in the century, new sources of oil were also being obtained from neighboring West Virginia and Ohio (see Figure 1.3). Production from these two states was rising rapidly even as production from Pennsylvania and New York started to fall. Ohio production would continue to rise before peaking in 1896, and West Virginia did not peak until 1900.

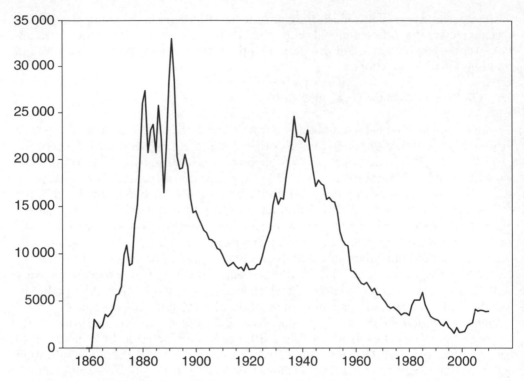

Data sources: see Appendix.

*Figure 1.2   Annual crude oil production (in thousands of barrels per year) from the
states of Pennsylvania and New York combined*

These four states together accounted for 90 percent of US production in 1896, with
the peak in production from the region as a whole coming that year (see Figure 1.4).
Overall US production declined for a few years with falling supplies from Appalachia,
but quickly returned to establishing new highs in 1900, thanks to growth in production
from new areas in the central USA, details of which are shown in Figure 1.5. Note the
difference in scale, with the vertical axes in Figure 1.5 spanning six times the magnitude
of corresponding axes in Figure 1.3. Each of the regions featured in Figure 1.5 would
eventually produce far more oil than Appalachia ever did. These areas began producing
much later than Appalachia, and each peaked much later than Appalachia. The combined production of Illinois and Indiana peaked in 1940, Kansas–Nebraska in 1957, the
southwest in 1960, and Wyoming in 1970.

Far more important for US total production were the four states shown in Figure 1.6,
which uses a vertical scale 2.5 times that for Figure 1.5. California, Oklahoma, Texas and
Louisiana account for 70 percent of all the oil ever produced in the USA. Production
from Oklahoma reached a peak in 1927, though it was still able to produce at 80 percent
of that level as recently as 1970 before entering a modern phase of decline that now leaves
it at 25 percent of the 1927 production levels. Texas managed to grow its oil production
until 1972, and today produces about a third of what it did then. California production

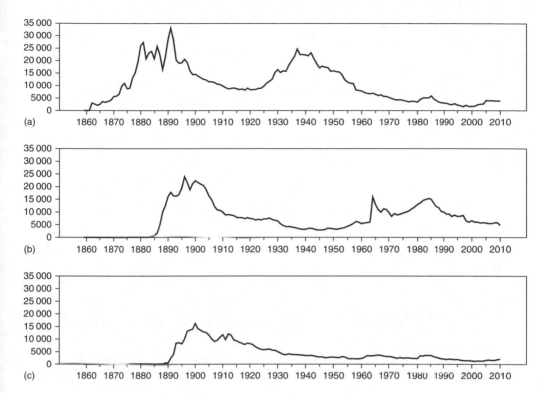

Figure 1.3   Annual crude oil production (in thousands of barrels per year) from the states of Pennsylvania and New York combined (a), Ohio (b), and West Virginia (c)

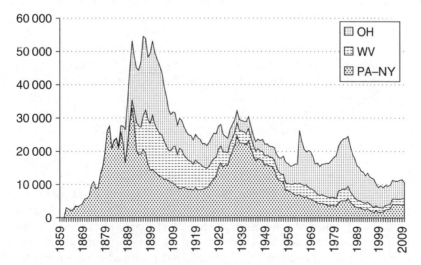

Figure 1.4   Combined annual crude oil production (in thousands of barrels per year) from the states of Pennsylvania, New York, West Virginia and Ohio

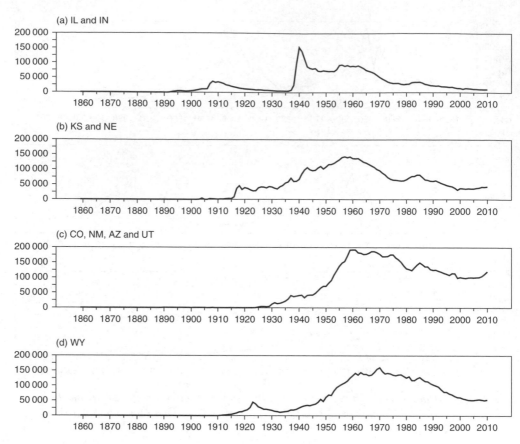

Figure 1.5   Annual crude oil production (in thousands of barrels per year) from assorted groups of states in the central USA

continued to grow until 1985 before peaking. The graph for Louisiana (bottom panel of Figure 1.6) includes all the US production from the Gulf of Mexico, growing production from which helped bring the state's indicated production for 2010 up to a value only 33 percent below its peak in 1971.

Figure 1.7 plots production histories for the two regions whose development began latest in US history. Production from Alaska peaked in 1988. North Dakota is the only state that continues to set all-time records for production, thanks in part to use of new drilling techniques for recovering oil from shale formations. To put the new Williston Basin production in perspective, the 138 million barrels produced in North Dakota and Montana in 2010 is about half of what the state of Oklahoma produced in 1927 and a fifth of what the state of Alaska produced in 1988. However, the potential for these fields looks very promising and further significant increases from 2010 levels seem assured.

The experience for the USA thus admits a quite clear summary. Production from every state has followed a pattern of initial increase followed by eventual decline. The feature that nonetheless allowed the total production for the USA to exhibit a seemingly uninterrupted upward trend over the course of a century was the fact that new, more

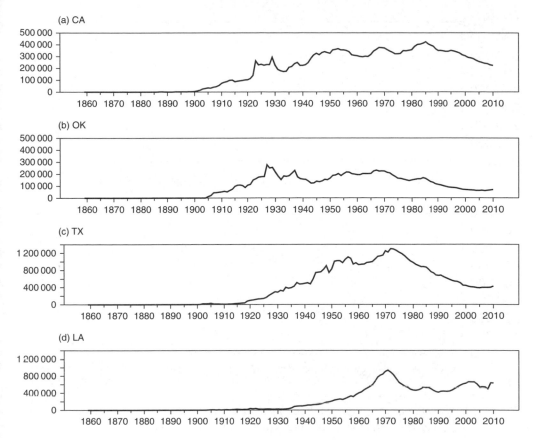

Note: California includes offshore and Louisiana includes all Gulf of Mexico US production.

Figure 1.6   Annual crude oil production (in thousands of barrels per year) from four leading producing states

promising areas were always coming into production at the same time that mature fields were dying out (see Figure 1.8). Total US production continued to grow before peaking in 1970, long after the original fields in Appalachia and the central USA were well into decline.

And the decline in production from both individual regions within the USA as well as the USA as a whole has come despite phenomenal improvements in technology over time. Production from the Gulf of Mexico has made a very important contribution to slowing the rate of decline over the most recent decade. Some of this production today is coming from wells that begin a mile below sea level and bore from there through up to a half-dozen more miles of rock – try doing that with three guys kicking a spring-pole down! The decline in US production has further come despite aggressive drilling in very challenging environments and widespread adoption of secondary and now tertiary recovery methods. The rise and fall of production from individual states seems much more closely related to discoveries of new fields and their eventual depletion than

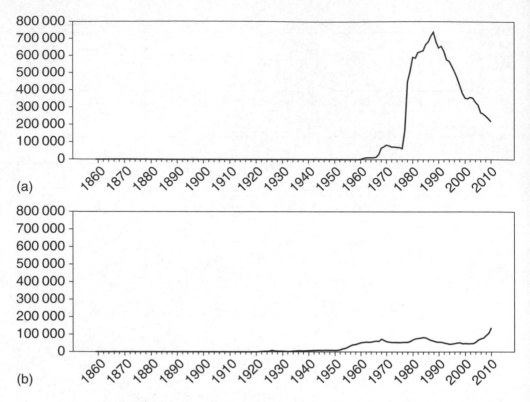

*Figure 1.7   Annual crude oil production (in thousands of barrels per year) from Alaska
(including offshore) (a) and North Dakota and Montana (b)*

to the sorts of price incentives or technological innovations on which economists are accustomed to focus.

Notwithstanding, technological improvements continue to bring significant new fields into play. The most important recent development has been horizontal rather than vertical drilling through hydrocarbon-bearing formations accompanied by injection of fluids to induce small fractures in the rock. These methods have allowed access to hydrocarbons trapped in rock whose permeability or depth prevented removal using traditional methods. The new methods have enabled phenomenal increases in supplies of natural gas as well as significant new oil production in areas such as North Dakota and Texas. Wickstrom et al. (2011) speculated that application of hydraulic fracturing to the Utica Shale formation in Ohio might eventually produce several billion barrels of oil, which would be more than the cumulative production from the state up to this point. If that indeed turns out to be the case, it could lead to a third peak in the graphs in Figure 1.3 for the Appalachian region that exceeds either of the first two, though for comparison the projected lifetime output from Utica would still only correspond to a few years of production from Texas at that state's peak.

Obviously price incentives and technological innovations matter a great deal. More oil will be brought to the surface at a price of $100 per barrel than at $10 per barrel,

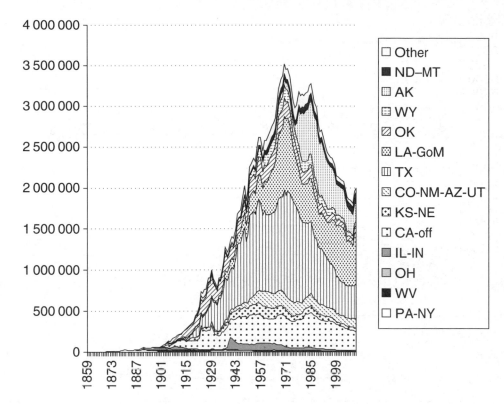

☐	Other
■	ND–MT
▦	AK
▤	WY
▨	OK
▦	LA-GoM
▥	TX
▨	CO-NM-AZ-UT
▣	KS-NE
▢	CA-off
▦	IL-IN
▢	OH
■	WV
☐	PA-NY

Note: Contributions from individual regions as indicated.

Figure 1.8   Annual crude oil production (in thousands of barrels per year) from the entire USA

and more oil can be produced with the new technology than with the old. But it seems a mistake to overstate the operative elasticities. By 1960, the real price of oil had fallen to a level that was one-third of its value in 1900. Over the same period, US production of crude oil grew to become 55 times what it had been in 1900. On the other hand, the real price of oil rose eight-fold from 1970 to 2010, while US production of oil fell by 43 percent over those same 40 years. The increase in production from 1900 to 1960 thus could in no way be attributed to the response to price incentives. Likewise, neither huge price incentives nor impressive technological improvements were sufficient to prevent the decline in production from 1970 to 2010. Further exploitation of offshore or deep shale resources may help put US production back on an upward trend for the next decade, but it seems unlikely ever again to reach the levels seen in 1970.

1.2   World Oil Production, 1973–2010

Despite the peak in US production in 1970, world oil production was to grow to a level in 2010 that is 60 percent higher than it had been in 1970. The mechanics of this growth are the same as allowed total US production to continue to increase long after production

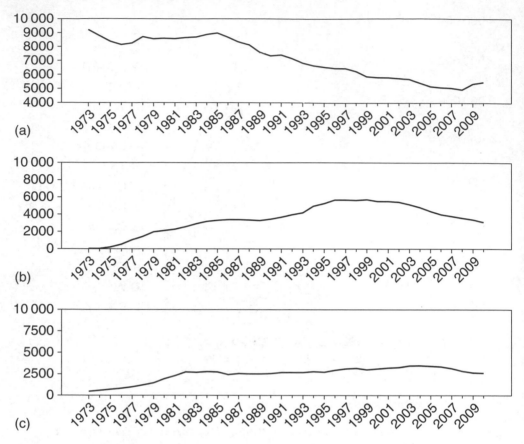

(a)

(b)

(c)

Data source: *Monthly Energy Review*, Sept. 2011, Table 11.1b (http://205.254.135.24/totalenergy/data/monthly/query/mer_data.asp?table=T11.01B).

Figure 1.9   Annual crude oil production, thousand barrels per day, for the USA (a), combined output of Norway and the UK (b) and Mexico (c) 1973–2010

from the initial areas entered into decline – increases from new fields in other countries more than offset the declines from the USA. For example, the North Sea and Mexico accounted for only 1 percent of world production in 1970, but had grown to 13 percent of total world output by 1999. But production from the North Sea peaked in that year, and in 2010 is only at 54 percent of the peak level (see Figure 1.9). Cantarell, which is Mexico's main producing field, also appears to have passed peak production, with the country now at 75 percent of its 2004 oil production.

Production from members of the Organization of Petroleum Exporting Countries (OPEC) must be interpreted from a very different perspective. The episodes of declining production one sees in the bottom panel of Figure 1.10 have little to do with geological depletion but instead often reflect dramatic geopolitical events such as the OPEC embargo of 1973–74, the Iranian Revolution and beginning of the Iran–Iraq War in 1978–81, and the first Persian Gulf War in 1990–91, events that will be reviewed in more

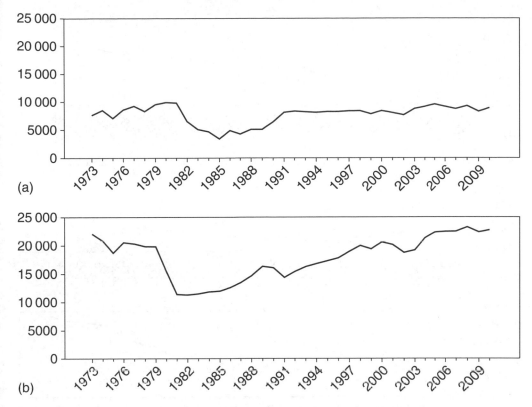

Data source: *Monthly Energy Review*, Sept. 2011, Table 11.1a (http://205.254.135.24/totalenergy/data/monthly/query/mer_data.asp?table=T11.01A).

Figure 1.10 *Annual crude oil production, thousand barrels per day, for Saudi Arabia (a) and the rest of OPEC (b)*

detail in the following section. In addition, Saudi Arabia in particular (top panel) has often made a deliberate decision to increase or reduce production in an effort to mitigate price increases or decreases. For example, Saudi Arabia cut production to try to hold up prices during the weak oil market in 1981–85 and recession of 2001, and boosted production to make up for output lost from other producing countries during the two Persian Gulf Wars. However, the decline in Saudi Arabian production since 2005 would have to be attributed to different considerations from those that explain the earlier historical data. The kingdom's magnificent Ghawar field has been in production since 1951, and in recent years had accounted for perhaps 6 percent of total world production all by itself. There is considerable speculation that Ghawar may have peaked, although this is difficult to confirm. What we do know is that, for whatever reason, Saudi Arabia produced 600 000 fewer barrels each day in 2010 than it did in 2005, and with growing Saudi consumption of their own oil, the drop in exports from Saudi Arabia has been even more dramatic.

A mix of factors has clearly also contributed to stagnating production from other

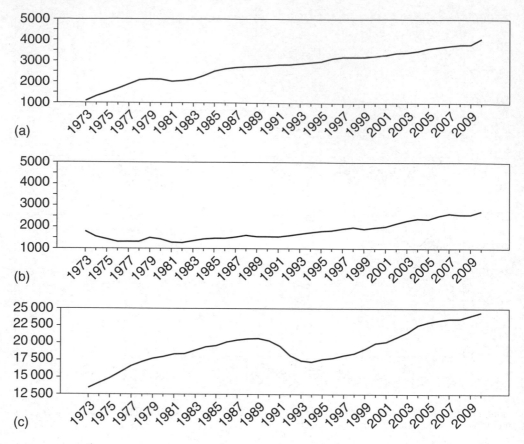

(a)

(b)

(c)

Data source: *Monthly Energy Review*, Sept. 2011, Table 11.1b (http://205.254.135.24/totalenergy/data/monthly/query/mer_data.asp?table=T11.01B).

Figure 1.11   *Annual crude oil production, thousand barrels per day. China (a) and Canada (b). Panel (c) combines all non-OPEC countries other than those in Figure 1.10 or top two panels*

OPEC members over the last five years. Promising new fields in Angola have allowed that country to double its production since 2003. In Nigeria and Iraq, conflicts and unrest have held back what appears to be promising geological potential. In Venezuela and Iran, it is hard to know how much more might be produced with better-functioning governments. But again, although there is a complicated mix of different factors at work in different countries, the bottom line is that the total production from OPEC has essentially been flat since 2005.

At the same time, some other countries continue to register increases in oil production (see Figure 1.11). China has doubled its oil production since 1982, although its three most important fields (Daqing, Shengli and Liaohe) peaked in the mid-1990s (Kambara and Howe, 2007). Canadian oil production continues to increase as a result of the contribution of oil sands. Unfortunately, exploitation of this resource is far more costly in terms

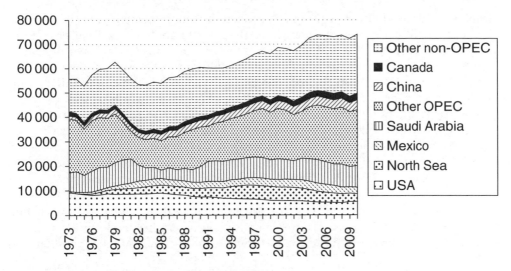

Note: Contributions from individual regions as indicated.

Data sources: As for Figures 1.9–1.11.

Figure 1.12 Annual crude oil production (in thousands of barrels per day) from entire world

of capital and energy inputs and environmental externalities relative to conventional sources, and it is difficult to see it ever accounting for a major fraction of total world oil production. Other regions such as Brazil, Central Asia and Africa have also seen significant gains in oil production (bottom panel of Figure 1.11). Overall, global production of oil from all sources was essentially constant from 2005 to 2010 (see Figure 1.12).

1.3 Reconciling Historical Experience with the Theory of Exhaustible Resources

The evidence from the preceding subsections can be summarized as follows. When one looks at individual oil-producing regions, one does not see a pattern of continuing increases as a result of ongoing technological progress. Instead there has inevitably been an initial gain as key new fields were developed, followed by subsequent decline. Technological progress and the incentives of higher prices can temporarily reverse that decline, as was seen for example in the impressive resurgence of Pennsylvanian production in the 1920s. In recent years these same factors have allowed US production to grow rather than decline, and that trend in the USA may continue for some time. However, these factors have historically appeared to be distinctly secondary to the broad reality that after a certain period of exploitation, annual flow rates of production from a given area are going to start to decline. Those encouraged by the 10 percent increase in US oil production between 2008 and 2010 should remember that the level of US production in 2010 is still 25 percent below where it had been in 1990 (when the real price of oil was half of what it is today) and 43 percent below the level of 1970 (when the real price of oil was one-eighth of what it is today).

Some may argue that the peaking of production from individual areas is governed by quite different economic considerations than would apply to the final peaking of total production from all world sources combined. Certainly, in an environment in which the market is pricing oil as an essentially inexhaustible resource, the pattern of peaking documented extensively above is perfectly understandable, given that so far there have always been enough new fields somewhere in the world to take the place of declining production from mature regions. One could also reason that, even if the price of oil has historically been following some kind of Hotelling path, fields with different marginal extraction costs would logically be developed at different times. Smith (2011) further noted that, according to the Hotelling model, the date at which global production peaks would be determined endogenously by the cumulative amount that could eventually be extracted and the projected time path for the demand function. His analysis suggests that the date for an eventual peak in global oil production should be determined by these economic considerations rather than by the engineering mechanics that have produced the historical record for individual regions detailed above.

However, my reading of the historical evidence is as follows. First, for much of the history of the industry, oil has been priced essentially as if it were an inexhaustible resource. Second, although technological progress and enhanced recovery techniques can temporarily boost production flows from mature fields, it is not reasonable to view these factors as the primary determinants of annual production rates from a given field. Third, the historical source of increasing global oil production is exploitation of new geographical areas, a process whose promise at the global level is obviously limited. The combined implication of these three observations is that at some point there will need to be a shift in how the price of oil is determined, with considerations of resource exhaustion playing a bigger role than they have historically.

A factor accelerating the date of that transition is the phenomenal growth of demand for oil from the emerging economies. Eight emerging economies – Brazil, China, Hong Kong, India, Singapore, South Korea, Taiwan and Thailand – accounted for 43 percent of the increase in world petroleum consumption between 1998 and 2005 and for 135 percent of the increase between 2005 and 2010 (the rest of the world decreased its petroleum consumption over the latter period in response to the big increase in price).[3] And, as Hamilton (2009a) noted, one could easily imagine the growth in demand from the emerging economies continuing. One has only to compare China's one passenger vehicle per 30 residents today with the one vehicle per 1.3 residents seen in the USA, or China's 2010 annual petroleum consumption of 2.5 barrels per person with Mexico's 6.7 or the USA's 22.4. Even if the world sees phenomenal success in finding new sources of oil over the next decade, it could prove quite challenging to keep up with both depletion from mature fields and rapid growth in demand from the emerging economies, another reason to conclude that the era in which petroleum is regarded as an essentially unlimited resource has now ended.

Some might infer that the decrease in Saudi Arabian production since 2005 reflects not an inability to maintain production flows from the mature Ghawar field but instead is a deliberate response to recognition of a growing importance of the scarcity rent. For example, Hamilton (2009a) noted the following story on 13 April 2008 from Reuters news service:

Saudi Arabia's King Abdullah said he had ordered some new oil discoveries left untapped to preserve oil wealth in the world's top exporter for future generations, the official Saudi Press Agency (SPA) reported.

'I keep no secret from you that when there were some new finds, I told them, "no, leave it in the ground, with grace from God, our children need it",' King Abdullah said in remarks made late on Saturday, SPA said.

If that is indeed the interpretation, it is curious that we would see the private optimizing choices predicted by Hotelling manifest by sovereign governments rather than the fields under control of private oil companies. In any case, it must be acknowledged that calculation of the correct Hotelling price is almost insurmountably difficult. It is hard enough for the best forecasters accurately to predict supply and demand for the coming year. But the critical calculation required by Hotelling is to evaluate the transversality condition that the resource be exhausted when the price reaches that of a backstop technology or alternatively over the infinite time horizon if no such backstop exists. That calculation is orders of magnitude more difficult than the seemingly simpler task of just predicting next year's supply and demand.

One could argue that the combined decisions of the many participants in world oil markets can make a better determination of what the answer to the above calculation should be than can any individual, meaning that if the current price seems inconsistent with a scenario in which global oil production will soon reach a peak, then such a scenario is perhaps not the most likely outcome. But saying that the implicit judgment from the market is the best guess available is not the same thing as saying that this guess is going to prove to be correct. The historical record surely dictates that we take seriously the possibility that the world could soon reach a point from which a continuous decline in the annual flow rate of production could not be avoided, and inquire whether the transition to a pricing path consistent with that reality could prove to be a fairly jarring event. For this reason, it seems worthwhile to review the historical record on the economic response to previous episodes in which the price or supply of oil changed dramatically, to which we now turn in the next section.

2 OIL PRICES AND ECONOMIC GROWTH

2.1 Historical Oil Price Shocks

There have been a number of episodes over the last half-century in which conflicts in the Middle East have led to significant disruptions in production of crude oil. These include closure of the Suez Canal following the conflict between Egypt, Israel, Britain and France in October 1956, the oil embargo implemented by the Arab members of OPEC following the Arab–Israeli War in October 1973, the Iranian Revolution beginning in November 1978, the Iran–Iraq War beginning in September of 1980, and the first Persian Gulf War beginning in August 1990. Figure 1.13 summarizes the consequences of these five events for world oil supplies. In each panel, the solid line displays the drop in production from the affected areas expressed as a percentage of total world production prior to the crisis. In each episode, there were some offsetting increases in production elsewhere in the world. The dashed lines in Figure 1.13 indicate the magnitude

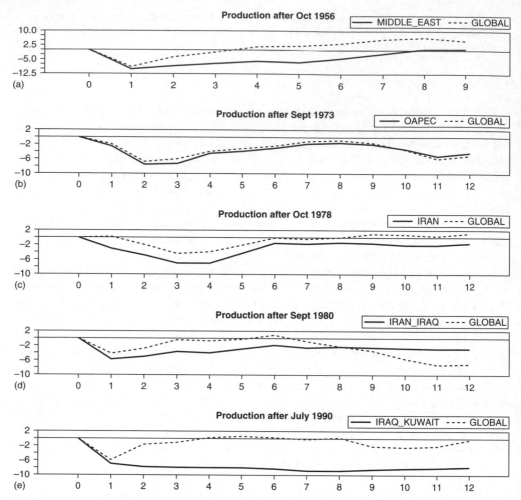

Notes: (a) After the Suez Crisis. Dashed line: change in monthly global crude oil production from October 1956 as a percentage of October 1956 levels. Solid line: change in monthly Middle East oil production from October 1956 as a percentage of global levels in October 1956. (b) After the 1973 Arab–Israeli War. Dashed line: change in monthly global crude oil production from September 1973 as a percentage of September 1973 levels. Solid line: change in monthly oil production of Arab members of OPEC from September 1973 as a percentage of global levels in September 1973. Horizontal axis: number of months from September 1973. (c) After the 1978 Iranian Revolution. Dashed line: change in monthly global crude oil production from October 1978 as a percentage of October 1978 levels. Solid line: change in monthly Iranian oil production from October 1978 as a percentage of global levels in October 1978. (d) After the Iran–Iraq War. Dashed line: change in monthly global crude oil production from September 1980 as a percentage of September 1980 levels. Solid line: change in monthly oil production of Iran and Iraq from September 1980 as a percentage of global levels in September 1980. (e) After the first Persian Gulf War. Dashed line: change in monthly global crude oil production from August 1990 as a percentage of August 1990 levels. Solid line: change in monthly oil production of Iraq and Kuwait from August 1990 as a percentage of global levels in August 1990. Horizontal axis: number of months from August 1990.

Source: Adapted from Figures 6, 10, 12, 13 and 15 in Hamilton (forthcoming).

Figure 1.13   Oil production after five world events

the actual decline in total global production following each event, again expressed as a fraction of world production. Each of these five episodes was followed by a decrease in world oil production of 4–9 percent.

There have also been some other more minor supply disruptions over this period. These include the combined effects of the second Persian Gulf War and strikes in Venezuela beginning in December 2002, and the Libyan Revolution in February 2011. The disruption in supply associated with either of these episodes was about 2 percent of total global production at the time, or less than a third the size of the average event in Figure 1.13.

There are other episodes since World War II when the price of oil rose abruptly in the absence of a significant physical disruption in the supply of oil. Most notable of these would be the broad upswing in the price of oil beginning in 2004, which accelerated sharply in 2007. The principal cause of this oil spike appears to have been strong demand for oil from the emerging economies confronting the stagnating global production levels documented in the previous sections (see Kilian, 2008, 2009; Hamilton, 2009b; Kilian and Hicks, 2011). Less dramatic price increases followed the economic recovery from the East Asian crisis in 1997, dislocations associated with post-World War II growth in 1947, and the Korean conflict in 1952–53. Table 1.1 summarizes a series of historical episodes discussed in Hamilton (forthcoming). It is interesting that of the 11 episodes listed, ten were followed by a recession in the USA. The recession of 1960 is the only US postwar recession that was not preceded by a spike in the price of crude oil.

A large empirical literature has investigated the connection between oil prices and real economic growth. Early studies documenting a statistically significant negative correlation include Rasche and Tatom (1977, 1981) and Santini (1985). Empirical analysis of dynamic forecasting regressions found that oil price changes could help improve forecasts of US real output growth (Hamilton, 1983; Burbidge and Harrison, 1984; Gisser and Goodwin, 1986). However, these specifications, which were based on linear relations between the log change in oil prices and the log of real output growth, broke down when the dramatic oil price decreases of the mid-1980s were not followed by an economic boom. On the contrary, the mid-1980s appeared to be associated with recession conditions in the oil-producing states (Hamilton and Owyang, forthcoming). Mork (1989) found a much better fit to a model that allowed for oil price decreases to have a different effect on the economy from oil price increases, although Hooker (1996) demonstrated that this modification still had trouble describing subsequent data. Other papers finding a significant connection between oil price increases and poor economic performance include Santini (1992, 1994), Rotemberg and Woodford (1996), Daniel (1997) and Carruth et al. (1998).

Alternative nonlinear dynamic relations seem to have a significantly better fit to US data than Mork's simple asymmetric formulation. Loungani (1986) and Davis (1987a, 1987b) found that oil price decreases could actually reduce economic growth, consistent with the claim that sectorial reallocations could be an important part of the economic transmission mechanism resulting from changes in oil prices in either direction. Ferderer (1996), Elder and Serletis (2010), and Jo (2011) showed that an increase in oil price volatility itself tends to predict slower GDP growth, while Lee et al. (1995) found that oil price increases seem to affect the economy less if they occur following an episode of high volatility. Hamilton (2003) estimated a flexible nonlinear form and found evidence for

Table 1.1   Summary of significant postwar events

Gasoline shortages	Crude oil price increase	Crude oil or gasoline price controls	Key factors	Business cycle peak
Nov 47–Dec 47	Nov 47–Jan 48 (37%)	No (threatened)	Strong demand, supply constraints	Nov 48
May 52	Jun 53 (10%)	Yes	Strike, controls lifted	Jul 53
Nov 56–Dec 56 (Europe)	Jan 57–Feb 57 (9%)	Yes (Europe)	Suez Crisis	Aug 57
None	None	No	–	Apr 60
None	Feb 69 (7%) Nov 70 (8%)	No	Strike, strong demand, supply constraints	Dec 69
Jun 73 Dec 73–Mar 74	Apr 73–Sep 73 (16%) Nov 73–Feb 74 (51%)	Yes	Strong demand, supply constraints, OAPEC embargo	Nov 73
May 79–Jul 79	May 79–Jan 80 (57%)	Yes	Iranian Revolution	Jan 80
None	Nov 80–Feb 81 (45%)	Yes	Iran–Iraq War, controls lifted	Jul 81
None	Aug 90–Oct 90 (93%)	No	Gulf War I	Jul 90
None	Dec 99–Nov 00 (38%)	No	Strong demand	Mar 01
None	Nov 02–Mar 03 (28%)	No	Venezuela unrest, Gulf War II	None
None	Feb 07–Jun 08 (145%)	No	Strong demand, stagnant supply	Dec 07

Source: Hamilton (forthcoming).

a threshold effect, in which an oil price increase that simply reverses a previous decrease seems to have little effect on the economy. Hamilton (1996), Raymond and Rich (1997), Davis and Haltiwanger (2001) and Balke et al. (2002) produced evidence in support of related specifications, while Carlton (2010) and Ravazzolo and Rothman (2010) reported that the Hamilton (2003) specification performed well in an out-of-sample forecasting exercise using data as they would have been available in real time. Kilian and Vigfusson (2011) found weaker (though still statistically significant) evidence of nonlinearity than reported by other researchers. Hamilton (2011) attributed their weaker evidence to use of a shorter data set and changes in specification from other researchers.

A negative effect of oil prices on real output has also been reported for a number of other countries, particularly when nonlinear functional forms have been employed. Mork et al. (1994) found that oil price increases were followed by reductions in real GDP growth in six of the seven OECD countries investigated, the one exception being the oil exporter Norway. Cuñado and Pérez de Gracia (2003) found a negative correlation between oil price changes and industrial production growth rates in 13 out of 14 European economies, with a nonlinear function of oil prices making a statistically significant contribution to forecast growth rates for 11 of these. Jiménez-Rodríguez and Sánchez (2005) found a statistically significant negative nonlinear relation between oil prices and real GDP growth in the USA, Canada, the euro area overall, and five out of six European countries, though not in Norway or Japan. Kim (2012) found a nonlinear relation in a panel of six countries, while Engemann et al. (2011) found that oil prices helped predict economic recessions in most of the countries they investigated. Daniel et al. (2011) also found supporting evidence in most of the 11 countries they studied. By contrast, Rasmussen and Roitman (2011) found much less evidence for economic effects of oil shocks in an analysis of 144 countries. However, their use of this larger sample of countries required using annual rather than the monthly or quarterly data used in the other research cited above. In so far as the effects are high frequency and cyclical, they may be less apparent in annual average data. Kilian (2009) has argued that the source of the oil price increase is also important, with increases that result from strong global demand appearing to have more benign implications for US real GDP growth than oil price increases that result from shortages of supply.

Blanchard and Galí (2010) found evidence that the effects of oil shocks on the economy have decreased over time, which they attributed to the absence of other adverse shocks that had historically coincided with some big oil price movements, a falling value of the share of oil in total expenses, more flexible labor markets, and better management of monetary policy. Baumeister and Peersman (2011) also found that an oil price increase of a given size seems to have a decreasing effect over time, but noted that the declining price elasticity of demand meant that a given physical disruption had a bigger effect on price and turned out to have a similar effect on output as in the earlier data. Ramey and Vine (2012) attributed the declining coefficients relating real GDP growth to oil prices to the fact that the oil shocks of the 1970s were accompanied by rationing, which would have magnified the economic dislocations. Ramey and Vine found that, once they correct for this, the economic effects have been fairly stable over time.

2.2 Interpreting the Historical Evidence

Equation (1.1) reports the regression estimates from equation (3.8) of Hamilton (2003), which is based on data from 1949:Q2 to 2001:Q3. Here y_t represents the quarterly log change in real GDP. The specification implies that oil prices do not matter unless they make a new high relative to values seen over the previous three years. If oil prices make a new high, $o_t^{\#}$ is the amount by which the log of the producer price index at the end of quarter t exceeds its maximum over the preceding three years, whereas $o_t^{\#}$ is zero if they do not. Standard errors appear in parentheses, and both y_t and $o_t^{\#}$ have been multiplied by 100 to express as percentage rates:

$$y_t = \underset{(0.13)}{0.98} + \underset{(0.07)}{0.22y_{t-1}} + \underset{(0.07)}{0.10y_{t-2}} - \underset{(0.07)}{0.08y_{t-3}} - \underset{(0.07)}{0.15y_{t-4}}$$

$$- \underset{(0.014)}{0.024o_{t-1}^{\#}} - \underset{(0.014)}{0.021o_{t-2}^{\#}} - \underset{(0.014)}{0.018o_{t-3}^{\#}} - \underset{(0.014)}{0.042o_{t-4}^{\#}} \tag{1.1}$$

Two aspects of this relation are puzzling from the perspective of economic theory. First, the effects of an oil price increase take some time to show up in real GDP, with the biggest drop in GDP growth appearing a full year after the price of oil first increases. Second, the size of the estimated effect is quite large. If the price of oil exceeds its three-year high by 10 percent, the relation predicts that real GDP growth will be 0.42 percent slower (at a quarterly rate) four quarters later, with a modest additional decline coming from the dynamic implications of $o_{t-4}^{\#}$ for y_{t-1}, y_{t-2} and y_{t-3}.

To understand why effects of this magnitude are puzzling,[4] suppose we thought of the level of real GDP (Y) as depending on capital K, labor N and energy E according to the production function,

$$Y = F(K, N, E)$$

Profit maximization suggests that the marginal product of energy should equal its relative price, denoted P_E/P:

$$\frac{\partial F}{\partial E} = P_E/P$$

Multiplying the above equation by E/F implies that the elasticity of output with respect to energy use should be given by γ, the dollar value of expenditures on energy as a fraction of GDP:

$$\frac{\partial \ln F}{\partial \ln E} = \frac{P_E E}{P Y} = \gamma \tag{1.2}$$

Suppose we believe that wages adjust instantaneously to maintain full employment and that changes in investment take much longer than a few quarters to make a significant difference to the capital stock. Then neither K nor N would respond to a change in the real price of energy, and

$$\frac{\partial \ln Y}{\partial \ln P_E/P} = \frac{\partial \ln F}{\partial \ln E} \frac{\partial \ln E}{\partial \ln P_E/P}$$

$$= \gamma \theta \tag{1.3}$$

for θ the price elasticity of energy demand.

The energy expenditure share is a small number – the value of crude oil consumed by the USA in 2010 corresponds to less than 4 percent of total GDP. Moreover, the short-run price elasticity of demand θ is also very small (Dahl, 1993). Hence it seems that any significant observed response to historical oil price increases could not be attributed to the direct effects of decreased energy use on productivity but instead would have to arise from forces that lead to underemployment of labor and underutilization of capital. Such effects are likely to operate from changes in the composition of demand rather than the physical process of production itself.[5] Unlike the above mechanism based on aggregate supply effects, the demand effects could be most significant when the price elasticity of demand is low.

For example, suppose that the demand for energy is completely inelastic in the short run, so that consumers try to purchase the same physical quantity E of energy despite the energy price increase. Then nominal saving or spending on other goods or services must decline by $E\Delta P_E$ when the price of energy goes up. Letting C denote real consumption spending and P_C the price of consumption goods,

$$\frac{\partial \ln C}{\partial \ln (P_E/P_C)} = \frac{P_E E}{P_C C} = \gamma_C$$

for γ_C the energy expenditure share in total consumption. Again, for the aggregate economy γ_C is a modest number. Currently about 6 percent of total US consumer spending is devoted to energy goods and services,[6] although for the lower 60 percent of US households by income, the share is closer to 10 percent (Carroll, 2011). And although the increased spending on energy represents income for someone else, it can take a considerable amount of time for oil company profits to be translated into higher dividends for shareholders or increased investment expenditures. Recycling the receipts of oil-exporting countries on increased spending on US-produced goods and services can take even longer. These delays may be quite important in determining the overall level of spending that governs short-run business cycle dynamics.

Edelstein and Kilian (2009) conducted an extensive investigation of US monthly spending patterns over 1970 to 2006, looking at bivariate autoregressions of measures of consumption spending on their own lags and on lags of energy prices. They scaled the energy price measure so that a one-unit increase would correspond to a 1 percent drop in total consumption spending if consumers were to try to maintain real energy purchases at their original levels. Figure 1.14 reproduces some of their key results. The top panel shows that, as expected, an increase in energy prices is followed by a decrease in overall real consumption spending. However, the same two puzzles mentioned in connection with (1.1) occur again here. First, although consumers' spending power first fell at date 0 on the graph, the decline in consumption spending is not immediate but continues to increase in size up to a year after the initial shock. Second, although the initial shock corresponded to an event that might have forced a consumer to cut total spending by 1 percent, after 12 months we see total spending down 2.2 percent.

The details of Edelstein and Kilian's other analysis suggest some explanations for both the dynamics and the apparent multiplier effects. The second panel in Figure 1.14 looks at one particular component of consumption spending, namely spending on motor vehicles and parts. Here the decline is essentially immediate, and quite large relative to normal expenditures on this particular category. The drop in demand for domestically manufactured motor vehicles could lead to idled capital and labor as a result of traditional Keynesian frictions in adjusting wages and prices, and could be an explanation for both the multiplier and the dynamics observed in the data. Hamilton (1988) showed that multiplier effects could also arise in a strictly neoclassical model with perfectly flexible wages and prices. In that model, the technological costs associated with trying to reallocate specialized labor or capital could result in a temporary period of unemployment as laid-off workers wait for demand for their sector to resume. Bresnahan and Ramey (1993), Hamilton (2009b) and Ramey and Vine (2012) demonstrated the economic importance of shifts in motor vehicle demand in the recessions that followed several historical oil shocks.

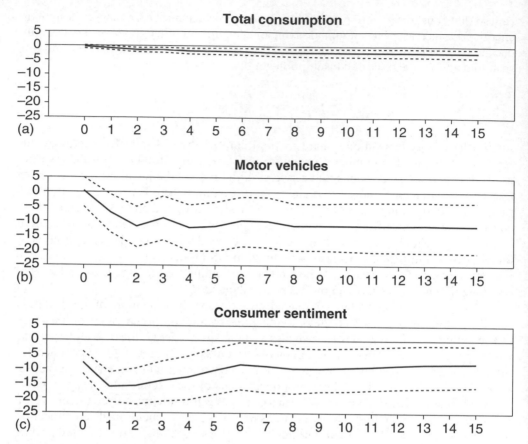

Notes: (a) Impulse–response function showing percentage change in total real consumption spending *k* months following an energy price increase that would have reduced spending power by 1%. (b) Percentage change in real spending on motor vehicles. (c) Change in consumer sentiment (measured in percentage points). Dashed lines indicate 95% confidence intervals.

Source: Adapted from Edelstein and Kilian (2009) and Hamilton (2009b).

Figure 1.14   US monthly spending patterns, 1970–2006

Another feature of the consumer response to an energy price increase uncovered by Edelstein and Kilian is a sharp and immediate drop in consumer sentiment (see the bottom panel of Figure 1.14). Again, this could produce changes in spending patterns whose consequences accumulate over time through Keynesian and other multiplier effects.

Bohi (1989) was among the early doubters of the thesis that oil prices were an important contributing factor in postwar recessions, noting that the industries in which one sees the biggest response were not those for which energy represented the biggest component of total costs. However, subsequent analyses allowing for nonlinearities found effects for industries for which energy costs were important both for their own production as well as for the demand for their goods (Lee and Ni, 2002; Herrera et al., 2011).

Bernanke et al. (1997) suggested that another mechanism by which oil price increases might have affected aggregate demand is through a contractionary response of monetary policy. They presented simulations suggesting that, if the Federal Reserve had kept interest rates from rising subsequent to historical oil shocks, most of the output decline could have been avoided. However, Hamilton and Herrera (2004) demonstrated that this conclusion resulted from the authors' assumption that the effects of oil price shocks could be captured by seven monthly lags of oil prices, a specification that left out the biggest effects found by earlier researchers. When the Bernanke et al. analysis is reproduced using 12 lags instead of seven, the conclusion from their exercise would be that even quite extraordinarily expansionary monetary policy could not have eliminated the contractionary effects of an oil price shock.

Hamilton (2009b) noted that what happened in the early stages of the 2007–09 recession was quite consistent with the pattern observed in the recessions that followed earlier oil shocks. Spending on the larger domestically manufactured light vehicles plunged even as sales of smaller imported cars went up. Had it not been for the lost production from the domestic auto sector, US real GDP would have grown 1.2 percent during the first year of the recession. Historical regressions based on energy prices would have predicted much of the falling consumer sentiment and slower consumer spending during the first year of the downturn. Figure 1.15 updates and extends a calculation from Hamilton (2009b), in which the specific parameter values from the historically estimated regression (1.1) were used in a dynamic simulation to predict what would have happened to real GDP over the period 2007:Q4 to 2009:Q3 based solely on the changes in oil prices. The pattern and much of the magnitude of the initial downturn are consistent with the historical experience.

Of course, there is no question that the financial crisis in the fall of 2008 was a much more significant event in turning what had been a modest slowdown up to that point into what is now being referred to as the Great Recession. Even so, Hamilton (2009b) noted that the magnitude of the problems with mortgage delinquencies could only have been aggravated by the weaker economy, and suggested that the oil price spike of 2007–08 should be counted as an important factor contributing to the early stages of that recession as well as a number of earlier episodes.

2.3 Implications for Future Economic Growth and Climate Change

The increases in world petroleum production over the first 150 years of the industry have been quite impressive. But given the details behind that growth, it would be prudent to acknowledge the possibility that world production could soon peak or enter a period of rocky plateau. If we should enter such an era, what does the observed economic response to past historical oil supply disruptions and price increases suggest could be in store for the economy?

The above analysis suggests that historically the biggest economic effects have come from cyclical factors that led to underutilization of labor and capital and drove output below the level associated with full employment. If we are asking about the character of an alternative long-run growth path, most economists would be more comfortable assuming that the economy would operate close to potential along the adjustment path. For purposes of that question, the relatively small value for the energy expenditure share

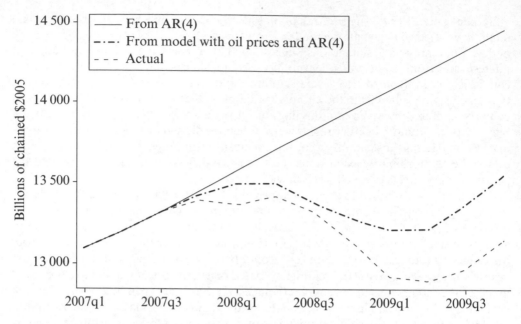

Figure 1.15   *Dynamic simulations of real GDP*

Notes: Dashed line: actual value for real GDP. Solid line: dynamic forecast (1 to 5 quarters ahead) based on coefficients of univariate AR(4) estimated 1949:Q2 to 2001:Q3 and applied to GDP data through 2007:Q3. Dotted line: dynamic conditional forecast (1 to 5 quarters ahead) based on coefficients reported in equation (3.8) in Hamilton (2003) using GDP data through 2007:Q3 and conditioning on the *ex post* realizations of the net oil price increase measure $o^{\#}_{t+s}$ for $t + s = 2007$:Q4 through 2009:Q3.

Source: Foote and Little (2011).

γ in equation (1.2) would seem to suggest a modest elasticity of total output with respect to energy use and relatively minor effects.

One detail worth noting, however, is that historically the energy share has changed dramatically over time. Figure 1.16 plots the consumption expenditure share γ_C since 1959. Precisely because demand is very price-inelastic in the short run, when the real price of oil doubles, the share nearly does as well. The relatively low share in the late 1990s and early 2000s, to which Blanchard and Galí (2010) attributed part of the apparent reduced sensitivity of the economy to oil shocks, basically disappeared with the subsequent price increases. If a peaking of global production does result in further big increases in the price of oil, it is quite possible that the expenditure share will increase significantly from where it is now, in which case even a frictionless neoclassical model would conclude that the economic consequences of reduced energy use would have to be significant.

In addition to the response of supply to these price increases discussed in Section 1, another key parameter is the long-run price elasticity of demand. Here one might take comfort from the observation that, given time, the adjustments of demand to the oil price increases of the 1970s were significant. For example, US petroleum consumption declined 17 percent between 1978 and 1985 at the same time that US real GDP increased

Note: Calculated as 100 times nominal monthly consumption expenditures on energy goods and services divided by total personal consumption expenditures, 1959:M1 to 2011:M8. Horizontal line drawn at 6%.

Data source: BEA Table 2.3.5U (http://www.bea.gov/national/nipaweb/nipa_underlying/SelectTable.asp).

Figure 1.16 Energy expenditures as a fraction of total US consumption spending

by 21 percent. However, Dargay and Gately (2010) attributed much of this conservation to one-time effects, such as switching away from using oil for electricity generation and space heating, which would be difficult to repeat on an ongoing basis. Knittel (2011) was more optimistic, noting that there has been ongoing technological improvement in engine and automobile design over time, with most of this historically being devoted to making cars larger and more powerful rather than more fuel-efficient. If the latter were to become everyone's priority, significant reductions in oil consumption might come from this source.

Knowing what the future will bring in terms of adaptation of both the supply and demand for petroleum is inherently difficult. However, it is not nearly as hard to summarize the past. Coping with a final peak in world oil production could look pretty similar to what we observed as the economy adapted to the production plateau encountered over 2005–09. That experience appeared to have much in common with previous historical episodes that resulted from temporary geopolitical conflict, being associated with significant declines in employment and output. If the future decades look like the last five years, we are in for a rough time.

Most economists view the economic growth of the last century and a half as being fueled by ongoing technological progress. Without question, that progress has been most impressive. But there may also have been an important component of luck in terms of finding and exploiting a resource that was extremely valuable and useful but ultimately finite and exhaustible. It is not clear how easy it will be to adapt to the end of that era of good fortune.

Let me close with a few observations on the implications for climate change. Clearly reduced consumption of petroleum by itself would mean lower greenhouse gas emissions. Moreover, since GDP growth has historically been the single biggest factor influencing the growth of emissions (Hamilton and Turton, 2002), the prospects for potentially rocky economic growth explored above would be another factor slowing the growth of emissions. But the key question in terms of climate impact is what we might do instead, since many of the alternative sources of transportation fuel have a significantly bigger carbon footprint than those we relied on in the past. For example, creating a barrel of synthetic crude from surface-mined Canadian oil sands may emit twice as much carbon dioxide equivalents as are associated with producing a barrel of conventional crude, while *in-situ* processing of oil sands could produce three times as much (Charpentier et al., 2009). This is not quite as alarming as it sounds, since greenhouse gas emissions associated with production of the crude itself are still dwarfed by those released when the gasoline is combusted in the end-use vehicle. The median study surveyed by Charpentier et al. (2009) concluded that, on a well-to-wheel basis, vehicles driven by gasoline produced from surface-mined oil sands would emit 17 percent more grams of carbon dioxide equivalent per kilometer driven compared to gasoline from conventional petroleum. Enhanced oil recovery and conversion of natural gas to liquid fuels are also associated with higher greenhouse gas emissions per kilometer driven than conventional petroleum, although these increases are more modest than those for oil sands. On the other hand, creating liquid fuels from coal or oil shale could increase well-to-wheel emissions by up to a factor of two (Brandt and Farrell, 2007).

In any case, if the question is whether the world should reduce combustion of gasoline produced from conventional petroleum sources, we may not have any choice.

NOTES

*   I thank Roger Fouquet and Lutz Kilian for helpful comments on an earlier draft.
1. See Fouquet and Pearson (2006, 2012) on the history of the cost of illumination.
2. See for example Williamson and Daum (1959, pp. 44–8).
3. Data source: total petroleum consumption, EIA (http://www.eia.gov/cfapps/ipdbproject/iedindex3.cfm?tid=5&pid=5&aid=2&cid=regions&syid=1980&eyid=2010&unit=TBPD).
4. The discussion in this paragraph is adapted from Hamilton (2011).
5. Other neoclassical models explore the possibility of asymmetric or multiplier effects arising through utilization of capital (Finn, 2000) or putty-clay capital (Atkeson and Kehoe, 1999). Related general equilibrium investigations include Kim and Loungani (1992) and Leduc and Sill (2004).
6. See BEA Table 2.3.5.u (http://www.bea.gov/national/nipaweb/nipa_underlying/SelectTable.asp).

REFERENCES

Arrow, Kenneth J. and Sheldon Chang (1982), 'Optimal pricing, use and exploration of uncertain natural resource stocks', *Journal of Environmental Economics and Management*, **9**, 1–10.

Atkeson, Andrew and Patrick J. Kehoe (1999), 'Models of energy use: putty-putty versus putty-clay', *American Economic Review*, **89**, 1028–43.

Balke, Nathan S., Stephen P.A. Brown and Mine Yücel (2002), 'Oil price shocks and the U.S. economy: where does the asymmetry originate?', *Energy Journal*, **23**, 27–52.

Baumeister, Christiane and Gert Peersman (2011), 'Time-varying effects of oil supply shocks on the U.S. economy', Working Paper, Bank of Canada.

Bernanke, Ben S., Mark Gertler and Mark Watson (1997), 'Systematic monetary policy and the effects of oil price shocks', *Brookings Papers on Economic Activity*, **1**, 91–142.

Blanchard, Olivier J. and Jordi Galí (2010), 'The macroeconomic effects of oil price shocks: why are the 2000s so different from the 1970s?', in Jordi Galí and Mark Gertler (eds), *International Dimensions of Monetary Policy*, Chicago, IL: University of Chicago Press, pp. 373–428.

Bohi, Douglas R. (1989), *Energy Price Shocks and Macroeconomic Performance*, Washington, DC: Resources for the Future.

Brandt, Adam R. and Alexander F. Farrell (2007), 'Scraping the bottom of the barrel: greenhouse gas emission consequences of a transition to low-quality and synthetic petroleum resources', *Climatic Change*, **84**, 241–63.

Bresnahan, Timothy F. and Valerie A. Ramey (1993), 'Segment shifts and capacity utilization in the U.S. automobile industry', *American Economic Review Papers and Proceedings*, **83**, 213–18.

Burbidge, John and Alan Harrison (1984), 'Testing for the effects of oil-price rises using vector autoregressions', *International Economic Review*, **25**, 459–84.

Carlton, Amelie Benear (2010), 'Oil prices and real-time output growth', Working Paper, University of Houston, TX.

Carroll, Daniel (2011), 'The cost of food and energy across consumers', *Economic Trends*, 14 March, Federal Reserve Bank of Cleveland (http://www.clevelandfed.org/research/trends/2011/0411/01houcon.cfm).

Carruth, Alan A., Mark A. Hooker and Andrew J. Oswald (1998), 'Unemployment equilibria and input prices: theory and evidence from the United States', *Review of Economics and Statistics*, **80**, 621–28.

Charpentier, Alex D., Joule A. Bergerson and Heather L. MacLean (2009), 'Understanding the Canadian oil sands industry's greenhouse gas emissions', *Environmental Research Letters*, **4**, 1–11.

Cuñado, Juncal and Fernando Pérez de Gracia (2003), 'Do oil price shocks matter? Evidence for some European countries', *Energy Economics*, **25**, 137–54.

Dahl, Carol A. (1993), 'A survey of oil demand elasticities for developing countries', *OPEC Review*, **17** (Winter), 399–419.

Daniel, Betty C. (1997), 'International interdependence of national growth rates: a structural trends analysis', *Journal of Monetary Economics*, **40**, 73–96.

Daniel, Betty C., Christian M. Hafner, Hans Manner and Léopold Simar (2011), 'Asymmetries in business cycles: the role of oil production', Working Paper, University of Albany, NY.

Dargay, Joyce M. and Dermot Gately (2010), 'World oil demand's shift toward faster growing and less price-responsive products and regions', *Energy Policy*, **38**, 6261–77.

Dasgupta, Partha and Geoffrey Heal (1979), *Economic Theory and Exhaustible Resources*, Cambridge: Cambridge University Press.

Davis, Steven J. (1987a), 'Fluctuations in the pace of labor reallocation', in Karl Brunner and Allan H. Meltzer (eds), *Empirical Studies of Velocity, Real Exchange Rates, Unemployment and Productivity*, Carnegie-Rochester Conference Series on Public Policy, 24, Amsterdam: North Holland.

Davis, Steven J. (1987b), 'Allocative disturbances and specific capital in real business cycle theories', *American Economic Review Papers and Proceedings*, **77**, 326–32.

Davis, Steven J. and John Haltiwanger (2001), 'Sectoral job creation and destruction responses to oil price changes', *Journal of Monetary Economics*, **48**, 465–512.

Derrick's Hand-Book of Petroleum: A Complete Chronological and Statistical Review of Petroleum Developments from 1859 to 1898 (1898), Oil City, PA: Derrick Publishing Company, obtained through Google Books.

Edelstein, Paul and Lutz Kilian (2009), 'How sensitive are consumer expenditures to Retail Energy Prices?', *Journal of Monetary Economics*, **56**, 766–79.

Elder, John and Apostolos Serletis (2010), 'Oil price uncertainty', *Journal of Money, Credit and Banking*, **42**, 1138–59.

Engemann, Kristie M., Kevin L. Kliesen and Michael T. Owyang (2011), 'Do oil shocks drive business cycles? Some U.S. and international evidence', *Macroeconomic Dynamics*, **15**, 498–517.

Ferderer, J. Peter (1996), 'Oil price volatility and the macroeconomy: a solution to the asymmetry puzzle', *Journal of Macroeconomics*, **18**, 1–16.

Finn, Mary G. (2000), 'Perfect competition and the effects of energy price increases on economic activity', *Journal of Money, Credit and Banking*, **32**, 400–16.

Foote, Christopher L. and Jane S. Little (2011), 'Oil and the macroeconomy in a changing world: a conference summary', Public Policy Discussion Paper, Federal Reserve Bank of Boston.

Fouquet, Roger and Peter J.G. Pearson (2006), 'Seven centuries of energy services: the price and use of light in the United Kingdom (1300–2000)', *Energy Journal*, **27**, 139–77.

Fouquet, Roger and Peter J.G. Pearson (2012), 'The long run demand for lighting: elasticities and rebound effects in different phases of economic development', *Economics of Energy and Environmental Policy*, **1**, 1–18.

Gisser, Micha and Thomas H. Goodwin (1986), 'Crude oil and the macroeconomy: tests of some popular notions', *Journal of Money, Credit, and Banking*, **18**, 95–103.

Hamilton, Clive and Hal Turton (2002), 'Determinants of emissions growth in OECD countries', *Energy Policy*, **30**, 63–71.

Hamilton, James D. (1983), 'Oil and the macroeconomy since World War II', *Journal of Political Economy*, **91**, 228–48.

Hamilton, James D. (1988), 'A neoclassical model of unemployment and the business cycle', *Journal of Political Economy*, **96**, 593–617.

Hamilton, James D. (1996), 'This is what happened to the oil price macroeconomy relation', *Journal of Monetary Economics*, **38**, 215–20.

Hamilton, James D. (2003), 'What is an oil shock?', *Journal of Econometrics*, **113**, 363–98.

Hamilton, James D. (2009a), 'Understanding crude oil prices', *Energy Journal*, **30**, 179–206.

Hamilton, James D. (2009b), 'Causes and consequences of the oil shock of 2007–08', *Brookings Papers on Economic Activity, Spring*, 215–61.

Hamilton, James D. (2011), 'Nonlinearities and the macroeconomic effects of oil prices', *Macroeconomic Dynamics*, **15**, 364–78.

Hamilton, James D. (forthcoming), 'Historical oil shocks', in Randall Parker and Robert Whaples (eds), *Handbook of Major Events in Economic History*, Routledge.

Hamilton, James D. and Ana Maria Herrera (2004), 'Oil shocks and aggregate macroeconomic behavior: the role of monetary policy', *Journal of Money, Credit, and Banking*, **36**, 265–86.

Hamilton, James D. and Michael T. Owyang (forthcoming), 'The propagation of regional recessions', *Review of Economics and Statistics*.

Herrera, Ana María, Latika Gupta Lagalo and Tatsuma Wada (2011), 'Oil price shocks and industrial production: is the relationship linear?', *Macroeconomic Dynamics*, **15** (53), 472–97.

Hooker, Mark A. (1996), 'What happened to the oil price–macroeconomy relationship?', *Journal of Monetary Economics*, **38**, 195–213.

Hotelling, Harold (1931), 'The economics of exhaustible resources', *Journal of Political Economy*, **39**, 137–75.

Jenkins, Gilbert (1985), *Oil Economists' Handbook*, London: British Petroleum Company.

Jiménez-Rodríguez, Rebeca and Marcelo Sánchez (2005), 'Oil price shocks and real GDP growth: empirical evidence for some OECD countries', *Applied Economics*, **37**, 201–28.

Jo, Soojin (2011), 'The effects of oil price uncertainty on the macroeconomy', Working Paper, UCSD.

Kambara, Tatsu and Christopher Howe (2007), *China and the Global Energy Crisis*, Cheltenham, UK and Northampton, MA, USA: Edward Elgar.

Kilian, Lutz (2008), 'The economic effects of energy price shocks', *Journal of Economic Literature*, **46**, 871–909.

Kilian, Lutz (2009), 'Not all oil price shocks are alike: disentangling demand and supply shocks in the crude oil market', *American Economic Review*, **99**, 1053–69.

Kilian, Lutz, and Bruce Hicks (2011), 'Did unexpectedly strong economic growth cause the oil price shock of 2003–2008?', Working Paper, University of Michigan.

Kilian, Lutz and Robert J. Vigfusson (2011), 'Are the responses of the U.S. economy asymmetric in energy price increases and decreases?', *Quantitative Economics*, **2**, 419–53.

Kim, Dong Heon (2012), 'What is an oil shock? Panel data evidence', *Empirical Economics*, **43**, 121–43.

Kim, In-Moo and Prakash Loungani (1992), 'The role of energy in real business cycle models', *Journal of Monetary Economics*, **29**, 173–89.

Knittel, Christopher R. (2011), 'Automobiles on steroids: product attribute trade - offs and technological progress in the automobile sector', *American Economic Review*, **101**, 3368–99.

Krautkraemer, Jeffrey A. (1998), 'Nonrenewable resource scarcity', *Journal of Economic Literature*, **36**, 2065–107.

Leduc, Sylvain and Keith Sill (2004), 'A quantitative analysis of oil-price shocks, systematic monetary policy, and economic downturns', *Journal of Monetary Economics*, **51**, 781–808.

Lee, Kiseok and Shawn Ni (2002), 'On the dynamic effects of oil price shocks: a study using industry level data', *Journal of Monetary Economics*, **49**, 823–52.

Lee, Kiseok and Shawn Ni and Ronald A. Ratti (1995), 'Oil shocks and the macroeconomy: the role of price variability', *Energy Journal*, **16**, 39–56.

Loungani, Prakash (1986), 'Oil price shocks and the dispersion hypothesis', *Review of Economics and Statistics*, **68**, 536–39.

Mork, Knut A. (1989), 'Oil and the macroeconomy when prices go up and down: an extension of Hamilton's results', *Journal of Political Economy*, **91**, 740–44.

Mork, Knut A., Øystein Olsen and Hans Terje Mysen (1994), 'Macroeconomic responses to oil price increases and decreases in seven OECD countries', *Energy Journal*, **15**(4), 19–35.

Pindyck, Robert S. (1978), 'The optimal exploration and production of nonrenewable resources', *Journal of Political Economy*, **86**, 841–61.

Ramey, Valerie A. and Daniel J. Vine (2012), 'Oil, automobiles, and the U.S. economy: how much have things really changed?', *NBER Macroeconomics Annual 2011*, Chicago, IL: University of Chicago Press.

Rasche, R.H. and J.A. Tatom (1977), 'Energy resources and potential GNP', *Federal Reserve Bank of St. Louis Review*, **59** (June), 10–24.

Rasche, R.H. and J.A. Tatom (1981), 'Energy price shocks, aggregate supply, and monetary policy: the theory and international evidence', in K. Brunner and A.H. Meltzer (eds), *Supply Shocks, Incentives, and National Wealth*, Carnegie-Rochester Conference Series on Public Policy, vol. 14, Amsterdam: North-Holland, pp. 9–94.

Rasmussen, Tobias N. and Agustín Roitman (2011), 'Oil shocks in a global perspective: are they really that bad?', Working Paper, IMF.

Ravazzolo, Francesco and Philip Rothman (2010), 'Oil and U.S. GDP: a real-time out-of-sample examination', Working Paper, East Carolina University.

Raymond, Jennie E. and Robert W. Rich (1997), 'Oil and the macroeconomy: a Markov state-switching approach', *Journal of Money, Credit, and Banking*, **29** (May), 193–213. Erratum **29** (November, Part 1), p. 555.

Rotemberg, Julio J. and Michael Woodford (1996), 'Imperfect competition and the effects of energy price increases', *Journal of Money, Credit, and Banking*, **28** (part 1), 549–77.

Santini, Danilo J. (1985), 'The energy-squeeze model: energy price dynamics in U.S. business cycles', *International Journal of Energy Systems*, **5**, 18–25.

Santini, Danilo J. (1992), 'Energy and the macroeconomy: capital spending after an energy cost shock', in J. Moroney (ed.), *Advances in the Economics of Energy and Resources*, vol. 7, Greenwich, CT: JAI Press, pp. 101–24.

Santini, Danilo J. (1994), 'Verification of energy's role as a determinant of U.S. economic activity', in J. Moroney (ed.) *Advances in the Economics of Energy and Resources*, vol. 8, Greenwich, CT: JAI Press, pp. 159–94.

Slade, Margaret E. (1982), 'Trends in natural resource commodity prices: an analysis of the time domain', *Journal of Environmental and Economic Management*, **9**, 122–37.

Smith, James L. (2011), 'On the portents of peak oil (and other indicators of resource scarcity)', Working Paper, Southern Methodist University.

Wickstrom, Larry, Chris Perry, Matthew Erenpreiss and Ron Riley (2011), 'The Marcellus and Utica Shale plays in Ohio', presentation at the Ohio Oil and Gas Association Meeting, 11 March (http://www.dnr.state.oh.us/portals/10/energy/Marcellus_Utica_presentation_ OOGAL.pdf).

Williamson, Harold F. and Arnold R. Daum (1959), *The American Petroleum Industry: The Age of Illumination 1859–1899*, Evanston, IL: Northwestern University Press.

APPENDIX

State-level production data (in thousands of barrels per year) were assembled from the following sources: *Derrick's Handbook* (1898, p. 805); *Minerals Yearbook*, US Department of Interior, various issues (1937, 1940, 1944 and 1948); *Basic Petroleum Data Book*, American Petroleum Institute, 1992; and Energy Information Administration online data set (http://www.eia.gov/dnav/pet/pet_crd_crpdn_adc_mbbl_a.htm). Numbers for Kansas for 1905 and 1906 include Oklahoma. The *Basic Petroleum Data Book* appears to allocate some Gulf of Mexico production to Texas but most to Louisiana. The EIA series (which has been used here for data from 1981 onward) does not allocate federal offshore Gulf of Mexico to specific states, and has been attributed entirely to Louisiana in Table 1A.1.

Table 1A.1   State-level oil production data, 1862–2010

Year	US total	PA–NY	WV	OH	IL–IN	CA	KS–NE	CO–NM–AZ–UT	TX	LA	OK	WY	AK	ND–MT	Other
1862	3056	3056	0	0	0	0	0	0	0	0	0	0	0	0	0
1863	2631	2631	0	0	0	0	0	0	0	0	0	0	0	0	0
1864	2116	2116	0	0	0	0	0	0	0	0	0	0	0	0	0
1865	2498	2498	0	0	0	0	0	0	0	0	0	0	0	0	0
1866	3598	3598	0	0	0	0	0	0	0	0	0	0	0	0	0
1867	3347	3347	0	0	0	0	0	0	0	0	0	0	0	0	0
1868	3716	3716	0	0	0	0	0	0	0	0	0	0	0	0	0
1869	4215	4215	0	0	0	0	0	0	0	0	0	0	0	0	0
1870	5659	5659	0	0	0	0	0	0	0	0	0	0	0	0	0
1871	5795	5795	0	0	0	0	0	0	0	0	0	0	0	0	0
1872	6539	6539	0	0	0	0	0	0	0	0	0	0	0	0	0
1873	9894	9894	0	0	0	0	0	0	0	0	0	0	0	0	0
1874	10927	10927	0	0	0	0	0	0	0	0	0	0	0	0	0
1875	8788	8488	0	0	0	0	0	0	0	0	0	0	0	0	0
1876	9133	8969	120	32	0	12	0	0	0	0	0	0	0	0	0
1877	13350	13135	172	30	0	13	0	0	0	0	0	0	0	0	0
1878	15396	15163	180	38	0	15	0	0	0	0	0	0	0	0	0
1879	19914	19685	180	29	0	20	0	0	0	0	0	0	0	0	0
1880	26286	26028	179	39	0	40	0	0	0	0	0	0	0	0	0
1881	27662	27377	151	34	0	100	0	0	0	0	0	0	0	0	0
1882	21073	20776	128	40	0	129	0	0	0	0	0	0	0	0	0
1883	23449	23128	126	47	0	143	0	0	0	0	0	0	0	0	5
1884	24218	23772	90	90	0	262	0	0	0	0	0	0	0	0	4
1885	21859	20776	91	662	0	325	0	0	0	0	0	0	0	0	5
1886	28065	25798	102	1783	0	377	0	0	0	0	0	0	0	0	5
1887	28283	22356	145	5023	0	678	0	76	0	0	0	0	0	0	5
1888	27612	16489	119	10011	0	690	0	258	0	0	0	0	0	0	5
1889	35163	21487	544	12472	34	303	1	317	0	0	0	0	0	0	5
1890	45824	28458	493	16125	65	307	1	369	0	0	0	0	0	0	6
1891	54293	33009	2406	17740	138	324	1	666	0	0	0	0	0	0	9
1892	50515	28422	3810	16363	699	385	5	824	0	0	0	0	0	0	7
1893	48431	20315	8446	16249	2336	470	18	594	0	0	0	0	0	0	3

Table 1A.1 (continued)

Year	US total	PA-NY	WV	OH	IL-IN	CA	KS-NE	CO-NM-AZ-UT	TX	LA	OK	WY	AK	ND-MT	Other
1894	49344	19020	8577	16792	3689	706	40	516	0	0	0	2	0	0	2
1895	52892	19144	8120	19545	4386	1209	44	438	0	0	0	4	0	0	2
1896	60960	20584	10020	23941	4681	1253	114	361	1	0	0	3	0	0	2
1897	60476	19262	13090	21561	4123	1903	81	385	66	0	1	4	0	0	0
1898	55367	15948	13618	18739	3731	2257	72	444	546	0	0	6	0	0	6
1899	57071	14375	13911	21142	3848	2642	70	390	669	0	0	6	0	0	18
1900	63621	14559	16196	22363	4874	4325	75	317	836	0	6	6	0	0	64
1901	69389	13832	14177	21648	5757	8787	179	461	4394	0	10	5	0	0	139
1902	88767	13184	13513	21014	7481	13984	332	397	18084	549	37	6	0	0	186
1903	100461	12518	12900	20480	9186	24382	932	484	17956	918	139	9	0	0	557
1904	117081	12239	12645	18877	11339	29649	4251	501	22241	2959	1367	12	0	0	1001
1905	134717	11555	11578	16347	11145	33428	0	376	28136	8910	12014	8	0	0	1220
1906	126494	11500	10121	14788	12071	33099	0	328	12568	9077	21718	7	0	0	1217
1907	166095	11212	9095	12207	29410	39748	2410	332	12323	5000	43524	9	0	0	825
1908	178527	10584	9523	10859	36969	44855	1801	380	11207	5789	45799	18	0	0	743
1909	183171	10434	10745	10633	33194	55472	1264	311	9534	3060	47859	20	0	0	645
1910	209557	9849	11753	9916	35303	73011	1128	240	8899	6841	52029	115	0	0	473
1911	220449	9201	9796	8817	33012	81134	1279	227	9526	10721	56069	187	0	0	480
1912	222935	8712	12129	8969	29572	87269	1593	206	11735	9263	51427	1572	0	0	488
1913	248446	8865	11567	8781	24850	97788	2375	189	15010	12499	63579	2407	0	0	536
1914	265763	9109	9680	8536	23256	99775	3104	223	20068	14309	73632	3560	0	0	511
1915	281104	8726	9265	7825	19918	86592	2823	208	24943	18192	97915	4246	0	0	451
1916	300767	8467	8731	7744	18483	90952	8738	197	27645	15248	107072	6234	0	0	1211
1917	335316	8613	8379	7751	16537	93878	36536	121	32413	11392	107508	8978	0	45	3110
1918	355928	8217	7867	7285	14244	97532	45451	143	38750	16043	103347	12596	0	100	4384
1919	378367	8988	8327	7736	12932	101183	33048	121	79366	17188	86911	13172	0	69	9305
1920	442929	8344	8249	7400	11719	103377	39005	111	96868	35714	106206	16831	0	90	8765
1921	472183	8406	7822	7335	11201	112600	36456	108	106166	27103	114634	19333	0	340	19510
1922	557530	8425	7021	6781	10470	138468	31766	97	118684	35376	149571	26715	0	1509	21707
1923	732407	8859	6358	7085	9750	262876	28250	86	131023	24919	160929	44785	0	2449	44705
1924	713940	8926	5920	6811	9016	228933	28836	543	134522	21124	173538	39498	0	2815	53458
1925	763743	9792	5763	7212	8692	232492	38357	2286	144648	20272	176768	29173	0	4091	84197

Year															
1926	770874	10917	5946	7272	8568	224673	41498	4434	166916	23201	179195	25776	0	7727	64751
1927	901129	11768	6023	7593	7846	231196	41069	4057	217389	22818	277775	21307	0	5058	47230
1928	901474	12559	5661	7015	7514	231811	38596	3717	257320	21847	249857	21461	0	4015	40101
1929	1007323	15197	5574	6743	7300	292534	42813	4188	296876	20554	255004	19314	0	3980	37246
1930	898011	16450	5071	6486	6730	227329	41638	11845	290457	23272	216486	17868	0	3349	31030
1931	851081	15255	4472	5327	5879	188830	37018	16772	332437	21804	180574	14834	0	2830	25049
1932	785159	15920	3876	4644	5479	178128	34848	13591	312478	21807	153244	13418	0	2457	25269
1933	905656	15805	3815	4235	4981	172010	41976	15035	402609	25168	182251	11227	0	2273	24271
1934	908065	18282	4095	4234	5317	174305	46482	18003	381516	32869	180107	12556	0	3603	26696
1935	996596	20046	3902	4082	5099	207832	54843	22043	392666	50330	185288	13755	0	4603	32107
1936	1099687	21733	3847	3847	5297	214773	58317	28873	427411	80491	206555	14582	0	5868	28093
1937	1279160	24667	3845	3559	8343	238521	70761	40459	510318	90924	228839	19166	0	5805	33953
1938	1214355	22471	3684	3298	25070	249749	60064	37171	475850	95208	174994	19022	0	4946	42828
1939	1264962	22480	3580	3156	96623	224354	60703	39041	483528	93646	159913	21454	0	5960	50524
1940	1353214	22352	3444	3159	152625	223881	66415	40755	493209	103584	156164	25711	0	6728	55187
1941	1402228	21935	3433	3510	139804	230263	85140	41719	505572	115908	154702	29878	0	7526	62838
1942	1386645	23200	3574	3543	113134	248326	98873	33743	483097	115785	140690	32812	0	8074	81794
1943	1505613	20816	3349	3322	87543	284188	106813	41216	594343	123592	123152	34253	0	7916	75110
1944	1677904	18815	3070	2937	82531	311793	99179	42638	746699	129645	124616	33356	0	8647	73978
1945	1713665	17163	2879	2838	79962	326482	96720	42387	754710	131051	139299	36219	0	8420	75535
1946	1733909	17829	2929	2908	82023	314713	97511	48670	760215	143669	134794	38977	0	8825	80846
1947	1856987	17452	2617	3108	72554	333132	105361	56628	820210	160128	141019	44772	0	8742	91264
1948	2020185	17288	2692	3600	71782	340074	111123	65847	903498	181458	154455	55032	0	9382	103954
1949	1841940	15799	2839	3483	74197	332942	102198	71869	744834	190826	151660	47890	0	9118	94285
1950	1973574	16002	2808	3383	72727	327607	109133	71898	829874	208965	164599	61631	0	8109	96838
1951	2247711	15599	2757	3140	71343	354561	117080	81847	1010270	232281	186869	68929	0	8983	94052
1952	2289836	15475	2602	3350	72126	359450	117467	90799	1022139	243929	190435	68074	0	11155	92835
1953	2357082	14449	3038	3610	71849	365085	120910	108653	1019164	256632	202570	82618	0	17103	91404
1954	2314988	12364	2902	3880	78002	355865	127100	122931	974275	246558	185851	93533	0	20220	91507
1955	2484428	11435	2320	4353	92411	354812	132872	137833	1053297	271010	202817	99483	0	26797	94983
1956	2617283	10978	2179	4785	93859	350754	140408	148875	1107808	299421	215862	104830	0	35255	102269
1957	2616901	10856	2215	5478	89745	339646	143200	154103	1073867	329896	214661	109584	0	40431	103214
1958	2448987	8235	2186	6260	92139	313672	140315	172074	940166	313891	200699	115572	0	42216	101562
1959	2574600	8140	2184	5978	88281	308946	142424	192115	971978	362666	198090	126050	187	47681	119879
1960	2574933	7822	2300	5405	89395	305352	137278	192515	927479	400832	192913	133910	559	52232	126940
1961	2621758	7301	2760	5639	88318	299609	136610	192503	939191	424962	193081	141937	6327	54558	128962
1962	2676189	6891	3470	5835	90873	296590	136970	182873	943328	477153	202732	135847	10259	56829	126539

Table 1A.1 (continued)

Year	US total	PA–NY	WV	OH	IL–IN	CA	KS–NE	CO–NM–AZ–UT	TX	LA	OK	WY	AK	ND–MT	Other
1963	2752723	6762	3350	6039	86698	300908	130953	181727	977835	515057	201962	144407	10740	55900	130385
1964	2786822	6987	3370	15859	81451	300009	125365	177257	989525	549698	202524	138752	11059	56378	128588
1965	2848514	6554	3530	12908	75189	316428	121949	178072	1000749	594853	203441	138314	11128	59128	126271
1966	3027762	6072	3674	10899	72278	345295	117588	181889	1057706	674318	224839	134470	14358	62506	121870
1967	3215742	6359	3561	9924	69223	359219	112573	187021	1119962	774527	230749	136312	29126	60274	116912
1968	3329042	5692	3312	11204	65083	375496	107688	187361	1133380	817426	223623	144250	66204	73500	114823
1969	3371751	5704	3104	10972	58565	375291	100822	183249	1151775	844603	224729	154945	73953	66657	117382
1970	3517450	5287	3124	9864	51234	372191	96304	178061	1249697	906907	223574	160345	83616	59877	117369
1971	3453914	4924	2969	8286	45742	358484	88594	170669	1222926	935243	213313	148114	79494	56252	118904
1972	3455369	4459	2677	9358	41004	347022	82449	170103	1301686	891827	207633	140011	72893	54528	129719
1973	3360903	4249	2385	8796	35981	336075	73467	171036	1294671	831524	191204	141914	72323	54855	142423
1974	3202585	4374	2665	9088	32472	323003	68302	176306	1262126	737324	177785	139997	70603	54251	144289
1975	3056779	4139	2479	9578	30699	322199	65226	176088	1221929	650840	163123	135943	69834	53296	151406
1976	2976180	3876	2519	9994	30902	326021	64896	165945	1189523	606501	161426	134149	63398	54539	162491
1977	3009265	3539	2518	10359	30922	349609	63464	160223	1137880	562905	156382	136472	169201	55953	169838
1978	3178216	3739	2382	11154	28051	347181	62448	151948	1074050	532740	150456	137385	448620	55279	172783
1979	3121310	3729	2406	11953	26508	352268	63063	140173	1018094	489687	143642	131890	511335	60871	165691
1980	3146365	3475	2336	12928	27680	356923	66391	130510	977436	469141	150140	126362	591646	69921	161476
1981	3128624	4570	3473	13551	28811	384958	72481	128088	932350	462097	154056	130563	587337	76237	150052
1982	3156715	5116	3227	14571	33273	401572	77397	124344	908217	475474	158621	118300	618910	78192	139501
1983	3170999	5113	3628	14971	34521	404688	77974	133990	882911	499334	158604	118303	625527	79915	131520
1984	3249696	5124	3524	15271	34394	412020	82181	143085	883174	536868	168385	124269	630401	82413	128587
1985	3274553	5922	3555	14988	35433	423877	82350	149743	869218	527852	162739	128514	666233	80625	123504
1986	3168252	4636	3145	13442	32004	406665	74132	144354	819595	532119	149105	121337	681310	72700	113708
1987	3047378	4012	2835	12153	27718	395698	65975	137049	760962	500544	134378	115267	715955	66410	108422
1988	2979126	3396	2621	11711	26141	386014	64802	136718	735495	464466	128874	113985	738143	62681	104079
1989	2778771	3196	2243	10215	23689	364250	61715	127921	688169	432222	117493	107715	683979	57700	98264
1990	2684679	3056	2143	10008	22954	350899	61317	125428	678478	417386	112273	103856	647309	56527	93045

1991	2707043	2958	1963	9156	22082	351016	62760	126377	682616	438825	108094	99928	656349	55470	89449
1992	2624631	2541	2068	9197	22319	348040	59087	122573	650623	443984	101807	96810	627322	51376	86884
1993	2499044	2371	2048	8282	20167	343729	54493	119714	619090	439791	96625	87667	577495	48363	79209
1994	2431483	2817	1918	8758	19640	343569	50948	115185	590735	440306	90973	79528	568951	44103	74052
1995	2394268	2243	1948	8258	18968	350686	47560	112544	559646	467203	87490	78884	541654	45865	71319
1996	2366021	2001	1680	8305	18098	346828	45330	108917	543342	505795	85379	73365	509999	48236	68746
1997	2354832	1597	1509	8593	18545	339307	43172	114850	536584	546302	83364	70176	472949	51358	66526
1998	2281921	2197	1471	6541	15940	329860	38715	113969	504662	582608	77578	64782	428850	52045	62703
1999	2146726	1677	1471	5970	14029	312719	31709	99164	449233	614072	70556	61126	383199	47819	53982
2000	2130720	1710	1400	6575	14304	306124	37420	101374	443397	628675	69976	60726	355199	48147	55693
2001	2117521	1786	1226	6051	12114	291766	36864	99832	424297	665095	68531	57433	351411	47611	53504
2002	2097121	2398	1382	6004	14013	287793	35500	98514	411985	661287	66642	54717	359335	47848	49703
2003	2073454	2569	1334	5647	13561	280000	36699	100382	405801	659242	65356	52407	355582	48726	46148
2004	1983300	2708	1339	5785	12739	267260	36365	101014	392867	615311	62502	51619	332465	55878	45448
2005	1890105	4144	1563	5652	11934	256848	36236	100184	387680	543259	62142	51626	315420	68515	44902
2006	1862259	3945	1749	5422	12054	249562	37964	101173	397220	547876	62841	52904	270486	76173	42890
2007	1848452	4033	1574	5455	11336	241378	38824	101631	396894	542763	60952	54130	263595	79887	46000
2008	1811819	3997	1593	5715	11281	238691	41976	105507	398014	494708	64065	52943	249874	94321	49134
2009	1956597	3880	1864	5834	10903	228994	41703	112443	403797	638004	67018	51333	235500	107428	47896
2010	1998138	3923	1992	4785	10901	223501	42672	120583	426700	633639	69513	53133	218762	138341	49693

2 Gas markets: past, present and future
Paul Stevens[1]

1. INTRODUCTION

The history of gas markets is both controversial and complex. This arises in large part because of the characteristics of natural gas as a source of primary energy. These characteristics differentiate gas from other hydrocarbon energies, in particular oil. The appendix outlines these characteristics and the consequent differences. This chapter attempts to lay out that history and consider the future role of gas in the future global primary energy mix. Section 2 provides a short history of gas markets. While specifics of the history of regional gas markets are explored in Section 2.2, certain generalities are considered in Section 2.1. Section 3 then considers how gas markets have generated conflict over the contractual terms by which gas is bought, sold and transported. Section 4 looks to the future, which is dominated by the possibilities for the development of shale gas. Section 5 draws some conclusions on the future of gas markets.

2. A SHORT HISTORY OF THE INTERNATIONAL GAS INDUSTRY

As the appendix explains, gas markets are different from oil markets. Specifically, while oil is now a truly international market (Stevens, 2008), gas markets remain regional and need to be treated historically in regional terms, as they will be in section 2.2. However, certain generalizations can be made.

2.1 General Issues of being a Constrained Industry

The starting point of the analysis is that gas, outside of the Former Soviet Union (FSU), had before the 1990s been a constrained industry. This was despite the fact that many saw gas as a premium fuel. It tended to have high conversion efficiencies at the burner tip. It was easy to handle once the infrastructure was in place and it was also seen as being cleaner to burn. Although in the 1970s there was a perception that there were limited gas resources, oil exploration by the international oil companies (IOCs) (encouraged by the nationalizations in OPEC in the 1970s) meant that reserves were growing, as can be seen from Figure 2.1, as more gas was discovered.[2]

Yet despite this long list of attractive characteristics of gas, as Figure 2.2 shows, if the FSU is excluded, gas's share in the global primary energy mix hardly changed in the 20 years after 1970.

The explanation for this is that gas had been a constrained fuel. The reasons for this go a long way to explain the generic history of the gas industry and for that reason are worth further explanation.

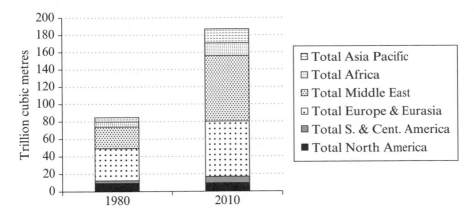

Figure 2.1   World gas reserves by region, 1980 and 2010

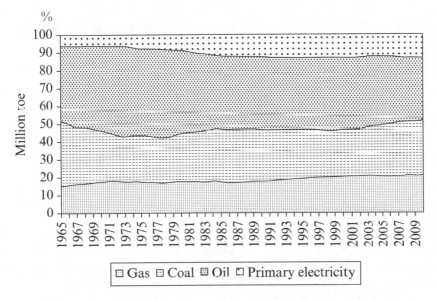

Figure 2.2   The share of gas in primary energy outside of the FSU, 1965–2010

The first source of constraint is the fact that gas, relative to the other hydrocarbons, because it is a high-volume, low-value commodity, is extremely expensive to transport, as Figure 2.3 illustrates.

Thus, if a country or region did not have physical access to gas, that is did not produce it, no gas would be consumed. For the European Union (EU) and the USA there was a further constraint. In the mid-1970s, a view gained ground that gas reserves were limited and that thanks to the characteristics of gas, it was considered too valuable a fuel to be simply burnt. It was argued that it should be reserved for 'special' uses. This 'premium fuel' argument led in both the EU and the USA to legislation in the mid-1970s that forbade the burning of gas in new power stations. In the OECD outside of the USA

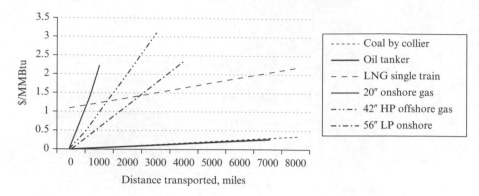

Source: Jensen (2004).

Figure 2.3   Comparative transport costs

there was yet a further constraint. In most cases national gas markets were dominated by monopoly and monopsonist state-owned utilities. By their nature they tended to use this dominant position to act as satisficers rather than as profit maximizers.

In the developing world there were further constraints to the entry of gas into the primary energy mix. For domestic use, there was the problem that the debt crisis that affected many developing countries in the 1980s meant that they could not afford the very large capital expenditure required to develop the gas infrastructure. Also there was what became known as the foreign company problem. Foreign companies discovered the majority of the gas reserves. Domestic use, say in local power stations, meant that the gas would be paid for in local currency. If this was non-convertible, then, although the projects were extremely profitable in terms of an internal rate of return criterion, the companies had no interest in pursuing them because they would be unable to remunerate their shareholders at home.[3]

In the context of developing countries with gas, an export option also faced significant constraints. Gas export projects need minimum levels of reserves to make them viable.[4] Often the reserves found were below this level and the foreign company problem outlined above inhibited further exploration.[5] This was reinforced because often the government was a monopsonist and therefore often offered very unattractive prices, whether in local currency or not.

There were also serious problems with negotiating gas export contracts. Historically there have been two types of gas market into which exporters could try to sell: a commodity gas supply market and a project gas supply market. In the former a large number of buyers and sellers of gas operated in a relatively transparent market. Thus there existed something that could be called a 'gas price' upon which to base the export contracts. There were few such markets.[6] Most markets were project supply markets where there were few buyers and sellers and poor transparency. Thus there was no 'gas price' upon which to base the contract price. This seriously complicated negotiations. Pricing was based upon some sort of formula in terms of potentially competing fuels (usually oil). However, there was still a requirement to fix a floor price to protect the supplier and a ceiling price to protect the buyer. Given that these were very long-term contracts, this was difficult to say the least and sowed the seeds of many of the subsequent conflicts over such export contracts.[7]

There were also problems with transit pipelines that, as will be developed below, suffered in many cases from serious and endemic conflict (Stevens, 2009). While it is true that most of these problems were associated with oil rather than gas transit pipelines[8] the poor reputation of oil transit pipelines in some cases inhibited plans for similar gas lines and acted as a deterrent. This was particularly relevant in the context of possible gas lines from the Persian Gulf region. LNG also had its problems, which until relatively recently were serious enough to constrain LNG projects.[9] The projects were complex and extremely expensive. LNG requires lowering the temperature of the natural gas to −161 degrees Celsius.[10] The engineering tolerances with such temperatures are extremely small, making the projects highly complex in terms of the technology. They were also energy intensive with somewhere between 9 to 12 per cent of the gas input simply going into the processing of the gas. Also, the projects had to involve all stages of the value chain from wellhead to final customer. The result was that in the 1970s and 1980s the projects were extremely expensive.

The projects also had extremely long lead times.[11] Because of the very high capital intensity, full capacity operation was crucial. Therefore, before the plant could even be designed, let alone built, all the output had to be contracted. Negotiating contracts for all the output could take years. Also accumulating the finance for such large projects could take a very long time. In the LNG projects that came on stream in the 1980s and early 1990s, lead times of 20 to 25 years were not unusual. A major consequence of the long lead times and the complexity of the contracts was that the trading links were very inflexible. As will be seen below, this made such contracts very vulnerable to conflict over the terms as market circumstances changed. Finally, if these problems were not enough, for the owner of the gas there was very little revenue in LNG. The effective gas price at the wellhead was extremely low and the economics of most LNG projects was highly dependent upon the value to the liquids stripped out of the wet gas.

However, as we begin to move through the 1990s, many of the constraints to the use of gas listed above began to erode. In 1990, thanks to the development of new ample reserves as well as the emergence of the new CCGT (combined cycle gas turbine) technology, both the USA and the EU dropped the legal restriction on gas in power generation. At the same time, growing concern about environmental issues started to make gas look an increasingly attractive option, especially in the context of the CO_2 targets set in Rio and Kyoto. In many countries the electricity sectors were undergoing reform in an attempt to attract private sector investment in generation. Such investment would invariably pick gas, if available, as the fuel of choice, given the huge benefits of CCGT.[12] At the same time countries are also trying to reform their gas sectors with the aim of moving them closer to a commodity supply market, which, other things being equal, would make it easier to negotiate export contracts.[13] Developments with respect to the Energy Charter Treaty also initially gave some hope that the problems with transit pipelines might have some form of collaborative solution.

Finally there have been major changes to LNG. This means that it might play a very much greater role in international gas markets. Between 1990 and 2004, the costs of projects came down, in large part because of the use of economies of scale. For example, in 1990 a 4.5 million tonne per year project would need two trains with a total cost of

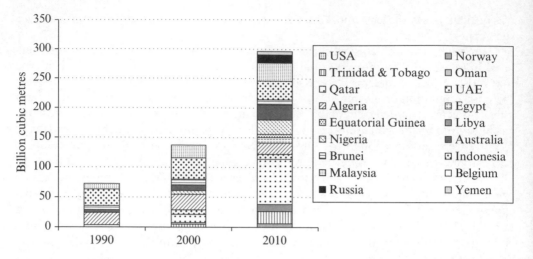

Figure 2.4   LNG exports, 1990–2010

$3.75 billion, while in 2004 with only one train the same-sized plant would cost $2.85 billion, with the bulk of the cost reduction coming in the stages of transportation and regasification (Jensen, 2009). These reductions in costs were crucial in making LNG projects look more attractive to private sector investors who were very conscious of their cost of capital. After 2004, like most projects in oil and gas, costs skyrocketed due to a combination of rising raw material prices and engineering cost increases due to tight service industry capacity.[14] At the same time, changes in the development of project financing methods meant it was much easier to raise the capital for the project. However, the really big change came because an increasing number of projects increased their flexibility. Figure 2.4 shows how the number of projects has increased between 1990 and 2007.

The significance of more projects is that, for the first time, there developed a spot trade in LNG. This would have been unthinkable only ten years ago. Thus the requirement that all of the output must be firmly contracted before the plant was built no longer applies.

These changes all gave rise to growing speculation that the regional gas markets may indeed become more global in the same way that the oil markets did following the dramatic reduction in the cost of transportation of crude oil.[15]

2.2   A History of Gas in Terms of Specific Regions

As has been indicated earlier, any historical narrative for gas must be on the basis of specific regions given the relatively little gas trade between the major regions.[16] In terms of the five regions identified below – North America, the EU, the FSU, the Far East and the rest of the world – only 26 per cent of pipeline movement is interregional trade and virtually all of that is from the FSU, North Africa and Norway into the EU. Obviously for LNG the interregional dimension of trade is much more important, but again, it is not large. In 2008, LNG trade accounted for only 7.5 per cent of global gas consumption. This lack of interregional trade is reflected in regional gas prices. As can be seen from

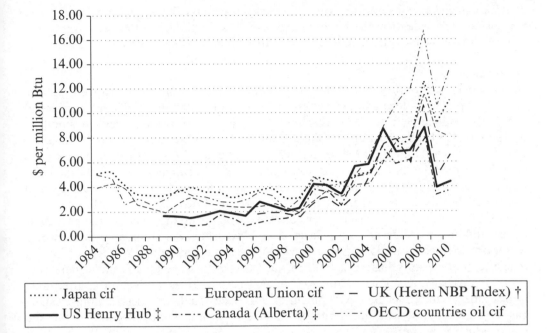

Figure 2.5   Regional gas prices

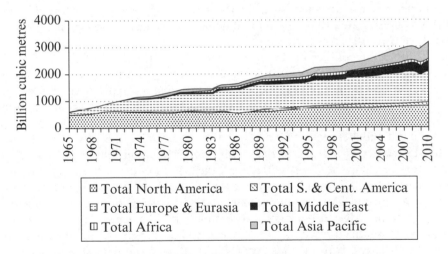

Figure 2.6   World gas consumption by region, 1965–2010

Figure 2.5, there is little or no coherence between regional prices. This clearly indicates the lack, at least to date, of any effective arbitrage in the global gas market, although the expansion of LNG trade described earlier may well change that in the future.

The development of consumption by region is shown in Figure 2.6. As discussed earlier, gas consumption was relatively slow to get going. In 1965, of global primary commercial energy consumption, only 22 per cent was in the form of natural gas.[17] There was also quite a wide disparity between regions. Thus, for the USA, the figure was 31 per cent, while for the EU (defined in terms of current membership) it was only 12 per cent.

North America
In the nineteenth century, consumption was dominated by 'town gas'. However, in the 1890s gas pipelines began to emerge from Texas and Oklahoma.[18] Thereafter natural gas consumption began to grow, reaching 44 per cent of primary energy consumption by 1965. In 1930, US consumption was around 5 billion cubic metres (bcm); by the early 1970s this reached a peak of 630 bcm (Darmstadter et al., 1971).[19] In 1966, except for a very few places, the manufacture of 'town gas' ceased. Throughout the 1960s, the USA was self-sufficient in gas but thereafter imports (predominantly from Canada) began to rise. In 1989, they were roughly 51 bcm (9.4 per cent of demand) and by the next peak in 2005 they were 112 bcm (18 per cent of demand). During the 2000s and in particular the second half of the decade many large-scale LNG trains were built all over the world to service the upcoming US import need. At the same time some 130 bcm of import regasification capacity has been built in the USA, with some 60 still under construction. Against all expectations, however, as will be developed in Section 4, after 2005 imports began to fall as the development of unconventional gas reserves, notably shale gas, began to reverse what most analysts had seen as an inevitable decline in domestic gas production in the Lower 48 (US states without Alaska) (Stevens, 2010). The effect has been particularly felt after 2008 with the economic crisis. The result was that many of the new LNG regasification plants that had been built/refurbished in anticipation of higher imports remain idle, creating a significant oversupply in the global LNG market. In 2010, it was estimated that less than 10 per cent of regasification capacity was being utilized (EPRINC, 2011).

A main source of conflict in the period before the oil shocks of the 1970s was the legacy of the nineteenth century, when consumers were constantly dissatisfied with prices charged by the local monopoly suppliers despite the presence of price regulation imposed by the various state governments. After the oil shocks of the 1970s, gas consumption fell sharply, reflecting the generally higher price of energy. By the early 1980s, consumption levels stabilized and between 1980 and 1988 averaged fairly steadily around 500 bcm annually. Thereafter they began to rise again from 541 bcm in 1989 to 653 bcm in 2007.

The history of gas in the USA is effectively the history of a regulated industry and the consequences of regulation and attempts to deregulate (Bradley, 1996). Before 1938, regulation was a mish-mash of disputes at local and state level, with the federal government hovering round the edges often significantly muddying the waters. The main targets of regulation were wellhead prices and also access to and the role of the growing pipeline network. In 1938, the Natural Gas Act (NGA) allowed the federal government (The Federal Power Commission (FPC)) to regulate the rates charged by the transmission companies managing interstate gas trade. In 1954, the Phillip's Decision by the Supreme Court classed gas producers involved in interstate trade as natural gas companies, thus bringing them under the regulatory umbrella of the NGA. Thus, between 1954 and 1978, wellhead prices were subject to price control. However, the result was relatively low prices that boosted demand but at the same time inhibited interstate supply,[20] leading to growing shortages in the context of higher energy prices linked to the oil price shocks.

To try to sort out the growing confusion in 1978, The Natural Gas Policy Act (part of President Carter's National Energy Act) began the process of deregulation. The aim was to try to create a single national gas market where market forces ultimately fixed prices. A series of other pieces of legislation between 1985 and 1992 completed the deregulation of US gas markets.[21] In particular, this effectively unbundled the gas supply industry so that while pipelines remained subject to regulation as natural monopolies, gas sales were turned over to the competition.

European Union

In Europe, the gas industry was dominated by 'town gas' until the 1960s. Then gas discoveries in the North Sea and in Holland saw the gradual rise of natural gas, with 'town gas' manufacture virtually ceasing in the 1980s. Between 1970 and 1980, gas production in the EU increased from 102 bcm to 197 bcm. As consumption rose and production pretty well stagnated after 1980, imports began to become more important. Pipelines began to bring gas from the Soviet Union and Algeria.[22] This began a debate that has been running ever since over vulnerability to imports from what were seen as potentially unreliable sources. In particular, there was intense opposition from the USA to the building of the line to bring Soviet imports, and huge pressure was put, especially upon Germany as the first landing point, not to go ahead. The argument was that the Soviet Union could use gas supplies to leverage influence in Europe, weakening NATO and the ability to manage the cold war. In the event, the record of supply through the pipeline has been until recently exemplary and arguably the recent problems arose because of problems between Russia and the Ukraine rather than between sellers and buyers.[23]

Concern over gas security of supply continues to be a dominant theme in European gas markets. Currently, gas import dependence in the EU is running at around 61 per cent of consumption. The view within Brussels is that the clear solution lies in diversifying supplies.[24] Currently around 95 per cent of EU gas imports come from three countries – Russia, Algeria and Norway. To this end the Commission has been very supportive of the proposed Nabucco gas line as a 'Fourth Corridor' to bring in Caspian and Middle East gas by pipeline and to compete with supplies from Russia and North Africa in Europe. However, this illustrates a key problem that affects gas markets concerning the issue of externalities and supply security. The benefits of a secure supply of gas accrue not only to the private consumer but also to the state. Gas outages would cause very considerable economic costs to any country. Thus many of the benefits of a line like Nabucco accrue to the state. A private investor would not be rewarded for this and therefore if such a pipeline were to be built, government funding would need to be involved. In a world where the Commission is looking for private funding exclusively, it is very difficult for pipelines such as Nabucco to be built. While large producing and importing companies promote South Stream and North Stream, mid-size and midstream companies promote the Nabucco pipeline. While Gazprom can carry out investments on strategic grounds, mid-sized and midstream companies cannot.

Another element of the gas story in Europe revolves around the attempts of the EU to create a unified and competitive gas market, thus converting what was a project supply market into a commodity supply market. After 1990, the UK embarked upon this route and, as can be seen from Figure 2.7, once the various constraints are lifted, as discussed earlier, the share of gas in primary energy rises dramatically. The story of the

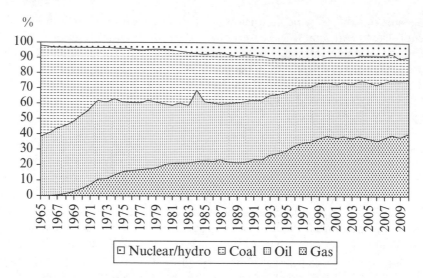

Figure 2.7   UK primary energy consumption by fuel, 1965–2010

Commission's attempts to achieve this and imitate the UK[25] is long and complex, but it revolves around the ability of national interests and quasi-monopoly positions of various gas companies that have succeeded in blocking attempts.

FSU
Before the Second World War there was barely anything that could be described as a gas industry in the Soviet Union (Wright, 1975). Two factors changed that – the strategic move of the country's industrial base to the East in the face of the German invasion in 1941 and the increasing discovery of gas reserves east of the Urals and especially in Western Siberia. Furthermore, these new gas reserves could be produced at a much lower resource cost than coal (especially) and even oil. The result was that in the 1950s, gas moved up Gosplan's (and its successor Commissions') agenda and became the highest priority in the Soviet energy sector. Between 1960 and 1972, gas production increased over five times and its share in primary energy rose rapidly. By 1970, 20 per cent of the primary energy mix was from gas (Chesshire and Huggett, 1975). This rising importance led to a huge expansion of the Soviet gas network. All of this was under the control of the Ministry of Gas Industry created in 1965. In 1989, the whole sector was put under the control of the first state-corporate enterprise – State Gas Concern Gazprom.

In the 1960s gas exports began to the Soviet Comecon partners in Eastern Europe via the Brotherhood pipeline. In 1969, three gas export contracts to Western Europe were signed to deliver some 5.3 bcm.[26] By 1980, the Soviet Union was earning around $15 billion from gas and oil exports, which represented the bulk of hard currency earnings.

Following the break-up of the Soviet Union in 1990, Russia's gas interests were all vested in Gazprom as a state-owned monopoly.[27] Subsequently, despite partial privatization and being listed abroad (via global depository receipts), there remained very close relationships between the company and the state. Gazprom became regarded as one of the 'national champions', a group of companies whose purpose was to promote state power and leverage Russian national interests. The ultimate example of this was when Dmitry Medvedev, who had been Gazprom's chairman, became Vladimir Putin's successor as President of Russia. However, this relationship goes both ways and it is believed that Russia's refusal to ratify the Energy Charter Treaty to date is because of opposition to the Treaty from Gazprom, which sees it as a threat to its monopoly status (Stevens, 2009). Since the 1990s, Gazprom has clearly indicated that it has ambitious targets to secure assets in the gas sectors of Western Europe. That has made many in Western Europe suspicious and concerned that Gazprom is seeking to extend monopoly powers into its downstream markets.

Far East
The Far East has always been an extremely important market for LNG, with Japan, South Korea and Taiwan initiating the growth in imports in the 1970s, followed by non-OECD Asia in the last 20 years. Gas consumption in the region is dominated by China, India, Indonesia, Japan, Malaysia, Pakistan, South Korea and Thailand, which together in 2009 accounted for 82 per cent of the total gas consumed. In terms of LNG imports, five countries dominate – China, India, Japan, South Korea and Taiwan collectively importing 152 bcm of LNG.[28] The imports are all on oil-indexed contracts and the markets are what are called 'firm market', which means they will effectively pay whatever the market demands.[29]

Rest of the world[30]
Here the main source of growth has been in the Middle East. In the 1970s, when the producer governments in the Middle East effectively nationalized the major oil companies, one of the first things that was attempted was to reduce associated gas flaring. This had been a major source of dispute between the governments and the companies. For the companies the associated gas had no commercial value. For the governments it was seen as part of the national patrimony to be preserved.

3. THE SOURCES OF DISPUTE OVER GAS CONTRACTS

Gas contracts and indeed gas transit agreements have tended to have a very mixed record. However, there have been many occasions when they have resulted in conflict between the parties. A major explanation of this is that, as is explained in the appendix, gas contracts have a tendency to be very long term. Fifteen-, 20- or even 25-year contracts are very common. The terms of any contract are an outcome of negotiations that take place in a specific market context in terms of supply, demand and prices. They are also the result of the relative bargaining power of the two parties. Inevitably, the market context changes, as does the relative bargaining power of the negotiating parties, especially once investments have been made and costs sunk. Once this happens, the obsolescing bargaining takes over and one or other signatory demands a renegotiation of the terms. It is this fundamental fact of gas markets, that is, long-term contracts, that threatens to generate conflict.

One interesting example of cooperation but with potential for conflict has been the

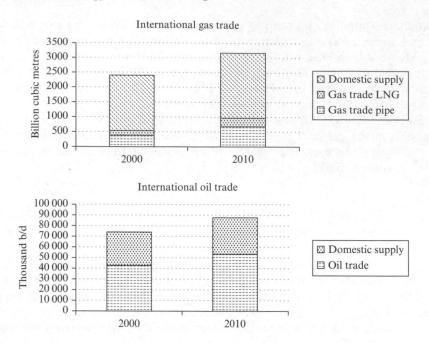

Figure 2.8   Comparative volumes of oil and gas traded internationally

development of the Gas Exporting Countries Forum (GECF). This held its first (infor-mal) meeting in Tehran in 2001. At the 7th Ministerial Meeting in Moscow at the end of 2008 a Charter was approved and a Secretariat created, to be located in Doha. There are currently 13 members that include all the major gas exporters with the exception of Norway, which is classed as an 'observer'. The official purpose of GECF is to promote exchanges between gas exporters in terms of information and research. However, inevi-tably, the Forum is seen by some as a potential gas-exporting cartel along the lines of OPEC. It is here where the potential conflict enters. In the early days, ministers attending the various meetings of the Forum were at pains to deny any such intention. However, in the last year or so, as gas prices have been weakening following the impact of uncon-ventional gas supplies in the USA, this has changed and there have been a number of statements, notably from Russia and Algeria, to the effect that GECF may well try to take action to prevent further declines in prices. How realistic this is remains debatable (Stevens, 2009), but the threat is sufficient to make gas consumers, especially in Western Europe, nervous.

Over time, as more and more gas has been found further away from the market, trans-port has been an increasing problem and in some cases (but by no means all) a major source of conflict in the industry. In reality there are only two obvious means of trans-port, pipelines and LNG.[31] As can be seen from Figure 2.8, much less gas is transported across international frontiers than is the case for oil.

Of the international trade in gas, in 2008 around 19.4 per cent was via pipelines and 7.5 per cent using LNG. A significant proportion of pipeline gas is forced to use transit pipelines. A transit pipeline can be defined as a pipeline that must pass through another 'sovereign' territory to get to market.[32] Unfortunately, while the history of transit pipelines has been very mixed, in general their operations have been characterized by conflict, although these problems are much more associated with oil than with gas pipelines.[33] Furthermore, these problems in recent years have increased simply because of the collapse of the Soviet Union, creating more jurisdictions.

A number of generalized explanations can be given for the poor record of transit pipelines. The one that tends to get the lion's share of the attention is politics. Because man by nature is an aggressive, greedy and selfish animal, it should come as no surprise that neighbouring countries have a long history of conflict. It should therefore also come as no surprise that these bad relations inevitably transfer into bad relations over transit pipelines between neighbours.[34] However, such explanations often miss an important source of conflict over pipelines. This relates to disputes arising over the level of transit fees paid to the transit entity.[35]

There are two problems that arise from the transit fee. First, there is no objective way to determine an appropriate transit fee. Thus the fee is set by the relative bargaining power of the parties. This changes over time and therefore there are constant pressures to renegotiate the terms of transit. A further problem compounds pressure for higher transit fees: the economist's 'bygones rule'. Pipelines attract very large economies of scale and therefore tend to be very large, capital-intensive projects (McLellan, 1992). Thus they are characterized by very high fixed costs and very low variable costs. The majority of costs are fixed, associated with securing 'rights of way' and building the pipeline and pumping stations. Variable costs relate only to maintenance and fuel for pumping; in the case of gas pipelines the latter is often supplied at below-market prices. The economist's 'bygones rule' explains that even if an operation is making a loss, the owner would be advised to continue operating so long as variable costs are covered and some contribution is being made to fixed costs. Thus if the loss exceeds the variable cost, closing the operation will minimize losses. However, if some contribution is being made to fixed costs, closing would not remove the fixed costs, and losses would be higher than continued operation. Thus the transit country can keep increasing transit demands even if it means that the pipeline begins to operate at a loss. It will not close if economic considerations are the only factor, since closure still entails fixed costs. This makes the transit agreement a very tempting target.

The second problem is that, given the nature of 'sovereignty' as defined, there is no overarching jurisdiction to ensure sanctity of contract. Thus the normal sanctions against unilateral behaviour over contracts are not available. Conflict is therefore greatly compounded. And, of course, if there is also a legacy of difficult political relations, this compounds the tendency to conflict.

The difficulty is that there are no obvious solutions to these problems that do not involve greater conflict. Several possibilities are available. An obvious if distinctly crude solution is to invade or to threaten to invade. However, in reality going to war would be an extremely expensive solution and is unlikely in the absence of other unconnected issues. There has been no case of this being used explicitly although it is interesting to speculate how far the threat – of either explicit military intervention or implicit subversion – may have muted transit country behaviour in some cases.

Another solution that certainly has been prevalent is to avoid 'bad' transit countries and use only 'good' ones. A number of examples can be cited, for example Russia's plans to bypass Ukraine. There are several problems with this solution. It assumes that an alternative route is available. Also determining in advance what makes a good or bad transit country is not always easy to establish (Stevens, 2009). If a pipeline already exists, building an alternative line significantly increases the cost of transport given the fact that pipelines are subject to very large economies of scale. It is much cheaper, other things being equal, to have one line of 1 million b/d capacity compared to two lines each with 500 000 b/d capacity. Finally, it is not always easy to identify what might make a 'good' or a 'bad' transit country. Various characteristics have been discussed in the literature but this is not a 'tick-box' exercise (ibid.).

A variant on this latter option is to develop an alternative route to avoid a country that has proved to be difficult over transit. Alternatives include another pipeline through an alternative country, but this suffers from the problems described in the previous paragraph. Other alternatives might include LNG, CNG and embodied gas. The main problem with this solution is that it assumes access to the high seas and it is also likely to involve high-cost solutions compared to a simple pipeline.

Historically, countries that have placed a high priority on attracting foreign investment have tended to be 'good' transit countries. The explanation is simple. Unilateral demand to renegotiate a bilateral contract will seriously damage the reputation of the country concerned and will act as a serious constraint on new inward investment. In this sense it might be argued that the process of increasing globalization that characterized the 1990s helped to reduce tensions over transit in many cases. For example, in the 1970s and 1980s Turkey had little interest in attracting FDI and its transit record in the period was very bad. However, since then FDI has been a central plank of economic policy and there are indications that Turkey's transit behaviour improved.

Other solutions that have been suggested include developing mutual dependence between countries such that 'bad' behaviour by the transit country can provoke effective retaliation by those hurt. For example, it has been suggested that if the Iran–Pakistan–India pipeline goes ahead, India could supply electricity to Pakistan so if that country played games with India's gas supply, India could retaliate by cutting off power. Another original solution (Stevens, 2009) is to make the transit fee a progressive fee linked to the price of the gas. Thus if prices rise, the transit fee automatically increases and vice versa.

Finally there is the option to try to create some form of overarching jurisdiction to manage disputes. The most obvious example of this is the Energy Charter Treaty. Unfortunately progress on this has been stalled by Russia's refusal to ratify it.

4.   THE FUTURE OF GAS MARKETS AND THE ROLE OF SHALE GAS

In recent years the USA has experienced what has become known as the 'shale gas revolution'. Shale gas is part of unconventional gas. Figure 2.9 illustrates the various types of unconventional gas. The simplest way to think about unconventional gas is that drilling is not enough. Something more must be done to allow the gas to flow in commercial quantities. The 'shale gas revolution' in the USA has been achieved by the application of horizontal drilling and hydraulic fracturing. Neither is a new technology, but their application to shale gas operations has produced results that are nothing short of spectacular.

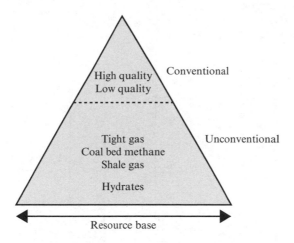

Source: IEA (2009).

Figure 2.9   Types of gas

Small entrepreneurial exploration companies using a dynamic and competitive service industry achieved this. While this might be seen as a triumph for the private sector, it was also underpinned by the US government spending very large sums of money on R&D into managing hydrocarbon production in a low-permeability environment.

The impact of the shale gas revolution is illustrated in Figure 2.10. Furthermore, the bulk of the increase in shale gas production has occurred since 2006. It is this that has justified the title 'revolution', although it is important to point out that this has been over 20 years in the making. It was not an overnight phenomenon.

A number of consequences have followed. Domestic gas prices have fallen, as can be seen from Figure 2.11, and LNG imports have also fallen, as can be seen from Figure 2.12.

Between 2005 and 2009, US regasification capacity increased by some 75 per cent in anticipation of falling US domestic gas production. In 2011, some 90 per cent of this is idle and US pipeline imports are the lowest since 1999 (EPRINC, 2011).

Shale gas clearly has the potential to transform the global energy scene. However, two questions are key. Can this revolution continue in the USA and can it be replicated elsewhere?

For the USA there are two main concerns. First, the lower gas prices threaten the economics of many shale gas operations, although to some extent this is offset by the fact that the large IOCs are beginning to buy into shale gas operations in the USA and they have deep pockets. Also, many of the small producers hedged their prices and so are not suffering from the current low prices. Second, there is concern over the possible negative environmental consequences of hydraulic fracturing. This involves injecting water and chemicals at very high pressure into the gas plays. Much of the concern arises because the 2005 Energy Act explicitly excluded hydraulic fracturing from the Environmental Agency's Clean Water Act. Therefore many shale gas operations have been undertaken with few or no proper environmental impact assessments. As concerns grow, drilling moratoria have been called at state and local levels while environmental impact studies are completed.

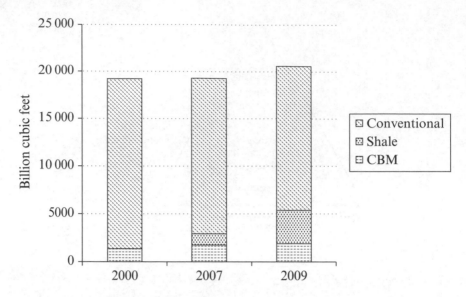

Note: The shale gas production data in the USA are very poor and much delayed. The latest data from the Energy Information Administration as of March 2012 are for 2009. Casual conversations by the author suggest that in 2011, shale gas accounted for around 25 per cent of US domestic gas production compared to 16.4 per cent in 2009.

Source: . US EIA (www.eia.gov/).

Figure 2.10   US gas production by source

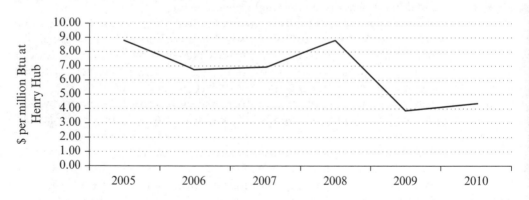

Source: US EIA (www.eia.gov/).

Figure 2.11   US domestic gas price

As for the issue of replication outside the USA, undoubtedly the resources exist. Potentially global unconventional gas resources (coal-bed methane, tight gas and shale gas) are estimated at five times conventional gas reserves. Figure 2.13 illustrates a recent estimate.

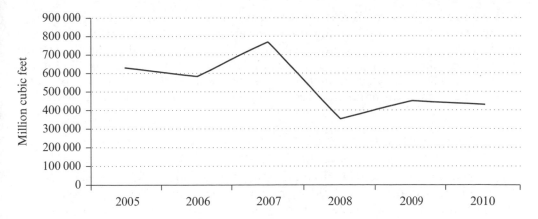

Source: US EIA (www.eia.gov/).

Figure 2.12   US LNG imports

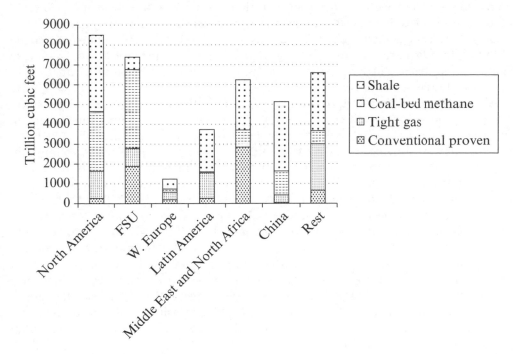

Sources: Conventional proven: BP (2008); others: NPC (2007).

Figure 2.13   Estimates of global gas resources

There are concerns, however, over the ability to translate these resources into proven reserves and production. The US shale gas revolution was triggered by favourable factors such as geology, tax breaks, a vibrant service industry and relatively easy access to pipeline networks by means of a legal regime based upon common carriage. However, in Western Europe, where there is much current interest, geology is less favourable, there are no tax breaks, the drilling industry is far behind the USA and pipeline options via third-party access are less favourable. Also in a densely populated Europe (compared to the USA), disruptions caused by shale gas developments will struggle to find public acceptance. This is driven by a fundamental difference in property rights. Unlike in the USA, benefits from gas production accrue to governments, not to local landowners.

5.   CONCLUSIONS

From the uncertainties deriving from these two key questions major problems arise. As the world recovers from global recession and as constraints on using gas continue to erode, gas demand will grow and probably gain ever-greater shares in the global primary energy mix. It is interesting to speculate in what ways GDP growth may affect the proportion of growth in gas compared to other primary energy sources. Electricity tends to have much higher income elasticities of demand than other energy sources. Given the increasing importance of gas in power generation as a result of the spread of CCGT technology, this implies that gas will gain ever-larger market shares of primary energy. Gas is also likely to gain ground in the transport sector as growing concerns about particulate pollution in urban areas encourage the growth of CNG in transport. Finally, if climate change issues remain high up policy agendas, more gas could be burnt in an effort to move to a lower-carbon-based economy.

   All this suggests that it is reasonable to assume that gas will increase its market share in global primary energy. However, given investor uncertainty, investment in future gas supplies will be lower than would have been required had the 'shale gas revolution' not happened or at least had not been so hyped up. If the 'revolution' in the USA continues to flourish and is replicated elsewhere in the world, this inadequate level of investment in gas will not matter. Consumers can look forward to a future, floating on unlimited clouds of cheap gas as unconventional gas fills the gaps. It might even be argued that cheaper gas could propel global economic growth in the same way that cheap oil fuelled the OECD economic miracle in the 1950s and 1960s. It could also presage an energy transition in the same way that the USA moved from wood to coal in the last 40 years of the nineteenth century and from coal to oil in the twentieth century. However, because of the growing complexity of energy markets and technology, such a transition is likely to take place over a longer rather than a shorter period. Clearly a key determinant of the speed of any possible transition will depend upon how the debate on climate change develops and where the issue will stand on the global policy agenda.

   If unconventional gas fails to deliver on current expectations – and we will not be sure of this for some time into the future – then in ten years or so, gas supplies will face serious constraints. Of course markets will eventually solve the problem as higher prices encourage a revival of investment in gas supplies. However, given the long lead times on most gas projects, consumers could face high prices for some considerable time. Failure to deliver on the hype might also mean that many economies would become increasingly at the mercy of fluctuations in gas prices.

   Another problem concerns investments in renewables for power generation. There is general agreement that the world must move to a low-carbon economy if climate change is to be controlled. Among other things, this requires much greater investment in renewables. The failure of the Copenhagen talks on climate change (and arguably in Durban) has already injected considerable uncertainty into the future investment climate in power generation, not least because of uncertainty over what the future price/cost of carbon might be. The uncertainties created by the 'shale gas revolution' have significantly compounded this investor uncertainty. In a world where there is the serious possibility of cheap, relatively clean gas, who will commit large sums of money to expensive pieces of equipment to lower carbon emissions? However, that said, any large-scale switching from coal and oil to gas will aid greenhouse gas mitigation. While not creating a transition to a zero-carbon economy, it would be better than many of the current projections for CO_2 emissions derived from a world dominated by coal into power.

Finally, it is interesting to consider how far the historical experiences of the relationship between the global economy and oil markets may offer lessons for the future of the relationship between the global economy and gas markets. For example, what are the prospects for the development of a gas exporters' cartel along the lines of OPEC? As the appendix outlines, gas is different from oil. Nonetheless it is tempting to draw analogies although there would have to be very significant changes in the cost of transporting gas before the experience of oil would have greater relevance.

NOTES

1. The author would like to thank Manfred Hafner for comments on an earlier draft of this chapter.
2. All statistics in this chapter, either in the figures or in the text, have been derived from BP (2011) unless otherwise stated. A major problem in dealing with international gas is that of differing units of measurement. Different sources swing between imperial and metric measures to the bewilderment of the reader as well as the analyst. For example, the BP *Statistical Review of World Energy* provides data in cubic feet, cubic metres and tonnes of oil equivalent (toe). Table 2A.1 provides conversions.
3. There was also the ever-present problem of high transport costs. This meant that the net back value of the gas was often low, which meant that the projects often were simply not profitable – a fact reinforced in markets where the domestic prices were kept artificially low for social reasons. By contrast, in countries that produced both oil and gas, gas often developed strongly because using it domestically released oil for export.
4. Estimates vary, but in rough terms a 2 billion cubic feet per day (BCFD) pipeline needs reserves of at least 15 trillion cubic feet (TCF). A 6 million tonne per year LNG (liquefied natural gas) project needs at least 8 TCF.
5. In the 1980s discovering gas rather than oil was regarded as bad news for the companies. Gas was seen as a problem rather than a blessing. Such views began to change in the 1990s as gas markets started (in some cases) to look more attractive, but the legacy remained.
6. Even today, only North America, the UK and perhaps Argentina (although there are significant price controls) can be described as real commodity supply markets.
7. There is a debate over whether gas-to-gas competition is superior to price indexation to a competitive fuel. It is argued that price indexation was the result of commercial negotiations between the parties. While there were commercial conflicts as circumstances changed, they were resolved usually without affecting the flow of gas. While in the absence of alternatives, given the stage of market development in the countries, this was acceptable, the use of market prices seems to be the simplest option.
8. Indeed, until relatively recently, the few gas transit lines which existed had a good record in terms of avoiding conflict.
9. As will be developed below, many of these problems and constraints are being removed, making LNG projects relatively more attractive.

10. To give an idea of how cold this is, if the temperature of a sheet of steel were lowered to this level, hitting the frozen sheet with a hammer would cause it to shatter like glass.
11. Defining lead times can be tricky. If timed from the final investment decision (FID), then they have not changed much. However, the earlier projects often took a very long time to get to the FID.
12. These include the fact that CCGT plants have a lower capital intensity than most other power plants, have a high energy efficiency, are environmentally friendly and have relatively short lead times. They can be small scale or large scale and, thanks to their low environmental impact, can be located much closer to demand loads. Also, they are subject to very quick payback for investors.
13. In particular, the European Commission has been trying for a long time to convert the European gas market to one resembling a commodity supply market. However, there are extremely powerful vested interests within the sector trying to resist these moves.
14. After 2008, in the aftermath of the global economic crisis, costs began to fall back again.
15. This occurred because of the economies of scale inherent in the development of very large tankers together with a large surplus of tanker capacity following the first oil shock of 1973–74 that lasted well into the early 1990s.
16. Much of the internationally traded gas identified in Figure 2.8 is in fact intraregional trade.
17. This figure actually understates 'gas' consumption because it fails to include 'town gas' since consumption of this would be subsumed in the data for coal consumption.
18. In fact, it was not until after the Second World War that a major network of gas pipelines began to emerge on any significant scale.
19. By far the largest consumer of gas has been the manufacturing sector. In many cases the boilers used in the sector are dual fired with oil, which creates significant competition between the two fuels, given the ease of switching at short notice.
20. Intrastate trade was not regulated to the same extent and therefore attracted higher prices. The result was very restricted interstate trade, with producers preferring to sell in state.
21. Specifically this involved the Federal Energy Regulatory Commission (FERC), the successor to FPC Order Number 436 1985, the Natural Gas Wellhead Decontrol Act of 1989 and the FERC Order 636 of 1992.
22. Russian gas started flowing in the late 1970s and Algerian gas in 1982.
23. Even here, the conflict was extremely limited. Supplies were disrupted only for a couple of days in January 2006. In January 2009, supplies were disrupted for several weeks, as the result of which there were deaths from hypothermia in South Eastern Europe where supplies were especially badly hit. Slovakia actually declared a State of Emergency.
24. There is also considerable optimism over the potential for unconventional gas but it is too early to determine how realistic an option this might be (see Section 4).
25. It is also important to note that, in the UK, gas development was based upon domestic production and there was a large number of producers of gas that, after the privatization of British Gas, allowed a competitive market to develop. This was not the case in continental Europe.
26. The contracts were with France (3 bcm), Italy (1.1 bcm) and West Germany (1.25 bcm).
27. This is in contrast with the oil sector, which was broken up and the assets distributed among a number of individual entities, some of which were then privatized.
28. Pipeline imports into Asia Pacific are very small, with Malaysia, Singapore and Thailand importing only 19 bcm in 2009.
29. Private communication with Manfred Hafner.
30. This also includes former members of Comecon who are not now members of the EU.
31. Other options are feasible but not really on a large scale. These include CNG (compressed natural gas); gas by wire and embodied gas.
32. In this context the term 'sovereign', which has very specific meaning in international law, simply means that the signatory can unilaterally rip up any contract without fear of legal redress. As will be developed, of course there may well be other penalties for such behaviour.
33. What follows draws heavily upon Stevens (2009). For further details of conflict over pipelines, see ESMAP (2003) and Stevens (1998).
34. It might be argued that in some cases transit pipelines can also contribute to stabilization of relations. Trade can be seen as a positive element to reduce conflict providing everyone gets what they perceive to be a reasonable share of the spoils.
35. Often the transit fee, normally based on a volumetric throughput, also includes access to offtake from the line at discounted terms as part of the 'fee'.

REFERENCES

BP (2008), *Statistical Review of World Energy 2008*, London: BP.

BP (2011), *Statistical Review of World Energy 2011*, London: BP.

Bradley, R.L. (1996), *Oil, Gas and Government: The US Experience*, Lanham, MD: Rowman & Littlefield for the Cato Institute.

Chesshire, J.H. and Huggett, C. (1975), 'Primary energy production in the Soviet Union: problems and prospects', *Energy Policy*, **3**(3), 223–44.

Darmstadter J., Teilelbaum, P.D. and Polach, J.G. (1971), *Energy in the World Economy*, Baltimore, MD: The Johns Hopkins Press for Resources for the Future.

EPRINC (2011), 'Natural gas industry fakes the moon landing', EPRINC Briefing Memorandum, 1 July.

ESMAP (2003), 'Cross border oil and gas pipelines: problems and prospects', ESMAP Technical Paper, UNDP/World Bank, Washington, DC.

IEA (2009), *World Energy Outlook*, Paris: International Energy Agency, OECD.

Jensen, J. (2004), 'The future of gas transportation in the Middle East: LNG, GTL and pipelines', presentation to the annual conference of the Emirates Center for Strategic Studies and Research, Abu Dhabi, 20 September 2004.

Jensen, J.T. (2009), 'LNG Its role in the internationalization of gas markets', presentation to Columbia School of International and Public Affairs, New York, 7 October.

McLellan, B. (1992), 'Transporting oil and gas – the background to the economics', *Oil and Gas Finance and Accounting*, **7**(2).

NPC (2007), 'Unconventional gas topic paper', Working Paper, 29 July.

Stevens, P. (1998), 'A history of transit pipelines in the Middle East: lessons for the future', in G.H. Blake, M.A. Pratt and C.H. Schofield (eds) *Boundaries and Energy: Problems and Prospects*, London: Kluwer Law International, pp. 215–32.

Stevens, P. (2008), 'National oil companies and international oil companies in the Middle East: under the shadow of government and the resource nationalism cycle', *Journal of World Energy Law and Business*, **1**(1), 5–30.

Stevens, P. (2009), 'The trouble with transit pipelines', Chatham House Report, Chatham House, London.

Stevens, P. (2010), 'The shale gas revolution: hype versus reality', Chatham House Report, Chatham House, London.

Wright, A.W. (1975), 'Contrasts in Soviet and American energy policies', *Energy Policy*, **3**(1), 38–46.

APPENDIX

Gas is different from oil. Several differences are key. Gas is essentially a regional rather than a truly global market. This arises simply because gas suffers from the 'tyranny of distance' which means that, because it is a high-volume, low-value commodity, it is expensive to transport.[1] Admittedly, in recent years, changes to the LNG market that will be outlined below have given rise to expectations that gas was moving from regional to more global markets but this trend may well have been halted by the sudden development of unconventional gas (Stevens, 2010). This regional dimension is strongly reinforced in the context of earlier periods. Early gas consumption was based upon 'town gas' manufactured from coal. Small-scale local companies invariably did the 'manufacturing'.[2] These were monopolies within relatively small areas for markets. Gradually, however, 'town gas' was increasingly replaced with natural gas, with 'town gas' production all but ceasing in the USA in 1966 and in Europe in the 1980s.[3]

There is much less economic rent in the gas price than in that of oil. This is because, at least to date, there is no gas cartel to fulfil the same role as OPEC, that is, restrain supply to ensure significantly higher prices than would exist in a competitive market.[4] Also, transporting gas is much more expensive than oil, therefore the netback, that is the rent, is much lower.

Security of gas supply is also far more complex than in the case of oil. A loss of oil supplies can obviously matter to an economy given the outage costs but once the disruption has been solved, supplies can easily be resumed. Also it is far easier to replace lost oil supplies given the flexibility of oil transport and trade. Gas has much less flexibility in terms of transport and trade. Also safety concerns and the integrity of the gas grid mean that it is difficult and dangerous to turn gas supplies off and on.[5]

The gas trade, unlike the oil trade, is mostly based on long-term contracts. The reason lies in the cost structure of gas projects and their specificity. Normally, producing gas and getting it to market requires very large projects. Such projects tend to be characterized by very high fixed costs and relatively low variable costs. This requires that the equipment be operated at full capacity. Less than full capacity means that the high fixed costs are spread over a smaller throughput and profits decline exponentially. Furthermore, because of the 'bygones rule', such losses will be borne by the operator for a long time before closure is a rational economic option. Also the very high initial cost needs to lock in future revenue streams to justify the project since the payback period is relatively long. Of course, such characteristics are by no means peculiar to gas. For example, upstream oil projects, especially those offshore, are very similar. However, because of the transport constraints facing gas, gas projects are highly specific. The end of a pipeline is the end of a pipeline. If nothing emerges, then finding alternative supplies of gas is very difficult simply in terms of the logistics let alone in terms of commercial issues. In a similar vein, LNG projects must have access to regasification plants and until recently there has been very limited if any flexibility in the trade. Thus, unlike the oil trade which has much greater flexibility, the gas trade depends upon long-term contracts. Also, as discussed above, due to lower rent for oil, gas projects have a much longer payback time than oil projects. Loss of supply or loss of customers presents a very serious threat to the economics of any gas project.

Gas transmission grids (and indeed single pipelines) are natural monopolies and therefore must either be in public ownership or, if privately owned, heavily regulated. This, together with the need for long-term contracts, has meant that gas has tended to have much greater state involvement than in the case of oil. Thus outside of the USA until the 1980s and 1990s, gas companies in most markets had a tendency to be state-owned utilities.

Table 2A.1   Gas conversions

From	To					
	billion cubic metres NG	billion cubic feet NG	million tonnes oil equivalent	million tonnes LNG	trillion British thermal units	million barrels oil equivalent
	Multiply by					
1 billion cubic metres NG	1	35.3	0.90	0.74	35.7	6.60
1 billion cubic feet NG	0.028	1	0.025	0.021	1.01	0.19
1 million tonnes oil equivalent	1.11	39.2	1	0.82	39.7	7.33
1 million tonnes LNG	1.36	48.0	1.22	1	48.6	8.97
1 trillion British thermal units	0.028	0.99	0.025	0.021	1	0.18
1 million barrels oil equivalent	0.15	5.35	0.14	0.11	5.41	1

Source: BP (2011).

Notes

1. In many ways this was similar to the oil markets before the advent of much cheaper transport following the development of large crude carriers in the later 1960s.
2. In Europe the municipalities themselves often owned these companies. In the USA they were largely private companies.
3. The calorific value of natural gas (methane) is very much higher than that of 'town gas'. This means that while both can use the gas supply networks, the gas-burning appliance must be adjusted for one or the other. This effectively precludes running both types of gas in the same system. When consumers switch to natural gas the whole of the gas-burning appliance stock must be refitted.
4. In many gas markets, the gas price is linked to oil prices and so there is a side-effect of some benefit from higher oil prices
5. In theory, to be absolutely safe each gas-burning appliance needs to have a gas engineer present before supplies can be reconnected following an outage. In the context of residential supplies this can be extremely time consuming. In the early 1980s, British Gas – the state gas monopoly – claimed that if Birmingham, Britain's second largest city, were cut off from supplies it would take around three years to reconnect all customers.

3 The likelihood and potential implications of a natural gas cartel

Steven A. Gabriel, Arild Moe, Knut Einar Rosendahl and Marina Tsygankova

INTRODUCTION

In 1973 the Organization of Petroleum Exporting Countries (OPEC) shocked the world by taking control of the oil market through price increases and production cutbacks. The history of OPEC, however, goes back to 1960, when Iran, Iraq, Kuwait, Saudi Arabia and Venezuela founded the organization. Can a comparable event take place in the natural gas market; that is, should the market be prepared for the appearance of a 'gas OPEC'? Already there exists a gas-exporting organization quite similar to the OPEC of the 1960s, the Gas Exporting Countries Forum (GECF), founded in 2001. The GECF member countries currently control almost two-thirds of global gas reserves. So far the Forum has lived up to its name, by acting more like a discussion forum than a forceful organization. The history of OPEC, however, clearly shows that organizations are dynamic by nature, and so the question remains whether a 'gas OPEC' will emerge from the GECF, or from another coalition of gas-exporting countries, in the foreseeable future.

The current chapter discusses the prospects for a natural gas cartel, taking the GECF as our point of departure. We consider the GECF countries' position in the current gas market, comparing them with the OPEC countries. Gas markets have certain characteristics that differ from the oil market, and these are important to assess as well. A prerequisite for turning any constellation of gas-exporting countries into a gas cartel is that it is sufficiently profitable to coordinate action. We discuss this in light of simulation results presented in a recent study (Gabriel et al., 2012), using the World Gas Model, a large-scale model for international gas markets. Moreover, the potential market implications of a gas cartel are considered.

The biggest gas exporter in the world is Russia, which also holds the largest share of global gas reserves. Russia is one of the key GECF members, and its position *vis-à-vis* cartelization will to a large degree determine whether or not, and in what form, a gas cartel will appear. Thus we discuss in a separate section Russian gas policies, speculating about its willingness to take part in an effective gas cartel.

The prospects for gas cartelization will obviously also depend on how supply and demand conditions evolve. The technological innovations in unconventional gas extraction in general, and shale gas production in particular, have substantially changed market expectations in the USA, and can potentially do so in other regions as well. The demand for gas will not least be affected by policies to curb emissions of greenhouse gases, and by policies that directly stimulate the use of renewable energy. Towards the end of this chapter we discuss in more detail these important drivers for supply and demand, including their potential effects on gas cartelization.

PROSPECTS FOR A NATURAL GAS CARTEL

The Gas Exporting Countries Forum (GECF) was founded at a ministerial meeting in Tehran in 2001, with 'the objective to increase the level of coordination and strengthen the collaboration between member countries' (www.gecf.org). As the GECF members together account for almost two-thirds of global gas reserves, it is no surprise that there have been regular speculations about a gas cartel *à la* 'gas OPEC' since 2001 (see e.g. Hallouche, 2006; Jaffe and Soligo, 2006). The GECF shares several similarities with OPEC, but there are also significant differences that we will come back to below. Some of these differences are related to the member countries of the two organizations, and their characteristics, but there are also differences related to the properties of oil and gas. In this section we will first discuss the likelihood of a future gas cartel, presumably originating from the GECF. Then we will consider potential implications for the gas market if such a cartel should become effective in optimizing its members' joint export profits.

GECF: A Future Gas OPEC?

The GECF currently has 13 members, including the world's three largest gas producers in terms of reserves, Russia, Iran and Qatar.[1] These three countries account for 53 per cent of global gas reserves (BP, 2010), and are therefore considered as the key members of the GECF. However, their collective share of global gas production is much more modest (25 per cent in 2009), and the importance of exports varies greatly between them. In fact, Iran has not been a net exporter of gas since the Iranian revolution in 1979 (except for a few single years). Thus, at the moment Iran has few incentives to cut back on supply in order to raise the price of gas exports. This could change in the future, though, if the country is able to develop its gas industry, for instance by attracting investments in LNG facilities. Qatar, on the other hand, with a population of fewer than two million, exports more than three-quarters of its gas production, and is the largest LNG exporter in the world with a market share of 20 per cent of global LNG trade. Russia, which we discuss in more detail in the next section, is the world's leading gas exporter with a share of 20 per cent of global gas trade (mainly through pipelines), but three-quarters of Russian gas production is consumed domestically.

According to the GECF website, one of the Forum's missions is to 'identify and promote measures and processes necessary to ensure that Member Countries derive the most value from their gas resources'. Taken literally, this mission seems to reflect the objective of a producer cartel, aiming to maximize the joint profits of its members. However, although some member countries have called for coordinated cuts of gas production, these suggestions have been voted down so far.[2] This may suggest either that the cited mission should not be taken too literally, or that there are too many difficult challenges to resolve before coordinated production cuts can be realized. The GECF members clearly have incentives to tone down expectations about production cuts that may not materialize, both because it may cause gas importers to diversify towards imports from non-GECF members as well as towards other energy sources, and because international companies may become more reluctant to invest in GECF countries.

One major challenge for any potential gas cartel is how to allocate production, or production cuts, between the cartel members. First of all, if we consider the GECF as a prospective gas cartel, the forum consists of 13 quite heterogeneous countries. This has already been touched upon above with respect to gas market characteristics, and to further exemplify we may add that GECF member Equatorial Guinea's gas production amounts to only 1 per cent of Russia's production level. In this respect, however, the GECF is not very different from OPEC. The effectiveness of a cartel is arguably largely dependent on the willingness and capability of its key members to restrict their supply. Thus a comparison between the GECF and OPEC should rather focus on their key members.

Saudi Arabia is clearly the key member and driving force of OPEC. Other important members are Kuwait, the United Arab Emirates (UAE), Iraq, Iran and Venezuela. All these countries export a majority of their oil production (in fact all OPEC countries do). On the other hand, as explained above, two out of three key GECF members (Russia and Iran) consume at least three-quarters of their gas production domestically. Furthermore, whereas five out of the six OPEC members just mentioned are located around the Persian Gulf, the most important GECF member is certainly Russia. We return to Russian gas policies in the next section, and just note here that Russia's position *vis-à-vis* cartelization seems to be decisive.

Another challenge for a prospective gas cartel lies in the characteristics of natural gas as an energy source. Due to the significant costs of transporting gas, the international gas markets have traditionally been regionalized, with for example one North American market, one European market and one East Asian market, and limited interregional trade. Over the last decade, however, trade across the Atlantic Ocean has increased, partly due to lower costs and increased capacity of LNG. Qatar's rising LNG exports, which can be directed to either Europe or South and East Asia, have led to larger market integration across regions.[3] Still, the prevalence of long-term contracts reduces the flexibility of the international gas market, and can make it difficult for a cartel to adjust supply and influence the price of gas (see e.g. Finon, 2007). Looking ahead, however, the share of spot market sales is growing, and is expected to further increase.

The substantial costs of transporting gas, as well as the huge dependence on infrastructure in the short to medium term (LNG facilities and pipelines),[4] further imply that production cuts in Russia may have quite different market effects than production cuts in Qatar. It thus follows that a profit-maximizing gas cartel not only has to consider the collective production cuts by the cartel members, but also the distribution of cutbacks among the cartel members. In addition, it matters whether for example Qatar cuts back on its exports to East Asia or to Europe. Consequently, it may be optimal for the cartel as a whole that country *i* cuts back substantially on its production, whereas country *j* only cuts very small amounts. This problem is less pronounced in the oil market, where transport costs represent a rather small share of the market price. However, as production costs may vary across countries, both for oil and gas extraction, similar tradeoffs can also be relevant in the oil market.

A well-functioning cartel can solve this problem by transferring money between members, so that for instance country *i* receives a transfer from country *j* to compensate for a large production cut. There are, however, a number of difficulties with such a scheme. First of all, independent countries may be reluctant to even consider money transfer to other countries. Second, it is not at all straightforward to calculate the gains and losses from production cutbacks, and country *i* would presumably understate its production costs and overstate the price effects from its cutbacks.

Potential Impacts of a Gas Cartel

Putting all these challenges aside, what are the potential impacts of a gas cartel? Here we will mainly rely on the findings in a recent study (Gabriel et al., 2012), which uses a large-scale partial equilibrium model of the international gas markets (World Gas Model) to explore the effects of gas cartelization. The study considers three cartel variants, all including the current GECF members.[5] Two variants also include the Caspian countries, and one of these two adds the rest of the Middle East as well. The model simulations abstract from all the difficulties with cartelization discussed above, and assume that the cartel maximizes joint discounted profits over the years 2010–30. Moreover, the market equilibrium is modelled as a Nash–Cournot equilibrium. That is, in equilibrium no player (including the cartel acting as one player) can gain from changing its supply, given other players' levels of supply to the different regions. An alternative model formulation could be to model the cartel as a Stackelberg leader, which takes into account the other players' response to the cartel's supply.

The first conclusion we may draw from this study is that a GECF-based gas cartel will have modest, but not insignificant, impacts on the international gas markets. Although it is optimal for the GECF to reduce joint production and exports, non-GECF countries will respond by expanding their production. Thus the impacts on prices and consumption of gas are mostly limited. Not surprisingly, the biggest effects are seen in Europe, where the wholesale prices of gas increase by around 15 per cent according to the simulations. Europe is the biggest export market for Russia, and eight of the GECF member countries exported gas to Europe in 2009 (BP, 2010).

The second conclusion we want to highlight is that the gains from cartelization are quite limited when the cartel consists (only) of the GECF countries. Changes in annual profits for the cartel reportedly vary between −3 per cent and +15 per cent when considering specific years. Although production cutbacks may give substantial price benefits in the short run, increased investments by non-GECF exporters may be detrimental for the cartel in the long run. Furthermore, several cartel members, including the key GECF member Russia, get lower profits than in the base case (i.e. without any cartel). Thus money transfers would be needed in order to make this outcome possible. Without money transfers, a suboptimal allocation of production cuts would be required, and the joint cartel profits would decline. Given the likely costs of implementing coordinated production cuts, in both economic and political terms, the reported simulations suggest that a GECF-based cartel is not very likely.

Adding more countries to the cartel makes it more powerful. The Caspian countries, located in between Russia and Qatar/Iran, reap substantial market shares when staying out of the cartel. Thus, by inviting these countries into the cartel, the gains from cartelization may improve notably. The simulated profits for the cartel members now rise by between 12 per cent and 30 per cent per year (*vis-à-vis* the base case). Moreover, the price increase in Europe is doubled compared to the GECF-only cartel scenario. Thus, if Russia is able to convince other Former Soviet Union countries to join the GECF, coordinated production cutbacks could be much more effective and thus tempting. The Caspian region consists of four significant gas producers, with Turkmenistan having the largest reserves (it ranks 4th in the world according to BP, 2010). If also the other Middle Eastern countries join the cartel (Saudi Arabia and the UAE are the two most important countries here), joint profits for the cartel members may increase by more than one-third (*vis-à-vis* the base case). Impacts on prices and consumption, especially in Europe but also to some degree in Asia, are intensified.

The last conclusion we will draw from the Gabriel et al. (2012) study is that cartelization of gas markets most likely will have insignificant effects for North and South American consumers. This is partly because only three of the smaller GECF members are located in the western hemisphere. Furthermore, both North and South America are expected to have little net trade over the next couple of decades. Only a few years back, the USA was expected to import gradually more gas over the coming years, with significant import needs expected in 2020–30. However, the rapid improvement in shale gas technologies over the last 3–5 years has drastically increased the level of profitable reserves in the USA. We come back to this issue in a separate section below, but note here that this development most likely has reduced the profitability of a future gas cartel inasmuch as a large market for the cartel members has more or less disappeared.

The discussion above seems to suggest that gas cartelization is not an imminent threat to gas consumers around the world. Even though GECF countries control about the same share of global gas reserves as OPEC's share of global oil reserves, more countries with significant gas resources must find it in their interest to join the GECF before this organization can have comparable impacts in the gas market as OPEC has in the oil market. In addition, GECF countries must be interested in *and* able to coordinate substantial production cuts.

The prospects for gas cartelization in the future obviously depend on how the gas market otherwise evolves. One important driver is the development of shale gas technologies, which we come back to below. More generally, technological progress can substantially alter the energy markets, and thus influence the potential for gas cartelization. Some kinds of technological improvements, for example in carbon-free technologies, can reduce the demand for gas and thus the gains from gas cartelization. Other kinds of technological improvements can have quite the opposite effects. For instance, technological progress in the LNG chain will reduce the costs of reaching more distant markets for gas exporters, possibly increasing the potential gains from gas cartelization. However, it can also lead to stronger competition in regions close to the cartel member countries, such as Europe.[6] Another important driver, which we also return to below, is energy and climate policies. It is not clear in general whether new policies will favour gas or not.

RUSSIAN GAS POLICIES – COHERENT WITH GAS CARTELIZATION?

Russia is endowed with huge energy resources. It controls 44 trillion cubic metres (tcm) of natural gas, which is about 23 per cent of the world's proven gas reserves. Russia is the world's largest natural gas exporter and second-largest gas consumer. Also, Russia has the world's second-largest coal reserves and eighth-largest oil reserves. The abundance of energy resources has shaped Russian economic and international policies around the use and sale of energy resources (especially natural gas and oil). It became common to refer to Russia as an 'energy superpower' when oil and gas production surged in the early 2000s and especially after energy became intertwined with Russian foreign policy, most notably in the relations with Ukraine, Belarus and Georgia during the second half of the last decade.

The Russian gas company Gazprom controls practically all activities in the Russian gas sector. In 2009, only 21 per cent of Russian gas production came from non-Gazprom producers, which are oil companies and smaller gas companies. But also they must work in accordance with Gazprom's interests. Gazprom owns all the high-pressure transmission pipelines in Russia and has an exclusive right to export natural gas. The Russian Federation is the main shareholder of Gazprom, and its current stake is 50 per cent. The company enjoys a close relationship with the Russian government beyond formal ownership rights.

Thus Russia exhibits characteristics of a vigorous cartel leader: it controls a substantial share of the cartel's resources – in fact 40 per cent of GECF countries production – and it has shown willingness to constrain gas supplies to achieve better commercial conditions, as witnessed in the gas conflicts with Ukraine in 2006 and 2009. Internal control over the gas sector is concentrated. The question is if Russia is ready to use its formidable role as a gas supplier on a larger scale, and in collaboration with other suppliers to enhance the price of gas.

Since the establishment of the GECF in 2001, Russia has several times supported the idea of closer coordination between producers. Still, Russian authorities have rejected allegations that they are aiming for a cartel.[7] Representatives of the other core members, Iran and Qatar, have been more aggressive on this point.[8] The appointment of Leonid Bokhanovsky as the first Secretary General of the GECF in 2008 formalized the leading position of Russia in the GECF. But this does not necessarily mean that Russia supports the idea of transforming the Forum into a cartel at some point. Seen from Russia, the impression of the oil cartel – OPEC – must be mixed. While as a whole that organization has been useful in achieving a higher oil price, it is also clear that Saudi Arabia as the dominant member at times has carried a disproportionate burden. In a gas cartel Russia would soon find itself in a similar position. Another interpretation of Russia's leadership in GECF is that the position lets Russia harness the organization to work according to Russia's priorities.

Exports of Russian Gas

In 2009 gas exports to Europe amounted to 69 per cent of Gazprom's foreign sales – the rest is sold mainly to Former Soviet Union (FSU) countries (Gazprom, 2010). The price of Russian gas in Europe is significantly higher than the regulated domestic gas price in Russia. In 2009 the average prices of natural gas sold by Gazprom to the domestic and the European markets were respectively US$62 and US$238 per thousand cubic metres. Even if the difference in net earnings is smaller, gas exports to Europe are the most lucrative source of Gazprom's revenue and are also very important for the Russian state budget.

Just as Gazprom and Russia rely on revenues from the gas sales to Europe, the

EU is highly dependent on imports of natural gas from Russia. In 2009 34 per cent of the EU's gas imports came from Russia. Russia's share of EU gas imports has recently been reduced somewhat due to the EU's efforts to diversify its gas imports and the rapid development of the LNG market. However, Russia is actively engaged in holding on to its position in the EU gas market. First, Russia has sought to contract a large part of the future gas production of Central Asian countries and the Caucasus, thereby impeding the emergence of Central Asian competitors in the European market. Second, Russia has recently built a new pipeline to Europe, Nord Stream under the Baltic Sea, to reduce its dependence on Belarus and Ukraine, the main transit countries for Russian exports to Europe. Russia has also proposed to build a new southern line, South Stream, partly to fend off competition from another project – Nabucco (Baev and Øverland, 2010). Third, Gazprom is becoming more active in downstream activities in the EU, such as gas storage, transmission and sales, particularly in Central Europe.

Such developments enhance Russia's potential leverage in dealings with European customers, but at the same time they tie Russia to the European market and increase its interest in the stability of the market. However, Russia makes no secret of its plans to expand its gas exports to Asia and thus reduce its dependence on Europe. Today Russia supplies gas to the Asian markets mainly through the Sakhalin-2 project, which ships LNG largely to Japan. These deliveries are very small compared to the Russian gas exports to Europe and the FSU countries. Gazprom also has ambitions to supply gas by pipeline to China, whose growing economy has a substantial need for imported energy. In 2006, Gazprom and China National Petroleum Corporation signed a protocol for gas deliveries to China starting in 2011. However, since then there has been little progress towards the commencement of Russian gas deliveries to China, due to problems in agreeing upon the gas price. In addition, a gas transportation system to China is still lacking. If anything, the negotiations with China demonstrate that there is not a world market price for gas, at least not piped gas, and that the spillover from price changes in one part of the world to another is limited.

Domestic Market: Room for Export Expansions?

Although Russia is the world's largest gas exporter, the country exports only about 25 per cent of its gas production; the rest is consumed domestically. Russia is unique in terms of the high share of gas in its domestic energy consumption. This share has been growing since the collapse of the Soviet Union and reached 55 per cent in 2008 (BP, 2010). Such a high level of gas consumption is connected to the pricing policy inherited from the Soviet era.

Gazprom has always been obliged to sell gas domestically at regulated prices, which have remained below long-run marginal costs. When Russia began moving towards a market economy in 1991, it was believed that low domestic gas prices would stimulate economic growth following the collapse of the planned economy. Besides there was no way the economy – or population – could handle a rapid price increase. The downside of this policy in the longer run has been an exceedingly high growth in domestic natural gas demand, lower efficiency in energy use, and heavy underinvestment in the gas sector. As a result, the ability of the Russian gas sector to meet future demand by domestic and foreign consumers has come under question; see, for example, Milov et al. (2006), IEA (2006), Goldthau (2008) and Tsygankova (2008).

Russia is planning to introduce netback pricing of gas in the domestic market starting from 2015. Netback pricing refers to the equalization of the gas price in Russia to the gas price in Europe after adjusting for export taxes, transportation costs and transit tariffs. A pricing policy that equalizes the returns from domestic and export markets has long been demanded by the IMF, the WTO, the EU and others, as low regulated Russian gas prices are claimed to be export subsidies for Russian manufacturing exporters whose products embody large amounts of energy. The Russian government and Gazprom consider netback pricing as a measure to increase domestic prices, secure investments into the Russian gas sector, and to stimulate reductions in domestic gas demand and thus obtain spare capacity for export. But Russian authorities do not want to lose control over price formation in Russia. Very high domestic prices are not in the interest of Russia, both for social reasons and with regard to the competitiveness of Russian gas-consuming industries. If a direct link between European and domestic prices is established, Russia's desire to increase gas prices in Europe may be reduced. Such considerations are also likely to affect the country's interest in a gas cartel aimed at higher prices.

The 'big three' super giant fields (Medvezhe, Urengoy and Yamburg) that have dominated Russian gas production for many years have peaked and are now in a declining phase. The fourth super giant field, Zapolyarnoe, which commenced production in 2001, is about to reach its production peak. Gazprom is bringing into operation smaller 'satellite' fields located around its super giant fields. But these fields can only compensate the fall in gas production from the super giants for a few years. With growing demand, there is considerable attention to new long-term supply alternatives. The most important region is the Yamal Peninsula. Gazprom produced its first Yamal gas in 2012, but the further build-up of capacities in that region is unclear due to the enormous capital expenditures required.

Pressure on Long-term Contracts

The development of the Yamal Peninsula fields and other prestigious projects such as construction of the Nord Stream and South Stream pipelines, and the development of the offshore Shtokman field, require huge investments. Gazprom needs guarantees that these investments will be covered by future revenues. During the many years with unprofitable gas sales to domestic Russian consumers, the European gas market remained the only source of profit for Gazprom and the long-term contracts have been important for Gazprom in order to secure its large and irreversible investments. Russia has sold practically all its gas to Europe at long-term contracts of 20–25 years, and has always argued against shorter contracts and the sale of gas at spot prices. An additional reason for this stance is the limited technical flexibility in Russian gas supplies, due to modest gas storage capacity. To benefit from a spot market, as well as being able to manipulate supplies within a cartel context, more storage is needed. Presently the Russian capacity is only 2 bcm, whereas plans call for an increase to 6.5 bcm by 2016.[9]

At the same time, the recent developments in the European gas market such as liberalization and the surge in LNG capacity is putting pressure on long-term contracts to make them more flexible and of a shorter duration. Gas demand in Europe shrank following the global economic crisis. In addition, increase of shale gas production in the USA has made LNG suppliers look towards Europe. This recently led to oversupply of gas to Europe and a drop in spot prices below the price of the Russian long-term contracts. In 2010 Gazprom, fearing that its traditional European buyers would switch to other more flexible gas suppliers, had to renegotiate its long-term contracts and sell about 15 per cent of its gas at lower, spot prices. Gazprom seems to struggle with a dilemma. Even if Gazprom, to hold on to its position in the market in the short term, is forced to adopt shorter and more flexible contracts, it also is dependent on predictable long-term contracts in order to secure development of its new gas deposits and, thus, to be able to retain its dominant position in Europe in the longer term.

Russian Interests in Cartelization

While Russia would like to see technical cooperation with other gas producers, it officially rejects the idea of a cartel. We believe this is based on a realistic assessment of Russian interests. A cartel exerting influence on prices through manipulation of supply would presuppose a market dominated by short-term or spot contracts. Such a market is not perceived to be in Russia's interest, both because long-term contracts are perceived as necessary for financing development of the industry and because a market dominated by spot will give less predictability. Besides, lack of storage may give Russia less potential to reap benefits of spot trading. Finally, the plans of netback pricing in the domestic market seem to be at odds with cartelization.

In a situation dominated by long-term contracts a cartel would presumably have to address prices more directly, through coordination of sellers' positions. Such coordination would raise big practical problems, however, since negotiations between sellers and buyers within the different gas supply contracts do not take place simultaneously.

In either case, Russia's interests with regard to cartelization must be compared to the interests of other cartel members. We will argue that Russia has much stronger interests in the European market, through its long-term investments, than other potential cartel members such as Qatar or Iran. Consequently, the negative fallout from a cartel decision to cut supplies or conspire about pricing might be much larger for Russia than for the others. Indeed, the disunity of interests in the export markets is likely to be a major factor holding Russia back from joining a cartel. The gas conflicts with Ukraine clearly showed a willingness to constrain supplies in order to reach a better deal. But this was a targeted action that supposedly was under Russian control. What the conflicts also demonstrated, however, was that the political and economic fallout was much wider and costlier than anticipated. The risk for Russia involving itself in a cartel not totally under its control should be daunting.

Despite such objections to a cartel, Russia has a clear interest in what we could call 'strategic cartelization'. By this we mean coordination of infrastructure investments to avoid oversupply. Such discussions may entail various forms of coordination and compensations with other gas suppliers, and the GECF may serve this purpose. The GECF framework may also serve as an additional justification for Russia's efforts to reintegrate exports to Europe, and thus limit competition from former soviet republics, notably Turkmenistan, Kazakhstan and Azerbaijan.

THE RISE OF SHALE GAS[10]

Within the last few years, shale gas has figured prominently in natural gas supply fore-casts, particularly in the USA. Along with gas from tight sands, and coal-bed methane, gas shales are classified as 'unconventional', with the terminology referring to more low-permeability reservoirs that produce mostly natural gas (no associated hydrocarbon liquids) (NPC, 2007). This increase in shale gas supply is due to engineering advances such as hydraulic fracturing and horizontal drilling. A new report by the US Department of Energy (DOE) based on data from Advanced Resources International, Inc. indicates that initial assessments of some 163 trillion cubic metres (tcm) of technically recoverable shale gas resources in 32 non-US countries compared with 24.4 tcm in the USA alone (EIA, 2011). According to the DOE, this shale resource estimate increases the technically recoverable resources for the total world by over 40 per cent to more than 623 tcm.

The USA is certainly at the forefront of shale gas production with already 23 per cent of its total natural gas production in 2010 coming from this source. This dependence is forecasted to rise to 46 per cent of US production in 2035 (EIA, 2011).

Only a few years ago, bringing Alaska gas down to the lower 48 US states was seen as a key supply initiative albeit a huge infrastructure project. These days this source of gas has been relegated to the back burner for the most part. Also, while discussion on having adequate capacity for LNG imports in the USA has mostly faded away due to the abundance of shale gas, the talk now is of having the USA export some of its shale gas to markets in Europe and Asia. Indeed, in May 2011 the DOE gave Cheniere Energy, Inc. permission to export up to 2.2 billion cubic feet of LNG per day for 20 years from its Sabine Pass plant (*WSJ*, 2011).

The upside to US producers of exporting gas could possibly be large. Since mid-2010, spot prices in Europe have well exceeded those in the USA. According to *WSJ* (2011), Barclays Capital analyst Biliana Pehlivanova argues that 'Given the current prices, exporting U.S. LNG looks extremely attractive'. From the European or Asian perspective, gas originating from the USA (shale or otherwise) could provide a counterbalance to dependence on Russia and other suppliers, especially given the low current prices and favourable environmental advantages of gas over other fossil fuels. In a recent study for the DOE, the University of Maryland considered several scenarios for future natural gas markets using its World Gas Model, that is, the same large-scale model of global gas markets as referred to earlier (UMD, 2010). With the quantity of shale gas rising from about 11 per cent of the US total in 2010 to 32 per cent in 2030, the results indicated that North American prices would drop about 10 per cent in 2030 relative to a base case due to this extra shale gas.

This lowering of prices is not seen in other regions such as Europe in part because major export of shale gas was not modelled at that point. These results do highlight a key issue, though, that of the potentially positive economic effects of using this abundant supply of shale gas. On the negative side, especially in the USA, recently there has been environmental concern related to the extraction processes for shale gas. One of the main issues is the composition of chemicals used in the hydraulic fracturing ('fracking') fluid, involving millions of litres. According to the industry, about 99.5 per cent of the typical fracking fluids are just water and sand with 'trace amounts of chemical thickeners, lubricants and other compounds added to help the process along' (*NYTimes*, 2010). However, in some cases, diesel fuel is used as part of the fracking fluid, with some of the chemical components including benzene, which is a carcinogen.

Some regulators and environmental groups have expressed alarm at the prospect of some of the chemicals leaving the well bore due to 'migration through layers of rock or spills and sloppy handling – and into nearby sources of drinking water' (*NYTimes*, 2011b). Indeed, due to lack of regulation at the state and federal levels, US Representative Henry A. Waxman of California and others indicated that this use of diesel fuel may be a violation of the Safe Drinking Water Act. The US Environmental Protection Agency (EPA) intends to have initial study results on hydraulic fracturing and environmental issues out by late 2012.[11] In the meantime, New York State, which is part of the large Marcellus shale gas region, has maintained a moratorium on hydrofracturing while they study the issue further.[12] Also, France for example has recently banned shale gas drilling through fracking.[13]

Clearly there are a number of parties interested in establishing a link or showing that no link exists between hydrofracturing in shale gas and potential detrimental environmental effects. Besides these issues, a recent study by Cornell Professor Robert Howarth concluded that relative to possible warming impact on the climate, shale gas is not only worse than conventional gas but in fact also worse than coal and oil. This is due to the methane that leaks into the atmosphere during the hydraulic fracturing process and the fact that methane is much more potent than carbon dioxide on a pound-for-pound basis.[14] Naturally industry representatives have not agreed with the findings, citing low-quality data and other concerns.[15] If the Cornell study is accurate and is related mostly to current drilling practices for shale gas, it appears that with improved hydrofracturing techniques the environmental benefits of gas over other fossil fuels can be re-established. Of course, this might require some regulation to incentivize companies to put in the necessary improvements (if actually needed). It seems reasonable to assume that new environmental regulations could lead to higher extraction costs for shale gas, with higher US gas prices as a result.

To what degree have the increased supply projections for shale gas affected the prospects of gas cartelization? Going back to the Gabriel et al. (2012) study, their conclusion is that the impacts of shale gas are in fact rather modest. In a counterfactual scenario with much lower shale gas reserves (consistent with earlier estimates) the gains from cartelization (i.e. increased profits for the cartel) are almost the same as in their main scenario. Although less shale gas supply would have led to more gas imports to North America in the coming decades, the GECF countries would not have benefited much from holding back on their supply in order to raise the US price. Nevertheless, it seems reasonable to assume that a further substantial increase in shale gas supply also in other parts of the world would imply fewer gains from cartelization in the gas market, and thus reduce the likelihood of cartelization even further.

In summary, shale gas has a huge potential upside in the USA as well as in other parts of the world due to its abundance and supply security features, especially for places like Europe that rely so heavily on imported gas. Indeed, given the recent nuclear power disaster, Japan could be looking for a safer energy supply such as natural gas to fill the gap in its energy supply. Other countries may perceive natural gas as a 'bridge fuel' to a cleaner energy future due to its lower carbon content than other fossil fuels (the Cornell study notwithstanding). Lastly, shale gas drilling, especially in the USA, has potentially a strong economic and employment benefit which should not be overlooked. For example, an industry-financed study published in 2010 indicated that up to $6 billion in government revenue and 280 000 jobs could be at stake in the Marcellus shale region of the USA alone (API, 2010).

IMPACTS OF NEW TECHNOLOGIES AND NEW POLICIES

Future gains from cartelization in the gas market obviously depend on how gas demand develops. Higher demand will increase the residual demand that a cartel will face, which increases the opportunities to combine high prices and substantial market shares. Future demand for natural gas will be affected by energy-related policies as well as the development of alternative energy technologies. In this section we will elaborate on two important issues. First, we discuss how climate and renewable policies may affect gas demand. Then we discuss whether the expected growth in renewable energy supply will be a competitor to natural gas, or a complement.

The Importance of Climate and Renewable Policies

The global community agreed upon the United Nations Framework Convention on Climate Change (UNFCCC) in 1992 and the Kyoto Protocol in 1997, setting out emissions targets for CO_2 and five other greenhouse gases (GHG). Although the commitments in the Protocol will not be fully met, mainly due to the withdrawal of the USA, policies to reduce GHG emissions have been introduced around the world. This is especially the case in Europe, where the EU Emissions Trading System (ETS) is perhaps the most important climate policy instrument introduced in the world so far.[16] A major share of global CO_2 emissions comes from combustion of the three fossil fuels coal, oil and natural gas. However, as CO_2 emissions from natural gas are lower than from the other fossil fuels (per energy content), it is not at all clear whether reduced emissions of CO_2 imply less or more use of natural gas. Moreover, this is also a question of policy choice – it is not only first-best policies, that is, policies that reduce emissions at lowest costs, that are on the table.

Let us first discuss how first-best policies, such as a uniform CO_2 tax or a comprehensive emission trading system with auctioning of quotas, may affect gas demand. In the electricity sector, gas power currently competes to a large degree with coal power. Thus, in the short to medium term we should probably expect increased demand for gas in this sector, which has also been the case in Europe as a consequence of the EU ETS (Ellerman and Buchner, 2008). However, this could change within a few decades as electricity from renewable energy is expected to play a bigger role, possibly taking market shares from natural gas due to CO_2 pricing (Aune et al., 2010). For the USA, MIT (2011) points out that 'a combination of demand reduction and displacement of coal-fired power by gas-fired generation is the lowest-cost way to reduce CO_2 emissions by up to 50%. For more stringent CO_2 emissions reductions, further de-carbonization of the energy sector will be required; but *natural gas provides a cost-effective bridge to such a low-carbon future.*' See also the discussion below about the links between renewable power and gas power. Demand for natural gas in the power market could also decline in the nearer future if very ambitious targets are implemented, leading to substantially higher prices and consumption reductions in the electricity market.

In other sectors, competition between energy goods is less pronounced, at least in the short to medium term when equipment and infrastructure limit the choice of energy carrier. Moreover, gas competes mostly with fuel oil and electricity in these other sectors, and not so much with coal. Hence a switch to natural gas due to CO_2 pricing is less likely than in the electricity sector.

For a number of reasons, first-best policies are rarely chosen, and so we should also consider the effects of alternative policies. For instance, in the EU ETS installations have been given substantial amounts of free quotas. This has not only distributional effects, but also affects prices and quantities such as the technology mix in the electricity sector. As shown by Golombek et al. (2013), the market share of gas power may highly depend on the choice of allocation mechanisms: for example, compared to first-best policies, grandfathering of quotas combined with so-called closure rules may impede gas power, whereas output-based allocation may lead to substantial increases in gas power production.

Climate policies are often supplemented with renewable policies. In the EU, the member countries have agreed to increase the joint share of renewable energy in total energy consumption to at least 20 per cent in 2020, with a burden sharing that partly reflects individual countries' potential for renewable energy production.[17] The USA also has similar goals but most of the renewable policies so far have been implemented by individual states.[18] Large developing countries such as China and Brazil are also promoting renewable energy. The most common policy instruments to stimulate electricity production from renewable energy are feed-in tariffs and green certificate markets (also called renewable portfolio standards).

Supplementing an existing emissions trading system with renewable policies will not reduce total emissions from the installations regulated by the ETS, and most likely lead to higher rather than lower costs of reaching a given emission target. More interestingly here, renewable policies in addition to emissions trading will shift market shares from the least emissions-intensive non-renewable energy sources towards the most emissions-intensive energy technologies, compared to the case with emissions trading only (see Böhringer and Rosendahl, 2010). In the electricity sector this will typically mean switching from gas power to coal power. The intuition here is that pushing more renewable power into the market depresses the price of electricity *and* the price of emissions (as long as the cap on emissions is fixed). The former mechanism has the same negative effect on all non-renewable energy, whereas the latter mostly benefits emissions-intensive energy. Thus, if policy makers around the world introduce or intensify policies to stimulate production of renewable energy, gas demand may be particularly hurt. This conclusion could change, however, if gas and renewable energy were complements in production. Similar arguments could be raised with respect to efforts to improve energy efficiency, which reduce overall energy demand and also reduce the incentives to switch from coal to gas for a given emissions target.

Natural Gas and Renewables: Substitutes or Complements?

A recent study of natural gas by the MIT Energy Initiative notes that natural-gas-fired power generation is the major backup to intermittent renewable supplies such as solar or wind power in most US markets (MIT, 2011). Thus, under assumptions of more intermittent renewable generation but no utility-scale storage, even more natural gas capacity is projected, but this may require some regulatory changes. From that perspective, natural gas and intermittent renewables would be complements to each other.

Nevertheless, this MIT report also notes that, in the short term, if there is a quick increase of renewable energy, all things being equal, this renewable power will force a decrease in gas-fired generation in the power sector. The study goes on to further note that, in the longer term, increases in renewable generation affect baseload generation such as coal, nuclear, or natural gas combined cycle generation so the relationship may depend on the assumptions and regions considered. Thus, what this study shows is that renewable energy can have a substitute effect for natural gas in the US power sector as well. This conclusion is also documented in the wind versus natural gas debate in Texas:

> The success of wind power in Texas has come at the expense of natural gas. If the wind build-out continues, by 2013 the amount of gas consumed to make electricity could fall by 18.5%, as gas plants sit idle for longer, according to Tudor Pickering & Holt, a Houston-based energy investment bank. (*WSJ*, 2010)

Another point to consider for the power sector is the amount of natural resources needed to convert solar and wind to electricity (as two prominent examples of renewable energy). A case in point is California, which recently mandated that one-third of its electricity must come from renewable energy by 2020. According to Robert Bryce (*NYTimes*, 2011a), 17 000 megawatts of renewable energy capacity would be needed to satisfy this new mandate and use up vast amount of land to accommodate the solar and wind power sources. Natural gas, while not without environmental issues of its own, has more of its infrastructure below ground (although some above ground of course) and thus might be easier to use (initially at least). So from the land-use perspective, gas is a competitor to these renewable energy sources.

The power sector may also interact with the transportation sector to make renewable energy and natural gas substitutes. Consider the Pickens plan (www.pickensplan.com), which is a proposal to displace gas from power generation with wind from the centre of the USA. In turn, gas would then be used for transportation purposes to displace the more polluting petroleum-based fuels. However, despite the currently favourable price of natural gas, there are a number of problems with its potential use as a vehicle fuel on a national scale (*NYTimes*, 2008). First, natural gas fuel pumps and fuelling stations are needed. Without a more extensive fuelling infrastructure, drivers will be stranded due to the limited vehicle range. Also, cars would need to be retrofit to use natural gas instead of petroleum-based fuel (e.g. gasoline). However, a recent bill in the US Congress (NatGas Act) will help by providing tax incentives for buying natural-gas-powered vehicles or tax credits for building the requisite infrastructure.[19] Moreover, US President Obama endorsed the Pickens plan for natural gas vehicles.

Despite the last observation, we think that anticipated climate and energy policies in Europe, the USA and elsewhere will not stimulate any substantial rise in gas demand,

and could possibly have the opposite effect. The same can be said with regard to the development of alternative energy technologies. Thus, in our view, the prospects of gas cartelization will probably not increase as a result of new policies or new technological developments.

CONCLUSIONS

Is cartelization in international gas markets likely in the foreseeable future? This chapter has attempted to shed light on this question. We have discussed the position of the GECF (Gas Exporting Countries Forum) and its key members in the current gas markets, and asked whether a gas cartel may emerge from this organization. A basic prerequisite for a gas cartel is that the cartel members as a group benefit from cartelization. In addition, a number of challenges must be overcome when a cartel consists of several countries with differing interests. In particular, key members must find it in their own interest to join a cartel instead of having sole control over their own market decisions.

Our discussion suggests that the likelihood of a gas cartel *à la* OPEC emerging from the GECF is rather limited in coming years. First of all, the joint benefits for a cartel seem to be rather modest unless more exporting countries join the cartel. Second, there are more hurdles to overcome for a cartel in the gas market than in the oil market, such as significant transport costs and long-term contracts. Last but not least, we believe that Russia, the most important player in international gas markets, does not see it in its interest to be part of a gas cartel. This does not rule out any coordinated action among GECF members, however, for example related to infrastructure investments or in situations with substantial excess supply. If a gas cartel should emerge from the GECF, the effects would be biggest in the European market, which is located closest to the main members of the GECF. European gas consumers could face higher prices, but the price increases would be rather modest unless most Middle East and Caspian countries also joined such a cartel.

It is difficult to make predictions about the future gas market, and we cannot completely rule out the possibility of gas cartelization in the longer term. Increased demand for natural gas, for example due to intensified climate policies around the world, could potentially increase the benefits from cartelization. However, as discussed above, it is not at all certain that new policies will stimulate gas demand. Moreover, additional gas supply from unconventional sources such as shale gas will make it more difficult for a cartel to raise the price of natural gas. Thus gas cartelization in the future will probably require some unexpected changes in the international markets for natural gas.

ACKNOWLEDGEMENT

We are grateful for financial support from the Petrosam programme of the Research Council of Norway.

NOTES

1. The current members of GECF are (in descending order of reserves): Russia, Iran, Qatar, UAE, Nigeria, Venezuela, Algeria, Egypt, Libya, Oman, Bolivia, Trinidad and Tobago, and Equatorial Guinea. In addition, there are four observing members, Iraq, Norway, Kazakhstan and the Netherlands. (Source: www.gecf.org, accessed 24 January 2013.)
2. For instance, Algeria called for coordinated cuts of gas production prior to the GECF meeting in April 2010 (WGI, 2010).
3. In 2009 Qatar exported significant amounts of gas to South and East Asia and to Europe, as well as small amounts to North and South America (BP, 2010). Several other countries also exported to all the three continents Europe, America and Asia (Nigeria, Egypt, Trinidad and Tobago, Equatorial Guinea and Yemen).
4. LNG is of course more flexible than pipelines, as LNG ships can be directed to any market with available LNG regasification capacity. It is, however, costly to ship gas over long distances, so a certain price arbitrage is needed in order to make it profitable.
5. Note that the study was undertaken before UAE and Oman joined the GECF.
6. As shown by Rosendahl and Sagen (2009), lower LNG costs could in fact lead to *higher* gas prices in Europe, as more gas may be shipped away from the European market towards more distant markets where prices are higher (e.g. in East Asia).
7. 'We are not setting up a cartel, and we do not intend to sign cartel agreements. None of us intends to cede even the slightest part of our independence in decision-making' (Prime Minister Vladimir Putin, 11 November 2008). http://www.premier.gov.ru/eng/events/news/2390,/official translation.
8. 'Rossiya voydet v gazovyj kartel' ('Russia will enter gas cartel'), *Vzglyad*, 21 October 2008. http://www.vz.ru/economy/2008/10/21/221324.html.
9. *RIA Novosti*, 25 June 2010, http://en.rian.ru/russia/20100625/159571853.html.
10. See also the discussion on shale gas in Chapter 2.
11. http://water.epa.gov/type/groundwater/uic/class2/hydraulicfracturing/index.cfm.
12. http://www.dec.ny.gov/energy/46288.html.
13. http://online.wsj.com/article/BT-CO-20110511-712242.html.
14. http://www.news.cornell.edu/stories/April11/GasDrillingDirtier.html.
15. http://www.ibtimes.com/articles/133706/20110413/cornell-shale-study-natural-gas-robert-howarth-shale-global-warming-climate-change-shale-natural-gas.htm.
16. Emissions trading has also been introduced to regulate CO_2 emissions from the electricity sector in parts of the USA (http://rggi.org/home and www.westernclimateinitiative.org), and several countries are planning or discussing implementation of an ETS. Greenhouse gas emissions have also been regulated through other policy instruments such as carbon taxes in some European countries.
17. http://ec.europa.eu/environment/climat/climate_action.htm.
18. http://apps1.eere.energy.gov/states/maps/renewable_portfolio_states.cfm#map.
19. http://www.green-energy-news.com/arch/nrgs2011/20110045.html.

REFERENCES

API (2010), 'New study finds natural gas in Marcellus Shale region worth 280,000 jobs, $6 billion in government revenue', 21 July, Eric Wohlschlegel, American Petroleum Institute, http://new.api.org/Newsroom/m-shale-econ-impact.cfm.

Aune, F.R., G. Liu, K.E. Rosendahl and E.L. Sagen (2010), 'Subsidising carbon capture: effects on energy prices and market shares in the power market', *Environmental Economics*, 1(1), 76–91.

Baev, P. and I. Øverland (2010), 'The South Stream versus Nabucco pipeline race: geopolitical and economic (ir)rationales and political stakes in mega-projects', *International Affairs*, 86(5), 1075–90.

BP (2010), *BP Statistical Review of World Energy 2010*, www.bp.com.

Böhringer, C. and K.E. Rosendahl (2010), 'Green serves the dirtiest. On the interaction between black and green quotas', *Journal of Regulatory Economics*, 37, 316–25.

EIA (Energy Information Administration) (2011), 'Today in energy: shale gas is a global phenomenon', 5 April 2011, http://www.eia.gov/todayinenergy/detail.cfm?id=811.

Ellerman, A.D. and B.K. Buchner (2008), 'Over-allocation or abatement? A preliminary analysis of the EU ETS based on the 2005–06 emissions data', *Environmental Resource Economics* 41, 267–87.

Finon, D. (2007), 'Russia and the "Gas-OPEC". Real or perceived threat', IFRI Russia/NIS Center, Russie.

NE.Visions, #24, November, http://www.ifri.org/files/Russie/ifri_RNV_ENG_Finon_opepdugaz_sept2007. pdf.

Gabriel, S.A., K.E. Rosendahl, R.G. Egging, H. Avetisyan and S. Siddiqui (2012), 'Cartelization in gas markets: studying the potential for a "gas OPEC"', *Energy Economics*, **34**, 137–52.

Gazprom (2010), *Gazprom in Figures 2005–2009*.

Goldthau, A. (2008), 'Rhetoric versus reality: Russian threats to European energy supply', *Energy Policy*, **36**(2), 686–92.

Golombek, R., S.A.C. Kittelsen and K.E. Rosendahl (2013), 'Price and welfare effects of emission quota allocation', *Energy Economics*, dx.doi.org/10.1016/j.eneco.2012.11.06.

Hallouche, H. (2006), 'The Gas Exporting Countries Forum: is it really a gas OPEC in the making?', Oxford Institute for Energy Studies, NG 13, Oxford.

IEA (2006), 'Optimizing Russian natural gas: reform and climate policy', International Energy Agency, Paris.

Jaffe, A.M. and R. Soligo (2006), 'Market structure in the new gas economy: is cartelization possible?', in D.G. Victor, A.M. Jaffe and M.H. Hayes (eds), *Natural Gas and Geopolitics: From 1970 to 2040*, Cambridge: Cambridge University Press, ch. 11.

Milov, V., L.L. Coburn and I. Danchenko (2006), 'Russian energy policy 1992–2005', *Eurasian Geography and Economics*, **47**(3), 285–313.

MIT (MIT Energy Initiative) (2011), 'The future of natural gas', http://web.mit.edu/mitei/research/studies/natural-gas-2011.shtml.

National Petroleum Council (2007), 'Hard truths facing the hard truths about energy', http://www.npchardtruthsreport.org.

NYTimes (*New York Times*) (2008), 'Pickens plan stirs debate, and qualms', 5 August, Kate Galbraith.

NYTimes (*New York Times*) (2010), 'EPA considers risks of gas extraction', 23 July, Tom Zeller Jr.

NYTimes (*New York Times*) (2011a), 'Gas is greener', 7 June, Robert Bryce.

NYTimes (*New York Times*) (2011b), 'Gas drilling technique is labeled violation', 31 January, Tom Zeller Jr.

Rosendahl, K.E. and E.L. Sagen (2009), 'The global natural gas market. Will transport cost reductions lead to lower prices?', *The Energy Journal*, **30**(2), 17–40.

Tsygankova., M. (2008), 'Netback pricing as a remedy for the Russian gas deficit', Discussion Papers No. 554, Statistics Norway.

UMD (University of Maryland) (2010), 'Final summary report of ARI and cartel cases using the World Gas Model', 15 September.

WSJ (*Wall Street Journal*) (2010), 'Natural gas tilts at windmills in power feud', 2 March.

WSJ (*Wall Street Journal*) (2011), 'Cherniere doubles down on its LNG bet', 28 May.

WGI (World Gas Intelligence) (2010), 'GECF falls short of spot sales deal', 21 April.

4 Global steam coal markets until 2030: perspectives on production, trade and consumption under increasing carbon constraints

Clemens Haftendorn, Franziska Holz, Claudia Kemfert and Christian von Hirschhausen

1 INTRODUCTION

The development of global coal markets has a major impact on CO_2 emissions and is therefore key to any climate policy effort. Economic development and the increase of energy demand have spurred the use of fossil fuels, and in particular coal, over recent years, mainly in emerging countries, while some industrial countries foresee the end of coal power in their national energy mix (e.g. the UK and Germany). Consequently, aggregate greenhouse gas emissions from the coal sector are increasing.

International coal trade has also undergone substantive change over recent years, with a shift of the focus from the Atlantic to the Pacific Ocean. The traditional production and trade patterns, for example with South Africa as the main supplier to Europe, are in the process of breaking up. At the horizon of 2020 or 2030, global steam coal flows are redirected, largely driven by energy demand in India and China.

This chapter provides an overview of the global market and drivers of steam coal, with a special focus on the changing structures of this market and the effect of different climate policies until 2030. To this end, we have developed a partial equilibrium model of the world's steam coal markets, which can be used to answer the questions raised. Our hypothesis is that a steam coal market will play an important role in future energy market structures, mainly in the Pacific basin, and that there are close interactions between climate policies and the global steam coal market.

The chapter is structured as follows. The next section introduces the 'COALMOD-World' model, and describes how its features can bear fruit on the questions raised. This section also sketches a generic base case for coal trading to 2030, absent of any specific climate policy consideration; we observe not only a shift of the focus of action from the Atlantic to the Pacific Ocean, but also an increasing importance of countries that are critical to climate policies, for example Indonesia, India and China. Section 3 then focuses on the interaction of climate policies and the global steam coal markets until 2030. We implement three different scenarios of varying climate policies and market environments: a unilateral European climate policy; an export-limited supply policy of Indonesia; and an accelerated CCTS (carbon capture, transport and storage) roll-out. The last section concludes.[1]

2   THE COALMOD-WORLD MODEL

2.1   The Value-added Chain of Steam Coal Markets

The COALMOD-World model setup follows the organization of the value-added chain of the steam coal sector. The value chain is complex and various types of players are involved at each stage. Producers can be large national and sometimes state-owned companies. There are a few large multinational coal companies as well as many smaller companies, usually operating in one country only. Transport infrastructure can be built by the mining company or by another entity. Often, it consists of rail infrastructure, but in some countries trucks or river barges are used. Export ports can be exclusively used by one company or by multiple companies, with a variety of possible ownership structures. Traders as intermediaries also play a role as they can be vertically integrated or contractually connected to every stage of the industry.

In Haftendorn et al. (2012a) we provide an analysis of the market structure for the global steam coal trade and simplify the value chain for the modeling purpose. There is some evidence that, contrary to the oil market, the international steam coal market tends to be competitive. This result allows us to make some simplifying assumptions for the COALMOD-World model: since in a competitive market prices equal marginal costs, we can simplify the role of the players in the value-added chain to obtain two types of model players, the producers and the exporters, shown in Figure 4.1: the stages of the real-world value-added chain that are included in the model are represented by the small rectangles included in the larger producer and exporter boxes. We exclude the

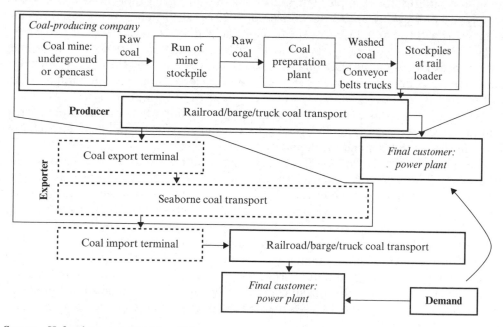

Source: Haftendorn et al. (2012a), p. 308.

Figure 4.1   Model players in the steam coal value-added chain

coal import terminals and the subsequent land transport link to the final consumer from our model representation because this capacity is assumed to be sufficient. *De facto*, we situate demand that cannot be reached by land close to the import port. The second type of demand node can be reached by a land link directly from the producer. The producer player oversees the coal-mining activity and also the land transport. The exporter operates the export terminal and pays for the sea transport. These players are aggregated at a national or regional (sub-national) level.

Research on international coal markets points out that the traditional separation of the Pacific and the Atlantic market has faded (see Zaklan et al., 2012; Li et al., 2010). In our model, we therefore consider the global market as one integrated market, albeit the spatial aspect of the market where transport costs play a role in determining the trade relations is not neglected.

2.2 Model Structure

The COALMOD-World model is a multi-period equilibrium model of the global steam coal market with two types of players: producers, f, and exporters, e, facing consumers, c, represented by a demand function. COALMOD-World's producers and exporters represent stylized players defined for aggregated production, export and consumption nodes, primarily determined using geographical parameters. A production node represents a geographically restricted area (mining basin) and aggregates the mining companies present in that area into one player called producer. In the model, production node and producer are equivalent terms. Production nodes are defined based on the following criteria: geography of reserves, type of coal, and production cost characteristics. An export node represents the coal export terminal of one region and aggregates the capacities of the real-world coal harbors present in that region into one model player called exporter. Here again, export node and exporter are used as equivalent terms. The export nodes are primarily defined based on geographic factors.

A demand node represents a geographic area where the coal is consumed. It aggregates the consumption by the coal-fired power plants in a region. It can have access to seaborne coal through a port or not. The demand nodes are primarily defined based on geographic factors, but other factors may come into play, such as the connection to a port or the presence of mine-mouth coal power plants. Figure 4.2 represents the model structure and the relationships between producers, exporters and demand. The model runs until 2040 and calculates yearly equilibria for the energy quantities sold in the years 2006, 2010, 2015, 2020, 2025, 2030, 2035 and 2040, which we call model years. Also, the players can make investments in each model year that will be available in the next model year. Thus the model not only calculates an equilibrium within each model year but also over the total model horizon regarding optimal investments. For the years between the model years we interpolate the produced quantities since these are necessary in order to model reserve depletion. We assume that production and other capacities will be made gradually available in the years between the model years to reach their new value in the following model year. Both producer and exporter problems are profit maximization problems over the entire model horizon. The players have perfect foresight, meaning that they choose the optimal quantities to be supplied in each model period and the

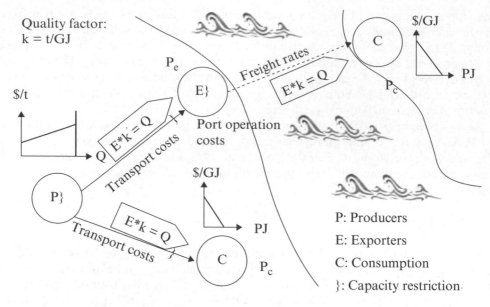

Source: Haftendorn et al. (2012a), p. 308.

Figure 4.2 COALMOD world model structure

investments between model periods under the assumption of perfect information about current and future demand. The market is assumed to be competitive following the results in Haftendorn and Holz (2010).

The detailed model including all equations is laid out in Haftendorn et al. (2012a), including the producer's profit optimization, production cost functions, the exporter's optimization problem, coal quality, final demand and the market-clearing condition. In the following we display and discuss the key elements of the model, the objective functions of the producer and the exporter:

Producers maximize their profits over the entire model horizon A for all model years $a < A$ (see equation 4.1). Producers extract coal and can then either sell it directly at their respective local demand node (x_{afc}) or to an exporter (y_{afe}); they have to bear production costs and overland transportation costs. In addition, the producers can invest in additional production capacities ($Pinv_{af}$) and/or in transportation capacities, either to their local demand mode ($Tinv\_c_{afc}$) or to the exporter ($Tinv\_e_{afe}$). The quality factor (κ_{af}) determines the conversion between different coal qualities from physical tons of energy units, and vice versa.[2]

The optimization of the producer is subject to production constraints, available reserves and transport capacities. Also, the investments in additional capacities are constrained in order to stick to reality and to avoid corner solutions.

The COALMOD-World producers' objective function equation is therefore

$$\max_{x_{afc};y_{afe};Pinv_{af};Tinv\_c_{afc};Tinv\_e_{afe}} \Pi_f^P(x_{afc};y_{afe};Pinv_{af};Tinv\_c_{afc};Tinv\_e_{afe})$$

$$= \sum_{a \in A} \left(\frac{1}{1 + r_f} \right)^a \cdot \left[\sum_c p_{ac} \cdot x_{afc} + \sum_c p_{ae} \cdot y_{afe} - C_{af}^p \left(\sum_c x_{afc} \cdot \kappa_{af} + \sum_e y_{afe} \cdot \kappa_{af} \right) \right.$$

$$- \sum_c trans\_c_{afc} \cdot x_{afc} \cdot \kappa_{af} - \sum_e trans\_e_{afe} \cdot y_{afe} \cdot \kappa_{af} - Pinv_{af} \cdot CPinv_{af}$$

$$\left. - Tinv\_c_{afc} \cdot CTinv\_c_{afc} - Tinv\_e_{afe} \cdot CTinv\_e_{afe} \right] \qquad (4.1)$$

An additional important feature of the production side in the model is the endogenous calculation of the yearly short-term cost functions. The calculation of future short-term cost functions is based on past extraction (depletion or loss of production capacity), investments, technology and geological factors of the production basin.

The exporter, too, maximizes its overall profit over the entire model period A. Each exporter is allocated a specific producing region. Profits are determined by overall sales minus the net coal price paid to producers (p_{ae}), the operating costs of the export terminal, and the investments into additional export post capacities. The exporter chooses an optimal quantity (z_{aec}) that it sells to each importing node c in each model year a. The exporter's optimization is also constrained by export capacities and a limitation of the amount of investments per period. Thus the exporter's objective function takes the form

$$\max_{z_{aec};Einv_{ae}} \Pi_e^E(z_{aec};Einv_{ae}) = \sum_{a \in A} \left(\frac{1}{1 + r_e} \right)^a \cdot$$

$$\left[\sum_c p_{ac} \cdot z_{aec} - \sum_c p_{ae} \cdot z_{aec} - \sum_c z_{aec} \cdot Cport_{ae} \cdot \kappa_{ae} \right.$$

$$\left. - \sum_c z_{aec} \cdot searate_{aec} \cdot \kappa_{ae} - Einv_{ae} \cdot CEinv_{ae} \right] \qquad (4.2)$$

2.3 Data and Calibration of a Base Case

We include all countries in our data set that were either consuming at least 5 Mtpa (million tons per annum) or producing and exporting at least 5 Mtpa in 2006. Some additional countries that are expected to become relevant players in the global market between 2010 and 2030 are included as well, for example Mongolia or Mozambique. We distinguish production and consumption nodes. For a country in which production takes place and that also consumes coal, we include at least one production node and one consumption node. For larger countries, there can be more than one production/demand node; this is the case for the USA, China, India, Russia and Australia. Reserve estimates are based on EIA, USGS and national data, which confirm the conventional wisdom of very high coal reserves, such as in China and the USA (about 114 and 245 Gt, respectively), Eurasia (253 Gt), India (92 Gt), South Africa (49 Gt) and Australia (78 Gt).

For the demand side of the model we rely on the IEA *World Energy Outlook* (IEA, 2010). These data are aggregated, so the demand projections of the IEA must be allocated to the model's demand nodes. To achieve this, we take a bottom–up approach based on national data and ensure consistency by checking with the IEA data.

The profit-maximizing players are 25 producers and 14 exporters serving a total of 41 demand centers. Virtually all worldwide steam coal demand is included as we model both domestic markets and the global seaborne market. The level of detail and disaggregation of the COALMOD-World model allows for a differentiated analysis of potential market adjustment effects as a reaction to climate policies.

For each model year, the COALMOD-World model delivers results for the inland and seaborne trade flows, the prices, the level of investments as well as the dual of the constraints.[3] Our results are based on the assumption of competitive and liberalized markets. We also assume that the markets are fully integrated. The base case results can be called 'ideal' results, as they tell us how future demand should be served optimally and in which countries investments should take place. We assume that there are restrictions on production capacity expansions for various geological, technical and economic reasons. The level of these restrictions is based on historical capacity data provided by the USGS in the country reports of the *Minerals Yearbook* and on historical production data. We also assume that there are restrictions on export capacity expansion for technical and economic reasons. These restrictions are based on historical experience as well as on planned and forecast expansions. They range from 5 to 30 Mtpa of additional capacity over a five-year period depending on the country.

The results for 2006 show a good fit with the observed trade pattern. The direction and relative amounts of the trade flows correspond to actual trade flows, with the only exception for flows from Australia to Japan.[4]

3   BASE CASE DEVELOPMENTS AND INTERNATIONAL TRADE (UNTIL 2030)

In this section we sketch a 'base case' to illustrate the application of the COALMOD-World model to a concrete policy setting. In accord with the 'new policies scenario' sketched out by IEA (2010), we assume a continuation of climate policies at the global level, and a stabilization of coal demand. More specific scenarios will be analyzed in the subsequent sections. Our base case relies on the demand projections of the new policies scenario from the *World Energy Outlook* (IEA, 2010). Underlying this scenario is some level of climate policy that leads to a stabilization of global steam coal demand after 2015. However, despite this stabilizing demand, trade increases its share until 2030 from 16 per cent today to 18 per cent of global consumption. This is due to the fact that India and China rely more and more on cheap imports to satisfy their increasing demand. See Haftendorn et al. (2012b) for a more detailed description and interpretation.

The international trade flows calculated in the COALMOD-World model base case are shown in Figure 4.3.

It is important to recall that these trade flows interact with the national trade flows that are five times more important and also calculated by the model but not represented here.

Regarding the global seaborne trade flows, we see significant changes in trade patterns as the traditional separation between the Atlantic and Pacific markets vanishes. Australia and Indonesia remain key players in the Pacific markets, their exports increasing to 179 Mtpa and 230 Mtpa by 2030, respectively. The rise of Indonesia is particularly spectacular, based on low-cost, but also low-quality, resources that make Indonesia by far the largest exporter worldwide.

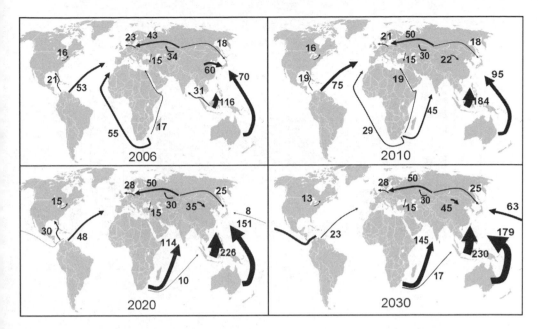

Figure 4.3 COALMOD-World model base case trade flow results in million tons

The most significant shift of trade flows is the shift eastwards of South African supplies. South Africa also has a high export potential, which is indicated by the fact that the export capacity investment restriction (15 Mtpa per five years) is constantly binding.

Traditionally a supplier to the European market, South Africa increasingly supplies Asian countries, especially India. This shift started between 2006 and 2010 and the model is able to reproduce the shift that actually took place. Further, the model projects that by 2020 South Africa will supply almost exclusively India, thus creating a new 'Indian Ocean' market. Indeed, one observes the emergence of this new 'sub-market', which is the market in the Indian Ocean, where South Africa will play a dominant role.

A second shift occurs westward after 2020. Asian demand, especially in China, is so strong that Colombia diverts trade flows from Europe to Asia, especially Japan. Europe loses another traditional supplier and by 2030 relies on Russia for most of its coal imports.

The Atlantic market changes significantly after 2010, from a swing market to a mainly land-based supplied market. By 2030 the only remaining overseas supplier is Colombia (23 Mtpa), whereas Poland and Russia become the most important suppliers to Europe. This confirms the most significant finding of the model – that is, the shift of overseas trade flows towards the Asian/Pacific markets.

With respect to prices, our results in the base case tend to confirm the secular trend of almost constant prices over the past century. Figure 4.4 shows the FOB (free on board) costs for all exporting countries in 2020, characterized by a relative homogeneity among the countries. Some significant increases in production costs are observed in Australia (Queensland), Russia, South Africa and China. Figure 4.4 also indicates that transport

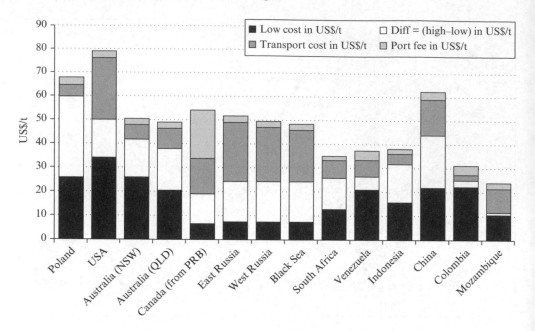

Figure 4.4   2020 FOB costs for all export countries calculated endogenously

costs and particularly inland transport costs continue to be an important element of total costs.

Global trade flows in the future will be influenced primarily by the growing demand from Asia. As we show in the next section, it is because of this fact that the interaction between the growing Asian market and future climate policies is crucial to assess the policies' success.

4   COMPARING CLIMATE POLICIES USING THE COALMOD-WORLD MODEL: SCENARIO SPACE AND REFERENCE ANALYSIS

4.1   The Partial Equilibrium Approach

Comparative static or multi-period scenario analyses have been widely used with partial equilibrium models, especially in the natural gas sector. Surprisingly, they have not up to now been much used for climate policy scenarios. One main focus has been on market power (e.g. Lise and Hobbs, 2008; Holz et al., 2008) and other scenarios including demand scenarios, supply modifications, investment constraints and disruptions (e.g. Lise et al., 2008; Huppmann et al., 2011). The numerical modeling literature for the coal market has focused on market power issues (Haftendorn and Holz, 2010; Paulus and Trüby, 2012) or infrastructure decisions (Paulus and Trüby, 2011).

We now turn to the representation of climate policies in the COALMOD-World model context. By doing so, and since we want to focus on the interaction of the coal

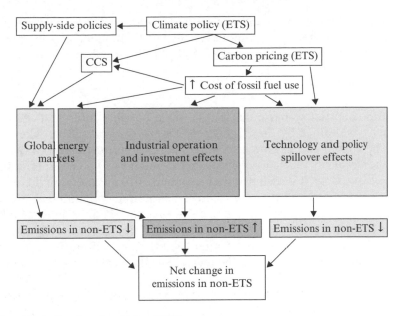

Source: Own illustration, based on Dröge (2009).

Figure 4.5   Carbon pricing and the sources of carbon leakage

market with climate policies only (and not with carbon leakage in general that would require CGE modeling), we concentrate on market effects.[5] The introduction of heterogeneous types of climate policies, geographically as well as in policy type, will affect the market supply and demand balance both locally and globally. The future use of coal in the world will be primarily influenced by climate policies, which will consequently affect the quantities of steam coal demanded. Quantity effects on the supply or demand side have price effects on the global market of steam coal and, in turn, as there is an elastic demand for coal, influence coal consumption. These 'market adjustments'[6] can affect the effectiveness of climate policy in a positive or negative way. Partial equilibrium models are able to assess these effects since detailed market effects are at the core of these models.

In this analysis, we focus on the global coal market and therefore do not investigate carbon leakage through relocation effects of industries. An overview of all possible channels of carbon leakage is given in Figure 4.5. The flowchart starts with one country or group of countries implementing climate policies to reduce their CO_2 emissions and then describes which effects this will have on the emissions of the rest of the world. We extend the framework proposed by Dröge (2009) and include a positive channel through the global energy markets. The beneficial and adverse channels of carbon leakage through the energy markets, which we call 'positive' and 'negative' market adjustments in the following, are explained and specified by the scenarios of this chapter.

We therefore concentrate our analysis on pure supply and demand market effects on the global steam coal market coming from market adjustments between demand centers. We use the COALMOD-World model and implement three different climate policy shocks in varying climate policy and market environments. The scenarios are: a unilateral European climate policy, an Indonesian export-limiting supply-side policy and a carbon capture, transport and storage (CCTS) fast roll-out policy.

4.2   The Scenario Space

We divide the potential scenario space into three possible futures of global climate policy intensity as defined by the IEA *World Energy Outlook* (2010). These scenarios are ordered here from the less intense climate policy implementation to the most: the current policies scenario, the new policies scenario and the 450 ppm scenario. In the IEA current policies scenario, it is assumed that as of mid-2010 no change in the current policies will be implemented and that the recently announced commitments are not acted upon. In the new policies scenario, the recently announced commitments and policies, for example from the 2009 Copenhagen Climate Conference, are fully implemented. Finally the 450 ppm scenario is named after the low CO_2 concentration in the atmosphere that is reached in order to keep the increase in global average temperatures below 2 °C.

The second division of the scenario space is made by an exogenous market constraint. We assume that there are restrictions on production capacity expansions for various reasons that can be geological, technical and economical (financial restrictions, lack of qualified labor force or equipment). For the constrained case, the level of this restriction is based on historical capacity data provided by the USGS in the country reports of the *Minerals Yearbook*[7] and on historical production data. This is a rather conservative assessment that can be regarded as the upper bound on the levels of exogenous market constraints. At the other extreme, these restrictions are completely lifted in the unconstrained case and the model producers can thus invest in production capacity as much as needed to maximize their profits.

Figure 4.6 shows all the scenarios implemented for this chapter as the little squares in their respective scenario space (policy framework and market condition) and, where it applies, with the additional policy shock implemented. We investigate the effects of three policy shocks: a unilateral climate policy by Europe, a restriction of Indonesian coal production and exports, and a fast roll-out of CCTS. To ensure scenario consistency and comparability the shocks are applied to the six reference scenarios (plain squares) and the results of the policy shock simulations are compared to their respective reference case. The reference cases are calibrated such that for most of the demand centers the consumed steam coal quantities are in a 10 percent range above or below the quantities defined by IEA (2010). As our demand functions are constructed using a reference demand, a reference price and a demand elasticity, we calibrate the references prices to fit the quantities. We calibrate such that at least 80 percent of the demand nodes for all model years are in the 10 percent range above and below the IEA (2010) quantities of their respective scenario.

A difficult issue in partial equilibrium analysis is the price elasticity of demand due to the lack of econometric studies. Paulus and Trüby (2012) give an overview of the results of econometric studies that estimate short-term price elasticities for coal. The range is between −0.05 and −0.57. The elasticities for the base year 2006 are based on our previous work (Haftendorn et al., 2012a). The elasticities of the following years are gradually

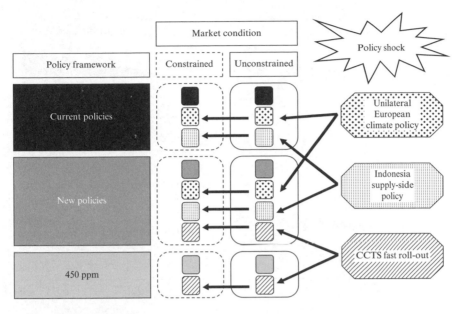

Figure 4.6   Scenario space

set higher as we assume that countries will have a more diverse energy mix and higher flexibility in their power systems in the future.

4.3   Reference Case: Worldwide Climate Policy

Before we describe the outcomes of the scenarios resulting from an additional policy shock on an already implemented level of global climate policy, we analyze the model outcomes from these global climate policies that represent our reference cases. We have three different levels of global climate policy: the current policies scenario, the new policies scenario and the 450 ppm scenario based on the projections of IEA (2010). Additionally to that policy framework we have to consider the market conditions as shown in Figure 4.6. In one case, investments in production capacity are constrained; in the other case they are not in order to represent two extremes of a continuum of exogenous market constraints.

Figure 4.7 shows the results from the different modeling runs of the reference cases in million tons of CO_2 emissions. Since the emissions are proportionally linked to the consumption of coal in energy units, we can also infer the development of consumption and trade from these figures. All the areas together show the emissions in the current policies scenario. Removing the areas 'Reduction from current policies', we obtain the emissions for the new policies scenario.

In the case of unconstrained investment possibilities in production capacity, coal consumption is significantly higher in the current policies scenario and slightly higher in the new policies scenario. Global seaborne trade remains important and will continue to grow. We see a reduction in global trade only in the 450 ppm scenario. In the case of a constrained market condition global seaborne trade is especially important for countries

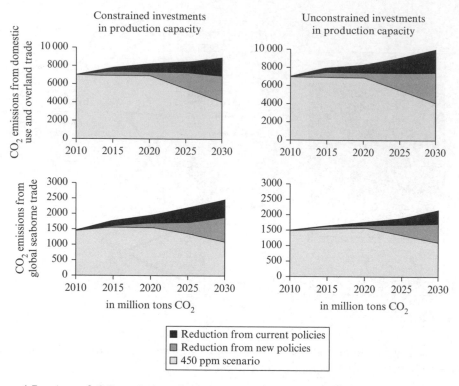

Figure 4.7   Annual CO_2 emissions from steam coal consumption in the six reference scenarios

like China and India to meet their coal demand as they might experience difficulties in expanding their domestic production base. The physical steam coal flows correspond to the 2010–30 scenarios sketched out in the previous section.

4.4   Scenario 1: Unilateral European Climate Policy

The unilateral European climate policy scenario is implemented in two global climate policy frameworks: the current policies and the new policies framework as shown in the scenario overview of Figure 4.6. In this scenario the EU goes a step further and aims at reducing CO_2 emissions by 30 percent (compared to the level of 1990) by 2020 with further reductions in the future. This goal is reached through a significantly lower coal consumption in the EU. In the IEA (2010) scenarios this is represented by the demand values of the 450 ppm scenario. The steam coal demand reductions compared to the reference scenarios are shown in Table 4.1.

4.5   Scenario 2: Yasuní-type Supply-side Policy in Indonesia

The Yasuní–ITT initiative proposed by the Ecuadorian government aims at combating global warming, protecting biodiversity and indigenous people as well as implementing

Table 4.1   EU demand reduction in the unilateral European climate policy scenario compared to the reference scenarios in percentage

	2020	2025	2030
Demand reduction compared to Current Policies	−0.24	−0.43	−0.64
Demand reduction compared to New Policies	−0.02	−0.20	−0.45

Source: Own compilation based on IEA (2010).

a sustainable social and energetic development by refraining indefinitely from exploiting the oil reserves of the Ishpingo–Tambococha–Tiputini (ITT) oil field within the Yasuní National Park (Larrea, 2010).[8] This field represents 20 percent of the Ecuadorian oil reserves and the initiative requires a capital contribution of at least half of the earnings Ecuador would receive from exploitation. Valuated at US$76.38 per barrel, this represents a sum of US$3.635 billion supplied by the international community to a fund managed by the United Nations Development Programme. The initiative represents 407 Mt CO_2 saved from not using the oil resource and an additional 820 Mt CO_2 mitigation potential over 20 years from avoided deforestation and forest management (Larrea, 2010).

For the Indonesian scenario, we transfer the idea and apply it to the case of steam coal in Indonesia: the bulk of coal exploitation in Indonesia takes place on the island of Kalimantan (formerly known as Borneo). This island is home to one of the greatest rainforests in the world and a treasure of biodiversity that is endangered by coal mining through deforestation and local air and water pollution. Fatah (2008) points out that coal mining has few to no beneficial effects on the local economy; the revenues and benefits go to private companies and the government. Thus one could imagine a supply-side climate policy mechanism similar to the Yasuní–ITT initiative to preserve the Indonesian forest and prevent the extraction, export and CO_2 emissions from coal albeit still allowing a local use of steam coal for power generation. In our particular case we modeled this policy as a gradual tightening of export restrictions, that is, a cap on the maximum quantity of coal that can be exported in a given year: 2006 to 2015, no restriction; 2020: 50 Mtpa; 2025: 25 Mtpa; 2030: 0 Mtpa.

4.6   Scenario 3: CCTS Fast Roll-out

Carbon capture, transport and storage (CCTS) is a set of technologies that aim at a reduction of CO_2 emissions into the atmosphere by separating and capturing the CO_2 at the power plant and transporting it to a geological sink where it will be compressed and stored underground (see IPCC, 2005). The CCTS technology is regarded by the IPCC and by the IEA as an important option for climate change mitigation. However, as of 2011 there are only about ten pilot CCTS plants operating in the world and not a single large-scale operation (22 are planned to start operating between 2014 and 2020).[9] In the IEA (2010) scenarios, CCTS plays a significant role in the 450 ppm scenario after 2025 and a smaller role in the new policies scenario but only after 2030. There are various technological and political barriers to the implementation that

Table 4.2   Assumed installed capacities of coal power plants with CCTS for the CCTS scenario in GW

	2020	2025	2030
World	150	286	423
OECD+ (incl. Europe, USA, Japan)	72	134	197
OME (incl. China, Russia, South Africa)	74	145	216
OC (incl. India, South-East Asia)	4	6	9

Source: Own calculations based on IEA (2010).

explain this late roll-out of the CCTS technology (see Gibbins and Chalmers, 2008; Herold et al., 2010).

In our CCTS fast roll-out scenario we assume that technological breakthroughs, a favorable regulatory framework as well as strong political support create the conditions for a fast CCTS roll-out with significant capacities coming in as early as 2020. Such a scenario makes sense only in an overall environment of ambitious climate policy; thus we apply this additional policy shock in the new policies and the 450 ppm policy framework only, as shown in Figure 4.6.

For this scenario we assume that the worldwide installed capacities of coal power plants with CCTS projected by the IEA (2010) in the 450 ppm scenario are put in place five years earlier. We assume that half of this additional capacity replaces existing older coal power plants, the other half is integrated in the power system as additional capacity, successfully competing with other technologies. Furthermore, for our coal demand calculations, we assume that CCTS power plants have a 38 percent efficiency and a capacity factor of 82 percent. Thus we compute two additional demand shocks: one coming from half of the CCTS capacity that is added to the coal demand and the other because the lower efficiency of CCTS power plants requires additional coal to produce the same amount of energy. The assumed capacities of CCTS for our modeling runs are shown in Table 4.2 divided into the following regions: OECD+, OME (other major economies) and OC (other countries).[10]

5   SCENARIO RESULTS AND INTERPRETATION

5.1   Negative Market Adjustments – Unilateral European Climate Policy

The modeling results of the unilateral European climate policy scenario are shown in Figure 4.8. We can see the actual reduction of emissions in the EU due to the EU unilateral climate policy. The remaining emissions are the actually targeted emissions and together with the emissions from market adjustments, they represent the actual global emissions. We can see that, given certain conditions, market adjustments can undermine a unilateral European climate effort. This is especially the case in the current policies framework with a constrained market condition. In that case global coal demand is high and the market is somewhat constrained so that a reduction in European coal demand allows the Asian countries to consume significantly more. In 2025, the market adjust-

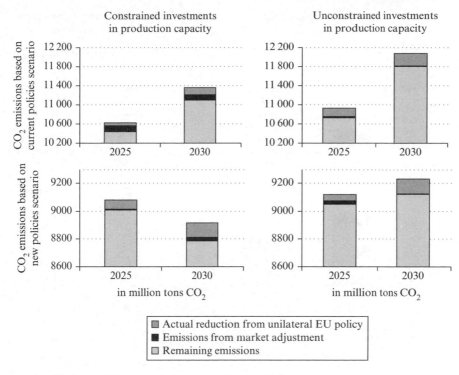

Figure 4.8   Unilateral European climate policy model scenario

ment nullifies 66 percent of the European reduction target and 29 percent in 2030. In the unconstrained case the market adjustment is negligible but global emissions are higher.

In the new policies framework the market adjustment is much lower. The market condition has a strong impact on the global level of emissions and on the market adjustment mechanism. In 2030, in the constrained case the mechanism works as described above for the current policies constrained case: the European reduction allows Asia to consume more through the global price mechanism. However, in the unconstrained case, in 2025, the market adjustment with considerably lower prices occurs because more quantities are potentially available when there is no constraint. The lower European consumption has a significant impact on prices that it does not have in the constrained case where global demand remains slightly restricted.

We can conclude that market adjustments are very likely to occur but that their adverse effect is generally low and will not overcompensate the emissions reductions from Europe. This is due to the relatively small size of EU demand in global steam coal demand. However, in the case of a low level of global climate policy as in the current policies scenario the adverse market adjustment effect can be very high. Thus there is an argument for the European Commission to link its aim of a 30 percent emissions reduction goal to other countries making binding commitments to higher reduction goals. In such a case, which can be described by the new policies scenario, Europe can always go an extra mile without expecting too many adverse market adjustments.

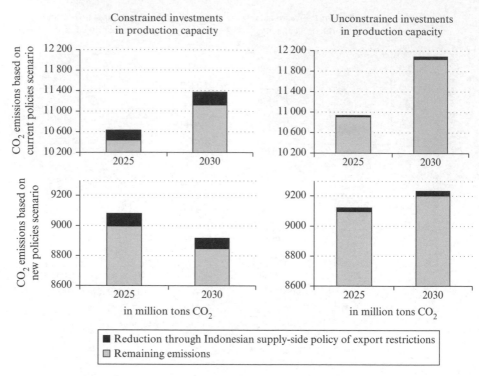

Figure 4.9   Worldwide emissions reduction in the Indonesian supply-side policy scenario

5.2   Indonesian Supply-side Policy: Positive Market Adjustment

In our reference cases for all policy environments Indonesia is the most important supplier to the global market with yearly export values that can be higher than 200 Mt.
The results of the Indonesia scenario run with export restrictions are summarized in
Figure 4.9. We see that the reduction effect is the strongest in the case of a constrained
global market because it is hard to find alternative suppliers on the world market that
could replace the Indonesian exports. In the unconstrained case the effect is lower as
Indonesian coal is substituted by other producers. Also, such an effect may be limited in
time as the supply gap may be covered by other producers over time.

5.3   CCTS Fast Roll-out Lowers Emissions

The results of the CCTS scenario are presented in Figure 4.10 for the new policies climate
policy framework. CCTS is insignificant in the reference scenario and therefore the additional CCTS capacity, half of which leads to new coal demand, has a strong effect on the
market. We see a market adjustment that is 'positive' for the climate as it lowers global
emissions. The higher demand leads to higher prices that lead to a reduction of demand
from conventional power plants. This effect is very strong in a constrained market
environment and still significant in the unconstrained environment.

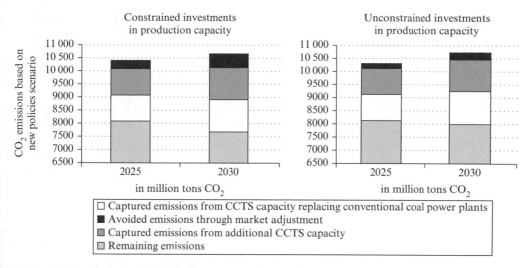

Figure 4.10   *Worldwide emissions in the CCTS scenario*

In the scenario based on the 450 ppm climate policy environment the global coal demand is so low that the market condition has very little effect on the scenario outcomes. We observe little market adjustment. In the case of a relatively intense global level of climate policy, such as in the new policies framework, a faster implementation of CCTS would be very beneficial. We expect that, additionally to the captured quantities of CO_2, a positive market adjustment will further reduce coal consumption and emissions.

5.4   Hedging of Market Adjustments

As shown in Section 4, negative market adjustments can have a significant impact and partly render ineffective an emissions reduction effort. In the case of a unilateral European climate effort the use of steam coal is reduced significantly. However, through pure market effects and price mechanisms, this lower consumption in Europe is compensated by a higher consumption in the rest of the world, especially Asia. One possibility to counteract this effect would be to accompany the demand reduction in Europe by another policy measure to induce beneficial market adjustments as a way to 'hedge' against adverse market adjustments effects. In the following modeling exercise we therefore present a combination of the unilateral European climate policy scenario with the Indonesian supply-side policy of export restriction.

Figure 4.11 shows the results of this combined model run in comparison to the results of the European unilateral climate policy scenario from Section 4. We concentrate on the case based on the current policies scenario with constrained investments in production capacity where we can expect the most drastic effects. We can see in Figure 4.9 that the additional emissions from negative market adjustments can be very strong in the original scenario. In the new scenario presented in Figure 4.11 we can see the avoided emissions in Europe and a positive market adjustment that is avoided emissions in the rest of the world.

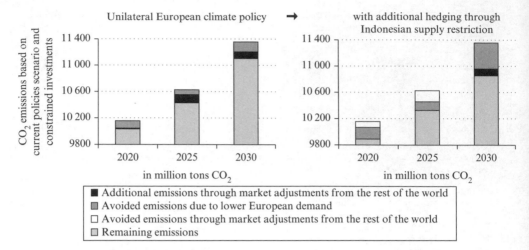

Figure 4.11   Unilateral European climate policy results with and without additional Indonesian supply-side policy

The right-hand graph shows the results of the combined scenarios: the addition of the Indonesian supply-side policy cancels out (or 'hedges against') and even overcompensates the negative market adjustment effect in 2020 and 2025 and leads to an overall emissions reduction. In 2020 and 2025, the Indonesian export restriction induces a lower consumption in Asia. In 2030, however, it affects Europe more, as the higher supply costs lead to an import reduction. In fact there is a beneficial market adjustment taking place in Europe that leads to an overachievement of the climate target. However, by 2030 we see the resurgence of some negative market adjustment effect in Asia, induced by an investment effect. The lacking supplies from Indonesia are compensated by investments in new mines, especially in China, that become economically available on a larger scale by 2030.

Despite the overall positive effect of this combined policy, we must be aware that its effects might be limited in time. Also the proposed hedging strategy may be politically too difficult to put in place. However, other 'hedging' strategies are possible to mitigate adverse market effects. These could be demand-reducing policies, for example through mechanisms already in place such as the Clean Development Mechanism (see Kemfert, 2002). The 'hedging' effect will appear when the build-up of renewable energy capacity replaces the use of steam coal for electricity generation.

6   CONCLUSIONS

In this chapter, we discuss likely trends in international steam coal markets, based on modeling exercises using the COALMOD-World model. The focus of coal trade is likely to move towards India and Asia, whereas traditional suppliers like South Africa, Australia or Indonesia will maintain and expand their role. Asian demand will be the main driver of the future markets. The COALMOD-World model also enables us to make differentiated conclusions about the efficiency of different types of climate policy alternatives through their interplay with the global coal market in various market conditions. We find that coal market adjustments can have a potentially adverse effect on climate policy effectiveness in the case of a unilateral European climate policy. However, on a global scale this effect never overcompensates saved emissions in Europe; even a unilateral European climate effort will be beneficial to the global climate. Thus the first priority of the European climate policy should be to reach a global level of climate policy that is at least at the level of the non-binding agreements in recent climate conferences in Copenhagen in 2009 and in Cancun in 2010. A global climate policy has the biggest effect on global carbon emissions from coal demand reduction. If a global climate consensus is reached, the EU can always go further in reducing its steam coal consumption without 'risking' much adverse effect from the global market.

A supply-side climate policy in Indonesia also has some significant CO_2 emissions reduction effects that are potentially in the same order of magnitude as the European unilateral climate policy. This supply-side policy has its best performance in the context where the European unilateral climate policy sees the most important negative market adjustment effects; that is, when there is a low intensity of global climate policy and when the market is constrained. Thus the EU could try to pursue such an unconventional climate policy as a way to hedge against adverse effects from its own domestic climate policy. Such a policy in Indonesia would also have additional beneficial effects for nature conservation, the protection of biodiversity as well as avoided CO_2 emission from deforestation.

The first advantage of a policy that aims at a faster roll-out of the CCTS technology is the emissions reduction through the capture of CO_2. Climate-beneficial market adjustment effects can also occur. A significant impact of CCTS can only be expected if this technology is implemented globally. Thus the strategy of the EU should be to support the roll-out of this technology in Europe but also abroad through international cooperation and knowledge spillovers.

NOTES

1. This chapter surveys previous work by the authors and builds on earlier work cited in the references. We thank Tim Winke for research assistance; the usual disclaimer applies.
2. The traded quantities are the energy quantities contained in the coal, expressed in petajoules. Whenever the model needs to deal with mass quantities in million tons of coal (for the costs, capacities and investments), these energy quantities are converted in mass using a conversion factor defined in tons per gigajoule κ_{af} that is different for every producer.
3. In constrained optimization the dual variable of a constraint indicates if it is binding or not. The dual is positive when a constraint, such as an infrastructure capacity constraint in COALMOD-World, is binding.
4. The underestimation of these trade flows by the model may be due to the strong relationship between Australia and Japan in the coal sector. Japanese firms are involved in coal production through joint ventures and long-term contracts are still common.
5. We do not address the issue of accelerated depletion of coal resources due to expectations of stricter future climate policy, since it is not evident that this effect plays an important role in competitive coal markets; for a more detailed treatment of this issue see Haftendorn et al. (2012b).
6. This mechanism has also been described as being part of the carbon leakage mechanisms in the literature (see Dröge, 2009) but is rarely analyzed. To avoid confusion we restrict the term leakage to relocation of manufacturing and introduce the term 'market adjustments' for pure market effects.

7. http://minerals.usgs.gov/minerals/pubs/country/.
8. http://yasuni-itt.gob.ec/.
9. Source: Carbon Capture and Sequestration Technologies Program at MIT (http://sequestration.mit.edu/tools/projects/index.html).
10. See IEA (2010) for an exact definition of these aggregates.

REFERENCES

Dröge, Susanne (2009), 'Tackling leakage in a world of unequal carbon prices: synthesis report', Climate Strategies.
Fatah, Luthfi (2008), 'The impacts of coal mining on the economy and environment of South Kalimantan Province, Indonesia', *ASEAN Economic Bulletin*, **25**(1): 85–98.
Gibbins, Jon and Hannah Chalmers (2008), 'Carbon capture and storage', *Energy Policy*, **36**(12): 4317–22.
Haftendorn, Clemens and Franziska Holz (2010), 'Modeling and analysis of the international steam coal trade', *The Energy Journal*, **31**(4): 201–25.
Haftendorn, Clemens, Franziska Holz and Christian von Hirschhausen (2012a), 'COALMOD-World: a model to assess international coal markets until 2030', *FUEL*, **102**: 305–25.
Haftendorn, Clemens, Claudia Kemfert and Franziska Holz (2012b), 'What about coal? Interactions between climate policies and the global steam coal market until 2030', *Energy Policy*, **48**: 274–83.
Herold, Johannes, Sophia Rüster und Christian von Hirschhausen (2010), 'Carbon capture, transport and storage in Europe: a problematic energy bridge to nowhere?', Brussels, CEPS Working Document No. 341 (November).
Holz, Franziska, Christian von Hirschhausen and Claudia Kemfert (2008), 'A strategic model of European gas supply (GASMOD)', *Energy Economics*, **30**(3): 766–88.
Huppmann, Daniel, Ruud Egging, Franziska Holz, Christian von Hirschhausen and Sophia Ruster (2011), 'The world gas market in 2030 – development scenarios using the world gas model', *International Journal of Global Energy Issues*, **35**(1): 64–84.
IEA (2010), *World Energy Outlook 2010*, Paris: OECD.
IPCC (2005), 'IPCC special report on carbon dioxide capture and storage', URL http://www.ipcc.ch/pdf/special-reports/srccs/srccs_wholereport.pdf.
Kemfert, Claudia (2002), 'Global economic implications of alternative climate policy strategies', *Environmental Science & Policy*, **5**(5): 367–84.
Larrea, Carlos (2010), 'Yasuni-ITT: an initiative to change history', URL http://yasuni-itt.gob.ec/wp-content/uploads/initiative_ change_history_sep.pdf, Government of Ecuador.
Li, Raymond, Roselyne Joyeux and Ronald Ripple (2010), 'International steam coal market integration', *The Energy Journal*, **31**: 181–202.
Lise, Wietze and Benjamin F. Hobbs (2008), 'Future evolution of the liberalised European gas market: simulation results with a dynamic model', *Energy*, **33**(7): 989–1004.
Lise, Wietze, Benjamin F. Hobbs and Frits van Oostvoorn (2008), 'Natural gas corridors between the EU and its main suppliers: simulation results with the dynamic GASTALE model', *Energy Policy*, **36**(6): 1890–1906.
Paulus, Moritz and Johannes Trüby (2011), 'Coal lumps vs. electrons: how do Chinese bulk energy transport decisions affect the global steam coal market?', *Energy Economics*, **33**(6): 1127–37.
Paulus, Moritz and Johannes Trüby (2012), 'Market structure scenarios in international steam coal trade', *The Energy Journal*, **33**(3): 91–123.
Zaklan, Aleksandar, Astrid Cullmann, Anne Neumann and Christian von Hirschhausen (2012), 'The globalization of steam coal markets and the role of logistics: an empirical analysis', *Energy Economics*, **34**(1): 105–16.

PART II

ELECTRICITY MARKETS

5 The future of the (US) electric grid

Henry D. Jacoby, John G. Kassakian and
Richard Schmalensee[1]

The first electric power system in the USA was installed by Thomas Edison in New York City in 1882. It served 59 customers in the Wall Street area at a price of about $5 per kilowatt hour (kWh). Over the decades since, the US electric system has grown into a vast physical and human network with thousands of electricity generators providing service to hundreds of millions of consumers. The grid tying it all together is a linked system of public and private enterprises operating within a web of government institutions: federal, regional, state and municipal. Over the years it has incorporated several generations of new technology, and has improved its performance accordingly.

Every expectation, though, is that change will be – and will need to be – more rapid in the next few decades than in the recent past. The grid will face new challenges, and new technologies will be available to meet them. Regulatory and other public policies will play a major role in determining the future of the grid, as will advances in technology. While some of the opportunities and challenges we discuss in this chapter are unique to the USA, we believe that many of the issues are being confronted as well by electric grids in other countries.

Like the grids of many other nations, for instance, the US grid faces the likelihood of increased reliance on generation from large-scale variable energy resources (VERs), chiefly wind and solar, the outputs from which are variable and imperfectly predictable. Many of the best sites for these generators are located far from the existing high-voltage grid. Further complicating matters, increased penetration of distributed generation, particularly roof-top solar units, and the charging of electric vehicles will pose challenges to distribution systems. Also, because of growth in air conditioning and reduction in base-load industrial use the difference between peak demand (which the system must be sized to accommodate) and average demand is increasing, reducing capacity utilization and thereby raising cost. Electric vehicles may accelerate this trend.

Much is expected in the USA and elsewhere of a future 'smart' grid that exploits the ongoing revolution in electronics and communications, and indeed a flood of new devices and analysis tools does offer more effective ways to observe and manage the grid. These tools can also be used to enhance incentives for efficient electricity supply and use. While some of these new technologies bear a broad resemblance to those that have supported the creation of the Internet, there is one very important difference: there are no 'killer apps' in prospect for the electric grid. New technologies promise to enable the same basic service – electric energy on demand – to be delivered more efficiently and reliably, but not accompanied by exciting new services or capabilities. Accordingly, there is no reason to expect great customer enthusiasm for new grid-related technologies. Moreover, the flip side of expanded communications and automated controls, with all their advantages, is an increase in avenues for cyberattack on the electric system and threats to consumer privacy.

We believe that new technologies will make it possible to meet the challenges described above while maintaining acceptable levels of reliability and electricity rates. But for this to happen, a number of policy changes are necessary to better align incentives. In addition, targeted research efforts are required, and the collection and sharing of critical data needs to be expanded. In this chapter we briefly explore these issues, with a focus on desirable directions in US federal policy. To provide background for that discussion we begin with a summary of some of the relevant, and in some ways unique, features of the US electric system.

THE ORGANIZATION OF THE US ELECTRIC SYSTEM[2]

Two related features of the US electric power system set it apart from systems in many other nations: the great diversity of entities involved in the supply of electric power, both in generation and distribution, and the lack of a comprehensive national policy toward the electric power sector. They are not unrelated, but are the joint product of a century of system evolution under the division of power and responsibility among levels of government that characterizes the US federal system.

Investor-owned firms account for about 84 percent of US electric generation. Another 4 percent is produced by the federal government, much of it sold on favorable terms (as so-called 'preference power') to cooperative and municipal distribution utilities. Cooperatives and publicly owned systems organized at the state or local level account for the remaining 12 percent of generation. Several hundred entities own parts of the transmission system, with government enterprises and cooperatives accounting for 27 percent.

Electric distribution is even more fragmented. About 3200 organizations provide electricity to retail customers. Nearly 2200 are publicly owned, mainly by municipalities, but they account for only 16 percent of electricity sales. Another 818 are cooperatives, accounting for an additional 10 percent. In some areas, particularly the Pacific Northwest, access to the favorably priced federal power is important for these municipal and cooperative utilities. Investor-owned utilities account for the rest, in some cases with intermediary services provided by retail power marketers who broker power but do not provide physical distribution. Investor-owned utilities are regulated at the state level, but cooperative and government-owned utilities are not subject to state regulation.

As this diversity in size and regulatory structure and the associated regional differences reflect, the USA lacks a comprehensive national electricity policy. The main federal agency, the Federal Power Commission (FPC), was established in 1920. It became the Federal Energy Regulatory Commission (FERC) in 1977. In the very early years of the industry, transmission was planned and built by vertically integrated utilities that met their native loads from their own generation, largely within state borders. But over time the promise of improved reliability led to increased interconnection among utilities, and in 1935 the FPC was empowered to regulate the wholesale sales of electric power, of which more below.

Over the industry's first century, most US electric service was provided by vertically integrated investor-owned utilities operating under cost-of-service regulation. Under this regime, state regulators set retail rates in order to cover utilities' costs, including a 'fair' rate of return on invested capital. In the 1980s, however, a new model of power system organization and governance began to emerge in the academic and policy literature (Joskow and Schmalensee, 1983). In this model, organized competitive markets set the price of wholesale electricity. Ownership of generation might be separated from the rest of the system, at least to some extent, and an independent entity would operate the transmission system and administer the wholesale markets. The provision of distribution services would remain a regulated monopoly, but there might be competition in the sale of electricity at the retail level. Overall, markets would perform some of the coordinating and cost-minimizing functions that traditionally took place within vertically integrated utilities. Originally discussed in the USA, this market-oriented model of electric system organization has been employed more completely and systematically in other countries, beginning with Chile in 1982 and the UK in 1990. It is now the model followed throughout the European Union (CEC, 2007).

To promote this use of competitive markets, in 1996 the FERC issued an order that required transmission owners to provide wholesale customers with open, non-discriminatory access to their systems, under regulated open access transmission tariffs. The order required that equal access be granted to both utility and non-utility generators. The FERC noted that one way to meet these requirements would be for the regional transmission system to be operated by an independent system operator (ISO), an independent, federally regulated entity without generation or distribution assets. The FERC later defined regional transmission operators (RTOs): ISOs that meet certain specified requirements and that have slightly greater responsibilities for system reliability.

Both Congress and the FERC have repeatedly expressed support for competitive wholesale markets and open, non-discriminatory access to the transmission system (Kelliher, 2007). And the FERC has strongly recommended that US utilities affiliate with RTOs or ISOs. But it has not required them to do so, lacking either the authority or the will (depending on whom you ask) to implement an effective, national pro-market policy. As a result the system has become a patchwork of forms of electricity market organization, as shown in Figure 5.1. Organized ISO/RTO wholesale markets now cover two-thirds of the US population and meet about two-thirds of electricity demand. In the Southeast, the traditional vertically integrated utility model remains dominant, while in the West, particularly the Pacific Northwest, federal, municipally owned and cooperative enterprises play an important role in the industry.

Where they have been established, RTOs or ISOs operate the wholesale electric market, dispatch generation to match load efficiently, and oversee the operation of the transmission system. They also generally are responsible for transmission system planning, a process that identifies and makes visible the need for strengthening the system to maintain reliability. In this new structure, as in vertically integrated utilities, central control of generators' outputs is necessary in the short run to minimize system cost, maintain security, and respond to unexpected changes in load and other events. In organized markets, the cost to be minimized is the cost of buying power from independent generators, at prices at which the generators are willing to supply.

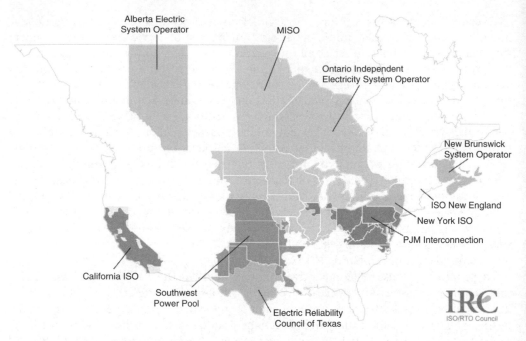

Source: ISO/RTO Council, http://www.isorto.org.

Figure 5.1   Regions with organized wholesale electricity markets

In organized wholesale markets, generators bid by offering a maximum amount of energy (MWh) for sale during specific periods of the next day at a specific price ($/MWh). These offers are arranged by the ISO/RTO in ascending order called the 'bid stack' and the generators are dispatched (told to generate) in this order until generation matches load. All the generators that are dispatched receive the same compensation called the 'clearing price' – the offer of the last generator dispatched. The actual process is more complicated than this simple explanation, incorporating such parameters as the time required to start the generator, out-of-economic-order dispatch due to congestion or reliability concerns, and security constraints. The detailed implementations of this new bulk power system structure have varied considerably over time and space, but they also have converged in important respects, and most observers agree that ISO/RTO systems are performing well.

At the retail level, most states were not eager to embrace competition. Nonetheless, 22 states and the District of Columbia have begun a variety of processes aimed at allowing competing vendors of electric power at the retail level, while continuing to treat the distribution of that power as a regulated monopoly.

In the remainder of this chapter we consider challenges, opportunities, and desirable policy and other changes in transmission and distribution separately, although the boundary between them is blurring. We then discuss some of the issues that arise because data communications are becoming increasingly important in modern electric power systems.

THE TRANSMISSION SYSTEM

The transmission network is the first link between large power generation facilities and electricity customers, supplying energy at high voltages to substations where it is then distributed to loads at lower voltages via the distribution network. The US network today operates reliably and efficiently, but various technologies offer the possibility of improvements in system performance. Sophisticated new monitoring systems may reduce the likelihood of cascading system failures, and other technologies may help solve problems associated with network expansion and help incorporate growing penetration of variable energy resources like wind and solar generation.

For example, developments in power electronics have enabled flexible alternating current transmission systems (FACTS) and other technologies that can expand the capacity of transmission systems and make them more flexible.[3] In addition, phasor measurement units (PMUs) can now provide frequent, time-stamped information on the state of the system at multiple locations that can be used to detect system stress and facilitate effective response to it.[4] And new high-voltage dc (HVDC) transmission technologies can provide efficient long-distance transmission with branching at multiple locations along the line, mitigating the point-to-point constraint that limits existing HVDC lines.

The main impediment to the use of FACTS and related power-electronics-based technologies is their costs, although these are declining. Stimulus spending in response to the 2008–09 US recession, particularly under the 2009 American Recovery and Reinvestment Act (ARRA), has funded initial expansion of PMU penetration into the grid, but new algorithms, software and communication systems need to be developed to translate the vast amount of data that PMUs produce into useful information and to integrate these new devices effectively into system operations. Proper interpretation of data on the actual state of the system requires an accurate model of the system under normal operating conditions, and development of such models will require more widespread sharing of PMU data than has thus far occurred.

As noted above, an important emerging challenge for the electric grid, in the USA and elsewhere, is the accommodation of increased generation from large-scale VERs, a goal of policies at both state and federal levels. The use of these technologies in the USA would be further stimulated by the adoption of policies aimed at substantial reductions in CO_2 emissions, although the advent of cheap natural gas could diminish this effect substantially in the near term. The greater the importance of VERs in any system, the greater the changes in system operation, with associated costs, that are required to adjust to the uncertainty and variability of supply as winds shift or clouds pass over solar panels. One much-discussed cost to power systems with substantial VER penetration is provision of reserve generating capacity and other sources of flexibility to enable the system to respond to output fluctuations on a wide range of time scales: seconds to hours to days.

Several improvements in system data and procedures that could help lower these costs are under discussion and development. For example, better wind forecasts would improve day-to-day planning of generating capacity, and more effective sharing of meteorological data at turbine hub height by existing wind operators would aid wind analysis and planning. Within-day management of generators could be improved by the provision of wind data on shorter forecast periods. And markets for ancillary services – spinning reserves, non-spinning reserves, and other measures that help maintain reliable operation of the network – may need to provide explicit incentives for the provision of flexibility to the system. It may also be necessary to provide incentives for non-VER generators to build in flexibility.

In addition, it is important to consider the performance and equipment standards (interconnection standards) that manage the connection of wind and solar sources to the transmission grid, to ensure that VERs play a proper role in maintaining system stability and reliability. Current US interconnection standards are adequate for current levels of VER penetration, but some tightening may be needed to insure safe and reliable operation at anticipated higher levels of wind and solar penetration in the future. Finally, the USA currently has 107 balancing authorities (the entities responsible for balancing generation and load in real time within a specified geographic region). Their number and distribution reflect the complex history of the electric power sector: there are six in Arkansas, but only one in New York, serving seven times the population. Virtual or full consolidation of some of these balancing areas could reduce the costs of VER integration by enabling wider geographic averaging of VER outputs and perhaps bring other efficiencies, but various institutional issues make this less than straightforward in some areas.

Increased penetration of VER generation will require expansion of the transmission system. Moreover, because the best wind and solar resources are often located far from major load centers, increasing the use of VER generation is likely to require construction of more transmission lines that cross state boundaries or land controlled by federal agencies.

As noted earlier, the US transmission system grew up largely within state borders, regulated by state authorities, and state approval is still required for the construction of new transmission facilities. Federal agencies must approve lines crossing land they control. Some lines crossing state boundaries and federal lands have been built over time, generally to enhance reliability, when the allocation of costs could be agreed upon by the actors involved and all relevant agencies granted approval. Over time the US transmission system has become progressively more integrated, so that today it is composed of three large 'interconnections' – east and west of the Rocky Mountains, and the area served by the Electric Reliability Council of Texas (ERCOT) – within each of which all generators are synchronized.[5]

But, particularly in the East, transmission planning is generally done on a regional basis, not interconnection-wide. Moreover, states' interests often conflict. For example, with cost allocation negotiated on a project-by-project basis, any involved state can block a multi-state project in hopes of getting a better deal. Finally, federal agencies with missions unrelated to energy manage 30 percent of the land area of the USA, and they can (and do) exercise the power to block or delay the construction of transmission lines across lands they control.

Recognizing that these barriers to building boundary-crossing transmission facilities constituted obstacles to the efficient expansion of the role of VER generation, the FERC issued Order 1000 in 2011 (FERC, 2011). This Order aims to significantly increase wide-area planning of transmission systems and rationalize the allocation of the costs of transmission facilities and base that allocation on the principle that costs should be allocated in proportion to benefits – the so-called 'beneficiary pays' principle. This

Order is a step toward a more coherent national transmission policy, but not a complete solution.

Order 1000 requires adjacent regions to coordinate planning, but does not mandate planning at the interconnection level. The public interest would be better served if the affected parties went beyond the Order's planning requirements and established permanent and collaborative processes for transmission planning at the interconnection level. To be effective, these processes should be hierarchical, combining top–down and bottom–up activities. In addition, research is needed to develop planning methods that can consider the optimal multi-period expansion of complex networks under uncertainty. In order to develop and test such methods more comprehensive data need to be available to researchers, particularly on the more complex Eastern Interconnection.

On the issue of the allocation of the costs of new facilities, Order 1000's endorsement of the 'beneficiary pays' principle is a step forward. Use of this principle will provide better locational signals than the commonly used alternative practice of 'socializing' the cost, that is, spreading it uniformly throughout the affected regions (Joskow, 2005). Moreover, if a project has overall benefits greater than its costs, allocating costs in proportion to benefits will generally make all affected parties better off and thus inclined to support the project. If it is not possible to allocate costs in a way that makes all parties better off, it is likely that total benefits fall short of total costs.

Unfortunately, benefits assessment is complicated by the fact that transmission affects not only electricity prices but also system reliability, jobs, local interests and the environment. And, as Order 1000 specifically recognizes, lines may be justified by policy objectives in addition to these system effects. As a result much work remains to develop procedures for allocation, which in practice likely will involve some combination of socialized costs and rough measures of benefit, along with more formal economic calculations (Hogan, 2011; MITEI, 2011, ch. 4).

Order 1000 only requires the development of standardized bilateral cost-allocation procedures for interregional projects. Since projects may involve multiple regions, it would be desirable for the parties involved to adopt a single interconnection-wide procedure in each interconnection.

Finally, even with effective planning and equitable cost allocation there remains the matter of granting permission to build a project that crosses state boundaries or federal lands. The FERC did not have the authority to address the siting of boundary-crossing lines in its Order 1000. The US federal government has addressed this type of problem before. In 1938, recognizing the growing importance of interstate natural gas pipelines, Congress gave FERC sole authority to site these facilities, including the power of eminent domain. In recognition of the increasing importance of interstate electricity transmission, the Energy Policy Act of 2005 contained a section apparently intended to give the FERC backup siting authority in the event states did not approve construction of proposed multistate electricity transmission facilities in areas experiencing transmission congestion. Unfortunately, a subsequent court decision has effectively annulled that section by holding that only delay, not denial of approval, triggered federal authority. Even though the administration has taken steps to streamline federal agency participation in siting decisions by creating an inter-agency Rapid Response Team for Transmission (CEQ, 2011), significant barriers remain in the way of efficient transmission expansion and exploitation of renewable energy.

One obvious solution, which has the merit of simplicity, is for the FERC to be granted sole siting authority over major interstate projects, as it has over interstate natural gas pipelines. Alternatively, giving FERC effective backstop authority, so that it could intervene only when existing procedures have failed to grant the necessary approvals, would create a slower process but one more sensitive to the legitimate concerns of individual states and federal agencies. While both approaches clearly have strengths and weaknesses, new legislation taking either approach would produce a significant and important improvement over the status quo.

THE DISTRIBUTION SYSTEM

In the USA and elsewhere, recent years have seen increasing investment in 'smarter' consumer meters, so-called advanced metering infrastructure (AMI), that can provide frequent, two-way communication between a distribution utility and its customers. These investments have been accelerated in the USA by ARRA expenditures under the US Department of Energy's Smart Grid Demonstration Program and companion Smart Grid Investment Grants. Most obviously, AMI installations facilitate the measurement of consumer use over short intervals of time and thus provide opportunities for improved pricing systems and other approaches to engaging customer demand, including the charging of plug-in-hybrid and pure electric vehicles, and, along with other new technologies, managing the distribution network itself.

Renewable energy mandates, subsidies to distributed generation and programs to spur the sales of electric vehicles have several important implications for distribution systems. First, while various types of small generators, including hydropower, gas-fired and oil-fired units already are connected to distribution networks, many are now used only as backup power supplies. Growth in residential- and commercial-scale solar and wind power generators that are in constant use are expected to have major effects on these systems in the future. Currently, such distributed power sources must meet a set of standards developed by the Institute for Electrical and Electronics Engineers (IEEE, 2008). Designed to limit potential negative effects of distributed sources on voltage regulation, system protection and worker safety, the current standards constrain the design of the power source and how it must respond to disruption elsewhere on the grid. Over the intervening years, technologies have become available to help manage these concerns, and a fundamental review and revision of these standards is needed to bring them up to date. Otherwise they may impose unneeded investment costs on utilities to accommodate the projected expansion, particularly of distributed solar and wind power, and possibly limit the benefits these technologies offer.

Second, adjustments are required to accommodate electric vehicles. Even at the upper end of projected market penetration over the next couple of decades at plausible subsidy levels, charging these vehicles is not likely to strain any regional distribution system as a whole. But local lines and transformers could become overloaded in neighborhoods where high income or local tastes may lead to a large number of electric vehicles on the same feeder. In addition, after-work charging could add significantly to system peak loads in some areas. Selective re-enforcement of the distribution network plus more peak-load generating capacity could accommodate this incremental load. Barring on-peak charging would eliminate the need for additional capacity but would likely be unacceptable to consumers. A more efficient – and probably more acceptable – solution would be to provide incentives for charging at off-peak times, when the electric power system has excess capacity and generation costs are relatively low.

More generally, facing customers with prices that reflect actual system marginal cost in real time, so-called dynamic pricing, can significantly reduce the costs of electric generation and the need for investments to expand system capacity. Most US customers are now billed for electricity at a price that changes at most with the season, and even among the largest commercial and industrial users dynamic pricing of electricity is far from universal. Constant prices tell consumers that the cost of delivering power is the same at all times in the billing period, when in fact it may vary twofold or more over the hours of a typical day, and by as much as a factor of ten or more at system peaks or when generation or transmission constraints are particularly binding. Some utilities have already offered residential (and other) customers programs in which they pay lower prices but give the utility the ability to reduce their demand (via, commonly, altering the operation of central air conditioning systems) when the grid is under particular stress. While expansion of these and related demand management systems would add some value, the ability that AMI provides to face consumers with prices that reflect actual costs is a substantial part of the opportunity for the electric power system as a whole to become 'smart'.

Of course, most residential consumers have no interest in watching and responding to real-time electricity prices, particularly since electricity in the USA and elsewhere accounts for only a few percent of average household budgets. There is much to be learned about how to overcome resistance to dynamic pricing, which we feel will involve automating response to price changes with a simple interface. For a variety of reasons, including federal project-specific subsidies under ARRA and state mandates, many utilities have installed or have committed to install AMI systems. We conclude that these utilities should begin a transition to dynamic pricing for all customers, a transition that will involve investment in ongoing consumer engagement and education, and publicly share data from their experiences. Consumers do not have a good understanding of dynamic pricing systems, technologies for automating response to price changes are still immature, and adverse reactions could derail program implementation and diffusion.

Reflecting uncertainty about consumer response, the benefits of dynamic pricing systems have been estimated only imprecisely, and other operational benefits may not always be sufficient in themselves to justify the cost of AMI deployment. Where commitments to AMI have not yet been made, deferring investment decisions to learn from the experience of early AMI adopters may prove a prudent strategy for many utilities. Much of this important information in the USA should come from the ARRA-funded investments, and it is thus important that detailed information about these projects and their results be made widely available.

In addition to AMI, several other new technologies are being introduced into distribution systems, some, in the USA, with support from the ARRA programs mentioned above. These 'distribution automation' technologies include sensors to provide instantaneous information on the state of the system, along with associated software systems and controls that can more efficiently handle consumer complaint calls, locate outages and allocate repair crews, and in some cases provide automatic 'self-healing' capability when faults occur. While these technologies are generally expensive, they have considerable promise, and their costs will come down. Sharing the results of early deployments and pilot studies will facilitate intelligent investments in the future.

These new technologies pose challenges for regulation. Unlike conventional distribution system investments to expand capacity or replace aging equipment, technologies that provide new capabilities or enhance the performance of a system may involve greater investment risk to the utility. Because both regulators and regulated monopolies are often punished politically for perceived failures but not rewarded for successes, traditional regulatory systems tend to encourage excessively conservative investment behavior. This is likely to become increasingly expensive for systems confronting the challenges of distributed generation, electric vehicles and energy efficiency programs, as well as the opportunities presented by AMI and distribution system automation technologies. Excessive risk aversion is thus an important problem, but one without an obvious solution. New approaches to investment oversight and rate setting may well be needed to give utilities appropriate incentives to make investments that are riskier – although perhaps on balance more attractive – than has been the norm.

Increased attention to measurement and reward of system performance should prove complementary to these approaches. The remuneration of most US distribution utilities is determined by an assessment of their capital and operation costs. Standardized measures of distribution system efficiency, reliability and service quality are not reported for most electric utilities in the USA. Some utilities do not report statistics on outages, for instance, while others employ a variety of definitions. Even where performance is measured, it is rarely linked directly to utility profits. We believe this may reflect the same risk aversion that characterizes investment behavior: no regulator wants to admit that the utility it supervises is performing poorly. Nonetheless, consumers would benefit if regulators adopted performance-based regulation. Whatever the regulatory regime, of course, greater efforts to develop and report comparable metrics of utility performance would improve decision making.

Finally, the growing penetration of distributed generation in response to subsidies and regulatory mandates of various sorts has increased the importance of the way network services are recovered. The costs of distribution and transmission networks are largely independent of usage, at least in the short run, but they are nonetheless generally recovered through volumetric charges – that is, charges per kWh of use – particularly for residential customers. This system reduces a distribution utility's incentive to accommodate distributed generation or to support energy-efficiency initiatives because both would reduce its sales and profits. And consumers who respond to volumetric charges for network costs by reducing their usage, through efficiency or installation of distributed generators, shift their share of fixed network costs onto other consumers.[6]

To correct these perverse incentives, state regulators and those who supervise government-owned and cooperative utilities should recover fixed network costs primarily through fixed charges that do not vary with kWh consumption. These charges should reflect each customer's contribution to the need for local distribution capacity, perhaps based on the individual consumer's historical pattern of use in peak demand periods or some consumer group load profile. Systems that continue to rely significantly on volumetric charges for cost recovery can improve utility incentives by decoupling utility revenues from short-run changes in sales, as some states have done.

COMMUNICATIONS, CYBERSECURITY AND PRIVACY

The introduction of new technologies for system control and customer metering is bringing a vast increase in data communications within the grid, as illustrated in Figure 5.2. These communications span an increasing collection of media – copper and fiber wires, radio and microwave, satellite systems, private networks, and the public Internet – to manage tasks ranging from the control of enormous generators to telling customers to adjust their air conditioners (or to adjust them directly). Increased grid connectivity is essential to make the grid more reliable and more efficient, but it also will open up vulnerabilities to accident and malfeasance that were not present in yesterday's grid. The threats range from disruption of generation and transmission facilities (at worst widespread, long-lasting blackouts) to the release of minute-to-minute data on business and home electricity use to anyone from home appliance marketers to career criminals. It is another area where the lack of coordinated national authorities and policies may create problems.

One key to smoothly flowing communication among the diverse grid components is standardization of equipment interfaces and protocols for communication and software applications. Developing these communication standards is a complex task, not least because the technologies are continually changing, and debates among alternative approaches are ongoing. Thus an important trade-off in standards design is between

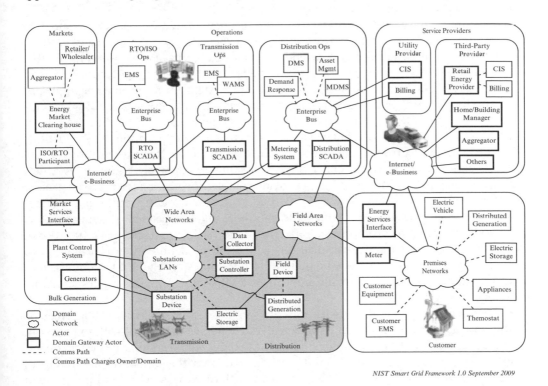

NIST Smart Grid Framework 1.0 September 2009

Source: NIST (2010).

Figure 5.2   Detailed communications flows in the future electric grid

early standardization, which may retard innovation, and delayed standardization, which could lead to future problems of interoperability. In the USA, the National Institute of Standards and Technology (NIST) oversees this inevitably messy process and is working on these issues with a broad-based consortium of utilities, consumer groups, public utility commissions and other federal agencies (NIST, 2010). While there seems general agreement that mandatory, detailed federal standards in this arena would not be productive, consumers would benefit if key standards could be adopted globally to avoid balkanization of equipment markets.

In the USA, the cybersecurity of the bulk power system is addressed in standards developed by the North American Electric Reliability Corporation, or NERC (NERC, 2011). NERC jurisdiction extends only to the bulk power system, however. For investor-owned utilities concern with cybersecurity at the distribution system level lies with state public utility commissions, and cooperative and municipal utilities are not subject to any regulatory authority on this issue. Given the increasingly interconnected nature of grid operations, from generation to distribution, such a fragmented approach is surely inefficient and, since the relevant expertise is not widely distributed, potentially dangerous. We believe it is imperative to give a single federal agency responsibility for cybersecurity throughout the US grid, along with the necessary regulatory authority. Its reach should encompass all interconnected transmission and distribution systems, and should cover prevention, response and recovery.

The White House has proposed that this agency should be the Department of Homeland Security, which has expertise in cybersecurity (White House, 2011). Legislation introduced in Congress would put the Department of Energy and the FERC, which have expertise in the electric grid, in the lead. Whichever architecture is chosen, the lead agency should, of course, take all necessary steps to insure that the appropriate expertise in cybersecurity and in grid operations and development is brought to the task by working with regulators, publicly owned and investor-owned utilities, as well as NERC, NIST and other federal agencies, and such expert organizations as the Institute of Electrical and Electronics Engineers and the Electric Power Research Institute.

Even with maximum effort under improved standards and regulations, it is widely acknowledged that it is not possible to provide complete protection from cyberattack to a system as complex as tomorrow's electric grid, with so many entry points and continually evolving technologies. Recognizing this, utilities and independent system operators need to be prepared to promptly and effectively recover from attacks when they occur. A risk management approach is called for, with appropriate attention to response and recovery, with a nationally coordinated effort by utilities, public utility commissions, and federal agencies to develop and implement a set of best practices for response and recovery from cyberattack.

Consumer anxiety over the implications for privacy of the installation of smart meters, with two-way communication of data, is real and will have to be dealt with to gain consumer acceptance of AMI technology. Failure to do this would complicate system operations and raise costs. As illustrated in Figure 5.3, data may flow not only back and forth to the local utility but potentially to suppliers of third-party services and government agencies – in and out of the jurisdiction of any single state.

Privacy is not a simple issue, and there are no generally accepted right answers. Nonetheless, nationwide coordination is needed among all involved parties to establish

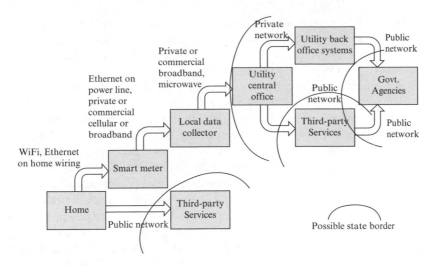

Source: MITEI (2011).

Figure 5.3   Potential consumer electricity data flows

consistent privacy policies. Failure to deal effectively with privacy issues could put at risk many of the advantages that could be gained from modern metering technology.

CONCLUDING OBSERVATIONS

Between now and 2030, the electric grids in the USA and many other nations will confront new challenges and inevitably undergo major changes. The current set of organizations that make up the US electric system – public and private utilities, state regulators and federal agencies – appears to be up to the challenge of taking advantage of a technology revolution while accommodating evolving objectives of federal energy and environmental policy. Fundamental reform of the electric power system is not necessary to keep the lights on, although nationwide implementation of competitive wholesale markets and an end to 'preference power' subsidies would certainly be desirable. Rather, continued adaptation, along with a few key reforms in the current fragmented US policy system, will produce an appropriate evolution of the grid.

The key reforms that we have discussed above fall into two categories: institutions and incentives. Institutional changes that are most important in the US context involve siting boundary-crossing transmission projects and national authority over cybersecurity at the distribution system level. The incentive issues we view as most important are those created by the opportunity to move to dynamic pricing, so that prices reflect actual costs over time, and to correct the economically unsound way fixed network costs are currently recovered.

To support these changes and meet other system challenges, additional research in a few key areas is likely to have substantial payoffs. These include the development of computational tools for new hardware that system operators can use to monitor and control the bulk power system, methods for wide-area transmission planning under uncertainty, processes for response to and recovery from cyberattacks, and careful social-science-based studies of customer response to alternative pricing regimes, perhaps supported by response automation. US electric utilities now generally lack the expertise necessary to perform this research internally, but the industry should be able to support the modest but sustained efforts required – although federal support would certainly be helpful. For this to happen, regulators will need to recognize that technical progress benefits consumers broadly and permit modest increases in utility R&D budgets. It will also likely be necessary for the industry to reverse the downward trend in cooperative R&D spending and make appropriate use of cooperative funding through the Electric Power Research Institute, one or more independent system operators, and project-specific coalitions.

In addition, better data need to be available to decision makers in both government and industry. Good data are critical inputs to good decisions regarding the grid, especially in the unfamiliar situations in which public and private actors will increasingly find themselves. Important needs include detailed data on the bulk power system (to be made available with careful attention to the need for security), standardized distribution utility performance metrics, the results of publicly funded 'smart grid' projects, and utility experiences with dynamic pricing.

Finally, despite all the challenges, opportunities and new technologies we have discussed, it is useful to recognize that from the point of view of most residential consumers, the US grid in 2030 is not likely to differ much from the grid today. Consumers will be able to get electricity on demand with a high degree of reliability. They may face prices that vary over time, but their household's response to those prices will be largely automated. Because the owners, operators and regulators of the grid will likely have responded to emerging challenges – increased penetration of VERs, distributed generation, electric vehicles and other policy-driven changes – by deploying new technologies and modifying operating procedures the system will look unchanged to its end users. The new grid will be much more data-intensive, but turning on a lamp will produce the same result as it does today.

NOTES

1. This chapter draws heavily on results of a study of the future of the US electric grid in which the authors participated (MITEI, 2011). We have learned a great deal from colleagues and graduate students involved in that study, and from members of a study advisory committee. More detailed discussion of, and support for, many of the points we make can be found in there. Nonetheless, opinions expressed here are not necessarily shared by anyone else, and any errors are entirely our responsibility.
2. Sources for the data presented here are given in MITEI (2011, ch. 1).
3. FACTS are a set of technologies employing power electronics that enable control of various transmission system operating parameters, including voltage-ampere-reactive support and power flow. For discussion of their use see MITEI (2011).
4. A PMU measures current, voltage, and frequency every 1/30th of a second or faster in synchronism with other such measurements across a wide area based on a GPS time signal. For detail on envisioned PMU applications, see NERC (2010) and EPRI (2007).
5. As a general matter, the FERC has authority over transmission pricing and related wholesale-level activities even though they only involve entities within a single state, since most transmission facilities are part of an integrated grid that crosses state lines. FERC does not have jurisdiction within Alaska, Hawaii, where this rationale clearly does not apply, or, by statute, within the ERCOT region, which covers most of the state of Texas.

6. Consider a zero-net-energy building that needs electricity from the grid only in emergencies. If all charges are per kWh, such a building would pay nothing for the insurance that connection to the grid provides.

REFERENCES

CEC (Commission of the European Communities) (2007), *An Energy Policy for Europe*, http://eur-lex. europa.eu/smartapi/cgi/sga_doc?smartapi!celexplus!prod!DocNumber&lg=en&type_doc=COMfinal&an_ doc=2007&nu_doc=1.

CEQ (Council on Environmental Quality) (2011), 'Interagency Rapid Response Team for Transmission', http:// www.whitehouse.gov/administration/eop/ceq/initiatives/interagency-rapid-response-team-for-transmission.

EPRI (Electric Power Research Institute) (2007), *Phasor Measurement Unit Implementation and Applications*, Technical Report 1015511, Palo Alto, CA.

FERC (US Federal Energy Regulatory Commission) (2011), *Transmission Planning and Cost Allocation by Transmission Owning and Operating Public Utilities, Docket No. RN 10-23-000, Order No. 1000*, 21 July, available at http://www.ferc.gov/industries/electric/indus-act/trans-plan/fr-notice.pdf.

Hogan, W. (2011), 'Transmission benefits and cost allocation,' White Paper, Harvard University, Cambridge MA, http://www.hks.harvard.edu/hepg/Papers/2011/Hogan_Trans_Cost_053111.pdf.

IEEE (IEEE Standards Association) (2008), *IEEE Standard for Interconnecting Distributed Resources With Electric Power Systems, IEEE Std. 1547-2003* (issued 2003, reaffirmed 2008), doi:10.1109/ IEEESTD.2003.94285.

Joskow, P. (2005), 'Patterns of transmission investment', Working Paper 2005-004, MIT Center for Energy and Environmental Policy Research, Cambridge, MA.

Joskow, P. and R. Schmalensee (1983), *Markets for Power: An Analysis of Electric Utility Deregulation*, Cambridge, MA: MIT Press.

Kelliher, Joseph T. (2007), 'Statement of Chairman Joseph T. Kelliher', Federal Energy Regulatory Commission, Conference on Competition on Wholesale Power Markets, AD07-7-000, 27 February; available at http://www.ferc.gov/media/statements-speeches/kelliher/2007/02-27-07-kelliher.asp#skipnav.

MITEI (Massachusetts Institute of Technology Energy Initiative) (2011), *The Future of the Electric Grid: An Interdisciplinary MIT Study*, Cambridge, MA: MIT Energy Initiative.

NERC (North American Electric Reliability Corporation) (2010), *Real-Time Application of Synchrophasors for Improving Reliability*, Princeton, NJ, http://www.nerc.com/docs/oc/rapirtf/RAPIR%20final%20101710.pdf.

NERC (North American Electric Reliability Corporation) (2011), 'Project 2008-06: Cyber Security Order 706 Phase II', 24 January, http://www.nerc.com/filez/standards/Project_2008-06_Cyber_Security_PhaseII_ Standards.html.

NIST (National Institute of Standards and Technology) (2010), *NIST framework and roadmap for smart grid interoperability standards, release 1.0*. Special Publication 1108, January, http://www.nist.gov/public_affairs/ releases/upload/smartgrid_interoperability_final.pdf.

White House (2011), *Cybersecurity Legislative Proposal, Fact Sheet*, 12 May, http://www.whitehouse.gov/ the-press-office/2011/05/12/fact-sheet-cybersecurity-legislative-proposal.

6 Increasing the penetration of intermittent renewable energy: innovation in energy storage and grid management

*Nick Johnstone and Ivan Haščič**

1. INTRODUCTION

Many governments have introduced policies to support the development and adoption of renewable energy technologies as a means of mitigating climate change. A significant barrier to the increased penetration of renewable energy arises from the 'intermittent' nature of the electricity produced. Output from individual plants can vary on a scale of seconds to minutes, as well as over several hours. The extent to which the grid as a whole can accommodate such variations is a function of its capacity to adjust to supply and demand shocks. As the penetration of intermittent renewable sources increases, the need for such capacity increases as well. This can be achieved through a variety of means, including increased capacity of 'dispatchable' power,[1] which can 'balance' intermittent renewable power, greater integration of grids within and across countries, and the use of a more diverse and dispersed mix of intermittent sources. However, flexibility can also be introduced into the system through increased energy storage capacity and improved grid management, both of which allow for improved matching of electricity supply and demand.

The motivation for the chapter arises out of a concern to provide policy makers with guidance on the targeting of public R&D support and other policy incentives. There may be greater benefits from targeting R&D expenditures at storage and grid management technologies rather than directly at intermittent generating technologies. At least three related reasons can be cited:

- it is more parsimonious with respect to information requirements for the government, since innovation in 'enabling' technologies (such as storage and grid management) is an important complement to innovation in all intermittent renewable generating technologies;
- the technologies are at a relatively early stage of development in comparison with the most important generating technologies, with greater returns on public R&D in the presence of learning curves; and
- such technologies are subject to important network externalities, implying that the rents from private investment will not be fully captured, and thus suboptimally provided.

In this chapter preliminary empirical results are presented on the factors that encourage innovation in system flexibility (energy storage and grid management technologies). The study draws upon a rich database of patent applications and IEA data on energy research expenditures, prices, capacity and production.

2.   INTERMITTENCY IN RENEWABLE ENERGY GENERATION AND SYSTEM FLEXIBILITY

A number of OECD (and other) governments have identified the increased penetration of renewable energy sources as a primary means of mitigating the emissions of greenhouse gases. Policy targets for renewable energy exist in at least 73 countries. Most national targets are for shares of renewable energy supply in total electricity production, typically 5–30 per cent, but ranging all the way from 2 to 78 per cent (REN21, 2008).

For instance, the European Union Directive of 2008, which succeeds the one from 2001, requires member states to increase their shares of renewable energies to meet a 20 per cent overall energy target by 2020. The Directive set a series of interim targets, known as 'indicative trajectories', in order to ensure steady progress towards the 2020 targets. EU countries are free to decide their own mix of renewables, allowing them to account for their different potentials, while Brussels reserves the right to enact infringement proceedings if states do not take appropriate measures towards their targets. Figure 6.1 gives the 2020 national targets for renewable use in EU as compared to the 2005 share of renewables in total energy supply.

In order to meet such targets, specific policies need to be introduced. By 2009, at least 64 countries had some type of policy to encourage renewable power generation (REN21, 2008). Such government policies target different stages of the industrial process, ranging from R&D through the investment in physical capital (plants and equipment) up to the production and sale/consumption of energy. However, the penetration of renewable energy remains relatively low. More than 80 per cent of produced electricity comes from coal, natural gas and nuclear power plants, while renewable energy sources rank fourth, and most of this comes from hydro (IEA, 2009a). Cost is, of course, the main reason. Despite many policy initiatives, the cost of renewable electricity generation remains prohibitively expensive, although there are exceptions. Moreover, since those renewables which are the most promising sources of future increases in capacity (e.g. offshore wind, tidal/stream, wave energy and solar photovoltaics) are intermittent in nature, the need to be able to adjust output levels and sources at short notice is likely to rise (see Infield and Watson, 2007; Sinden, 2007).

In a sense, all plants have 'variable' output, in so far as there is some probability of a breakdown which puts the plant off line for a period of time. Since power outages can impose significant economic costs, most regulators have a target 'loss of load probability' (LOLP). For instance, in the UK this is set at nine (i.e. nine outages per century). This is met by building in a system margin, allowing the system to meet unexpected decreases in supply from some plants and/or unexpected increases in demand.

The introduction of intermittent renewable energy plants increases the required system margin in order to meet the target LOLP since they are able to contribute less to peak demand (Neuhoff, 2005).[2] This can be measured as a capacity credit – that is, the amount of electricity (expressed in terms of conventional thermal capacity) that can be served by intermittent plant without increasing the LOLP. For instance, while wind plants generally have a capacity factor in the region of 20 per cent to 40 per cent (relative to 80 per cent–90 per cent for conventional fossil-fuel-fired plants), the

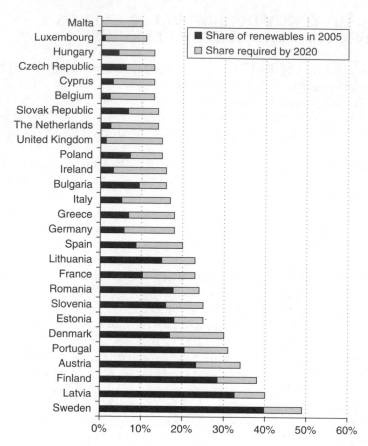

Source: See 'Europe 2020 targets' at http://ec.europa.eu/commission_2010–2014/president/news/speeches-statements/pdf/20110623_2_en. pdf.

Figure 6.1   *Renewable energy targets in Europe (share in final energy by 2020 as compared to share of renewables in 2005)*

capacity credit is less, reflecting the high variability of output through time. At 20 per cent wind power penetration, Gross et al. (2007) estimate a capacity credit of 19–26 per cent for a plant with an average annual capacity factor of 35 per cent. On the basis of a formula developed by Gross et al. (2007), this can be converted into a 'reliability cost'[3] of approximately £4/MWh, which can be considered as an 'externality cost' arising out of intermittency.

In the face of such vulnerability, some governments have even sought to cap the penetration of intermittent renewable energy sources in the grid (see IEA, 2008). Moreover, as penetration levels of intermittent renewables rise still further, the ratio between capacity credit and capacity factor falls, reflecting increased vulnerability of the system. The extent to which the penetration of intermittent renewables increases LOLP is a function of the flexibility of the system. Flexibility can be introduced into the system in six ways:

- *Improved weather forecasting*   Improved forecasting of meteorological and other ecological conditions (ocean wave activity, solar radiation etc.) can help system operators efficiently balance the amount of dispatchable and different intermittent power sources in the grid (see, e.g., IEA, 2008; APS, 2010).
- *Geographic dispersal of intermittent renewable energy plants*   Since spatial dispersion will probably reduce the extent of correlation among the output of different plants, this will 'smooth' system-wide output (see, e.g., Inage, 2009).
- *Diversity in the portfolio of renewable energy sources*   Since output variations between types of renewable is unlikely to be strongly correlated, this will also smooth output (see, e.g., Sinden, 2007; Gross et al., 2007; Infield and Watson, 2007).
- *Trade in electricity supply services*   The reasoning is analogous to the point made above, but with imperfect grid connections between countries it is worth highlighting. Even within continental Europe there are significant lacunae (e.g. France↔Spain) (see Milborrow, 2007; IEA, 2008).[4]
- *Improvements in grid management*   The development of 'intelligent' grids allows for improved balancing of demand and supply, and more flexible transfer between sources of supply (see, e.g., IEA, 2008; Duff and Green, 2008).
- *Energy storage*   Historically the primary back-ups (or reserves) have been fossil fuel plants which can come on line relatively quickly. However, this can be costly and as a consequence pumped hydro plants have been commonly used as a reserve source of energy. In recent years there have been significant innovations with respect to different types of energy storage of sufficient scale to serve as back-up for the grid (see, e.g., Hall and Bain, 2008; IEA, 2005, 2008).

It is developments in grid management and storage that are the primary focus of much ongoing work, and this chapter focuses on the case of innovation in these areas. Both improved storage and grid management have beneficial consequences for the delivery of electricity services more generally, irrespective of the supply mix. For instance, they can increase the reliability of supplies in the face of shocks (i.e. due to extreme weather events or earthquakes). They can also reduce transmission losses and increase efficiency more generally. However, by allowing for improved spatial and temporal balancing of supply and demand, improved storage and grid management have particular benefits for renewable energy. Exploiting such possibilities is, of course, dependent upon the implementation of institutional and regulatory measures which allow for such balancing in the market.[5]

Inage (2009) and IEA (2005) highlight a number of different storage technologies (in addition to pumped hydro) which are 'efficient' at scales of 1 MW and above. These include compressed air energy storage, superconducting magnetic energy storage, advanced lead-acid batteries, lithium-ion batteries and flow batteries. However, it remains the case that efficient energy storage is a significant constraint on the penetration of renewable energy sources in the market. The increased availability of energy storage at reasonable cost is, therefore, one of the strategies which a government can pursue in order to increase penetration of renewables in the electricity supply industry. Publicly supported innovation efforts which reduce cost of storage may be a cost-effective strategy to bring about reduced CO_2 emissions.

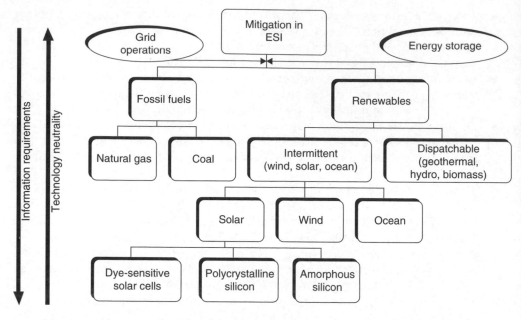

Figure 6.2   Policy incidence and technology neutrality

As noted above, improved grid management becomes increasingly important as the penetration rises to 15–20 per cent of the grid. Technologies such as flexible AC transmission systems (FACTS) enhance transmission networks. High-voltage DC systems are particularly valuable for remote generating facilities (i.e. offshore wind). High-temperature superconductors reduce transmission losses and improve generation efficiency. In addition, various information and communication technologies enable more efficient management and use of the grid. Advanced metering infrastructure (AMI) that allows for two-way communication will enable the transfer of information (and incentives) from users to suppliers (see IEA, 2011a).

Moreover, there are good reasons to believe that public research efforts targeted at storage and grid management are likely to be more cost-effective. Why? The information requirements (for governments) are more limited than would be the case in allocating public resources across different generating technologies. By targeting innovation 'upstream', flexibility with respect to the choice of generating technology is retained downstream. This point is represented in Figure 6.2. In effect, energy storage technologies and grid operations can serve as an 'enabling' technology, allowing for increased power system flexibility. This increases the competitiveness of all intermittent renewable energy sources relative to dispatchable sources (renewable and other), without specifically favouring one intermittent source over another (i.e. neutrality). It is a form of 'local' general purpose technology (GPT) within the basket of intermittent renewable energy sources.[6]

Moreover, Popp et al. (2010) have argued that improved electricity transmission systems benefit all technologies and will 'typically not reap great rewards for the innovator'. As a consequence there is less likely to be crowding out of private R&D investment

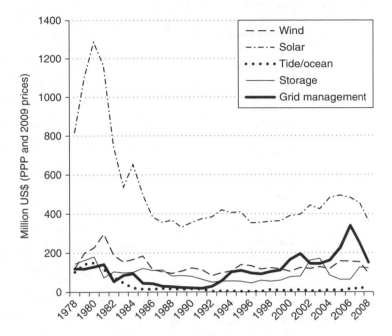

Source: IEA (2011b).

Figure 6.3   *Public RD&D on energy storage, grid management and electricity generation from intermittent renewables (1978–2009, OECD countries)*

if support is targeted at energy storage and grid management than if intermittent generating technologies are targeted directly.

For these reasons, targeted support for innovation in storage and grid management technologies (rather than generating technologies) may avoid some of the potential pitfalls associated with public R&D subsidies. Moreover, redirecting support from generating technologies to storage technologies is likely to have a significant impact since the scale of public R&D going to storage and grid management technologies is small relative to generating technologies (see Figure 6.3). However, R&D on grid management has been increasing recently.

The percentage of total energy RD&D expenditures devoted to both intermittent generating and storage and grid management technologies varies widely across countries. Switzerland, Greece and Austria target more than 8 per cent of energy R&D at storage and grid management (Figure 6.4). France and Japan have very low investment in both intermittent renewable and storage and grid management R&D. Only three countries (Austria, Czech Republic and Finland) have greater investment in storage and grid management R&D than intermittent renewable R&D. It is interesting that the two countries with the highest relative level of investment in intermittent renewable R&D (Denmark and Greece) have very different levels of investment in storage and grid management R&D. This may due in part to differences in their capacity to use other strategies to increase system flexibility (e.g. grid interconnections).

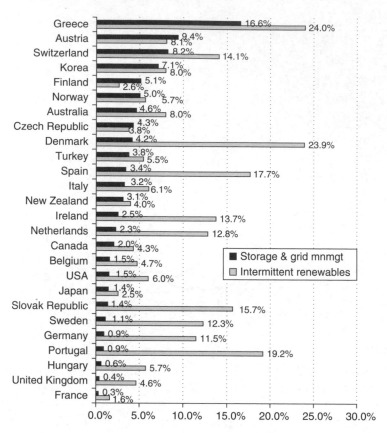

Source: IEA (2011b).

Figure 6.4   *Public R&D on energy storage, grid management and electricity generation from intermittent renewables (% of total energy RD&D, 1978–2008)*

3.   INNOVATION IN ENERGY STORAGE AND GRID MANAGEMENT

As a measure of innovation in energy storage (and generating) technologies, patent counts have been developed. Patents are a set of exclusionary rights (territorial) granted by a state to a patentee for a fixed period of time (usually 20 years) in exchange for the disclosure of the details of a given invention. Patents are granted by national or regional patent offices on invention (devices, processes) that are judged to be new (not known before the application date of the patent), involving a non-obvious inventive step and that are considered useful or industrially applicable. The use of patent data as proxy for innovation has a long history in the field of innovation economics. Griliches (1990) argues that patents are imperfect but useful indicators of inventive activity. Their main limitation is linked to the facts that not all innovations are patented, not all patented

Table 6.1　Patent classification codes for energy storage and grid management

ENERGY STORAGE	
Battery technology	Y02E60/12
Lithium-ion batteries	Y02E60/12B
Alkaline secondary batteries, e.g. NiCd or NiMH	Y02E60/12D
Lead-acid batteries	Y02E60/12F
Hybrid cells	Y02E60/12H
Ultracapacitors, supercapacitors, double-layer capacitors	Y02E60/13
Thermal storage	Y02E60/14
Sensible heat storage	Y02E60/14B
Latent heat storage	Y02E60/14D
Cold storage	Y02E60/14F
Pressurized fluid storage	Y02E60/15
Mechanical energy storage, e.g. flywheels	Y02E60/16
Pumped storage	Y02E60/17

GRID MANAGEMENT	
Flexible AC transmission systems	Y02E40/12-16
Active power filtering	Y02E40/22-34
Arrangements for reducing harmonics	Y02E40/40
Arrangements for eliminating or reducing asymmetry in polyphase networks	Y02E40/50
Superconducting generators	Y02E40/62
Superconducting or hyperconductive transmission lines or power lines or cables or installations thereof	Y02E40/64
Superconducting transformers or inductors	Y02E40/66
Superconducting energy storage for power networks, e.g. SME, superconducting magnetic storage	Y02E40/67
Protective or switching arrangements for superconducting elements or equipment	Y02E40/68
Current limitation using superconducting elements	Y02E40/69
Methods and systems for the efficient management or operation of electric power systems	Y02E40/7
. . . characterized by remote operation, interaction, monitoring or reporting system, e.g. smart grids	Y02E40/72

innovations have the same economic value, and that propensity to patent may vary across countries and technological fields.

The patent data used in this chapter have been extracted from the European Patent Office's (EPO) Worldwide Patent Statistical Database, or PATSTAT (EPO, 2010). PATSTAT is unique in that it contains data from more than 90 patent offices and on over 70 million patent documents. Patent documents are categorized using the international patent classification (IPC) and some national and regional classification systems, including the European classification scheme (ECLA). In addition to the basic bibliometric and legal data, the database also includes patent descriptions (abstracts) and citation data.

Data have been extracted on patent applications filed from 1975 to 2008. The relevant inventions have been identified using specially developed tagging codes for climate change mitigation technologies (see Table 6.1). This tagging scheme (Y02) has been developed by a team of patent examiners at the European Patent Office, with inputs

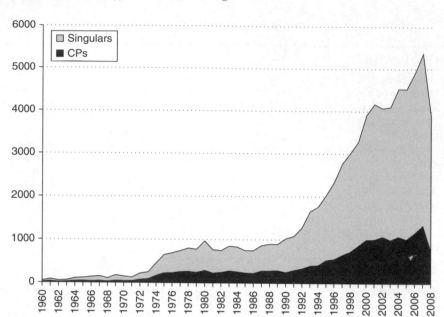

Figure 6.5   Invention in energy storage technologies (claimed priorities (CP) and singular patent applications by in-sample countries)

from collaborating researchers (including the authors) and has now been implemented into the ECLA system.

Figures 6.5 and 6.6 present data on patented inventions in the area of energy storage. Data are presented for both 'claimed priorities' (CP) (i.e. inventions for which protection is sought in at least two countries) and 'singulars' (i.e. inventions protected only in one country). This distinction is useful because it indicates different market value of the inventions patented (see OECD, 2011). With respect to energy storage, there was significant 'take-off' around 1990, with a six-fold increase in singulars since that point. Claimed priorities have tripled in the same period.

Over the period 1978–2008 the most important inventor countries in terms of 'claimed priorities' were Japan, the USA, Germany and Korea, which together are responsible for approximately three-quarters of total applications. Switzerland is amongst the top ten. Figure 6.6 breaks down the energy storage data in terms of technology type. Lithium-ion batteries are the most prevalent technology type, followed by alkaline batteries. While pumped hydro storage is very important in terms of system flexibility, there are few patents in this area since it is a very mature area and most of the inventions are likely to have more general applications. Pressurized fluid storage (often referred to as compressed air energy storage) represents a very small percentage of the total.

In terms of grid management innovations, Figure 6.7 indicates that the absolute level of patented inventions is more than an order of magnitude lower than in storage. There has been no obvious acceleration in the rate of invention. CPs represent a somewhat higher percentage of total counts than is the case in energy storage.

In terms of countries, Korea is much less prevalent than in the case of energy storage.

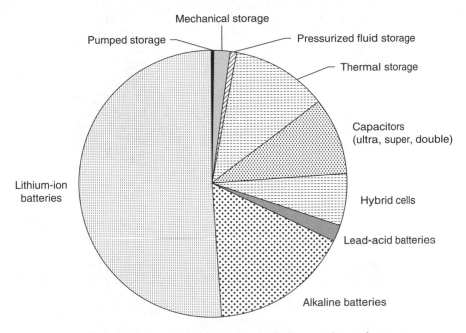

Figure 6.6   Invention in energy storage by technology type (claimed priorities 1978–2008)

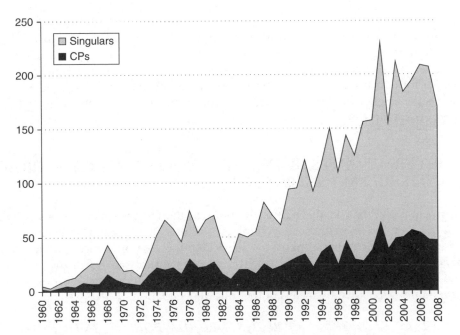

Figure 6.7   Invention in electricity grid management (CP and singular patent applications by in-sample countries)

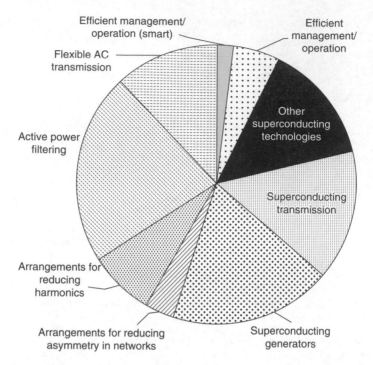

Figure 6.8   *Invention in grid management by technology type (claimed priorities 1998–2008)*

The converse is true of France. Both Sweden and Switzerland have relatively higher counts than is generally found in other fields. In terms of technology types, superconducting generators, transformers and other devices make up almost half of the total of claimed priorities (Figure 6.8). Currently, the one significant omission in the Y-tagging scheme is the absence of high-voltage DC transmission patents, which are important for isolated generating sources.

4.   DETERMINANTS OF INNOVATION IN ENERGY STORAGE AND GRID MANAGEMENT

Our principal objective in this section is to assess the factors which affect innovation in storage and grid management technologies. This is a first (and necessary) step in determining whether public support for RD&D in the area of energy storage and grid management will lead to increased penetration of renewable energy in the grid. Based on the discussion in Section 2, invention in system flexibility (energy storage and grid management) is specified by means of the following equation:

$$CNT\_EP\_STORGRID_{it} = f(R\&D\_STORGRID_{it}, POLICY_{it}, INTR\_PERC_{it},$$

$$INTR\_CONC_{it}, ELEC\_TRADE_{it}, CNT\_EP\_TOTAL_{it}, \omega_i) + \varepsilon_{it} \qquad (6.1)$$

where i indexes country and t indexes year. The dependent variable (*CNT_EP_STORGRID*) represents the number of patent applications deposited at the European Patent Office, classified by inventor country[7] and priority year.[8] We have used EPO patent applications because the search strategy is based on the Y-tags developed by the EPO, and alternative counts would be biased since the coverage of Y-tags for applications from other offices is uneven. (The data source is PATSTAT, as discussed above.)

As a control, a variable reflecting the propensity to invent and patent technologies in general (*CNT_EP_TOTAL*) is included as an explanatory variable. This variable is constructed in a manner analogous to the dependent variable (a count of patent families by inventor country and priority year), with the difference that all types of technologies (not only storage and grid management) are covered.[9] Ideally, we would estimate the model using a two-stage procedure where total patenting activity is first estimated. This approach was followed in Johnstone et al. (2011) and it was found that results from the two-stage estimation were closely comparable with those from a reduced-form model. Since many observations would be lost with such an approach in this case we have decided to adopt this strategy.[10]

Public sector expenditure on R&D in energy storage and grid management (*R&D_STORGRID*) is included as an explanatory variable, expressed in million US$ using 2008 prices and PPP (purchasing power parity). Assuming that such expenditures either result in patented innovations by the public sector, or contribute to patented innovations by the private sector, the sign is expected to be positive. The data source is IEA (2011b).

The data for energy storage cover government expenditures for RD&D on batteries, super-capacitors, superconducting magnetic, water heat storage, sensible/latent heat storage, photochemical storage, kinetic energy storage, and other means (excluding hydrogen and fuel cells). The data for grid management cover RD&D on 'electricity transmission and distribution'. They cover solid state power electronics, load management and control systems, superconducting cables, AC and DC high-voltage cables and transmission, other transmission and distribution R&D related to integrating distributed and intermittent generating sources into networks, and all high-temperature superconducting research not covered elsewhere.[11]

We also include a variable which reflects policy incentives. This is based on a database of public policies aimed at developing renewable energy sources compiled at the IEA. A composite policy dummy variable is constructed equal to one when any of the six policy types were in place and zero otherwise (these include investment incentives, tax measures, feed-in tariffs, voluntary programmes, obligations and tradable permits; see also Johnstone et al., 2010).

Trade in electricity services (*ELEC_TRADE*) is included as a binary variable reflecting whether or not there was any trade (exports or imports) in electricity in the year in question. This variable is intended to capture the extent to which countries are able to compensate for intermittency by importing or exporting electricity services. In the absence of trade, the incentives for innovation in system flexibility are greater. An alternative variable in which trade is constructed as the maximum of imports or exports over total generation was also used, but the results remain qualitatively the same. The source of the data is IEA (2009a).[12]

The percentage of 'intermittent' energy sources (wind, solar, ocean/tidal) in total electricity supply (*INTR_PERC*) is included to reflect the vulnerability to intermittency.

Table 6.2 Descriptive statistics

Variable	Obs	Mean	Std dev.	Min.	Max.
CNT_EP_STORGRID	674	22.167	56.337	0	435.667
R&D_STORGRID	674	9.778	22.861	0	244.139
POLICY	674	0.616	0.487	0	1
INTR_PERC	674	0.593	2.027	0	18.319
INTR_CONC	674	0.546	0.454	0	1
ELEC_TRADE	674	0.838	0.368	0	1
CNT_EP_TOTAL (in 000s)	674	3.344	6.476	0	35.465

Note: The estimation sample includes data on 28 OECD countries (AT, AU, BE, CA, CH, CZ, DE, DK, ES, FI, FR, GB, GR, HU, IE, IT, JP, KR, LU, NL, NO, NZ, PL, PT, SE, SK, TR, US) for the period 1974–2007.

The expected sign is positive. In addition, since the vulnerability to intermittency may be obviated through a more diverse portfolio of sources, a variable is constructed to reflect this variation. The variable *INTR_CONC* is constructed as the squared sum of the differences between the percentage dependence on each intermittent source and mean dependence for all intermittent sources. This is then normalized by percentage of intermittent sources in the grid to isolate the effect of concentration. The expected sign is positive. The source of the data is IEA (2009b).[13] Descriptive statistics are set out in Table 6.2.

Finally, country fixed effects account for omitted country-variant effects that influence the dependent variable in a time-invariant manner. Most notably, this would capture the effect of geographical area, and thus potentially smoothing of intermittent sources. All the residual variation is captured by the error term.

Our dependent variable represents the number of patent applications – patent counts. Count data models, such as the Poisson and negative binomial, have been suggested for estimating the number of occurrences of an event, or event counts (see, e.g., Wooldridge, 2002; Cameron and Trivedi, 1998; Hausman et al., 1984; Maddala, 1983: 51). Formally, the Poisson model is derived by assuming that a random variable y is Poisson-distributed with the conditional density of y equal to $(y|x) = (e^{-\theta}\theta^y)y!$, where $\theta = E[y|x]$. The log of the mean θ is assumed to be a linear function of a vector of independent regressors x: $\ln\theta = x'\beta$, where β is a parameter vector. This specification ensures non-negativity of θ (Cameron and Trivedi, 1998). However, the Poisson specification imposes a heavy restriction on the data – the equality of conditional mean and conditional variance, $E[y|x] = V[y|x] = \theta$, referred to as the equidispersion property of the Poisson. Indeed, as with most empirical data, casual inspection of the sample mean and sample variance indicates that their conditional counterparts are likely to be different for both dependent variables, indicating overdispersion.

One way to account for overdispersion is the negative binomial model suggested by Cameron and Trivedi (1998). They derive a negative binomial model from a Poisson–gamma mixture distribution (ibid.: 100–102). In addition to y being conditionally Poisson-distributed, parameter θ is assumed to be the product of a deterministic term and a random term, $\theta = e^{x'\beta+\varepsilon} = e^{x'\beta} + e^{\varepsilon} = \mu\nu$. In other words, the unobserved error

Table 6.3   *Estimated coefficients of the system flexibility innovation models*

	Model 1 (base model)	Model 2 (lagged policy)	Model 3 (without USA)	Model 4 (just storage)	Model 5 (just grid mgmt)
R&D_STORGRID	0.004***		0.008***		
Lag1_R&D_STORGRID		0.006***			
R&D_STOR				0.006**	
R&D_GRID					0.003
POLICY	0.804**		0.720***	0.789***	0.761***
Lag1_POLICY		0.650***			
INTR_PERC	0.055***	0.055***	0.058**	0.038**	0.126***
INTR_CONC	0.335**	0.368***	0.362***	0.253***	0.778***
ELEC_TRADE	−0.516***	−0.362	−0.466**	−0.510**	0.172***
CNT_EP_TOTAL	0.087**	0.088***	0.086***	0.094***	0.044
Number of observations	674	666	641	676	674
Log pseudo-likelihood	−1617.8	−1612.3	−1449.7	−1579.8	−617.7

Note: * $p < 0.050$, ** $p < 0.025$, *** $p < 0.010$ based on robust standard errors. Fixed effects are included in all models but are not reported here. The estimation sample includes data for 28 OECD countries and 34 years (1974–2007).

parameter (v) introduces heterogeneity in the variance, and the intensity parameter (μ) is explained (in log) by a vector of explanatory variables (x). Therefore, by assuming a gamma distribution for v (mean 1, variance α), Cameron and Trivedi show that the marginal distribution of y is the negative binomial with the first two moments $E[y|\mu,\alpha] = \mu$ and $V[y|\mu,\alpha] = \mu + \alpha\mu^2$ (for the NB2 variance function; ibid.: 63). It follows that as $\alpha \to 0$ the NB model converges to the Poisson distribution with intensity μ. The dispersion parameter α is to be estimated.

The results of the estimation are presented in Table 6.3. The total sample is a maximum of 676 observations, with an unbalanced sample of 28 countries and 34 years. In total five models are estimated. Model 2 includes lagged R&D and policy variables. Model 3 excludes the US sample since the pattern of R&D expenditures for the USA is quite distinct. Models 4 and 5 estimate on storage and grid management separately.

As expected, public RD&D has a positive and significant impact. Based on the estimated elasticities (Table 6.4), a 10 per cent increase in public sector R&D results in approximately 0.4–0.5 per cent increase in patents in the base model. The effect is relatively small, and further work will be undertaken to determine whether such investments result in significant increases in the efficiency of storage and grid management technologies, as well as increases in renewable energy penetration. Indeed, it is interesting to note that the coefficient reflecting the effect of the introduction of renewable energy support measures (*POLICY*) is highly significant and much larger. Countries which are encouraging renewable energy penetration in their grid are also innovating in storage and grid management. Moreover, as the penetration of intermittent renewable energy in the supply mix rises, innovation in storage and grid management increases. However, other findings indicate that this is conditioned by other factors. For instance, greater diversity in the mix of wind, solar and ocean-tide energy reduces innovation rates in storage and

Table 6.4   Estimated elasticities for base model

R&D_STORGRID	0.040
POLICY	0.451
INTR_PERC	0.027
INTR_CONC	0.172
ELECT_TRADE	−0.121
CNT_EP_TOTAL	0.299

Note: Based on conditional marginal effects evaluated at sample means.

grid management. In addition, greater trade will have the same effect. Interestingly, when disaggregated one can see that this is only true of storage (Model 4) and not grid management (Model 5). This is probably due to the fact that many aspects of grid management are complementary with improved capacity to exploit trade in electricity services.

5.   CONCLUSIONS AND FURTHER RESEARCH

Insufficient system 'flexibility' is a significant constraint on the penetration of intermittent renewable energy sources in the grid. Imperfect and uncertain matching of supply and demand means that the 'real' contribution of increased capacity of wind, solar and ocean/tide power is less than the nominal capacity, and sometimes very much less. Improved energy storage and grid management can overcome this constraint by increasing system flexibility.

There are good reasons to believe that there are benefits of targeting public R&D expenditures at storage technologies rather than directly at intermittent generating technologies (see above).

In this chapter preliminary results have been presented on the determinants of innovation in energy storage and efficient grid management. The results are consistent with our expectations, with an important role played by both underlying technological factors (i.e. penetration and concentration of renewables) and policy factors (i.e. policy incentives and public R&D expenditures). However, this is just a first step. Further work is required to assess whether public support for R&D is better targeted directly at the specific intermittent generating technologies or at supporting technologies which are of wider benefit to system flexibility. Such an evaluation would require an assessment of the impact that such innovation has on generating costs, and thus penetration of renewable energy in the grid.

NOTES

*   The authors wish to acknowledge the inputs of Victor Veefkind and colleagues at the European Patent Office for the development of the search strategy upon which the patent data used in this report is based. The contributions of Fleur Watson in data preparation are also acknowledged. The views expressed in this chapter are those of the authors, and do not necessarily reflect those of the OECD or its member countries.

1. IEA (2008) prefers the term 'firm' and 'variable'. Sinden (2007) uses the term 'dispatchable' and 'non-dispatchable'. The key factors are: the source of generating capacity; whether there is significant variation in the potential output of the source; and whether this variation is a consequence of exogenous factors beyond direct control (i.e. ecological conditions).
2. However, there are some 'intermittent' variables which correlate with peak demand (e.g. solar photo-voltaic and air conditioning). See Heal (2009) and Gross et al. (2007).
3. The difference between the fixed cost of energy-equivalent thermal plant minus the fixed cost of thermal plant displaced by capacity credit of the intermittent plant.
4. While 20 per cent of Danish electricity is generated by wind, only 9 per cent is consumed domestically, with the balance exported to Norway and Sweden. Since the power exported to Norway and Sweden displaces power that is itself partly carbon neutral (e.g. hydro), the benefits in terms of carbon reduction may be limited (see CEPOS (Center for Politisker Studier), Wind Energy: The Case of Denmark).
5. For instance, some (e.g. Swift-Hook, 2010) have argued that storage does not facilitate penetration of intermittent renewable energy since as low-marginal cost sources they will be the last to be stored by operators. However, as Wilson et al. (2011) point out, this would not be the case if storage operators were able to exploit price differential through intertemporal trading.
6. With 'locality' restricted to the basket of intermittent renewable energies. Such a strategy would favour intermittent over dispatchable renewable (e.g. hydro, geothermal) energy. Moreover, it would favour renewable energy in the electricity supply industry over other means of carbon abatement.
7. 'Fractional' counts are generated in cases when inventors from multiple countries are listed.
8. 'Priority date' indicates the earliest application date worldwide (within a given patent family).
9. This is achieved by extracting data on all EPO patent applications with an (any) ECLA code assigned.
10. In the sample used for econometric analysis, storage patents represent on average only 0.2 per cent of total patents. Nevertheless, in order to avoid any concern over possible endogeneity, regressions are esti-mated on the difference between the patent total and the dependent variable.
11. http://wds.iea.org/wds/pdf/documentation_RDD.pdf.
12. http://wds.iea.org/wds/pdf/doc_Electricity_2009.pdf.
13. http://wds.iea.org/WDS/tableviewer/document.aspx?FileId=1315.

REFERENCES

APS (2010), 'Integrating renewable electricity on the grid: a report by the APS Panel on Public Affairs', American Physical Society, Washington, DC.

Cameron, A.C. and Trivedi, P.K. (1998), *Regression Analysis of Count Data*, Cambridge: Cambridge University Press.

Duff, D. and Green, A. (2008), 'A comparative evaluation of different policies to promote the generation of electricity from renewable sources', in S. Bernstein et al. (eds), *A Globally Integrated Climate Policy for Canada*, Toronto: University of Toronto Press, pp. 222–46.

EPO (2010), Worldwide Patent Statistical Database (PATSTAT), September, European Patent Office.

Griliches, Z. (1990), 'Patent statistics as economic indicators: a survey', *Journal of Economic Literature*, **28**: 1661–707.

Gross, R., Heptonstall, P., Leach, M., Skea, J., Anderson, D. and Green, T. (2007), 'The UK Energy Research Centre review of the costs and impacts of intermittency', in D. Boyle (ed.), *Renewable Electricity in the Grid*, London: Earthscan, pp. 73–94.

Hall, P.J. and Bain, E.J. (2008), 'Energy storage technologies and electricity generation', *Energy Policy*, **36**: 4352–5.

Hausman, J., Hall, B.H. and Griliches, Z. (1984), 'Econometric models for count data with an application to the patents–R&D relationship', *Econometrica*, **52**: 909–38.

Heal, G. (2009), 'The economics of renewable energy', NBER Working Paper No. 15081, Cambridge, MA: NBER.

IEA (2005), 'Variability of wind power and other renewables: management options and strategies', Paris: IEA.

IEA (2008), *Empowering Variable Renewables: Options for Flexible Electricity Systems*, Paris: OECD/IEA.

IEA (2009a), *Electricity Information*, Paris: OECD/IEA.

IEA (2009b), *Renewables Information*, Paris: OECD/IEA.

IEA (2011a), 'Technology roadmap: smart grids', Paris: OECD/IEA.

IEA (2011b), 'Energy technology research and development budgets', Paris: OECD/IEA.

Inage, S.I. (2009), 'Prospects for large-scale energy storage in decarbonised power grids', Paris: IEA.

Infield, D. and Watson, S. (2007), 'Planning for variability in the longer term: the challenge of a truly sustainable energy system', in D. Boyle (ed.), *Renewable Electricity and the Grid: The Challenge of Variability*, London: Earthscan, pp. 201–10.

Johnstone, N., Haščič, I. and Popp, D. (2010), 'Renewable energy policies and technological innovation: evidence based on patent counts', *Environmental and Resource Economics*, **45**(1): 133–55.

Johnstone, N., Haščič, I., Poirier, J., Hemar, M. and Michel, C. (2011), 'Environmental policy stringency and technological innovation: evidence from survey data and patent counts', *Applied Economics*, **44**: 2157–70.

Maddala, G.S. (1983), *Limited-Dependent and Qualitative Variables in Econometrics*, Cambridge: Cambridge University Press.

Milborrow, D. (2007), 'Wind power on the grid', in D. Boyle (ed.), *Renewable Electricity and the Grid: The Challenge of Variability*, London: Earthscan, pp. 31–54.

Neuhoff, K. (2005), 'Large-scale deployment of renewables for electricity generation', *Oxford Review of Economic Policy*, **2**(1): 88–100.

OECD (2011), 'Methodological issues in the development of indicators of innovation and transfer in environmental technologies', in OECD (ed.), *Invention and Transfer of Environmental Technologies*, Paris: OECD, Annex A.

Popp, D., Haščič, I. and Medhi, N. (2010), 'Technology and the diffusion of renewable energy', *Energy Economics*, **33**(4): 648–62.

REN21 (2008), 'Renewables 2007: global status report', Renewable Energy Policy Network for the 21st century (REN21), Paris: REN21 Secretariat and Washington, DC: Worldwatch Institute.

Sinden G. (2007), 'Renewable resource characteristics and network integration', in D. Boyle (ed.), *Renewable Electricity and the Grid: The Challenge of Variability*, London: Earthscan, pp. 55–72.

Swift-Hook, D.T. (2010), 'Grid-connected intermittent renewables are the last to be stored', *Renewable Energy*, **35**: 1967–9.

Wilson, I.A.G. et al. (2011), 'Grid-connected renewable, storage and the UK electricity market', *Renewable Energy*, 1–5.

Wooldridge, J.M. (2002), *Econometric Analysis of Cross Section and Panel Data*, Cambridge, MA and London: MIT press.

7 Electric vehicles: will consumers purchase them?

Henry Lee and Grant Lovellette

Governments around the world are becoming aware that exclusive reliance on petroleum to power the rapidly expanding transportation sector may become very costly in terms of financial, environmental and security impacts. As per capita incomes in China, India and other emerging countries grow, so too will the worldwide demand for automobiles and the fuels that power them. In 2008, the USA had 256 million highway-registered vehicles, compared with 128 million in China and 40 million in India. Globally, there were 750 million vehicles, and this figure is predicted to increase to 1.1 billion by 2030. Is it realistic to assume that a passenger fleet of this size will be populated exclusively by gasoline-powered vehicles? Even if the answer is yes, how long will the gasoline supply be priced low enough to sustain this scenario?

Countries are becoming increasingly aware of the environmental and security implications of their present transportation systems and are looking at alternatives to petroleum, including biofuels, natural gas and electricity. President Obama has called for the deployment of one million electric cars on US roads by 2015. Others advocate a goal of having 20 percent of the passenger fleet consist of electric cars by 2030 – approximately 30 million electric vehicles nationwide. European countries are looking at electric vehicles as the urban car of the future, ameliorating the problems of both conventional and unconventional air pollution.[1]

The technology itself is not in question – many of the global automobile companies are planning to sell electric vehicles by 2012. The key question is, will consumers buy them?

This chapter attempts to answer this key question by examining three additional questions: (1) Will electric vehicles be cost competitive with cars powered by internal combustion engines? (2) Will the electric infrastructure provide sufficient power to charge the vehicles? (3) Will electric vehicles have attributes that are comparable with those of gasoline-powered cars? This chapter examines these questions from a distinctly US perspective, but its basic arguments pertain to many countries.

There are two basic categories of electric vehicle – battery electric vehicles (BEVs), which run solely on the electric energy stored in the battery, and plug-in hybrid electric vehicles (PHEVs), which operate on both a rechargeable battery and a gasoline-powered engine. With both types of vehicles, the major incremental expense compared to a conventional vehicle is the cost of the battery. While the industry is working hard to reduce these costs, a battery in a BEV with an average range of 60–80 miles costs between $10 000 and $15 000. Hence, this chapter compares the net present lifetime cost of electric vehicles with that of conventional cars, both at today's costs and at projected future costs. The chapter also runs comparison scenarios with different assumptions about gasoline and electricity costs, battery costs, consumer discount rates and vehicle efficiency levels.

1.   ELECTRIC VEHICLES

What is an electric vehicle (EV)? There are several varieties. The two most common are battery electric vehicles (BEVs) and plug-in hybrid electric vehicles (PHEVs), and there are several varieties of the latter. In this chapter, a conventional hybrid (HEV), such as a Toyota Prius, which is capable of drawing some of its energy from an electric battery, but not from the electric grid, will not be considered an electric vehicle.

BEVs run solely on chemical energy stored in rechargeable electric battery packs. Most planned production BEVs have a theoretical average range of around 100 miles (although some have ranges of up to 250 miles), and their performance can mirror that of a conventional vehicle.

Highway-capable BEVs are not yet in widespread use anywhere in the world, although the Israeli and Danish governments have announced aggressive programs to accelerate their penetration. Gasoline prices in both countries as of January 2011 were above $7.70 per gallon,[2] and both countries are small and have high population densities.[3] While several companies produce BEVs in small numbers – for example, BMW, Mitsubishi, Tesla and Smart – only Nissan Motors has presented a detailed roadmap for selling BEVs to a wide customer base. The Nissan Leaf debuted in the fall of 2010, and the company hopes to sell 5000 vehicles in the USA in 2011. There are several smaller companies, as well as a couple of larger manufacturers, that are planning to market BEVs by the end of 2012.

All of the BEVs that will enter mass production in the near term will follow the traditional model of automobile ownership – the consumer pays the purchase price for the car and is responsible for maintenance and energy costs. Alternative purchasing agreements are possible, such as the consumer buying the car and leasing the battery separately. Such alternative arrangements are being actively considered in China, as well as in a few other smaller countries.

Most companies, however, are betting that there will be a much stronger market for the second EV option – plug-in electric hybrids (PHEVs). The trade-off between BEVs and PHEVs will be between cost and range. BEVs involve fewer parts and thus are predicted to be less expensive, but their reduced range – about 70–100 miles – is considered by many to be a showstopper.

Similar to BEVs, PHEVs have batteries that can be recharged by connecting a plug to an electric source. Unlike BEVs, when its battery is depleted, a PHEV is capable of running on a small conventional motor. Hence consumers' range anxiety is substantially reduced. Probably the most familiar PHEV is the Chevrolet Volt, which went on sale in the fall of 2010. The Volt is a PHEV-40, which means that it can drive up to an estimated 40 miles on power from the battery alone. After depleting the battery, it can drive an additional 350 miles on gasoline.

No one doubts that the auto industry can build electric cars, but the key question is: will consumers buy them? The answer depends in part on their comparative costs. If an electric car costs more than a conventional car, what additional benefit(s) will the consumer obtain for the extra cost? A disproportionate percentage of the value of today's electric cars will be in the form of reduced environmental or energy security externalities – public benefits that the individual consumer has difficulty capturing. Additionally, for many consumers, a BEV's lack of range when compared with conventional cars is a significant detriment.

Further, recharging an EV will require an infrastructure that is readily available (including the recharging equipment and outlets), an upgraded electric distribution grid, and sufficient generation capacity to meet the additional demand. As with so many aspects of the electric car, the availability of this infrastructure depends on a number of uncertainties. How fast will electric vehicles penetrate the fleet? If slowly, then the market will not want to invest in charging equipment and wire upgrades that are subsequently stranded for many years. Will electric car sales be evenly distributed across the country, or disproportionately located in certain areas, such as the two coasts? Since the conditions of the grid and the adequacy of generating capacity depend on regional variables, one will need to look at this issue from a state perspective, if not from that of individual utility franchises. Finally, when will consumers be recharging their vehicles? There is a big difference between scenarios in which a high percentage of consumers charge their vehicles at 7:00 p.m. and ones in which a majority wait until midnight.

We will now look at these three factors, beginning with costs.

2. COST COMPARISON

We have developed a simple costing model that incorporates a number of key assumptions.[4] The model's results are quite sensitive to these assumptions, and there is substantial uncertainty surrounding most of them. Hence any definitive statements on the comparative price of electric vehicles in 2025 do not rest on a robust foundation. We compare the economics of four types of vehicles – conventional internal combustion engine cars (ICEs), conventional hybrids (HEVs), battery-only electric vehicles (BEVs), and plug-in hybrid vehicles (PHEVs), and we consider seven scenarios: (1) approximate cost estimates as they exist in 2010–11; (2) a conservative estimate of likely (real) costs in 2025–30; (3) a low battery cost of $150 per kWh; (4) a high oil price scenario; (5) a scenario in which the discount factor is 30 percent; (6) a scenario in which electricity prices rise to $0.24 per kWh; and (7) a scenario in which the efficiency of conventional cars reaches 55 MPG and the efficiency of HEVs reaches 80 MPG.

Before looking at the results of our calculations, it is worth examining the three factors battery costs, fuel costs and discount rates – that have the greatest impact on our results.

Batteries

In EV batteries, power refers to the rate of energy transfer from the battery to the wheels, measured in kilowatts. Greater power equals better acceleration and performance. As a way to compare with an ICE, 100 horsepower is equivalent to 75 kilowatts.[5] Energy, measured in kilowatt-hours (kWh), is a measure of the storage capacity of an EV's battery. Hence, all else being equal, the higher the kWh capacity of a battery pack, the farther a vehicle can be driven between charges. Comparing kWh of electricity with gasoline, a car with a 15-gallon tank has usable energy capacity equivalent to a BEV with a battery pack of 135 useable kWh,[6] which would be equivalent to a pack 15 times larger than that contained in the Chevrolet Volt.[7]

In over a year of interviewing authorities and reading studies, we found no consensus on the cost of batteries in the future or even the cost today. Estimates ranged from as high as $875 per nominal kWh[8] to below $200 per kWh. In our calculations, we begin with an estimate of $600 per kWh as today's cost, which is the figure contained in the Electrification Coalition's study.[9] The coalition consists of electric vehicle advocates, and their estimates are slightly lower than those we have seen in other formal studies, but we believe them to be within the range of reasonableness. If this number is correct, then the 16 kWh battery pack in the Chevy Volt costs approximately $10 000, and the Leaf's 24 kWh pack costs $14 400, while the Tesla Roadster's 53 kWh pack costs more than $30 000. We do not have actual cost data from the companies, but using the numbers published by one of the industry's own coalitions seems like a practical starting point.

The batteries in today's electric vehicles use lithium and are similar to the batteries that power a laptop computer. The battery costs include (1) obtaining the lithium; (2) building the battery pack to meet rigorous safety and reliability standards, and (3) constructing the plant where the batteries are manufactured. Advocates often forget to include this third element in their estimates – a category that can comprise about 30 percent of the total cost of the battery.

Some critics worry about the future availability of lithium. We did not find evidence to support this concern. A typical lithium-ion battery is between 3 and 5 percent lithium,[10] which means that each electric vehicle will need between 6 and 10 kg of lithium per car. The current world reserve of lithium is sufficient to power 1.7 billion vehicles, or about 500 million more vehicles than the total projected to be on the road in 2030. Furthermore, much of the lithium used in EVs will be recyclable.

Can lithium-ion battery costs be significantly reduced below the estimated $600 per kWh figure? The Electrification Coalition has set a goal of $150 per kWh. While battery manufacturers keep actual costs a closely guarded secret, at least one has hinted that their costs are already below $300 per kWh.

Deutsche Bank has projected that the cost of lithium-ion batteries will decline 7.5 percent a year as production increases and will reach $250 per kWh by 2020.[11] Research on alternative battery types has increased dramatically in the last few years, and there are several promising technologies, including thin-film methods for producing lithium-ion batteries that could dramatically reduce the weight of the battery, as well as lithium-air batteries, which might be able to produce five times the energy per kg of battery mass as current lithium-ion batteries. However, both technologies are at least six years away from penetrating the electric car market.

One of the ongoing difficulties that battery manufacturers must face is enormous pressure to meet safety, reliability and lifetime performance constraints. A lithium battery is a complex series-parallel electrical machine made of many small-voltage batteries. It must be carefully made and managed. A single bad or low-resistance cell can cause thermal/electric runaway, damaging or destroying the battery.[12]

Batteries rely on chemical reactions to absorb and discharge electricity; therefore short circuits and leakage are a concern for the large-scale units used in electric vehicles. Batteries can be damaged by overcharging, exposure to extreme heat or cold, or by impacts and collisions. Manufacturers have to reassure buyers that the batteries in their vehicles will operate under a wide range of weather and driving conditions and that passengers will be safe in the event of an accident. These constraints affect the construction and location of the battery packs, and thus their cost.

In addition, because battery packs are so expensive, manufacturers seek to reassure motorists that they will have a useful and lengthy lifespan. The industry standard seems to be 100 000 miles, which is about ten years. Since battery performance degrades over time as the battery is discharged and recharged, this constraint has forced manufacturers to make only a portion of the battery available for charging/discharging in order to minimize the stress that comes from deep discharging and recharging and to keep some of the battery's capacity as spare reserve. For example, the Chevy Volt only charges its battery to 90 percent of its nominal capacity and does not discharge the last 25 percent of its nominal capacity.[13] While this lengthens the useful life of the battery pack, it also greatly increases the cost per useful kWh, and thus the price of the battery. In the case of the Volt, it takes 1.6 kWh of nominal battery capacity to produce 1 kWh of useable energy capacity. The battery industry believes that it can dramatically improve this ratio and will eventually be able to increase usable capacity to more than 90 percent of nominal capacity while still meeting safety and reliability standards.

Fuel Costs

The consumer will purchase an electric vehicle only if he or she perceives that the purchase will provide greater benefits than purchasing a conventional vehicle. One of the strongest private benefits may be the relative lifetime cost of operating an electric vehicle as compared to a conventional passenger car. (We note that this is different from the societal life-cycle cost of the vehicle, but this chapter examines the cost from the consumer's perspective because, for most consumers, this will be their primary motivation in deciding which vehicle to purchase.) Since battery costs, even under the most optimistic scenarios, will ensure that the upfront capital costs are higher for EVs than for conventional vehicles, the only way EVs will be less expensive over their lifetimes is if their operating costs are lower than those for ICEs. Fuel costs dominate the cost of operating a conventional ICE vehicle, and thus gasoline prices are a critical factor in any comparative analysis. If one assumes $6.55 per gallon gasoline, the net present cost of purchasing a BEV, even under a $600 per kWh battery scenario, is equal to that of a conventional vehicle, at a 15 percent consumer discount rate. (Our costing model is discussed in detail below.)

One might argue that gasoline prices might not increase to $7 or even $5 per gallon. At lower gas prices, electric vehicles are less attractive. We realize that future oil prices are fraught with uncertainty, but in a world where the global vehicle fleet is 60 percent larger than today's, we expect that demand will place significant upward pressure on oil prices, and thus a $5 per gallon scenario seems more likely than a $3 scenario. The EIA's 'Annual Energy Outlook' for 2011 projects a 2035 price of $3.89 per gallon and a high estimate of $5.85 (in 2010 dollars).[14]

A strong possibility remains that between now and 2030 governments will place a value on carbon emissions, either through taxes or policies that will have an equivalent effect. This will push the gasoline price even higher. The bottom line is that the future marketplace for passenger vehicles will probably be characterized by significantly higher oil prices.

Discount Factor

Electric vehicles will cost more to purchase. However, since they are primarily fueled by electricity, not gasoline, they will have lower operating costs over their lifetimes. To compare operating savings against higher front-end capital costs, analysts need to discount the annual operating savings. Hence the choice of discount rate becomes important in comparing the total lifetime costs of a conventional vehicle with those of an EV.

While one might argue that it is in a consumer's interest to pay more in the form of upfront costs in order to capture the longer-term benefits of reducing operating costs, consumers do not necessarily value the savings in operating costs over the life of a vehicle as highly as theory might lead one to expect. In fact, studies suggest that consumers are reluctant to accept a differential in upfront costs for efficiency gains greater than the difference in fuel costs over the first three years. In other words, they want a payback period that is no longer than three years. Given the volatility in fuel markets, this may be quite rational behavior, but it means that the implicit discount rate used by consumers may be closer to 30–40 percent than the 5–10 percent one sees in many studies.

David Greene surveyed the economics literature and found that estimates of consumer discount rates varied significantly. He found that many estimates were under 10 percent, but an equally large number were over 20 percent.[15] Most fell between 4 percent and 40 percent, which is a wide range, but it is not implausible if one examines discount rates for energy savings from building or appliances.[16] The choice of discount rate will have a substantial effect on our comparison of the net present costs of the differently fueled vehicles.

Results

In order to test our assumptions, we constructed a simple model that calculates the net present cost differential between HEVs, PHEVs, BEVs and ICEs.[17]

Our model does not account for financing arrangements such as leases or loans, or for battery degradation over time. It assumes a uniform number of miles driven per year and does not account for consumer preferences such as performance, size and range. Perhaps most importantly, the model does not capture pricing decisions made by the automakers. Nevertheless, we believe it is a helpful tool in illustrating the main drivers of the cost differentials between various vehicle technology types and for demonstrating the sensitivity of cost outcomes to key assumptions and variables.

In our initial case, we simply wanted to determine the comparative costs in 2011. Unlike all the other cases, this case uses existing numbers and assumptions. We assume a gasoline price of $3.75 per gallon (which is approximately equal to pump prices as of summer 2011) and battery costs of $600 per kWh. We assume that all vehicles in our comparison are driven 12 000 miles per year and that the average price of electricity is $0.12 per kWh. We use a discount rate of 15 percent and assume there are no government subsidies. Finally, we assume that consumers pay $1500 to install a Level II 220–240 volt outlet to charge their EV and that the plug-in hybrid is driven 85 percent of the time in an all-electric mode. Some of these assumptions may be optimistic, but all are within the range of reasonableness.

We compare the lifetime costs to the consumer of four different vehicles – a conventional ICE vehicle, a conventional hybrid (HEV) using technology similar to that used by Toyota's Prius, a plug-in hybrid with an all-electric range of 40 miles (PHEV-40) and an all-electric vehicle (BEV) with a rated range of 100 miles. The comparison consists of looking at all the costs – both upfront capital costs and operation and maintenance costs – over a ten-year timeframe. To compare the lifetime costs of the four vehicles, we discount the savings in the later years to derive a net present cost for the upfront capital costs and the discounted operation and maintenance costs.

Table 7.1   Today's costs

	Conventional	HEV	PHEV	BEV
Total net present cost	$32 861	$33 059	$38 239	$37 680
Cost differential with conventional car	–	$197	$5 377	$4 819

Table 7.2   Future costs – base case

	Conventional	HEV	PHEV	BEV
Total net present cost	$34 152	$32 680	$34 601	$30 674
Cost differential with conventional car	–	($1 472)	$449	($3 478)

Scenario 1
In the initial case, we find that the net present cost of a conventional car over its lifetime is $32 861 (assuming a purchase price of $21 390); the net present cost of an HEV is $33 059 (assuming a purchase price of $22 930); the net present cost of a PHEV is $38 239 (assuming a purchase price of $30 235); and the net present cost of a BEV is $37 680 (assuming a purchase price of $33 565).

Thus, in 2011, a conventional car is $4819 less costly over its lifetime than a battery-powered electric car, and is $5377 less than a PHEV-40 (see Table 7.1).

Scenario 2
We then constructed what we believe is a reasonable future base case scenario. We assume that technology breakthroughs reduce battery costs (to $300 per kWh of nominal battery capacity), that gasoline prices increase to $4.50 per gallon, and that electricity prices will be 30 percent higher ($0.15 per kWh). Under these assumptions, BEVs are less costly than both conventional ICEs and HEVs. PHEVs remain higher priced than the other three options (see Table 7.2).

We used this 'future scenario' as the base case for the construction of five additional scenarios, in each of which we varied one of the assumptions.

Scenario 3
The assumptions are the same as in Scenario 2, except that gasoline costs increase to $6 per gallon. Electric vehicles, both BEVs and PHEVs, are now less expensive than conventionally fueled vehicles, with BEVs enjoying an advantage of over $6000 (see Table 7.3).

Table 7.3   Future costs – high gasoline prices

	Conventional	HEV	PHEV	BEV
Total net present cost	$36733	$34323	$34847	$30674
Cost differential with conventional car	–	($2411)	($1886)	($6059)

Table 7.4   Future costs – high discount rate

	Conventional	HEV	PHEV	BEV
Total net present cost	$29251	$28475	$31349	$28940
Cost differential with conventional car	–	($776)	$2097	($312)

Table 7.5   Future costs – low battery costs

	Conventional	HEV	PHEV	BEV
Total net present cost	$34152	$32080	$32549	$26971
Cost differential with conventional car	–	($2072)	($1603)	($7181)

Table 7.6   Future costs – high electricity prices

	Conventional	HEV	PHEV	BEV
Total net present cost	$34152	$32680	$35624	$31897
Cost differential with conventional car	–	($1472)	$1472	($2273)

Scenario 4
The assumptions are the same as in Scenario 2, except that the consumer's discount rate is 30 percent. BEVs are now only slightly less costly that conventional vehicles, illustrating the sensitivity of the cost calculations to the estimates of discount rates (see Table 7.4).

Scenario 5
The assumptions are the same as in Scenario 2, except that battery prices (for both BEVs and PHEVs) are $150 per kWh as opposed to $300 per kWh. Electric vehicles' cost advantage is larger than in any other scenario, and BEVs become almost $6000 less expensive than plug-in hybrids (see Table 7.5).

Scenario 6
The assumptions are the same as in Scenario 2, except that electricity prices are $0.24 per kWh instead of $0.15 per kWh. Despite the higher electric price, BEVs are still less costly than other transport options (see Table 7.6).

Scenario 7
The assumptions are the same as in Scenario 2, except that the fuel efficiency of conventional cars reaches 55 miles per gallon, while the fuel efficiency of HEVs and PHEVs

Table 7.7   Future costs – higher fuel efficiency

	Conventional	HEV	PHEV	BEV
Total net present cost	$31 336	$31 140	$34 370	$30 679
Cost differential with conventional car	–	($196)	$3 033	($662)

when burning fuel reaches 80 miles per gallon. Not surprisingly, the price advantages of HEVs and BEVs relative to conventional vehicles decrease, but even in this scenario, BEVs remain the cheapest transport option (see Table 7.7).

3.   COMPARING ATTRIBUTES

There are two schools of thought about how consumers' purchasing behavior will affect the electric car market. The first asserts that consumers look to purchase new vehicles whose attributes are superior to those of their existing vehicles. Few people go to the car dealer looking to purchase a vehicle that has worse performance, fewer accessories and less room than the car that they are replacing. For electric vehicles to be competitive, they must be able to perform at the same level, possess the same attributes, and be approximately the same size as the conventional cars that they are replacing. For example, the Nissan Leaf is rated up to 90 kW of power – equivalent to a 120 horsepower in a conventional car – which is in line with a large number of compact and intermediate vehicles, such as a VW Jetta, a Toyota Corolla or a Ford Focus.

However, performance is not the only measure that consumers use to compare vehicles. The two attributes that present the greatest challenge to PHEVs and BEVs are range and reliability. The average conventional vehicle has a range of between 350 and 450 miles per tank of gasoline. Relatively long-range travel in a BEV requires very large batteries. For example, the Tesla Roadster sports car achieves a maximum rated range of 244 miles through a massive 450 kg (992 pounds) battery pack consisting of 6831 lithium cells at an estimated cost of $30 000.[18] (For a sense of scale, the Toyota Prius HEV has a battery pack of 168 cells that operates at around 36 horsepower.) PHEVs can continue being driven after the electricity in the battery pack is exhausted, which will occur after 20 to 40 miles, depending on the driving conditions and the size of the PHEV's battery pack. Since they are more complicated to build and run at least partially on gasoline, PHEVs will have a higher lifetime cost than comparable BEVs, and this difference may be a good proxy for the range anxiety exhibited by consumers. That is, if consumers pay $4000 more for PHEVs than comparable BEVs, that figure may represent the value of eliminating the range anxiety connected with the ownership of a BEV.

The second school of thought argues that range anxiety is overhyped. The proponents of this thesis point out that there are 80 million motorists living in cities or nearby suburbs. Surveys by the National Highway Administration show that these motorists drive fewer than 20 miles per day on average.[19] Thus, on an average day, they would not come close to exhausting the power stored in a BEV with a range of 100 miles. In fact, they would only have to recharge their batteries every two to four days. Furthermore, if charging stations existed at their workplaces or in the parking lots where they shopped, there would be only a few days per year when the range of an electric car would be insufficient. Under these conditions, concern over range might shrink to such a level that consumers would purchase electric cars with smaller batteries at lower prices, improving the comparable cost estimates. In fact, the consulting firm McKinsey released a study asserting that electric cars will make major inroads into the passenger vehicle fleet in large cities, such as New York, Berlin, Paris and Shanghai, by 2015, primarily because range will not be an issue for many motorists with daily commutes of less than 30 miles.[20]

The actual range of a BEV may be less than the maximum rated range per charge, however, exacerbating range anxiety. For example, actual mileage of a Nissan Leaf may be significantly less than 100 miles per charge. If the driver uses the air conditioner, entertainment equipment or even headlights, electricity will be drawn from the battery, leaving less available for driving the vehicle. Even if the actual mileage available were closer to 60 miles per charge, most commuters would not actually be inconvenienced, because average daily mileage is below 20 miles per day. Nevertheless, most automobile buyers are risk-averse and might envision being stranded on the side of the highway with no power, despite the small likelihood of this actually happening.

Some might point out that if electric car owners average only 20 miles per day, they would be likely to drive 6000 miles or less per year. Therefore the numbers in our earlier calculations might misstate the net present cost advantage of EVs. It is true that an electric car that is driven 6000 miles per year will use less electricity than one that is driven 12000 miles. However, electricity is not very expensive, so the additional net present savings of cutting one's driving in half in a BEV would be about $1000 under our future baseline scenario. By contrast, if the owner of a conventionally fueled car drives it 6000 miles per year instead of 12000, the net present cost of his or her vehicle would decrease by just over $4000. Thus conventional cars would benefit more than electric vehicles if they were driven less, since the marginal cost of driving a conventional car (the price of gasoline) is far greater than the marginal cost of driving a BEV (the price of an equivalent amount of electricity).

While range anxiety may be the greatest challenge, it is not the only attribute that concerns potential BEV purchasers. BEVs and PHEVs are unfamiliar technologies to the average consumer, who has been driving some version of gasoline-powered car his or her entire life. Since a new vehicle is usually the most expensive purchase (other than a new home) that consumers make, they are generally risk-averse when purchasing vehicles. Consequently, they will likely purchase vehicles that closely resemble the ones they currently own.

Admittedly, very little data exists on the reliability of electric vehicles, and for this reason we emphasize consumer perceptions as opposed to factual realities. The industry is aware of this problem and has invested billions of dollars to ensure that EVs operate as reliably as conventional vehicles. However, it will take time for the public to adjust its perceptions. To accelerate this process, the government offers tax credits to early adopters who are willing to take on the perceived technology risks. These credits expire either over time (as in Europe) or after a certain number of cars are sold (as in the USA).

Some argue that recharging a car at home will prove to be more convenient than having to drive to the local gasoline station to fill up. Again, this may be the case, but is it worth paying an additional $50, $250, $1000 or $4500+? No actual data exist to answer this question, but we suspect that most consumers will not be willing to pay much to exchange the hassle of weekly trips to the gas station for the anxiety that their car's battery may fail during their commute to work. Further, the possibility exists that PHEV owners will simply 'forget' or stop charging their vehicle regularly and instead run it on gasoline most of the time, thus counteracting the environmental benefits.

Other benefits asserted by EV advocates include reduced pollution, greater vehicle efficiency, improved national energy security and reduced carbon emissions. All these are real benefits and may prove to be large, but they are all public benefits (or positive externalities), which are not captured by the private individual. Admittedly, there will be individuals who place a large value on being environmentally responsible, and they are likely to be less dissuaded by high initial prices or limited range than the majority of consumers.[21] However, these consumers account for a small segment of the population.

The bottom line is that electric vehicles currently do not provide private attributes and benefits that exceed, or even equal, those of most conventional gasoline-powered cars. Range deficiency is the single largest technical barrier separating electric vehicles from their gasoline counterparts, compounded by long recharge times when compared to filling up at a gas station. As we have discussed, the cost of range anxiety depends on the individual. The most obvious measure would be the net present cost difference between a PHEV and a comparable BEV, which in our scenarios ranges from about $2500 to $5500. As mentioned earlier, there may be other measures. As more EVs show up in company showrooms, we will begin developing a database from which analysts can answer this question.

4. ADEQUACY OF INFRASTRUCTURE

Electric cars are fueled wholly or partly by electricity, which presumes access to a reliable source of power. Equipment for connecting an EV to a source of electricity is required, at home and/or outside the home. Proponents are concerned that adequate electric distribution and transmission infrastructure might not exist when and where it is needed. This concern goes to the heart of the debate in Congress over the question of whether to subsidize installation of public charging stations in five to 15 EV deployment communities.[22] Our initial conclusion is that a market failure justifying a strong federal presence is not evident. While there are regional or national differences in the adequacy of the existing electric distribution, transmission and generating systems, there is no evidence to conclude that investments will not be forthcoming from private companies to meet those needs, if and when they manifest themselves.

We anticipate that most EV owners will want to charge their vehicles at home, so appropriate charging infrastructure is needed at each household. A standard wall outlet will provide power at 110 volts and 20 amps. In the simplest case, owners can run an electric cord from the house to the EV. The electric load will be equivalent to adding a new clothes dryer to the house, which means that an electrician may be needed to install a dedicated outlet, although many houses would be able to manage the extra electricity pull without any additional wiring. This type of connection is labeled a Level I system and is relatively inexpensive, assuming that the vehicle comes with the equipment to receive the electric charge. The downside is that the charging process is excruciatingly slow. For example, charging a Tesla Roadster from 'empty' on a Level I system will take over 30 hours, and fully charging a Nissan Leaf from 'empty' will take about 20 hours. Charging the Chevy Volt (a PHEV) from 'empty' will take about seven hours, so if the owner plugged in her Volt at 10:00 p.m., it would be ready by early morning.

In practice, most EV owners will find Level I charging unacceptable as a sole energy source and will install a Level II charging system, which operates at 220 volts and between 30 and 40 amps. With this system, a Nissan Leaf could be charged in seven hours.[23] Such a system would equal the electric pull of about two clothes dryers, and thus installing such a system would be approximately the same as adding one-half a house to a normal residential area. A Level II system can be installed by any licensed electrician and includes a special connector. The cost estimates vary and are likely to decrease over time, but are now approximately $1500–$2200 per household.[24]

A popular topic among EV enthusiasts is 'quick charging', which is charging an electric car in approximately the same time it takes to fill up a gasoline-powered car at your local gasoline station. These Level III systems draw about 210 kW for ten minutes, or about the same draw as 140 houses, if one assumes that each home draws about 1.5 kW of power. The current could be up to 500 volts at 200 amperes, which would require very expensive conductors and sophisticated safety systems. This system has been demonstrated and is technically feasible, but for safety reasons would only be available at dedicated service stations. Finally, Level III charging would likely subject the battery pack to significantly greater wear and tear than Level II charging and might cause the EV battery to degrade more quickly.

Level III charging would also put an enormous strain on the existing electric distribution systems and would require an industrial-sized substation to handle the power surges at each individual location. The cost of this system would be substantial, although we were not able to identify estimates that we could confidently embrace.

US utility companies have taken a relatively passive investment position. Many companies see no advantage of investing in advance of demand. On the other hand, Chinese electric grid companies have identified the emergence of electric cars as a potentially large opportunity. They are proposing to build a network of large battery recharging operations located near transmission substations that would feed a network of stations where BEV owners could quickly exchange their old batteries for new ones. It is still too early to determine the effectiveness of this scheme, but it demonstrates that penetration of electric cars into the market could provide economic opportunities for electricity transmission and distribution companies. For this model to work, car manufacturers would have to be willing to warranty their electric vehicles even though battery packs would be transferred from one EV owner to another.

If one is skeptical and believes that governments must intervene and subsidize the upfront costs of non-residential charging stations, a demonstrated market failure must justify government involvement. Will the local utilities fail to make the investments, and will they prevent others from investing? The former scenario is quite possible, given that utilities are risk-averse investors, but there is no reason that they would not welcome new paying customers, especially if they themselves do not have to take any of the front-end risk. Given that Level I and II charging stations are relatively inexpensive and can be erected in a matter of hours, the system can expand to meet demand quite easily.

Finally, proponents argue that urban residents are more likely to buy electric cars than those living in rural or suburban neighborhoods, but many urban dwellers do not have private garages and therefore park on the street. Lack of on-street charging will be a major deterrent to owning an electric car in dense city environments. Hence on-street charging stations and/or battery exchange stations may be deemed essential for densely populated cities.

If the demand for electric vehicles increases significantly, additional electricity will have to be generated and distributed. Projections of possible impacts of EVs on the grid vary greatly according to location, time of day and the technology used for charging. Kintner-Meyer et al. and Denholm and Short examine the potential impact of PHEV adoption on the US grid (the impact of which would be, by definition, smaller than an equivalent number of BEVs) and conclude that if the PHEVs were charged at lower voltages at optimal times of day (i.e. night), then between 50 percent and 73 percent of the US light-duty vehicles could be converted to PHEVs and be supported by the existing electric infrastructure.[25] These numbers assume that the grid can remain relatively stable while constantly running at or near 100 percent capacity, which may not be realistic, but their analyses suggest that a substantial number of EVs could be absorbed by the current grid without adding generating capacity.

In the same way that generation capacity to supply electric vehicles varies by region, so too does the need to enhance the distribution systems. In some localities, there will be a need for significant new investment. For example, if every home in a ten-house cul-de-sac installed a Level II charging system, the pull on the transformers would be equivalent to adding approximately five new homes. Hence the utility would probably have to upgrade its distribution system. On the other hand, there will be areas where the addition of electric cars will not make much of a difference and will not require much investment.

In a very optimistic scenario, let us assume that 30 percent of the vehicles in Southern California shift to electric cars by 2030. This would be over 2 million vehicles. Let us further assume that all of their owners install a Level II system at their homes and that an additional 300000 charging stations are installed in commercial locations. If the ratio of households to cars is 1.25 to 1, then the draw on the distribution system would be equivalent to about 4 million new clothes dryers – a substantial increase in demand. However, the utility would have 20 years to make the necessary investments, assuming linear growth in electric vehicles and no major change in battery technology. Thus the utility would only have to upgrade a small percentage of its system every year. The actual percentage and cost would be specific for each utility.

If the need for additional power emerges rapidly and existing siting and permitting hurdles for new power plants remain, utilities will be under pressure to keep their older plants in use, since they will not be able to permit, finance and build new facilities fast enough. In many parts of the country, these plants are coal-fired. The irony is that a rapid electrification campaign, justified on environmental and climate grounds, may extend the life of many older and dirtier coal plants, diluting the environmental benefits of electric vehicle adoption.

If the argument in favor of government intervention to ensure that charging stations are available is not compelling, why is it being made and, in many cases, accepted? There are two possibilities. In the first case, one might argue that our confidence that private parties will respond to an emerging demand for public charging stations is misplaced, and unlike their counterparts in China, they will ignore this market opportunity. We do not find this argument compelling. In the second case, perhaps the demand for EVs will not emerge unless public charging stations are widely available. People will purchase fewer electric vehicles than they otherwise would due to worries about the absence of charging opportunities beyond their homes. Since one charge per night will provide about 50–70 miles per day and most people living in urban areas drive less than 20 miles, this issue may be more of a perception problem than a real one, especially if most people opt to buy PHEVs as opposed to BEVs. Nevertheless, even if it is primarily a problem of perception, it will certainly retard some investments in electric vehicles.

Thus government subsidization of public charging is aimed at spurring consumers at the margin to purchase new electric cars. A logical question would be: if the government wants to provide an additional subsidy to spur EV sales, is subsidizing the installation of public charging stations the best way to do so? Perhaps increasing tax credits to consumers from $7500 per vehicle to $10 000 or extending the life of the credit beyond the first 250 000 vehicles produced would be more effective and more economical.

CONCLUSION

The future of electric cars is fraught with uncertainties. Will oil prices increase? Will battery technologies improve? Will the efficiency of conventional cars reach 50–60 mpg? Will biofuels emerge as a cost-competitive alternative to gasoline? Will the electric infrastructure be sufficient to supply increasing demand stemming from the penetration of electric cars? The answers to these questions will have enormous bearing on the future viability of electric vehicles. Despite these uncertainties, our analysis suggests the following:

1. At gasoline prices of $4.50 or greater, BEVs will be not only cost-competitive with conventional ICE vehicles, but in many cases less costly. Plug-in electric hybrids, which are in vogue today, will probably be more expensive than BEVs in the future and, under many scenarios, will even be more expensive than conventional hybrid vehicles.
2. Despite the cost advantage, limited range will greatly hamper consumer enthusiasm for BEVs, even though most car owners drive 20 miles per day or less, on average. Increasing range at competitive prices will probably require advances in battery technology beyond the current generation of lithium-ion batteries. China may be an exception, as Chinese consumers are less wedded to commonly desired vehicle attributes such as size, range, performance, and even reliability. Further, the Chinese government, automobile industry and electric utility companies see electric vehicles as both a compelling economic opportunity and as a solution to urban transportation and environmental problems, so they may be more willing than US actors to promote EV adoption aggressively.
3. The infrastructure needed to generate, transmit and distribute power to charge millions of electric vehicles has yet to be built. However, there is no reason to believe that private investments in such will not be forthcoming if demand emerges. Densely populated urban areas may be an exception, especially those that offer little or no off-street parking. Yet new infrastructure models, such as battery exchange net-

works like those being developed in Israel and Denmark, may prove to be a viable option.

In the short term, electric cars will gain market share in jurisdictions where gasoline and car purchase prices are high (especially if governments waive tax levies on new EV purchases) and where alternatives to traveling moderate and long distances are readily available (or where the demand for such travel is very limited). In the next five to ten years, lower consumer discount rates and strong government and industry engagement may stimulate consumers, private and public, to purchase electric vehicles in a few countries, such as China. In the longer term, however, improvements in battery technology will be needed to make electric vehicles attractive to consumers in the USA and other industrialized countries. The policy question is: will the notoriously impatient American consumer be willing to wait for improvements in battery technology, or will improvements in conventional automobile technology remove many of the incentives to purchase electric vehicles? Alternatively, will extremely high oil prices force American consumers into electric vehicles? We cannot be sure, but worldwide, the future for EVs appears bright – although perhaps brighter elsewhere than in the USA.

NOTES

1. '9 Million EV's in Berlin', *European Energy Review*, 29 September 2011: China faces the same challenges and has announced an aggressive program to stimulate the deployment of electric vehicles with an emphasis on its cities.
2. 'Europe's energy portal', accessed at http://www.energy.eu/#domestic.
3. 'Our gas costs 25% more than in Europe', *The Jerusalem Post*, 25 January 2011, accessed at http://www.jpost.com/Business/Globes/Article.aspx?id=205075.
4. For details on the model, see Henry Lee and Grant Lovellette, 'Will electric cars transform the U.S. vehicle market?', Belfer Center for Science and International Affairs, Harvard University Discussion Paper 2011-08, http://belfercenter.ksg.harvard.edu/publication/21216/will_electric_cars_transform_the_us_vehicle_market.html.
5. Electrification Coalition, 'Electrification roadmap', p. 74. November 2009, accessed at http://www.electrificationcoalition.org/sites/default/files/SAF_1213_EC-Roadmap_v12_Online.pdf.
6. ICEs have poor efficiency levels relative to electric cars, but not all energy stored in a BEV's battery pack is available for use.
7. We assume that only 65 percent of the nominal battery capacity is discharged, similar to the Volt.
8. *Transitions to Alternative Transportation Technologies: Plug-in Hybrid Electric Vehicles* (Washington, DC: The National Academies Press, 2010).
9. Electrification Coalition, 'Electrification Roadmap', p. 79.
10. Karen Fisher et al., 'Better waste management life cycle assessment', Environmental Resources Management, 18 October 2006, accessed at http://www.epbaeurope.net/090607_2006_Oct.pdf.
11. 'Deutsche Bank revises li-ion battery cost forecasts downward to $250/kWh by 2020', Green Auto Blog, accessed at http://green.autoblog.com/2011/01/06/deutsche-bank-li-ion-battery-cost-forecast-per-kwh/.
12. Comment received from Robert A. Frosch, 10 March 2011.
13. 'Electrical energy consumption in the Chevy Volt', GM Volt company webpage, 3 December 2010, accessed at http://gm-volt.com/2010/12/03/electrical-energy-consumption-in-the-chevy-volt/.
14. '2011 Annual Energy Outlook', US Department of Energy, 26 April 2011, (Washington, DC, Report #DOE/EIA 0383 ER 2011).
15. David Greene 'Why the market for new passenger cars generally undervalues fuel economy', (Joint Transport Center, OECD), ITE, January 2010, Discussion Paper 2010-6.
16. K.E. Train 'Discount rates in consumers' energy-related decisions: a review of the literature', *Energy*, **10**(12) (1985): 1243–53.
17. For details on the model, see Lee and Lovellette, 'Will electric cars transform the U.S. vehicle market?'

18. Gene Berdichevsky et al., 'The Tesla Roadster battery system', Tesla Motors web archive, 16 August 2006, accessed at http://teslamotors.com/display_data/TeslaRoadsterBatterySystem.pdf.
19. Hu and Reusher, 'Summary of travel trends: 2001 National Household Travel Survey', table 5, page 15.
20. Russell Hensley et al., 'The fast lane to adoption of electric cars', *McKinsey Quarterly*, February 2011.
21. Kelly Sims Gallagher and Erich Muehlegger, 'Giving green to get green: incentives and consumer adoption of hybrid vehicle technology', Working Paper, John F. Kennedy School of Government, Harvard University, February 2008.
22. The Electric Vehicle Deployment Act of 2010, US Congress (S.3442, H.R. 4442), accessed at http://www.govtrack.us/congress/bill.xpd?bill=s111-3442.
23. 'Nissan Leaf: charging FAQ', Nissan USA company website, accessed at http://www.nissanusa.com/leaf-electric-car/faq/list/charging#/leaf-electric-car/faq/list/charging.
24. Nick Bristow, '2011 Leaf US pricing officially announced', Autoblog Green website, accessed at http://green.autoblog.com/2010/03/30/2011-nissan-leaf-us-pricing-officially-announced-as-low-as-25/.
25. M. Kintner-Meyer et al., 'Impacts assessment of plug-in hybrid vehicles on electric utilities and regional U.S. power grids', November 2007, accessed at https://www.ferc.gov/about/com-mem/wellinghoff/5-24-07-technical-analy-wellinghoff.pdf; and P. Denholm and W. Short, 'An evaluation of utility system impacts and benefits of optimally dispatched plug-in hybrid electric vehicles', October 2006, accessed at http://www.nrel.gov/docs/fy07osti/40293.pdf.

PART III

ENERGY POLICY

8 The contribution of energy efficiency towards meeting CO_2 targets*

Joanne Evans, Massimo Filippini and Lester C. Hunt

1. INTRODUCTION

The Kyoto Protocol set an agenda for GHG emission reductions (relative to the 1990 emission levels) in participating countries between the years of 2008 and 2012. Despite emission reduction measures and strengthening political will internationally,[1] global CO_2 emissions reached their highest ever level in 2010[2] (IEA, 2010a), with an estimated 40 per cent of global emissions coming from OECD countries. Unsurprisingly non-OECD countries, led by China and India, saw much stronger increases in emissions as their economic growth accelerated. However, on a per capita basis, OECD countries collectively emitted 10 tonnes, compared with 5.8 tonnes for China, and 1.5 tonnes in India (IEA, 2010a). This emissions profile is informative as international discussions in Copenhagen (2009) and Cancun (2010) focused on countries contributing in line with 'common but differentiated responsibilities and respective capabilities' (Article 3 of the UNFCCC; United Nations, 1992) to 2020 economy-wide emissions reduction targets where developed countries should commit to emissions targets and countries party to Kyoto would strengthen their targets. Developing nations would 'implement mitigation actions' that are nationally appropriate to slow growth in emissions (UNFCCC, 2010). In 2011 in Durban, South Africa, the spirit of the negotiations changed and in conjunction with extending the life of the Kyoto Agreement by between five and eight years[3] to at least 2017, the so-called Durban Platform deal (UNFCCC, 2011) commits the world to negotiating a new legally binding climate treaty by 2015 for implementation by 2020. The emissions levels specified in this new treaty would be legally binding on all nations, including the USA and China.

These commitments will require a mix of instruments to be employed; however, improving energy efficiency has often been assumed to be one of the most cost-effective ways of reducing CO_2 emissions, increasing security of energy supply, and improving industry competitiveness. It is against this backdrop that the contribution of improvements in energy efficiency alone might make to national CO_2 emissions targets of 29 OECD countries is considered here. During the last 20 years, there has been considerable debate within energy policy about the possible contribution from an improvement in energy efficiency and about the effectiveness of ecological tax reforms in the alleviation of the greenhouse effect and in the decrease of dependence on fossil fuels. Many of the national and international 'think tanks' and policy agencies[4] suggest a major role for improvements in energy conservation and efficiency. In order to design and implement effective energy policy instruments to promote an efficient and parsimonious utilization of energy, it is necessary to have information on energy demand price and income elasticities in addition to sound indicators of energy efficiency.

In practical energy policy analysis, the typical indicator used at the country level is energy intensity, defined as the ratio of energy consumption to GDP. The IEA (2011) analysis is based on this simple traditional measure. However, according to an earlier report (IEA, 2009), 'Energy intensity is the amount of energy used per unit of activity. It is commonly calculated as the ratio of energy use to GDP. *Energy intensity is often taken as a proxy for energy efficiency, although this is not entirely accurate* since changes in energy intensity are a function of changes in several factors including the structure of the economy and energy efficiency' (p. 15, emphasis added). This highlights the weakness of this simple aggregate energy consumption to GDP ratio in that it does not measure the level of 'underlying energy efficiency' that characterizes an economy;[5] hence it is difficult to make conclusions for energy policy based upon this simple measure.

As the IEA (2011) highlights, energy intensity has decreased in many countries. This decrease, sometimes justified by the 'dematerialization' of the economies of these countries (e.g. Medlock, 2004), has allowed GDP growth to be decoupled to a certain extent from the growth of energy demand, although there may be other explaining factors. Richmond and Kaufmann (2006), for example, argue that energy prices explain the evolution of energy intensity in most countries, so that the dematerialization hypothesis should be rejected when prices are considered. This view of the role of energy prices, which partly drive greater efficiency of processes and structural shifts, is supported by the work of Metcalf (2008) and Sue Wing (2008), although these papers come to different conclusions, with the former suggesting a major role for energy efficiency and the latter underscoring the role of structural shifts.

In the energy economics literature some approaches have been proposed in order to overcome the problems related to the use of simple monetary-based energy efficiency indicators like the energy–GDP ratio, such as index decomposition analysis (IDA) and frontier analysis. IDA is basically a bottom–up framework that can be used to create economy-wide energy efficiency indicators.[6] Whereas frontier analysis is based on the estimation of a parametric, as well as a non-parametric, best-practice production frontier for the use of energy where the level of energy efficiency is computed as the difference between the actual energy use and the predicted energy use.[7]

An example of the use of parametric frontier analysis at the sectoral level is Buck and Young (2007), who used a parametric approach to estimate a stochastic energy use frontier function for a sample of Canadian commercial buildings, with energy use per square foot as a function of several variables pertaining to the activities and physical characteristics of the building. Another example (Boyd, 2008) estimated an energy use frontier function for a sample of wet corn-milling plants, where energy use is a function of four output variables and the capacity utilization. Both of these studies utilize the stochastic frontier function approach introduced by Aigner et al. (1977). An example of a non-parametric approach is Zhou and Ang (2008), who measured the energy efficiency performance of 21 OECD countries over five years (1997–2001) using a DEA (data envelopment analysis) model that consisted of four energy inputs, two non-energy inputs, a desirable output, GDP, and an undesirable output, CO_2 emissions.[8] Furthermore, in a developing strand of the literature, Filippini and Hunt (2011) introduced the idea of a frontier energy demand relationship (discussed below) as a way of estimating 'underlying energy efficiency' for 29 OECD countries.

The next section of this chapter therefore discusses what energy efficiency is. It begins with a discussion of the various measures of energy efficiency as well as the IDA approach and then introduces the definition of 'underlying energy efficiency' advocated in this chapter. Section 3 introduces the empirical framework to estimate the 'underlying energy efficiency' for 29 OECD countries. In Section 4 these estimates are used to calculate the contribution that improvements in 'underlying energy efficiency' alone might make to national CO_2 targets. Section 5 summarizes and concludes.

2. WHAT IS ENERGY EFFICIENCY?

Energy demand is derived from the demand for energy services, for example a warm house, cooked food, hot water, lighting, transport or the demand for other goods. Households and firms combine energy, labour and capital to produce a composite energy service. Therefore, behind any energy service a production process and associated production function can be identified. From an economic perspective it is important to produce the energy service in an efficient way – that is, by minimizing the amount of inputs used in the production of a given output and by choosing the combination of inputs that minimizes the production cost. In general, improvements in the level of efficiency of the use of energy are assumed to reduce energy consumption below where it would have been without those improvements (Sorrell, 2009, pp. 204–5). Thus the relationship between energy service demand, energy demand and improvements in energy efficiency is not straightforward but is important over the long run (see, e.g., Fouquet, 2008, 2011).

Energy efficiency is often discussed with regard to the so-called 'rebound effect' (see, for example, Greening et al., 2000; Sorrell, 2009). The rebound effect occurs when an engineering improvement in technology suggests that less energy is needed to produce a certain level of energy service; however, the reduction is not as great as that predicted by the technological improvement as consumers adjust their behaviour according to the new implicit prices of energy services that come about due to the technological improvement.[9] The rebound effect normally discussed in the literature is based on consumers of energy and energy services operating at the economically optimal point on their indifference curves, isoquants, production functions, cost curves and so on; therefore the rebound effect is their 'rational' adjustment to the change in implicit prices and thus moves them to a new economically optimal point on the associated curve. However, this is not what is considered here, because, as explained in more detail in Section 2.2, the analysis here is based on the theory of productive efficiency. This chapter attempts to unpick exactly what is meant by the term energy efficiency and redefine it in terms of productive economic efficiency and inefficiency. The focus is on where consumers of energy and energy services are 'away' from their economically optimal position on the isoquant (i.e. they are inefficient) and from this develop a measure of the 'underlying energy efficiency' based on economic principles (as explained below). However, it is worth noting that if consumers of energy and energy services were to reduce this inefficiency and become more efficient, then a rebound effect could also result given that the implicit price of the energy service would fall. Nevertheless, it is important to be clear that the cause of this is an improvement in the economically efficient use of energy and not an engineering improvement in technology; hence it is rather different from the rebound effect normally discussed in the literature. Hence the rebound effect is not the focus of this chapter.

As briefly discussed in the introduction, the energy economics literature tends to use definitions of energy efficiency based on the simple ratio of output to energy consumption, where the output and inputs can be measured in energy/thermodynamic units, physical units or economic monetary units (Patterson, 1996; Bhattacharyya, 2011). Engineers normally employ definitions based on thermodynamics, where energy efficiency is defined as the ratio of the heat content of both the output and the one input, energy. In contrast, economists tend to use a combination, where energy efficiency is defined as the ratio of output measured in monetary terms to the one input, energy measured in thermodynamic units. Both of these can be considered partial productivity indicators. However, although the simple ratio method to measure energy efficiency is pervasive within the energy economics field, there is also a definition based on the microeconomic theory of production (Huntington, 1994), which is advocated in this chapter. In particular, the importance of utilizing a framework where energy is used in a production function to produce an energy service in order to analyse the level of (energy) efficiency is demonstrated. The simple ratio indicators are therefore considered further in the next subsection, before introducing the concept of energy efficiency based on the microeconomic production theory in the following subsection.

2.1   Ratio Indicators

According to Bhattacharyya (2011), the simple ratio definitions of energy efficiency found in the literature all have the following basic origin:

$$Energy\, efficiency = \frac{Useful\, output\, of\, a\, process}{Energy\, input\, into\, a\, process}$$

Furthermore, Patterson (1996) identifies a number of ways in which the outputs and inputs for this ratio can be quantified, as follows:

- *Thermodynamic indicators* rely on measurements derived from the science of thermodynamics. Some are simple ratios and some are more sophisticated measures that relate actual energy usage to an 'ideal' process. Thermodynamic quantities for measuring energy efficiency are calculated in terms of 'state functions' of the process providing unique and objective measures for a given process in the context of a particular environment (prescribed by temperature, pressure, concentration, chemical formula, nuclear species and magnetization). Thus, for any change in physical conditions that results from some dynamic process, there is an associated change in the values of the state functions. However, according to Bhattacharyya (2011), these measures 'find limited use outside engineering design' (p. 143).
- *Physical–thermodynamic indicators* are hybrid indicators where the energy input is measured in thermodynamic units and output is measured in physical units. These physical units attempt to measure the service delivery of the process. One advantage of using these physical measures is that they can be objectively measured, just as thermodynamic measures can, but they directly reflect what consumers are

actually requiring in terms of an end-use service, and can readily be compared. However, their use in energy economics analysis appears to be very limited and they are partial productivity indicators for energy that consider only physical measures and do not consider all factors of production.

- *Economic indicators* measure both the outputs and the inputs purely in terms of monetary values. However, Patterson (1996) states that 'these types of indicators have not to date been developed for monitoring energy efficiency' (p. 383).
- *Economic–thermodynamic indicators* are also hybrid in nature as the output is measured in terms of market prices (such as £s) while the energy input is measured in terms of conventional thermodynamic units. These indicators can be applied at product, sectoral or national levels of economic activity and for primary energy consumption and secondary final energy consumption.

However, one of the most often used ratios in energy analysis is the energy–GDP ratio, which is in fact the inverse of the economic–thermodynamic indicator of energy efficiency identified by Patterson (1996). This is the ratio of total national energy consumption to GDP (or GNP) and is normally calculated for annual data and is a measure of the energy intensity of an economy at the aggregate level.[10] A decrease in the energy intensity ratio signifies, on average, a reduction in energy requirements to generate a unit of national output. It is suggested that this measure is useful for cross-country comparisons as GDP is measured in a common unit. However, system boundary and measurement problems mean that the application of the energy–GDP ratio is of limited use; for example in many developing countries, non-commercial fuels such as biomass form a significant portion of their total energy consumption but reliable data are generally unavailable and these fuels are used with very low efficiencies compared to modern fuels. This leads to complications as to how to include them with modern energy sources. Indeed, different approaches used could lead to very different energy–GDP ratio estimates (Ang, 1987).[11]

Furthermore, GDP represents many diverse activities, whose energy intensities may differ widely, and consequently a structural change, which is unrelated to energy efficiency, can cause significant variations in the energy–GDP ratio over time regardless of changes in the energy intensities. As an energy efficiency indicator, the energy–GDP ratio may not be a good measure of the efficiency with which energy is used at the end-use level.

The above illustrates the different measures that conceptually could be utilized. However, the predominant measure in the energy economics literature is the inverse of what Patterson (1996) calls the economic–thermodynamic indicator, that is, the simple energy–GDP ratio – or energy intensity. One way in which this has been developed in an attempt to analyse energy efficiency is to 'decompose' the energy intensity ratio.

- The *index decomposition analysis (IDA)* derives an economy-wide energy efficiency indicator by aggregating the effects of energy intensity changes at energy end-use or subsector level (Ang and Zhang, 2000; Ang, 2004). This bottom–up framework for developing an economy-wide composite energy efficiency index has been widely employed on a sectoral basis and with some minor modification the technique can be applied to economy-wide analysis (Tedesco and Thorpe, 2003). However, it should be noted that this approach is not directly related to the microeconomic production theory.

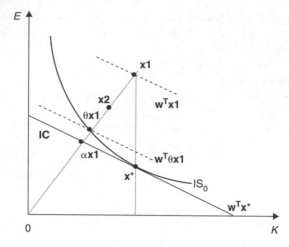

Figure 8.1   Productive efficiency

2.2   Productive Efficiency and Efficiency in the Use of Energy

In order to present the concept of 'energy efficiency' within the microeconomics theory of production, the definition of technical, allocative and cost efficiency or overall productive efficiency used here is based on the work of Farrell (1957). Within this theoretical framework, the expression 'energy efficiency' is imprecise. In fact, in order to reduce energy consumption for the production of a given output, the level of technical and/or allocative efficiency generally has to improve, which implies a change in the quantity and combination of the inputs it is possible to consume.[12] Figure 8.1 presents the situation of an economic agent that is using capital (K) and energy (E) in order to produce an energy service or an output (y).[13] The situation is illustrated using an isoquant (IS_0) and some isocost (IC) lines. A technically efficient economic agent uses combinations of E and K that lie on the isoquant IS_0.

If an economic agent uses quantities of inputs defined by point **x1** in Figure 8.1, it is technically inefficient as the point lies above IS_0. The technical inefficiency of the economic agent is represented by the distance between points **x1** and θ**x1**, which is the amount by which all inputs could be proportionally reduced without a reduction in output. Technical efficiency θ can be expressed as the ratio between the distance from the origin to technically efficient input vector θ**x1** and the distance from the origin to input vector **x1**.

If the input price ratio, represented by the slope of isocost line w^T**x1**, is known, then a cost-efficient input combination can be identified. An economic agent that uses a cost-minimizing input vector is presented by point **x***, where isocost line w^T**x*** is a tangent to input isoquant IS_0. Thus the minimum costs that can be achieved for the production of a given output y are w^T**x***. From Figure 8.1 the economic agent operating at θ**x1** is technically efficient but allocatively inefficient since it operates with higher costs (isocost line $w^T\theta$**x1** lies above the line w^T**x***). The distance between α**x1** and θ**x1** measures the allocative inefficiency of the economic agent. The allocative efficiency is defined as the ratio between the distance from the origin to α**x1** and the distance from the origin to θ**x1**, whereas the total cost efficiency α can be calculated as the ratio between the distance from the origin to αx1 and the distance from the origin to x1. To reach the optimal input combination and thus become cost-efficient, the economic agent would have to change its relative input use in the direction of increasing the use of input K and reducing the use of input E.

It is now possible to identify several cases that can improve the level of productive efficiency and, therefore, reduce the energy consumption by keeping the level of production of energy services or output constant:

- *Case A*   The economic agent is producing the level of energy service y using the input combination θx1. In this case, the economic agent could improve the level of allocative efficiency by moving to the optimal inputs combination x*. In this case the energy consumption will decrease; that is, energy is substituted with capital, allowing the economic agent to consume less energy. In order for the economic agent to reach x* it improves the use of capital stock, for example installing a device on a cooling system to improve the function of the system.

- *Case B*   The economic agent is producing the level of energy service y using the input combination x1. In this case, the economic agent could improve the level of overall productive efficiency by moving to the optimal inputs combination x*. In this case the energy consumption will decrease and there is no substitution with capital. So, for example, a household optimizes the amount of time that the windows in the house are opened during the day so as to reach x* or a firm optimizes the use of a cooling system. This is a special case and reflects the concept of input-specific technical efficiency introduced by Kopp (1981), defined as the ratio of minimum feasible to observed use of energy, conditional on the production technology and the observed levels of outputs and other inputs.

- *Case C*   The economic agent is producing the level of energy service y using the input combination x2. In this case, the economic agent could improve the level of overall productive efficiency by moving to the optimal input combination x*. In this case the energy consumption will decrease, as energy is substituted with capital, allowing the economic agent to consume less energy. This occurs when a household or a firm improves the insulation of the building in order to reach x*.

When technological change allows the economic agent to produce the same level of the energy service y, with less energy and capital, such technical progress shifts the isoquant.[14] For instance, this occurs when the temperature of rooms in a house is maintained at say 20°C, with less energy and capital – maybe due to new insulation technology or a new heating system. In this case, the technological progress will move the isoquant, IS$_0$, to the left as depicted in Figure 8.2.[15] In this case, the amount of energy and capital used to produce the energy service has decreased and the economic agent reaches point xt1*. From this discussion it is clear that the level of energy consumption for the production of a predefined level of output can change over time because of an increase in the level of productive efficiency and/or technical progress.

In order to estimate the level of overall productive efficiency it is possible to use the frontier analysis approach that is based on the estimation of production, distance, cost and input demand frontier functions using both parametric and non-parametric

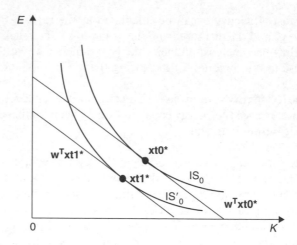

Figure 8.2   Technical progress

approaches. The non-parametric approaches include data envelopment analysis (DEA) and the free disposal hull (FDH). In these approaches the cost frontier is considered as a deterministic function of the observed variables. These methods are non-parametric in that they do not impose any specific functional form or distribution assumption. Apart from a few exceptions, all parametric methods have a stochastic element in their frontier function. Thus this group of methods is also labelled stochastic frontier analysis (SFA). The main exception to a deterministic frontier is the COLS (corrected ordinary least squares) method.

From a theoretical point of view, the estimation of a production frontier function or a frontier distance function allows the estimation of the level of technical efficiency, while the estimation of a cost frontier function allows the estimation of the level of cost efficiency or overall productive efficiency. It is also possible to estimate simultaneous-equation cost frontier models, that is, to estimate the cost frontier function together with the input demand frontier functions. In this case, as illustrated by Schmidt and Lovell (1979), the self-duality of the Cobb–Douglas functional form is used to derive a system of log-log stochastic cost-minimizing input demand equations which are also known as input frontier equations. In this context, actual input demands differ from the input frontier demands due to the presence of both allocative and technical inefficiency.[16]

Estimation of production and distance functions for energy services requires information on outputs, labour, capital and energy, whereas for the estimation of the cost and input demand functions, information on outputs and input prices is required. Sometimes, when there is a lack of data mainly on some inputs or input prices, it is also possible to estimate just one input frontier demand function, for instance the energy demand frontier function. This approach, which can be considered an *ad hoc* approach, does not completely consider the theoretical restriction imposed by the production theory; however, it does allow, in an approximate way, the estimation of the difference between the actual input demand and the input frontier demand. In order to consider the effect of technological change on production, cost or input demands, all frontier models can be estimated by introducing a time trend or a set of time dummy variables. Generally, these variables capture the shift in the frontier functions due to change in the technology.

A new strand of literature has developed where the stochastic frontier approach (SFA) is employed to try to estimate energy efficiency across OECD countries and US states (Filippini and Hunt, 2011, 2012; Zhou et al., 2012). The analysis by Filippini and Hunt (2011, 2012) is based upon a frontier energy demand relationship, where an *ad hoc* approach is used to estimate the difference between the actual energy demand and the frontier energy demand. In contrast, Zhou et al. (2012) estimate energy-specific technical efficiency using a frontier distance function with the economy-wide energy efficiency performance measured in 2001 (cross-sectional analysis) for a sample of 21 OECD countries.

These new developments for assessing energy efficiency using parametric frontier techniques are interesting and are therefore discussed further below. In particular, the frontier aggregate demand approach advocated by Filippini and Hunt (2011) is outlined in the next section, given the attempt to estimate possible energy and CO_2 emissions savings that might be achieved by the more efficient energy consumption of OECD countries.

3. A PANEL 'FRONTIER' WHOLE-ECONOMY AGGREGATE ENERGY DEMAND FUNCTION USING PARAMETRIC STOCHASTIC FRONTIER ANALYSIS

Filippini and Hunt (2011) use a parametric frontier approach to estimate an energy demand frontier function in an attempt to isolate 'underlying energy efficiency'. This is done by explicitly controlling for income and price effects, country-specific effects, climate effects and a common underlying energy demand trend (the UEDT), capturing 'exogenous' technical progress and other exogenous factors. Hence it allows for the impact of 'endogenous' technical progress through the price effect and 'exogenous' technical progress (and other factors) through the UEDT.

As stated above, their aim, and the aim here, is to analyse economy-wide energy efficiency; hence the estimated model introduced below is for aggregate energy consumption for the whole economy. Economy-wide aggregate energy demand is derived from the demand for energy services such as heat, illumination, cooked food, hot water, transport services, manufacturing processes and so on. To produce the desired services it is generally necessary to use a combination of energy fuels and capital equipment, including household appliances, cars, insulated walls and machinery. This implies that the demand for energy is influenced by the level of energy efficiency of the equipment and, generally, of the production process. For instance, some relatively new equipment and production processes are able to provide the same level of services and products using less energy than old equipment. This comes from R&D that improves the thermodynamic efficiency of appliances and the capital stock, as well as production processes – there is a technical improvement. Of course, in reality, apart from the technological and economic factors, there is a range of exogenous institutional and regulatory factors that are important in explaining the level of energy consumption; furthermore, these exogenous changes are unlikely to have an impact at a consistent rate over time. Hence it is important that the UEDT is specified in such a way that it is 'non-linear' and could increase and/or decrease over the estimation period, as advocated by Hunt et al. (2003a, 2003b). Therefore, given that a panel data set is used, this is achieved by time dummies as proposed by Griffin and Schulman (2005) and Adeyemi and Hunt (2007).[17]

In order to try to tease out these different influences, a general energy demand relationship found in the standard energy demand modelling literature, relating energy consumption to economic activity and the real energy price, is utilized for the estimation of an aggregate energy demand function for a panel of OECD countries. Moreover, in order to control for other important factors that vary across countries and hence can affect a country's energy demand, some variables related to the climate, size and structure of the economy are introduced into the model. Thus the framework adopted here attempts to isolate the 'underlying energy efficiency' for each country after controlling for income, price, climate effects, technical progress and other exogenous factors, as well as effects due to difference in area size and in the structure of the economy.[18] The estimated model therefore isolates the level of underlying energy efficiency, defined with respect to a benchmark, for example a best-practice economy in the use of energy by estimating a 'common energy demand' function across countries, with homogeneous income and price elasticities, and responses to other factors, plus a homogeneous UEDT. This is seen as important, given the need to isolate the different underlying energy efficiency across the countries.[19] Consequently, once these effects are adequately controlled for, it allows for the estimation of the underlying energy efficiency for each country showing (i) how efficiency has changed over the estimation period and (ii) the differences in efficiency across the panel of countries.

In the case of an aggregate energy demand function, the frontier gives the minimum level of energy necessary for an economy to produce any given level of goods and services. The distance from the frontier measures the level of energy consumption above the baseline demand, that is, the level of energy inefficiency. Energy efficiency measures the ability of a country to minimize the energy consumption given a level of GDP.

3.1 An Aggregate Frontier Energy Demand Model

Given the discussion above, and the model proposed by Filippini and Hunt (2011), it is assumed that there exists an aggregate energy demand relationship for a panel of OECD countries, as follows:

$$E_{it} = E(P_{it}, Y_{it}, POP_{it}, DCOLD_i, DARID_i, A_i, ISH_{it}, SSH_{it}, D_t, EF_{it}) \qquad (8.1)$$

where E_{it} is aggregate energy consumption, Y_{it} is GDP, P_{it} is the real price of energy, $DCOLD_i$ is a cold climate dummy, $DARID_i$ is a hot climate dummy, POP_{it} is population, A_i is the area size, ISH_{it} is the share of value added of the industrial sector and SSH_{it} is the share of value added for the service sector all for country i in year t.[20] In contrast to the model estimated by Filippini and Hunt (2011), equation (8.1) includes an extra dummy variable for extreme high temperatures $(DARID_i)$. D_t is a variable representing the UEDT that captures the common impact of important unmeasured exogenous factors that influence all countries. Finally, EF_{it} is the unobserved level of 'underlying energy efficiency' of an economy. This could incorporate a number of factors that will differ across countries, including different government regulations as well as different social behaviours, norms, lifestyles and values. Hence a low level of underlying energy efficiency implies an inefficient use of energy (i.e. 'waste energy'), so that, in this situation, awareness of energy conservation could be increased in order to reach the 'optimal' energy demand function. Nevertheless, from an empirical perspective, when using OECD aggregate energy data, the aggregate level of energy efficiency of the capital equipment and of the production processes is not observed directly. Therefore this underlying energy efficiency indicator needs to be estimated. Consequently, in order to estimate this economy-wide level of underlying energy efficiency (EF_{it}) and identify the best-practice economy in term of energy utilization, the stochastic frontier function approach introduced by Aigner et al. (1977) is used.[21]

A cost frontier function gives the minimum level of cost attainable by a firm or a household for any given level of output. For an input demand function (as utilized here) the frontier gives the minimum level of input used by an economic agent for any given level of output; hence the difference between the observed input and the cost-minimizing input demand represents, as mentioned before, both technical as well as allocative inefficiency. In the case of an aggregate energy demand function, used here, the frontier gives the minimum level of energy necessary for an economy to produce any given level of energy services. In principle, the aim is to apply the frontier function concept in order to estimate the baseline energy demand, which is the frontier that reflects the demand of the countries, that is, firms and households that utilize highly efficient equipment and manage production process efficiently.[22]

3.2 Econometric Specification

This frontier approach allows the possibility to identify if a country is, or is not, on the frontier. Moreover, if a country is not on the frontier, the distance from the frontier measures the level of energy consumption above the baseline demand, that is, the level of energy inefficiency. The approach used in this study is therefore based on the assumption that the level of the economy-wide energy efficiency can be approximated by a one-sided non-negative term, so that a panel log-log functional form of equation (8.1) adopting the stochastic frontier function approach proposed by Aigner et al. (1977) can be specified as follows:

$$e_{it} = \alpha + \alpha^y y_{it} + \alpha^p p_{it} + \alpha^{pop} pop_{it} + \delta_t D_t + \alpha^C DCOLD_i + \alpha^R DARID_i$$

$$+ \alpha^a a_i + \alpha^I ISH_{it} + \alpha^S SSH_{it} + v_{it} + u_{it} \qquad (8.2)$$

where e_{it} is the natural logarithm of aggregate energy consumption (E_{it}), y_{it} is the natural logarithm of GDP (Y_{it}), p_{it} is the natural logarithm of the real price of energy (P_{it}), pop_{it} is the natural logarithm of population (POP_{it}), $DCOLD_i$ is a cold climate dummy variable, $DARID_i$ is a hot climate dummy, a_i is the natural logarithm of the area size of a country (A_i),[23] ISH_{it} is the share of value added of the industrial sector, SSH_{it} is the share of value added for the service sector and D_t is a series of time dummy variables.

Furthermore, the error term in equation (8.2) is composed of two independent parts. The first part, v_{it}, is a symmetric disturbance capturing the effect of noise and

Table 8.1   Descriptive statistics

Variable		Mean	Std dev.	Min.	Max.
Description	Name				
Energy consumption (toe)	E	117472	260780	2214	1581622
GDP (1000 US$2000 PPP)	Y	832.26	1571.90	8.56	11693.2
Population in millions	POP	38.40	53.52	0.36	301.75
Real price of energy (2000 = 100)	P	89.08	15.89	12.63	149.33
Area size in km²	A	1241662	2755333	2590	9984670
Share of industrial sector as % of GDP	ISH	31.2	5.35	15.40	46.20
Share of service sector as % of GDP	SSH	64.25	6.81	45.40	84.30
Climate dummy	$DCOLD$	0.30	0.46	0	1
Arid country dummy	$DARID$	0.30	0.46	0	1

as usual is assumed to be normally distributed. The second part, u_{it}, which represents the underlying energy level of efficiency EF_{it} in equation (8.1), is interpreted as an indicator of the inefficient use of energy, for example, the 'waste energy'.[24] It is a one-sided non-negative random disturbance term that can vary over time, assumed here to follow a half-normal distribution.[25] An improvement in the energy efficiency of the equipment or in the use of energy through a new production process will increase the level of energy efficiency of a country. The impact of technological, organizational, and social innovation in the production and consumption of energy services on the energy demand is therefore captured in several ways: the time dummy variables, the indicator of energy efficiency and through the price effect. In summary, this is a slightly modified version of equation (2) in Filippini and Hunt (2011) which is estimated in order to estimate the underlying energy efficiency for each country in the sample. The data and the econometric specification of the estimated equations are discussed in the next subsection.

3.3   Data

The study is based on an unbalanced panel data set for a sample of 29 OECD countries ($i = 1, \ldots 29$)[26] over the period 1978 to 2008 ($t = 1978–2008$). This data set is based on information taken from the International Energy Agency database (IEA, 2010b) and from the general OECD database 'Country profile statistics' available at www.oecd.org. E is each country's aggregate energy consumption in thousand tonnes of oil equivalent (ktoe), Y is each country's GDP in billion US$ 2000 PPP, P is each country's index of real energy prices (2000 = 100), and POP is each country's population in millions. The climate dummy variables, $DCOLD$ and $DARID$, indicate whether a country's climate might best be characterized as cold or hot respectively (according to the Köppen–Geiger climate classification[27]) and A is the area size of a country measured in squared kilometres. Finally, the value added of the industrial and service sectors is measured as a percentage of GDP (ISH and SSH). Descriptive statistics of the key variables are presented in Table 8.1.

3.4 Panel Data Estimation

From the econometric specification perspective, the literature on the estimation of stochastic frontier models using panel data needs to be considered.[28] A first approach that can be used for the estimation of model (8.2) is the panel data version of the Aigner et al. (1977) half-normal model proposed by Pitt and Lee (1981). In this 'pooled' model specification the error term is composed of two uncorrelated parts: the first part, u_{it}, is a one-sided non-negative disturbance reflecting the effect of inefficiency (including both allocative and technical inefficiencies), and the second component, v_{it}, is a symmetric disturbance capturing the effect of noise. Usually the statistical noise is assumed to be normally distributed, while the inefficiency term u_{it} is assumed to follow a half-normal distribution. A shortcoming of this model is that the unobserved, time-invariant, country-specific heterogeneity is not directly considered in the estimation. Therefore this approach can suffer from the 'unobserved variables bias', because the unobserved characteristics may not be distributed independently of the explanatory variables. This heterogeneity bias can be reduced to some extent by introducing several explanatory variables and by considering a relatively long period. A second approach, also proposed by Pitt and Lee (1981), assumes the inefficiency effects, u_i, to be constant over time.[29] A major shortcoming of this model is that any unobserved, time-invariant, country-specific heterogeneity is considered as inefficiency.[30] Moreover, the level of efficiency does not vary over time. Therefore this approach is not ideal for an analysis that considers a relatively long period. In order to solve this problem using panel data, Greene (2005a, 2005b) extended the SFA panel data version of Aigner et al. (1977), also employed by Pitt and Lee (1981), by adding a fixed or random individual effect to the model.[31] It should be noted that these models produce efficiency estimates that do not include the persistent inefficiencies that might remain more or less constant over time. In fact, the time-invariant, country-specific energy inefficiency is captured by the individual random or fixed effects. Therefore, to the extent that there are certain sources of energy inefficiency that result in time-invariant excess energy consumption, the estimates of these models could provide relatively high and imprecise levels of energy efficiency. Of course, one advantage of the approaches proposed by Greene (2005a, 2005b) with respect to the panel data version of Aigner et al. (1977) is the reduction of the potential so-called 'unobserved variables bias' – for example a situation where correlation between observables and unobservables could bias some coefficients of the explanatory variables. However, by introducing several explanatory variables such as the climate, the area size, population and some variables on the structure of the economy it is possible to reduce this problem, at least in part. In summary the choice of model is not straightforward. However, given the relatively long period considered here, it could be assumed that in each country part of the inefficiency changed over time, whereas part of the inefficiency remained constant over time. For this reason the Aigner et al. (1977) model is believed to be the most suitable for the estimation of equation (8.2).[32]

3.5 Efficiency Estimation

The country's efficiency is estimated using the conditional mean of the efficiency term $E[\langle u_{it}|u_{it} + v_{it}\rangle]$, proposed by Jondrow et al. (1982).[33] The level of energy efficiency can be expressed in the following way:

Table 8.2   Estimated frontier energy demand for 29 countries (1978–2008)

Coefficient	Estimate (*t*-values in parentheses)
Constant (α)	1.835
	(3.11)
α^y	0.807
	(27.46)
α^p	−0.245
	(−5.97)
α^{pop}	0.109
	(4.03)
α^a	0.074
	(12.70)
α^C	0.083
	(4.16)
α^R	−0.311
	(−16.07)
α^I	0.028
	(8.78)
α^S	0.028
	(8.46)
Time dummies	Yes
Lamda (λ)	0.723
	(7.79)

$$EF_{it} = \frac{E_{it}^F}{E_{it}} = \exp(-\hat{u}_{it}) \qquad (8.3)$$

where E_{it} is the observed energy consumption and E_{it}^F is the frontier or minimum demand of the ith country in time t. An energy efficiency score of one indicates a country on the frontier (100 per cent efficient); while non-frontier countries, for example countries characterized by a level of energy efficiency lower than 100 per cent, receive scores below one. This therefore gives the measure of underlying energy efficiency estimated below.[34]

This approach therefore proceeds by estimating equation (8.2) and then utilizing equation (8.3) to estimate the efficiency scores for each country for each year. Given the econometric specifications discussed above, and given the approach chosen to take into account the effect of the impact of exogenous technical change and other exogenous factors, a pooled frontier energy demand model can be estimated. The discussion of the results are given in the next subsection.

3.6   Estimation Results

The estimation results for the frontier energy demand model, equation (8.2), are given in Table 8.2.[35] This shows that the estimated coefficients (alphas) and lambda (λ) have the expected signs and are statistically significant.[36]

For the variables in logarithmic form, the estimated coefficients can be directly interpreted as elasticities. The estimated income elasticity and the estimated own price elastic-

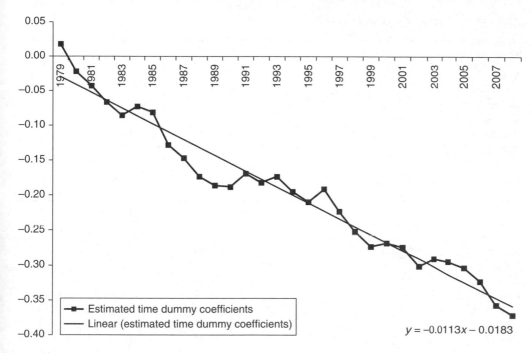

Figure 8.3 Estimated time dummy coefficients (relative to 1978)

ity are about 0.8 and −0.25 respectively. These values are not too out of line with previous estimates. The estimated population elasticity is about 0.1 and the estimated area elasticity is about 0.07. The estimated coefficients for the climate variables, DCOLD and DARID, are 0.08 and −0.31 respectively. These coefficients indicate that, as expected climate has an important influence on a country's energy demand. Further, the impact of high temperatures is larger than the impact of low temperatures. Similarly, larger shares of a country's industrial and service sectors will also increase energy consumption.

The time dummies in the pooled model, as a group, are significant and, as expected, the overall trend in their coefficients is negative, as shown in Figure 8.3; however, they do not fall continually over the estimation period, reflecting the 'non-linear' impact of technical progress and other exogenous variables.

3.7 Underlying Energy Efficiency Estimates

Table 8.3 provides descriptive statistics for the overall underlying energy efficiency estimates of the countries obtained from the econometric estimation, showing that the estimated mean average efficiency is about 91 per cent (median 91 per cent); nonetheless, there is a fair degree of variation around the average. Table 8.4 presents the average energy efficiency score for every country over the whole sample with their ranking and Figure 8.4 illustrates the rankings by reordering according to the estimated efficiency. These show that, when judged over the whole period the most efficient country was Switzerland followed by Denmark. At the other end of the spectrum the two countries

Table 8.3   Energy efficiency scores

Summary measure	
Min	0.740
Max	0.953
Mean	0.906
Median	0.909
Std dev.	0.027

Table 8.4   Estimated average energy efficiency scores and rankings (1978–2008)

Country	Score	Rank
Australia	0.898	18
Austria	0.932	4
Belgium	0.891	21
Canada	0.879	26
Czech Rep.	0.881	25
Denmark	0.934	2
Finland	0.879	27
France	0.927	7
Germany	0.911	15
Greece	0.917	11
Hungary	0.896	19
Ireland	0.921	10
Italy	0.913	14
Japan	0.883	24
Korea	0.895	20
Luxembourg	0.848	29
Mexico	0.929	6
Netherlands	0.888	22
New Zealand	0.926	8
Norway	0.933	3
Poland	0.899	17
Portugal	0.903	16
Slovak Rep.	0.884	23
Spain	0.914	12
Sweden	0.923	9
Switzerland	0.950	1
Turkey	0.913	13
UK	0.931	5
USA	0.875	28

Note: A score of 1 indicates that a country is energy efficient (i.e. it is on the energy demand frontier) and a score below 1 indicates the degree to which the country is inefficient.

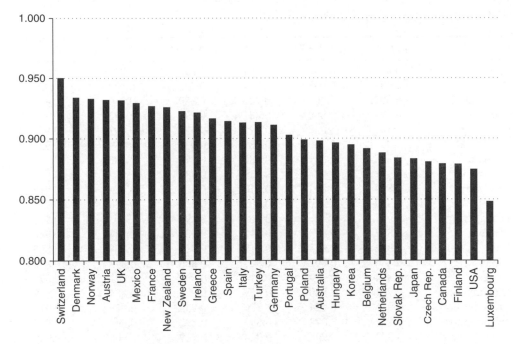

Figure 8.4   Estimated average underlying energy efficiency (1978–2008)

found to be the relatively least efficient over the whole period are Luxembourg and the USA.[37]

Table 8.5 presents the average energy efficiency score for every country for six sub-periods of the estimation period considered in the analysis. Figure 8.5 shows the estimated underlying energy efficiency scores for each country over the estimation period (relative to energy intensity). It should be noted that, although presented individually for each country, the estimated efficiencies of each country should not be taken as the precise position of each country given the stochastic technique used in estimation. However, they do give a good relative indication of a country's change in efficiency over time and a country's relative position *vis-à-vis* other countries. Bearing this in mind, Table 8.5 and Figure 8.5 show that the estimated underlying energy efficiency generally increased over the whole estimation period for some countries, such as Denmark, Finland, France, Ireland, Luxembourg, Netherlands, Norway, Sweden, the UK and the USA.[38] However, for some countries the opposite is the case, with the estimated underlying energy efficiency generally decreasing, such as Italy, Japan, Mexico, New Zealand, Portugal, Spain and Turkey. Moreover, taking Italy as an example, it is the third-least energy-intensive country over the period 1978 to 2008 with its energy to GDP ratio generally falling throughout the period, suggesting that it has been becoming more 'efficient'. However, the estimated measure of underlying energy efficiency advocated here suggests that in fact Italy has become less efficient over the period, highlighting that energy intensity gives a poor indication of Italy's change in energy efficiency over time.

Figure 8.5 also illustrates that the estimated underlying energy efficiency would appear to be negatively correlated with energy intensity for most countries (i.e. the level

Table 8.5   Average energy efficiency scores

	1978–82	1983–87	1988–92	1993–97	1998–2002	2003–07
Australia	0.900	0.899	0.897	0.898	0.895	0.899
Austria	0.934	0.933	0.934	0.938	0.933	0.923
Belgium	0.897	0.901	0.903	0.889	0.880	0.882
Canada	0.877	0.880	0.881	0.878	0.882	0.875
Czech Rep.	n/a	n/a	0.827	0.937	0.892	0.897
Denmark	0.922	0.930	0.935	0.937	0.939	0.939
Finland	0.865	0.881	0.882	0.879	0.885	0.881
France	0.921	0.924	0.929	0.930	0.927	0.929
Germany	0.894	0.893	0.910	0.925	0.925	0.917
Greece	0.939	0.920	0.907	0.912	0.908	0.913
Hungary	n/a	n/a	0.863	0.890	0.903	0.909
Ireland	0.897	0.899	0.911	0.930	0.943	0.948
Italy	0.922	0.923	0.919	0.917	0.907	0.895
Japan	0.886	0.892	0.891	0.886	0.873	0.872
Korea	0.894	0.917	0.914	0.892	0.873	0.878
Luxembourg	0.761	0.840	0.859	0.873	0.884	0.868
Mexico	0.946	0.924	0.926	0.933	0.929	0.918
Netherlands	0.873	0.872	0.893	0.897	0.901	0.892
New Zealand	0.943	0.940	0.924	0.919	0.906	0.920
Norway	0.932	0.930	0.930	0.935	0.935	0.936
Poland	n/a	n/a	0.858	0.869	0.914	0.918
Portugal	0.929	0.912	0.903	0.902	0.891	0.883
Slovak Rep.	n/a	n/a	n/a	0.868	0.878	0.900
Spain	0.932	0.930	0.922	0.915	0.898	0.892
Sweden	0.919	0.923	0.924	0.917	0.923	0.930
Switzerland	n/a	n/a	0.951	0.952	0.950	0.948
Turkey	0.943	0.923	0.921	0.915	0.901	0.894
UK	0.923	0.926	0.929	0.931	0.936	0.941
USA	0.834	0.861	0.877	0.891	0.895	0.889

Note: n/a represents the situation where the average is not available over the sub-period. Due to the unbalanced panel, some averages are calculated over a slightly shorter period than indicated.

of energy intensity decreases with an increase of the level of energy efficiency), but with some exceptions (discussed further below). This is to be expected in one sense. However, for this technique to be a useful procedure for extracting the underlying energy efficiency, a perfect, or even near perfect, negative correlation would not be expected since all the useful information would be contained in the standard energy to GDP ratio. This is confirmed, given that the average correlation coefficient between the estimated underlying energy efficiency and energy intensity across all countries is −0.48. Within this, there is a relatively high negative correlation for some countries, such as the Czech Republic, Denmark, Germany, Greece, Hungary, Ireland, Luxembourg, the Netherlands, New Zealand, Poland, Portugal, Slovak Republic, the UK and the USA, whereas for some countries the (negative) correlation is somewhat less, such as Canada, Finland, France, Korea, Norway, Spain and Sweden. Furthermore, for Australia, Austria, Belgium, Italy,

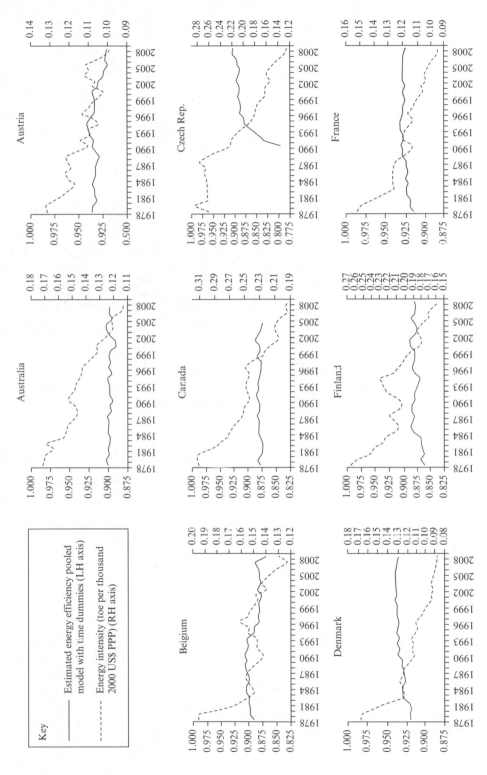

Figure 8.5   Comparison of estimated underlying energy efficiency with energy intensity

193

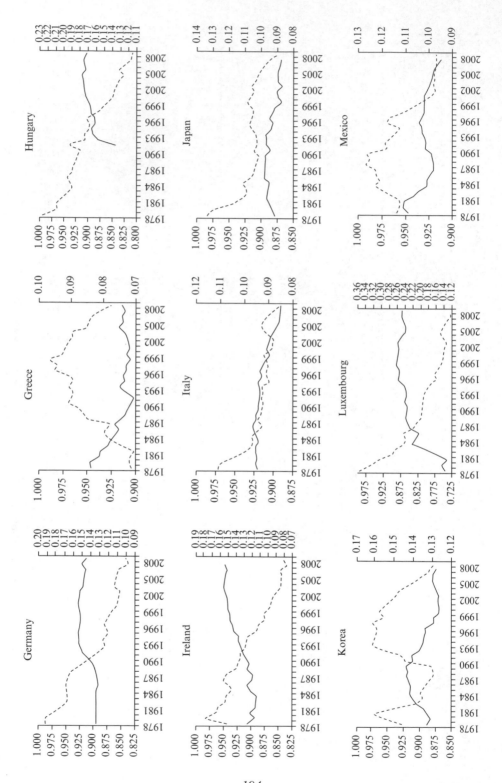

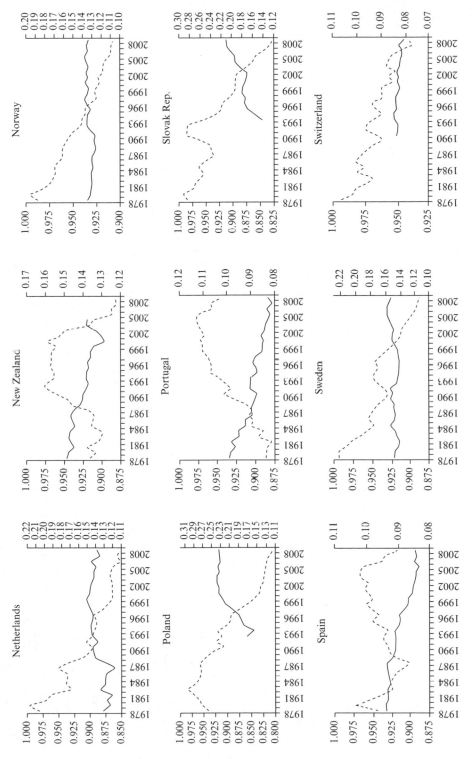

Figure 8.5 (continued)

195

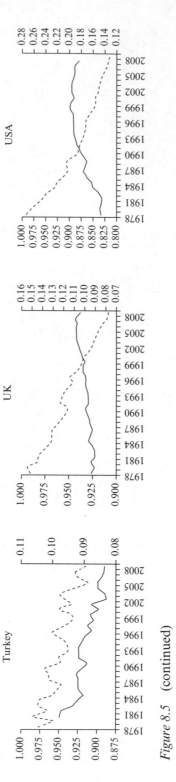

Figure 8.5 (continued)

196

Table 8.6   *Comparison of the rankings for average estimated underlying energy efficiency and energy intensity (2003–07)*

	Estimated underlying energy efficiency ($\overline{EF_i}$)		Energy intensity (energy–GDP ratio, toe per 1000 US$2000 PPP)	
	Level	Rank	Level	Rank
Australia	0.899	16	0.118	16
Austria	0.923	8	0.108	12
Belgium	0.882	24	0.134	22
Canada	0.875	27	0.203	29
Czech Rep.	0.897	17	0.148	26
Denmark	0.939	4	0.091	5
Finland	0.881	25	0.173	28
France	0.929	7	0.101	11
Germany	0.917	12	0.109	13
Greece	0.914	13	0.085	3
Hungary	0.909	14	0.126	19
Ireland	0.948	2	0.083	1
Italy	0.895	18	0.091	6
Japan	0.872	28	0.100	10
Korea	0.878	26	0.139	23
Luxembourg	0.868	29	0.139	23
Mexico	0.918	10	0.097	8
Netherlands	0.892	20	0.119	17
New Zealand	0.920	9	0.129	20
Norway	0.936	5	0.114	15
Poland	0.918	11	0.130	21
Portugal	0.883	23	0.110	14
Slovak Rep.	0.900	15	0.150	27
Spain	0.892	21	0.099	9
Sweden	0.930	6	0.124	18
Switzerland	0.948	1	0.084	2
Turkey	0.894	19	0.093	7
UK	0.941	3	0.085	3
USA	0.889	22	0.141	25

Note: A rank of 29 for underlying energy efficiency represents the least efficient country by this measure, whereas a rank of 1 represents the most efficient country. A rank of 29 for energy intensity represents the most energy-intensive country whereas a rank of 1 represents the least energy-intensive country.

Japan, Mexico, Switzerland and Turkey, there appears to be a positive relationship between the energy to GDP ratio and estimated energy efficiency. This suggests that for some countries energy intensity is a reasonable proxy for energy efficiency, whereas for others it is a very poor proxy. Consequently, unless the kind of analysis advocated here is conducted it is not possible to identify for which countries energy intensity is a good proxy and for which it is a poor proxy.

Focusing now on the 2003–07 period, Table 8.6 compares the estimated underlying energy efficiency with energy intensity. This shows that Luxembourg, Japan, Canada

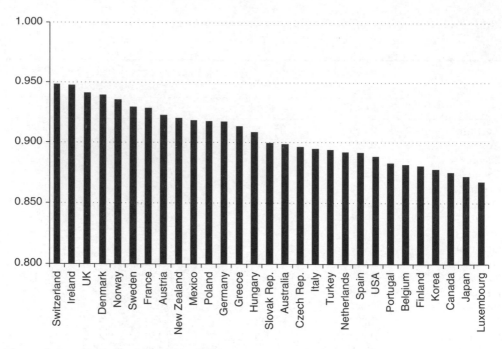

Figure 8.6a   Estimated average underlying energy efficiency (2003–07)

and Korea are estimated to be the four least efficient countries, with Switzerland, Ireland, the UK and Denmark estimated to be the four most efficient countries.[39] This is further shown in Figure 8.6a, with the countries reordered from the most efficient to the least efficient. However, Table 8.6 also shows that Ireland, Switzerland, Greece and the UK are the least energy intensive whereas Canada, Finland, the Slovak Republic and the Czech Republic are the most energy intensive; this is further shown in Figure 8.6b, with the countries reordered from the least energy intensive to the most energy intensive. Thus, although there would appear to be a generally negative relationship between the rankings of the estimated underlying energy efficiency and energy intensity, there is not a one-to-one correspondence. For example, for the period 2003–07 according to the energy intensity measure, Italy, Turkey and Japan are ranked 6th, 7th and 10th respectively whereas they are estimated to be 18th, 19th and 28th respectively according to the estimated energy efficiency measure; thus for these countries the simple energy intensity ratio would appear to overestimate their relative efficiency position. On the other hand, according to the simple energy intensity ratio Sweden, New Zealand, Poland and the Slovak Republic are ranked 18th, 20th, 21st and 27th respectively, whereas they are estimated to be 6th, 9th, 11th and 15th respectively according to the estimated energy efficiency measure; thus for these countries the simple energy intensity ratio would appear to underestimate their relative efficiency position. This relationship between the two measures is further illustrated in Figure 8.7.[40]

The discussion above illustrates the importance of attempting to adequately define and model 'energy efficiency' rather than just relying on the simple energy to GDP

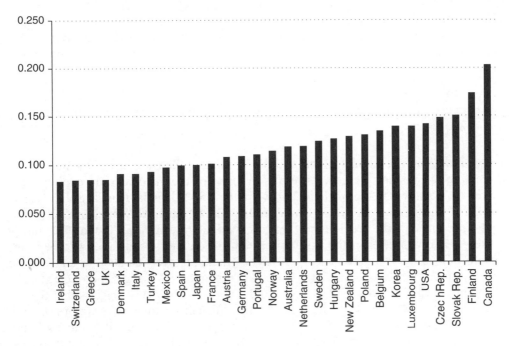

Figure 8.6b   Energy intensity (toe per US$1000 PPP, 2003–07)

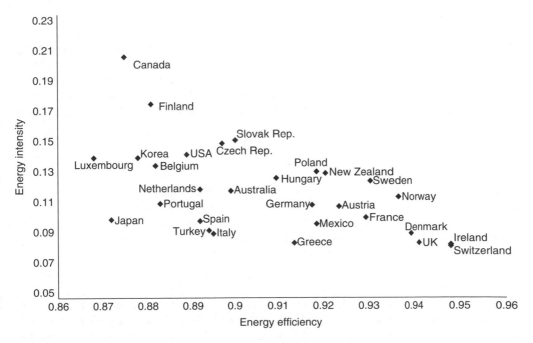

Figure 8.7   Energy efficiency and energy intensity, 2003–07

ratio – energy intensity. Furthermore, the estimated levels of efficiency give an indication of the possible savings in energy consumption that countries could make if they were all efficient. This is further analysed, along with the potential CO_2 savings, in the next section.

4.   THE CONTRIBUTION OF ENERGY EFFICIENCY TO CO_2 TARGETS

Before analysing the impact of the estimated energy efficiency of the countries, it is interesting to consider the general results in terms of the major carbon emitters. The top ten CO_2 emitters in the world in 2007 were the USA, Canada, Russia, South Korea, Germany, Japan, the UK, Iran, China and India, as illustrated in Figure 8.8.[41] The results presented in the previous section reveal an interesting result in that four out of the six OECD countries in the world's top ten carbon emitters are found to be in the bottom half of the estimated energy efficiency rankings over the 2003 to 2007 period with Japan at 28th, Canada at 27th, Korea at 26th and the USA at 22nd (see Table 8.6 and Figure 8.6a); whereas the UK is 3rd and Germany 12th. This suggests that there is some scope for CO_2 savings from more energy-efficient behaviour. Therefore the efficiency measures calculated in Section 3 are employed to determine the contribution of energy efficiency alone to the reduction of CO_2 emissions by the OECD countries so as to contribute towards national Kyoto targets. The next subsection sets out how this is achieved.

4.1   Method for Calculating CO_2 Savings

In order to determine the impact of improvements in energy efficiency of each country on their respective emissions, a CO_2 coefficient is constructed which, when multiplied by the level of energy demand assuming the country is efficient, gives an estimate of the emissions that might be saved if each country were on the efficient frontier. Therefore, λ_i, the average CO_2 coefficient for country i over the period 2003 to 2007, is calculated as follows:

$$\lambda_i = \frac{\overline{CO2}_i}{\overline{E}_i} \tag{8.4}$$

where $\overline{CO2}_i$ represents average CO_2 emissions for country i over the period 2003 to 2007 and $\overline{E}_i$ average energy consumption for country i over the period 2003 to 2007.

The notional energy consumption for each country i that would be consumed if it were operating efficiently ($\overline{E}_i^*$) is therefore estimated by:

$$\overline{E}_i^* = \overline{E}_i \times \overline{EF}_i \tag{8.5}$$

where $\overline{EF}_i$ is the average level of energy efficiency for country i over the period 2003 to 2007 estimated in Section 3 above (see Table 8.6). This is then used to estimate the amount of notional CO_2 for each country i that would be produced if it were operating efficiently ($\overline{CO2}_i^*$) given by:

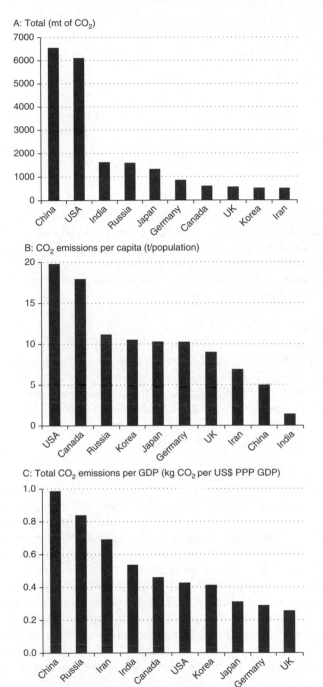

Source: UNSD (2009).

Figure 8.8 Top ten world CO₂ emitters, 2007

$$\overline{CO2}_i^* = \lambda_i * \overline{E}_i^* \tag{8.6}$$

Thus the average potential savings of energy and CO_2 for each country if it were being energy efficient is given by:

$$Esav_i = \overline{E}_i - \overline{E}_i^* \tag{8.7}$$

and

$$CO2sav_i = \overline{CO2}_i - \overline{CO2}_i^* \tag{8.8}$$

respectively.

Finally, for each of the OECD nations that are party to the Kyoto Agreement, the implications of consuming energy efficiently on achieving their emissions targets are considered. This is achieved by comparing the Kyoto target level of CO_2 emissions (which varies between countries for the period) with the actual level of CO_2 emissions and the estimated level of notional CO_2 emissions that would have been produced had the annual aggregate energy consumption been efficient.

4.2   Additional Data

The average energy consumption and the average energy efficiency estimates for each country over the period 2003–2007 obtained in Section 3 are used along with each country's CO_2 emissions obtained from fuel consumption (sectoral approach mt of CO_2)[42] for 29 OECD countries between 2003 and 2007 from the IEA (2010c) database.

4.3   Estimated Potential Energy CO_2 Savings

The results from the above calculations are presented in Table 8.7, showing the predicted energy savings attributed to improvements in efficiency alone and the estimated potential CO_2 emissions reductions for each individual country.[43] Although presented individually, the estimated CO_2 emissions reduction of each country should not be taken as the precise value given the stochastic technique used in estimation of the level of efficiency. However, they do give a good approximation of the potential direction of a country's change in efficiency and CO_2 emissions over time.

Unsurprisingly, the countries with the relatively lower energy efficiency rankings are, broadly, among those that stand to make the most potential CO_2 savings were they to operate on their efficient energy demand frontiers as seen in Figures 8.9 and 8.10. It can be seen in Figure 8.9 that Korea, Germany, Canada, Japan and the USA, which are among the top ten world emitters, are among those countries with the largest estimated potential to consume less energy if they operated efficiently. That said, the ten countries shown in Figure 8.9 to have the estimated least potential to reduce energy consumption (Luxembourg, Ireland, Denmark, New Zealand, Switzerland, Slovak Rep., Norway, Hungary, Greece and Belgium) would collectively reduce their energy consumption by just under 28 mtoe, being more than the potential savings for Canada of about 25 mtoe. Furthermore, as Figure 8.10 illustrates, the ten countries with the

Table 8.7   Estimated potential emissions and energy savings if energy efficient (2003–07)

Country	Average energy cons. (ktoe)	Average CO$_2$ emissions mt	CO$_2$ coefficient (kt/toe)	Estimated notional energy cons. (ktoe)	Estimated notional CO$_2$ emissions mt	Estimated potential energy savings (ktoe)	Estimated potential CO$_2$ savings mt
i	$\overline{E_i}$	$\overline{CO2_i}$	λ_i	E_i^*	$\overline{CO2_i^*}$	$Esav_i$	$CO2sav_i$
Australia	73 537	380.54	0.005175	66 089	342.00	7 448	38.54
Austria	27 016	72.57	0.002686	24 930	66.97	2 086	5.60
Belgium	41 157	112.83	0.002742	36 285	99.48	4 871	13.35
Canada	200 696	556.62	0.002773	175 622	487.08	25 073	69.54
Czech Rep.	27 405	120.97	0.004414	24 570	108.46	2 834	12.51
Denmark	14 884	52.71	0.003541	13 982	49.52	901	3.19
Finland	26 380	65.23	0.002473	23 234	57.46	3 146	7.78
France	167 871	382.51	0.002279	155 888	355.20	11 983	27.30
Germany	241 188	824.25	0.003417	221 208	755.97	19 980	68.28
Greece	20 985	94.76	0.004515	19 169	86.56	1 816	8.20
Hungary	19 107	55.95	0.002928	17 361	50.84	1 746	5.11
Ireland	11 821	43.05	0.003642	11 200	40.79	621	2.26
Italy	139 417	452.23	0.003244	124 746	404.65	14 671	47.59
Japan	344 870	1218.76	0.003534	300 806	1063.05	44 063	155.72
Korea	141 005	470.31	0.003335	123 791	412.89	17 214	57.42
Luxembourg	3 902	10.77	0.002761	3 385	9.35	516	1.43
Mexico	106 175	386.60	0.003641	97 495	355	8 679	31.60
Netherlands	59 766	181.37	0.003035	53 309	161.78	6 456	19.59
New Zealand	12 420	33.02	0.002659	11 427	30.38	993	2.64
Norway	20 478	37.28	0.001820	19 158	34.88	1 320	2.40
Poland	62 001	297.53	0.004799	56 902	273.06	5 099	24.47
Portugal	20 126	58.40	0.002902	17 772	51.57	2 354	6.83
Slovak Rep.	11 461	37.62	0.003282	10 310	33.84	1 150	3.78
Spain	100 119	330.64	0.003302	89 260	294.78	10 858	35.86
Sweden	34 568	50.64	0.001465	32 135	47.07	2 432	3.56
Switzerland	20 634	43.57	0.002112	19 566	41.32	1 067	2.25
Turkey	67 634	226.08	0.003343	60 456	202.09	7 177	23.99
UK	147 331	530.83	0.003603	13 866	499.61	8 666	31.22
USA	1 568 233	5731.59	0.003655	1 393 429	5092.71	174 804	638.88

least estimated potential for CO$_2$ savings individually (Luxembourg, Switzerland, Ireland, Norway, New Zealand, Denmark, Sweden, Slovak Rep., Hungary and Austria) have an estimated potential to save less than 6 mt of CO$_2$ emissions; however, added together, their saved CO$_2$ emissions would be about 32 mt. This is similar to the estimated potential savings for the UK and Mexico, suggesting that the relatively small emitters and the savings that they can make if operating efficiently should not be ignored.

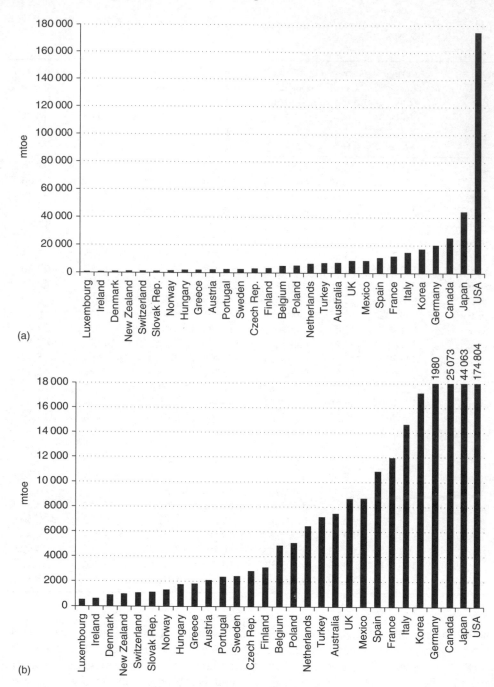

(a)

(b)

Note: (b) is an adjusted version of (a) in order to show the countries with the smaller estimated potential energy savings.

Figure 8.9   Estimated potential energy savings for the 29 OECD countries (2003–07)

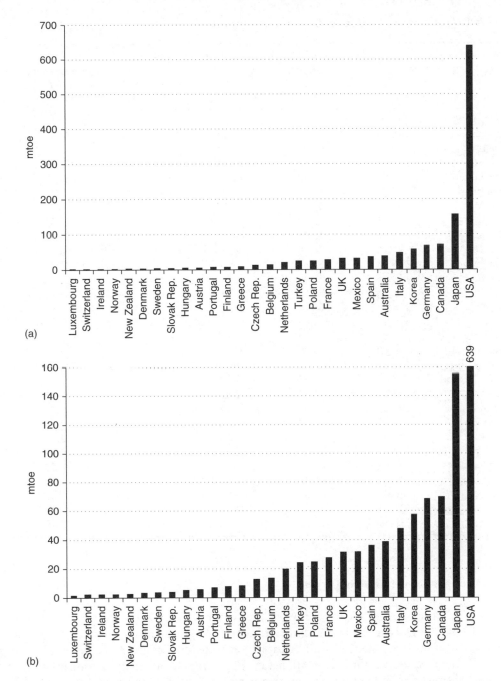

(a)

(b)

Note: (b) is an adjusted version of (a) in order to show the countries with the smaller estimated potential energy savings.

Figure 8.10   Estimated potential CO$_2$ savings if energy efficient for the 29 OECD countries (2003–07)

Table 8.8   Comparison of potential CO_2 savings with Kyoto targets

Country	Average CO_2 emissions (mt) $CO2_i$	Notional CO_2 emissions (mt) $CO2_i^*$	Potential CO_2 savings (mt) $CO2sav_i$	Kyoto target as % of 1990 emissions %	Emission target based on 1990 emissions and Kyoto obligations (mt)
i					
Australia	380.54	342.00	38.54	8	280.59
Austria	72.57	66.97	5.60	−8	52.04
Belgium	112.83	99.48	13.35	−8	101.47
Canada	556.62	487.08	69.54	−6	406.29
Czech Rep.	120.97	108.46	12.51	−8	142.68
Denmark	52.71	49.52	3.19	−8	46.35
Finland	65.23	57.46	7.78	−8	50.05
France	382.51	355.20	27.30	−8	323.94
Germany	824.25	755.97	68.28	−8	874.39
Greece	94.76	86.56	8.20	−8	64.52
Hungary	55.95	50.84	5.11	−6	64.40
Ireland	43.05	40.79	2.26	−8	28.18
Italy	452.23	404.65	47.59	−8	365.97
Japan	1218.76	1063.05	155.72	−6	1007.14
Korea	470.31	412.89	57.42	n/a	
Luxembourg	10.77	9.35	1.43	−8	9.63
Mexico	386.60	355.00	31.60	n/a	
Netherlands	181.37	161.78	19.59	−8	144.06
New Zealand	33.02	30.38	2.64	0	21.37
Norway	37.28	34.88	2.40	1	28.73
Poland	297.53	273.06	24.47	n/a	
Portugal	58.40	51.57	6.83	−8	36.14
Slovak Rep.	37.62	33.84	3.78	−8	52.19
Spain	330.64	294.78	35.86	−8	189.38
Sweden	50.64	47.07	3.56	−8	48.53
Switzerland	43.57	41.32	2.25	−8	37.45
Turkey	226.08	202.09	23.99	n/a	
UK	530.83	499.61	31.22	−8	508.73
USA	5731.59	5092.71	638.88	−7	4522.86

4.4   Comparison with Kyoto Targets

When comparing the emissions savings of each country had it been operating efficiently with the Kyoto emissions targets that parties to the agreement have agreed to for the 2008–12 first phase of the Kyoto obligations illustrated in Table 8.8, it can be seen that in most instances improvements in efficiency alone are not sufficient to contribute fully to eliminating the gap between the emissions and the emissions targets (based on 1990 emissions levels). This implies that energy efficiency improvements in conjunction with changes in the fuel mix are necessary to enable many countries in this analysis to achieve their emissions targets under the Kyoto Agreement. Many were emitting above the level of the Kyoto targets and, even assuming efficiency, would still have done so.

There are, however, a number of EU member countries in the group of OECD countries including the Czech Republic, Hungary, Poland, Slovak Republic and Germany that were all emitting below the Kyoto target during the period, as shown in Figure 8.11.[44] This is of particular interest in the case of Germany given its total level of emissions and suggests that rather than employing a flat EU reduction target of a reduction of 8 per cent on 1990 emission levels, these countries could have been challenged by a higher emissions reduction target. This was an initiative encouraged by the Copenhagen Accord (2009) and the UNFCCC meeting in Cancun (2010), although it delivered little improvement on a possible compromise. On emission reductions the text of the Cancun document says that countries could either cut emissions by a specified percentage or simply implement their chosen target without regard to how ambitious it is.[45] The thinking at Cancun was along the lines of forming regional arrangements, where similar economies via 'coalitions of the willing' established binding agreements to cover the spectrum of climate problems limiting their own emissions, helping others to limit theirs and promoting low carbon development. The results suggest that regional agreements, like the 8 per cent reduction offered by many of the EU countries under the Kyoto Protocol, will not necessarily be as challenging for all regional members and indeed some with the appropriate incentive could make further significant reductions in their emissions.

There are interesting exceptions from the group in Figure 8.11. In each of the six countries in Figure 8.12 energy efficiency improvements alone would have taken each country below their emissions for at least some of the period. In the case of Denmark and Finland energy efficiency alone would have led to the attainment of the target in 2005 and 2008, in Belgium energy efficiency improvements alone would have taken it below its environmental target from 2004 on and in Sweden it would have meant achieving the target about a year (to 18 months) ahead of when the target emission barrier was broken. In Luxembourg there are relatively small levels of emissions but nevertheless for the entire sample period energy efficiency alone would have meant that emissions targets were met. This is also the case for the UK; however, the magnitude of the emissions level is significantly larger than that of Luxembourg and indeed approximately ten times larger than that of Sweden, Denmark and Finland. The UK could have produced CO$_2$ emissions below its Kyoto targets for the entire period had it been operating on the efficient frontier; this is significant for one of the world's top ten CO$_2$ emitters.

Of the remaining countries in the sample, Figure 8.13 presents those that are still exceeding their emissions target (in some cases quite significantly). This pattern of emissions is of concern given the undertaking of many of these nations at Cancun to seek to strengthen their existing targets. The Australians' commitment to only increase emissions by 8 per cent from 1990 levels is still way above this 'generous' target and efficiency measures alone will not be enough. Figure 8.13 shows that the US target of a reduction of 7 per cent and Canada 6 per cent of 1990 emissions levels are not close to being met and both countries have committed to targets of a 17 per cent reduction of 2005 emissions levels by 2020. Energy efficiency improvements will assist but alone will not be sufficient to attain such ambitious targets. Japan and Korea have committed to CO$_2$ reduction targets of 25 per cent and 30 per cent respectively by 2020 (UNFCCC, 2010). Germany along with other EU member states has committed to a 30 per cent reduction in emissions from 1990 levels by 2020 and the UK in its 4th Green Budget of May 2011

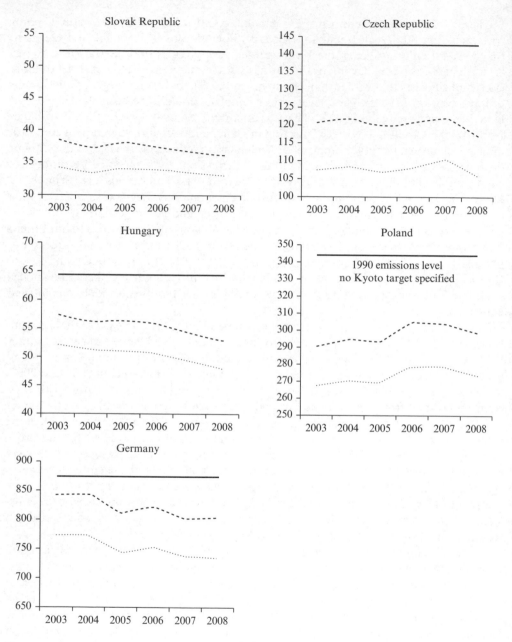

Note:
Solid line represents emission target based on 1990 figures.
Dashed line represents actual CO_2 emissions.
Dotted line represents the estimated efficient level of CO_2 emissions.

Figure 8.11   OECD countries for whom the Kyoto targets were above emissions (2003–07)

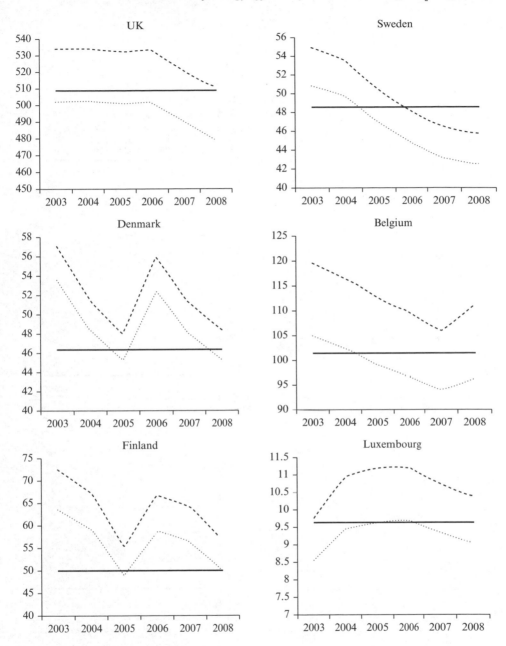

Note: As for Figure 8.11.

Figure 8.12   OECD countries where energy efficiency makes a difference to attaining target (2003–07)

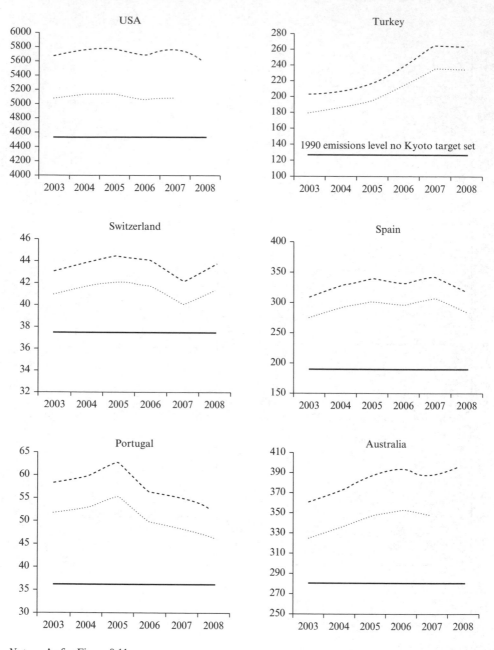

Note: As for Figure 8.11.

Figure 8.13   OECD countries for which target below emissions (2003–07)

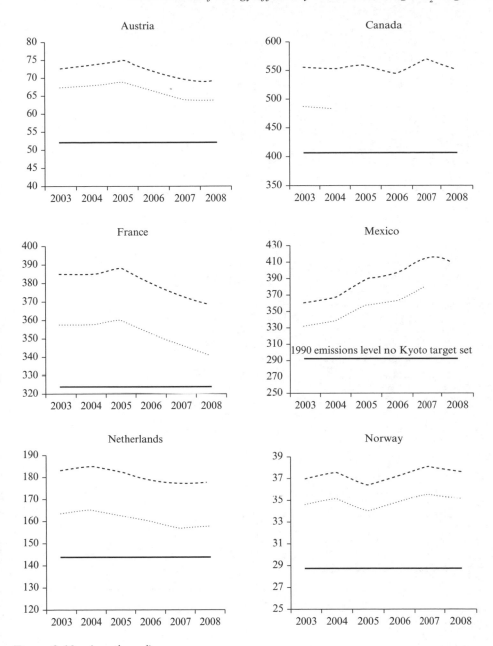

Figure 8.13   (continued)

Figure 8.13   (continued)

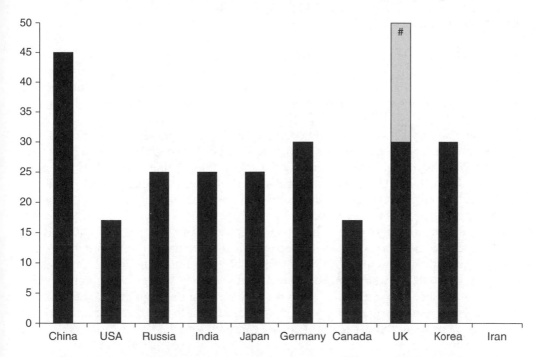

Notes: * USA and Canada base year is 2005. # The extension for the UK commitment represents the additional 20 per cent committed by the UK in its fourth Green Budget in May 2011 (HM Government, 2011). This budget covers the five years from 2023 to 2027 and commits the UK to a 50 per cent reduction of CO_2 emissions on 1990 emissions levels by 2025 (with a 34 per cent emissions reduction by 2020). On this basis emissions reductions by 2050 are targeted at 80 per cent of 1990 levels. The carbon budget sets out emissions targets for the UK as required under the Climate Change Act 2008.

Source: UNFCCC (2010).

Figure 8.14   National commitments under Copenhagen Accord emission reduction pledges by 2020 (% of 1990 emissions levels*)

(as shown in Figure 8.14) has committed to a 50 per cent reduction in CO_2 emissions from 1990 levels.

When considering the national benefits of improvements in energy efficiency, with notable exceptions the benefits nationally appear modest and in the short term are possibly outweighed by the costs of doing so. However, the sum of the energy that would have been saved on average over the period if all the 29 OECD countries had operated on their efficient frontiers is 24.8 per cent of the USA's actual average energy consumption for the same period. That is, if all 29 OECD countries operated on the efficient energy demand frontier, this would have saved average annual energy consumption equivalent to 24.8 per cent of US average annual energy consumption. This saving in average annual energy consumption of the 29 OECD countries would allow for a reduction in CO_2 emissions equivalent to just over 12 per cent of US CO_2 emissions for the period 2003–07 and about 5.5 per cent of the total average CO_2 emissions over the same period for the 29 OECD countries included in this study.

4.5   Beyond Kyoto

At UNFCCC meetings in Copenhagen (2009) and again in Cancun (2010) measures for action beyond Kyoto started to take shape. A new round of more ambitious commitments was sought from countries party to the Kyoto Agreement and those that operated outside the agreement. The commitments of the top emitters are presented in Figures 8.14 and 8.15.

For the majority of the 29 OECD countries analysed, improvements in efficiency of energy consumption alone were not sufficient to meet emissions targets of the six countries shown in Figure 8.15 that are among the top ten OECD emitters for 2007 and 2010, the USA, Japan, Canada and South Korea were over target, Germany under target and the UK is over target but energy efficiency would have put the it under target. Therefore championing energy efficiency policies alone will be unlikely to yield the target results. Indeed the common target across the EU of a reduction of 8 per cent on 1990 emission levels appears to have been relatively easily attainable for the likes of a number of EU countries including Germany, but not others. There are the six notable exceptions (as shown in Figure 8.12) where improvements in efficiency of energy consumption alone were, and are likely to be, the difference between meeting the emissions target and not. In the UK, the results suggest that without any adjustment to the fuel mix employed, but with only increases in the efficiency of energy consumption, the Kyoto targets should be within reach.

As national governments face up to their environmental obligations and consider the most appropriate measures to take beyond Kyoto (2008–12),[46] they have been guided by the text of the Cancun documentation (December 2010), which suggests that countries could either cut emissions by a specified percentage or simply implement their chosen target without regard as to how ambitious it might be nationally. The formation of regional arrangements via 'coalitions of the willing' was the suggested way forward. This established binding agreements to cover the spectrum of climate problems limiting domestic emissions, while helping others to limit theirs and promoting low carbon development. The 2020 emissions targets of some of the top ten emitters are in some instances very ambitious, as set out in Figures 8.14 and 8.15. This is despite the text of the Cancun document specifying that countries could either cut emissions by a specified percentage or simply implement their chosen target without regard to how ambitious it is. In all instances it would seem that energy efficiency measures alone will be insufficient to meet CO_2 emissions targets and a mix of policy initiatives as suggested by the IEA (2008) is necessary to achieve the targets set.

There have been a variety of CO_2 reduction measures and legislation in process. In the USA, the 2009 Waxman–Markey bill (US Congress, 2009a)[47] was prompted by the then upcoming UNFCCC meeting in Copenhagen, the need for a framework for climate change legislation beyond 2012 and the desire to introduce a cap-and-trade system. It was considered to be the most comprehensive and ambitious of the climate change bills proposed during this period.[48] As well as a cap-and-trade scheme that would cover 84.5 per cent of the US emissions by 2016, the bill included provisions that mandate emission performance standards for new coal-fired power stations, nationwide renewable energy standards and funding provisions, energy efficiency and clean transport technologies. The bill capped emissions from large sources at 17 per cent below 2005 levels

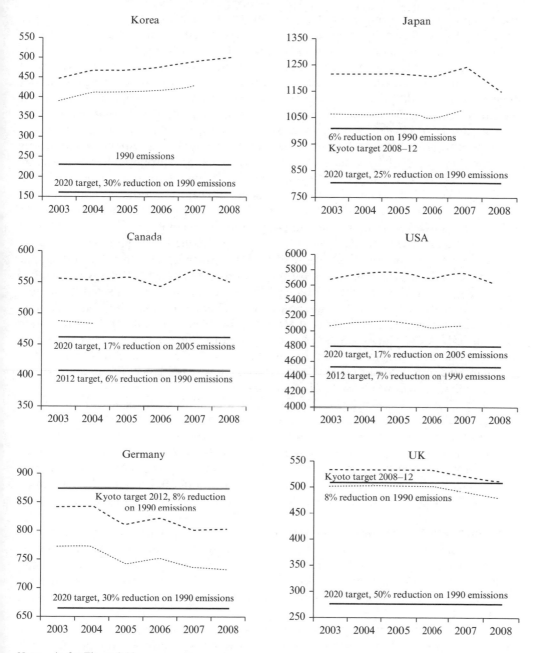

Note: As for Figure 8.11.

Figure 8.15   *2020 emission reduction targets for six countries listed among the 2007 and 2010 top ten OECD emitters*

in 2020 and at 83 per cent below 2005 in 2050. The bill also mandated generation of 20 per cent of total electricity from renewable sources by 2020 with the possibility to use energy efficiency measures to cover 5 per cent of the target (rising to 8 per cent in some circumstances) – that is, energy efficiency improvements of 220 TWh by 2020 equivalent to half of the UK's annual electricity generation. A number of other energy efficiency provisions are made in the bill regarding lighting products and appliances as well as funding of energy efficiency technologies. However, even with these provisions the EPA has estimated that US energy intensity will fall by less than 12 per cent by 2050.[49] This bill never became law.[50] The American Power Act (US Congress, 2009b) was subsequently sponsored in May 2010 by Senators Kerry and Lieberman. It proposed to put a price on carbon emissions from roughly 7500 power plants and other industrial facilities. The bill proposed to establish a market to trade emissions allowances in order to reduce carbon emissions 17 per cent below 2005 levels by 2020, and 80 per cent below 2005 levels by 2050. While stirring debate, this bill did not progress to become law.

Following the 2010 Cancun meetings, the US CO_2 reduction commitment lodged to the UNFCCC was that the USA would commit to reducing CO_2 emissions in the range of 17 per cent (base year 2005) in conformity with the anticipated US energy and climate change legislation; the final target reported would be in light of the enacted legislation. There is a pathway in the pending legislation for a 30 per cent reduction in economy-wide emissions by 2025, a 42 per cent reduction by 2030 and an 83 per cent reduction by 2050 (UNFCCC, 2010). Canada has committed to aligning its CO_2 reductions with those committed by the USA in specifying a target of 17 per cent emissions reduction of economy-wide emissions below 2005 emissions levels in accordance with paragraph 4 of the Copenhagen Accord (2009) (UNFCCC, 2010).

In July 2011 the Korean government submitted a bill to parliament with the aim of cutting CO_2 emissions by 30 per cent below that expected in 2020. The bill includes plans to establish an ETS from January 2015 and sets out cuts of 244 mt of CO_2 below the projected 2020 figure, with 83 mt coming from industry, 68.2 mt from electricity production, 48 mt from the building and construction sector and 36.8 mt from the transport sector. The South Korean parliament had until the end of 2011 to pass this legislation (UNFCCC, 2010). In contrast, Japan has committed to a 25 per cent reduction on 1990 levels.

A number of EU members including Germany and the UK have committed to a 30 per cent emissions reduction target from 1990 levels by 2020. However, the UK has gone beyond this target in its 4th Green Budget in May 2011 by committing the most ambitious emissions reduction target of all (to date) of a 50 per cent emissions reduction on 1990 emissions levels by 2020, as shown by the extension to the UK's commitment in Figure 8.14.

The EU has also begun to consider measures beyond 2020 via the European Commission's Roadmap of March 2011, which sets out a pathway to reach the EU's objective of cutting greenhouse gas emissions by 80–95 per cent of 1990 levels by 2050. In conjunction with the Roadmap a European Energy Efficiency Plan has also been set out (EC, 2011). The Roadmap suggests that the most cost-effective way of achieving the 2050 target requires a 25 per cent emissions cut by 2020 (5 per cent higher than current European targets) via domestic measures because by 2050 it is predicted that international credits for offsetting emissions will be less available. The Energy Efficiency Plan is a set of proposed measures aimed at creating benefits for households and businesses through lowered emissions targets of 20 per cent improvement in energy and efficiency and goes some way to meeting the reduction targets set out by the Roadmap. The impact of the Energy Efficiency Plan will be reviewed in 2013 and legally binding targets introduced if insufficient progress has been made towards the Roadmap targets. An additional annual investment of 1.5 per cent of EU GDP over the next 40 years is estimated to be required.

It might have been expected that the negotiations at the UNFCCC meetings in Durban, South Africa in December 2011 would continue towards voluntary national climate commitments in line with the Bali Action Plan and the agreements made in Copenhagen and Cancun. However, the Durban Platform raises the stakes by refocusing countries on the negotiation of a legally binding agreement to be in place by 2020 (UNFCCC, 2011). To this end the EU has promised to register existing emissions pledges under the extended Protocol, as have a number of other countries.[51] In conjunction with these measures, progress was also made towards establishing a Green Climate Fund to help developing countries meet their emissions reduction commitments.

5. SUMMARY AND CONCLUSIONS

This chapter attempts to estimate a panel 'frontier' whole-economy aggregate energy demand function for 29 countries over the period 1978 to 2008 using parametric stochastic frontier analysis (SFA). Consequently, unlike standard energy demand econometric estimation, the energy efficiency of each country is also modelled and it is argued that this represents a measure of the underlying efficiency for each country over time, as well as the relative efficiency across the 29 OECD countries.[52] This shows that energy intensity is not necessarily a good indicator of energy efficiency, whereas the measure of energy efficiency obtained via this approach, by controlling for a range of economic and other factors is. This is particularly relevant in a world dominated by environmental concerns with the subsequent need to conserve energy and/or use it as efficiently as possible. Moreover, the results show that although for a number of countries the change in energy intensity over time might give a reasonable indication of efficiency improvements, this is not always the case. Therefore, unless this analysis is undertaken, it is not possible to know whether the energy intensity of a country is a good proxy for energy efficiency or not. Hence it is argued that this analysis should be undertaken to avoid potentially misleading advice to policy makers.

As national governments face up to their environmental obligations and participate in the negotiation of the new legally binding treaty to replace the Kyoto Agreement beyond 2020, they must ensure that the legally binding emission reduction targets are nationally challenging, requiring energy efficiency improvement as well as other measures. The results presented here suggest that, due in part to different levels of energy efficiency, common regional targets will significantly challenge some countries, while not challenging others at all. For the majority of the 29 OECD countries analysed, improvements in efficiency of energy consumption alone were not sufficient to meet emissions targets. Therefore championing energy efficiency policies alone will be unlikely to yield the target results.[53]

As shown, the common 8 per cent reduction target across the EU on 1990 emission levels appears to have been relatively easily attainable for a number of EU countries including Germany, but not for others, with six notable exceptions where improvements in efficiency of energy consumption alone could make the difference between meeting the emissions target and not. Further, for the UK, it suggests that increases in the efficiency of energy consumption alone should be enough to meet the Kyoto targets, even without any adjustment to the fuel mix employed. Therefore a uniform reduction target, of the same fixed percentage of a given year's CO_2 emissions for all countries, is not likely to produce the most efficient outcome. Indeed, for some countries the target set on this basis will be too ambitious (and not realistically attainable) while for others not challenging enough to incentivize the drive for optimal energy efficiency.

Instead the legally binding requirements of the treaty could usefully include national emissions reduction targets which are determined in two parts. The first part of the target should be calculated on the basis of desirable national energy efficiency improvements, which could be specified in the treaty. That is, the first part of the target should determine the emissions reductions possible when each nation is energy efficient. Beyond this, the second part of the target could require emissions reductions via additional measures and might indeed be set as a flat-rate reduction target or on a sliding scale (to accommodate developing countries). Such a measure would ensure that each nation had a challenging emissions reduction target and the incentive to improve both energy efficiency and undertake other measures to reduce CO_2 emissions. The results suggest that measures to improve energy efficiency alone have been and therefore are likely to continue to be insufficient to meet CO_2 emissions targets and therefore a mix of policy initiatives will be required to achieve the emissions reductions levels necessary.

NOTES

*   We gratefully acknowledge the comments of participants at the 34th International IAEE Conference, Stockholm, Sweden, June 2011, the editor of this volume and an anonymous referee. The authors are, of course, responsible for all errors and omissions.
1.   Although the Accord was not legally binding, countries representing over 80 per cent of global emissions engaged with Copenhagen Accord in December 2009, with 114 parties agreeing to 'take note of' it. The Accord does not require countries to agree to a binding successor to the Kyoto Protocol; however, it does endorse the continuation of Kyoto-like measures.
2.   After a dip in 2009 caused by the global financial crisis, emissions in 2010 were estimated to be 30.6 gigatonnes (Gt), a 5 per cent increase on 2008, when levels reached 29.3 Gt. In terms of fuels, 44 per cent of the estimated CO_2 emissions in 2010 came from coal, 36 per cent from oil, and 20 per cent from natural gas. The top 10 CO_2 emitting countries of 2008 accounted for two-thirds of the world CO_2 emissions (top 10 total 19.1 Gt CO_2, world total 29.4 Gt CO_2). The combined share of electricity and heat generation (41 per cent of global CO_2 emissions in 2008) and transport (22 per cent of global CO_2 emissions in 2008) represented two-thirds of global emissions in 2008 (IEA, 2010a, p. 11, Figure 4).
3.   The final decision as to the exact term of the extension would be specified at the UNFCCC meeting in Qatar in December 2012.
4.   Such as European Climate Change (Action) Programmes (ECCP) 2000 and 2005; IPCC (2007), IEA (2008), EC (2005, 2006a and 2006b), European Roadmap 2050; EC (2011) and IEA (2011).
5.   Energy intensity can vary between countries for a whole range of reasons including the level of industrialization, the mix of services and manufacturing, the climate, the level of energy efficiency of the appliance and capital stock and production processes and the organization of the production and consumption processes in space.
6.   See Ang (2006) for a general discussion and application of this method.
7.   See Huntington (1994) for a discussion on the relation between energy efficiency and productive

efficiency using the production theory framework. One of the first studies that made use of the frontier approach was Ferrier and Hirschberg (1992).

8. During the last few years several studies have also been published on the measurement of environmental performance of OECD countries using an environmental DEA approach that also considers energy as an input into the production process. See, for instance, Zofio and Prieto (2001) and Zhou et al. (2008).

9. This could result in more energy being consumed (often referred to as 'backfire'), depending on the strength of the changes in behaviour by energy consumers.

10. Thus the inverse of the energy intensity ratio is a measure of the energy efficiency at the aggregate level without an adjustment for changes in factors unrelated to efficiency.

11. Comparisons were made between primary energy consumption and the exchange-rate-converted GDP in the 1970s and early 1980s. The exchange-rate-converted GDP tends to exaggerate the real income differences between the developing countries and industrial countries. In the mid-1980s the availability of GDP data given in purchasing power parity (PPP) helped overcome this problem (Summers and Heston, 1991). Another problem in using GDP as a measure of output which is difficult to overcome is the existence of a substantial and difficult-to-quantify non-monetized sector in the low-income developing countries.

12. However, it is worth noting that an improvement in the level of technical and/or allocative efficiency could actually result in an increase in energy consumption in certain circumstances.

13. An economic agent could be a firm or a household. Moreover, Figure 8.1 could also represent the economy-wide aggregate production function for a country.

14. Note that this is a technical relationship, and the amount of service actually chosen by consumers (such as the temperature of the rooms in a house) will depend upon the size of the rebound effect discussed briefly above.

15. See also Gillingham et al. (2009) for a discussion of this effect.

16. See Kumbhakar and Lovell (2000, p. 148) for a discussion on the interpretation of the efficiency in an input demand function.

17. It is worth noting that Kumbhakar and Lovell (2000) highlight that the use of a large number of time dummies in a parametric frontier framework can create estimation problems. Thus, although not done here, Filippini and Hunt (2011, 2012) also consider a time trend for the specification of the UEDT, but there is no discernible difference in the efficiency rankings.

18. Note that previous studies by Buck and Young (2007) and Boyd (2008) did not base their estimation on an energy demand function, in that they did not consider the energy price as an explanatory variable, hence omitting this important control variable.

19. The UEDT includes exogenous technical progress and it could be argued that even though technologies are available to each country they are not necessarily installed at the same rate; however, it is assumed that this results from different behaviour across countries and reflects 'inefficiency' across countries; hence it is captured by the different (in)efficiency terms for all countries.

20. Unfortunately, it is not possible to obtain more sectoral disaggregated data (e.g. data on energy-intensive sectors) on a consistent basis for all 29 countries for all the years.

21. The frontier function approach suggested by Aigner et al. (1977) was developed within neoclassical production theory and the main goal of this literature has been to estimate production and cost frontier in order to identify the level of productive inefficiency (allocative and technical inefficiency). In this study, the neoclassical production theory is discarded and instead the concept of a stochastic frontier within the empirical approach traditionally used in the estimation of economy-wide energy demand function is employed. Of course, behind the concept of underlying energy inefficiency developed here is still a 'production process'.

22. The energy demand function estimated in this chapter can be considered an input demand function derived through a cost-minimizing process from an aggregate production function. Of course, theoretically the demand for energy might also depend on the price of other inputs, but in line with previous energy demand studies, data limitations make it impossible to include these variables. For this reason this equation is specified, similar to the general energy demand literature, in a relatively *ad hoc* way with an indirect reference to production theory.

23. The size of the country should reflect the extension of the transport system and therefore, indirectly, should consider the use of energy in the transport sector. Two countries could have the same GDP but consume a different amount of energy because of the differences in size resulting in different transport activities.

24. The assumption that the error term is composed of two independent parts, a two-sided classical 'noise' component, v_{it}, and the non-negative inefficiency component u_{it}, is crucial in the stochastic frontier approach. This assumption is needed in order to identify the level of efficiency for each country separately. The two-sided classical 'noise' component captures unobserved random factors affecting the energy demand, including exogenous shocks and measurement errors.

25. Based on this assumption, the empirical results here show a good fit. For a discussion of alternative distributions such as the truncated normal or the exponential, see Kumbhakar and Lovell (2000).
26. Australia, Austria, Belgium, Canada, Czech Republic, Denmark, Finland, France, Germany, Greece, Hungary, Ireland, Italy, Japan, Korea, Luxembourg, Mexico, Netherlands, New Zealand, Norway, Poland, Portugal, Slovak Republic, Spain, Sweden, Switzerland, Turkey, the UK and the USA. For some countries, data for some of the explanatory variables are not available for the whole sample period; for this reason the data set is unbalanced.
27. See Peel et al. (2007) for a discussion of this classification.
28. For a presentation of several approaches for the estimation of frontier models in the energy sector see Farsi and Filippini (2009).
29. Battese and Coelli (1992) propose a model where the variation of efficiency with time is considered as a deterministic function that is commonly defined for all firms.
30. In order to address this problem, Filippini and Hunt (2012) employ a Mundlak version of the model proposed by Pitt and Lee (1981) to a relatively short period of data. In the Mundlak version of the model the correlation of the individual specific effects and the explanatory variables is considered in an auxiliary equation.
31. For a successful application of these models in network industries, see Farsi et al. (2005) and Farsi et al. (2006).
32. In this model, the unobserved inefficiency that remains constant over time should be absorbed by the inefficiency term or by the classical error term. In the latter case, of course, the inefficiency term would underestimate the level of inefficiency.
33. See also Kumbhakar and Lovell (2000) and Battese and Coelli (1992).
34. This is in contrast to the alternative indicator of energy inefficiency given by the exponential of u_{it}. In this case, a value of 0.2 indicates a level of energy inefficiency of 20 per cent.
35. Note that, in comparison with Filippini and Hunt (2011), in this study the data set contains information for an extra year and the model specification includes an extra explanatory variable for climate.
36. Lambda (λ) gives information on the relative contribution of u_{it} and v_{it} on the decomposed error term e_{it} and shows that, in this case, the one-sided error component is relatively large.
37. However, one of the reasons for the estimated poor performance of Luxembourg could be the presence of 'tank tourism', which is not captured in the aggregate model.
38. The Czech Republic, Hungary, Poland and the Slovak Republic also generally increased over the shorter period that data were available for these countries.
39. However, it should be noted that, given the unbalanced panel used in estimation, the figures for Switzerland are over a much shorter period.
40. We are grateful to Dermot Gately, who suggested presenting the results in this way.
41. It is worth noting that although the magnitude of the CO_2 emissions has changed over the period 2003 to 2007, the ranking of the countries contributing the most emissions, as shown in Figure 8.8, saw little movement over the period.
42. CO_2 emissions in the IEA database are measured in two different ways: the sectoral approach and the reference approach. The CO_2 reference approach data contain total CO_2 emissions from fuel combustion; they are based on the supply of energy in a country and as a result include fugitive emissions from energy transformation and for this reason are likely to overestimate national CO_2 emissions. The difference between the sectoral approach and the reference approach includes statistical differences, product transfers, transformation losses and distribution losses (IEA, 2010c, p. 8). The sectoral approach contains total CO_2 emissions from fuel combustion including emissions only when the fuel is actually combusted (IEA, 2009, p. 3). Consequently, the sectoral approach data are used here since by definition this measure provides the most accurate measure of emissions. Nevertheless, there is no discernible difference between the results generated using each of the two measures.
43. It should be noted that the estimates for Canada and New Zealand given in Tables 8.7 and 8.8 and Figure 8.13 are more uncertain given that the available data for the estimation for these countries only go up to 2004.
44. Assuming Poland was to have a target consistent with the other EU target of a reduction of 1990 by 8 per cent, in which case it too would be polluting less than this target.
45. The first week of talks was dominated by tension following Japan's unwillingness to accept the continuation of the Kyoto Protocol, a position condemned by developing countries and Britain alike. However, there is underlying concern in many quarters that the Protocol covers less than a third of global emissions. Britain, for one, will not allow the Protocol to continue unless China signs an agreement to cut emissions.
46. They must consider the relative roles of improvements in efficiency of energy consumption as one means of reducing energy consumption or the alternative options of reducing carbon emitted per unit of energy by the fuel mix employed so as to be able to maintain and grow economic output.
47. The Democrat authors of this global warming bill were Representative Henry A. Waxman of California, chairman of the House Energy and Commerce Committee and Representative Edward J. Markey of

Massachusetts, a subcommittee chairman. The bill was passed in June 2009 and on 7 July 2009 was read the second time and placed on Senate Legislative Calendar under General Orders, Calendar No. 97.

48. Obama proposed a bill, as did Lieberman and Warner.
49. This is small when compared to the drop of 46 per cent in US energy intensity that was observed between 1975 and 2005.
50. At the end of each two-year session of Congress all proposed bills and resolutions that have not passed are cleared from the books.
51. Brazil, South Africa, China and the USA have indicated that they would accept binding commitments under a new treaty. India is holding to the position of the original 1992 Framework Convention on Climate Change, which stated that countries have 'common but differentiated responsibilities'. Canada, however, has been unable to meet its Kyoto commitments, and announced on 12 December 2011 that it would formally withdraw from the protocol (UNFCCC, 2011).
52. It should be noted that the estimated underlying energy efficiency and associated potential CO$_2$ savings for each country should not be taken as precise values given the stochastic technique used for the estimation. However, they do give a good approximation of each country's direction of change in efficiency and CO$_2$ emissions over time. Moreover, no account is made for any possible rebound effect resulting from an improvement in a country's energy efficiency and possible CO$_2$ reductions since it is outside the scope of this work. However, this might mitigate some of the possible gains highlighted here.
53. This is consistent with the IEA findings. While energy efficiency is still important in limiting increases in energy use in IEA countries, the rate of energy efficiency improvement since 1990 has been slower than previously and will need to speed up to make a more significant contribution. This will require government action to promote the deployment of energy efficiency technologies (IEA, 2008).

REFERENCES

Adeyemi, O.I. and L.C. Hunt (2007), 'Modelling OECD industrial energy demand: asymmetric price responses and energy saving technical change', *Energy Economics*, **29**(4): 693–709.
Aigner, D.J., C.A.K. Lovell and P. Schmidt (1977), 'Formulation and estimation of stochastic frontier production function models', *Journal of Econometrics*, **6**(1): 21–37.
Ang, B.W. (1987), 'Structural change and energy demand forecasting in industry with applications to two newly industrialized countries', *Energy*, **12**(2): 101–11.
Ang, B.W. (2004), 'Decomposition analysis for policy making in energy: which is the preferred method?', *Energy Policy*, **32**(9): 1131–39.
Ang, B.W. (2006), 'Monitoring changes in economy-wide energy efficiency: from energy–GDP ratio to composite efficiency index', *Energy Policy*, **34**: 574–82.
Ang, B.W. and F.Q. Zhang (2000), 'A survey of index decomposition analysis in energy and environmental studies', *Energy*, **25**(12): 1149–76.
Battese, G.E. and T.J. Coelli (1992), 'Frontier production functions, technical efficiency and panel data: with application to paddy farmers in India', *Journal of Productivity Analysis*, **3** (1/2): 153–69.
Bhattacharyya, S.C. (2011), *Energy Economics: Concepts, Issues, Markets and Governance*, London: Springer.
Boyd, G.A. (2008), 'Estimating plant level manufacturing energy efficiency with stochastic frontier regression', *The Energy Journal*, **29**(2): 23–44.
Buck, J. and D. Young (2007), 'The potential for energy efficiency gains in the Canadian commercial building sector: a stochastic frontier study', *Energy – The International Journal*, **32**: 1769–80.
EC (2005), *Doing More with Less: Green Paper on Energy Efficiency*, Luxembourg: Office for Official Publications of the European Communities, European Commission.
EC (2006a) 'A European strategy for a reliable, competitive and sustainable energy supply', Green Paper, European Commission, (SEC(2006) 317), COM(2006) 105 final, March.
EC (2006b), 'Action Plan for Energy Efficiency: realising potential', Communication from the Commission COM(2006)545 final, European Commission, (SEC(2006)1173) (SEC(2006)1174) (SEC(2006)1175), October.
EC (2011), 'A Roadmap for moving to a competitive low carbon economy in 2050', Communication from the Commission to the European Parliament, The Council, The European Economic and Social Committee and the Committee of the Regions, COM (2011)112 final, European Commission, (SEC(2011) 287 final) (SEC(2011) 288 final) (SEC(2011) 289 final), March.
Farrell, M. (1957), 'The measurement of productive efficiency', *Journal of the Royal Statistical Society*, Series A, General, **120**(3): 253–81.
Farsi, M. and M. Filippini (2009), 'Efficiency measurement in the electricity and gas distribution sectors', in

J. Evans and L.C. Hunt (eds), *International Handbook on the Economics of Energy*, Cheltenham, UK and Northampton, MA, USA: Edward Elgar, pp. 598–623.

Farsi, M., M. Filippini and W. Greene (2005), 'Efficiency measurement in network industries: application to the Swiss railway companies', *Journal of Regulatory Economics*, **28**(1): 69–90.

Farsi, M., M. Filippini and M. Kuenzle (2006), 'Cost efficiency in regional bus companies: an application of alternative stochastic frontier models', *Journal of Transport Economics and Policy*, **40**(1): 95–118.

Ferrier, G.D. and J.G. Hirschberg (1992), 'Climate control efficiency', *The Energy Journal*, **13**: 37–54.

Filippini, M. and L.C. Hunt (2011), 'Energy demand and energy efficiency in the OECD countries: a stochastic demand frontier approach', *The Energy Journal*, **32**(2): 59–80.

Filippini, M. and L.C. Hunt (2012), 'US residential energy demand and energy efficiency: a stochastic demand frontier approach', *Energy Economics*, 34(5): 1484–91.

Fouquet, R. (2008), *Heat, Power and Light: Revolutions in Energy Services*, Cheltenham, UK and Northampton, MA, USA: Edward Elgar.

Fouquet, R. (2011), 'Divergence in long run trends in the prices of energy and energy services', *Review of Environmental Economics and Policy*, **5**(2): 196–218.

Gillingham, K., R.G. Newell and K. Palmer (2009), 'Energy efficiency economics and policy', *Annual Review of Resource Economics*, **1**: 597–619.

Greene, W.H. (2005a), 'Reconsidering heterogeneity in panel data estimators of the stochastic frontier model', *Journal of Econometrics*, **126**: 269–303.

Greene, W.H. (2005b), 'Fixed and random effects in stochastic frontier models', *Journal of Productivity Analysis*, **23**(1): 7–32.

Greening, L.A., D.L. Greene and C. Difiglio (2000), 'Energy efficiency and consumption – the rebound effect – a survey', *Energy Policy*, **28**(6–7): 389–401.

Griffin J.M. and C.T. Schulman (2005), 'Price asymmetry in energy demand models: a proxy for energy-saving technical change?', *The Energy Journal*, **26**(2): 1–21.

HM Government (2011), 'Implementing the Climate Change Act 2008: The Government's proposal for setting the fourth carbon budget', Policy Statement, May, available at http://www.decc.gov.uk/emissions/carbonbudgets.

Hunt, L.C., G. Judge and Y. Ninomiya (2003a), 'Underlying trends and seasonality in UK energy demand: a sectoral analysis', *Energy Economics*, **25**(1): 93–118.

Hunt, L.C., G. Judge and Y. Ninomiya (2003b), 'Modelling underlying energy demand trends', in L.C. Hunt (ed.), *Energy in a Competitive Market: Essays in Honour of Colin Robinson*, Cheltenham, UK and Northampton, MA, USA: Edward Elgar, pp. 140–74.

Huntington, H.G. (1994), 'Been top down so long its looks like up to me', *Energy Policy*, **22**: 833–9.

IEA (2008), 'Energy efficiency policy recommendations in support of the G8 Action Plan', Paris, available at www.iea.org.

IEA (2009), 'Progress with implementing energy efficiency policies in the G8', International Energy Agency Paper, available at http://www.iea.org/Textbase/publications/free_new_Desc.asp?PUBS_ID_2127.

IEA (2010a), 'Energy statistics', available at www.iea.org.

IEA (2010b), 'Energy balances and energy statistics of OECD countries', International Energy Agency, ESDS International, University of Manchester.

IEA (2010c), 'CO_2 emissions from fuel combustion', International Energy Agency, ESDS International, University of Manchester.

IEA Scorecard (2011), 'Implementing energy efficiency policy: Progress and challenges in IEA member countries', OECD/IEA, www.iea.org.

IPCC (2007), *Climate change 2007: synthesis report. Contribution of Working Groups I, II and III to the Fourth Assessment Report of the Intergovernmental Panel on Climate Change* [Core Writing Team, R.K Pachauri and A. Reisinger (eds)], Geneva, Switzerland: IPCC.

Jondrow, J., C.A.K. Lovell, I.S. Materov and P. Schmidt (1982), 'On the estimation of technical efficiency in the stochastic frontier production function model', *Journal of Econometrics*, **19**(2/3): 233–8.

Kopp, R.J. (1981), 'The measurement of productive efficiency: a reconsideration, *The Quarterly Journal of Economics*, **96**(3): 477–503.

Kumbhakar, S.C. and C.A. Lovell (2000), *Stochastic Frontier Analysis*, Cambridge: Cambridge University Press.

Medlock, K.B. (2004), 'Economics of energy demand', *Encyclopedia of Energy*, **2**: 65–78.

Metcalf, G.E. (2008), 'An empirical analysis of energy intensity and its determinants at the state level', *The Energy Journal*, **29**: 1–26.

Patterson, M.G. (1996), 'What is energy efficiency? Concepts, indicators and methodological issues', *Energy Policy*, **24**(5): 377–90.

Peel, M.C., B.L. Finlayson and T.A. McMahon (2007), 'Updated world map of the Köppen–Geiger climate classification', *Hydrology and Earth System Sciences*, **11**: 1633–44.

Pitt, M. and L. Lee (1981), 'The measurement and sources of technical inefficiency in the Indonesian weaving industry', *Journal of Development Economics*, **9**: 43–64.

Richmond, A.K. and R.K. Kaufmann (2006), 'Energy prices and turning points: the relationship between income and energy use/carbon emissions', *The Energy Journal*, **27**: 157–78.

Schmidt, P. and C.A.K. Lovell (1979), 'Estimating technical and allocative inefficiency relative to stochastic production and cost frontiers', *Journal of Econometrics*, **9**: 343–66.

Sorrell, S. (2009), 'The rebound effect: definition and estimation', in J. Evans and L.C. Hunt (eds), *International Handbook on the Economics of Energy*, Cheltenham, UK and Northampton, MA, USA: Edward Elgar, pp.199–233.

Sue Wing, I. (2008), 'Explaining the declining energy intensity of the U.S. economy', *Resource and Energy Economics*, **30**: 21–49.

Summers, R. and A. Heston (1991), 'The Penn World Table (Mark 5): an expanded set of International comparisons, 1950–1988', *The Quarterly Journal of Economics*, May: 327–68.

Tedesco, L. and S. Thorpe (2003), 'Trends in Australians energy intensity 1973–74 to 2000–01', ABARE Report 03.9 for the Ministerial Council of Energy, Canberra.

United Nations (1992), *UN Framework Convention on Climate Change* (UNFCCC), New York: United Nations.

UNFCCC (2010), 'Communications received from Parties in relation to the listing in the chapeau of the Copenhagen Accord', Appendix I – Quantified economy-wide emissions targets for 2020, available at http://unfccc.int/meetings/copenhagen_dec_2009/items/5264.php.

UNFCCC (2011), 'Report of the Conference of the Parties on its seventeenth session, held in Durban from 28 November to 11 December 2011', 15 March 2012, available at http://unfccc.int/resource/docs/2011/cop17/eng/09a01.pdf.

UNSD (2009), Millennium Development Goals Indicators Database, UN, Department of Economic and Social Affairs, Population Division, World Population Prospects: The 2008 Revision, New York, 2009, UNSD Demographic Yearbook, available at http://mdgs.un.org/unsd/mdg/Data.aspx.

US Congress (2009a), H.R. 2454: American Clean Energy and Security Act of 2009, 111th Congress 2009–2010, available at http://www.govtrack.us/congress/bill.xpd?bill=h111-2454.

US Congress (2009b) S. 1733: Clean Energy Jobs and American Power Act, 111th Congress 2009–2010, available at http://www.govtrack.us/congress/bill.xpd?bill=s111-1733.

Zhou, P. and B.W. Ang (2008), 'Linear programming models for measuring economy-wide energy efficiency performance', *Energy Policy*, **36**: 2911–16.

Zhou, P., B.W. Ang and K.L. Poh (2008), 'A survey of data envelopment analysis in energy and environmental studies', *European Journal of Operational Research*, **189**: 1–18.

Zhou, P., B.W. Ang and D.Q. Zhou (2012), 'Measuring economy-wide energy efficiency performance: a parametric frontier approach', *Applied Energy*, **90**(1): 196–200.

Zofio, J.L. and A.M. Prieto (2001), 'Environmental efficiency and regulatory standards: the case of CO_2 emissions from OECD industries', *Resource and Energy Economics*, **23**: 63–83.

9 Economic analysis of feed-in tariffs for generating electricity from renewable energy sources

G. Cornelis van Kooten

INTRODUCTION

The process accompanying the UN's 1992 Framework Convention on Climate Change (FCCC) culminated in an agreement at the Third Conference of the Parties (COP3) held at Kyoto, Japan, in December 1997. Industrialized nations agreed that, by the first commitment period 2008–12, they would reduce their collective emissions of carbon dioxide and equivalent greenhouse gases (hereafter just CO_2) to an average of 5.2 percent below what they were in 1990. Since then, countries have sought to build on Kyoto and reduce CO_2 emissions even further in order to address predicted climate change. Given the perceived urgency of addressing global warming, the leaders of the G8 countries agreed at a July 2009 meeting in L'Aquila, Italy, to limit the increase in global average temperature to no more than 2 °C above pre-industrial levels. They would do this by reducing their own greenhouse gas emissions by 80 percent or more, and global emissions by 50 percent, by 2050 relative to 1990 or some more recent year. The European Union (EU) already has in place a '20–20–20 target' – a 20 percent reduction in CO_2 emissions from 1990 levels by 2020, with 20 percent of energy to be produced from renewable sources.[1] At COP15 in Copenhagen in late 2009, and again at COP16 in Cancun, Mexico in 2010, the EU was prepared to impose a 30 percent reduction in CO_2 emissions by 2020, if there had been some sort of climate agreement.

The USA also appeared ready to reduce CO_2 emissions by a significant amount: the House of Representatives passed the American Clean Energy and Security Act (also known as Waxman–Markey) on 26 June 2009. The Act required large emitters to reduce their aggregate CO_2 emissions by 3 percent below 2005 levels in 2012, 17 percent below 2005 levels in 2020, 42 percent in 2030, and 83 percent in 2050. The Senate has yet to pass legislation, but had been contemplating major reductions in emissions. The American Power Act (2009) proposed by Senators Kerry and Lieberman added to the Waxman–Markey cap-and-trade scheme a carbon tax on large emitters. The tax would have a floor of $12 per ton CO_2 that would rise by the rate of inflation plus 3 percent, and a ceiling of $25 indexed to inflation plus 5 percent. This bill was subsequently replaced by a 2010 bill (S.3813) sponsored by Senator Bingham to create a national 'Renewable Electricity Standard' (RES). It required that, by 2021, 15 percent of the electricity sold by an electric utility be generated from wind or certain 'other' renewable energy sources (presumably solar, etc., and not hydro); up to four of the 15 percent points could, theoretically, be achieved by actions that improve energy efficiency, although these were tightly defined.

To date, no legislation has actually been passed by the Senate and, given Republican gains at the mid-term elections in November 2010 (including a majority of the seats in the House), it will be difficult but perhaps not impossible to pass climate legislation, especially legislation that involves some sort of economic instrument, cap-and-trade or carbon taxes. At the same time, the US Environmental Protection Agency was granted power to regulate CO_2 emissions as a result of a 2007 Supreme Court ruling that CO_2 is a pollutant.

To achieve the kinds of emission reduction targets envisioned, it is necessary to radically transform the fundamental driver of global economies – the energy system. The main obstacle to so doing is the abundance and ubiquity of fossil fuels, which can be expected to power the industrialized nations and the economies of aspiring industrial economies into the foreseeable future. Realistically, global fossil fuel use will continue to grow and remain the primary energy source for much of the next century (Bryce, 2010; International Energy Agency, 2009; Duderstadt et al., 2009; Smil, 2003).

ECONOMIC INSTRUMENTS FOR REDUCING CO_2 EMISSIONS

Several options are available to the authority for reducing reliance on fossil fuels – for reducing emissions of CO_2. These are regulation (also known as mandates) and economic incentives, namely, a carbon tax, a cap-and-trade scheme or subsidies of one form or other. Each is discussed below.

Regulation

The government can choose to regulate emissions of CO_2 from fossil fuels, and other sources, in a variety of ways. In the transportation sector, fleet fuel economy standards can be mandated: these require that an automobile manufacturer's sales of vehicles in a particular market achieve a specified average fuel economy. Coal-fired power plants may be required to install equipment to capture CO_2 from the smokestack (or new plants must be able to do so). Electric system operators or utilities may be required to derive a specified proportion of their power from renewable generating sources. In some cases, the authority may even specify the extent to which the operator must rely on wind-generated power.

Most environmental economics textbooks provide a simple demonstration as to why economic incentives are more efficient than regulation. Hence it is surprising that policy makers still rely on regulation as a principal means of tackling market failures caused by unwanted emissions.

Aside from transaction costs, which include monitoring compliance, direct intervention leads to economic inefficiencies because economic agents seek only to comply with the regulations, but not lower emissions at least cost. CO_2 emissions are not necessarily reduced by those firms that can do so at the lowest cost. In addition, as new firms enter or new plants are built, emissions can expand even while mandates, such as requirements to adopt the best available technology, are met. Thus there is no guarantee that regulations will actually reduce CO_2 emissions. Yet, by failing to pass legislation to address climate change, the US Congress has chosen to rely, through the Environmental Protection Agency, on regulation as the vehicle for lowering CO_2 emissions.

Carbon Taxes and Emissions Trading

Carbon taxes target prices, while cap-and-trade targets quantity. A carbon tax raises the cost of emitting CO_2, thereby increasing the price of energy produced by fossil fuels and, if correctly applied, the costs of energy from biomass burning (as it also releases CO_2).[2] With cap-and-trade, emissions of CO_2 are restricted; this causes them to take on value, thereby raising costs of releasing CO_2 into the atmosphere. By permitting economic agents to trade the limited quantity of emissions (the cap), the cost of a permit falls to its lowest possible value. In principle, the state can choose the tax level (price) or the number of emission permits to auction (quantity), but if all is known the outcome will be the same – the targeted level of CO_2 emissions reduction will be achieved.

Given a choice between carbon taxes and cap-and-trade, economists generally prefer carbon taxes for three reasons. First, the transaction costs associated with emissions trading are likely much larger than with a tax.

Second, a carbon tax and cap-and-trade scheme are identical in theory, but when abatement costs and/or benefits are uncertain, picking a carbon tax can lead to the 'wrong' level of emissions reduction, while choosing a quantity can result in a mistake about the forecasted price that firms will have to pay for auctioned permits (Weitzman, 1974). Such errors have social costs. If the marginal cost of abatement is steep while the marginal benefit (marginal damages avoided) curve is relatively flat, then a small increase in the number of permits that are issued can have a large impact on their price (Pizer, 1997; Weitzman, 1974, 2002). 'Uncertainty about compliance costs causes otherwise equivalent price and quantity controls to behave differently and leads to divergent welfare consequences . . . [so] that price controls are more efficient [than quantity controls]' (Pizer, 2002, p. 409). On economic grounds, a carbon tax is preferred over cap-and-trade.

Finally, large income transfers are involved. With a tax, the authority drives a wedge between the supply and demand curves that causes the price of energy from fossil fuels to rise above the marginal cost of providing that energy by the amount of the tax. The government collects the difference as revenue. With a quantity restriction (a cap), the difference between price and marginal cost of provision constitutes a large rent that is 'up for grabs'. Large industrial emitters can capture this rent if all or a significant proportion of the permits are grandfathered rather than auctioned. Rent seeking occurs and, thus, grandfathering of permits is likely to be required for a cap-and-trade scheme to be politically acceptable. Likewise, large financial intermediaries will lobby for cap and trade as they gain from trading permits. Further, there is the potential for corruption if permits can be purchased abroad through such devices as the Clean Development Mechanism. If that is the case, we have emissions trading but not a true cap-and-trade scheme, and emission reduction targets are unlikely to be met.

Politicians have generally eschewed carbon taxes as these are seen as just another means to raise overall taxes. Resistance in the USA to cap-and-trade has also come about because it too is increasingly viewed as another form of taxation. This most likely explains the failure of the US Congress to pass climate legislation.

Subsidies

Subsidies are also a form of economic incentive. Governments can subsidize everything from research and development of new technologies that substitute for or reduce reliance on fossil fuels to the construction of energy-efficient buildings and manufacturing plants; states can even subsidize the purchase of end products such as eco-friendly vehicles. Governments have subsidized biofuel production facilities, research into electric, hybrid and hydrogen vehicles, and the construction of biomass power-generating plants, wind turbines and solar photovoltaic panels. Needless to say, firms prefer subsidies over taxes and emissions trading (unless they can capture large rents from the grandfathered emission permits); sometimes the public even appears to prefer subsidies, but only as long as they are unaware of the tradeoffs in public spending on other programs and/or assume the funds spent for this purpose will reduce expenditures on things they oppose.

In the USA, federal government subsidies for energy totaled $37.2 billion in Fiscal Year 2010, a 108 percent increase from FY 2007 (US Energy Information Administration, 2011). Subsidies for renewable energy accounted for 60 percent of the FY 2010 total, compared with 7.6 percent for natural gas and petroleum (mainly exploration tax breaks). Coal received less than 4 percent of energy subsidies, a decline of more than 80 percent from the support provided in 2007. The nuclear energy sector only received some 7 percent of federal subsidies, mainly to decommission assets. Clearly, subsidies and their allocation are an important component of US climate policy.

In practice, one finds many of the above economic instruments operating simultaneously. For example, a government might subsidize farmers for growing energy crops and energy companies for building biofuel processing plants, while at the same time regulating ethanol content in gasoline and imposing carbon taxes on gasoline from petroleum. Indeed, they might even at the same time be subsidizing exploration for new sources of petroleum or natural gas.

Feed-in tariffs are a particular type of subsidy to the electricity sector. They guarantee power producers a fixed price for their electricity for a specified period of time. The electricity sector is important because it already accounts for nearly one-fifth of the world's final energy consumption, power can be generated from a great variety of energy sources, and electricity could possibly play a large role in future transportation, whether directly to recharge electric vehicles or indirectly by producing hydrogen fuel. Hence we turn our attention to the electricity sector.

ELECTRICAL POWER GENERATION

Fossil fuels are the most important source of energy in the global generation of electricity (Figure 9.1). Approximately two-thirds of electricity is produced from fossil fuels, while the remainder comes primarily from hydro and nuclear sources. Geothermal, biomass, solar, wind and other sources contribute a meager 2.6 percent of the energy required to produce electricity.

To obtain some notion regarding which countries generate the most electricity and the importance of coal in the global electricity generating mix, consider Table 9.1. Nearly 20 000 terawatt hours (TWh), or 20 petawatt hours (PWh),[3] of electricity

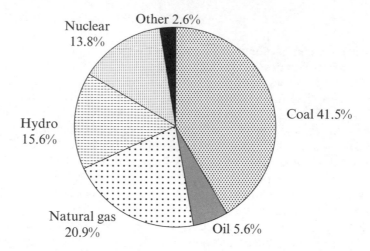

*Figure 9.1   Global electricity production by energy source, 2007, percent; total = 19 771
TWh*

*Table 9.1   Largest electricity producers, total and by selected fossil fuel energy source,
2007, TWh*

TOTAL		Coal/Peat		Gas	
USA	4 323	China	2 656	USA	915
China	3 279	USA	2 118	Russia	487
Japan	1 123	India	549	Japan	290
Russia	1 013	Japan	311	Rest of world	2 435
India	803	Germany	311	Total	4 127
Canada	640	South Africa	247		
Germany	630	Australia	194	Oil	
Rest of world	7 960	Korea	171	Total	1 114
Total	19 771	Russia	170		
		Poland	148		
		Rest of world	1 353		
		Total	8 228		

Source: International Energy Agency (2010b).

were generated in 2007, the latest year for which statistics are available from the
International Energy Agency (IEA, 2010a, 2010b). Notice that the USA and China are
the largest producers of electricity and also the largest producers of coal-fired power.
Other large industrial nations generate large amounts of electricity, with many relying
on coal (Figure 9.1). Canada is the sixth-largest power producer, but much of it comes
from hydro sources and a significant amount (about 25 TWh annually) is exported
to the USA. Clearly, rich countries are rich because they consume large amounts of
energy, especially electricity.

Table 9.2   Lifetime generation costs by generating type, $ per MWh[a]

Generating type[b]	Midpoint	Low	High
Wind onshore	68.08	36.39	168.71
Wind offshore	78.54	59.09	144.38
Solar thermal	193.64	193.64	315.20
Solar PV	192.21	141.10	2195.39
Run-of-river/small hydro	108.28	46.45	283.02
Large-scale hydro	53.12	53.12	99.33
Nuclear	30.71	24.34	80.26
Coal (lignite)	39.35	34.40	75.35
Coal (high-quality)	31.90	30.30	80.85
Coal (integrated coal gas)	44.73	31.94	69.15
Gas (CCGT)	54.62	44.69	73.24
Gas (open)	54.64	54.64	57.33
Waste incineration	11.39	−4.68	61.19
Biomass	48.74	43.64	117.59

Notes:
[a] Costs include capital, operating and maintenance, and fuel costs over the lifetime of a power plant,
   discounted to the present and 'levelized' over the expected output of the generating source over its lifetime.
   Values are in 2008 US dollars. The midpoint and low values are based on a 5 percent discount rate, as is the
   low value; the high value is derived using a 10 percent discount rate.
[b] Open-cycle gas turbines lose exhaust heat but can respond quickly to changes in demand; closed-cycle gas
   turbines (CCGT) recycle exhaust heat, which makes them suitable as base-load plants but makes it more
   difficult for them to ramp up and down.

Source: van Kooten and Timilsina (2009).

Although oil dominates total global consumption of energy because of its use in transportation, very little oil is used to generate electricity. With the exception of a few large, base-load power plants that rely on oil, petroleum is used mainly in diesel generators that power small grids such as those found in remote communities, and in much of sub-Saharan Africa where few alternatives to diesel generation currently exist. While energy security is often cited in the USA as a reason to subsidize wind, solar and other renewable sources of power generation, electricity generation is not reliant on imports of energy from offshore (e.g. the Middle East, Nigeria and elsewhere). As noted, the USA imports (hydro) electricity from Canada but the remainder is generated from domestic sources of energy.

An indication of the costs of producing electricity from various sources is provided in Table 9.2. The costs of producing electricity from wind and solar sources have fallen dramatically, while costs of geothermal, tidal, wave and some other renewable energy technologies are not yet known because they are in various stages of development. Advances in nuclear power generation technology and experience also continue, particularly with regard to performance and safety (Ansolabehere et al., 2003; Deutch et al., 2009). Yet most renewable energy programs tend, in practice and probably realistically, to ignore geothermal, tidal and wave energy in favor of wind and solar power. But they also exclude from consideration the substitution of natural gas for coal and greater reliance on nuclear energy, two important and proven low-carbon technologies. In essence,

Table 9.3   Energy densities of selected energy sources

Energy source	Energy density (W/m²)
Corn ethanol	<0.1
Biomass-fueled power plant	0.4
Wind turbines	1.2
Solar PV	6.7
Small oil well (10 barrels/day)	27.0
Average natural gas well (3300 m³/day)	287.5
Nuclear power plant[a]	56.0

Note: [a] Based on a 4860 ha location in Texas, although the power plant occupies only a very small area within the property.

Source: Bryce (2010, pp. 91–3).

therefore, the objective of reducing CO_2 emissions is confused with encouraging renewable energy in the generation of electricity (Deutch et al., 2009, p. 9).

Consider the future prospects of renewable energy sources in generating electricity, especially their near-term prospects as many developed countries have ambitious greenhouse gas emission targets, some of which are supposed to come into force within a decade. A major problem facing renewable energy relates to its low energy density. As indicated in Table 9.3, the energy density of renewable energy sources is simply too low compared with that of fossil fuels and nuclear power to make them sufficiently competitive with there sources. Therefore subsidies are required. Nearly 40 countries, and many more jurisdictions if provinces, states and cities are counted separately, provide potential generators of renewable power with feed-in tariffs. The four most common renewable sources of energy that qualify for feed-in tariffs are biomass, small-scale (usually run-of-river) hydro, solar and wind. The prospects of these four energy sources are briefly discussed in order below. With the exception of biomass, these renewable energy sources tend to be intermittent and therefore non-dispatchable.

Biomass for Generating Electricity

Increasing electrical power production from forest biomass, sawmill residue and 'black liquor' from pulp mills is constrained by high transportation costs and competition for residual fiber. This makes forest biomass an expensive source of energy.

Because of the extent of mountain pine beetle damage to forests in the interior of British Columbia, an obvious use of beetle-killed trees was considered to be power generation. While early studies suggested that this could be done without reliance on large feed-in tariffs (Kumar et al., 2008; Kumar, 2009), later studies indicated that this optimistic conclusion was based on average past costs of harvesting and hauling timber from the forest to sawmills.[4] Niquidet et al. (2012) found that, when account is taken of the rising costs of hauling timber as more remote beetle-damaged sites need to be harvested, marginal costs rise rapidly with truck cycle times (the time required to travel to and from the harvesting site). An electrical generating facility turns out to be only a marginally attractive option for reducing CO_2 emissions when feedstock costs are low; but, as feedstock costs alone rise from an equivalent of 4¢/kWh to 8.5¢/kWh, biomass power is no longer an economically viable option.

Producing char from biomass through a process known as pyrolysis (a form of incineration that chemically decomposes organic matter by heat without oxygen) suffers from similar problems, although high transportation costs might be mitigated somewhat by producing char on site. Nonetheless, the amount of char available for generating electricity will be negligible in comparison to what is needed and there are concerns that the process produces hazardous wastes.

Perhaps the best option for generating electricity from wood biomass is wood pellets. Wood pellet production plants are relatively inexpensive to construct and can, in some instances, be moved quite easily to new locations (although they are not mobile enough to be located at the harvesting site). Wood pellets can be used directly in coal-fired power plants with little or no adjustments to the burners – pellets can be pulverized much like coal and pellets are preferred over wood chips (which are used for pulp).

Because of their flexibility, relatively low production costs and government subsidy programs, demand for pellets has increased sharply. European demand for wood pellets has grown rapidly since about 2005 because of subsidies. As a result, British Columbia's wood pellet production capacity had risen to about one million tons by 2010. But as demand for other energy uses of wood biomass increase, prices will rise.

Using a regional fiber allocation and transportation model, Stennes et al. (2010) demonstrate a major drawback of timber feedstocks. As noted, hauling costs make it costly to employ timber for biomass generation of electricity. Indeed, in British Columbia and other jurisdictions where logging and hauling are important cost components of wood supply, wood residuals and other wood waste are available at a reasonable cost only as a result of timber harvests for sawmilling and production of lumber. Without lumber production, it is often too expensive to access biomass to support a bioenergy sector. In British Columbia, chips from sawmilling operations form the mainstay of the province's pulp industry. Other sawmill residues (bark, sawdust etc.) are already allocated by mills to on-site space heating and power generation, with some excess chips and residues used in the production of such things as wood pellets, oriented strand board and other products.

Competition for sawmill residuals occurs between pulp mills and other wood product manufacturers as well as heating and electricity. While there is some leeway to increase available wood waste by hauling roadside and other waste from harvest operations to power-generating facilities, competition for residual fiber raises prices. That is, when account is taken of the supply and demand of wood fiber for all its different purposes, there is little excess fiber available for generating power at reasonable cost. Feed-in tariffs and other subsidies for electricity production would harm existing users of fiber, such as pulp mills or wood pellet producers (Stennes et al., 2010). While pulp producers can outbid energy producers for wood fiber if pulp prices are high, they would have a harder time competing at lower prices for pulp, especially if bioenergy producers are subsidized. Thus feed-in tariffs for biomass energy in a jurisdiction such as British Columbia might well be politically unacceptable.

In other forest jurisdictions, there might be more leeway for fast-growing tree species to provide power, but similar problems are encountered. Competition for fiber implies that subsidies are required if the fiber is to be used for generating electricity. For example, the EU requires that 20 percent of total energy come from renewable sources by 2020, although only 7 percent came from renewable sources in 2009. To meet these targets, many countries will rely primarily on wind and energy from biomass. As a result, a wood deficit of 200 to 260 million m^3 is forecast for the EU by 2020, which is greater than Canada's annual harvest. Globally, an ECE/FAO report estimates that there will be a wood deficit of 320 to 450 million m^3 annually simply to satisfy planned biomass energy needs plus a growing wood-based industry.[5] Global wood fiber prices will certainly increase, resulting in potentially detrimental changes in land use (see below).

What is often neglected in discussions of biofuels and biomass-fired power generation is the fact that bioenergy is not carbon neutral as is often claimed. The combustion of biomass releases CO_2, more than that released from fossils fuels to generate an equivalent amount of energy. It is only when plants and trees grow that CO_2 is removed from the atmosphere, and this can take quite a long time in the case of trees. The timing of CO_2 emissions and sequestration matters: if CO_2 released by burning takes 20 years to be sequestered, for example, there is a CO_2 penalty associated with biomass burning that cannot be ignored – burning biomass releases more CO_2 per kWh than coal, while trees can take a long time to grow (van Kooten, 2009).

Further, greenhouse gas emissions related to harvests and hauling, and nitrogen fertilizers, may offset any CO_2 benefits of biomass as a fuel (Crutzen et al., 2008). Finally, using biomass for energy can result in land-use changes that largely offset gains from burning biomass in lieu of fossil fuels (Searchinger et al., 2008, 2009). More CO_2 is released in gathering biomass across a large landscape than is the case with coal, for example, as coal deposits are concentrated near a particular location. To mitigate the time that trees take to grow (upwards of 80 years), fast-growing tree species such as hybrid poplar or plants such as switchgrass can be used. While this tilts emissions more favorably towards the use of biomass, nitrogen fertilizer is often required to spur growth, and nitrogen oxides are a more potent greenhouse gas than CO_2.

From a policy perspective, energy crops (including trees) are not an efficient means of addressing climate change, although there may be potential to source energy from various biological organisms in the future. However, energy from biological organisms does not appear to be a major component of governments' policy arsenals for combating climate change. Landfill gas generated from solid waste is also a potential source of electricity, but even if it is employed on a large scale, its contribution to the globe's electricity needs would necessarily be extremely small. The same holds for the incineration of municipal wastes.

Hydraulics, Storage and Run-of-river Hydro

A number of countries have developed their hydraulic resources to build large-scale hydropower facilities. With the so-called 'three gorges' dam (affecting the Upper Mekong, Yangtze and Salween Rivers), China now has the greatest hydro capacity in the world (Table 9.4). Yet, in 2007, hydro production only accounted for 14.8 percent of China's consumption of electricity. This is much less than the proportions accounted for by hydro in Norway (98 percent), Brazil (84 percent), Venezuela (72 percent) and Canada (57 percent). India relied on hydropower to a greater extent than China, as did Russia despite its relatively abundant fossil fuel resources.

Table 9.4 Hydroelectric power production and capacity, 2007

Country	Production (TWh)	Capacity (GW)[a]	% of domestic consumption
China	485	126	14.8
Brazil	374	73	84.0
Canada	369	73	57.6
USA	276	99	6.3
Russia	179	46	17.6
Norway	135	29	98.2
India	124	35	15.4
Japan	84	47	7.4
Venezuela	83	n.a.	72.3
Sweden	66	n.a.	44.5
Rest of world	987	n.a.	n.a.
WORLD	3162	889	15.9

Notes:
[a] Data for 2006.
n.a. not available.

Source: International Energy Agency (2010b).

Large-scale hydro remains one of the best options for generating 'clean' electricity, but its main drawbacks relate to inadequate runoff for power generation (especially in regions where water is scarce, intermittent and/or unreliable) and negative environmental externalities (changes in the aquatic ecosystem, impediments to fish migration, land inundation by reservoirs etc.). Environmentalists oppose large-scale hydro development, particularly in developing countries because of the ecological damage it causes, while even small-scale, run-of-river projects have been opposed in rich countries on environmental grounds. Because of strong environmental opposition to hydroelectric developments, hydropower's future contribution to increases in overall generating capacity will inevitably remain limited in scope. Expansion of water power is not expected to be a large contributor to the mitigation of climate change.

Although unlikely to contribute much in the way of additional clean power, existing large-scale hydro and strategic expansions of reservoir storage capacity (which raise generating capacity) might serve an important purpose when combined with intermittent sources of energy, particularly wind and solar sources. For example, wind-generated power is often available at night, when base-load power plants are able to supply all demand. Wind energy would then need to be curtailed (wasted) or, where possible (and it may not always be possible), base-load plants would need to reduce output, causing them to operate inefficiently. If a base-load plant is coal fired, inefficient operation implies that CO_2 emissions are not reduced one for one as wind replaces coal. In some cases, the tradeoff is so poor that CO_2 emissions are hardly reduced at all. This problem can be overcome if adequate transmission capacity exists so that the excess wind-generated power could be stored behind hydro dams by displacing electricity demand met by hydropower. This is the case in Northern Europe, where excess wind power generated at night in Denmark is exported to Norway, with hydropower imported from Norway during peak daytime hours.

Similar relationships are found elsewhere. In Canada, for example, the provinces of Quebec and British Columbia rely almost exclusively on hydropower, while the respective neighboring provinces of Ontario and Alberta generate significant base-load power from coal (or nuclear in Ontario's case). Ontario and Alberta are both expanding their installed wind capacity. During nighttime, off-peak hours, excess wind and/or base-load power from Ontario (Alberta) is sold to Quebec (British Columbia), with hydropower sold back during peak periods. Given that the rents from these transactions have accrued to the provinces with hydro assets, Ontario and Alberta have nonetheless been less than keen to upgrade the transmission interties, preferring to look at other possible solutions to the intermittency and/or storage problems.

In all three cases, there are net economic and climate benefits from the development of higher-capacity transmission interties; or, in the case of Northern Europe, simply more interties between jurisdictions where wind power is generated (northern Germany, other parts of Denmark) and those with hydro resources (Norway and Sweden). The main obstacle is the lack of incentives for the wind-generating region to 'dump' power into the region with storage, as the latter captures all the rents from such an exchange. This is a game-theory problem: if institutions can be developed that facilitate the sharing of both the economic rents and the climate benefits (emission reduction credits), the jurisdictions have the incentive to better integrate the operations of their electricity grids (including construction or upgrading of transmission interties) so that overall CO_2 emissions are minimized.

Wind and Solar Energy: Generating Electricity from Intermittent Energy Sources

There are a number of promising renewable energy sources that could at some time in the future make a significant contribution to global electrical energy needs. However, the likelihood that these will have a major impact in the short or medium term (five to 50 years) is small. It is evident from Figure 9.1 that non-conventional sources of energy constitute only about 3 percent of global electricity production. Raising that to 20 percent or more constitutes an enormous challenge, especially in a world where energy demand is rapidly increasing as a result of economic development in countries such as India and China. Simply expanding the use of renewable energy and then incorporating renewable energy sources into energy systems will prove difficult, not least because an expansion in the use of renewables will lead to increases in their prices (as we noted with regard to wood biomass).

Among alternative energy sources, solar and wind energy are especially promising. The energy or irradiance from the sun averages some 1.366 kW/m², or 174 PW for the entire globe, but it is difficult to convert to usable energy. Other than through plant photosynthesis, there are two ways to harness this solar energy: (1) solar photovoltaic (PV) converts the sun's energy directly into electricity, while (2) solar heaters provide energy to warm a fluid such as water (swimming pools, water tanks etc.). Solar heaters convert up to 60 percent of the sun's energy into heat, while PV cells convert only 12 percent to 15 percent of the energy into electricity, although PV laboratory prototypes are reaching 30 percent efficiency. One problem with solar electricity is its prohibitive capital costs, which amount to some $13000 to $15000 per kilowatt (kW) of installed capacity (IEA, 2005; also see Table 9.2), although costs have fallen in the past several years. In addition, solar power is intermittent (e.g. output is greatly reduced on cloudy days), unavailable at night, and, at high latitudes, less available in winter when demand is high than in summer (due to shorter days). Nonetheless, for remote locations that receive plenty of sunshine and are not connected to an electrical grid, the costs of constructing transmission lines to bring in outside power might make solar PV and solar heaters a viable option.

Given the current drawbacks of many other renewable sources of energy, wind energy appears to be the renewable alternative of choice when it comes to the generation of electricity. As a result, global wind-generating capacity has expanded rapidly from only 10 megawatts (MW) of installed capacity in 1980 to 157899 MW by the end of 2009, an average annual rate of increase of some 49 percent (GWEC, 2010). Such a high average annual rate of increase is a result of starting from a low base, and is driven largely by feed-in tariffs and other government incentives.

ECONOMICS OF FEED-IN TARIFFS

Because electricity can be produced from any conceivable source of energy, renewable sources of energy can most easily be integrated into an economy through the electricity grid. Consequently, many jurisdictions have set renewable electricity targets, using regulation, feed-in tariffs or other forms of subsidy to encourage the generation of electricity from non-fossil-fuel, non-nuclear sources. Regulation takes the form of mandates, the best examples of which are renewable electricity standards that require electrical utilities to produce some proportion of their power from renewable sources by certain dates. Such requirements are being adopted in many developed countries, with some even having mandated the elimination of all coal-fired power plants.

With the exception of biomass, which has its own demons, wind, solar, run-of-river, wave and tidal energy suffer from intermittency – output is erratic and capacity factors are usually well below 30 percent.[6] This has serious complications for the way that electricity markets operate. In this section, we consider this complication in more detail as it pertains to feed-in tariffs (FITs). While the focus is on intermittent energy sources, some of the discussion also applies to FITs for biomass or, for that matter, diesel and other forms of power generation.

Electricity Markets

If the prices consumers pay for electricity are fixed, the demand function is essentially a vertical line – demand for electricity is completely inelastic and does not respond to changes in wholesale prices. Time-of-use pricing at the retail level affects demand directly (giving the vertical demand function a downward slope), but to implement time-of-use pricing requires a 'smart grid' – something beyond just smart meters. Smart meters can only detect how much electricity a customer uses at each time of the day; a smart grid enables the customer to adjust electricity use in response to price changes. Thus smart meters can be used to implement a tiered pricing scheme (such as daytime and nighttime pricing differentials), which can tilt the demand curve slightly; the smart grid enables off-site control of large appliances in response to changing prices.

Even if some degree of real-time pricing can be implemented, it is likely that the demand for electricity will remain highly inelastic. Based on cross-section and time-series analyses, the short-run elasticity of demand is about −0.3 (US Energy Information Administration, 2010, p. 26), while it is between −1.5 and −0.5 in the long run.[7] This implies that a 1 percent increase in the price of electricity results in a 0.3 percent reduction in demand in the short run, and a reduction of 0.5 percent to 1.5 percent in the long run.

To examine the supply side, assume an electricity system that is deregulated at the wholesale level. The electricity system operator (ESO) requires owners of generating facilities to commit to produce electricity at a given hour one day (24 hours) ahead of actual delivery. Each generator will offer to produce a certain amount of electricity at a particular price, knowing that the final price received is the market-clearing price for that hour. In essence, a power plant will offer units of electricity at a single price (or variety of prices if costs of producing electricity differ across units) to be produced and delivered on a specified hour the next day. This is known as day ahead unit commitment. Of course, as the hour approaches for which an owner of a generating facility has committed to supply power, more information about the status of generators and the evolution of prices becomes known. Therefore generators are able to make changes to their offers up to two hours before delivery. The extent of permitted changes is increasingly constrained by penalties as the hour nears.

What do the offers to supply electricity look like? Base-load nuclear and coal-fired power plants, and for some grids base-load hydropower dams, will bid in lowest. Indeed, for base-load facilities that cannot readily change their power output, or can do so only at high cost, the optimal strategy is to provide very low (even zero) price bids to ensure that they can deliver power to the grid. Open-cycle, natural gas peaking plants will want to bid in at their true marginal cost of production, which is essentially determined by the price they have to pay for fuel. The facilities that provide the highest bids are those that wish to export electricity to another system, regardless of the energy source used to generate the power; by setting their price high, their output is unlikely to be chosen by the system operator and can thus be exported. (Importers will want to set their prices low to guarantee that the imported power will be chosen.) In between the extreme prices are found a variety of generating facilities, such as biomass plants, combined-cycle gas plants (CCGT), and sub-units of extant plants that are at different levels of readiness, maintenance and so on. Once the ESO has all the information regarding the amounts of electricity that the various components of the generating system are willing to supply, and their associated prices, a 'market merit order' is developed to allocate power across the generators depending on demand. An example is provided in Figure 9.2.

In Figure 9.2, the supply curve is given by the market order. Base-load nuclear and coal facilities bid in at the lowest price, followed by CCGT and other generating facilities as indicated.[8] The market-clearing price is determined by the location of the demand curve at that hour. Assuming the demand curve on the right, the market price P is given by the marginal open-cycle natural gas plant (NG 2). If the transmission infrastructure somehow impedes NG 2 (or some other plant) from delivering power, then NG 3 deter-

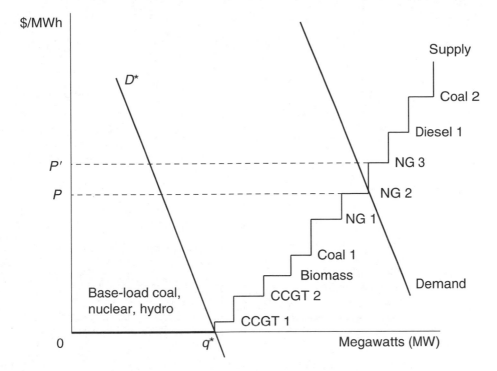

Figure 9.2   Market merit order

mines the market-clearing price, which becomes *P'*. All generators get paid *P* for the period in question (or *P'* if transmission capacity results in NG 3 coming on line instead of some other generator).

Base-load facilities bid in at zero price to avoid incurring the high costs of curtailing output, but also knowing they will receive *P* or *P'*. If the demand curve is *D** then the wholesale price is zero and only base-load facilities generate electricity. Assuming that investments in base-load capacity were determined from the load duration curve, demand would never be less than *D**, with *q** representing the system's minimum load. To reiterate, base-load plants would bid in at a zero price despite potentially earning no revenue; this is to avoid high costs of ramping production or, worse, dumping power in an emergency-like situation (i.e. instantaneously reducing pressure in the boiler).

Suppose a feed-in tariff for biomass-generated power increased biomass generating capacity. In terms of Figure 9.2, biomass would drop in the merit order because of the subsidy and more would be available. This could result in moving CCGT 2 or even CCGT 1 and CCGT 2 'higher up' in the merit order – essentially the bid prices would remain the same but biomass would be chosen before these generating sources. All other generators would be chosen later in the merit order, with NG 1 or even 'coal 1' becoming the marginal power plant. The price of electricity would fall, *ceteris paribus*. If biomass generation becomes base-load, it will be necessary to displace some nuclear and/or coal base-load capacity. This might be desirable except that, as noted earlier, there may be constraints on wood fiber availability.

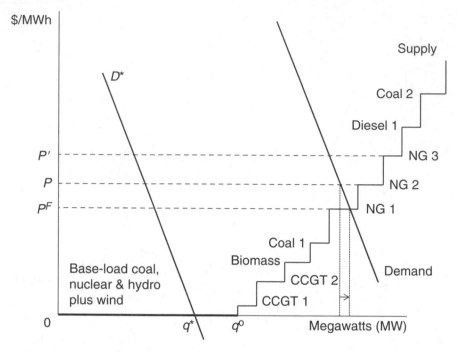

Figure 9.3   Market merit order with wind added

The picture changes completely when wind, solar, run-of-river or other variable generating capacity is introduced into the electricity system as a result of a feed-in tariff. The situation can be illustrated with the aid of Figure 9.3. The only difference between Figures 9.2 and 9.3 is the addition of $q*q^0$ electricity from variable generating sources (hereafter referred to as wind). This shifts the supply curve in Figure 9.2 to the right by amount $q*q^0$. Now, with the original demand curve (the one on the right in Figure 9.3), it is no longer NG 2 that is the marginal producer of electricity; rather, it is the plant with a lower marginal cost, NG 1. The market-clearing price of electricity for that hour falls from P to P^F. The feed-in tariff lowers the price of electricity, which will induce consumers to purchase more of it (as indicated by the arrow).

What does one do with the wind energy $q*q^0$ if the demand in a given hour is $D*$? Clearly, either the wind must be curtailed (wasted) or base-load output reduced. Base-load hydropower can easily be reduced, as discussed below, so consider only a system with base-load thermal generating capacity. If $q*q^0$ could be reliably produced in every period, so that it can be considered part of the base load, then some coal or nuclear base-load capacity becomes redundant and can be eliminated – an ideal outcome.[9] However, wind-generated power is not reliable and thus cannot replace thermal base-load capacity, except at some cost.

Suppose base-load capacity is reduced by the amount $q*q^0$. Then, whenever wind power is less than $q*q^0$, this is the same as shifting the supply curve in Figure 9.2 to the left, which would raise the market price for every hour that wind is less than $q*q^0$, while lowering price if wind output exceeds $q*q^0$. Thus the effect of a feed-in tariff for wind (or solar, wave, tidal etc.) is to increase price volatility if thermal base-load capacity is driven from the system;[10] if thermal base-load capacity is not driven from the system, electricity prices will generally be lower, but base-load plants will need to ramp up or down if wind energy is non-dispatchable (i.e. considered to be 'must run'), which will increase their operating costs (van Kooten, 2010). Alternatively, if wind is considered dispatchable, wind will need to be curtailed or wasted.

The situation is somewhat different in a system with significant hydropower-generating capacity, because hydropower can provide base-load power and serve the peak load and reserve markets. The presence of significant hydro capacity enables a system to absorb wind power that might overwhelm the ability of a system with a high thermal capacity in the generating mix to absorb it, or raise system costs by too much in doing so. That is, the existence of hydro reservoirs enables a system to store wind energy that would be wasted in systems lacking hydro-generating capacity in the mix. However, there must be times when this stored wind energy is required to meet load, perhaps at peak load times.

Because demand and supply of electricity must balance at all times, there is one further aspect to electricity markets: the need for operating reserves. These consist of regulating reserves that adjust supply continually to meet small changes in demand (load) and supply over a time frame of several seconds to 10–15 minutes. Thermal power plants have the ability to adjust output very quickly over a small range in less than a minute. Some open-cycle gas and diesel generators are operating below capacity or on standby and, by adjusting the fuel received (in essence apply more or less pressure to the 'gas pedal'), can readily adjust output. Some generators will simply be idling in standby mode, not delivering electricity to the grid; these are referred to as 'spinning reserve', as distinguished from units that are operating below capacity. For example, generator NG 1 in Figure 9.3 is not operating at full capacity and can easily adjust supply (e.g. by the amount indicated by the arrow).

Storage devices, such as batteries and flywheels, might also be used in a regulatory capacity, as might hydropower. Automated generation control, which is also known as regulation, is used to manage small fluctuations in the supply-load balance.

Contingency reserves, on the other hand, are required to meet a situation where power from any given generator is suddenly lost for whatever reason. They are designed to handle emergencies – the contingency that a power plant goes 'off line' and is unable to provide the electricity that it had committed. For example, the Western Electricity Coordinating Council (WECC) requires that contingency reserves be sufficient to cover the most severe potential loss (loss of the largest generating unit) plus some proportion of the total production from hydro and thermal sources.[11] The market for contingency reserves is indicated in Figure 9.4.

Suppose that the merit-order demand for contingency reserves in a given hour is denoted D^C, which is determined by the conditions set out by WECC. The various units bid their reserves much as they do in the establishment of the merit order in Figure 9.2, although the bid also includes a capacity payment needed to maintain the reserve position. If the offer is accepted, the capacity payment would be made by the ISO regardless of whether any power was dispatched to the grid.

In the ancillary market, the open-cycle and diesel peakers will now want to bid in at a low price because they are the ones that can get off the mark the quickest. Likewise, the bid price of the hydro contingent reserves will be low, perhaps even zero (in which

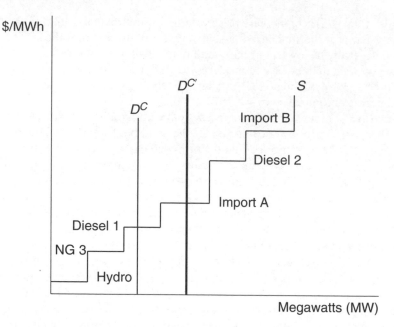

Figure 9.4   Market for ancillary contingent reserves

case they obtain the marginal market-clearing price P in Figure 9.2), but the prices bid by peakers NG 3 and Diesel 1 will also be low because they know that, when there is a demand for ancillary services, they will receive at least the price determined by the marginal generator (NG 2 in Figure 9.2) plus their own bid in the ancillary market. Base-load plants, on the other hand, will bid in very high, if at all, because they can only ramp up output at great expense. The actual bid price will depend on the strategy of the owners of the various units, which will depend on the anticipated state of the units at the time. A market for ancillary contingent reserves is established in a fashion similar to that of the real-time market, except that units will also receive a capacity payment for their reserve position, while receiving the market-clearing price for any electricity they are asked to dispatch.

In the previous discussion of intermittent renewable sources of energy, it was assumed that wind output was predictable. When wind enters an electricity system, however, there is a real risk that output from this source falls (or rises) dramatically and unexpectedly during the course of an hour. This means that the system operator must not only meet its normal reserve requirements (e.g. as set out by the WECC), but must also have additional regulating and contingency reserves that address the variability in wind. In terms of Figure 9.4, this is seen by the shift of demand for contingent reserves from D^C to $D^{C'}$. When wind capacity is installed, contingency reserves could increase by upwards of 10 percent and regulating reserves by even more (Gross et al., 2006, 2007).This often requires additional investments in operating reserves of various types.

The additional costs of reserves must be added to the operating costs that intermittent energy imposes on an electricity grid. These are well documented and consist of costs related to more frequent ramping up and down of base-load facilities, operating power plants below their optimal operating range, and more frequent stops and starts of diesel and gas peaking generators (Maddaloni et al., 2008; Prescott and van Kooten, 2009; van Kooten, 2010; le Pair and de Groot, 2010).

THE REAL WORLD OF FEED-IN TARIFFS

Although many jurisdictions provide direct subsidies to the construction of renewable electrical generating facilities, transmission lines and R&D related to generation of electricity from wind, solar and other sources, they have also implemented feed-in tariffs (FITs) as an incentive to increase power generation from various renewable sources. FITs guarantee developers of renewable generating capacity a preferred price for any power they deliver to the grid. For example, if the market price of electricity is $0.10 per kilowatt hour (kWh), the FIT might be $0.15 per kWh, which amounts to a subsidy of $0.05 per kWh. The supplier of renewable power is often guaranteed the higher rate for a period of five or more years. Along with the FIT, the authority might also have to guarantee the renewable generator of power access to the grid through newly constructed or extant transmission lines.

While subsidies might help in the short run, they are not sustainable in the long run because they distort production decisions, resulting in inefficiencies. Suppose only some countries provide FITs that result in greater reliance on wind and solar energy, thereby displacing fossil fuels in power generation. This does not mean that the fossil fuels are no longer burned. After all, the mining of fossil fuels creates economic rents that owners will continue to exploit. Thus, for example, eliminating coal power in the USA does not necessarily prevent the mining of coal and its sale to China, India or elsewhere to be burned for power generation. If the USA were to close its coal-fired power plants, owners of coal mines in Appalachia would more than likely sell coal overseas for the production of electricity (*The Economist*, 29 January 2011, pp. 64–5). Indeed, curtailing coal use in the USA and Europe could reduce global coal prices, at least in the short term. This in turn will incentivize those countries that continue to rely on coal to use it less efficiently, and overseas power producers to invest more in coal-fired capacity than they would otherwise. Once such investments have been made, it could be 40 or more years before such capacity becomes redundant. In this manner, the climate benefits of the original subsidies are offset. This is one aspect of what has been termed 'leakage' in some of the climate change literature – the offsetting increase in emissions that accompany programs designed to reduce emissions in the first place (Jenkins et al., 2011).

As we will see, FIT programs are primarily but not exclusively a rich-country phenomenon. Given that developing countries are increasing electricity output from all sources, feed-in tariffs and other subsidies in some of those countries are used to encourage investments in all forms of power generation, as the desire is for more electrical generating capacity and not for renewable power *per se*. Thus it is FIT program benefits in rich countries that are inevitably countered to some extent by rising CO_2 emissions in developing countries, which is due to their improved access to fossil fuels. This is another example of a leakage.

Table 9.5　Feed-in tariffs for onshore and offshore wind projects, including length of programs, various jurisdictions, as of November 2010

Jurisdiction	Years	$/kWh
Large-scale offshore wind		
Germany (without bonus)	20	0.174
(with bonus)	20	0.201
France	15	0.174
Ontario	20	0.190
Large-scale onshore wind		
Germany	20	0.123
Ontario	20	0.135
France	15	0.110
Spain	25	0.105
South Africa	20	0.179
Vermont	20	0.128
Small onshore wind projects (examples)		
Portugal <3.68 kW (Microgenerator)	15	0.578
UK 1.5 kW–15 kW	20	0.424
UK >15 kW<100 kW	20	0.383
Italy <200 kW	15	0.402
Israel <15 kW	20	0.347
Israel <50 kW	20	0.444
Switzerland <10 kW	20	0.204
Vermont <15 kW	20	0.204
Washington State out of state	6	0.123
Washington State fully in state	6	0.419
Wisconsin, Xcel Energy	10	0.067
Wisconsin, Madison Gas & Electric	10	0.062
Indianapolis Power & Light >50 kW<100 kW	10	0.143

Notes:　Feed-in tariffs for solar photovoltaic (PV) are much larger than those for wind power, as indicated in Table 9.6. Again Ontario's feed-in tariffs for solar power are the most lucrative. The State of Washington is the only jurisdiction to offer more money to solar power producers, but only if the solar panels are manufactured within the state.

Source:　Calculations by Paul Gipe, Bakersfield, CA. Available at http://www.wind-works.org/articles/feed_laws.html.

Existing Feed-in Tariff Programs: A Comparison[12]

Many countries employ feed-in tariffs as a means of encouraging private development of renewable generating capacity. In Table 9.5, we summarize the extent of FIT programs for wind projects in various jurisdictions. The Ontario FIT program that is discussed in the next subsection is one of the more lucrative programs, paying the highest tariffs for large onshore and offshore projects. The only exceptions are South Africa for onshore projects and Germany for offshore projects, but then only as a bonus if a wind project is grid compatible (so there is no need for increasing transmission line capac-

Table 9.6   Feed-in tariffs for solar photovoltaic, including length of programs, various jurisdictions, as of November 2010

Jurisdiction	Application	Years	$/kWh
Italy[a]	Rooftop small	20	0.524
	Rooftop large	20	0.443
	Ground-mounted small	20	0.465
	Ground-mounted large	20	0.368
South Korea	>3 kW	15	0.758
India	< 1 MW	25	0.435
France	Building Integrated	20	0.562
Germany[b]	Rooftop small (<30 KW)	20	0.576
	Rooftop large (> 1 MW)	30	0.442
	Ground-mounted	20	0.428
Czech Republic		15	0.664
Spain (2007 RD)	<100 kW	15	0.455
Austria	<5 kW	12	0.616
Ontario		20	0.802
Washington State[c]	Manufactured in state	8	1.175
California[c]	Commercial	5	0.673
South Australia[c]	Residential	5	0.431

Notes:
[a] After May 2011, but declining by 6 percent beginning in 2012.
[b] Data for 2009.
[c] Form of net metering.

Source: Calculations by Paul Gipe, Bakersfield, CA. Available at: http://www.wind-works.org/articles/feed_laws.html.

ity). As indicated in the appendix, there are many other jurisdictions that offer FITs for wind power generation, but the scale of projects is limited to at most 10 MW, with 500 kW a more frequent maximum. A 10 MW project constitutes perhaps five large wind turbines or 40 very small turbines. In some countries, FIT payments are of short duration and programs are subject to arbitrary cancelation. An indication of the extent to which small wind projects are subsidized in various places is provided in the lower half of Table 9.5 (see also Table 9.6 for solar examples).

It is important to note, however, that the feed-in tariff is nothing more than a guaranteed payment to the electricity producer. The producer receives the wholesale price as determined from the merit order and load in each hour plus a premium determined by the value of the FIT. Thus the actual subsidy paid to the producer varies from one hour to the next and from one jurisdiction to another. In jurisdictions that are not deregulated, it is impossible to determine the actual value of the subsidy, while it requires significant effort to determine the subsidy in the case of a deregulated market such as that described in the previous section.

Many other countries offer feed-in tariffs to producers of electricity generated from wind, hydro, solar, biomass, and even geothermal, tidal and wave sources. A list of countries providing FITs for the main renewable energy sources is provided in the appendix.

It should be noted, however, that power producers employing renewable energy are not the only ones to benefit from FITs. In Tanzania, for example, producers of diesel-generated electricity in rural areas are eligible for a feed-in tariff of $0.253/kWh, while small power producers in general receive between $0.068 and $0.091 per kWh depending on the season, with power provided during the dry season receiving a premium. In all cases, contract lengths for FITs in Tanzania are 15 years.

Some countries are now rethinking their FIT structures. For example, Germany's FITs for rooftop solar PV have fallen in 2011 to $0.385/kWh for units with a capacity less than 30 kW and to $0.290/kWh for units with a capacity greater than 1 MW, while tariffs for freestanding units have been reduced to $0.296/kWh. At the same time, the length of the German FIT program for large-scale rooftop PV has been reduced from 30 to 20 years. The German government is in discussions with solar power producers to reduce FITs even further because

> consumer energy prices for those not using photovoltaic have risen noticeably to cover the subsidy costs for those using solar energy ... [The government] is facing growing resistance from non-solar electricity bill payers. Energy providers also have difficulties incorporating any surplus solar power into their networks, as almost half the world's solar panels are mounted on German roofs, and not linked directly to the grid.[13]

Thus it is a combination of ratepayer dissatisfaction and difficulties that system operators have in managing generating assets to accommodate increasing amounts of intermittent sources of power that is leading the government to back off its subsidies.

The Netherlands is reconsidering its feed-in tariffs for wind power because it is finding that much of the presumed reduction in CO_2 emissions is offset through leakages. In essence, the installation of wind, run-of-river hydro and solar energy capacity does not result in a one-to-one reduction in fossil fuel capacity, nor is there a one-to-one reduction in CO_2 emissions for each kW of wind that replaces a kW of fossil-fuel-generated power. The reasons are well known. With energy sources that are intermittent or erratic, traditional thermal-generating facilities cannot be taken off line; rather, they often produce below their optimal efficient operating range, requiring more fuel and releasing more CO_2 per unit of output. Even if large-scale storage is available (e.g. hydro reservoirs), it cannot entirely prevent base-load and back-up generation from remaining on line and releasing CO_2. And, as noted earlier, when it comes to burning biomass for electricity, there may be little in the way of CO_2 savings.

It is not leakages of various kinds that are the only problem, however. As noted, subsidies must be paid by taxpayers or ratepayers. As noted in the theoretical discussion, a feed-in tariff for wind energy causes the price received by producers to be somewhat reduced but also more volatile. Thus, in order to pass the subsidy to ratepayers, the ISO or regulator has to add to the market-clearing wholesale price an amount to cover the subsidy, which assumes inelastic demand. It is partly the backlash from electricity consumers that is causing governments to rethink feed-in tariffs. Lucrative feed-in tariffs lead to high electricity rates because someone must eventually pay for the subsidization of higher-cost alternatives for generating electricity – pay the costs of reducing CO_2 emissions.

Another problem with FITs is that the authority chooses the technology to be used for reducing CO_2 emissions, and such choices have a social cost. Alternative technologies might be de-emphasized or overlooked, unanticipated costs related to the existing generating mix and transmission constraints may have been ignored, and externalities may have been disregarded (e.g. transmission lines from a wind site to a load center may need to bypass a wilderness area thereby greatly increasing costs). Often the authority bears these costs because it has done more than simply provide a FIT: it has also guaranteed access to transmission lines. As a result, the costs of generating electricity are higher than the amount of the subsidy, which equals the FIT minus the market price, and reductions in CO_2 emissions are lower than anticipated, sometimes much lower. Further, the FITs lock the authority into long-term payments as contracts average 20 years and tariffs ratchet upwards as a result of inflation clauses.

Feed-in Tariffs: The Case of Ontario

Ontario probably has the world's most lucrative feed-in tariffs for renewable energy. For this reason, the Ontario program is considered in somewhat more detail as an example of the potential costs and pitfalls. The province of Ontario is committed to reducing its reliance on thermal generation of electricity, especially coal and nuclear. It aims to eliminate coal-fired power generation entirely and, especially in light of events in Japan in March 2011, much nuclear capacity as well; this is quite a challenge because together they account for well over half of Ontario's generating capacity. Thus the Ontario government launched one of the most ambitious attempts to affect power generation from renewable sources when it passed the Green Energy and Green Economy Act on 14 May 2009.

The FIT schedule under the Act is provided in Table 9.7. The important thing to note about the FIT schedule is that feed-in tariffs are indexed to inflation, with the exception of solar power. Solar power is not indexed to inflation because the subsidy is high to begin with and prices of solar panels are expected to fall dramatically in the future. There is a significant feed-in tariff for electricity produced from biomass as the government seeks to replace coal-fired power with wood pellets.[14] For biomass generators exceeding 10 MW capacity the FIT is 13.0¢/kWh, while it is 13.8¢/kWh for smaller generators. Contracts are 20 years in length and tariffs are indexed to the Ontario Consumer Price Index. The tariff is also increased by a factor of 1.35 during peak hours (7:00 a.m. to 11:00 a.m. and 5:00 p.m. to 9:00 p.m.), but reduced by 0.90 for all off-peak hours. Already wood producers in British Columbia and Ontario are investing in wood pellet production for domestic use and export (see Stennes et al., 2010).

Recall that the feed-in tariff is a guaranteed price that a power producer receives. Therefore, to estimate the potential size of the subsidy associated with Ontario's FIT program it is necessary to have some idea about the (wholesale) market price. Assuming that the retail and wholesale prices are nearly identical, it is possible to make some crude calculations regarding the subsidy using information about electricity rates. Residential customers with smart meters pay 9.9¢ per kWh at peak times, 8.0¢/kWh during mid-peak periods (11:00 a.m. to 5:00 p.m.) and 5.3¢/kWh during off-peak times (9:00 p.m. to 7:00 a.m.). Customers without smart meters pay 6.5¢/kWh for the first 600 kWh (in summer the first 1000 kWh) and 7.5¢/kWh thereafter.

Ontario's average electrical load was some 16 000 MW during 2007, although it has

Table 9.7 *Ontario power authority's feed-in tariff (FIT) program for renewable energy projects, base date: 30 September 2009*

Renewable type	Size (capacity of generating plant)[b]	Contract price (¢/kWh)	Percentage escalated[a]
Biomass			
	≤ 10 MW	13.8	20
	> 10 MW	13.0	20
Landfill gas			
	≤ 10 MW	11.1	20
	> 10 MW	10.3	20
Biogas			
on-farm	≤ 100 kW	19.5	20
on-farm	> 100 kW, ≤ 250 kW	18.5	20
biogas	≤ 500 kW	16.0	20
biogas	> 500 kW, ≤ 10 MW	14.7	20
biogas	> 10 MW	12.2	20
Wind			
onshore	Any size	13.5	20
offshore	Any size	19.0	20
Solar			
roof/ground	≤ 10 kW	80.2	0
rooftop	> 10 kW, ≤ 250 kW	71.3	0
rooftop	> 250 kW, ≤ 500 kW	63.5	0
rooftop	> 500 kW	53.9	0
ground mount	> 10 kW, ≤ 10 MW	44.3	0
Water power[a]			
	≤ 10 MW	13.1	20
	> 10 MW, ≤ 50 MW	12.2	20

Notes:
[a] Performance factor: 1.35 peak, 0.90 off peak.
[b] Generally a 20-year contract with 2–3-year lead time; for hydro, 40-year contracts.
[c] Indexed by the Ontario CPI.

Source: http://fit.powerauthority.on.ca/Storage/99/10863_FIT_Pricing_Schedule_for_website.pdf (viewed 21 April 2010).

fallen somewhat since then as a result of the financial crisis, which caused some major demanders of power to shut down. Coal- and gas-generating capacities are both about 4000 MW; nuclear-generating capacity amounts to some 10 000 MW, while hydro capacity is nearly 6000 MW. To provide some indication of the costs and benefits of Ontario's FIT program, assume that only 30 percent of the load is satisfied by fossil fuels, or 4800 MW per hour, and the objective is to eliminate that production. Further, assume that, despite the capacities of coal and natural gas generation, coal-generated power accounts for half or more of fossil-fuel-generated power. Finally, assume that biomass- and wind-generated power substitute for fossil fuel power – biomass accounts for either half or one-quarter of the required substitute power with onshore and offshore wind accounting for two-thirds and one-third, respectively, of the remainder.

Table 9.8 *Costs and benefits of Ontario's feed-in tariff program: hourly CO_2 flux (tCO_2) and cost (US\$) of reducing CO_2 emissions, various scenarios*

	Biomass 50%; wind 50%			Biomass 25%; wind 75%		
Coal to NG →	1 : 0	¾ : ¼	½ : ½	1 : 0	¾ : ¼	½ : ½
CO_2 flux						
Coal saving	1749.2	1311.9	874.6	1749.2	1311.9	874.6
NG saving	0	247.9	495.8	0	247.9	495.8
Sequestered[a]	665.8	665.8	665.8	332.9	332.9	332.9
Biomass emission	2058.2	2058.2	2058.2	1029.1	1029.1	1029.1
Net flux	*356.9*	*167.5*	*−21.9*	*1053*	*863.7*	*674.3*
Subsidy	\$272 000	\$272 000	\$272 000	\$300 000	\$300 000	\$300 000
Subsidy per tCO_2	\$762	\$1624	n.a.	\$285	\$347	\$45

Notes:
[a] Carbon sequestered in tree growth over 25 years using growth function $v(t) = ¼\, t^2 e^{-0.2t}$, including all above ground biomass, and with carbon discounted at 2 percent.
n.a. indicates not applicable because eliminating fossil fuel generation results in a net release of CO_2 – there is no climate change benefit whatsoever in this scenario.

For every metric ton (tonne or t) of coal that is burned, 7506 kWh of energy are generated and 2.735 tonnes of CO_2 are released.[15] Thus it takes 320 tonnes of coal to burn half of the 4800 MWh of electricity supplied by coal-fired generation, releasing 874.6 tCO_2 hourly or 7.660 $GtCO_2$ per year. At the same time, natural gas plants will release 495.8 tCO_2 each hour or 4.343 $GtCO_2$ annually if they generate 2400 MW of electricity each hour.

The costs to the government of the FIT program depend on the extent to which various renewables substitute for fossil fuel generation and the average amount that final consumers pay for electricity. In Table 9.8, it is assumed that consumers pay an average of 8.5¢/kWh. Using various biomass and wind combinations and fossil fuel displacement scenarios, and FIT data from Table 9.7, it is possible to calculate carbon fluxes and costs to the public treasury of reducing CO_2 emissions. Results provided in Table 9.8 suggest that costs to the treasury could amount to \$2.4–\$2.6 billion annually, which will put a severe strain on the provincial treasury. In essence, by substituting fossil fuel energy with renewable sources in the generation of electricity, Ontario will pay a subsidy ranging from some \$45 per tCO_2 to well over \$1000, depending primarily on the extent of biomass generation. Greater reliance on biomass compared to wind leads to higher costs.

Several points are worth mentioning. First, there are much cheaper ways to reduce CO_2 emissions, as indicated by the low price at which certified emission reduction credits are traded on the EU carbon market (approximately €13 or \$17.50 per tCO_2 in August 2011). Second, the analysis in Table 9.8 is crude, focuses only on the costs to the public treasury and excludes any other costs, some of which can be quite high. For example, it is assumed that wind energy can substitute directly, one for one for fossil fuels, which is certainly not the case (van Kooten, 2010). Third, as noted above, Ontario is not the only jurisdiction to employ feed-in tariffs. Germany subsidies to wind, solar and hydro generation amounted to $7.3 billion in 2009 and were forecast to rise to $11.3 billion by the end of 2010.[16]

Finally, it is important to consider how FIT subsidies are paid. Currently, Ontario has not raised electricity rates to reflect the cost of its FIT program partly because it fears this will make Ontario firms less competitive and partly to avoid any political fallout. By not allowing rates to rise, however, there is also no incentive for consumers to reduce their electricity consumption; indeed, unless the costs of the FITs are passed along to consumers as a tax on market-determined wholesale rates, prices might even fall (as discussed previously), potentially leading to an offsetting increase in CO_2 emissions. In all likelihood, the Ontario government will eventually need to shift program costs onto rate-payers because the budgetary burden of large subsidies (Table 9.8) cannot be sustained in an era of fiscal restraint. At that point, political opposition to renewable energy subsidies is likely to increase.

DISCUSSION

Unlike carbon taxes or emissions trading, feed-in tariffs distort the playing field towards the authority's preferred electrical generating option(s). In effect, the authority (e.g. a regulator such as the US EPA) selects those technologies that it feels will best accomplish its objectives. One objective might be to reduce CO_2 emissions, but, given the questionable nature of intermittent sources of energy and the high costs of wood fiber, there clearly exist other objectives that FIT programs seek to address. These objectives might include a desire to create jobs, develop a wind turbine and/or solar panel production sector, eliminate coal burning for non-climate-related reasons (replacing coal with biomass), dramatically increase harvests from domestic forests, diversify sources of energy, appease environmental lobby groups and so on. However, these are anything but climate-related objectives – they are the domain of industrial and/or macroeconomic policy and not climate policy.

As industrial policy, renewable energy subsidies have been a disappointment. Subsidies for biomass power generation redirect wood fiber away from pulp and engineered wood products. Feed-in tariffs designed to promote wind and solar generation of electricity have benefited primarily Chinese manufacturers. Because wind turbines and photovoltaic panels require inputs of rare earth minerals that are available almost exclusively from China (*The Economist*, 2 October 2010, p. 64) and because production costs are so much lower in Asia, no jurisdiction has yet been able to become a powerhouse in the production of turbines or solar panels despite efforts to do so. It is true that costs of turbines and panels are falling, but primarily as a result of subsidized R&D in the developed countries. But these improvements have not made wind or solar more competitive with nuclear, coal or natural gas, while manufacturing continues to be centered in Asia.

As seen in this chapter, most FIT programs provide subsidies that last for a decade or more, with many programs also providing inflation protection to those investing in electrical generating capacity based on alternative energy sources that are eligible for FIT payouts. As a result, an electrical grid might be locked into generating assets that are not compatible with existing generating assets, while reducing incentives to invest in generating assets that might lower overall CO_2 emissions in the future. For example, in a study of the Ontario system, Fox (2011) finds that, under a carbon tax (or an equivalent emissions cap), it is optimal to decommission coal and invest more in natural gas capacity, although when the tax is set very high (the cap is very stringent) it is optimal to invest only in nuclear capacity. In all his scenarios, the role of wind is limited by its intermittency.

At the same time, FIT programs can lock a government into the subsidization of power from alternative energy sources for an extended period. This can impose a long-term burden on the treasury and taxpayers. If costs can be passed along to consumers of electricity in the form of higher rates, then consumers are forced to pay for politically motivated programs, ones that may not directly target CO_2 emission reduction or do so in an inefficient fashion (as noted by Fox). Higher electricity rates disadvantage industry relative to that in other jurisdictions where rates are lower, transfer income from general ratepayers to recipients of FIT subsidies, and harm those least able to pay higher rates for heating or cooling their homes.

If the government is unable to transfer the costs of FIT subsidies to ratepayers (or simply desires not to), the burden on the treasury could be unsustainable. For Ontario, which has an average load of 16 GW, the annual cost was calculated to be some $2.5 billion or more than $32 billion over a 20-year project life (discounted at 7.5 percent), a not insignificant sum. Meanwhile, as demonstrated in this chapter, the feed-in tariff leads to a reduction in electricity prices at the wholesale level, while the deployment of intermittent generating capacity causes prices to fluctuate to such an extent that it reduces incentives to invest in new capacity. Assuming that part or all of the reduction in wholesale prices gets passed along to (at least some) consumers, there is a rebound effect that could offset any reduction in CO_2 emissions by 60 percent or more (Jenkins et al., 2011).

Given a desire to promote development of alternative energy sources for generating electricity, what policies might a government employ? If the sole objective is to reduce CO_2 emissions, the economist would favor a carbon tax as such a tax would address the issue at hand. The tax would tilt the playing field against fossil fuels, particularly coal but also biomass, and give non-CO_2-emitting renewable energy sources a leg up. However, the required carbon tax may need to be quite high to encourage investment in wind and solar energy; while it adversely affects extant generators, it might not be enough to close them down. The authority might simply reject a tax in favor of feed-in tariffs because it has in mind objectives over and above that of reducing greenhouse gas emissions. The analysis in this chapter suggests that, if this is the case, the authority needs to consider alternative policies (e.g. construction subsidies to wind, solar etc., tax holidays, capital cost allowances) that lead to investments in desired renewable energy alternatives. These might be more effective and less costly.

NOTES

1. This target is directed particularly at the countries of Western Europe, or the EU-25, although more recent entrants to the EU are also expected to make significant gains towards its achievement.
2. Proponents of biomass energy argue that it is CO_2 neutral, but this is not the case, as pointed out below in the discussion pertaining to the use of biomass for power generation.
3. A watt (W) equals 1 joule (J) per second. A kilowatt (kW) equals 1000 W; a megawatt (MW) = 10^6 W; a gigawatt (GW) = 10^9 W; a terawatt (TW) = 10^{12} W; a petawatt (PW) = 10^{15} W.

4. Using average harvest and transportation data, Stennes and MacBeath (2006) found it was more economical to transport wood fiber from the BC interior to coal-fired power plants in Alberta than to construct a biomass power-generating facility locally. The reason had to do with the adequacy of wood fiber supply over the life of a power plant, something that could have been addressed with a combined wood–coal-fired plant.
5. Results reported by Don Roberts, CIBC, in presentations given in early 2010.
6. The capacity factor of a generator refers to the actual power output over a period compared to what it could potentially produce. For example, a 2 MW capacity wind turbine could conceivably produce 17.52 GWh of electricity in one year (= 2 MW × 8760 hours). But this depends on perfect wind conditions. With variable wind, the turbine might only produce 4200 MWh, and its capacity factor would be 24.0 percent. The capacity factor of a base-load, coal-fired power plant, on the other hand, might be 85 percent or more.
7. Price elasticities between 0 and −1 indicate inelastic demand. In a meta-regression analysis of studies of US residential demand for electricty, Espey and Espey (2004) concluded that the best estimates of short-run and long-run elasticities were −0.28 and −0.81, respectively.
8. CCGT power plants are often base-load facilities but they have a little more wiggle room in ramping production than base-load coal and nuclear power plants. The reason is that available heat from the fuel can be adjusted more quickly for gas than coal or nuclear fuel rods. Likewise, biomass-fueled plants are often base-load; only their capacity tends to be much smaller than that of coal, nuclear and CCGT plants.
9. Of course, with concern about climate change, the optimal solution would be to reduce coal-fired capacity.
10. Using a grid model, this is precisely what a major European consulting firm found: as installed European wind capacity increased to the levels required to meet 2020 renewable energy targets in electricity, wholesale prices fluctuated wildly, making investments in electrical generating capacity riskier (see Pöyry, 2011). This conclusion held even if transmission interties and capacity increased to facilitate wind entering a Europe-wide grid.
11. Information related to the WECC is available at www.wecc.biz.
12. All values provided in this chapter are in Canadian dollars. The FIT values discussed in this subsection have been converted to a Canadian dollar basis by Paul Gipe, who assumed an exchange rate of €1 = $C 1.33878 (see http://www.wind-works.org/articles/feed_laws.html). However, the reader can interpret the values as being US dollars as the Canadian dollar has been trading at or slightly above par against the US dollar for the past several years.
13. Article in *Earth Times* entitled 'German solar power producers agree to subsidy cuts', available at: http://www.earthtimes.org/articles/news/363164,producers-agree-subsidy-cuts.html (viewed 15 March 2011).
14. Wood pellets are easy to transport and can readily be used in lieu of coal in power plants; wood pellet production facilities are also simple to construct, and require relatively little capital investment.
15. From http://bioenergy.ornl.gov/papers/misc/energy_conv.html (viewed 16 March 2011), coal releases 25.4 tonnes of carbon per terajoule (TJ) compared to 14.4 for natural gas.
16. See http://www.upi.com/Science_News/Resource-Wars/2010/10/05/Solar-boom-drives-up-German-power-price/UPI-74351286299555/ (viewed 11 October 2010).

REFERENCES

Ansolabehere, S., J. Deutch, M. Driscoll, P.E. Gray, J.P. Holdren, P.L. Joskow, R.L. Lester, E.J. Moniz, N.E. Todreas, N. Hottle, C. Jones and E. Parent (2003), *The Future of Nuclear Power. An Interdisciplinary MIT Study*, Cambridge, MA: Massachusetts Institute of Technology. http://web.mit.edu/nuclearpower/ (accessed 3 March 2011).

Bryce, R. (2010), *Power Hungry: The Myths of 'Green' Energy and the Real Fuels of the Future*, New York: Public Affairs.

Crutzen, P.J., A.R. Mosier, K.A. Smith and W. Winiwarter (2008), 'N_2O release from agro-biofuel production negates global warming reduction by replacing fossil fuels', *Atmospheric Chemistry and Physics*, **8**: 389–95.

Deutch, J.M., C.W. Forsberg, A.C. Kadak, M.S. Kazimi, E.J. Moniz, J.E. Parsons, Y. Du and L. Pierpoint (2009), *Update of MIT 2003 Future of Nuclear Power. An Interdisciplinary MIT Study*, Cambridge, MA: Massachusetts Institute of Technology. http://web.mit.edu/nuclearpower/ (accessed 3 March 2011).

Duderstadt, J., G. Was, R. McGrath, M. Muro, M. Corradini, L. Katehi, R. Shangraw and A. Sarzynski (2009), *Energy Discovery-Innovation Institutes: A Step toward America's Energy Sustainability*, Washington, DC: Brookings Institution. http://www.brookings.edu/~/media/Files/rc/reports/2009/0209_energy_innovation_muro/0209_energy_innovation_muro_full.pdf.

Espey, J.A. and M. Espey (2004), 'Turning on the lights: a meta-analysis of residential electricity demand elasticities', *Journal of Agricultural and Applied Economics*, **36**(1): 65–81.

Fox, Conrad (2011), 'The effects of CO_2 abatement policies on power system expansion', unpublished M.A.Sc. thesis, Department of Mechanical Engineering, University of Victoria, Victoria, Canada, September, available at http://www.iesvic.uvic.ca/.

Gross, R., P. Heptonstall, D. Anderson, T. Green, M. Leach and J. Skea (2006), 'The costs and impacts of intermittency: an assessment of the evidence on the costs and impacts of intermittent generation on the British electricity network', March, London, Energy Research Centre, Imperial College, viewed 19 April 2011 at: http://www.ukerc.ac.uk/Downloads/PDF/06/0604Intermittency/0604IntermittencyReport.pdf.

Gross, R., P. Heptonstall, M. Leach, D. Anderson, T. Green and J. Skea (2007), 'Renewables and the grid: understanding intermittency', *ICE Proceedings, Energy*, **160**(1): 31–41.

GWEC (2010), 'Global wind report: annual market update 2010', retrieved from http://www.gwec.net/fileadmin/images/Publications/GWEC_annual_market_update_2010_2nd_edition_April_2011.pdf.

International Energy Agency (IEA) (2005), *Projected Costs of Generating Electricity. 2005 Update*, Paris: OECD/IEA.

International Energy Agency (IEA) (2009), *World Energy Outlook 2008*, Paris: OECD/IEA.

International Energy Agency (IEA) (2010a), *World Energy Outlook 2009. Executive Summary*, Paris: OECD/IEA.

International Energy Agency (IEA) (2010b), *Key World Energy Statistics 2009*, Paris: OECD/IEA.

Jenkins, J., T. Nordhaus and M. Shellenberger (2011), 'Energy emergence. Rebound & backfire as emergent phenomena', February, Oakland, CA: Breakthrough Institute.

Kumar, A. (2009), 'A conceptual comparison of bioenergy options for using mountain pine beetle infested wood in Western Canada', *Bioresource Technology*, **100**(1): 387–99.

Kumar, A., P. Flynn and S. Sokhansanj (2008), 'Biopower generation from mountain pine infested wood in Canada: an economical opportunity for greenhouse gas mitigation', *Renewable Energy*, **33**: 1354–63.

le Pair, C. and K. de Groot (2010), 'The impact of wind generated electricity on fossil fuel consumption', Nieuwegein and Ledischendam, Netherlands, at http://www.clepair.net/windefficiency.html (viewed 20 April 2011).

Maddaloni, J.D., A.M. Rowe and G.C. van Kooten (2008), 'Wind integration into various generation mixtures', *Renewable Energy*, **34**(3): 807–14.

Niquidet, K., B. Stennes and G.C. van Kooten (2012), 'Bio-energy from mountain pine beetle timber and forest residuals: the economics story', *Canadian Journal of Agricultural Economics*, **60**(2): 195–210.

Ontario Power Authority (2009), *Feed-in Tariff Program. Program Overview Version 1.1*, 30 September, http://fit.powerauthority.on.ca/Storage/97/10759_FIT-Program-Overview_v1.1.pdf (viewed 5 November 2009).

Pizer, W.A. (1997), 'Prices vs. quantities revisited: the case of climate change', Resources for the Future Discussion Paper 98–02, Washington, DC, October.

Pizer, W.A. (2002), 'Combining price and quantity controls to mitigate global climate change', *Journal of Public Economics*, **85**: 409–34.

Pöyry, (2011), 'The challenges of intermittency in North West European power markets. The impacts when wind and solar deployment reach their target levels', Oxford, UK: Pöyry Management Consulting (UK) Ltd, March (www.poyry.com).

Prescott, R. and G.C. van Kooten (2009), 'The economics of wind power: destabilizing an electricity grid with renewable power', *Climate Policy*, **9**(2): 155–68.

Searchinger, T., R. Heimlich, R.A. Houghton, F. Dong, A. Elobeid, J. Fabiosa, S. Tokgoz, D. Hayes and T. Yu (2008), 'Use of U.S. croplands for biofuels increases greenhouse gases through emissions from land-use change', *Science*, **319**: 1238–40.

Searchinger, T.D., S.P. Hamburg, J. Melillo, W. Chameides, P. Havlik, D.M. Kammen, G.E. Likens, R.N. Lubowski, M. Obersteiner, M. Oppenheimer, G.P. Robertson, W.H. Schlesinger and G.D. Tilman (2009), 'Fixing a critical climate accounting error', *Science*, **326**: 527–8.

Smil, V. (2003), *Energy at the Crossroads. Global Perspectives and Uncertainties*, Cambridge, MA: MIT Press.

Stennes, B. and A. MacBeath (2006), *Bioenergy Options for Woody Feedstock: Are Trees Killed by Mountain Pine Beetle in British Columbia a Viable Bioenergy Resource?* Information Report BC-X-405. Victoria: Natural Resources Canada, Canadian Forest Service, Pacific Forestry Centre.

Stennes, B., K. Niquidet and G.C. van Kooten (2010), 'Implications of expanding bioenergy production from wood in British Columbia: an application of a regional wood fibre allocation model', *Forest Science*, **56**(4): 366–78.

US Energy Information Administration (2010), *Assumptions to the Annual Energy Outlook 2010*. Report #DOE/EIA-0554(2010). http://www.eia.doe.gov/oiaf/aeo/assumption/pdf/0554%282010%29.pdf. (viewed 16 September 2010).

US Energy Information Administration (2011), *Direct Federal Financial Interventions and Subsidies in Energy in Fiscal Year 2010*, Washington, DC: US Department of Energy, July.

van Kooten, G.C. (2009), 'Biological carbon sequestration and carbon trading re-visited', *Climatic Change*, **95**(3–4): 449–63. doi: 10.1007/s10584-009-9572-8.

van Kooten, G.C. (2010), 'Wind power: the economic impact of intermittency', *Letters in Spatial & Resource Sciences*, **3**: 1–17.

van Kooten, G.C. and Govinda R. Timilsina (2009), 'Wind power development: economics and policies', Policy Research Working Paper 4868, Washington, DC: The World Bank, Development Research Group, Environment and Energy Team, March.

Weitzman, M.L. (1974), 'Prices vs quantities', *The Review of Economic Studies*, **41**(October): 477–91.

Weitzman, M.L. (2002), 'Landing fees vs harvest quotas with uncertain fish stocks', *Journal of Environmental Economics and Management*, **43**: 325–38.

APPENDIX

Table 9A.1 Countries with feed-in or other price subsidies for renewable energy

Country	Wind	Hydro	Solar	Biomass
Algeria	✓	✓	✓	
Argentina	✓			
Australia			✓	
Austria	✓		✓	
Bosnia Herzegovina	?	?	?	?
Brazil	✓	✓		✓
Bulgaria	✓	✓	✓	✓
Canada	✓	✓	✓	✓
China	✓		✓	✓
Croatia	✓	✓	✓	✓
Czech Republic	✓	✓	✓	✓
Denmark	✓			✓
Dominican Republic	?	?	?	?
Finland	✓			✓
France	✓	✓	✓	✓
Germany	✓	✓	✓	✓
Greece	✓		✓	
Hungary	?	?	?	?
India			✓	
Iran	?	?	?	?
Ireland	✓	✓		✓
Israel	✓		✓	
Italy	✓		✓	✓
Japan	✓			
Luxembourg			✓	
Malaysia	✓	✓	✓	✓
Malta	?	?	?	?
Mongolia	✓	✓	✓	
Portugal	✓	✓	✓	
Serbia	✓	✓	✓	✓
Slovakia	✓		✓	✓
Slovenia	?	?	?	?
South Africa	✓	✓	✓	
South Korea	✓	✓	✓	
Spain	✓	✓	✓	✓
Switzerland	✓	✓	✓	✓
Taiwan	✓	✓	✓	✓
Thailand	✓	✓	✓	✓
The Netherlands	✓	✓	✓	✓
Turkey	✓	✓	✓	✓
Ukraine	✓	✓	✓	✓
UK	✓	✓	✓	✓
USA	✓	✓	✓	✓
Vietnam	?	?	?	?

Notes:
Some countries such as Canada and the USA have separate jurisdictions that have implemented their own FITs. A ? indicates that a country has a renewable subsidy but little is known about it.

Source: Derived from http://www.wind-works.org/articles/feed_laws.html (Paul Gipe, Bakersfield, CA).

10 A renewable energy future?
Michael Jefferson

INTRODUCTION

By 2008 the world's dependence on the fossil fuels for its primary energy supply remained at 85 per cent, despite at least 20 years of appeals and purported action to reduce this dependence greatly as a matter of urgency. Also by 2008 renewable energy sources accounted for 12.9 per cent of primary energy supply. However, traditional biomass accounted for most of this, at 10.2 per cent. Of the remaining 2.7 percentage points, hydropower accounted for 2.3 per cent. This left all other forms of renewable energy accounting for just 0.4 per cent – of which wind energy accounted for 0.2 per cent; geothermal 0.1 per cent; direct solar energy also 0.1 per cent; and ocean energy 0.002 per cent.

The contribution of renewable energy sources to world electricity supply in 2008 was more encouraging. Renewable energy contributed 19 per cent of supply, made up of: hydro electricity 16 per cent; 3 per cent other renewable sources. Traditional biomass contributed 17 per cent of the world's demand for heating; modern biomass 8 per cent; solar thermal and geothermal each about 1 per cent, to make a total contribution of 27 per cent. Biofuels contributed 2 per cent of the world's total road transport fuel supply.[1]

In 2009 the rate of capacity expansion of some forms of renewable energy supply was indeed impressive. Grid-connected solar photovoltaics (PV) expanded capacity over 50 per cent, wind energy capacity increased over 30 per cent, solar hot water/heating capacity expanded just over 20 per cent, and biofuels increased their contribution to the world's road transport use from 2 per cent in 2008 to 3 per cent in 2009. But these percentage increases are from a very low level, and it takes 40 or more years of such rates to make big inroads into global infrastructure and patterns of use. Moreover, it is actual output that counts, not installed capacity, when it comes to those forms of renewable energy where the source is variable – wind energy is a classic example. And where variability is an important feature then so too, usually, is the need for a 'spinning reserve' from conventional sources to make good those periods when the renewable energy source's contribution falls dramatically. For electricity generation, this means that large 'spinning reserves' of natural gas, or coal or nuclear (or some combination of these) are essential to meet demand. In a world where a low-carbon future and greater efficiency in the provision and use of energy are widely desired, the accelerated expansion of sound renewable energy developments is the most attractive option. As the UK's Minister of Energy has said: 'It is vital that our support for renewable electricity both encourages investment and represents value for money for consumers.'[2] The issue is how much renewable energy investment has represented, and is likely to represent, good value for consumers.

There are firm grounds for considering it desirable, as a precautionary measure, to shift towards greater use of renewable energy sources even where there is scepticism that anthropogenic greenhouse gas emissions and their increased atmospheric concentration will raise near-surface temperatures as much as most of the models now in circulation suggest. This is particularly the case for conventional oil, where the 'peak oil' hypothesis strongly suggests that by the 2030s world supplies will be moving sharply down the right-hand side of the Hubbert-type 'bell curve'. There are concerns about the environmental impacts of 'fracking' for shale gas, notably water pollution resulting from the introduction of noxious chemicals, and the overall wisdom of separating and sequestering the carbon dioxide (CO_2) emissions from coal exploitation. In the wake of the Fukushima Daiichi disaster in March 2011, renewed concerns were expressed about the future of nuclear power and some countries decided to opt out of this form of energy provision.

This chapter, however, raises questions as to how far renewable energy sources can (and will over the next forty years or so) realistically make a substantial contribution to primary and secondary energy supply and use. It explores why there are more severe limitations than is commonly acknowledged. There are questions raised about technical potential, performance to date, exaggerated and false claims, and the unintended (or perverse) effects of subsidies. In the process the defects, not least the internal contradictions, of policies and measures are examined. Finally, two scenarios are offered to outline future prospects.

PREDETERMINED ELEMENTS

In October 2011, the world's population reached 7 billion. It is not only the absolute number which is bringing additional pressures to bear on available food, water, energy and other resources. In a growing number of countries populations are ageing, while younger members of the population are being required or asked to take on heavier financial burdens and work commitments. There are greater pressures on population movement – from rural areas to urban centres (75 per cent of the world's population is expected to live in cities by 2050), and from one country to another. Migration of people across borders – increasing because of food and water shortages, lack of job and income-generating opportunities, threats to personal security, social aspirations or simply wish for change – is raising controversy and even hostility.

Among the invasive elements of population increase and movement are increased population density and loss of contact with nature. These have considerable significance for the prospects for some forms of renewable energy. Thus the visual intrusion of large structures, especially in fine landscapes or near residences (not least historic assets), arouses considerable concern in rural areas and is likely to prove a constraint on future renewable energy development. This is particularly the case with large wind turbines and large-scale solar thermal plants unless situated in fairly remote areas. The erection, for example, of thousands of 125 metre high wind turbines in a densely populated country, affecting through aerodynamic modulation the sleeping patterns and hence health of significant numbers of residents within 1.5 km of their nearest wind turbine; and adversely affecting residential property values, are matters that have caused widespread concern in rural areas. The combined effects of loss of contact with our rural environments and the impositions of our urban environments are having consequences that may be considered contrary to the basic elements of sustainable development.

Another aspect of population growth is reflected in pressures on food and water availability, and on prices of raw materials – including food. Many of those living in relatively deprived conditions, in richer economies as well as in poorer ones, are particularly exposed to shortages of, and/or price increases in, food, water and commodities. Commodity price increases are having a heavy impact on energy availability, and on the goods and services processed and manufactured from such commodities as oil and copper. These pressures are of particular concern in non-OECD countries, where over a 90 per cent increase in the demand for energy services is expected in the period 2010 to 2050. Non-OECD countries are expected to account for some 90 per cent of global energy demand increases over the next 40 years, but even this figure will prove unattainable unless urgent and successful efforts are made to improve energy efficiency globally. The subject of improving the efficiency of energy supply and use has scarcely featured in international negotiations over the past 30 years. It was confined to one short chapter, along with all other energy matters, in Agenda 21 (chapter 9, 'The protection of the atmosphere') at the Rio Earth Summit in 1992. Reference to the word 'energy' was completely omitted from the eight Millennium Development Goals in 2000 (and the document behind them). There have been initiatives by UN agencies to correct this oversight, in 2005 and 2010, and only time will tell whether lost time will be made up.

Discussion of the role and nature of some renewable energy resources also often makes one curious assumption, and overlooks one important fact. The curious assumption is summarized in the following statement: 'The meaning of efficiency is a redundant concept to apply to wind energy, where the fuel is free.'[3] But the harnessing of wind energy involves investments and, usually, substantial subsidies from users and costs of impacts. It is also important to site wind turbines optimally in relation to mean wind speeds, resistance factors and other technicalities. What we should all be concerned about is the *efficacy* of renewable energy developments: is the technology applied and location optimal, or near optimal?

The important fact is that all renewable energy sources have lower power densities than fossil fuels or nuclear power. Among the renewable energy resources, flat-plate solar heat collectors in areas of high solar irradiation can achieve a power density of up to 80 W per square metre. Solar PV can achieve up to 50 W per square metre in areas of high solar irradiation. Wind power, geothermal and the more efficient hydroelectric schemes can achieve 5–15 W per square metre. Biomass is closer to 1 W per square metre. By contrast, the fossil fuels as primary energy have power densities in the range 1–10 kW per square metre, although for coal-fired plants this can fall to around 600 W per square metre taking the whole cycle of fuel, transportation, generation and transmission into account.[4]

Another useful concept is that of energy return on (energy) invested, or EROI. For thousands of years cultures have relied upon transforming nature – through agriculture, exploiting woodland, using natural resources for heating and cooking – and latterly for large-scale recovery of minerals, industrial processing and transportation. But, of course, the energy return from renewable energy is heavily dependent upon location. If close to the equator, then solar irradiation is likely to be high, and therefore exploitation of solar energy makes good sense. The further one goes from the equator, the more likely it is that solar irradiation weakens and the less plausible solar energy becomes. Similarly, placing wind turbines where mean wind speeds are relatively high makes better sense than locating them where mean wind speeds are low. Thus the EROI concept can usefully be applied to discriminate between better and poorer locations for renewable energy developments, and hence the efficacy of a development or subsidy towards it. However, it serves little purpose to claim that the EROI for solar photovoltaic is 6.8:1 or wind turbines is 18:1 when actual performance can vary greatly from location to location, development to development – and when, indeed, the purpose of this measure should be to define the efficacy of a development rather than the technical efficiency of, say, a particular wind turbine.[5]

The prospect is for rapidly rising demands for energy to provide the energy services people require (or aspire to) from modern energy forms. Given this prospect, it may be considered, the brilliant future of a '100 per cent renewable energy (world) by 2050' seems to beckon.[6] The reality is somewhat different.

BARRIERS TO THE DEPLOYMENT OF SOME KEY RENEWABLE ENERGY RESOURCES

The reality is that the problems associated with the deployment of key renewable energy sources are substantial and real. In particular, modern biomass (in addition to the longstanding concerns about the collection and use of traditional biomass), biofuels, wind and solar energy (unless sourced in areas of relatively high solar irradiation, and/or successfully brought to areas of lower solar irradiation by ultra high voltage direct current – UHVDC – transmission) face huge challenges and create huge problems. In a nutshell, the challenges are of high costs and detrimental effects on the environment and reliability of supplies.

This is not, of course, either a fashionable or a popular view except among those who have been adversely affected or have grounds to fear they will be. The fashionable and popular view has been sustained in part because of concerns about greenhouse gas emissions from human activities and their possible climatic impacts, in part because of the local and regional impact of emissions from fossil fuel use, in part because of concerns about nuclear safety, and in part because of political and industrial perceptions and interests.

In focusing on biomass/biofuels, wind and solar, it is not intended to suggest that no challenges are faced by the future exploitation of other forms of renewable energy – for example, hydro (large hydro schemes have come in for their fair share of criticism in the past) or tidal range–estuarine barrages (the La Rance scheme in France has been described as destroying the local ecology by its developer, Électricité de France[7]). But the challenges covered here are currently the most pressing.

It also needs to be stressed that the purpose of what follows is to highlight the ways in which some renewable sources, or particular developments, are being exploited suboptimally. It is not, for example, being suggested that modern biomass using wastes, and/or drawing on crops which are not diverting food and water from where they may be better used should be discouraged. It is not suggested that wind turbines should not be erected where mean wind speeds are high, visual impacts moderate (partly a function of subjective views about landscape quality), and effects on people and their homes are minimal. Nor is it suggested that solar PV systems located where solar irradiation is adequate to justify their placement free of significant subsidy should be discouraged.

MODERN BIOMASS AND BIOFUELS

Concerns about the implications of modern biomass for food and water availability (and for food prices) are not quite as straightforward as sometimes suggested. Clearly, where wastes are used – including municipal wastes, manufacturing wastes and wood residues – then this should be advantageous. Care needs to be taken, however, to ensure that when anaerobic digesters are involved, the same or very similar feedstocks are employed; otherwise the microbes breaking down the wastes can be exterminated. Where the feedstock comes from land not suited to food production, due to soil quality or severe gradients, for example, then again there may be a clearer case for modern biomass exploitation – although water availability may qualify some of these alternative considerations. However, there are sufficient concerns about the impacts of modern biomass to require caution.

Jean Ziegler, then the UN Special Rapporteur on the Right to Food, expressed the view that the expansion of modern biomass – and especially of biofuels – was a 'crime against humanity'.[8] Within months food riots had broken out in 47 different countries. In the early months of 2011 further food riots broke out in Tunisia, Algeria, Morocco and Egypt due to rising prices – leading some commentators to conclude that these significantly promoted the onset of the so-called 'Arab Spring'. A World Bank policy research study concluded that the expansion of biofuels and their impact on low grain stocks, together with large land shifts including increases in tropical deforestation, speculative activity and bans on exports, were responsible for 70–75 per cent of total food price rises.[9] By the end of 2008 food prices were again falling, but double-digit food price rises in 2010–11 in many countries – not least for grains – have reawakened concerns about the links to modern biomass and biofuels activity. Much of this activity is the direct result of subsidies, and the rate of expansion of ethanol fuel production (up five-fold between 2003 and 2011) indicates the scale and speed of change. Yet serious questions have been raised about the implications. For example, it has been concluded that if all US corn and soybean production were to be devoted to biofuels, then this would meet only 12 per cent of gasoline demand and 6 per cent of diesel demand.[10] There has also been serious questioning of the claims whether, once a life-cycle assessment of biofuels has been undertaken (including land-use changes), modern biomass and biofuels reduce carbon emissions. There is a considerable body of evidence to suggest that the opposite is true. Corn and cellulosic ethanol have been found to increase carbon emissions compared to gasoline by up to 93 per cent and 50 per cent, respectively.[11] In a remarkable (but not unique) case in the UK, a developer put forward a proposal for a 6 MW electricity generating plant in the middle of rural England, where he wished to burn palm oil as his 'preferred feedstock'. The planning inspector – Alan D. Robinson – permitted the proposal to go ahead because (in his judgement) palm oil was defined under UK Planning Guidance as 'a renewable source of energy'. This conclusion was reached although it was pointed out that it had been found, on average, that exploitation of Indonesian palm oil involved the emission of 33 tonnes of CO_2 for every tonne of palm oil produced – before being shipped to the UK and then road-hauled from port to site. Concern was also expressed about tropical deforestation, habitat loss and impacts on rare species such as – specifically – the orangutan. Other scientific evidence was brought to the planning inspector's notice and to local authority planning officers, to no avail. This case, from Chelveston, Northamptonshire, England, is a classic of its absurd type.[12]

More recent research results and publications serve to reinforce the concerns that modern biomass can, and biofuels more surely do, constrain food and water availability and raise food prices, while encouraging large-scale changes in land use, tropical deforestation and habitat loss for rare species.[13]

There are aspirations to produce biofuels from second- and third-generation technologies, which rely on sustainable feedstocks – cellulosic ethanol, biomethanol, wood diesel, biohydrogen diesel and algae. Each of these technologies is, however, in an early stage of development. Even producing ethanol from cellulose is no easy matter, and the more encouraging technical advances suggest that the use of high-temperature conversion of cellulose into sugars will be required. Much is made in the specialist literature of the use of algae in third-generation technologies, but again these are early days. Above all, it will take many generations to produce large volumes of biofuels from these technologies (as would be required), even where successful.

For a sound sustainable future, modern biomass and biofuels developments will need to pay more heed to their impacts on food availability and prices, water requirements, deforestation and habitat loss than in recent years. These are all issues that many commentators have highlighted – for example in discussion of the EU Directive on renewable energy – but few authorities have taken sufficient notice of them.

WIND ENERGY

Wind energy is clearly a variable resource. Despite this unavoidable fact, instead of being treated as a useful adjunct to mainstream energy supplies, it has been treated in some countries as the main 'saving grace' against fossil fuel use and greenhouse gas emissions. In the UK, for example, wind energy has been claimed to be able to provide 10 per cent of electricity generation by 2010 and that 'the lion's share of (35% to 45% of electricity coming from green sources by 2020) will have to be wind'. 'The UK is the windiest country in Europe, so much so that we could power our country several times over using this free fuel.'[14] The realities are very different. Even the claim that the UK is the windiest country in Europe has to be placed in context: there are many days during which there is insufficient wind (under 4 m/s) to turn 90 per cent of turbines; there are already 25 days per year when there is too much wind and output has had to be reduced by the National Grid, and National Grid anticipate this will rise to 38 days per year.[15] In May 2011, UK wind energy operators were paid £2.6 million to keep their turbines idle (a cost borne by UK households), £613 000 on 24 May alone. However, it is the claims that wind energy is a reliable source, with no complicating factors, that the resource rarely fails, that it achieves a 'typical' capacity factor (per cent of installed capacity) of 30 per cent, that reference to capacity (or load) factors is 'bizarre pseudo-science' and 'absolute nonsense', and that the UK's Renewable Obligation scheme 'is not a subsidy' that need to be challenged. All these claims are at best misleading, and can be shown to be incorrect. Although it may be argued that many recognize that industry lobby groups traditionally make exaggerated claims to extract additional funding from governments (i.e. taxpayers) or subsidies from consumers, and therefore the details need not concern us, the precise detail of such exaggerated claims should be examined in order that the extent and nature of the exaggerations are clearly understood.

To give one example: in the UK official planning policy guidance claims that wind energy developments typically achieve a capacity factor of 30 per cent (Office of the Deputy Prime Minister, 2004, p. 165). The reality is that none of the constituent parts of the UK have achieved that – although Scotland has come close in two recent years. For the rest of the UK barely two-thirds of that figure have been achieved. The same official guidance claims that capacity factors range between 20 per cent and 50 per cent, whereas in recent years the actual range has been around 4 per cent to 45 per cent.

There is, of course, a substantial global wind resource. However, even this is widely overstated. Instead of a potential of 900 terawatts (TW), the actual likely range is 18–68 TW.[16] Then it is widely assumed that if the wind is not blowing strongly enough onshore, or just offshore, then it will be blowing sufficiently strongly further afield – if not offshore the UK, for example, then off the Irish, or Spanish or German coasts – and the resultant power could make up for any UK shortfalls via a supergrid (of the sort discussed below). Disappointingly, research evidence indicates that this 'does not seem justified as neighbouring countries are seen to experience a simultaneous fall in wind power'.[17] Moreover, claims made about a nation's wind resource[18, 19] have been found to be greatly exaggerated.[20]

Regrettably, exaggerated claims have rarely been effectively challenged to date. For example, the UK Advertising Standards Authority was content to allow a developer to claim that, at a relatively low mean wind speed site in England they would achieve a higher capacity factor (34.03 per cent) than that achieved anywhere onshore in the country in 2009 or 2010. Under the higher mean wind speeds operating in 2007 and 2008, the proposed development would have ranked 4th and 9th onshore, respectively. The ASA claimed to have taken 'expert advice', but refused to reveal from whom.[21] Even the simplest distortions of reality may be attempted. For example, a developer claimed to regard as viable sites where the mean wind speed, according to the UK government's wind speed database, exceeds 6.5 m/s at 45 metres above ground level. This claim took seconds to expose: half the proposed turbines were in a location (grid square) where the mean wind speed is 6.3 m/s at 45 metres above ground level.[22] These efforts, seen by many as attempts to hoodwink the public, may well lead to scepticism about this renewable source of energy and constrain its future contribution.

Another area of concern surrounding wind energy developments concerns the 'noise' from wind turbines (amplitude modulation or 'fluctuating swish'). For example, the official UK guideline – ETSU-R-97 – was secretly 'modified' to raise permitted decibel limits well above those advised by consultants. The consultants, Hayes McKenzie Partnership, had recommended a limit of 33 decibels if turbines were to emit a 'beating' noise as they spun but found the official guideline of 43 decibels retained in ETSU-R-97.[23] A member of the Noise Working Group, Dick Bowdler, stated in his resignation letter:

> Looking at the Government Statement it is clear that the views of the Noise Working Group (that research into aerodynamic modulation to assist the sustainable design of windfarms in the future) have never been transmitted to government and so the Statement is based on misleading information.[24]

Certainly research on aerodynamic modulation carried out in other countries has shown it to have adverse effects on the sleep patterns and health of some nearby residents (e.g. the writings of Frits van den Berg, Eja Pederson, Erik Rudolphi and Kerstin Persson Waye). There is no adequate justification for the manipulation that has gone on in the UK on this subject. The recent IPCC *Special Report on Renewable Energy Sources* appears to express greater concern about this issue (IPCC, 2011, ch. 7, pp. 54–5). Unless the impact of aerodynamic modulation can be satisfactorily resolved, there are likely to be increasingly severe constraints on the contribution onshore wind energy developments will make in future.

There are concerns, perhaps but not solely related to impacts of aerodynamic modulation, arising from the impact of proposed and actual wind energy developments on residential property prices. The UK industry, for example, has relied upon a small, and flawed, research project in the South West of England. More robust evidence indicates that the problem is real and extensive.[25] Again, this concern is likely to constrain future contributions from onshore wind developments unless sited in remote locations – which may be far from existing grid systems and major areas of electricity demand.

Earlier the claim that the UK is the windiest country in Europe was quoted. At one level this is probably true. Further examination of the evidence points to the claim being deeply flawed. The Risø National Laboratory, in Roskilde (Denmark), produces the European wind resources Atlas and the European wind resources over open sea Atlas.[26] From these two maps the following features are plain to see: mean wind speeds onshore are relatively high in Scotland and close to the northern and western coasts of Northern Ireland. For almost the whole of England, all of Wales, and most of Northern Ireland mean wind speeds are modest. For all of Central England mean wind speeds are low – under 6.5 m/s at 50 metres above ground level. This feature is held in common by inland areas of Northern France, Belgium, The Netherlands, Germany and Sweden (although in most of the rest of France, Southern Germany, almost all of Portugal and Spain, and virtually all of Italy mean wind speeds are even lower – under 5.5 m/s except on hills). This is why capacity factors achieved vary so much, and are so often low. Thus in Germany the average capacity factor in most years has been close to 16 per cent, but is more often close to 12 per cent in inland *Länder*; in Italy, although most turbines have been placed on hills and ridges, the average capacity factor has generally been below 16 per cent; and in France it has been below 15 per cent.

Offshore, the Risø National Laboratory map shows wind speeds east and west of a latitude well into the North of England as exceeding 9 m/s at 50 metres above sea level, and this area extends to the northern half of Denmark's west coast. South of this line the mean wind speed is between 8 m/s and 9 m/s at 50 metres above sea level, still very satisfactory. The point here is that a particular turbine model located at a site where the mean wind speed is 6 m/s will only produce half as much energy as at a site where the mean wind speed is 8 m/s (the cubed effect).

Data for the USA (California and Texas) suggest that there are lengthy periods when demand for electricity is high (the month of July) and capacity factors achieved are low – under 17 per cent in Texas and they can be as low as 4 per cent in California.[27] In China and India, for comparison, average capacity factors achieved in recent years have been around 12 per cent and 16 per cent, respectively, although IPCC (2011) produced a higher figure for China – the same source also providing a global wind 'resource' map which indicates why over most of the USA, India, China, Africa and all of Latin America (excepting Patagonia) the wind resource is poor.[28]

Three further matters concerning wind energy need consideration. First, the importance of linking wind turbine developments to the grid. This has been a concern for many developments in Scotland and offshore generally, and is always a point for consideration by developers. In Denmark the problems, primarily associated with linking into the two grid systems there, have resulted in only about half the wind-generated electricity in Denmark being available to Danish consumers. Half, so some 10 per cent of the total Danish electricity requirements, has had to be exported – frequently at a loss – to Norway, Sweden or Germany. Thus claims that Denmark generates some 20 per cent of its electricity from wind energy have to be qualified: only about 10 per cent is available for use in Denmark. IPCC (2011) states that Denmark has the highest wind electricity penetration in the world, equating to 20 per cent of total electricity demand, and then states the benefits of Denmark having access to (foreign) markets and strong transmission interconnections to neighbouring countries, so no serious reliability issues arise as the result of having 'two different electric systems'. It is not made clear at any point in the report that the result is that half the wind-generated electricity has to be exported, usually at a loss.[29]

Second, due to the variability of wind it is necessary to maintain a 'spinning reserve' of electricity generating capacity using traditional sources – 'spinning' 24 hours per day – as back-up in case wind power fails. In the UK, for instance, this means that 17 gas-fired stations will be required to back up wind power – at an estimated cost of £10 billion. It is not quite true to claim that wind capacity carries 'no fuel costs'. The issue is perhaps more serious than IPCC (2011) appears to suggest when it remarked: 'this additional flexibility is not free, as it increases the amount of time that fossil fuel-powered units are operated at less efficient part-load conditions' (ch. 7, p. 40).

Third, although many observers (including the present writer) hope that some day substantial storage capacity for wind-generated electricity will enable it to cover for shortfalls in wind, there is a strong body of opinion within the industry that takes another view:

> Many enthusiasts for large-scale storage see intermittent sources of renewable energy such as wind and solar as a splendid new opportunity to press their case. Unfortunately, this thinking is misguided and, with storage available on a power system, wind and solar are in fact, literally, the last types of energy to be stored.[30]

In the event that this last opinion proves to be well founded, it will raise even more serious questions over the maximum contribution a variable renewable energy source can make without causing major disruptions in the event of a sudden drop in that contribution. Some have considered 20 per cent the maximum contribution, but if – say – wind energy's contribution to a country's electricity generation falls from over 20 per cent to barely 2 per cent in a matter of hours, is this a satisfactory maximum contribution? Hence the need for costly back-up in the form of a 'spinning reserve'.

Given this lengthy catalogue of issues, what should the future of wind energy be? First, it should not be the principal plank on which to build sustainable energy or electricity supplies. It is a matter of concern that instead of encouraging the careful planning analysis of proposed wind energy developments, for rejection or acceptance according to stiff criteria, there is an attitude of impatience in some quarters which shows insensitivity to the potential adverse impacts of schemes if approved. IPCC (2011) suggests this mind set when referring to 'cumbersome and slow planning, siting and permitting procedures that impede wind energy deployment' (ch. 7, p. 32).

Second, the hurdles over which wind energy applications need to go should be higher in terms of being focused only on sites where mean wind speeds are relatively high (e.g. not less than 7 m/s at 45 metres above ground level). This would eliminate consideration of many proposals for inland developments, thereby saving rural landscapes from severely adverse visual impacts, save a great deal of local planning resources now tied up in evaluating suboptimal proposals, and focus more attention and resources on where mean wind speeds are relatively high and benefits in terms of electricity generation and carbon emissions avoidance greater.

At present, economic, social and environmental aspects of renewable energy development proposals are not being addressed satisfactorily in many countries; appropriate environmental safeguards are not being built in; planning policies often fail to provide sufficient reasoned justification as to why constraints *should* be placed on specific types of renewable energy at specific locations; the wider environmental and economic costs and benefits of proposals are insufficiently analysed; the technical and commercial feasibility of proposals (with, and in the absence of, consumer subsidies) are not properly examined by the planning authorities; would-be developers all too often fail to engage in active consultation and discussion with local communities from an early stage of the planning process; and development proposals frequently fail to demonstrate adequately their environmental, economic and social benefits. In this regard proposals fail to conform – for example – to seven, out of the eight in total, Key Principles of the UK's official planning guidance for renewable energy just listed.

Third, due weight should be given to population density and quality of remaining rural landscapes, to concerns about aerodynamic modulation and the risks to sleep patterns and health which have been posed, and to compensation for loss of residential property values by developers.

Fourth, it is widely recognized that once wind energy achieves a 20 per cent share of an electricity generation market (on a very windy day!), the need to cope with variability and bring on reserves becomes a major issue. In fact there have been problems at much lower levels of penetration. This is despite the recent IPCC *Special Report on Renewable Energy Sources* mentioning a number of significant integration studies from Europe and the USA that have pointed out that accommodating wind electricity penetrations of up to 20 per cent is technically feasible, though not without challenges.[31]

Finally, here, there needs to be not only greater awareness of the extent to which an industry can – and in the case of the wind energy industry in some countries clearly has done – engage in political pressure, rhetoric and attacks on legitimate (soundly based) contrary views, but a firmer resolve on the part of those responsible for public policy to follow an objective and soundly rational path. There is a long history of sectors of the energy industry, in common with many other sections of industry and society, seeking to influence policy and attract funding. Governments and international organizations, especially since the Rio Earth Summit in 1992 and the various subsequent conferences under the UN Framework Convention on Climate Change, have seen a need to influence energy market outcomes. This is understandable. That these institutions have so often encouraged suboptimal outcomes is less readily understandable. More finely tuned subsidy programmes would help promote a more sustainable future.

DIRECT SOLAR POWER

Some of the same concerns as those expressed about wind energy are applicable to solar power, but can be addressed briefly.

Passive solar – the positioning of buildings, their layout, the materials from which they are built – has been a focus in architecture since the Ancient Greeks. Efforts to maximize the availability of sunlight, protect against cold winds, avoid intense heat from the sun or provide shelter from it, permit heat to absorb materials used, circulate water effectively, were all employed. In the first century before the Christian era the Roman architect, Vitruvius, took further the advice he had imbibed from the earlier Greeks. In many parts of the world, traditional, vernacular architecture sought to cope with local weather conditions. Many modern architects and planners appear to have lost their way in maximizing the benefits to be gained from passive solar, as can be seen in many high-rise buildings around the world – and not least in the Arabian Gulf. Yet ancient houses – in Yemen, southern Iran, and in a few examples in Bahrain, Sharjah and Oman, for example – show that our forebears had a sound understanding of the means of harnessing or avoiding the sun and the wind (the use of *barjeel*, or wind towers), and water. In the interests of energy efficiency, more attention needs to be paid to exploiting passive solar.

Solar PV (photovoltaics) technology goes back to 1839 and the experiments of the Frenchman Edmund Becquerel, and the introduction of the first solar cells a half-century later by the American Charles Fritts, followed by the leap forward achieved by Frank Shuman.[32] The efficiency of the cells and the materials they are made of has improved greatly in recent years. Solar PV has powerful attractions when used in areas of relatively high solar irradiance (or insolation). Thus in California, New Mexico and Florida, in the Middle East and elsewhere in tropical and near sub-tropical regions in the western and eastern hemispheres, solar PV is a very attractive renewable energy option. However, there have been considerable efforts to subsidize the diffusion of solar PV in areas where solar irradiance is relatively low. IPCC (2011) has claimed: 'Although not all countries are equally endowed with solar energy, a significant contribution to the energy mix from direct solar energy is possible for almost every country.'[33] This proposition cannot be taken seriously unless confined to the future application of concentrating solar power (CSP), transmitted long distances by ultra high voltage direct current transmission (UHVDC) to countries and areas where solar irradiance is low. Certainly the use of solar PV in countries where solar irradiance is relatively low needs serious questioning.

For example, in the year 2000 the German government introduced a scheme to support solar PV in Germany. By 2010 it had cost over €50 billion, and produced 0.7 per cent of Germany's electricity. The International Energy Agency estimates the cost per tonne of carbon emissions avoided at €1000 per tonne. A German academic institution, Ruhr University, has produced a somewhat lower estimate of €716 per tonne. There has been widespread dissatisfaction at such a high cost, and many have taken the view that such expenditure is not the most effective route to reducing carbon emissions and expanding renewable energy use.

Notwithstanding the German experience, the current UK government introduced two parallel schemes to encourage solar PV: one for larger-scale developments and the other for domestic schemes. The former was quickly withdrawn, the latter was curtailed in December 2011. Again, questions have reasonably been raised about the efficacy of solar PV in a country not known for its high solar irradiance.

More interesting, and with far greater technical potential, is CSP. The technology is well understood (the first comparable plant was in operation under the guidance of Frank Shuman at Meadi, near Cairo in 1912 (see Butti and Perlin, 1980, pp. 106–10)). Long-distance transmission technology is also well understood, and recent innovations in UHVDC have meant that transmissions losses have fallen below 3 per cent per 1000 km. In recent years CSP plants have been developed, first in the USA and more recently in Spain, Morocco and Algeria. The laying down of parabolic trough mirrors on a large scale to capture the sun in North Africa and transmit it to Europe has long been the goal of the Trans-Mediterranean Energy Co-operation or Desertec concept (http://www.trecers.net/concept.html). Alternatively, central-receiver or dish systems have been considered. There is some scope for thermal energy storage (currently up to 7.5 hours). The impacts on the grid system of CSP are modest, due to the thermal mass of the collector system and the spinning mass of turbines, as well as alleviation through the thermal storage system. Some concerns have arisen from land take-up of CSP systems, but their usual desert location suggests that these may not be severe. There are ongoing concerns about the political stability in North Africa and some technical concerns – water needs, and costs and availability of materials required. However, with the sun providing some 10000 times the amount of energy the world currently uses in a year, this technology for employing direct solar energy over a wide area has unparalleled attractions.

WHERE ARE WE HEADED? SOME CONCLUDING COMMENTS

There is widespread, though not universally accepted, concern that there is an urgent need to reduce the world's reliance on fossil fuels, reduce or constrain the growth of greenhouse gas emissions from human activities and their atmospheric concentration, and to pursue the sound accelerated diffusion of renewable energy use. This chapter has supported all these goals. However, it has also concluded that in far too many instances – not least relating to wind, modern biomass and biofuels, and solar PV developments – efforts to accelerate the diffusion of renewable energy use have not always been sound.

The world's 85 per cent reliance, still, on fossil fuels for energy provision is a matter of great concern. So is the fact that modern renewable energy sources still account for only 3 per cent of the world's primary energy use. Carbon emissions from fossil fuel use have risen 46 per cent since 1990 (to 2010).[34] Yet not only have unsound policies been resorted to in the pursuit of renewable energy expansion. The internationally accepted standard of greenhouse gas emissions accounting is distorting an even worse reality by failing to include emissions 'embedded' in imported goods. Thus manufactured goods imported from, say, China and India (mostly produced using large volumes of coal-fired electricity generation), because the importing country no longer manufactures them itself, are not included in the emissions of the importer. Thereby the importing country can maintain and increase its standard of living while pointing to the rapidly increasing emissions of some developing and emerging economies. The chief scientific advisor to the UK's Department of the Environment, Food, and Rural Affairs has drawn attention to this.[35] In September 2010, he estimated that instead of the UK's carbon emissions having gone down 16 per cent they had in fact gone up about 12 per cent since 1990. By end-2010, instead of these emissions having gone down some 20 per cent they had risen 16 per cent. The same phenomenon has been at work in many other countries. Thus the large declines in carbon emissions since 1990 officially claimed by Germany, Sweden, Switzerland and Denmark are overstated. The increases of the USA, Canada, Australia New Zealand, Spain, Portugal, Greece, Ireland, Austria and The Netherlands are even greater.

Currently, if one were to outline two energy supply scenarios for the world over the next 50 years or so, two archetypes seem to present themselves. The first is a world in which there is failure to pursue rational policies effectively, with measures and lack of sufficient technological innovation and diffusion over the next half-century (at least) to achieve the results claimed to be desirable. The move to a low-carbon economy proves to be slow and full of contradictory elements, including suboptimal decisions on the support for and siting of renewable energy schemes. Among the consequences are interrupted energy services and wider chaos. We can call this scenario 'Complexity and Chaos Rule'. The other is a world where sound policies and measures are rapidly and widely introduced, that support needed technologies and their widespread diffusion at all desired scales. Markets prove effective, lifestyles shift where required to accommodate an accelerated move towards a fossil-fuel-free future. Renewable energy schemes are optimized according to investments required, location, reliability and useful energy produced. Brains and institutions focus on how to handle complexity and perversity (such as 'the Law of Unintended Consequences'). Currently, the emphasis would have to be on the rule of 'Complexity and Chaos'. The chances of a 'rational transition' becoming feasible at present seem remote.[36] This is, not least, the result of unrealistic expectations and targets for renewable energy use, suboptimal investment for society at large, and poor value for consumers' money.

Over the next few decades, if there is to be 'rational transition', many observers consider better and faster results will come from abandoning suboptimal renewable energy projects, being more selective about those which remain, and placing greater emphasis on improving the efficiency of energy provision and use. A better balance should make a sustainable future more readily achievable. This is not an original idea.

Maurice Strong, when opening the 1992 Rio Earth Summit, said:

> Nowhere is efficiency more important than in the use of energy. The transition to a more energy-efficient economy that weans us off our over-dependence on fossil fuels is imperative to the achievement of sustainable development.[37]

Over the ensuing years this focus has been largely lost.

NOTES

1. The data in this section are from IPCC (2011), ch. 1, pp. 15–16.
2. UK Energy Minister Charles Hendry introducing a report prepared by Arup (Arup 2011) for the UK Department of Energy & Climate Change, 10 June 2011, http://www.decc.gov.uk/en/content/cms/news/pn11_47/pn11_47.aspx.
3. RenewableUK (formerly British Wind Energy Association), http://www.bwea.com/ref/faq.html.
4. See Smil (2008), and 'Power Density' Primer, Parts I to V, May 2010, at www.vaclavsmil.com.
5. See Hall and Klitgaard (2012), Table 14.1, p. 313.
6. See WWF International, Ecofys and OMA (2011).
7. See, for example, Rodier (1992), p. 308. In Canada's Bay of Fundy their Annapolis estuarine barrage facility has been closed due to some (relatively minor) ecological problems and developments at Cumberland and Minas Basins withdrawn. In the UK a few vested interests have supported an estuarine barrage across the River Severn, with other UK estuaries having even greater ecological (especially ornithological) sensitivity.
8. 'ONU diz que biocombustíveis são crime contra a humanidade', http://www1.folha.uol.com.br/folha/dinheiro/ulto91u391866.shtml; and 'UN Rapporteur calls for biofuel moratorium', http://www.swissinfo.org/eng/front/detail/UN_rapporteur calls_for.biofuel_moratorium/html?siteSect105&sid=8305080.
9. See Mitchell (2008), p. 17.
10. Hill et al. (2006), pp. 11206–10.
11. See Searchinger et al. (2008). See also OECD/IEA (2008); OECD (2007); IEA (2009); and Eisentraut (2010). See also Fargione et al. (2008).
12. Robinson (2009). References cited included: Childs and Bradley (2008) and Fargione et al. (2008). The UK Planning Guidance referred to is Planning Policy Statement 22: 'Renewable Energy', 2004. Despite its severe flaws, not least in respect of wind energy development applications, this 'Guidance' remains in operation as of November 2012.
13. For example: Fargione et al. (2008), Fischer et al. (2009); Delucci (2010) and FAO et al. (2011).
14. RenewableUK: 'Onshore Wind', http://www.bwea.com/onshore/index.html, accessed 27 July 2011.
15. National Grid announcement, 13 June 2011, 'Operating the electricity transmission networks in 2020 – update June, 2011', p. 9, para. 2.18.
16. Miller et al. (2011) and extensive list of references.
17. Oswald et al. (2008), p. 3211.
18. Sinden (2007b), p. 119.
19. Sinden (2007a).
20. For example: Mackay (2009), pp. 32–34; http://inference.phy.cam.ac.uk/mackay/presentations/WIND/mgp00041.html; http://withouthotair.blogspot.com/2009/05/wind-farm-per-unit-area-data.html; Jefferson (2008), pp. 4120–21; Ian Fells, Written evidence to the House of Lords Science and Technology Committee, Report No. HL Paper 126–11.
21. In December, 2009, the UK Advertising Standards Authority (ASA) – following protests from opponents – cleared RWE/npower renewables of illegitimately claiming that their proposed development at Nun Wood, Northamptonshire, England would achieve a 34.03 per cent capacity factor (AO9-8370). A curious claim, which the ASA followed up with the claim that its view was supported by 'expert advice', as this was higher than any onshore wind energy development in England in 2009 or 2010.
22. Michael Jefferson, Proof of Evidence to Bicton Appeal Inquiry APP/H0520/A/11/2146394, re: Appeal by Broadview, August 2011.
23. See Jonathan Leake and Harry Byford in *The Sunday Times*, 13 December 2009, for details of Freedom of Information request and Hayes McKenzie (http://www.timesonline.co.uk/tol/news/environment/article6954565.ece...); and http://www.semantise.com/-lewiswindfarms/OpenItemURL=S000BFC2F for Dick Bowdler's resignation.
24. Bowdler (2005, 2008), and Stigwood (2009).
25. See Etherington (2009), ch. 7, pp. 112–28. For a report on two Cornish villages, see Dent and Sims (2007). An earlier report from the RICS, based upon responses from its professional members, found that 76 per cent of respondents in the South West of England (within which Cornwall lies) considered wind energy developments have negative effects on residential property prices. The figure was 74 per cent for respondents in the Midlands and Eastern Regions of England. IPCC (2011) is more equivocal on this subject, and cites Dent and Sims without further comment (ch. 7, p. 55).
26. See: http://www.windatlas.dk/europe/EuropeanWindResource.html and http://www.windatlas.dk/europe/oceanmap.html.
27. See: http://www.windaction.org/documents/4032.

28.   http://lightbucket.wordpress.com/. For global wind resource map see IPCC (2011) ch. 7 ('Wind energy'), pp. 15 and 17.
29.   Ibid., pp. 37 and 104.
30.   Swift-Hook (2010).
31.   IPCC (2011), ch. 7, p. 45.
32.   See Butti and Perlin (1980). It is encouraging to note that the recent IPCC (2011), ch. 3, 'Direct solar energy', starts from the same point and reference, (p. 7).
33.   Ibid., p. 4.
34.   BP (2011).
35.   BBC News: 'Openness urged on UK's emissions', Report by Roger Harrabin on Robert Watson's (former Chairman of the IPCC) view, 3 September 2010.
36.   Jefferson (2011).
37.   Maurice Strong, quoted in Johnson (1993), p. 55.

REFERENCES

Arup, O. (2011), *Review of the Generation Costs and Deployment Potential of Renewable Electricity Technologies in the UK*, UK Department of Energy & Climate Change.
Bowdler, D. (2005), 'ETSU-R-97: Why it is wrong', *New Acoustics*, July, www.newacoustics.co.uk.
Bowdler, D. (2008), 'Amplitude modulation of wind turbine noise', *Acoustics Bulletin*, July–August.
BP (2011), *Statistical Review of World Energy*, Carbon Dioxide Emissions Spreadsheet, June.
Butti, K. and John Perlin (1980), *A Golden Thread: 2500 Years of Solar Architecture and Technology*, New York: Van Nostrand.
Childs, Britt and Rob Bradley (2008), *Plants at the Pump: Biofuels, Climate Change and Sustainability*, Washington, DC: World Resources Institute.
Delucci, M. (2010), 'Impacts of biofuels on climate change, water use, and land use', *Annals of the New York Academy of Sciences*, **1195**, 28–45.
Dent, Peter and Sally Sims (2007), 'What is the impact of wind farms on house prices?', Oxford Brookes University report funded by a Royal Institute of Chartered Surveyors educational trust, March.
Eisentraut, A. (2010), *Sustainable Production of Second-Generation Biofuels*, Paris: IEA.
Etherington, J. (2009), *The Wind Farm Scam: An Ecologist's Evaluation*, London: Stacey International.
FAO et al. (2011), 'Price volatility in food and agricultural markets: policy responses', Rome: FAO.
Fargione, Joseph, Jason Hill, David Tilman, Stephen Polasky and Peter Hawthome (2008), 'Land clearing and the biofuel carbon debt', *Science*, **319**(5867), 1235–8.
Fischer, G., E. Hizsnyik, S. Prieler, M. Shah and H. van Velthuizen (2009), *Biofuels and Food Security*, Laxenburg, Austria: International Institute of Applied Systems Analysis.
Hall, C. and K. Klitgaard (2012), *Energy and the Wealth of Nations: Understanding the Biophysical Economy*, New York: Springer.
Hill, J. et al. (2006), 'Environmental, economic, and energetic costs and benefits of biodiesel and ethanol bio-fuels', *Proceedings of the National Academy of Sciences*, **3**(30).
IEA (2009), *Report on 1st–2nd Generation Biofuel Technologies*, Paris: IEA.
IPCC (2011), *Special Report on Renewable Energy Sources and Climate Change Mitigation*, Geneva, Switzerland: IPCC.
Jefferson, M. (2008), 'Accelerating the transition to sustainable energy systems', *Energy Policy*, **36**(11), 4116–25.
Jefferson, M. (2011), 'Energy supply scenarios: building blocks from 2011', presentation to Centre for International Business and Sustainability Conference, London Metropolitan Business School, April.
Johnson, S. (1993), *The Earth Summit*, London/Dordrecht/Boston: Graham & Trotman/Martinus Nijhoff.
Mackay, D. (2009), 'Sustainable energy – without the hot air', UIT, Cambridge, England.
Miller, L.M., F. Gans and A. Kleidon (2011), 'Estimating maximum global land surface wind power extract-ability and associated climatic consequences', *Earth System Dynamics*, **2**, 1–12.
Mitchell, D. (2008), 'A note on rising food prices', World Bank Policy Research Working Paper 4682.
OECD (2007), *Biofuels for Transport: Policies and Possibilities*, Paris: OECD.
OECD/IEA (2008), *From 1st to 2nd Generation Biofuel Technologies: An Overview of Current Industry and RD&D activities*, Paris: OECD. HMSO.
Office of the Deputy Prime Minister (2004), *Planning for Renewable Energy: A Companion Guide to Planning Policy Statement 22*, London: HMSO.

Oswald, J., M. Raine and H. Ashraf-Ball (2008), 'Will British weather provide reliable electricity?', *Energy Policy*, **35**(1), 3202–15.

Robinson, A. (2009), Planning Inquiry Decision Reference APP/G2815/A/08/2088102, Planning Inspectorate, Bristol, England.

Rodier, M. (1992), 'The Rance Tidal Power Station: a quarter of a century', in Clare, R. (ed), *Tidal power: trends and developments*, Thomas Telford, London.

Searchinger, T. et al. (2008), 'Use of US croplands for biofuels increases greenhouse gases through emissions from land-use change', *Science*, **319**(5867), 1238–1240.

Sinden, G. (2007a), 'Wind power and the UK wind resource', Environmental Change Institute, University of Oxford.

Sinden, G. (2007b), 'Characteristics of the UK wind resource: long-term patterns and relationships to electricity demand', *Energy Journal*, **35**(1), 112–27.

Smil, V. (2008), *Energy in Nature and Society*, Cambridge, MA: MIT Press.

Stigwood, M. (2009), 'Large wind turbines – are they too big for ETSU-R-97?', *Wind Turbine Noise*, Institute of Acoustics, 16 January.

Swift-Hook, D. (2010), 'Grid connected wind and solar energy will never be stored', *Renewable Energy*, **35**(1967–9).

WWF International, Ecofys and OMA (2011), *The Energy Report: 100% Renewable Energy by 2050*, New York: WWF.

11 Energy policy: a full circle?
Colin Robinson

1. INTRODUCTION

Almost always and almost everywhere the energy industries are regulated by governments that attempt to steer those industries in directions they claim are conducive to the 'public interest'. Since the end of the Second World War, most governments have had 'energy policies' most of the time.[1]

However, the balance between state regulation and voluntary action has varied over time. Governments have sometimes intervened extensively in energy markets and at other times they have partially withdrawn, leaving a greater role to market forces. Looking back over nearly 70 years, three sub-periods in energy policy can be identified. In the aftermath of the war, in the economies of the wartime participants, central planning of energy was in vogue for about 35 years. Then, in the last 20 years or so of the twentieth century, as liberal market economics staged a revival, energy planning became less fashionable and governments tended to give freer rein to market forces.[2] That revival proved very short-lived and in the early years of the twenty-first century government energy planning began a comeback, as governments claimed that, without their intervention, the free operation of energy markets would have significant adverse effects on the natural environment and on energy security.

Reasons for these variations are complex. Changes in the prevailing view in the economics profession form part of the background. From the 1940s to the late 1970s that view tended to favour significant government intervention in markets, but in the last quarter of the twentieth century there was a short-lived revival of classical liberal ideas, to be followed in the twenty-first century by a renewed, if more moderate, belief in government action. No doubt energy policies, like other government policies, were affected by the views of economic advisers.

Nevertheless, economists often exaggerate the extent to which their professional views influence policy-making. One particularly important, if neglected, factor in the energy field is the changing influence of pressure groups. Vote-seeking governments are susceptible to the attentions of rent-seeking interest groups because they are chronically short of information on which to base decisions and the ensuing information vacuum is often filled by organized groups which act in the interests of their members. Public choice analysis points out that citizens' interests as producers are concentrated in a particular activity whereas their interests as consumers are dispersed. Consequently organized producer groups have an incentive to invest in lobbying, the prospective returns from which seem to be high: any gains made will be concentrated on the members of a group and losses to society as a whole are dispersed, usually affecting any individual very little. For example, if a government can be induced to protect a particular industry from overseas competition, those in the industry gain much more as producers than they lose as consumers since the costs of protection, even if very large in aggregate, are spread thinly over society as a whole.

Producer and related groups have always been strong in the energy industries and, in recent times, other organized interests – such as groups of scientists and environmental associations – have been active and apparently influential in the energy field. All such groups tend to flourish at times when governments are intervening extensively because the market becomes politicized and it is obvious that government is ready to support interests it favours. The potential gains from lobbying therefore appear substantial. In periods when competitive markets are allowed to operate and intervention is less, the scope for pressure group activity diminishes. Thus the activities of organized interest groups may well amplify changes in policy that would be taking place anyway.

This chapter describes and analyses energy policy over this 70-year period in the UK. At the same time, it points to ways in which pressure groups have influenced policy in each of the sub-periods. Although the analysis concerns only one country, it is a useful one to consider because the changes in policy were probably more extreme than elsewhere: the UK seems to have moved further than any other country from energy planning to energy liberalization, and it is now reverting to planning.

2.   CENTRALIZED PLANNING, 1945–80

Immediately after the Second World War, the UK, in common with other European countries, continued the centralized planning of energy which had originated in the 1930s but had been much enhanced in wartime. Coal, gas and electricity were among the 'commanding heights' of the economy that were nationalized in the 1940s and the whole of the energy sector, including the international oil companies, was subject to substantial government intervention. The principal immediate postwar objective of policy was to restore to its prewar level the production of coal, the predominant primary fuel. By 1952 that objective was achieved, but thereafter coal production was on a downward trend, as relatively high-cost indigenous coal could not compete with products refined from imported oil and, from the 1970s onwards, with natural gas and crude oil discovered in the UK's offshore fields. In 1950, coal's share of UK primary energy consumption was 90 per cent, but by 1980, despite two oil 'shocks' in the 1970s which drastically raised crude oil prices, it was down to 37 per cent. The percentage shares of oil, gas and nuclear power were, respectively, 37, 22 and 4.

The 'energy policies' pursued during this 35-year period were essentially protectionist measures intended to aid British coalmining, with the subsidiary aim of promoting British-designed nuclear power stations.[3] This was a period in which the policies of successive governments, of both major political parties, were heavily influenced by producer pressure groups. The classic conditions for successful pressure group activity existed. Governments were known to be ready to tax, subsidize or take administrative measures in order to promote particular sources of energy and, not surprisingly, producer groups tried to obtain for their members the favours that government was ready to provide. Rent-seeking was rife and the energy market was highly politicized.

The influence of particular pressure groups stemmed from the circumstances of the time. Coalmining employed nearly three-quarters of a million men in the early postwar years, it had a strong union and workers in mining enjoyed considerable public sympathy. Management and unions in mining constituted a very effective political force and it is hardly surprising therefore that the wall of protection that surrounded British-mined coal rose higher and higher as governments tried, without great success, to moderate the forces that were turning consumers away from indigenous coal. But coal was not the only powerful energy lobby. Civil nuclear power was favoured by powerful scientific interests that saw it as a rich source of future research funding and which blinded governments with science, persuading them that a programme of British-designed nuclear power stations would produce electricity 'too cheap to meter'. Moreover, an unintended consequence of a government policy that enhanced the market power of the coalmining industry was that governments came to regard nuclear energy as a counterweight to that power that would help to keep the lights on if, for example, the miners went on strike. Whereas coal was a long-established energy source on which Britain's industrial revolution had been based, civil nuclear energy was a newcomer in the postwar years, conjured into existence by the state on the basis of wartime programmes and then maintained by it.[4]

One of the principal ways in which government policies were pursued was by using the nationalized electricity supply industry to support favoured fuels.[5] In effect, governments determined the fuel choices of the monopoly generator, the Central Electricity Generating Board (CEGB). The industry became a captive market for indigenous coal and it was induced to embark on two nuclear power programmes using British-designed reactors, first Magnox and then advanced gas cooled reactors, followed by a third that moved away from British designs to a Pressurized Water Reactor but which resulted in only one power station (Sizewell B). The CEGB was also prevented from burning in its power stations the natural gas that had been discovered in the British North Sea. Governments interfered less in the development of North Sea oil and gas than they did in the rest of the energy sector,[6] but they were concerned that a switch to natural gas in power generation could displace considerable amounts of British-mined coal and so they made it clear to the oil companies that such a switch would not be permitted. The CEGB had little incentive to resist the government's influence since it was compensated by electricity consumers (to whom the costs of support could readily be passed on because of the industry's statutory monopoly) and by taxpayers.

Whether the extensive government intervention in the energy industries during this period can properly be labelled 'policy' is questionable. An 'energy policy' is generally regarded as a carefully considered long-term plan, formulated and implemented by government or its agencies and intended to improve on what would otherwise be the market outcome. But in the period in question at no time did such a policy exist. Government actions were essentially short-termist reactions to perceived problems, usually under the influence of producer pressure groups: only later were these *ad hoc* interventions rationalized and described as 'policy'. Instead of policy being formulated, to be followed by action to implement it, the process was reversed: actions were taken, later to be set out in a White Paper[7] or a speech and dignified as 'policy'.[8]

Inevitably, there were unintended consequences. Measures aimed at protecting indigenous coalmining and favouring nuclear power spilled over into a protected fuel market in which imports were limited and competition was severely constrained. In electricity generation in particular, the effect of government policy was to allocate shares to the favoured fuels, limiting the use of oil and banning the burning of natural gas in power stations. In some parts of the market outside power generation there was sufficient competition to permit oil and later natural gas to displace coal, but elsewhere cartel-like arrangements prevailed because of government action. In the protected fuel market that then existed, all energy suppliers benefited from the high level of fuel prices, and consumers, especially residential consumers, were disadvantaged compared with what their position would have been in a competitive energy market. It was surely not the kind of energy market that any reasonable person would have planned.

3. A LIBERALIZED MARKET

Significant changes in policy towards the energy industries began quite soon after the election of the first Thatcher government in 1979. The first step came when Nigel (now Lord) Lawson was appointed Secretary of State for Energy in 1981 and began to change the department's culture of planning. For years the department had used a 'predict and provide' approach, which involved making energy demand and supply projections that invariably showed an 'energy gap' which it was assumed it was the government's job to fill. Lawson recognized that this approach, producing fictional gaps which apparently demonstrated that there was a role for government, was incompatible with the Thatcher government's emphasis on the role of markets. He argued that such blueprints for the future have no value and that it was more important to price fuel realistically.[9]

Privatization and Liberalization

Soon afterwards, energy privatization started a process that led eventually to energy market liberalization. British Gas was privatized in 1986, to be followed in 1989 by electricity. Privatization had the beneficial effect of establishing transferable property rights so that energy suppliers entered the market for corporate control and were subject to efficiency pressures from their shareholders. Nevertheless, in their early stages these privatizations had little liberalizing effect on product markets because the newly privatized companies retained much of their market power. Producer pressure groups again had a powerful effect at the time of privatization. The managements of the nationalized industries combined with their unions and with the City to press for monopoly privatizations, and the government, anxious for expeditious and smooth privatizations and substantial revenues, generally conceded. Thus, although efficiency gains could be expected (and did occur), there was no mechanism for passing these benefits on to consumers because of the lack of marketplace rivalry. Even so, privatization was as an important enabling step on the road to liberalization because it lifted the previous prohibition on entry to gas and electricity markets, allowing competitive markets gradually to appear through the actions of entrepreneurs, regulators and the competition authorities.

There were some novel features of the UK regulatory regime for the utilities, established in the 1980s, that distinguished it from previous regulatory systems, in particular giving it a pro-competition emphasis quite different from traditional regulation.[10] First, the regulatory office for each industry was given a degree of independence from the political process, which provided markets with confidence that the industries would no longer be used by government to pursue political objectives, as they had been under nationalization. Of course, governments can always override a regulator if they are determined to do so, but the UK system provided safeguards that regulation would be independent and, again unlike nationalization as it had been practised, based on clear rules.

Second, price cap regulation, based on an RPI-X formula, provided better efficiency incentives for parts of the utilities that had to be price-regulated than had regulatory systems based on rate of return.[11] Third, separation of networks from the rest of the utilities was a crucial factor in liberalization. Separation was carried out only partially, or not at all, at the time of privatization. At privatization, the gas pipeline network was left in the hands of British Gas, leading to a large vertically integrated private monopoly which made entry extremely difficult, though in electricity there was some network separation since the long-distance transmission system was placed in a separate company. Much of the separation that occurred was carried out post privatization by regulators and the competition authorities, isolating 'natural monopoly' sectors of the utilities so that they could have their prices and service standards regulated while the other (potentially competitive) parts of the utilities could be opened to competition.

Fourth, the most fundamental component of the UK regulatory regime was to give regulators a duty to promote competition. Traditional regulation had been based on the view that a regulator should aim at achieving the results of a competitive market. The view underlying the UK regime was that, since it is impossible for a regulator to guess what the outcome of a competitive market will be, the task of the regulator in potentially competitive markets should be to start a competitive process. This process will be beneficial to consumers since it will promote entrepreneurship and innovation, the gains from which will be passed on to consumers because of marketplace rivalry.

Regulation thus had a different emphasis from what had been traditional. Instead of trying to control prices or profits, regulators could be pro-active, taking the initiative to stimulate entry to any potentially competitive markets they supervised. In gas and electricity, the regulators[12] embarked on a determined drive to introduce effective competition, defining the 'natural monopoly' areas narrowly. They stripped away and opened to competition activities that had previously been considered parts of natural monopoly networks (such as meter provision, meter reading storage and connections), permitted competitive extensions to networks and generally minimized the area to which price regulation applied. In sum, energy regulators liberalized markets where possible and only regulated in the traditional sense where a competitive market was either infeasible (natural monopoly) or was some way off (pre-competitive markets).

The new regime seemed to deliver positive results. As it evolved from uncertain beginnings, government interference declined, competition increased, the old monopolies lost their power, rent-seeking declined and efficiency improved. In the energy industries, where consumers were in 1998 given freedom of choice of supplier and wholesale markets were subsequently established so that gas and electricity became more like other commodities with functioning spot and forward markets, a healthy injection of competition ensued. Falling prices for both domestic and business consumers, improved service and better supply security followed as regulation was replaced by competition across much of the energy market.[13]

The Effects on Energy Policy

Pre-1979-style energy policy could not, of course, survive under the new, more liberal regime, though it took some time to die. The government continued to protect indigenous coal for a time, mainly via contracts with the electricity generators and until the late 1980s it also continued to favour nuclear power. But after the miners' strike of 1984–85, which the government was deemed to have won, British coalmining was never again a powerful pressure group and, in 1994, a much-diminished coal industry was privatized. Nuclear power fell out of favour after failure of the first attempt to privatize it, with the rest of electricity supply, in 1989 when investors were startled by revelations about the unexpectedly high and uncertain costs of some of its activities.[14] By the late 1990s, UK energy policy was much less protectionist than it had previously been. Indeed, the opening up to market forces of electricity and gas, together with the freeing of imports and the decline of protection for indigenous fuels, transformed a previously heavily protected energy market into probably the most competitive energy market in the world. For a time, producer pressure groups no longer had much influence: with competitive markets and less interventionist government, there was no longer such scope for producers to obtain favours from government, and rent-seeking diminished.

4. RECENT CHANGES TO THE REGIME

However, the liberalized energy market, despite being generally regarded as highly successful, lasted a remarkably short time. By the early years of the twenty-first century it was clear that government intervention in energy was re-emerging, justified mainly on the grounds that looming environmental problems and supply insecurity made such intervention desirable. Governments began to formulate 'strategic objectives' for the energy sector that affected all engaged in the market and that could not readily coexist with independent, pro-competition regulation of the energy utilities.

The 1997–98 Review and the Formulation of Strategic Objectives

The effect on the recently liberalized utilities was particularly marked. The seeds of change of a different approach to utility regulation can be detected in a 1997–98 regulatory review,[15] and subsequent legislation (beginning with the Utilities Act, 2000) that implemented the conclusions of the review. The review – which in general supported the existing regulatory regime and its emphasis on competition-promotion – concluded nevertheless that regulators had enjoyed too much discretion in social and environmental matters: government should therefore in future give statutory guidance to regulators on social and environmental objectives. These statutory guidance provisions, incorporated in the legislation, had the potential gradually to restore government control of the utilities, even though they would be implemented by regulators.[16]

Another significant change that occurred with the Utilities Act was the replacement of individual regulators by boards so that responsibility for carrying our regulatory objectives became more dispersed. The entrepreneurial spirit shown by the early regulators, especially in their efforts to introduce competition into gas and electricity markets, seemed likely to be curbed by the new committee structures, and the greater number of appointments to regulatory positions left more scope for governments to appoint people with views that it favoured, thus potentially politicizing the appointments process.

Nevertheless, in the first few years after the Utilities Act, there was little sign that government control was increasing. Indeed, the liberalization of the electricity and gas markets, begun in the 1990s, continued in the early years of this century.[17] Only in the last few years of the 1997–2009 Labour government did central control of the energy industries significantly increase. Evidently in pursuit of two principal strategic objectives – promoting security of energy supply and avoiding damaging climate change – the government began to intervene extensively. It introduced a range of measures, explained below, centred on energy conservation and the promotion of non-fossil energy sources (especially 'renewables'), which towards the end of Labour's term had begun to have pervasive effects across the energy sector, impinging on and constraining the energy regulator's activities.[18] The coalition government, elected in May 2010, continued along broadly the same path, beginning with a statement in July 2010 that adopted a highly interventionist stance towards the energy market.[19] Subsequent statements continued this interventionist theme, culminating in a programme of 'reform' for the electricity market which has a theme familiar from earlier times – ensuring that the electricity generators conform with the government's views about what fuels should be used to generate electricity.[20]

Changes in the Energy Regulator's Objectives and Role

As a consequence of the changed approach, the energy regulator is operating in an environment in which its independence is increasingly constrained by government instructions or guidance. The Office of Gas and Electricity Markets (Ofgem), the energy regulator, has had its objectives modified on several occasions and they have become more complex. Instead of a straightforward competition-promoting goal, it now has to protect the interests of consumers by promoting effective competition 'wherever appropriate' and (under the 2008 Energy Act) to have regard to the need to 'contribute to the achievement of sustainable development'. The 'sustainable development' qualification must be a particularly difficult issue for a regulator since it is very much a matter of subjective opinion whether any given action is or is not consistent with such a vague objective. In January 2010, the previous government issued revised social and environmental guidance to Ofgem, to which the latter must have regard, which made clear that the interests of consumers, which Ofgem must protect, include *inter alia* their interests in the reduction of greenhouse gas emissions and in security of supply.

The incoming coalition government set up a review of Ofgem to consider whether changes are required to 'align the regulatory framework with the government's strategic policy goals'. The review concluded that Ofgem would continue to regulate independently of government but that government would set out clear strategic roles in a new 'Strategy and Policy Statement'.[21] It seems clear that, although Ofgem may still have operational independence to pursue objectives set for it by government, the original 'hands-off' approach has now gone as governments intervene more and more in energy markets and change the objectives of the 'independent' regulator so that they are in accord with actions the government wishes to take. Ofgem was established to promote liberalized energy markets, with minimal government intervention. Now, in a very significant if little-remarked change of approach, it is cooperating in a revived interventionist approach and appears to place little emphasis on market liberalization.

Reversion to Earlier Practices

Seen in historical perspective, the interventionist trend of recent years appears as a reversion to practices characteristic of the period from the end of the Second World War to the early 1980s when, as explained in Section 2 above, governments pursued 'energy policies' in which they supported certain favoured forms of energy production (in those days coal and nuclear power), leading to a protected home market for energy. At that time, of course, gas (except for offshore production), electricity and coal were nationalized and there was no energy regulator. The new manifestation of the old policy is the setting by government of two 'strategic objectives', relating to climate change and energy security, which assume that the energy market is likely to fail to provide both adequate protection of the natural environment and also adequate security of supply. The principal measures intended to achieve these objectives aim to reduce emissions of carbon dioxide (CO_2) and other 'greenhouse gases' by means of substituting non-fossil for fossil energy sources and by improving the efficiency with which energy is used.

Some of the measures are at the instance of the EU, but UK governments have gone further than would be required by EU membership: indeed, the coalition government has (so far unsuccessfully) been pressing other EU members into increasing the EU's greenhouse gas emission reductions target from 20 per cent to 30 per cent by 2020. Already, under the 2008 Climate Change Act, the UK now has 'carbon budgets' (claimed to be the first in the world), which aim to cut carbon emissions from 1990 levels by 34 per cent by 2020 and 'at least' 80 per cent by 2050. A range of subsidies is available for those suppliers that take steps to reduce customers' energy consumption and those that avoid using fossil fuels (gas, coal and oil). In general, obligations are imposed by government on suppliers, which then pass the costs on to consumers.

For energy consumers, there have been two principal consequences of renewed energy market intervention. First, energy prices have increased and will increase further in future, compared with what they would have been had consumers and producers interacted freely in a competitive market, because government policy increases costs, especially in electricity generation. Generating costs have risen, and will rise further under current policy, because of the higher costs of using renewable rather than fossil fuels. There has also been an increase in electricity transmission and distribution costs because of the tendency to locate renewable generation in rather remote places. Second, energy prices have moved further out of alignment with costs because some energy consumers are cross-subsidizing other consumers: for example, when suppliers encourage their customers to take subsidized energy-saving measures under government schemes, the subsidy is paid for by consumers who do not undertake such action (or had taken it already).

A particularly important feature of present energy policy, which is a consequence of the attempt to force renewable and other non-fossil energy sources into electricity generation, is that the government is once again assuming a major role in determining what types of power stations should be built, even though the electricity supply industry has now been in private hands for over 20 years. This intervention has repoliticized the energy market, which is converging on the old nationalized regime under which the investment programme of the Central Electricity Generating Board (CEGB) was largely under the control of government.

Section 2 above explained that the CEGB was used to support indigenous energy producers and equipment suppliers and was told by government to build coal and nuclear power stations.[22] Today, coal is of course out of favour. The government is now insisting on a big programme of wind turbines, intended to produce 30 per cent of electricity from wind and other renewables by 2020 (but requiring extensive back-up from fossil fuel generation). Despite differences in the coalition about nuclear power, the government also appears to be trying to promote a new nuclear power programme which, if it goes ahead, will involve a subsidy in the form of a minimum price for carbon (from 2013) which would benefit nuclear (and other non-fossil fuels) at the expense of fossil fuels. At the same time, some coal and oil plants are being closed early because of EU regulations, on the grounds that they are excessively polluting.

5.   PROBLEMS IN THE RETREAT FROM LIBERALIZATION

This revival of an interventionist centralized energy policy is a significant change from the previous 'hands-off' policy of both Conservative and Labour administrations. It raises the question of whether the likely benefits of government action will outweigh the likely costs.

Centralized Planning of Energy

If a centralized 'energy policy' were being pursued by an altruistic and omniscient government of the sort typical of neoclassical textbooks in economics, the outcome might well be an improvement in social welfare. Such a government would be able to do two important things and be willing to undertake a third. First, it would be able to see reasonably well into the future, so it would have a good idea what the market outcome would be if it took no action; second, it would be able to make reasonable predictions of the consequences of various actions it might take and so assess what it should do to improve on the market outcome; and third, it would take welfare-improving actions, having regard not to its own interests but only to the 'public interest' (which it would be able to define).

Real-world governments can do neither of the first two things and may be unwilling to undertake the third. As to the first two, the record of centralized predictions – both of outcomes based on unchanged policies and on the effects on outcomes of various policies – is so poor in the energy field that it suggests that such predictions provide no useful basis for policy. Indeed, there is a serious underlying problem, which is that the suppression of markets which is implied by central planning inevitably suppresses much of the information that governments require to plan. The UK government is, by definition, ignorant of most of the information it needs if it is successfully to engage in the future planning of the UK energy sector. For example, it has no means of knowing at what level it should set the minimum price for carbon it intends to impose to promote the use of non-fossil fuels (nor even, as explained below, of whether any such minimum can be justified). Inevitably, its efforts to intervene will have all manner of unintended consequences, most likely including the weakening of the market incentives that would otherwise bring about adaptation to incipient problems. Finally, as regards willingness, it cannot be assumed that, even if welfare-enhancing actions can be identified, they will be undertaken by a government if they appear to be against its own interests.

Moreover, a serious complication in the real world is that planning from the centre invariably encourages pressure group activity, for the reasons that were set out earlier – the government is seen to be willing to provide support for favoured activities so everyone will want to be on the favoured list and, once there, will press for as much support as possible. The 35-year period after the Second World War, as explained in Section 2 above, provides a striking example of the kind of market that appears in such circumstances where producers flourish and consumers are disadvantaged. In the new interventionist environment of today, some of the interest groups that are flourishing are familiar from earlier times – many energy producers are happy that competitive forces have been blunted and the scientific establishment continues to encourage the promotion of nuclear power. However, the scientific establishment now has another cause: it promotes the view that centralized action is essential if the dire consequences it predicts will result from human-induced climate change are to be avoided.[23] Pressure groups that favour measures to protect the natural environment are also, of course, vocal in support of action against the possible effects of climate change. Some contrary views are now appearing that may eventually be influential: in particular there are protests from business about the costs of government programmes. But, as in earlier times, the bulk of residential consumers, unorganized as they are, have little influence except at the times of (infrequent) elections when many other issues are under discussion. Despite the fundamental problems of centralized planning and despite the lessons that might be learned from earlier periods, the UK government is proceeding with a central plan for the energy industries which centres on setting the two 'strategic objectives' explained above, reflecting the view that there are severe market failures which can and should be corrected by government. What substance is there in this view?

Security of Supply

As regards the security of supply objective, the idea that government can enhance the security that would otherwise be provided by the energy market is superficially attractive. However, both theory and experience suggest that it is very difficult for government to do so[24] and that there is a significant risk that the attempt will have the perverse effect of reducing security (as past British governments have often done) because it removes responsibility from market participants and blunts their incentives to provide security. For example, government intervention in decisions about new electricity-generating plant is likely to affect the behaviour of actual and potential electricity generators in ways that will make supplies less secure. Electricity generators contemplating new plant in a system in which government is determining a large proportion of what is built, as it now is, must try to assess how political whims might change and to what extent existing plans might fail. A particular issue concerns investment in types of generating plant not presently favoured by government (such as gas, coal and oil). If the government is insisting on a large quantity of wind, other renewable and nuclear capacity, trying to estimate the future residual demand for these other types of power stations is guesswork indeed. For example, hardly anybody believes that the present very ambitious government targets for wind energy will be achieved but there is no way of knowing what the shortfall will be. Nor is there any way of assessing to what extent future governments might change the plans. Consequently, one of the unintended consequences is likely to be to increase uncertainty, deter investment and reduce security. Responsibility for security is, in effect, removed from producers (which are the main instruments for achieving it) and handed over to government. Market-based security has much firmer foundations, based as it is on the incentives which producers have to diversify sources of supply and supply technologies, and on supplier incentives to purchase energy which is likely to be in continuous supply. The popular idea that there is a major market 'failure' that requires government to intervene to ensure energy security has little basis: security is a valuable attribute of any energy product which is incorporated in the prices that appear in market transactions. Government 'failure' is a more likely problem in security provision.

Climate Change Policy

There are also serious problems underlying pursuit of the climate change objective.[25] There is room for reasonable people to differ about the influence of human beings on the climate, the likely extent of any future climate change, the associated economic and social consequences and the need for government action. Clearly, the conventional view is that governments must act because the looming problems are so serious. But, because of information failures and government failures in policy implementation, it is very difficult to translate that view into action that will in practice improve social welfare. In particular, in formulating policy, it is essential to show some awareness of the uncertainties surrounding projections of climate change and its effects.

Information failure is a particular problem in climate change policy in that not only is there huge scientific uncertainty about how much (if any) global warming there will be as a consequence of human activities, there is also extreme uncertainty about the economic and social consequences of a given amount of warming. The economic and climate models that are used as a basis for the conventional conclusion – that drastic human-induced climate change with dire economic and social consequences is virtually certain in the rest of this century – go out into the far distant future, typically a century or more, and well beyond the range of any experience. It is difficult to believe that they would be taken seriously in any field other than environmental policy where the scientific and economic establishments, and their followers in the media, have encouraged an uncritical attitude towards the relevant issues, with their dismissive claim that 'the science is settled'. Such models seem a wholly inadequate basis for the government to conclude, as it evidently does, that it knows the future direction of climate change, its approximate magnitude and its likely effects sufficiently well to formulate a highly interventionist policy to deal with the problems it foresees.

Like its immediate predecessor, the present UK government shows no awareness of the extent of the uncertainties surrounding future climate change and its possible effects. It seems determined to plan centrally, with all the rigidities that implies, in circumstances in which central planning is inappropriate. In doing so, it is committing one of the commonest errors of central planners – to view the future as though it contained only a narrow range of possible outcomes – leaving little flexibility to deal with the inevitable surprises. As explained below, policy towards climate change should reflect the huge uncertainties of the future, retaining a degree of flexibility, rather than taking a fixed position.

6. TIME FOR A CHANGED APPROACH?

In UK energy, the story of the last 70 years is one of changing views about the roles of markets and governments which have been translated into practice via considerable input from producer and other interest groups that have been successful in ensuring that, in periods of significant government intervention, it has worked to their advantage. After a brief period of liberalization in which the power of the organized groups was undermined, a rent-seeking energy market has reappeared and the direction of change is now towards a market like that from the 1950s to the early 1980s when government was heavily involved and a cartel-like structure developed among suppliers, to the advantage of incumbent producers and to the detriment of consumers. All UK energy producers learned to accept and profit from that structure, which gave them a significant amount of protection from competition. They were happy to be regulated by government, in an 'orderly' market protected from disruptive entry in which they enjoyed a quiet life at the expense of energy consumers. Rent-seeking was encouraged by the politicization of the market.

A similar kind of market has reappeared in the last few years. Competition in the gas and electricity markets has diminished, the regulator is following government-set objectives and is no longer concentrating on promoting competition, and producers find it easy to raise prices since they are confident their competitors will follow suit. 'Policy' is once again being heavily influenced by pressure groups. Potential suppliers of renewable and nuclear power are queuing up to prise out of the government the subsidies available to those who invest in lobbying.

The revived centralized approach seems destined to end in failure. Like previous such ventures, it is beset by innate information failures and rigidities. The case for government intervention to improve energy security is very weak and indeed any such action is likely to have perverse consequences. As for action on anticipated climate change and its effects, the future is so uncertain and the degree of ignorance is so great that there should be a premium on flexibility rather than on measures that assume, as does present policy, that the future is foreseeable.

The climate change objective has become so dominant in energy policy that it is worth pointing to alternative ways of dealing with it. There seem to be three possible policy approaches to the perceived problem.

1. The overwhelming favourite, as far as politicians are concerned, is centralized action, by some variant of picking winners. Adopting this approach necessarily means looking a long way into a very uncertain future and settling now on means of dealing with apparent problems. Yet, given the well-known problems of central planning, it seems unlikely that governments and international bodies have the information and the incentives to use this approach successfully. Such action runs the risk of serious

errors, even in the direction of policy, and the major investment programmes that are involved would be difficult to change, even marginally, let alone reverse, if circumstances required. Another serious difficulty in relying on political action to solve any climate change problems that may emerge is that enthusiasts for winner-picking assume that, if a genuine climate change problem occurred, governments would act. However, in such circumstances the political calculus might mean that governments continued to concentrate on the appearance of taking action. For that reason, and because of practical experience so far of climate change policy and experience during other 'crises', there are legitimate doubts about whether, if effective action were necessary, such action would ensue.

2.   A second approach, popular with neoclassical economists, is to try to correct the apparent 'market failure' by internalizing the relevant externality, in this case introducing a carbon tax or carbon trading. Although this approach seems more attractive than picking winners, in practice it is very difficult to apply successfully. One major problem is that no government is willing to rely solely on it: those that favour a tax or trading (as in the EU) want to pick winners as well. More fundamentally, the information requirements for the successful use of a carbon tax or carbon trading are huge. Given all the uncertainties about not just the magnitude but the direction of any future climate change, setting the necessary tax rate or the traded volumes is guesswork and so it is uncertain whether a carbon tax or carbon trading would be welfare-improving.

3.   The third approach is to leave it primarily to the market. Hardly anyone seems to favour it, perhaps because most economists are so impressed with the magnitude of the apparent market 'failure'. But it is a constructive approach which recognizes the huge uncertainties that exist, and seeks flexible means of dealing with the problems that may arise. There is, after all, already a spontaneous market reaction in favour of 'green' products and services. That reaction is based on some very imperfect information, much of it emanating from governments and international bodies, but consumers are used, in most markets, to filtering distorted information (such as that from producer advertising).

Economists who propose reliance on markets to deal with natural resource and environmental problems are sometimes accused of being Panglossian. But the true descendants of Dr Pangloss may well be those who have faith in very long-term forecasts of climate change and its effects and who think that, in the face of recent experience and examples they can see all round them, centralized action is the answer. In practice, if damaging man-made climate change is in prospect, the only real hope of avoiding the damage is probably through the problem-solving mechanisms of markets: experience so far suggests that the chances of effective action by governments and international bodies are very low.

A big advantage of relying primarily on markets is their flexibility and adaptability. It is unnecessary to peer many decades ahead into a very murky future and make long-term commitments to massive investments to deal with supposed problems. Nor is it necessary to wait for politicians to act.

Markets will start to deal with problems as soon as they are perceived as such, and they will adapt as views change. Global warming may appear a more serious issue than now, in which case markets will enhance the profitability of 'greenery', so reacting in the 'right' direction. Or it may seem less serious, so that 'greenery' starts to go out of fashion. Can anyone be so confident that the big programmes now being urged by climate change activists will show a similar degree of adaptability to changing circumstances? They are all too likely to set communities on courses that are very difficult to change in response to the changing views of climate scientists. A market-based, decentralized approach to climate change would allow adaptation to changes in climate that scientists cannot predict and in ways that central authorities cannot envisage.

NOTES

1. Colin Robinson and Eileen Marshall, 'The regulation of energy: issues and pitfalls', in Michael Crew and David Parker (eds), *International Handbook on Economic Regulation*, Cheltenham, UK and Northampton, MA, USA: Edward Elgar, 2006, pp. 325–49. The reasons why governments regulate energy are complex. Presumably democratically elected governments believe there are votes to be won by such regulation.
2. Colin Robinson, 'Energy economists and economic liberalism', *The Energy Journal*, 21(2), 2000, 1–22.
3. For some history of the period see PEP, *A Fuel Policy for Britain*, London: Political and Economic Planning, 1966; William G. Shepherd, *Economic Performance under Public Ownership – British Fuel and Power*, London: Yale University Press, 1965; Dieter Helm, *Energy, the State and the Market*, Oxford: Oxford University Press, 2003; and Colin Robinson, 'Gas, electricity and the energy review', in Colin Robinson (ed.), *Successes and Failures in Regulating and Deregulating Utilities*, Cheltenham, UK and Northampton, MA, USA: Edward Elgar, 2004, pp. 184–205.
4. Colin Robinson, *The Power of the State: Economic Questions over Nuclear Generation*, London: Adam Smith Institute, 1991.
5. The industry was used also to provide support to British manufacturers of power-generating equipment.
6. Colin Robinson and Jon Morgan, *North Sea Oil in the Future*, London: Macmillan, 1978.
7. There were two White Papers on Fuel Policy, Cmnd 2798 in 1965 and Cmnd 3438 in 1967.
8. Colin Robinson, 'A policy for fuel?', Institute of Economic Affairs Occasional Paper 31, 1969 and 'Energy policy: errors, illusions and market realities', Institute of Economic Affairs Occasional Paper 90, 1993.
9. Nigel Lawson, *The View from Number 11*, London: Bantam Press, 1992, especially ch. 15.
10. Robinson and Marshall, 'The regulation of energy', and Eileen Marshall, 'Electricity and gas regulation in Great Britain: the end of an era', in Lester C. Hunt (ed.), *Energy in a Competitive Market*, Cheltenham, UK and Northampton, MA, USA: Edward Elgar, 2003, ch. 1.
11. The classic analysis of the problems of the US rate of return regulatory regime is in H.A. Averch and L.L. Johnson, 'Behavior of the Firm under Regulatory Constraint', *American Economic Review*, 52(5), 1962.
12. Initially OFFER and OFGAS, later OFGEM.
13. The National Audit Office examined the impact of the reforms implemented by the gas and electricity regulators in the late 1990s and early 2000s in a number of reports and concluded that, in general, they had been beneficial. For example, National Audit Office, *Giving Customers a Choice – the introduction of competition into the domestic gas market*, HC 403, Session 1998–99, May 1999; Comptroller and Auditor General, *Giving Domestic Customers a Choice of Electricity Supplier*, HC 85, Session 2000–01, December 2000; and Comptroller and Auditor General, *The New Electricity Trading Arrangements in England and Wales*, HC 624, May 2003.
14. Robinson, *The Power of the State*. The industry was eventually privatized in 1996.
15. Department of Trade and Industry, *A Fair Deal for Consumers: Modernising the Framework for Utility Regulation*, Cm 3898, March 1998 and *A Fair Deal for Consumers: Modernising the Framework for Utility Regulation – The Response to Consultation*, July 1998.
16. Colin Robinson, 'After the Regulatory Review', in M.E. Beesley (ed.), *Regulating Utilities: A new era?*, London: Institute of Economic Affairs, Readings 49, 1999.
17. Marshall, 'Electricity and gas regulation in Great Britain'.
18. Colin Robinson, 'The rise and decline of UK utility regulation', *The Business Economist*, 41(3), 2010. There had been some (lesser) support for nuclear and renewables from the early days of privatization.
19. DECC, *Annual Energy Statement*, July 2010.

20. DECC, *Planning our electric future: a White Paper for secure, affordable and low-carbon electricity*, July 2011.
21. DECC, *Ofgem Review Final Report*, July 2011.
22. At the time of privatization, almost 90 per cent of the fuel used in electricity generation was coal and nuclear. Relevant statistics are in DECC, *Digest of UK Energy Statistics*, Table 5.1.1.
23. Colin Robinson, 'Economics, politics and climate change: are the sceptics right?' The Julian Hodge Lecture, University of Cardiff Business School and Julian Hodge Bank, 2008.
24. Colin Robinson, 'The economics of energy security: is import dependence a problem?', *Competition and Regulation in Network Industries*, December 2007, p. 425, and Eileen Marshall, 'Energy regulation and competition after the White Paper', in Colin Robinson (ed.), *Governments, Competition and Utility Regulation*, Cheltenham, UK and Northampton, MA, USA: Edward Elgar, 2005, pp. 118–40.
25. Robinson, 'Economics, politics and climate change'.

PART IV

CLIMATE AGREEMENTS

12 Anthropogenic influences on atmospheric CO_2

*David F. Hendry and Felix Pretis**

1 INTRODUCTION

We identify anthropogenic contributions to atmospheric CO_2 measured at Mauna Loa using the statistical automatic model selection algorithm *Autometrics*. Estimating the determinants of atmospheric CO_2 is traditionally a challenge due to the complex systems of data involved. CO_2 is a highly autocorrelated, non-stationary time series, and globally there exist a large number of potential carbon sources and sinks. There is mixed evidence in the literature on anthropogenic contributions to atmospheric CO_2: the long-term trend is widely attributed to human factors, while the main seasonal fluctuations are thought to be driven by the biosphere. However, the statistical measures applied are often somewhat unsatisfactory due to the complexities of dealing with large numbers of variables.

Over the long run of geological time, evidence of repeated glaciations, and of coal and oil deposits from extinct tropical forests, reveals that atmospheric CO_2 has varied greatly, and manifestly without any anthropogenic influence, including very low levels and levels as high as 1000 parts per million (ppm): see, for example, Hoffman and Schrag (2000); Hendry (2011) provides a summary. In the more recent half million years of 'ice ages', natural fluctuations include highs and lows of 300 and 180 ppm from Antarctic ice sheet drilling (see Juselius and Kaufmann, 2009). Finally, in the last 10000–12000 years, humanity has transformed planet Earth, replacing forests by agriculture and creating an industrial world (see, e.g., Ruddiman, 2005). Against the background of such movements, it is important to establish that recent levels of atmospheric CO_2 are not merely a natural event, but have an anthropogenic signature.[1]

We introduce a new approach to modeling changes in atmospheric CO_2 using a model selection algorithm which allows for a larger number of potential variables than observations without *a priori* forcing any to be significant or to be excluded. Using this method, the main relevant explanatory variables are determined and their magnitudes estimated while irrelevant factors are dropped from the model.

The model controls for a number of natural carbon sources and sinks: vegetation measured by the Normalized Difference Vegetation Index (NDVI); temperature (measured as anomalies in the northern hemisphere); weather phenomena (measured through the Southern Oscillation Index); as well as accounting for dynamic transport by including seasonal interaction terms. This allows an estimate of the anthropogenic contribution to CO_2 as measured by industrial output indices and fossil fuel use for different geographical areas. The resulting estimates describe the direct effects on CO_2 growth and the proportional contribution of each factor. We find that vegetation, temperature and other natural factors alone cannot explain either the trend or all the variation in CO_2 growth. Industrial production components, driven by business cycles and shocks, are highly significant contributors.

Section 2 provides an overview of related literature, and Section 3 discusses model selection, impulse indicator saturation (IIS) – which we use to detect multiple breaks in the models – and the *Autometrics* algorithm. Section 4 describes the data used, Section 5 outlines the estimation procedures and Section 6 reports the main results. Section 7 concludes.

2   LITERATURE REVIEW

There is a plethora of literature on atmospheric CO_2 and its link to anthropogenic factors. Key aspects in the literature are finding an appropriate measure for anthropogenic activity and sufficient controls for natural effects such as vegetation and oceanic absorption. Atmospheric CO_2 has been measured consistently and regularly since 1958, mostly due to the effort of Charles Keeling, who initiated and supported the measurement at Mauna Loa, Hawaii and later other measurement stations (see Sundquist and Keeling, 2009). This led to the now well-established and often cited 'Keeling curve', showing the increasing trend and highly seasonal pattern in atmospheric CO_2 (Figure 12.1). We are primarily interested in identifying the anthropogenic contribution to atmospheric CO_2, so the following section reviews existing evidence and factors that need to be included in models.

Sources and Sinks

Anthropogenic sources
While atmospheric CO_2 has been consistently measured at multiple sites for a long time (see Keeling et al., 1976; Sundquist and Keeling, 2009), the choice of anthropogenic

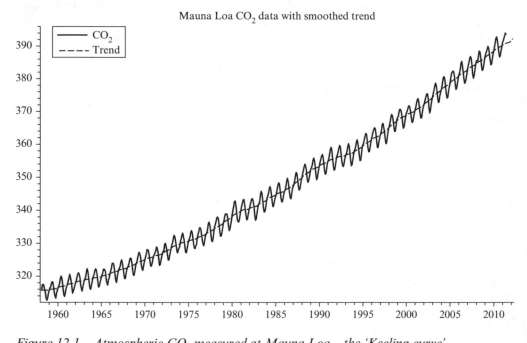

Figure 12.1   Atmospheric CO_2 measured at Mauna Loa – the 'Keeling curve'

variables is less straightforward. Three variables are regularly used: estimates of fossil fuel emissions; population; and cement production. In most cases, variables are measured on an annual basis and interpolated to monthly frequencies.

A standard measure for estimating fossil fuel emissions is the product of the amount of fuel produced, the proportion of the fuel that is oxidized, and the fuel carbon content (Marland and Rotty, 1984). Variations of these data are used in Erickson et al. (2008), Jones and Cox (2005), Randerson et al. (1997), and Nevison et al. (2008). Recent data using this methodology are available at an annual frequency from Marland et al. (2011). In contrast, Hofman et al. (2009) as well as Newell and Marcus (1987) focus on population as a measure of human industrial output. Granados et al. (2012) extend the model of population by including a measure of global GDP. Cement production is a further major component of anthropogenic emissions. CO$_2$ emissions in production through limestone calcination, kiln operation and power generation are estimated to make up approximately 5 percent of anthropogenic emissions (Worrell et al., 2001).

These measures provide a good starting point to capture anthropogenic emissions in the long run, but due to the annual measurement frequency do not capture short-run fluctuations and seasonality. Population and GDP are too broad as measures to capture variability other than a trend. Measurement could be improved through supplementing overall fuel emissions by disaggregate individual fuel consumption. However, most importantly, a higher-frequency (monthly) anthropogenic measure is required to capture seasonal and short-term effects.

Terrestrial biosphere and transport

Aside from anthropogenic emissions, the terrestrial biosphere (vegetation) is a major factor in the carbon cycle's sources and sinks. When trying to model anthropogenic contributions to CO$_2$ it is therefore important to account for the most important factors in the natural world.

Atmospheric CO$_2$ falls and rises seasonally each year due to photosynthetic activity (during summer) and respiratory release (during winter) of CO$_2$ in the terrestrial biosphere (Keeling et al., 1996; Buermann et al., 2007). The intensity of these effects depends on the length of the growing season, a fertilization effect (feedback to plant growth from increased CO$_2$), and shifts in seasonal patterns (Kaufmann et al., 2008). The literature on atmospheric CO$_2$ suggests a strong link between the earth's biosphere and the level of CO$_2$.

The Normalized Difference Vegetation Index (NDVI) provides a direct measure of photosynthetic activity. As Myeni et al. (1995) describe, chlorophyll found in plants absorbs visible light for photosynthesis and reflects near infra-red light. The more active a plant is (indicated by higher density of green leaves), the more visible radiation is absorbed and the more near infra-red is reflected. Thus the difference between the two measures increases with higher leaf density. Using satellite-based remote sensing, the intensity of reflected visible and infra-red light can be measured. Using the ratio of the difference between the two measures, the NDVI is defined by

$$\text{NDVI} = \frac{\rho_{NearIR} - \rho_{Visible}}{\rho_{Visible} + \rho_{NearIR}} \qquad (12.1)$$

where $\rho_{Visible}$ and ρ_{NearIR} are measures for visible and near infra-red light respectively. NDVI is therefore an index ranging from -1 to $+1$, with values around 0 denoting non-vegetation objects, and values around > 0.7 indicating dense vegetation (Tucker et al., 2010). The NDVI provides a direct measure for vegetation activity with high values close to 1 during the growing season (summer in the northern hemisphere), and lower values closer to 0 during the less active season (winter in the northern hemisphere). Kaufmann et al. (2008) investigate the link between NDVI and atmospheric CO_2 using econometric methods and find evidence that NDVI values 'Granger-cause' CO_2. There is also evidence of a feedback effect of increased CO_2, leading to enhanced vegetation activity.

Due to transport airflows, the primary influence of vegetation on measured atmospheric CO_2 depends on the location of the measurement station. In the case of Mauna Loa, Hawaii, the seasonal variation due to the biosphere is driven by long-range transport from Eurasia during winter and short-range transport from North America during summer (see Buermann et al., 2007; Taguchi et al., 2003). Narrowing down the time frame, Levin et al. (1995) suggest that Eurasian airflows dominate from October to June while North American airflows are dominant from July to September.

In terms of long-term development of terrestrial vegetation, there has been a greening trend – an increase of the growing season in the northern hemisphere (Lucht et al., 2002). This trend, however, was interrupted by the eruption of Mount Pinatubo in June 1991, which led to a decline in vegetation during 1992–93. This poses the question whether volcanic influence on atmospheric CO_2 needs to be controlled for separately from the biosphere. Lucht et al. (2002) state that the main channel through which volcanoes affect atmospheric CO_2 is indirect through temperature, while Hofman et al. (2009) propose that the Pinatubo eruption enhanced photosynthesis through scattered sunlight. Overall, Gerlach (2011) finds the direct effect of volcanic activity on measured atmospheric CO_2 to be small, thus accounting for vegetation, and temperature is expected to be sufficient without considering volcanoes separately. However, sudden breaks due to large eruptions will be detected through our IIS procedure.

Oceanic absorption and El Niño

A large amount of CO_2 is transferred between the ocean and the atmosphere, where the ocean acts as both a source and a sink of atmospheric CO_2. There are two main factors determining oceanic carbon absorption and release (Bacastow, 1976). First, generally ocean absorption of CO_2 is governed by the difference in partial pressure between the atmospheric and oceanic CO_2:

$$\frac{dp}{dt} = -k \cdot (p - P) \tag{12.2}$$

where p is the partial pressure of atmospheric CO_2, P is the partial pressure of oceanic CO_2 beneath a surface layer and k is a variable capturing layer thickness and wind. These values are affected primarily by temperature and wind speed. Temperature is a key factor in oceanic CO_2 transfer, as higher sea surface temperature reduces uptake and increases outgassing (Watson et al., 1995). Increased wind speed increases k, so leads to higher oceanic CO_2 uptake. The second effect is upwelling – dense cold water driven to the surface releases CO$_2$ stored in the ocean. Absorption and upwelling play opposite roles; which dominates is debated and depends on the geographical region.

The atmospheric fluctuations of air pressure differences known as Southern Oscillation affect oceanic absorption through the two channels described above. Southern Oscillation describes the change of air pressure differences between Tahiti and Darwin, Australia (see Troup, 1965; Bacastow, 1976; Keeling and Revelle, 1985). It is measured as an index (SOI) from the Australian Bureau of Meteorology (2011), and defined as

$$\text{SOI} = 10 \cdot \frac{\Delta P_t - \overline{\Delta P_t}}{\sigma_{\Delta Pt}} \tag{12.3}$$

where ΔP_t is the difference in the average of mean sea level pressure between Tahiti and Darwin for month t. $\overline{\Delta P_t}$ is the long-run average of ΔP_t and $\sigma_{\Delta Pt}$ is the long-run standard deviation of ΔP_t for the given month respectively. Negative values of the SOI are generally referred to as El Niño years, while positive values correspond to episodes of La Niña. However, the effect on oceanic absorption is not so clear cut. Episodes of La Niña (SOI > 0) are associated with increased wind speeds (increase in k in equation (12.2)), thus making uptake easier. Nevertheless, increased wind also increases upwelling, which leads to a release of oceanic CO$_2$. Bacastow et al. (1985) suggest that easier absorption outweighs upwelling during episodes of La Niña (SOI > 0), resulting in higher absorption of CO$_2$ by the ocean when the SOI is positive. In turn, this implies less absorption during El Niño (SOI < 0) years.

On the contrary, Francey et al. (1995) find the opposite – during La Niña years (SOI > 0) oceanic absorption is relatively lower because of the large upwelling effect. Keeling and Revelle (1985) side with Francey on the theoretical model that upwelling should outweigh increased absorption, but empirically find that less atmospheric CO$_2$ is absorbed during El Niño episodes (SOI < 0), which agrees with Bacastow's (1976) findings.

Another factor that is not often considered in the literature is CO$_2$ use by oceanic algae, as Ritschard (1992) mentions. Nevertheless, data on algae are limited as they are not covered by the NDVI satellite measures, and consequently are not considered in our study.

Looking at the bigger picture, most evidence suggests that the ocean has become a carbon sink for anthropogenic emissions (Christopher et al., 2007). As increased CO$_2$ emissions into the atmosphere increased atmospheric partial pressure of CO$_2$, absorption by the ocean increased due to the pressure difference. However, the absolute magnitude of this effect is not known, and they estimate that approximately 48 percent of fossil fuel emissions are absorbed in the ocean; and atmospheric CO$_2$ would be approximately 55 ppm (parts per million) higher if there were no oceanic uptake. Orr et al. (2001) similarly find the ocean to be a net carbon sink but, as well as Nevison et al. (2008), suggest that most models overestimate the proportion of CO$_2$ emissions absorbed by the ocean.

In order to capture the key factors determining oceanic uptake when modeling atmospheric CO$_2$, it is important to account for temperature (which crucially affects partial pressure) and control for Southern Oscillation, even though the effects of El Niño and La Niña are not fully understood.

Modeling Methodology

The models for atmospheric CO_2 can broadly be classed into two categories: atmospheric transport and statistical models. Data for both are often decomposed into a long-run trend, cycle and noise using Fourier decompositions or the Hodrick–Prescott (HP) filter.

Atmospheric transport describes spatial three-dimensional models with vertical levels based on solving the fundamental equations for conservation of mass, momentum and energy. For different latitude and longitude grid resolutions, these models simulate carbon emissions and global transport (for the set-up and methodology of these models see Hansen et al., 1983; Kawa et al., 2004). Within this group of models, Erickson et al. (2008), Nevison et al. (2008), Randerson et al. (1997) and Keeling et al. (1995) use annual emissions data to analyze anthropogenic effects.

Statistical approaches also vary in methodology. Thoning and Tans (1989), Keeling et al. (1976) and Enting (1987) use Fourier decompositions to study trend and seasonal cycle. Granados et al. (2012) and Granados et al. (2009) use cointegration and time-series regression to study links between population, GDP and HP-filtered CO_2 growth. However, no actual measure of anthropogenic emissions is used in these studies. Hofman et al. (2009) use regression and graphical comparisons of carbon and population, while Jones and Cox (2005) regress growth rates of CO_2 on global emissions and cement production. Newell and Marcus (1987) look at the simple correlation between levels of CO_2 and global population.

Concerning modeling methods, atmospheric transport models and many statistical techniques are widely applied and well documented. However, they suffer from similar problems. Often inappropriate statistical techniques are applied without considering the time-series properties of the data. Correlation of time series alone is not an appropriate measure of dependence between them. The low frequency of measurement of emissions data is problematic and models are restricted by an initial choice of a small number of independent variables. Original data are rarely used: instead series are decomposed. This step is not necessary *a priori*, especially when explanatory variables that are seasonal themselves are available. Additionally, the regression analyses applied in many papers are not robust to outliers or structural breaks, do not always handle non-stationarity, and present few tests for mis-specification.

Summary of the Main Findings

The long-run trend in increasing CO_2 is clearly driven by anthropogenic factors, whereas the short-run seasonal fluctuations and changes in amplitude are mostly attributed to changes in the biosphere.

Long-term trend

The long-term trend in atmospheric CO_2 is fossil fuel induced. Pre-industrial levels of CO_2 are estimated to be around 260–280 ppm (see Wigley, 1983; Hofman et al., 2009), based on ice core, tree ring and oceanic data. Consistent and repeated measurement, starting with Keeling's work in 1958, has documented the rise in CO_2 to a current level of approximately 390 ppm measured at Mauna Loa. The rate of increase of CO_2 is proportional to combustion of fossil fuels (see Keeling, 1973; Keeling et al., 1976; Keeling et al., 1995; Thoning and Tans, 1989). Using population as a proxy measure for emissions yields similar results (Hofman et al., 2009; Newell and Marcus, 1987; Granados et al., 2009; Granados et al., 2012).

Given the large departure from pre-industrial levels, Hansen et al. (2008) investigate the question of a target level CO_2. They suggest that (at the time of their writing) the level of 385 ppm is too high to maintain climate conditions that current life has adapted to. Levels of 450 ppm in the Caenozoic era were associated with near ice-free conditions. Consequently, they propose a target level of at most 350 ppm.

Seasonal variation and amplitude
Seasonal fluctuations and changes in amplitude are mainly attributed to factors in the biosphere rather than industrial emissions. There are two effects described in the literature: one is the general pattern of seasonality; the second is an increase in the amplitude of this seasonality. In particular a perceived increase in the growing season is alleged to be the driving force behind increases in amplitude.

Many studies propose that the seasonal component of atmospheric CO_2 reflects the inter-annual uptake by plants. This is supported by the fact that the amplitude of this seasonality for a given season decreases towards the equator (Keeling et al., 1976). In particular, Enting (1987) argues that vegetation is sufficient to account for most of the inter-annual variation and that economic data do not show the required seasonality. While the peak to trough ratio measured at Mauna Loa was approximately 0.8 for the time period Enting investigates, he suggests that industrial emissions are not sufficient to cause this seasonal change. However, as is obvious from many economic time series, there is high seasonality in production and therefore in emissions.

The amplitude of the seasonal effect has been increasing over time. Keeling et al. (1996), Randerson et al. (1997), Kohlmaier et al. (1989), and Bacastow et al. (1985) characterize the increase as a result of a lengthening growing season with only a very small effect directly from fossil fuel emissions. The effect from anthropogenic emissions in these studies ranges from 0.01 to 0.2 percent on the change of amplitude. Additionally, Keeling et al. (1995) find that changes in the overall growth rate of CO_2 are driven by changes in vegetation and temperature rather than changes in industrial emissions.

A major issue with many of the above-mentioned studies is that anthropogenic emissions and production data are measured annually and therefore do not have the required frequency to be able to account for seasonal fluctuations. In a recent paper, Erickson et al. (2008) investigate this issue and find that economic data would suggest that the highest anthropogenic fluxes occur at the same time as the respiration phase of plants (winter in the northern hemisphere). Once models account for seasonality in fuel consumption, this will then lead to a diminished effect of seasonality from the biosphere.

Contribution of this Chapter

There are recurring problems with existing models of anthropogenic contributions to CO_2. Climate and atmospheric carbon fluxes are complex systems; nevertheless, many of the models are restricted by *a priori* selections of explanatory variables. The data used to account for anthropogenic emissions are often measured at too low a frequency to capture any seasonality. The main series of CO_2 is often decomposed into cycles and trends, something that is not necessary if the explanatory data are measured at a reasonable frequency. On the one hand, a significant number of the papers that approach the problem from a statistics or economics point of view do not sufficiently control for the biosphere or other natural factors. On the other hand, many models coming from a natural science background use statistical methods that are ill fitted given the time-series characteristics of the data. Modern econometric methods can provide an interdisciplinary solution to these problems.

To address these issues, we introduce an extended general-to-specific (*Gets*) modeling approach based on automatic model selection and the theory of reduction. This allows for a large number of candidate explanatory variables; in particular, models can be estimated with more explanatory variables than observations. It is therefore possible to include many lags to capture time dynamics as well as a wide range of controls for natural factors and industrial output measures. As our main measure of anthropogenic productivity and emissions is industrial production (measured at monthly intervals), the data are analyzed as a whole without requiring prior decomposition into trends and cycles. Models are also not restricted to a tight *a priori* selection of variables. Using IIS, the methods are robust to outliers and structural breaks, handle unit roots reasonably well, and provide a straightforward method of testing for mis-specification.

Overall, the literature indicates a clear necessity to control for the biosphere, temperature, El Niño effects, and long- (as well as short-) run anthropogenic measures. Intuitively, our approach is to utilize a large number of potential determinants controlling for the above-mentioned factors, and then use automatic model selection techniques to determine which forces are significant. Starting with a theory-based, but very broad, general unrestricted model (GUM), the initial system is reduced to a specific model. This is a comparatively agnostic and data-driven approach that imposes few restrictions on explanatory variables while being robust to sudden shifts (structural breaks).

3   METHODOLOGY

The carbon cycle, with many potential sinks and sources, is a complex system, which makes it near-impossible to correctly specify an appropriate model *a priori*. To model complex equations, we rely on general to specific (*Gets*) modeling approaches (see Campos et al., 2005). The unknown data-generating process (DGP) is the underlying structure that creates the data. Empirical modeling will always deal with a subset of variables of the DGP; thus an important factor is the local data-generating process (LDGP) – the generating process in the space of the variables under analysis: see Hendry (2009). The approach, therefore, is to construct a set of data based on broad theoretical assumptions, which nests the LDGP; then, within this set, reduce the model from its general form down to a specific representation. This is a two-step procedure. One: define a set of N variables that includes the LDGP as a sub-model. Two: starting with that general model as a good approximation of the overall properties of the data, reduce its complexity by removing insignificant variables, while checking that at each reduction the validity of the model is preserved. This is the basic framework of *Gets* modeling.

This section introduces theoretical concepts of model selection, their use in mis-specification testing, followed by the introduction of impulse indicator saturation (IIS) and its generalized version. All these concepts are then united and applied through the automatic search algorithm *Autometrics*. The algorithm combines these features through automated selection based on *Gets* while handling more variables than observations with IIS for detecting breaks and outliers, and mis-specification testing.

Model Selection

The theory of reduction characterizes the operations implicitly applied to the DGP to obtain the local LDGP. Choosing to analyze a set of variables, denoted $\mathbf{y}_t$, $\mathbf{x}_{1,t}$, determines the properties of the LDGP, and hence of any models of $\mathbf{y}_t$ given $\mathbf{x}_{1,t}$ (with appropriate lags, non-linear transforms thereof etc.). A congruent model is one that matches the empirical characteristics of the associated LDGP, evaluated by a range of mis-specification tests (see, e.g., Hendry, 1995, and the following section). A model is undominated if it encompasses, but is not encompassed by, all other sub-models (see, e.g., Mizon and Richard, 1986; Hendry and Richard, 1989; and Bontemps and Mizon, 2008).

Mis-specification Testing

Using a large number of variables with IIS (discussed in the next section) also provides a new view of model evaluation: to avoid mis-specification and non-constancy, start as general as possible within the theoretical framework, using all the available data unconstrained by $N > T$ (where N is the number of variables and T the number of observations), retaining the theory-inspired variables and only selecting over the additional candidates.

Even so, this approach does not obviate the need to test the specification of the auxiliary hypotheses against the possibility that the errors are not independent, are heteroskedastic (non-constant variance) or non-normal, that the parameters are not constant, that there is unmodeled non-linearity, and that the conditioning variables are not independent of the errors. When $N \ll T$, the first five are easily tested in the initial general model; otherwise their validity can be checked only after a reduction to a feasible sub-model. Congruence is essential not only to ensure a well-specified final selection, but also for correctly calibrated decisions during selection based on Gaussian significance levels, which IIS will help ensure.

More Variables than Observations: $N > T$

The model selection approach introduced here allows for more variables than observations to be used in modeling ($N > T$). For *Autometrics* this was first introduced through impulse indicator saturation, and has recently been extended to the general case.

Impulse indicator saturation
The numbers, timings and magnitudes of breaks in models are usually unknown, and are obviously unknown for unknowingly omitted variables, so a 'portmanteau' approach is required that can detect location shifts at any point in the sample while also selecting over many candidate variables. To check the null of no outliers or location shifts in a

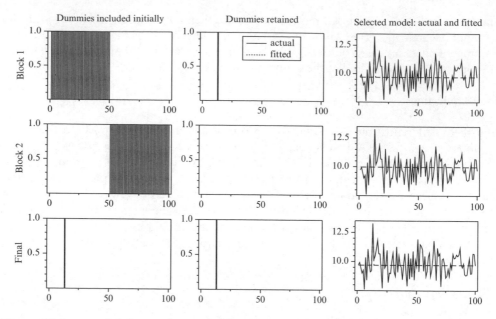

Figure 12.2   Impulse indicator saturation under the null of no break

model, impulse indicator saturation (IIS) creates a complete set of indicator variables $\{1_{\{j=t\}}\}$ equal to unity when $j = t$ and equal to zero otherwise for $j = 1, \ldots, T$, and includes these in the set of candidate regressors. Although this creates T variables when there are T observations, in the 'split-half' approach analyzed in Hendry et al. (2008), a regression first includes $T/2$ indicators. By dummying out the first half of the observations, estimates are based on the remaining data, so any observations in the first half that are discrepant will result in significant indicators. The location of the significant indicators is recorded, then the first $T/2$ indicators are replaced by the second $T/2$, and the procedure repeated. The two sets of significant indicators are finally added to the model for selection of the significant indicators. The distributional properties of IIS under the null are analyzed in Hendry et al. (2008), and extended by Johansen and Nielsen (2009) to both stationary and unit-root autoregressions.

Figure 12.2 illustrates the 'split-half' approach for $y_t \sim \text{IN}[\mu, \sigma_y^2]$ for $T = 100$ selecting indicators at a 1 percent significance level (denoted α). The three rows correspond to the three stages: the first half of the indicators, the second half, then the selected indicators combined. The three columns respectively report the indicators entered, the indicators finally retained in that model, and the fitted and actual values of the selected model. Initially, although many indicators are added, only one is retained. When those indicators are dropped and the second half entered (row 2), none is retained. Now the combined retained dummies are entered (here just one), and selection again retains it. Since $\alpha T = 1$, that is the average false null retention rate.

We next illustrate IIS for a location shift of magnitude λ over the last k observations:

$$y_t = \mu + \lambda 1_{\{t \geq T-k+1\}} + \varepsilon_t \qquad (12.4)$$

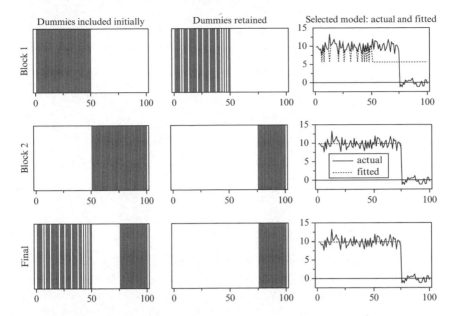

Figure 12.3   IIS stages for a location shift

where $\varepsilon_t \sim IN\ [0, \sigma_\varepsilon^2]$ and $\lambda \neq 0$. The optimal test in this setting would be a t-test for a break in (12.4) at $T - k + 1$ onwards, but requires precise knowledge of the location-shift timing, as well as knowing that it is the only break and is the same magnitude break thereafter. Figure 12.3 records the behavior of IIS for a mean shift in (12.4) of $10\sigma_\varepsilon$ occurring at $0.75T = 75$. Initially, many indicators are retained (top row), as there is a considerable discrepancy between the first-half and second-half means. When those indicators are dropped and the second set entered, all those for the period after the location shift are now retained. Once the combined set is entered (despite the large number of dummies), selection again reverts to just those for the break period. Under the null of no outliers or breaks, any indicator that is significant on a subsample would remain so overall, but for many alternatives, subsample significance can be transient, due to an unmodeled feature that occurs elsewhere in the data set. Thus there is an important difference between 'outlier detection' procedures and IIS.

While IIS is perhaps surprising initially, many well-known statistical procedures are variants of IIS. The Chow (1960) test corresponds to subsample IIS over $T - k + 1$ to T, but without selection, as Salkever (1976) showed, for testing parameter constancy using indicators. Recursive estimation is equivalent to using IIS over the future sample, and reducing the indicators one at a time. Johansen and Nielsen (2009) relate IIS to robust estimation, and show that under the null of no breaks, outliers or data contamination, the average cost of applying IIS is the loss of αT observations. Thus, at $\alpha = 0.01$, for $T = 100$ one observation is 'dummied out' by chance despite including 100 'irrelevant' impulse indicators in the search set and checking for location shifts and outliers at every data point. Retention of theory variables is feasible during selection with IIS, as is jointly selecting over the non-dummy variables, and IIS can be generalized to multiple splits of unequal size. While IIS entails more candidate variables than observations as $N + T > T$, selection is feasible as *Autometrics* undertakes expanding as well as contracting block

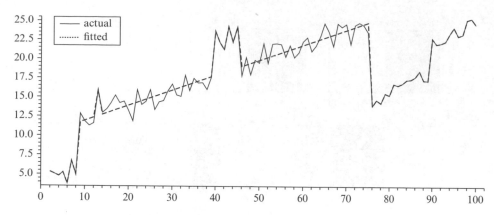

Figure 12.4   IIS outcomes for four location shifts

searches (see next section). Non-linear model selection (including threshold models) is examined in Castle and Hendry (2011).

For a single location shift, Hendry and Santos (2010) show that the detection power is determined by the magnitude of the shift τ; the length of the break interval $T - T^*$, which determines how many indicators need to be found; the error variance of the equation σ_η^2; and the significance level, α, where a normal-distribution critical value, c_α, is used by the IIS selection algorithm. Castle et al. (2012b) establish the ability of IIS to detect multiple location shifts and outliers, including breaks close to the start and end of the sample, as well as correcting for non-normality. Figure 12.4 shows the application of the general autometrics algorithm to a trending process with four breaks of varying magnitudes over $1, \ldots, 10; 40, \ldots, 45; 75, \ldots, 90;$ and $90, \ldots, 100$, to illustrate the ability of IIS to capture multiple breaks, at both the start and end of the sample.

General case of $N > T$

The idea of generalizing using more variables than observations from IIS to all forms of independent variables is introduced by Hendry and Krolzig (2005) as well as Hendry and Johansen (2013). Suppose there are N total regressors partitioned into J blocks of n_j, where $N = \sum_{j=1}^{J} n_j$ such that $N > T$ and $n_j < T$ for all j. Consequently the total number of variables N exceeds the number of observations T but total variables can be partitioned into J blocks n_j, each smaller than T. Their approach suggests randomly partitioning the set of variables into blocks of n_j, applying *Gets* to each block retaining the selected variables and crossing the groups to mix variables. The next step is to use the union of selected variables from each block to form a new initial model and repeat the process until the final union of selected variables is sufficiently small. *Autometrics* implements a variant of this algorithm to handle the general case of $N > T$.

Autometrics

Autometrics (see Doornik, 2009) is the latest installment in the automated *Gets* methodology and is available in the *OxMetrics* software package. The algorithm is based on the following main components:

1. GUM: the general unrestricted model (GUM) is the starting point of the search. The GUM should be specified based on broad theoretical considerations to nest the LDGP.
2. Pre-search: prior to specific selection, a pre-search lag reduction is implemented to remove insignificant lags, speeding up selection procedures and reducing the fraction of irrelevant variables selected (denoted the gauge of the selection process). Pre-search is only applied if the number of variables does not exceed the number of observations ($N < T$).
3. Search paths: *Autometrics* uses a tree search to explore paths. Starting from the GUM, *Autometrics* removes the least significant variable as determined by the lowest absolute *t*-ratio. Each removal constitutes one branch of the tree. For every reduction, there is a unique sub-tree which is then followed; each removal is back-tested against the initial GUM using an *F*-test. If back-testing fails, no sub-nodes of this branch are considered (though different variants of this removal exist). Branches are followed until no further variable can be removed at the pre-specified level of significance α. If no further variable can be removed, the model is considered to be terminal.
4. Diagnostic testing: each terminal model is subjected to a range of diagnostic tests based on a separately chosen level of significance. These tests include tests for normality (based on skewness and kurtosis), heteroskedasticity (for constant variance using squares), the Chow test (for parameter constancy in different samples), and residual autocorrelation and autoregressive conditional heteroskedasticity. Parsimonious encompassing of the feasible general model by sub-models both ensures no significant loss of information during reductions, and maintains the null retention frequency of *Autometrics* close to α: see Doornik (2008). Both congruence and encompassing are checked by *Autometrics* when each terminal model is reached after path searches, and it backtracks to find a valid less-reduced earlier model on that path if any test fails. This repeated re-use of the original mis-specification tests as diagnostic checks on the validity of reductions does not affect their distributions (see Hendry and Krolzig, 2003).
5. Tiebreaker: as a result of the tree search, multiple valid terminal models can be found. The union of these terminal models is referred to as the terminal GUM. As a tiebreaker to select a unique model, the likelihood-based Schwarz (1978) information criterion (SIC) is used, although other methods are also applicable, and terminal models should be considered individually.

In simulation experiments, models are primarily evaluated based on three concepts: gauge, potency and the magnitudes of the estimated parameters' root mean-square errors (RMSEs) around the DGP values (see Doornik and Hendry, 2009). Gauge describes the retention of irrelevant variables when selecting (i.e. variables that are selected but do not feature in the DGP). Potency measures the average retention frequency of relevant variables (variables that are selected and feature in the DGP). Low gauge (close to zero) and high potency (close to 1) are preferred, as are small RMSEs.

The main calibration decision in the search algorithm is the choice of significance level α at which selection occurs. Selection continues until retained variables are significant at α, although it can be the case that variables in the final model are also retained at a level above α if removal leads to diagnostic tests failing. α is approximately equal to the gauge of selection. Further, the choice of diagnostic tests and lag length selection for residual autocorrelation and autoregressive conditional heteroskedasticity need to be set.

In the general case of $N > T$ and IIS, *Autometrics* groups variables into two categories: selected and not selected (Doornik, 2010). Not currently selected variables are split into sub-blocks and the algorithm proceeds by alternating between two steps: first, the expansion step – selection is run over not-selected sub-blocks to detect omitted variables. Second, the reduction step – a new selected set is found by running selection on the system augmented with the omitted variables found in step one. This is repeated until the dimensions of the terminal model are small enough and the algorithm converges, so the final model is unchanged by further searches for omitted variables.

Autometrics has been applied successfully in a range of fields: see, for example, Hendry and Mizon (2011) on US food expenditure, Bårdsen et al. (2010) on unemployment in Australia, and Castle et al. (2011) for a comparison with other selection methods.

Nevertheless, overall selections should be interpreted carefully. Successful identification of the underlying LDGP can be adversely affected by collinearity of the independent variables. Most simulations of *Autometrics* with large numbers of variables use orthogonal regressors, which makes selection easier. Furthermore, when $N > T$ in the block selection algorithm of *Autometrics*, adding or dropping a variable from the initial GUM may change the block partitioning of variables, so the selection is not invariant to the initial specification.

The next section covers the data used to construct the GUM in an attempt to nest the LDGP for atmospheric CO_2. In the section following, *Autometrics* is then used to determine the anthropogenic contributions to CO_2.

4   DATA

CO_2

The atmospheric CO_2 data used here are taken from Keeling's measurements at Mauna Loa, available from Tans and Keeling (2011) (Scripps Institution of Oceanography). The time series of CO_2 in monthly averages runs from 1958:3 until 2011:7 at the time of writing. Simple inspection of the data shows that both the level (see Figure 12.1) and the annual change (see Figure 12.5) are increasing over time.

The seasonal fluctuations are apparent in the data, and as Buermann et al. (2007) poetically describe it, the regular seasonal cycle of CO_2 at Mauna Loa 'records the breathing of the Northern Hemisphere biosphere'. In economic terms the level of atmospheric CO_2 can plainly be described as a stock variable. The total level of CO_2 can be approximated by the integral of the netflow to the atmosphere:

$$CO_{2_t} = \int \text{Netflow } dt \qquad (12.5)$$

where Netflow = carbon sources – carbon sinks. Therefore any analysis of the impact of anthropogenic emissions should be based on the change in atmospheric CO_2:

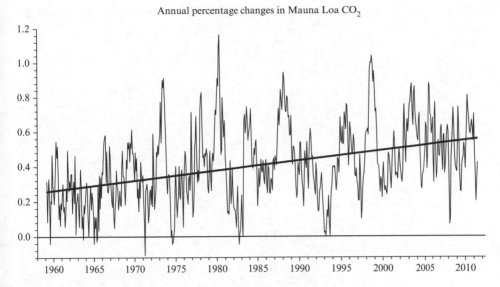

Figure 12.5 Annual percentage changes in Mauna Loa CO_2

Figure 12.5   Annual changes in atmospheric CO_2

$$\frac{dCO_2}{dt} = \text{Netflow} \tag{12.6}$$

Approximating this relationship from continous to discrete time:

$$\frac{dCO_2}{dt} \approx \Delta CO_{2,t} = CO_{2,t} - CO_{2,t-1} = \text{Netflow}_t \tag{12.7}$$

The dependent variable modeled is therefore $\Delta CO_{2,t}$. Equations (12.5)–(12.7) suggest that any relationship between the change in atmospheric CO_2 and netflow should be modeled in levels rather than any non-linear transformation thereof. The following section identifies variables that make up the netflow, both anthropogenic as well as natural sources and sinks.

Terrestrial Biosphere

We use NDVI data to account for vegetation effects on CO_2. Data are available for the NDVI from the Oak Ridge National Laboratory Distributed Archive Center (see Tucker et al., 2010), ranging from 1981:7 until 2002:12 at spatial resolutions of 0.25, 0.5 and 1.0 degrees latitude and longitude.

CO_2 measured at Mauna Loa is driven by North American airflow during the summer and airflow from Eurasia during the winter. Therefore the NDVI data are split into two main regions: North America and Eurasia. Figure 12.6 shows the NDVI measure in detail for North America in August and January respectively. Using 1° spatial resolution, an algorithm then takes the average of every 3 × 3 observation grid on land within the two regions (excluding water, permanent ice and missing observations). To capture

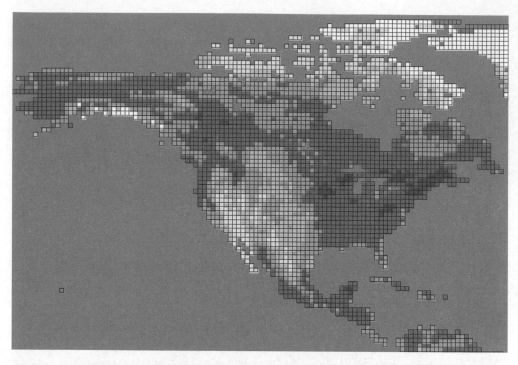

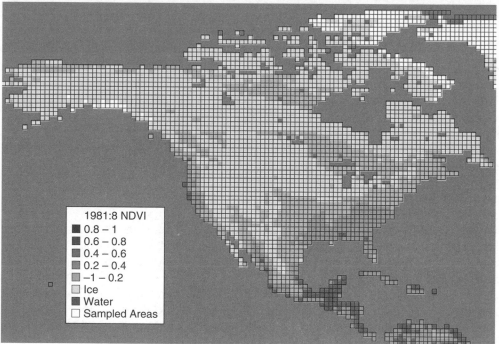

Figure 12.6   NDVI at 1° North America in 1981: August (top) and January (bottom)

Table 12.1   Principal components for vegetation, 1981:7–2002:12

North America	Proportion of variance	Cumulative
Principal component 1	0.831	
Principal component 2	0.064	0.895
Principal component 3	0.042	0.937
Eurasia		
Principal component 1	0.735	
Principal component 2	0.108	0.843
Principal component 3	0.042	0.885

the main variation of vegetation, similar to the spatial analysis in Buermann et al. (2007), the North American region is defined by the rectangle ranging from 86°N/167°W to 14°N/48°W, and the Eurasian region by the rectangles ranging from 75°N/9°E to 36°N/51°E and 76°N/52°E to 7°N/358°E. This generates 198 time-series variables for North America and 567 for Eurasia.

Due to the nature of a common growing season in the northern hemisphere, the generated time series are highly collinear. Principal components (PCs) are used to reduce the number of variables, while retaining most of the variation in the data. Since this process captures the overall variation in vegetation, it should also reduce the problem of random noise due to cloud cover at the time of satellite measurement. Although PCs are just linear transformations of the original times series, they have two potential advantages. First, PCs are mutually orthogonal, so adding or eliminating any one PC has little effect on the coefficient estimates of others, making the model more robust. Second, linear combinations of 'small' effects can be statistically significant (so retained during model selection) when individual time series would not be: see Castle et al. (2012a) for a more detailed discussion. The contributions of individual variables can be disentangled if needed.

For the following analysis, the first three principal components are entered for both North America and Eurasia. Cumulatively they explain 93.6 percent of the variation in North America and 88.5 percent of Eurasian variation. Table 12.1 summarizes the principal components that account for variation in the biosphere, and Figure 12.7 shows the highly seasonal variation present in the biosphere as measured by PCs. The first principal component shows a higher amplitude in Eurasia compared to North America but the seasonal pattern is nearly identical. NDVI implicitly covers changes in land use since it is a measure of photosynthetic activity for a particular area. A forest that is cut down would result in a change of NDVI from around +1 to closer to zero for that particular region. However, once NDVI is calculated for large regions and reduced in dimensionality (by PCs), changes in land use would have to occur on a grand scale to be identified in the time series. The principal components of NDVI should, therefore, be interpreted primarily as the variation in plant activity of photosynthesis and respiratory release.

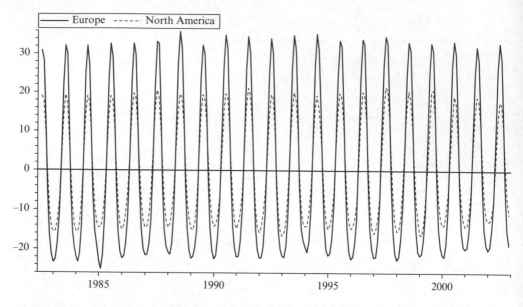

Figure 12.7   First principal components for Eurasian and North American NDVI

Oceanic Indicators: Temperature and Southern Oscillation

The measure of temperature used here is the anomaly in land and sea surface temperature for the northern hemisphere. The temperature anomaly measure is expected to capture the main factors of ocean CO_2 absorption and is available from the NASA Goddard Institute for Space Studies (GISS) (2011) Surface Temperature Analysis from 1880:1 to 2011:9. The data are measured as an index in 0.01 degrees Celsius of deviations from the 1951–80 base period (see Hansen and Lebedeff, 1987; Hansen et al., 2010 for the detailed measurement methodology). Land measures are taken from multiple stations and are combined and corrected for urban and other non-climatic factors. Sea surface temperature measures are restricted to ice-free regions. As Hansen et al. (2010) describe, temperature in the northern hemisphere has been increasing despite recent El Niño effects. Figure 12.8 shows temperature anomalies from 1958:3 to 2010:9.

The feedback effect of CO_2 is one of the main concerns in climate change. The level of atmospheric CO_2 feeds into temperature which, in turn, affects the rate of growth of CO_2, particularly through oceanic absorption. There would be a potential problem of endogeneity if the level of CO_2 were modeled by temperature. However, lagged temperature measures should be predetermined for the growth of CO_2.

To take account of weather phenomena through the Southern Oscillation we include the Southern Oscillation index (SOI). Data on the SOI are available from the Australian Bureau of Meteorology (2011) from 1876:1 until 2011:9. Figure 12.9 shows the SOI for 1958 until 2011, with a noticeably strong episode of El Niño in 1997–98 (SOI < 0).

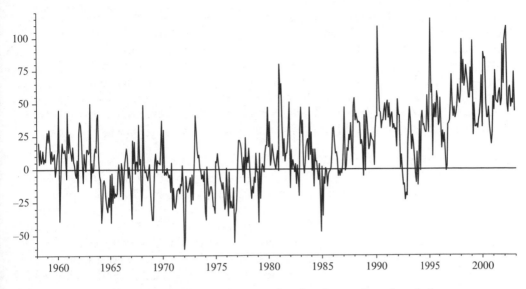

Figure 12.8   Land and sea temperature anomalies for the northern hemisphere

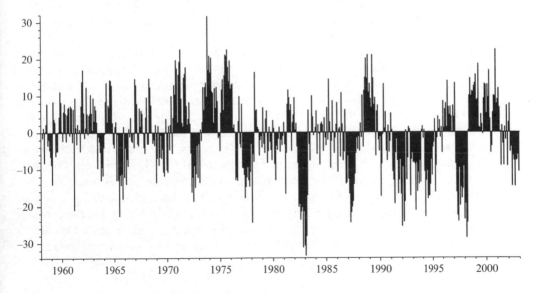

Figure 12.9   Southern Oscillation index

Economic Indicators

Anthropogenic contributions to atmospheric CO_2 are normally approximated by low-frequency economic indicators; for example annual GDP, population, an estimate of total CO_2 emissions or cement production (see Section 2). This works reasonably well when trying to explain the long-run dynamics of carbon. However, using only low-frequency annual measures does not allow for estimation of any effect of anthropogenic emissions on the seasonal variation of CO_2. High-frequency (monthly) measures of anthropogenic output permit a richer analysis. Here we use a combination of multiple low-frequency (annual) and high-frequency (monthly) indicators. The annual data are included to provide a robustness check and potentially account for long-run growth. Monthly measures are included for short-run dynamics, which could explain the seasonal fluctuations as well as long-run growth. To capture atmospheric transport, variables are chosen to reflect North America as well as Europe/Asia.

High-frequency (monthly) measures

The main high-frequency indicators for anthropogenic contribution to CO_2 used here are monthly industrial production indices for multiple regions. Industrial production for North America is given by the US Industrial Production Index (2005 = 100) available from the Federal Reserve (2011). The data are not seasonally adjusted and range from 1919:1 until 2011:5. The index measures real output in the sectors covering manufacturing, electric and gas utilities, and mining, and thus accounts for a large share in business-cycle fluctuations. To cover Europe and Asia, industrial production indices for the UK, Germany, India and Japan are included. These measures function as a proxy for business-cycle fluctuations in the Eurasian region. Ideally Chinese and Russian production should also be included; however, no data are available on industrial production for both countries at the required frequency and time span.[2] UK industrial production is measured as an index (2005 = 100) of non-seasonally adjusted manufacturing activities. The data are obtained from the Office of National Statistics (2011) and range from 1968:1 to 2011:6. German and Japanese industrial production is measured by the industrial production index (2005 = 100) covering manufacturing, mining and electricity, gas and water supply. The series is available only in seasonally adjusted format from the OECD (2011) from 1960:1 until 2011:2, and for Germany after October 1990 the data account for the accession of the German Democratic Republic to West Germany. The Indian industrial production index (2005 = 100) covers manufacturing, mining and electricity (Ministry of Statistics, Government of India 2011) and ranges from 1981:4 to 2011:5. Table 12.2 and Figure 12.10 summarize the high-frequency measures.

The seasonal adjustment of German and Japanese industrial production is visible in their dampened seasonal cycles. The higher seasonal variation in UK industrial production likely stems from it covering primarily manufacturing, which is more volatile to business cycles and seasonal factors than mining and energy production included in the other indices.

Table 12.2   High-frequency (monthly) variables

Measure	Description	Range	Source
US industrial production	Index 2005 = 100	1958:3–2011:5	US Federal Reserve
UK industrial production	Index 2005 = 100	1968:1–2011:6	ONS
Germany industrial production	Index 2005 = 100	1960:1–2011:2	OECD
India industrial production	Index 2005 = 100	1981:4–2011:5	Govt. of India
Japan industrial production	Index 2005 = 100	1960:1–2011:2	OECD

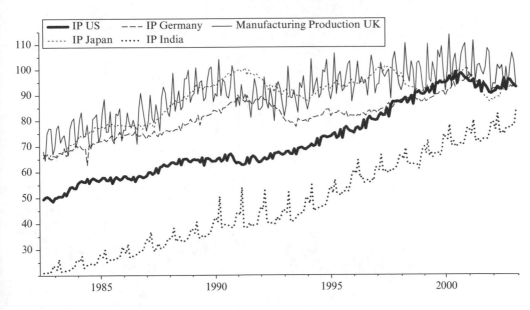

Figure 12.10   Industrial production indices, 2005 = 100

Table 12.3   Principal components for industrial production, 1981:4–2011:2

IP	Proportion of variance	Cumulative
Principal component 1	0.809	
Principal component 2	0.119	0.928
Principal component 3	0.047	0.975

These industrial production series are highly collinear, so we again employ principal components to reduce the dimensionality and work with orthogonal variables, which improves robustness in selection as discussed above. Table 12.3 summarizes the first three industrial production components, which cumulatively explain approximately 97 percent of variation in production. Figure 12.11 compares US industrial production with the three components used in selection.

These are the first three anthropogenic high-frequency components included in the GUM. While industrial production reflects the intensity of economic activity associated with higher emissions, it does not account for changing emission intensity. More efficient processes could lead to an increase in industrial production without an equivalent increase in emissions. This is a crucial missing measure and difficult to control for: an attempt is made by including overall long-run emissions estimates in addition to production.

Low-frequency (annual) measures

Low-frequency anthropogenic measures are variables reported on an annual basis, and capture the long term of human-driven CO$_2$ emissions. These low-frequency variables include total estimated CO$_2$ emissions (in thousand metric tons of carbon) for North

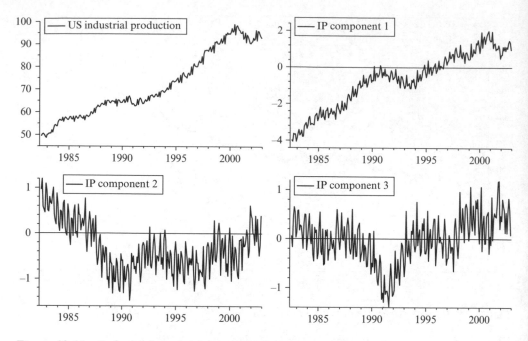

Figure 12.11   Industrial production principal components

Table 12.4   Low-frequency (annual) variables

Measure	Description	Range	Source
N. America CO_2 emissions	000s tons carbon	1950–2007	US Dep. of Energy
W. Europe CO_2 emissions	000s tons carbon	1950–2008	US Dep. of Energy
E. Europe CO_2 emissions	000s tons carbon	1950–2008	US Dep. of Energy
Central Asia CO_2 emissions	000s tons carbon	1950–2008	US Dep. of Energy
Far East CO_2 emissions	000s tons carbon	1950–2008	US Dep. of Energy

America, Western Europe, Eastern Europe, Central Asia and the Far East.[3] Emissions are estimated based on the burning of fossil fuels, cement manufacture and gas flaring by the US Department of Energy (see Marland et al., 2011) from 1950 to 2007. Table 12.4 summarizes the annual measures.

In order to usefully combine these annual variables with the monthly measures listed above, all annual variables are linearly interpolated to monthly observations. In the case of moving from annual to monthly observations, this method estimates a straight line over 12 months between each annual observation. While this may be restrictive, if the seasonal structure of the variables is not known there is no *a priori* reason to prefer a different interpolation algorithm. In any case, seasonal dummy variables and IIS can 'pick up' any systematic or large deviations. As before, the low-frequency measures are reduced in dimensionality by calculating their principal components. The first three components capture 99 percent of variation in the emissions series. Table 12.5 shows

Table 12.5   Low-frequency (annual) components

CO_2 emissions	Proportion of variance	Cumulative
Principal component 1	0.733	
Principal component 2	0.222	0.955
Principal component 3	0.039	0.994

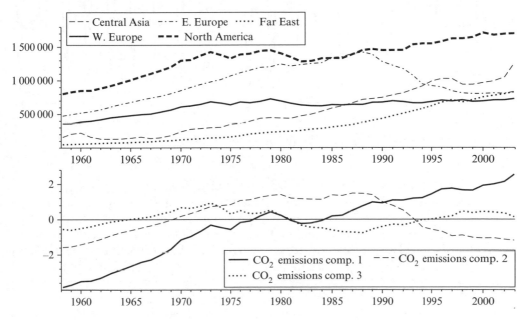

Figure 12.12 CO_2 emissions and principal components

explained variation of each component and Figure 12.12 displays interpolated CO_2 emissions as well as the three low-frequency PCs.

The components of the interpolated annual variables are included to account for potential low-frequency movements, the industrial production short-term indicators are expected to be sufficient to capture inter-annual dynamics. These are the first three low-frequency anthropogenic components included in the GUM.

5   ESTIMATION

Overview

The dependent variable ΔCO_2 is modeled as a finite autoregressive-distributed lag model (ADL) (see, e.g., Hendry, 1995):

$$\Delta CO_{2,t} = a(L)\Delta CO_{2,t-1} + \sum_{i=1}^{p} b_i(L)x_{i,t} + \mathbf{z}_t'\phi + \varepsilon_t \tag{12.8}$$

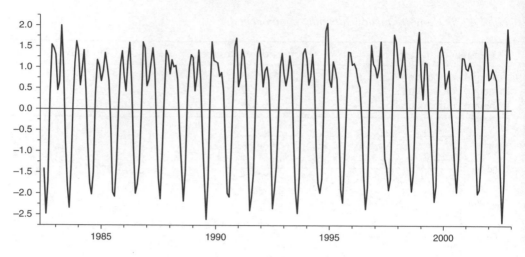

Figure 12.13   Dependent variable $\Delta CO_{2,t}$ over 1982:7–2002:12

where L denotes the lag operator, and p is the number of explanatory variables x_i. The vector z consists of non-lagged deterministic terms such as a linear time trend, centered seasonal variables and impulse indicators. Let q denote the total number of explanatory variables that appear on the left-hand side of equation (12.8). Then the change in atmospheric CO_2 is modeled as a function of past values of the change in CO_2, current and past values of selected independent variables x_i, and deterministic components. Figure 12.13 graphs $\Delta CO_{2,t}$ for 1982:7–2002:12. The seasonal variation is so large that it is difficult to discern the slow but persistent growth.

There is a large number of potential explanatory variables x_i in modeling atmospheric CO_2. The model needs to account for all the above-mentioned anthropogenic and natural factors as well as their lag reactions, as CO_2 is a highly autocorrelated time series. Adding IIS quickly moves the general unrestricted model to a situation with more explanatory variables than observations. This used to be a major difficulty in modeling; however, as outlined in Section 3, it can now be handled by estimation in blocks using *Autometrics* to select variables to retain in the final model in the form of (12.8). The estimation procedure operates as follows: first the theory-motivated GUM is specified, then estimation in blocks following the *Autometrics* algorithm selects down to individual terminal models. The union of terminal models is captured by the final GUM. Formally, the selection for the final model can use the likelihood-based SIC, although, as each terminal model represents a valid representation, final model selection can be based on other theoretical considerations.

Formulation of the GUM

The dependent variable that is being modeled is ΔCO_2 (the change in atmospheric CO_2 measured at Mauna Loa). While *Autometrics* has been shown to be effective at recovering the data-generating process in large models, the algorithm is not perfect and is sensitive to initial specification. As a robustness check, therefore, we estimate multiple variations of initial sample specifications. The subsample GUMs 1 and 2 are selected

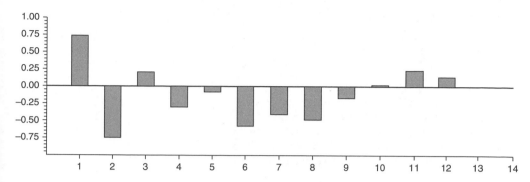

Figure 12.14   Partial-autocorrelation function for ΔCO_2

for different initial variables, although they always include the main variables of interest: short-term dynamics captured by monthly industrial production and the control variables for natural factors – temperature, SOI and vegetation. Non-anthropogenic emissions are captured through potential oceanic release of CO_2 (SOI and temperature) and the respiratory release phase in vegetation (NDVI close to zero). The models include general control variables of a constant, a linear time trend and 11 centered seasonal dummy variables (annual frequency −1 dummies with mean zero in the long run), which are subject to selection so not automatically included in terminal models.

To capture interseasonal transport dynamics, interaction terms for winter/summer are included with each vegetation measure. Summer is defined as May–October and winter is defined as November–April. Thus binary variables (winter and summer weights) are interacted with all region-specific NDVI variables and included in the GUM. This could be extended to smoothed weights following a sine/cosine pattern, but binary weights are expected to capture the main seasonal effect of different atmospheric transport.

Sample 1 includes all variables measured at a monthly frequency: natural control variables as well as the first three components of industrial production. Sample 2 includes all short-term dynamics as in sample 1, and adds the first three low-frequency interpolated fossil fuel emission components.

Since ΔCO_2 is a highly autocorrelated series when measured monthly, a long lag length is allowed. Based on the partial-autocorrelation function (PACF) of ΔCO_2, the longest lag length is selected to be 12, as longer lags fall below the 95 percent critical level. Figure 12.14 shows the PACF for ΔCO_2. Additionally, 12 lags of each independent variable are added to GUM 1. Lag lengths for GUM 2 are specified based on selection from GUM 1 to compensate for a larger number of independent variables (see Section 6).

As controls for the terrestrial biosphere are included in every GUM, it is the NDVI measure that defines the maximum number of available observations. CO_2 is measured at Mauna Loa from 1958:3 until the present, but NDVI data are only available from 1981:7 until 2002:12, so the vegetation control limits the maximum number of observations to 258. A maximum lag length of 12 then leads to the estimated sample size being 246 observations ($T = 246$).

Impulse indicator saturation includes one binary variable for each observation. In the case of a model with 12 lags, this means that 246 individual binary variables are added

Table 12.6   GUM: sample 1

Variables included	Lag length
Temperature	12
NDVI PC1 Eurasia (Eur) + winter interaction	12
NDVI PC2 Eurasia (Eur) + winter interaction	12
NDVI PC3 Eurasia (Eur) + winter interaction	12
NDVI PC1 North America (NA) + summer interaction	12
NDVI PC2 North America (NA) + summer interaction	12
NDVI PC3 North America (NA) + summer interaction	12
SOI	12
Industrial production comp. 1	12
Industrial production comp. 2	12
Industrial production comp. 3	12
Constant	yes
Trend	yes
Centered seasonal variables	yes
Impulse indicators	yes
Total variables	492

to the GUM. Tables 12.6 and 12.7 provide an overview of the variables making up the sample-specific GUMs.

Non-stationarity

Non-stationarity is a common feature of time-series data. Broadly, a non-stationary process is defined as a process whose distribution changes over time, for example the mean or variance of the process is non-constant (see Hendry and Juselius, 2000, for a detailed discussion). The trending level of CO_2 measured at Mauna Loa is non-stationary as its mean is increasing with time. There are various forms of non-stationarity. The time series could be integrated processes (a time series is said to be integrated of order r, or $I(r)$, if differencing the series r times yields a stationary process), or alternatively there could be structural breaks/shifts in coefficients or levels. Non-stationarity through structural breaks need not be removed by differencing. IIS is used to detect structural breaks (as well as mis-specification): if a large number of impulse indicators is selected, the model may be mis-specified, data badly mis-measured; or there are breaks in the data.

Non-stationary data are not a problem in automatic model selection so long as it is approached correctly. Selection in *Autometrics* is primarily based on t- and F-statistics that under non-stationarity can follow non-standard distributions. Sims et al. (1990) show that the limit distributions of these test statistics are standard if the original equation can be rewritten in terms of coefficients on mean-zero stationary variables. That is to say, the actual re-parametrization is not required, merely the existence of a linear re-parametrization in mean-zero variables is sufficient for the test statistics in the original equation to have standard distributions. So long as the equation can be written that way, selection based on t- and F-statistics will be valid. However, there are potential problems

Table 12.7 GUM: sample 2

Variables included	Lag length
Temperature	6
NDVI PC1 Eurasia (Eur) + winter interaction	12
NDVI PC2 Eurasia (Eur) + winter interaction	12
NDVI PC3 Eurasia (Eur) + winter interaction	12
NDVI PC1 North America (NA) + summer interaction	12
NDVI PC2 North America (NA) + summer interaction	12
NDVI PC3 North America (NA) + summer interaction	12
SOI	6
Industrial production comp. 1	6
Industrial production comp. 2	6
Industrial production comp. 3	6
CO$_2$ emissions comp. 1	6
CO$_2$ emissions comp. 2	6
CO$_2$ emissions comp. 3	6
Constant	yes
Trend	yes
Centered seasonal variables	yes
Impulse indicators	yes
Total variables	483

during selection when a path is considered in which this re-parametrization is not possible. Such a difficulty is hard to avoid, so is handled here by using tight significance levels that allow for possible non-standard distributions. As Figure 12.5 showed, $\Delta_{12}CO_2$ manifests a strong stochastic trend so is clearly not stationary: in monthly data, that trend is 'swamped' by the seasonal variation. Due to this strong seasonality, augmented Dickey–Fuller (Dickey and Fuller, 1981) type tests would not successfully identify a unit root. Seasonal unit-root tests are somewhat unreliable as other determinants, such as breaks, are not included, but based on the apparent stochastic trend in $\Delta_{12}CO_2$, it is safe to say that ΔCO_2 is integrated of order one, I(1). Unit roots can also not be rejected for most of the independent time-series variables that are included as potential determinants. The order of integration of the dependent variables is then the same as the order of integration of the independent variables. Given that the model can be written as coefficients on stationary mean-zero variables (see Banerjee et al., 1993), we proceed by estimating the model in levels with tight significance levels to account for selection effects where an I(0) re-parametrization is not possible.

Algorithm Specification

The *Autometrics* algorithm is used to estimate and select within the GUMs described in Tables 12.6 and 12.7. The algorithm is calibrated to the following parameters. Selection is done at a 0.1 percent significance level, considerably tighter than the conventional 5 percent or 1 percent used in the literature. In models starting with K irrelevant variables, this implies that on average $0.001K$ irrelevant variables are retained in the terminal model. The current model is reduced by removing the least significant variable until variables cannot be dropped at the 0.1 percent level. At the termination of each path, models are backtested to the initially specified GUM when feasible. Diagnostic tests, defined formally below (see Doornik and Hendry, 2009), are conducted at a 1 percent level, for normality, heteroskedasticity, coefficient constancy (set to a 70 percent sample split), residual autocorrelation and autoregressive conditional heteroskedasticity, both based on eight lags. The specified GUM includes more variables than observations with IIS, so diagnostic tests are only applied to terminal models, and, if satisfactory, conventional standard errors are used.

6   RESULTS

Overview

The specifications given in Tables 12.6 and 12.7 are selected from general unrestricted models using *Autometrics* by the algorithm outlined in Section 5. Sample 1 includes all natural controls and monthly anthropogenic components; sample 2 includes all natural controls as well as monthly and annual anthropogenic components. The crucial feature of *Autometrics* is determining the selection of variables, rather than their estimated coefficients, although bias corrections have been implemented. Below, we also note the relative importance of individual variables through decomposition of explained variance.

The selection algorithm results in 14 terminal models for sample 1, and 16 terminal models for sample 2. It may surprise that so many congruent, undominated different representations can be found at such a tight significance level as 0.1 percent, but this merely reflects the high collinearity both between the different series and over time as represented by their lagged values. Most of the terminal models are minor variations on others as the final GUMs had 21 and 30 variables respectively.[4] The final models are selected from the set of terminal models by the smallest SIC value. Equations (12.9) and (12.10) show the selected final models for sample 1 and sample 2 respectively.

$$\Delta CO_{2,t} = \underset{(0.053)}{0.24}\, \Delta CO_{2,t-1} - \underset{(0.049)}{0.67}\, \Delta CO_{2,t-2} + \underset{(0.05)}{0.20}\, \Delta CO_{2,t-3} \qquad (12.9)$$

$$- \underset{(0.037)}{0.32}\, \Delta CO_{2,t-4} + \underset{(0.057)}{0.24}\, IP_{1,t-1} - \underset{(0.056)}{0.30}\, IP_{1,t-4}$$

$$- \underset{(0.034)}{0.20}\, IP_{2,t-4} + \underset{(0.035)}{0.15}\, IP_{3,t} + \underset{(0.0005)}{0.003}\, Temp_{t-4}$$

$$- \underset{(0.0013)}{0.006}\, SOI_{t-5} - \underset{(0.007)}{0.042}\, NDVI_{1,Eur_{t-1}} + \underset{(0.003)}{00.019}\, NDVI_{1,Eur_{t-10}}$$

$$- \underset{(0.006)}{0.020}\, NDVI_{1,Eur_{t-12}} + \underset{(0.007)}{0.026}\, w\_NDVI_{3,Eur_{t-8}}$$

$$\hat{\sigma} = 0.212 \ T = 246 \ n = 14 \ \text{SIC} = -0.006 \ F_{ar}(8, 224) = 1.55$$

$$\chi^2_{nd}(2) = 1.47 \ F_{reset}(2, 230) = 1.42 \ F_{arch}(8, 230) = 1.07 \ F_{het}(28, 217) = 1.42$$

Let F_{name} denote an approximate Lagrange-multiplier F-test, then F_{ar} tests for autocorrelation of order k (Godfrey, 1978), F_{het} tests for heteroscedasticity (White, 1980), F_{arch} tests for k^{th}-order autoregressive conditional heteroskedasticity (ARCH: Engle, 1982), F_{reset} tests for functional-form mis-specification (White, 1980), and $\chi^2_{nd}(2)$ tests for non-normality (Doornik and Hansen, 2008).

$$\Delta CO_{2,t} = -\underset{(0.051)}{0.57} \ \Delta CO_{2,t-2} - \underset{(0.050)}{0.22} \ \Delta CO_{2,t-4} + \underset{(0.062)}{00.27} \ IP_{1_{t-1}} \tag{12.10}$$

$$-\underset{(0.063)}{0.34} \ IP_{1_{t-4}} - \underset{(0.039)}{0.28} \ IP_{2_{t-4}} + \underset{(0.035)}{0.16} \ IP_{3_t} - \underset{(0.21)}{0.74} \ I_{1990(7)}$$

$$+\underset{(0.0006)}{0.004} \ Temp_{t-4} - \underset{(0.001)}{0.007} \ SOI_{t-5} - \underset{(0.008)}{0.047} \ NDVI_{1,Eur_t}$$

$$-\underset{(0.006)}{0.044} \ NDVI_{1,Eur_{t-3}} + \underset{(0.006)}{0.033} \ NDVI_{1,Eur_{t-11}} - \underset{(0.009)}{0.042} \ NDVI_{1,Eur_{t-12}}$$

$$+\underset{(0.008)}{0.030} \ NDVI_{2,Eur_{t-11}} - \underset{(0.007)}{0.029} \ w\_NDVI_{2,Eur_{t-7}} - \underset{(0.006)}{0.022} \ w\_NDVI_{2,Eur_{t-8}}$$

$$+\underset{(0.007)}{0.035} \ w\_NDVI_{3,Eur_{t-8}} - \underset{(0.008)}{0.039} \ s\_NDVI_{1,NA_{t-8}} + \underset{(0.007)}{0.034} \ s\_NDVI_{1,NA_{t-9}}$$

$$\hat{\sigma} = 0.199 \ T = 246 \ n - 19 \ \text{SIC} = -0.048 \ F_{ar}(8, 219) = 1.62$$

$$\chi^2_{nd}(2) = 0.044 \ F_{reset}(2, 225) = 0.042 \ F_{arch}(8, 230) = 0.57 \ F_{het}(36, 208) = 1.03$$

First: as is to be expected from the theory, controls for natural factors are selected in both final models. Temperature anomalies enter the model with a positive coefficient, likely capturing the effect of oceanic uptake such that CO_2 increases with higher temperatures. Vegetation controls through the principal components of NDVI are selected in both models, as is the control for Southern Oscillation. However, a key finding is that in both terminal models, natural controls are insufficient to account for the variation in the change of atmospheric CO_2. Anthropogenic factors captured through components of industrial output indices are consistently selected in both models. Selection of these components is highly consistent over the two models as the selected production components are identical in models 1 and 2. Selection of these is robust to the addition of emissions components which are not selected in sample 2, suggesting that the high-frequency measures provide a better approximation for anthropogenic emissions measured at Mauna Loa.

Second: most selected variables enter the model in lagged form. Only the third principal component of production (in samples 1 and 2) and the first component of Eurasian NDVI (in sample 2) have an estimated immediate effect on the growth of CO_2. Most anthropogenic emissions and vegetation growth 'lead' measured atmospheric CO_2 by suggested time periods of 1 to 12 months. Relative to the initial sizes of the GUMs, few variables are retained, yet relative to the tight significance levels, many more are retained than could be attributed to chance (less than 1 on average).

Third: the final models appear to be well specified. As a result of the algorithm, all models pass tests for normality, heteroskedasticity, residual autocorrelation and autoregressive conditional heteroskedasticity. The number of selected indicators from IIS is low. There is only one indicator selected for 1990:7. This suggests that the model is correctly specified and there appear to be no real structural breaks or shifts in the change of atmospheric CO_2. No deterministic variables are selected: no constant, time trend or centered seasonals appear in the final models. This suggests that changes in CO_2 are well approximated by the selected final variables covering anthropogenic and natural factors. If constants are included post-selection (both not statistically different from zero in models 1 and 2), R^2 can be used as a rough measure of goodness of fit. Both final models exhibit a high goodness of fit ($R^2 = 0.974$ and $= 0.978$ respectively). This is not a straight result of selection as *Autometrics* does not directly maximize the goodness of fit. Moreover, R^2 measures should not be attributed much weight when assessing models, as there are preferred likelihood-based measures that also account for the number of parameters included.

GUM: Sample 1

Sample 1 in equation (12.9) covers all variables measured at a monthly frequency. *Autometrics* in sample 1 with 246 observations estimated 345 models reducing the number of explanatory variables from an initial 492 (split into initial six blocks) down to 14 in the final model. The final model passes the stationarity test on the residual, unit roots ranging from lags 1 to 12 are rejected at the 1 percent level using an ADF test. There are no impulse indicator variables selected in the final GUM. Together this provides evidence for a well-specified model that forms a stationary relationship.

Selection: neither the constant, the linear time trend nor centered seasonal variables feature in the final model, so that the seasonality and increase in the growth of CO_2 are explained by the anthropogenic and natural factors. All selected variables (apart from $IP_{3,t}$) enter the model as lags, suggesting a delay in the effect of CO_2 emissions/absorption and measurement at Mauna Loa. The longest lag on an anthropogenic component is four months.

The anthropogenic sources that are selected are all three principal components for industrial production at lag lengths ranging from immediate t to $t - 4$. Component 1 is selected at $t - 1$ and $t - 4$ with opposite signs, suggesting that it enters the model mainly as a difference. As these variables are principal components of production indices, the coefficients are not straightforward to interpret. The key result is the consistent selection, relative importance and lag length of these, rather than the signs of individual estimated coefficients.

The non-seasonally weighted principal components for NDVI are mostly selected, but only the Eurasian region is included. Given that the growth cycle is relatively similar in North America and Eurasia, this should not be over-interpreted. It is likely that the PCs for North American and Eurasian vegetation capture a very similar pattern and are to a considerable extent interchangeable. The negative coefficient on the first PC of vegetation (at $t - 1$, as well as in the long-run solution below) supports the theory that increased vegetation activity slows down CO_2 growth. The near equal magnitude, opposite signs on $t - 10$ and $t - 12$ suggest a difference, a pattern also seen in (12.10). The coefficient

Table 12.8   GUM: sample 1 relative importance

Variable	Partial R^2	Hierarchical partitioning
$\Delta CO_{2,t-2}$	0.4472	0.0661
$\Delta CO_{2,t-4}$	0.2461	0.0643
$NDVI_{1,Eur_{t-10}}$	0.1583	0.0257
$IP_{2_{t-4}}$	0.1259	0.0036
$NDVI_{1,Eur_{t-1}}$	0.1259	0.2795
$Temp_{t-4}$	0.1225	0.0067
$IP_{1_{t-4}}$	0.1105	0.0117
SOI_{t-5}	0.0836	0.0014
$\Delta CO_{2,t-1}$	0.0796	0.1954
IP_{3_t}	0.0756	0.0074
$IP_{1_{t-1}}$	0.0727	0.0157
$\Delta CO_{2,t-3}$	0.0619	0.0372
$w\_NDVI_{3,Eur_{t-8}}$	0.0578	0.0200
$NDVI_{1,Eur_{t-12}}$	0.0483	0.2648

on temperature is positive at a lag of $(t-4)$, likely capturing reduced oceanic absorption under higher temperature. Southern Oscillation enters the model at a lag of five months with a negative coefficient. Thus, during El Niño episodes ($SOI < 0$), growth in CO_2 appears to increase, in line with findings of other papers (see Bacastow, 1976; Keeling and Revelle, 1985).

To quantify and assess the relative importance of each regressor, we decompose the total explained variation in ΔCO_2. Decomposition is not straightforward when independent variables are correlated. We use two measures, partial R^2 and hierarchical partitioning. The partial R^2 provides a measure of the marginal contribution to explained variation for a given variable, while holding other factors constant. In hierarchical partitioning the explained variance is decomposed by calculating the average contribution of each variable over all potential orderings of the variables (see Kruskal, 1987, and method LMG in Groemping, 2007).[5] Individual variance contributions sum to unity and can be interpreted as percentages. This yields values for individual variables that sum to the total explained variance (R^2). Table 12.8 ranks the selected variables by the partial R^2 and also reports relative importance based on hierarchical partitioning.

Based on this decomposition, the single largest (non-autoregressive) contribution comes from the Eurasian NDVI principal component of vegetation followed by industrial production. The anthropogenic components cumulatively explain a large fraction of the variation in atmospheric CO_2. Perhaps surprisingly, both the SOI and temperature are ranked low based on hierarchical partitioning.

GUM: Sample 2

Sample 2 in (12.10) covers all variables measured at a monthly frequency as well as interpolated annual components for long-term anthropogenic emissions. Lag selection for sample 2 is based on selection in sample 1. Selection in model 1 results in a maximum lag of 4 on anthropogenic components and 5 for temperature and SOI. The longest lag for vegetation was selected at $t - 12$. Therefore GUM 2 starts with a maximum lag of 6 for anthropogenic components, and a maximum lag of 12 for vegetation. *Autometrics* in sample 2 with 246 observations estimated 571 models, reducing the number of explanatory variables from an initial 483 (split into initial six blocks) down to 19 in the final model. Unit roots are rejected for the model residuals at the 1 percent level using ADF tests covering up to 12 lags. There is only a single impulse indicator selected (1990:7), suggesting no major breaks or mis-specification.

Selection: sample 2 results in a slightly higher number of selected variables in the final model relative to sample 1. In terms of robustness, selection is highly consistent relative to sample 1. Anthropogenic components, as well as temperature and Southern Oscillation, are selected identically to model 1. No deterministic terms are selected.

Anthropogenic emissions are captured solely by the principal components for industrial production: the low-frequency fossil fuel measures are not selected. These findings suggest that high-frequency industrial production provides a better measure for anthropogenic factors than interpolated fuel emissions. This may appear surprising given that anthropogenic emissions directly measure emitted CO_2: however, this result likely stems from the annual frequency of CO_2 emissions that miss any seasonal component.

In terms of natural controls, the estimated coefficient on temperature is positive and that on Southern Oscillation is negative, both as in the previous model. The selection of vegetation variables moves towards seasonally weighted Eurasian indicators but also adds North American NDVI measures. The selection of seasonally weighted vegetation measures supports the theory of atmospheric transport. The shift in selection of vegetation variables is likely due to the collinearity in the NDVI measure resulting from a similar growth cycle in North America and Eurasia. The selected indicator for 1990:7 (July) suggests there was a reduction in growth relative to other years. Although the cause for this is not apparent in the data, it should be noted that such an indicator creates a step shift in the level of CO_2, so may correspond to China's resurgent growth thereafter.

Table 12.9 ranks the selected variables by relative importance based on the partial R^2. Anthropogenic components rank similarly to model 1, with IP_{2t-4} being the second largest (non-autoregressive) contributor to explained variation after Eurasian vegetation. However, the measure of relative importance matters – based on hierarchical partitioning, anthropogenic measures are ranked lower than vegetation, while still above temperature and SOI.

Comparisons between Sample 1 and Sample 2

GUM sample 1 and GUM sample 2 are estimated as robustness checks. Comparing *Autometrics* selection in samples 1 and 2, there is highly consistent selection of key variables. Industrial production, temperature, vegetation and Southern Oscillation are selected identically in both models. Anthropogenic emissions are consistently selected in the form of high-frequency industrial production rather than low-frequency annual emissions. Conversely, selection of regional NDVI varies slightly between models, which could be the result of a similar growing season and pattern in North America and Eurasia as measured through NDVI data. Differences in selection from sample 1 to sample 2 may seem surprising given that all vegetation variables in sample 1 are identi-

Table 12.9 GUM: sample 2 relative importance

Variable	Partial R^2	Hierarchical partitioning
$\Delta CO_{2,t-2}$	0.3582	0.0298
$NDVI_{1,Eur_{t-3}}$	0.2147	0.0290
IP_{2t-4}	0.1814	0.0040
$NDVI_{1,Eur_t}$	0.1376	0.1750
$Temp_{t-4}$	0.1349	0.0055
SOI_{t-5}	0.1149	0.0013
IP_{1t-4}	0.1109	0.0067
$NDVI_{1,Eur_{t-11}}$	0.1098	0.0950
$w\_NDVI_{3,Eur_{t-8}}$	0.0974	0.0122
$s\_NDVI_{1,NA_{t-8}}$	0.0936	0.0550
$s\_NDVI_{1,NA_{t-9}}$	0.0869	0.0464
IP_{3t}	0.0840	0.0034
$NDVI_{1,Eur_{t-12}}$	0.0813	0.1740
$\Delta CO_{2,t-4}$	0.0805	0.0778
IP_{1t-1}	0.0792	0.0086
$w\_NDVI_{2,Eur_{t-7}}$	0.0770	0.0780
$w\_NDVI_{2,Eur_{t-8}}$	0.0605	0.1125
$NDVI_{2,Eur_{t-11}}$	0.0555	0.0810
$I_{1990(7)}$	0.0507	0.0018

cal to those in sample 2. However, the PCs are only orthogonal within regions, and are highly correlated between regions, as Figure 12.7 shows. The result of different selection may be due to computational short cuts, namely dropping branches in the tree search after a model failed backtesting, and block partitioning in *Autometrics*, which adversely affect consistent selection.

Overall, based on *Autometrics* selection, natural factors such as vegetation, temperature and Southern Oscillation are necessary but not sufficient to explain changes in atmospheric CO_2 measured at Mauna Loa. Industrial production variables are consistently selected. Most estimated effects from selected variables affect ΔCO_2 with a lag, and there seem to be few or no structural breaks in the relationships being modeled. As a further robustness check, future work will involve applying the estimation method to other measurement stations such as Barrow, Alaska (Keeling et al., 2008). Atmospheric CO_2 at Barrow has been measured from 1974:2 until 2007:12 and displays higher amplitude and higher autocorrelation due to its location relative to Mauna Loa.

Using relatively few assumptions, automatic model selection with *Gets* modeling can provide a tool to successfully model complex relationships. Starting from a broad GUM that nests the LDGP theoretically (accounting for natural and anthropogenic sinks and sources), the analysis proceeds with an agnostic approach to determine the key factors in changes in atmospheric CO_2. First, automatic model selection reduces the GUM to terminal models. Shortcomings of automatic model selection are computational issues in *Autometrics* selection because of short cuts and block partitioning. However, the terminal models appear congruent (acceptable diagnostic tests, few to no indicators or seasonal dummies selected) and are also supported by theoretical conclusions from the

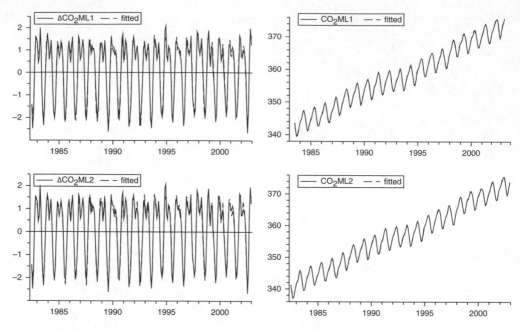

Figure 12.15   Graphs of fitted and actual values in differences and levels

broader literature – natural effects are selected with the expected signs on coefficients. Key in the results is that additionally to the natural determinants selected, all terminal models include a large number of anthropogenic factors. The explained variance is decomposed to establish the relative importance of the variables. Using hierarchical partitioning and ranking based on partial correlations, the anthropogenic contribution to explained variation is quantifiable and found to be high.

From Changes to Levels

The models estimated as GUM sample 1 and GUM sample 2 describe the change in atmospheric CO_2, ΔCO_2. While most of the analysis has been focused on the change in CO_2, it is possible to recover the estimated level from the models 1 and 2. Level estimates for model 1 and model 2, based on $CO_{2,t} = \Delta CO_{2,t} + CO_{2,t-1}$ are given in Figure 12.15 and show the close fit. The next step is to attribute the components of the long-run explanation. To do so, we derive the relation after all dynamics from lagged variables have been solved out (the 'long-run solution': see Hendry, 1995). In a simple ADL(1) of the form

$$y_t = \lambda_1 x_{1,t} + \lambda_2 y_{t-1} + e_t \tag{12.11}$$

where $|\lambda_2| < 1$, the long-run conditional expected value is

$$E[y|x_t] = \lambda_1 x_1/(1 - \lambda_2) = \beta_1 x_1 \tag{12.12}$$

Then, based on the theoretical specifications given in equations (12.5)–(12.7), the solved estimated model is in the approximate form given in equation (12.13):

$$\Delta CO_{2,t} = \beta_1 x_{1,t} + \beta_2 x_{2,t} + \cdots + \beta_q x_{q,t} + \varepsilon_t \tag{12.13}$$

This can be rewritten as

$$CO_{2,t} = CO_{2,t-1} + \beta_1 x_{1,t} + \beta_2 x_{2,t} + \cdots + \beta_q x_{q,t} + \varepsilon_t \tag{12.14}$$

Recursive substitution for $CO_{2,t-1}, CO_{2,t-2}, \ldots$ in equation (12.14) yields

$$CO_{2,t} = CO_{2,0} + \beta_1 \sum_{j=1}^{t} x_{1,j} + \beta_2 \sum_{j=1}^{t} x_{2,j} + \cdots + \beta_q \sum_{j=1}^{t} x_{q,j} + \bar{e}_t \tag{12.15}$$

where $\bar{e}_t = \sum_{j=1}^{t} \varepsilon_j$

We divide the variables into two different groups: let s be the number of variables $x_{i,t}$ that have a stationary cumulative sum (non-trending) $\sum_{j=1}^{t} x_{i,j} \sim I(0)$, so that $q - s$ variables have a non-stationary cumulative sum (trending) $\sum_{j=1}^{t} x_{i,j} \sim I(r)$, where $r > 0$. Equation (12.15) can then be expressed as

$$CO_{2,t} = CO_{2,0} + \mathbf{x}'_{s,t} \boldsymbol{\beta}_s + \mathbf{x}'_{q-s,t} \boldsymbol{\beta}_{q-s} + \bar{e}_t \tag{12.16}$$

where $\mathbf{x}'_{s,t}$ and $\mathbf{x}'_{q-s,t}$ are $s \times 1$ and $(q - s) \times 1$ column vectors respectively with $\sum_{j=1}^{t} x_{i,j}$ as their row elements. Equation (12.16) implies that the trending level of CO_2 is a function of the cumulative sums of the stationary ($\mathbf{x}'_{s,t}$) and non-stationary ($\mathbf{x}'_{q-s,t}$) variables in our model. Stationary variables in $\mathbf{x}'_{s,t}$ by nature cannot drive the trend. Only explanatory variables with non-stationary cumulative sums, $\mathbf{x}'_{q-s,t}$, determine the trend. This provides a straightforward method of evaluating the underlying factors of the long-run trend.

Out of the selected variables in models 1 and 2, only a subset exhibits trending cumulative sums, which are all the anthropogenic factors and the temperature anomaly. Both natural controls of NDVI and SOI remain approximately stationary around zero over time. Importantly, neither final model includes a deterministic intercept or trend, which on summation would become a linear or a quadratic time trend. However, summed variables do not have a straightforward interpretation in the case of PCs of industrial output and temperature. The trending temperature anomaly is likely a mutually supporting feedback effect, as mentioned in Section 4. Overall, the trend in the levels of CO_2 is derived from the trends in the independent variables in both estimated models, so is driven primarily by the PCs and partly by temperature. Specifically, the empirical equivalents of (12.16) for model 1 are

$$\mathbf{x}'_{s,t} \boldsymbol{\beta}_s = -0.0038 \sum_{j=1}^{t} SOI_j + \left(-0.0275 \sum_{j=1}^{t} NDVI_{1,Eur_j} + 0.0169 \sum_{j=1}^{t} w\_NDVI_{3,Eur_j} \right) \tag{12.17}$$

$$\mathbf{x}'_{q-s,t} \boldsymbol{\beta}_{q-s} = -0.037 \sum_{j=1}^{t} IP_{1_j} - 0.127 \sum_{j=1}^{t} IP_{2_j} + 0.097 \sum_{j=1}^{t} IP_{3_j} + 0.0018 \sum_{j=1}^{t} Temp_j \tag{12.18}$$

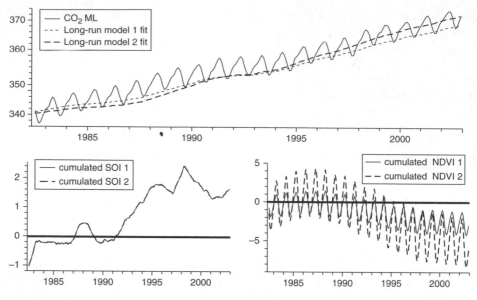

Figure 12.16   Level of CO_2 and cumulative sums of anthropogenic and natural factors

Figure 12.16 shows the resulting coefficient-weighted cumulative sums of vegetation ($NDVI_{EUR}$ principal components), SOI, and the combined anthropogenic components (IP_1 to IP_3 and temperature trend) for both models 1 and 2, together with the recorded level of CO_2. The industrial production components and temperature approximate the level of CO_2 well, marking a slight slowdown in the trend around 1991–93. Both cumulative stationary components vary over a small range, so contribute little to the long-run changes. Even though the model is estimated in net inflows to atmospheric CO_2, in re-parametrized form it can explain the long-run trend – and attributes it primarily to anthropogenic emissions. This is an outcome of the data analysis and is not enforced.

7   CONCLUSIONS

We identified anthropogenic contributions to atmospheric CO_2 measured at Mauna Loa using an automatic model selection algorithm. Traditionally, estimation of anthropogenic effects on CO_2 relied on *a priori* selection of variables, which may not be appropriate in complex relations, low-frequency measures of anthropogenic emissions, and decompositions of time series. Using *Autometrics* in a general to specific modeling approach allows for model selection with more variables than observations, stringent mis-specification testing and a more agnostic way of modeling complicated interactions. The algorithm is applied to model changes in atmospheric CO_2 controlling for natural as well as anthropogenic sinks and sources without *a priori* restrictions of the determinants. While they are not completely robust to initial specification, we find that natural factors such as vegetation, temperature and the Southern Oscillation are necessary, but not sufficient in explaining variation of atmospheric CO_2. Industrial production components measured monthly are highly significant and consistently selected in the estimated models. A higher proportion of variation should be attributed to anthropogenic sources than has been previously the case in the literature. Producing congruent models, our methodology introduces *Gets* modeling through *Autometrics* as a useful tool in modeling complicated climate relationships.

NOTES

*   This research was supported in part by grants from the Open Society Foundation and the Oxford Martin School. We thank Roger Fouquet for useful comments and Qin Duo for helpful notes on Chinese production.

1. 1 ppm by volume of CO_2 in the atmosphere is approximately equal to 2.13 gigatons of carbon: see Clark (1982) and http://cdiac.ornl.gov/pns/convert.html.
2. Chinese gross industrial output is only available from 1979 to 1999 with a change in measure thereafter. Inclusion of this series did not improve selection at the cost of a reduction in observations.
3. See Marland et al. (2011) for a detailed listing of which countries are included. Eastern Europe includes Russia, Central Asia includes China, and India is covered by the Far East category.
4. Detailed results are available from the authors on request.
5. An intercept is forced to be retained in selection of these models, though it is not statistically different from zero for models 1 and 2.

REFERENCES

Australian Bureau of Meteorology (2011), 'Southern Oscillation index archive', http://www.bom.gov.au/climate/current/soihtm1.shtml.

Bacastow, R.B. (1976), 'Modulation of atmospheric carbon dioxide by the Southern Oscillation', *Nature*, **261**, 116–18.

Bacastow, R.B., Keeling, C.D. and Whorf, T.P. (1985), 'Seasonal amplitude increase in atmospheric CO_2 concentration at Mauna Loa, Hawaii 1959–1982', *Journal of Geophysical Research*, **90**, 10529–40.

Banerjee, A., Dolado, J.J., Galbraith, J.W. and Hendry, D.F. (1993), *Co-integration, Error Correction and the Econometric Analysis of Non-Stationary Data*, Oxford: Oxford University Press.

Bårdsen, G., Hurn, S. and McHugh, Z. (2010), 'Asymmetric unemployment rate dynamics in Australia', Creates Research Paper 2010-2, Aarhus University, Denmark.

Bontemps, C. and Mizon, G.E. (2008), 'Encompassing: concepts and implementation', *Oxford Bulletin of Economics and Statistics*, **70**, 721–50.

Buermann, W., Lintner, B.R., Koven, C.D., Angert, A., Pinzon, J.E., Tucker, C.J. and Fung, I.Y. (2007), 'The changing carbon cycle at Mauna Loa observatory', *Proceedings of the National Academy of Sciences of the United States of America*, **104**, 4249–54.

Campos, J., Ericsson, N.R. and Hendry, D.F. (2005), Editors' introduction, in J. Campos, N.R. Ericsson and D.F. Hendry (eds), *Readings on General-to-Specific Modeling*, Cheltenham, UK and Northampton, MA, USA: Edward Elgar, pp. 1–18.

Castle, J.L., Doornik, J.A. and Hendry, D.F. (2011), 'Evaluating automatic model selection', *Journal of Time Series Econometrics*, 3(1), DOI: 10.2202/1941–1928.1097.

Castle, J.L., Doornik, J.A. and Hendry, D.F. (2012a), 'Model selection in equations with many "small" effects', *Oxford Bulletin of Economics and Statistics*, doi:10.1111/j.1468-0084.2012.00727.x.

Castle, J.L., Doornik, J.A. and Hendry, D.F. (2012b), 'Model selection when there are multiple breaks', *Journal of Econometrics*, **169**, 239–46.

Castle, J.L. and Hendry, D.F. (2011), 'Automatic selection of non-linear models', in L. Wang, H. Garnier and T. Jackman (eds), *System Identification, Environmental Modelling and Control*, New York: Springer, pp. 229–50.

Castle, J.L. and Shephard, N. (eds) (2009), *The Methodology and Practice of Econometrics*, Oxford: Oxford University Press.

Chow, G.C. (1960), 'Tests of equality between sets of coefficients in two linear regressions', *Econometrica*, **28**, 591–605.

Christopher, S.L., Feely, R.A., Gruber, N., Key, R.M., Lee, K., Bullister, J.L., Wanninkhof, R., Wong, C.S., Wallace, D.W.R., Tilbrook, B., Millero, F.J., Peng, T.H., Kozyr, A., Ono, T. and Rios, A.F. (2007), 'The oceanic sink for anthropogenic CO_2', *Science*, **305**, 367–71.

Clark, W.C. (ed.)(1982), *Carbon Dioxide Review*, New York: Oxford University Press.

Dickey, D.A. and Fuller, W.A. (1981), 'Likelihood ratio statistics for autoregressive time series with a unit root', *Econometrica*, **49**, 1057–72.

Doornik, J.A. (2008), 'Encompassing and automatic model selection', *Oxford Bulletin of Economics and Statistics*, **70**, 915–25.

Doornik, J.A. (2009), 'Autometrics', in Castle and Shephard (2009), pp. 88–121.

Doornik, J.A. (2010), 'Econometric model selection with more variables than observations', Working Paper, Economics Department, University of Oxford.

Doornik, J.A. and Hansen, H. (2008), 'An omnibus test for univariate and multivariate normality', *Oxford Bulletin of Economics and Statistics*, **70**, 927–39.

Doornik, J.A. and Hendry, D.F. (2009), *Empirical Econometric Modelling using PcGive: Volume I*, London: Timberlake Consultants Press.

Engle, R.F. (1982), 'Autoregressive conditional heteroscedasticity, with estimates of the variance of United Kingdom inflation', *Econometrica*, **50**, 987–1007.

Enting, I.G. (1987), 'The interannual variation in the seasonal cycle of carbon dioxide concentration at Mauna Loa', *Journal of Geophysical Research*, **92**, 5497–504.

Erickson, D.J., Mills, R.T., Gregg, J., Blasing, T.J., Hoffmann, F.M., Andres, R.J., Devries, M., Zhu and Kawa, S.R. (2008), 'An estimate of monthly global emissions of anthropogenic CO_2: impact on the seasonal cycle of atmospheric CO_2', *Journal of Geophysical Research*, **113**.

Federal Reserve (2011), 'Industrial production and capacity utilization', http://www.federalreserve.gov/releases/g17.

Francey, R.J., Tanis, P.P., Allison, C.E., Enting, I.G., White, J.W.C. and Trolier, M. (1995), 'Changes in oceanic and terrestrial carbon uptake since 1982', *Nature*, **373**, 326–30.

Gerlach, T.M. (2011), 'Changes in oceanic and terrestrial carbon uptake since 1982', *EOS, Transactions, American Geophysical Union*, **92**, 201–2.

Godfrey, L.G. (1978), 'Testing for higher order serial correlation in regression equations when the regressors include lagged dependent variables', *Econometrica*, **46**, 1303–13.

Granados, J.A.T., Ionides, E.L. and Carpintero, O. (2009), 'A threatening link between world economic growth and atmospheric CO_2 concentrations', Working Paper, University of Michigan.

Granados, J.A.T., Ionides, E.L. and Carpintero, O. (2012), 'Climate change and the world economy: short-run and long-run determinants of amtospheric CO_2', *Environmental Science and Policy*, **21**, 50–62.

Groemping, U. (2007), 'Estimators of relative importance in linear regression based on variance decomposition', *American Statistician*, **61**(2), 139–47.

Hansen, J. and Lebedeff, S. (1987), 'Global trends of measured surface air temperature', *Journal of Geophysical Research*, **92**(13), 345–72.

Hansen, J., Ruedy, R., Sato, M. and Lo, K. (2010), 'Global surface temperature change', *Review of Geophysics*, **48**, 2010RG000345.

Hansen, J., Russel, G., Rind, D., Stone, P., Lacis, A., Lebedeff, S., Ruedy, R. and Travis, L. (1983), 'Efficient three-dimensional global models for climate studies: Models I and II', *Monthly Weather Review*, **111**, 609–62.

Hansen, J., Sato, M., Kharecha, P., Beerling, D., Berner, R., Masson-Delmotte, V, Pagani, M., Raymo, M., Royer, D.L. and Zachos, J.C. (2008), 'Target atmospheric CO_2: where should humanity aim?', *The Open Atmospheric Science Journal*, **2**, 217–31.

Hendry, D.F. (1995), *Dynamic Econometrics*, Oxford: Oxford University Press.

Hendry, D.F. (2009), 'The methodology of empirical econometric modeling: applied econometrics through the looking-glass', in T.C. Mills and K.D. Patterson (eds), *Palgrave Handbook of Econometrics*, Basingstoke: Palgrave Macmillan, pp. 3–67.

Hendry, D.F. (2011), 'Climate change: possible lessons for our future from the distant past', in S. Dietz, J. Michie and C. Oughton (eds), *The Political Economy of the Environment*, London: Routledge, pp. 19–43.

Hendry, D.F. and Johansen, S. (2013), 'Model discovery and Trygve Haavelmo's legacy', *Econometric Theory*, forthcoming.

Hendry, D.F., Johansen, S. and Santos, C. (2008), 'Automatic selection of indicators in a fully saturated regression', *Computational Statistics & Data Analysis*, **33**, 317–35.

Hendry, D.F. and Juselius, K. (2000), 'Explaining cointegration analysis: Part I', *Energy Journal*, **21**, 1–42.

Hendry, D.F. and Krolzig, H.-M. (2003), 'New developments in automatic general-to-specific modelling', in B.P. Stigum (ed.), *Econometrics and the Philosophy of Economics*, Princeton, NJ: Princeton University Press, pp. 379–419.

Hendry, D.F. and Krolzig, H.-M. (2005), 'The properties of automatic Gets modelling', *Economic Journal*, **115**, C32–C61.

Hendry, D.F. and Mizon, G.E. (2011), 'Econometric modelling of time series with outlying observations', *Journal of Time Series Econometrics*, **3**(1), DOI: 10.2202/1941-1928.1100.

Hendry, D.F. and Richard, J.-F. (1989), 'Recent developments in the theory of encompassing', in B. Cornet and H. Tulkens (eds), *Contributions to Operations Research and Economics. The XXth Anniversary of CORE*, Cambridge, MA: MIT Press, pp. 393-440.

Hendry, D.F. and Santos, C. (2010), 'An automatic test of super exogeneity', in M.W. Watson, T. Bollerslev and J. Russell (eds), *Volatility and Time Series Econometrics*, Oxford: Oxford University Press, pp. 164-93.

Hoffman, P.F. and Schrag, D.P. (2000), 'Snowball Earth', *Scientific American*, **282**, 68-75.

Hofman, D.J., Butler, J.H. and Tans, P.P. (2009), 'A new look at atmospheric carbon dioxide', *Atmospheric Environment*, **43**, 2084-86.

Johansen, S. and Nielsen, B. (2009), 'An analysis of the indicator saturation estimator as a robust regression estimator', in Castle and Shephard (2009), pp. 1-36.

Jones, C.D. and Cox, P.M. (2005), 'On the significance of atmospheric CO₂ growth rate anomalies in 2002-2003', *Journal of Geophysical Research*, **32**.

Juselius, K. and Kaufmann, R. (2009), 'Long-run relationships among temperature, CO₂, methane, ice and dust over the last 420,000 years: cointegration analysis of the Vostok ice core data', *IOP Conference Series: Earth and Environmental Science*, **6**.

Kaufmann, R.K., Paletta, L.F., Tian, H.Q., Myeni, R.B. and D'Arrigo, R.D. (2008), 'The power of monitoring stations and a CO₂ fertilization effect: evidence from causal relationships between NDVI and carbon dioxide', *Earth Interactions*, **12**, 1-23.

Kawa, S.R., Erickson, D.J., Pawson, S. and Zhu, Z. (2004), 'Global CO₂ transport simulations using meteorological data from NASA data assimilation system', *Journal of Geophysical Research*, **109**, D18312.

Keeling, C.D. (1973), 'Industrial production of carbon dioxide from fossil fuels and limestone', *Tellus*, **25**(2), 174-98.

Keeling, C.D., Bacastow, R.B., Brainbridge, A.E., Ekdahl, C.A., Guenther, J.P.R. and Waterman, L.S. (1976), 'Atmospheric carbon dioxide variations at Mauna Loa observatory, Hawaii', *Tellus*, **6**, 538-51.

Keeling, C.D., Chin, J.F. and Whorf, T.P. (1996), 'Increased activity of northern vegetation inferred from atmospheric CO₂ measurements', *Nature*, **382**, 146-49.

Keeling, C.D. and Revelle, R. (1985), 'Effects of El Niño/southern oscillation on the atmospheric content of carbon dioxide', *Meteoritics*, **20**(2), 437-50.

Keeling, C.D., Whorf, T.P., Whalen, M. and van der Plicht, J. (1995), 'Interannual extremes in rate rise of atmospheric carbon dioxide since 1980', *Nature*, **375**, 666-70.

Keeling, R.F., Piper, S.C, Bollenbacher, A.F. and Walker, S.J. (2008), 'Atmospheric CO₂ records from sites in the SIO air sampling network', in *Trends: A Compendium of Data on Global Change*, Carbon Dioxide Information Analysis Center, US Department of Energy, Oak Ridge, TN, USA, http://cdiac.ornl.gov/trends/co2/sio-bar.html.

Kohlmaier, G.H., Sire, E.O., Janecek, A., Keeling, C.D., Piper, S.C, and Revelle, R.(1989), 'Modelling the seasonal contribution of a CO₂ fertilization effect of the terrestrial vegetation to the amplitude increase in atmospheric CO₂ at Mauna Loa observatory', *Tellus*, **41B**, 487-510.

Kruskal, W. (1987), 'Relative importance by averaging over orderings', *American Statistician*, **41**(1), 6-10.

Levin, I., Graul, R. and Trivett, N.B.A. (1995), 'Long-term observations of atmospheric CO₂ and carbon isotopes at continental sites in Germany', *Tellus*, **47B**, 23-34.

Lucht, W., Prentice, I.C., Myeni, R.B., Sitch, S., Friedlingstein, P., Cramer, W., Bousquet, P., Buermann, W. and Smith, B. (2002), 'Climate control of the high-latitude vegetation greening trend and Pinatubo effect', *Science*, **296**, 1687-9.

Marland, G., Boden, T.A. and Andres, R. (2011), 'Global, regional, and national fossil fuel CO₂ emissions', in *Trends: A Compendium of Data on Global Change*, Carbon Dioxide Information Analysis Center, Oak Ridge National Laboratory, US Department of Energy, Oak Ridge, TN, USA, http://cdiac.ornl.gov/trends/emis/overview.html.

Marland, G. and Rotty, R.M. (1984), 'Carbon dioxide emissions from fossil fuels: a procedure for estimation and results for 1950-1982', *Tellus*, **36B**, 232-61.

Ministry of Statistics Government of India (2011), 'Index of industrial production (IIP)', http://mospi.nic.in/.

Mizon, G.E. and Richard, J.F. (1986), 'The encompassing principle and its application to non-nested hypothesis tests', *Econometrica*, **54**, 657-78.

Myeni, R.B., Hall, F.G., Sellers, P.J. and Marshak, A.L. (1995), 'The interpretation of spectral vegetation indexes', *Transactions on Geoscience and Remote Sensing*, **33**(2), 481-6.

NASA Goddard Institute for Space Studies (GISS) (2011), 'GISS surface temperature analysis', http://data.giss.nasa.gov/gistemp/.

Nevison, C.D., Mahowald, N.M., Doney, S.C, Lima, I.D., van der Werf, G.R., Randerson, J.T., Baker, D.F., Kasibhatla, P. and McKinley, G.A. (2008), 'Contribution of ocean, fossil fuel, land biosphere, and biomass burning carbon fluxes to seasonal and interannual variability in atmospheric CO_2', *Journal of Geophysical Research*, **113**.

Newell, N.D. and Marcus, L. (1987), 'Carbon dioxide and people', *Society for Sedimentary Geology*, **2**(1), 101–3.

OECD (2011), 'Composite leading indicators: MEI', http://stats.oecd.org/Index.aspx.

Office of National Statistics (2011), 'Primary production', http://www.statistics.gov.uk/hub/business-energy/.

Orr, C.J., Maier-Reimer, E., Mikolajewicz, U., Monfray, P., Sarmiento, J.L., Toggweiler, J.R., Taylor, N.K., Palmer, J., Gruber, N., Sabine, C.L., Le Quere, C., Key, R.M. and Boutin, J. (2001), 'Estimates of anthropogenic carbon uptake from four three-dimensional global ocean models', *Global Biogeochemical Cycles*, **15**(1), 43–60.

Randerson, T.J., Thompson, M.V., Conway, T.J., Fung, I.Y. and Field, C.B. (1997), 'The contribution of terrestrial sources and sinks to trends in the seasonal cycle of atmospheric carbon dioxide', *Global Biogeochemical Cycles*, **11**(4), 535–60.

Ritschard, R.L. (1992), 'Marine algae as a CO_2 sink', *Water, Air and Soil Pollution*, **64**, 289–303.

Ruddiman, W. (ed.)(2005), *Plows, Plagues and Petroleum: How Humans took Control of Climate*, Princeton, NJ: Princeton University Press.

Salkever, D.S. (1976), 'The use of dummy variables to compute predictions, prediction errors and confidence intervals', *Journal of Econometrics*, **4**, 393–7.

Schwarz, G. (1978), 'Estimating the dimension of a model', *Annals of Statistics*, **6**, 461–4.

Sims, C.A., Stock, J.H. and Watson, M.W. (1990), 'Inference in linear time series models with some unit roots', *Econometrica*, **58**, 113–44.

Sundquist, E.T. and Keeling, R.F. (2009), 'The Mauna Loa carbon dioxide record: lessons for long-term earth observations', *Geophysical Monograph Series*, **183**, 27–35.

Taguchi, S., Murayama, S. and Higuchi, K. (2003), 'Sensitivity of inter-annual variation of CO_2 seasonal cycle at Mauna Loa to atmospheric transport', *Tellus*, **55B**, 547–54.

Tans, P. and Keeling, R. (2011), 'Mauna Loa, monthly mean carbon dioxide', Scripps Institution of Oceanography (scrippsco2.ucsd.edu/) and NOAA/ESRL http://www.esrl.noaa.gov/gmd/ccgg/trends/.

Thoning, K.W. and Tans, P.P. (1989), 'Atmospheric carbon dioxide at Mauna Loa observatory 2. Analysis of the NOAA GMCC data, 1974–1985', *Journal of Geophysical Research*, **94**, 8549–65.

Troup, A.J. (1965), 'The "Southern Oscillation"', *Quarterly Journal of the Royal Meteorological Society*, **91**, 490–506.

Tucker, C.J., Pinzon, J., Brown, M. and GIMMS/GSFC/NASA (2010), 'ISLSCP II GIMMS monthly NDVI, 1981–2002', in Forrest G. Hall, G. Collatz, B. Meeson, S. Los, E. Brown de Colstoun and D. Landis (eds), ISLSCP Initiative II Collection. Data set, http://daac.ornl.gov/ from Oak Ridge National Laboratory Distributed Active Archive Center, Oak Ridge, TN, USA.

Watson, A.J., Nightingale, P.D. and Cooper, D.J. (1995), 'Modeling atmosphere ocean CO_2 transfer', *Philosophical Transactions of the Royal Society of London Series B – Biological Sciences*, **353**, 41–51.

White, H. (1980), 'A heteroskedastic-consistent covariance matrix estimator and a direct test for heteroskedasticity', *Econometrica*, **48**, 817–38.

Wigley, T.M.L. (1983), 'The pre-industrial carbon dioxide level', *Climate Change*, **5**, 315–20.

Worrell, E., Price, L., Martin, N., Hendriks, C. and Meida, L.O. (2001), 'Carbon dioxide emissions from the global cement industry', *Annual Review of Energy and the Environment*, **26**, 303–29.

13 International cooperation on climate change: why is there so little progress?

Bjart Holtsmark*

1. INTRODUCTION

It is more than two decades since the problem of climate change became a major item on the political agenda. According to the reports from the Intergovernmental Panel on Climate Change (IPCC), the potentially dangerous and irreversible consequences of the rapidly rising atmospheric concentration of greenhouse gases have become no less urgent over the years (IPCC, 2007).

Nevertheless, little is being done to limit emissions at present, at least in comparison with the deep cuts recommended by IPCC (2007). What action is being taken is largely cosmetic, more likely to give people a clear conscience than make a real difference. The price of carbon on the EU market has for example slumped to very low levels, and the national quotas under the first commitment period of the Kyoto Protocol were far too generous (Böhringer, 2002; Wara and Victor, 2008).[1] In any case, the Kyoto Protocol regulated less than 30 per cent of global emissions, and the planned second commitment period will involve fewer countries and a correspondingly smaller share of global emissions, probably around 15 per cent. Moreover, negotiations on establishing a more effective, comprehensive agreement to follow on have shown little progress in recent years. The USA is not intending to sign up to a new agreement unless leading developing countries like India and China do so. India and China, for their part, pointed to the far higher per capita emissions in the developed countries. They were also emitting substantial quantities of greenhouse gases long before developing countries.

In other words, although there is general agreement that we are facing a problem, the international community has not been able to get together behind a strategy capable of effecting substantial cuts in emissions.

This chapter employs simple games to illustrate why it is so difficult to rally support behind effective agreements to reduce greenhouse gas emissions and why there is reason to be pessimistic about the outcome of future negotiations. The main problem appears to be that because free-riding is such a favourable option, it is difficult to establish a coalition of countries that agree on deep cuts in emissions (Barrett, 1994, 2003, 2005; Hoel, 1992).

At the same time the chapter illustrates that the world is unlikely to make real progress in addressing climate change without an international agreement with broad participation. The current political debate, especially in Europe, reflects a belief in the efficacy of countries 'setting a good example' by cutting their own emissions, and that others will then follow their lead.

It is a weakly founded hypothesis, and I shall use game theory to show the likely effect on international climate cooperation of one country signalling readiness to make unilateral cuts, whether an international agreement is concluded or not. I shall conclude that global emissions are unlikely to fall as a result of such unilateral actions. On the contrary, it can be argued that such unilateral commitments or actions are more likely to increase global emissions, by increasing the likelihood of a less ambitious international agreement.

This conclusion echoes Hoel (1991), who found that in a non-cooperative equilibrium and given a Nash bargaining solution, unilateral emission reductions may well push up global emissions, although Hoel's framework does allow for falling global emissions in the wake of unilateral cuts. What I want to demonstrate in this chapter is that if one country *announces* an ambitious national climate policy plan before the negotiations, and irrespective of the eventual negotiated outcome, it will, given the applied theoretical framework, always reduce the scope and ambition of the final agreement and, at the end of the day, give higher global emissions than if the single country had made its emission cuts contingent on other countries doing likewise.

The standstill in the international negotiations has prompted a proposal: let a *set* of agreements, each including a subset of countries, be the goal, rather than a single agreement with global participation (Osmani and Tol, 2009). If each of these subsets of countries were small enough, all countries, in principle at least, could sign up to so-called internally stable agreements, that is, agreements where all signatories would be worse off if they left the coalition (Carraro and Siniscalco, 1993). However, in Section 6 of this chapter a numerical example suggests that although many countries might endorse the idea, it would not provide for emission reductions at a satisfactory level.

A great deal of research is being done on international climate agreements, and I do not intend to review every contribution. I base my approach on non-cooperative game theory and the concept of stable coalitions as originally conceived by d'Aspremont et al. (1983). I do not discuss the results of a strand of literature, which is also based on non-cooperative game theory, that has an international permit market explicitly at centre stage, where the initial allocation of permits results from a non-cooperative game, and where no group of agents ever engages in maximizing its joint objectives (Carbone et al., 2009; Crampton and Stoft, 2010a, 2010b; Godal and Holtsmark, 2011b; Helm, 2003; Holtsmark and Sommervoll, 2012). Nor does the chapter apply results from the strand of literature that uses cooperative game theory (Chander, 2007; Chander and Tulkens, 1995, among others). Moreover, I shall simply be analysing agreements that give each country a national emission quota. I do not discuss agreements requiring countries to commit to specific instruments, such as a carbon tax (Hoel, 1997; Hovi and Holtsmark, 2006; Stiglitz, 2006; Hoel and Karp, 2001, 2002). In recent years, there has also been more focus on the need to consider how policy instruments, agreements and technological progress interact; see for example Golombek and Hoel (2005). However, this is also outside the scope of this chapter.

Sections 2–5 of this chapter introduce simple games to illustrate the importance of international agreements. The games also demonstrate the delicate balance these agreements need to achieve, since an individual country will have much to gain from free-riding on the back of others who do cut their emissions. The games also illustrate the system's dependence on large coalitions to achieve adequate cuts (depth of cooperation). Section 6 presents a numerical example that illustrates that when we take country differences and side payments into consideration, governments may be more willing to join in and reduce emissions. A stable coalition on these terms might result in more abatement than the examples dealing with identical countries.

2. THE 'TRAGEDY OF THE COMMONS'

As an introduction this section revisits a famous article of Garrett Hardin, professor of biology at University of California, Santa Barbara. He is renowned for writing 'The tragedy of the commons', an article published by the journal *Science* in 1968. Hardin drew and expanded on a parable set forth in an 1833 lecture published by William Forster Lloyd, then professor of political economy at Oxford.[2] The parable goes as follows.

Several cattle-owners let their livestock graze open pastures – the commons. The cattle-owners are allowed to let as many cows graze as they like, and do so without encountering problems. The capacity of the land is limited, however, and as the populations grow a point will inevitably be reached when 'the inherent logic of the commons remorselessly generate(s) [a] tragedy' (Hardin, 1968: 1244). The land becomes over-grazed, with the collective activity of the herders leading to diminishing returns for each of them separately.

The question each cattle-owner has to ask is 'What is there to be gained from adding an extra cow to my herd?' The positive component comes from the sale of the additional quantities of beef, milk and hides provided by the additional cow. The negative component is the added pressure on the land, causing a general decline in the productivity of the owner's original livestock.

The 'tragedy of the commons' results therefore from the failure of each individual cattle-herder to take into account how adding to his herd will affect the productivity of all the other farmers' livestock that are grazing the same pasture. The result will be overgrazing and all lose out.

As Harding puts it, 'Each man is locked into a system that compels him to increase his herd without limit – in a world that is limited.'

The analogy with the current climate situation is simple. The atmosphere is a global commons, into which we all want to discharge our industrial CO_2 and other greenhouse gases. The approach worked for a long time, but according to IPCC (2007) the system is evidently straining under the load. As more greenhouse gases enter the atmosphere, the greater the adverse impact on the earth's climate.

Just as the individual cattle-herder will have few incentives to reduce the number of cattle grazing the commons, the individual state will find few incentives to downsize its emissions.

3. A NUMERICAL ILLUSTRATION OF THE EFFECTS OF UNILATERAL ACTIONS

As a follow-up to the preceding section, this section presents a few model simulations showing how tenuous the incentives are for major economies like the USA and China to cut emissions. And even the entire group of industrialized countries, together with China, will not achieve much unless the rest of the world joins in. Table 13.1 presents

Table 13.1   *The slowdown in global warming when China, India and the USA*
             unilaterally cut their emissions and if all developed countries or the whole
             world do the same[a]

	Emission reductions from BAU[a](%)	Temperature change compared to no action °C[b]				
		China	India	USA	All developed countries	Global action
2025	−15	−0.01	−0.002	−0.01	−0.02	−0.04
2050	−65	−0.07	−0.03	−0.06	−0.12	−0.36
2100	−95	−0.23	−0.12	−0.16	−0.30	−1.56

Notes:
[a] In the assumed BAU scenario China's CO_2 emissions are set to rise from 5.8 $GtCO_2$ in 2010 to 11.0 $GtCO_2$ in 2100. By assumption, India's and the US BAU emissions will rise from 1.6 to 5.5 and from 6.0 to 7.0 over the period, respectively.
[b] The temperature changes caused by emission changes are calculated using an impulse response function (IRF) derived from the carbon cycle model Bern 2.5CC (IPCC, 2007). This IRF was selected in the IPCC Fourth Assessment Report as their preferred model.

simulations that illustrate the incentives problem in mitigating global warming. The table shows the estimated reduction in global warming in 2025, 2050 and 2100 resulting from emission cuts by the world's three biggest countries, all developed countries, and by the whole world.

For example, the third column in Table 13.1 shows a case where China follows a path implying emissions cuts of 15, 65 and 95 per cent to the business-as-usual levels in 2025, 2050 and 2100 respectively. The numbers measure the temperature effect of China's emission reductions only. The result would be a relatively modest slowdown in global warming; 0.01 °C, 0.07 °C and 0.23 °C lower global temperature in 2025, 2050 and 2100, respectively. The corresponding numbers are similar or smaller for the USA and India. Table 13.1 also shows the relatively limited effect of common abatement efforts in all developed countries.

We see that a single country's effort has a disappointingly small impact and might be too small to motivate the public in these countries to accept end-user price hikes on fossil fuels and/or other expensive abatement policies. Without arguments for a climate policy that could motivate the public to accept high energy prices and other impacts of an effective climate policy, policy makers will find it difficult to argue for and implement such policies.

It follows that only a joint effort by all or most countries in the world will make sense to the public and policy makers in the respective countries. And this is the basic problem that is studied from a theoretical perspective in the subsequent sections. I want to show there why there are grounds to be pessimistic about the likelihood of a comprehensive and effective international climate agreement.

The numbers presented in Table 13.1 point to another characteristic of the climate problem – its long-term nature, that is, the time lag between the start of serious emission cuts and the result, a significant reduction in global warming.

In 2100, not many people living today will still be around. But our great-grandchildren and their children will. Hence, if emissions are cut substantially in the coming decades, only future generations will benefit from the result. This might well be considered an argument against tough emission cuts today, especially by poor countries, because future generations are likely to be considerably better off than current generations in these countries.

4. A CLIMATE CHANGE GAME INVOLVING TWO COUNTRIES

Consider a world of two identical countries. Denote the abatement in country 1 and 2 as a_1 and a_2 respectively. The cut in global emissions is given by $a_1 + a_2$. Emission reduction is a public good; in other words, each country benefits from the overall emission reduction. Assume a linear relationship between emission abatement and each country's benefits, expressed by $b(a_1 + a_2)$, where b is a positive parameter.[3]

The benefit of emission abatement is less damage from drought, warmer weather and so forth, and lower costs of adaptation to impacts, such as a rising sea level.

Assume quadratic abatement costs.[4] If country i is to cut its emissions by a_i units, the cost is given by $(c/2)(a_i)^2$, where c is a positive parameter. This entails the assumption that the marginal costs of abatement rise linearly with rising abatement.

The payoff for each country is given by the following expression:

$$v_i = b(a_1+a_2) - (c/2)(a_i)^2, \; i = 1,2 \tag{13.1}$$

In the analysis below, the results are not affected by the values chosen for b and c. To simplify the calculations, let us therefore assume that $b = c = 1$. The payoffs for country 1 and 2 are then as follows:

$$v_i = a_1 + a_2 - \tfrac{1}{2}(a_i)^2, \; i = 1,2 \tag{13.2}$$

Figure 13.1 shows the consequences of the different choices available to the players. The first number in each cell shows the payoff for country 1, and the second number the payoff for country 2.

If both countries choose not to reduce their emissions, there are no costs, but nor are there any benefits. Thus the payoff to both countries is 0, as shown in the top left-hand cell.

But this is not a good choice. Assume that country 2 chooses not to reduce its emissions. What is the best strategy for country 1 in this case? If country 1 reduces its emissions by 1 unit, equation (13.1) shows that the abatement cost is 0.5 and the benefit is 1, so the payoff to country 1 is 0.5. If country 1 reduces its emissions by 2 units, its payoff drops back to 0. This is because the marginal abatement costs rise and exceed the benefits of the additional abatement. Thus country 1 will benefit most by reducing its emissions by 1 unit. Further inspection of Figure 13.1 shows that, regardless of the strategy chosen by country 2, country 1 will always receive the largest payoff if it chooses abatement level 1. For country 1, the *dominant* strategy is therefore to reduce its emissions by 1 unit.

Since countries 1 and 2 are identical, the same argument must also apply to country 2.

Country 2 – abatement level chosen

		0		1		2		3	
	0	0	0	1	0.5	2	0	3	–1.5
Country 1 –	1	0.5	1	**1.5**	**1.5**	2.5	1	3.5	–0.5
abatement	2	0	2	1	2.5	**2**	**2**	3	0.5
level chosen	3	–1.5	3	–0.5	3.5	0.5	3	1.5	–1.5

Figure 13.1

We can therefore conclude that regardless of what country 1 does, country 2 will receive the largest payoff by reducing its emissions by 1 unit.

From this, we can conclude that absent cooperation between the countries, they will both choose abatement level 1, and both will receive a payoff of 1.5. This is the only Nash equilibrium in this game.

Having said that, however, Figure 13.1 also shows that both countries will get a bigger payoff if they are able to cooperate and both reduce their emissions by 2 units. If this is a one-shot game, that is, a game played in only one period, such cooperation is usually not considered a possibility. The incentives to free-ride are too strong, and there is no mechanism by which countries can punish the free-rider for not cooperating. Why should either government then not choose abatement level 1 if the game ends after the first period? However, if the game is repeated, the question is then whether it is possible to design a compliance mechanism by which both countries are better off if they stick to abatement level 2. This was the question Asheim and Holtsmark (2009) posed. They showed, in a game with n players, that with a sufficiently low discount rate, such a punishment mechanism does exist. In this two-country case, the following compliance mechanism will make compliance the most attractive strategy if the discount rate is smaller than 0.5: if player 1 defects by abating only 1 in period t, player 2 will punish the defector by abating only 1 unit in the subsequent period, and the other way round. To see how this makes compliance attractive, assume that player 1 reverts to an abatement level of 2 from period $t + 1$ onwards, after defection in period t. With the described punishment rule, player 1 will then receive the payoffs 2.5, 1 and 2, . . . in the periods t, $t + 1$ and $t + 2$, . . . , respectively. The alternative would have been a payoff of 2 in each of these periods. It can easily be calculated that if the discount rate is lower than 100 per cent, it is worthwhile for the parties to comply with the agreement. An agreement on these terms can therefore function satisfactorily. Note also that Heitzig et al. (2011) found that if emissions trading were introduced within the same game, the punishment rule described could be designed to deter compliance irrespective of the number of countries and the size of the discount factor.

From Figure 13.1 we can also see that neither country would benefit from an agreement on further cuts in emissions beyond 2 units. If, for example, both countries agreed to reduce their emissions by 3 units, their payoff would drop to 1.5 because the additional costs of abatement would exceed the additional climate-related benefits.

5. A CLIMATE CHANGE GAME INVOLVING THREE OR FOUR COUNTRIES

The game in the previous section showed that if the world consisted of only two countries, the prospect of achieving an agreement on emission reductions should be good. In this section, we will see that it may not be as easy to reach agreement in a world consisting of more than three countries.

Assume there are n identical countries, each of which will receive the following payoff:

$$v_i = b(a_1 + a_2 + \ldots + a_n) - (c/2)a_i^2 \qquad (13.3)$$

for $I = 1, \ldots, n$, where a_i is the abatement level in country i. We continue to assume that $b = c = 1$. The results below are not affected by the choice of parameter values. The payoff for each country can be expressed by the following equation:

$$v_i = a_1 + a_2 + \ldots + a_n - \tfrac{1}{2} a_i^2 \qquad (13.4)$$

It can easily be shown that if all n countries agree to reduce their emissions by n units, all countries receive the payoff $\tfrac{1}{2} n^2$. If the abatement level is set at $n - 1$ or $n + 1$, the payoff for each country is $\tfrac{1}{2} n^2 - \tfrac{1}{2}$.

To illustrate how difficult it can be to conclude an agreement between all the countries on such an ambitious abatement target, I first examine the possibility of three or four countries succeeding and look at the depth of cooperation that is achieved.[5]

Figure 13.2 illustrates a situation where $n = 3$ and countries 1 and 2 constitute a coalition and have already concluded an agreement on adopting the same emission cuts. The left-hand number in each cell shows the payoff to countries 1 and 2 individually, and the right-hand number shows the payoff to country 3. I have omitted the option of no abatement since this is not a relevant strategy. The numbers in the bottom-right corner of each cell show the sum of the payoffs to the three countries.

Let us first assume that countries 1 and 2 agree to cut emissions by 2 units. In principle, country 3 can be a free-rider to their agreement, and will then receive a payoff of 4.5 if it chooses abatement level 1. However, countries 1 and 2 would like country 3 to join their coalition. If an abatement level of 2 is maintained, the payoff to country

Country 3 – abatement level chosen

		1		2		3		4	
Country 1 and 2 – abatement level chosen	1	2.5 2.5	7.5	3.5 2	9	4.5 0.5	9.5	5.5 −2	9
	2	3 4.5	10.5	4 4	12	5 2.5	12.5	6 0	12
	3	2.5 6.5	11.5	3.5 6	13	4.5 4.5	13.5	5.5 2	13
	4	1 8.5	10.5	2 8	12	3 6.5	12.5	4 4	12

Figure 13.2

3 will drop to 4 if it joins the agreement. To make it attractive for country 3 to join the agreement, the abatement level must be increased to 3. In this case, Figure 13.2 shows that country 3 will receive the same payoff as it would by free-riding under the original terms of the agreement. Unless country 3 is very unreasonable, the three countries should therefore find it possible to conclude an agreement on abatement level 3 (assuming again that the game is repeated and a sufficient compliance mechanism is in place).

However, an agreement by the three countries to cut emissions by more than 3 units would not be a good solution for the parties. Figure 13.2 shows, for example, that if the abatement level is increased to 4, the payoff drops to 4 for all three parties.

5.1   The Consequence of Unilateral Pledges

Figure 13.2 also tells us whether it would be a sensible move for a country to announce large cuts in emissions independently of international commitments. Assume the government of country 3 announces its intention to reduce emissions by 3 units regardless of what countries 1 and 2 do. Figure 13.2 shows that the payoff for countries 1 and 2 will be 4.5 if they follow suit and choose abatement level 3. However, since they know that country 3 will cut its emissions by 3 units regardless of what they do, their best option will be to conclude a bilateral agreement to reduce their emissions by 2 units. The payoff to countries 1 and 2 will then be 5.

Thus a pledge by the government of country 3 to make deep cuts in emissions regardless of what is agreed internationally may make it difficult to persuade countries 1 and 2 to do the same. The unilateral policy followed by country 3 will result in an overall global reduction in emissions of 7 units, whereas it would have been possible to achieve a reduction of 9 units under an international agreement.

5.2   Is Cooperation between Four Countries Possible?

Next, assume the world consists of four countries ($n = 4$), and countries 1, 2 and 3 constitute a coalition with a common abatement level. Figure 13.3 illustrates the situation for the parties to that agreement and for country 4.

Let us first assume that the three coalition members have chosen abatement level 3,

Country 4 – abatement level chosen

	1		2		3		4		5	
1	3.5	3.5 ₁₄	4.5	3 ₁₆.₅	5.5	1.5 ₁₈	6.5	−1 ₁₈.₅	7.5	−4.5 ₁₈
2	5	6.5 ₂₁.₅	6	6 ₂₄	7.5	4.5 ₂₅.₅	8	2 ₂₆	9	−1.5 ₂₅.₅
3	5.5	9.5 ₂₆	6.5	9 ₂₈.₅	7.5	7.5 ₃₀	8.5	5 ₃₀.₅	9.5	1.5 ₃₀
4	5	12.5 ₂₇.₅	6	12 ₃₀	7	10.5 ₃₁.₅	8	8 ₃₂	9	4.5 ₃₁.₅
5	3.5	15.5 ₂₆	4.5	15 ₂₈.₅	5.5	13.5 ₃₀	6.5	11 ₃₀.₅	7.5	7.5 ₃₀

Countries 1, 2 and 3 – abatement level chosen (rows 1–5)

Figure 13.3

while country 4 chooses abatement level 1, which is this country's dominant strategy. The payoffs to each of the coalition members will be 5.5, and the payoff to country 4 will be 9.5. The sum of all countries' payoffs is 26. This is a good solution for country 4, but there are better solutions for the world as a whole. If all four countries choose abatement level 4, the payoff to each of them will be 8, giving a global payoff of 32. Any increase in the abatement level beyond this will give a lower total payoff and is therefore of less interest.

The problem is that country 4 loses out by joining this agreement, since its payoff drops from 9.5 to 8. Since all four countries are identical, the same applies to each of them: for each country, the ideal solution would be for the other three to adopt an agreement on abatement level 3, while it chose abatement level 1 for itself. Thus a coalition of all four countries is not stable.

In this situation, a form of 'chicken game' can easily develop, where the aim is to be the country that is not party to an agreement. This makes it rational for the authorities in each country to give the impression that they are not very interested or willing to join a climate change agreement. It is impossible to say whether this could explain the real-world situation today, where key actors such as the USA, China and India have so far been reluctant to join negotiations on binding agreements. Their behaviour may also be the result of other strategic considerations.

In any case, we now know that a coalition of three countries is stable, whereas a coalition of four is unstable. Below, I show that this finding is true regardless of the total number of countries involved in the game, as long as we stick to the quadratic-linear model (LQ, constant marginal benefits from abatement and quadratic abatement cost functions).

5.3 The Size of Stable Coalitions in the *n* Country Case

Let us now assume there is an unspecified number (n) of countries in the world, of which k have agreed to reduce their emissions by k units each. The remaining ($n - k$) countries reduce their emissions by 1 unit each. Let v_{sk} be the payoff to a party to the agreement, while v_{nk-1} is the payoff to non-parties. From equation (13.4), we obtain the following payoffs to the parties:

$$v_{sk} = k^2 + (n - k) - \tfrac{1}{2}k^2 \tag{13.5}$$

Next, assume one country withdraws from the agreement. The $k - 1$ remaining parties to the agreement will maximize their joint welfare if they adjust their agreed abatement level to $k - 1$, while the withdrawn country will choose its dominant strategy, which is abatement level 1. The payoff to the withdrawing country will be given by the following equation:

$$v_{nk-1} = (k - 1)^2 + (n - (k - 1)) - \tfrac{1}{2} \tag{13.6}$$

From (13.5) and (13.6) we can obtain

$$v_{sk} - v_{nk-1} = \tfrac{1}{2}(k - 1)(3 - k) \tag{13.7}$$

Hence the gain (loss) resulting from participation (withdrawal) is given by the right-hand side of (13.7). This expression is negative if $k > 3$. It follows that if *an agreement includes four parties or more*, a signatory will benefit from withdrawing from it. This confirms the result of the discussion about Figure 13.3, which showed that with the LQ model the maximum stable coalition had three members.

The critical reader is likely to ask how far this result can be generalized. As I have pointed out, the values of parameters b and c do not affect the result.[6] On the other hand, the result is sensitive to choice of functional forms. In this chapter, I have chosen to use a quadratic cost function and linear benefits from abatement. Other functional forms will give different results. However, within a reasonable range of assumptions regarding functional forms and parameter values small stable coalitions will follow.[7]

6. NUMERICAL ILLUSTRATION USING 20 COUNTRIES

The world's 20 largest countries account for almost 80 per cent of global greenhouse gas emissions. Let us therefore, as a further illustration, assume there are 20 identical countries, and that the payoff function for each of them is given by

$$v_i = a_1 + a_2 + \ldots + a_{19} + a_{20} - \tfrac{1}{2} a_i^2, \text{ where } i = 1, \ldots, 20 \tag{13.8}$$

An efficient agreement in a world of 20 countries would entail all players reducing their emissions by 20 units. Each country would receive a payoff of 200, and the global emission abatement would total 400 units. If the agreed abatement level is higher than 20, the payoff to each country will be lower than 200. If the abatement level is set at 21, for example, the payoff drops to 199.5. The payoff to each country is also lower than 200 if the abatement level is lower than 20.

Without any form of cooperation, it would be profitable for each country to abate only 1 unit, giving each country a payoff of 19.5, while the global emission abatement would drop to 20 units.

Assume one of the countries, for example country 1, announces that regardless of what the other countries do, it will reduce emissions by 20 units. The other 19 countries meet to negotiate a joint abatement level, knowing that country 1 will choose abatement level 20 regardless of the outcome.

If the 19 countries decide to rise to the challenge from country 1 and also choose abatement level 20, the payoff to each of them is 200, as we saw above. However, if they instead agree on abatement level 19, the 19 parties to the agreement each receive a payoff of 200.5, and the payoff for country 1 drops to 181. So a decision by country 1 to lead the way may lessen the ambition of the international agreement, and in this case to a drop in the global abatement level from 400 to 381.

To return to the problem of stable coalitions, I showed above that it can be difficult for more than three parties to achieve an agreement because a fourth country would benefit from being a free-rider to the agreement. And an agreement between three countries will not have much effect on a global problem. Another solution might be to go for a *set* of three-country agreements. Given that $n = 20$, it is possible to draw up six three-country agreements and one two-country agreement and thus include all countries in stable coalitions. However, from the discussion of Figure 13.2 we know that each of the three-party agreements would commit the parties to reduce emissions by 3 units and the single two-party agreement would commit its two parties to reduce emissions by 2 units, cf. Figure 13.1, making the global emission abatement from these agreements 56, much lower than the efficient abatement level of 400, and only 36 units more than the Nash abatement level of 20.

For each country, the ideal situation would be to be the free-rider to an agreement between the other 19 countries. For the 19 parties to the agreement, it would be profitable to adopt abatement level 19, resulting in a payoff of 181.5. The free-rider would choose abatement level 1 and collect a payoff of 361.5. The payoff to the free-rider would therefore be about twice that to each of the 19 parties to the agreement.

However, the 19 countries do not form a stable coalition either. If one country defects, the 18 remaining parties will reduce their abatement level to 18, and the payoff to each of them will be 164. The payoff to each of the two free-riders will be 325.5.

We can continue the process: it is only when three countries are left in the agreement that the coalition is stable.

At first sight, this may appear to be a hopeless situation. However, in the next section, I illustrate the concept of stable agreement with some numerical examples. These examples indicate that in the real world, where countries are different and some countries are relatively large, things may look somewhat more promising.

7. STABLE COALITIONS WHEN COUNTRIES ARE DIFFERENT – A NUMERICAL EXAMPLE

In the preceding game analysis the countries were all identical. In the real world countries are very different, with a few large countries and a large number of smaller countries. For example, the USA, China and India account for 42 per cent of the global population. Larger countries are likely to shoulder more responsibility for the global environment than small countries because what they do matters; their large populaces will collect the benefits. Small countries, on the other hand, are more likely to conclude that whatever they do, it will not matter. So given a few stable coalitions of the world's largest countries, cooperation could have significant gains. Correspondingly, McGinty (2006) found that with asymmetric countries and side payments, stable coalitions may contain a larger number of countries and achieve larger emission reductions than coalitions of identical countries. This section illustrates this point with a simple but empirically based model.

First, a few new concepts need to be introduced. Let the n countries from now on have the following payoff functions:

$$v_i = b_i \sum_{j \in N} a_j - \frac{c_i}{2} a_i^2 , i = 1,2, \ldots, n \qquad (13.9)$$

Assume that a subset $C \subseteq N$ of countries are members (signatories) to a coalition, while the countries outside C are singletons (non-signatories). As before, a coalition maximizes its members' joint objectives. Signatories therefore abate at a level such that

$$a_i = \frac{\sum\limits_{j \in C} b_j}{c_i} \tag{13.10}$$

while non-signatories abate such that $a_i = b/c_i$.

As the countries' behaviours are determined by participation in coalition C, the payoffs could be defined as functions of C. As in the preceding sections, coalition C is said to be *internally stable* (IS) if

$$v_i(C) \geq v_i(C\backslash\{i\}) \tag{13.11}$$

for all $i \in C$ so that a signatory does not prefer to leave the coalition. However, when countries are no longer identical – as in the preceding sections – an additional stability concept is relevant. Following Carraro et al. (2006), a coalition C is said to be *potentially internally stable* (PIS) if

$$\sum_{i \in C} v_i(C) \geq \sum_{i \in C} v_i(C\backslash\{i\}) \tag{13.12}$$

So if a coalition is PIS, it generates a surplus that is large enough, by the use of side payments, to be allocated among the signatories such that no signatory can benefit from leaving the coalition. It follows that the set of IS coalitions is a subset of the set of PIS coalitions.

As mentioned above, McGinty (2006) considered the size of stable coalitions when countries are different. However, rather than using an empirically based model, McGinty (2006) considered random parameter values. In the following, a set of 16 countries/regions is studied numerically assuming the LQ setting introduced above. The parameter values are estimated by Osmani and Tol (2009). Their values are given in Table 13.2.

At this point, a note on the data is appropriate. Unfortunately the units in Osmani and Tol's (2009) model include not only single countries, but regions or large groups of countries; see Table 13.2. This is in contrast to international environmental agreements, where *single* countries represented by their governments usually are the signatories, although the EU is an exception. The level of aggregation matters and for many purposes it would give misleading results to discuss games using aggregates of countries when single countries' behaviour is to be studied. To see this, consider six identical countries with parameters $b_i = c_i = 1$, $i = 1, \ldots, 6$. This group of countries would be able to establish two internally stable coalitions of three members each, and each signatory abate 3 units. Total abatement is 18. Consider instead aggregates of countries (1, 2), (3, 4), as well as (6, 6), and give these 3 units indexes A, B and C, respectively. It follows that $b_i = 2$, and $c_i = \frac{1}{2}$, $i = A, B, C$. These 3 units could establish a single stable coalition; cf. the discussion in the preceding sections. It follows that all 3 units now will abate 12. After aggregation, then, the result changes such that a stable coalition now achieves total abatement of 36.

What we learn from this is that the level of aggregation matters. If the three pairs of countries 1 and 2, 3 and 4, and 5 and 6 *actually* act as three unions, then aggregation is appropriate. However, if the reality is that each of these six governments acts independently, the aggregation leads to a misleading and far too optimistic result.

Now consistent sets of empirically founded estimates of both abatement cost and

Table 13.2   The players and parameter values in the numerical example

			b_i	c_i	E_i
1	USA	USA	2.1965	116.19	1.647
2	Canada	CAN	0.0932	1592.12	0.124
3	Western Europe	WEU	3.1572	679.16	0.762
4	Japan and South Korea	JPK	−1.4209	967.07	0.525
5	Australia and New Zealand	ANZ	−0.0514	2159.11	0.079
6	Central and Eastern Europe	EEU	0.1013	380.70	0.177
7	Former Soviet Union	FSU	1.2724	26.43	0.811
8	Middle East	MDE	0.0474	98.00	0.424
9	Central America	CAM	0.0665	872.18	0.115
10	South America	LAM	0.2684	822.46	0.223
11	South Asia	SAS	0.3557	76.41	0.559
12	Southeast Asia	SEA	0.7316	291.24	0.334
13	China	CHI	4.3569	33.52	1.431
14	Northern Africa	NAF	0.9663	609.69	0.101
15	Sub-Saharan Africa	SSA	1.0738	419.19	0.145
16	Small Island States	SIS	0.0555	1092.86	0.038

Source: Osmani and Tol (2009).

damage cost functions are not easily retrieved. I will therefore, for illustrative purposes, use the parameter set provided by Osmani and Tol (2009), although we are dealing here with aggregates of countries.[8] The results should be interpreted with care, because the aggregation level is important for the results, as the above numerical example showed. While the numerical example might be useful in illustrating the importance of side payments and differences between countries when forming coalitions, it cannot tell us much about the real prospects of international cooperation on climate issues.

Given 16 players, there are 65 519 possible coalitions containing at least two countries. In Figure 13.4 all these coalitions are plotted, each marker representing the outcome of a coalition after applying the model described above. The efficiency index is calculated on the basis of the countries' aggregate welfare, with the index set to 1 in the efficient solution and zero with business as usual.

A total of 397 coalitions are IS when side payments between coalition members are excluded. The best-performing IS coalition when side payments are excluded is the five-country coalition of the USA, JPK, CHI, NAF and SSA. There are IS coalitions with up to nine signatories (without side payments). Note, however, that these coalitions provide small or even negative efficiency gains compared to the non cooperative equilibrium.[9]

When side payments are included, the number of (potentially) internally stable coalitions increases dramatically and some of them perform much better compared to the best-performing coalitions in cases where side payments are excluded. There are 33 107 PIS coalitions. The best-performing PIS coalition has 13 signatories and an efficiency index of 0.81, more than three times as high as the non-cooperative equilibrium (0.24). Hence the numerical example supports McGinty's (2006) finding: given asymmetries and side payments, internally stable coalitions could have considerably more than three countries and provide correspondingly larger efficiency gains.

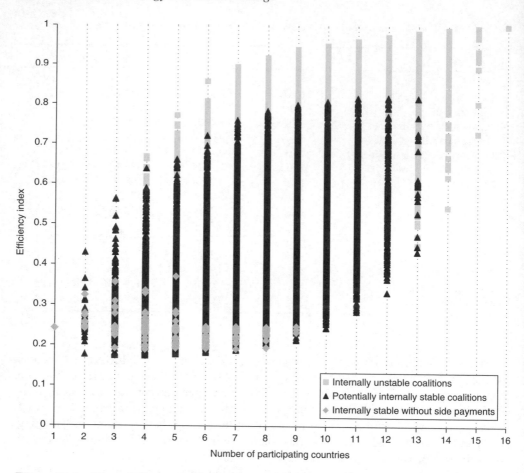

Figure 13.4   Potentially internally stable coalitions, internally stable coalitions and unstable coalitions with respect to number of signatories and efficiency

It should, however, be noted that the side payments necessary to make internally unstable coalitions stable might be difficult to implement in practice. That is because the typical receiver of side payments in these agreements is a country with low damage costs, while a country supposed to pay side payments typically is a country with high damage costs. For example, not only does China not receive side payments in any of the agreements; it has to pay other countries that are less affected by climate change to keep them on board.

Hence, as Godal and Holtsmark (2011a) show, the stability of these coalitions depends on the willingness of countries mostly affected by climate change, which most likely are developing countries, to pay the less affected countries, which most likely are richer developed countries. Unless the climate change victims are willing to make such side payments, participation in the agreement would not be in the interest of the less affected countries.

8. CONCLUSIONS

In this chapter, games have thrown light on whether the international community can get its act together to address the problem of climate change. The basic obstacle is the strong incentives to free-ride at the expense of parties to an international agreement on emission abatement. A single country is in a favourable situation if it does nothing or very little while as many other countries as possible agree to reduce their emissions substantially. We should therefore not be surprised if political leaders develop creative arguments explaining why precisely their country should not commit to cuts. And we may be able to recognize strategies of this kind in the real world today.

Because free-riders are in such a favourable situation, there is a high probability that one country after another will defect from an ambitious international agreement and fail to meet its commitments. The larger the number of parties to an agreement and the deeper the specified emission cuts, the greater the benefits of defecting and becoming a free-rider.

However, others are following a very different approach. For example, the Norwegian government has announced that it will cut emissions substantially and unilaterally, bringing the country to carbon neutrality by 2030. Similarly, the EU has announced that it intends to cut emissions significantly irrespective of cuts by other countries.

It is paradoxical that, according to the modelling framework used in this chapter, such well-meant unilateral pledges in fact make it *more difficult* to achieve a comprehensive international agreement, thus facilitating a global rise in emissions, the exact opposite of what is intended.

Another conclusion is that international cooperation and international agreements will be the key to dealing with climate change. If each country acts separately, the impact on emission levels will be far too small. And even if small groups of countries establish a set of separate agreements in order to get around the problem of small stable coalitions and in that way involve the majority of countries in an agreement, emission reductions are likely to be small.

A key underlying assumption of this chapter is that states will always act to protect their own interests. However, I do not therefore assume that governments are ignoring the problem of climate change. On the contrary, addressing climate change through emission abatement is modelled as a public good in the payoffs to all countries. Barrett (2003) includes a thorough discussion of the relevance of the assumption that countries act in their own best interests, and describes several examples showing that, in practice, it is difficult for governments to base their policies on anything else than their country's own interests.

NOTES

* I would like to thank Geir B. Asheim, Roger Fouquet, Katinka Holtsmark and Jon Hovi for their valuable comments to different draft versions, and Chris Saunders for editing.
1. At the time of writing, the carbon price for the EU market's second period, 2008–12, was about €6 per tonne CO_2.
2. As Copeland and Taylor (2009) have noted, Hardin primarily popularized and raised awareness of the problems attending on resource management. Hardin (1968) did not provide a complete analysis by any

means of the problems arising from free access to a resource. That analysis was conducted by H. Scott Gordon (1954), however. For a recent critical comment on Hardin (1968), see Walker (2009).

3.   A linear relationship between the damage caused by emissions (and thus the benefits of emission abatement) and the volume of emissions is unlikely. Nevertheless, the approach may provide a reasonable approximation.

4.   This is a frequently used functional form in the relevant literature; see for example Barrett (1994, 2003), Osmani and Tol (2009), Godal and Holtsmark (2011a).

5.   Cooperation depth means the abatement level the countries agree to.

6.   If we do not assign numerical values to b and c, equation (13.7) will read as follows: $v_{sk} = v_{nk-1} + \frac{1}{2}b^2(k - 1)(3 - k)/c$. Hence the gain resulting from participation is equal to $\frac{1}{2}b^2(k - 1)(3 - k)/c$. This expression is non-negative if $1 \leq k \leq 3$. A coalition of more than three countries will therefore be unstable irrespective of the size of b and c, provided they are positive; see Barrett (1994, 2005) and Hoel (1992).

7.   A commonly applied theoretical model in the literature has a type of abatement cost functions where the countries have only two choices, no abatement or abatement at a predetermined level (usually set to 1). In this type of binary abatement choice model, the maximum size of the stable coalition depends on the parameter values, and very large stable coalitions may exist, in principle containing all countries in the world; see for example Barrett (2003). These results do not carry over to continuous-choice models with increasing marginal abatement costs.

8.   Osmani and Tol (2009) did not mention whether their choice of aggregation level necessarily influences their results, and if so, to what extent.

9.   The reason we have negative efficiency gains is related to the negative marginal damage costs of JPK and ANZ. Because these countries collect net benefits from climate change, according to Osmani and Tol (2009), they will subsidize emissions. If JPK and ANZ are able to influence the design of an agreement, it is in their interest to minimize any abatement efforts.

REFERENCES

Asheim, G.B., and Holtsmark, B. (2009), 'Renegotiation-proof climate agreements with full participation – conditions for Pareto efficiency', *Environmental and Resource Economics*, **43**, 519–33.

Barrett, S. (1994), 'Self-enforcing international environmental agreements', *Oxford Economics Papers*, **46**, 878–94.

Barrett, S. (2003), *Environment & Statecraft – The Strategy of Environmental Treaty-Making*, New York: Oxford University Press.

Barrett, S. (2005), 'The theory of international environmental agreements', in *Handbook of Environmental Economics*, **3**: 1457–516.

Böhringer, C. (2002), 'Climate politics from Kyoto to Bonn: from little to nothing?', *The Energy Journal*, **23** (2), 51–72.

Carbone, J.C., Helm, C. and Rutherford, T.F. (2009), 'The case for international emission trade in the absence of cooperative climate policy', *Journal of Environmental Economics and Management*, **58**, 266–80.

Carraro, C. and Siniscalco, D. (1993), 'Strategies for the international protection of the environment', *Journal of Public Economics*, **52**, 309–28.

Carraro, C., Eyckmans, J. and Finus, M. (2006), 'Optimal transfers and participation decisions in international environmental agreements', *Review of International Organizations*, **1**, 379–96.

Chander, P. (2007), 'The gamma-core and coalition formation', *International Journal of Game Theory*, **35**, 539–56.

Chander, P. and Tulkens, H. (1995), 'A core-theoretic solution for the design of cooperative agreements on transfrontier pollution', *International Tax and Public Finance*, **2**, 279–93.

Copeland, B.R., and Taylor, M. Scott (2009), 'Trade, tragedy, and the commons', *American Economic Review*, **99**, 725–49.

Crampton, P. and Stoft, S. (2010a), 'International climate games: from caps to cooperation', Research Paper 10-07, Global Energy Policy Center, USA (http://www.global-energy.org/lib).

Crampton, P. and Stoft, S. (2010b), 'Price is a better climate commitment', *The Economists' Voice*, **7**(1), Article 3.

d'Aspremont, C., Jaquemin, A., Gabszewicz, J.J. and Weymark, J. (1983), 'On the stability of collusive price leadership', *Canadian Journal of Economics*, **16**, 17–25.

Godal, O. and Holtsmark, B. (2011a), 'On the efficiency gains of emissions trading when climate deals are non-cooperative', SNF Working Paper 17/11.

Godal, O. and Holtsmark, B. (2011b), 'Emissions trading: merely an efficiency neutral redistribution away from climate change victims?', *Scandinavian Journal of Economics*, **113**, 784–97.

Golombek, R. and Hoel, M. (2005), 'Climate policy under technology spillovers', *Environmental & Resource Economics*, **31**, 201–27.

Gordon, H. Scott (1954), 'The economic theory of a common-property resource: the fishery', *Journal of Political Economy*, **62**, 124–42.

Hardin, G. (1968), 'The tragedy of the commons', *Science*, **162**, 1243–48.

Heitzig, J., Lessmann, K. and Zou, Y. (2011), 'Self-enforcing strategies to deter free-riding in the climate change mitigation game and other repeated public good games', *Proceedings of the National Academy of Sciences*, **108**, 15739–44.

Helm, C. (2003), 'International emissions trading with endogenous allowance choices', *Journal of Public Economics*, **87**, 2737–47.

Hoel, M. (1991), 'Global environmental problems: the effects of unilateral actions taken by one country', *Journal of Environmental Economics and Management*, **20**, 55–70.

Hoel, M. (1992), 'International environment conventions: the case of uniform reductions of emissions', *Environmental and Resource Economics*, **2**, 141–59.

Hoel, M. (1997), 'International coordination of environmental taxes', in C. Carraro (ed.), *New Directions in the Economic Theory of the Environment*, Cambridge: Cambridge University Press.

Hoel, M., and Karp, L. (2001), 'Taxes and quotas for a stock pollutant with multiplicative uncertainty', *Journal of Public Economics*, **82**, 91–114.

Hoel, Michael, and Karp, Larry (2002), 'Taxes versus quotas for a stock pollutant', *Resource and Energy Economics*, **24**(4), 367–84.

Holtsmark, B. and Sommervoll, D.E. (2012), 'Welfare implications of emissions trading when endowments are endogenous', Discussion Paper 542, Statistics Norway, Oslo, Norway.

Hovi, J. and Holtsmark, B. (2006), 'Cap-and-trade or carbon taxes? The feasibility of enforcement and the effects of non-compliance', *International Environmental Agreements: Politics, Law and Economics*, **6**(2), 137–55.

IPCC (2007), *The Physical Science Basis. Contribution of Working Group I to the Fourth Assessment Report of the International Governmental Panel on Climate Change*, Cambridge and New York: Cambridge University Press.

Lloyd, W.F. (1833), *Two Lectures on the Checks to Population*, Oxford: Oxford University Press.

McGinty, M. (2006), 'International environmental agreements among asymmetric nations', *Oxford Economic Papers*, **45**, 45–62.

Osmani, D. and Tol, R. (2009), 'Toward farsightedly stable international environmental agreements', *Journal of Public Economic Theory*, **11**, 455–92.

Stiglitz, J.E. (2006), 'A new agenda for global warming', *The Economists' Voice*, **3**(7), Art. 3, available at http://www.bepress.com/ev/vol3/iss7/art3.

Walker, P.A. (2009), 'From "tragedy" to commons: how Hardin's mistake might save the world', *Journal of Natural Resources Policy Research*, **1**, 283–6.

Wara, Michael W. and Victor, David G. (2008), 'A realistic policy on international carbon offsets', Working Paper No.74, Program on Energy and Sustainable Development. Stanford University.

14 Long live the Kyoto Protocol!
Richard S.J. Tol

1. INTRODUCTION

The international climate negotiations are widely considered to have stalled, if not failed. The main piece of international legislation on climate policy, the Kyoto Protocol, seems to have had little impact on emissions (Pielke, 2010). The 15th Conference of the Parties (COP15) to the United Nations Framework Convention on Climate Change, in Copenhagen in December 2009, almost collapsed. The Copenhagen Accord was negotiated at the last minute by a handful of countries; it contains a long-term pledge but no concrete commitments. The main result of COP16, in Cancun in 2010, was that negotiations would continue in Durban in 2011. Expectations for COP17 are low.

I take exception to this view. In this chapter I argue that, as it stands, international legislation for climate policy, here referred to as the Kyoto Protocol, is roughly what is needed. Obviously, many people are disappointed by the lack of progress but the Kyoto Protocol is, in my opinion, the best way forward given the current circumstances.

I build up that argument as follows. Climate change is an international and inter-generational externality. It would be great if there were global, intertemporal planners to impose the appropriate carbon tax. That is not the world we live in. I review these arguments in Sections 2 and 3. Unfortunately, although well rehearsed, these arguments have fallen on deaf ears with the international negotiators, who are trying to agree what cannot be agreed and to enforce what cannot be enforced. I then turn, in Section 4, to the three key elements of a feasible international climate regime: data sharing, pledge and review, and flexibility mechanisms. In Section 5, I argue that these three elements are there already. The international climate policy regime is thus complete. It is not perfect, but it is complete.

2. GAMES OF INTERNATIONAL ENVIRONMENTAL AGREEMENTS

The failure of Copenhagen and Cancun was widely anticipated. A cynic would wonder why COP15 or COP16 would succeed where COP1 to 14 failed.

A game theorist would point out that it is difficult to provide a global public good (such as greenhouse gas emission reduction) and that international environmental agreements cannot be stringent (Carraro and Siniscalco, 1992, 1993). In 1994, Scott Barrett showed that stable treaties have either a large number of signatories who commit to very little, or a small number of signatories who commit to more ambitious targets (Barrett, 1994). Nordhaus and Yang (1996) confirm this using a numerical model. Despite the best efforts of a fair number of smart people to prove Barrett wrong, his result still stands. Indeed, Brazil, China, India, South Africa and the USA ignored the other 188 countries in Copenhagen and agreed their own accord. Similarly, the Kyoto Protocol commits only a small number of countries to greenhouse gas emission reduction.

Finus and Rundshagen (1998) and Yang (2003) study the effect of renegotiation on climate treaties. Chen (1997) and Eyckmans and Tulkens (2003) analyse the impact of transfers. Dellink et al. (2008) examine the stability of climate coalition for a wide range of parameter choices. Escapa and Gutierrez (1997) discuss alternative solution concepts. Asheim et al. (2006) and Osmani and Tol (2010) consider multiple coalitions. Kemfert et al. (2004) study the impact of international emissions and costs spillovers. Lise and Tol (2004) and Osmani and Tol (2009) analyse farsightedly stable coalitions. These studies reach the same conclusion: it is unlikely that there will an international treaty that effectively caps greenhouse gas emissions.

The reason for these results is simple. Climate change is an international externality. Greenhouse gas emission reduction is therefore a public good. The costs of abatement fall, by and large, on individual economies, while the benefits are shared by all. It is therefore preferable if others reduce their emissions, but domestic emission reduction is less attractive.

3. TIME

Most greenhouse gases stay in the atmosphere for a long time. Methane is the exception, with an e-folding time of only about 11 years. For nitrous oxide, it is 120 years. The situation for carbon dioxide is more complicated. A fair share (10 per cent) is removed from the atmosphere within a few years, but some 13 per cent of carbon dioxide emissions are removed by rock weathering, that is at a geological time scale. For human purposes, a fraction of the carbon dioxide emitted stays in the atmosphere for ever and will continue to affect the climate in perpetuity.

We call our planet 'Earth' but 70 per cent of the surface is covered in water. On average, the ocean is some three kilometres deep. Water has a much greater heat capacity than air and land. The temperature of the atmosphere is therefore largely driven by the temperature of the ocean. Greenhouse gases warm the atmosphere. Most of the additional heat dissipates into the ocean. This will continue until ocean and atmosphere reach a new equilibrium. That is a process that takes hundreds of years, if not longer.

For these reasons – long atmospheric residence times and slow response rates – climate change is a slow problem that plays out over centuries and millennia.

Climate policy is slow too. Capital is long-lived. For example, power generation is a major source of emissions. Power plants have an economic lifetime of 25 years or more. Basic technologies change only slowly. For example, transport is another major source of emissions. The internal combustion engine is still much as it was at the turn of the twentieth century. Demand for fundamental services evolves slowly too. Transport demand is largely determined by where people live and work. Settlement patterns do not change very rapidly. As another example, dietary preferences do not change much either. Meat, dairy and rice consumption are responsible for a large amount of emitted methane. In fact, methane emissions are intrinsic to meat and dairy production. Methane is the evolutionary response of ruminants to their grass-based diet. Similarly, carbon dioxide is fundamental to the chemical process that generates energy from the combustion of fossil fuels.

For these reasons, climate policy will only gradually reduce emissions. Both the climate problem and its solution are slow.

This has two implications. First, climate policy is an investment in the future. Costs are incurred in the short run, and benefits accrue in the long run. The discount rate is thus one of the most important parameters in any climate policy analysis. Although many people have argued that future generations are as worthy as the current generation (Azar and Sterner, 1996; Broome, 1992; Cline, 1993; Fearnside, 2002; Koopmans, 1967; Ott, 2003; Quiggin, 2008; Ramsey, 1928; Stern, 2008, 2010; Stern et al., 2006; Stern and Taylor, 2007), and some have argued that future generations are more worthy (Davidson, 2006, 2008), all empirical evidence shows that people are impatient (Arrow et al., 1996; Beckerman and Hepburn, 2007; Birdsall and Steer, 1993; Bradford, 2001; Cole, 2008; Evans and Sezer, 2005; Frederick et al., 2002; Lind, 1995; Lind and Schuler, 1998; Pearce et al., 2003; Schelling, 1995, 2000). It will therefore be hard to justify a substantial investment in climate policy.

The second implication is that the success of climate policy will depend on the actions of future policy makers. Future policy makers will make their own trade-offs and set their own priorities. Current policy makers cannot commit future policy makers to any action, but they can of course influence their behaviour. Policy makers genuinely interested in improving the climate should therefore focus on reducing the costs of emission abatement, and avoid policies that are expensive show-offs. Policies that create sustained revenues are more likely to be perpetuated than policies that support a small lobby group. Polarization is unwise. In a democracy, the opposition is likely to gain power sooner or later. A modest climate policy supported by sober advice is more likely to obtain cross-party support than an assertive policy based on exaggerated claims.

4.   IN SEARCH OF ALTERNATIVES

The question is what to do now that COP1 to 16 have failed to deliver an international treaty on greenhouse gas emission reduction. The fundamental positions of the main countries involved in climate policy are unlikely to change much over the next few years, so the next rounds of the negotiations will probably be a repetition of the moves in Copenhagen and Cancun – with a similar non-result. Indeed, I argue in Sections 2 and 3 that the problems with the climate negotiations are structural.

There is a plan B. The Kyoto Protocol was initiated in 1997, completed in 2001, and entered into force in 2005. There are two components to the Kyoto Protocol. First, some countries have emissions targets for the period 2008–12. Second, there are 'international flexibility mechanisms'. There is no sunset clause in the Kyoto Protocol. The targets will become obsolete in 2013, but the mechanisms will remain in force. The Kyoto Protocol is therefore the default option should future climate negotiations fail to deliver. The USA is the only substantial country that has not ratified the Kyoto Protocol. However, the USA did not reject it either: the Kyoto Protocol was signed by the president but never put to a vote in the Senate. The USA primarily objects to the targets in the Kyoto Protocol, so that a ratification after 2012 would be uncontroversial.

5.   FEASIBLE CLIMATE POLICY

Is the default any good? In order to answer that question, we need to look at the best international climate regime that is feasible. The best, but infeasible, option is a global carbon tax (Nordhaus and Yang, 1996).

How would one structure international climate policy? There are three crucial ingredients – domestic support, international reassurance and flexibility – discussed in turn below.

5.1   Domestic Support for Emission Abatement

It is futile to try to negotiate legally binding targets for sovereign nations (cf. Section 2). This is why the Kyoto Protocol has targets for a minority of countries only (the rich ones); and why some of these countries did not ratify the treaty (USA) or withdrew after ratification (Canada). The basis for climate policy lies in a domestic demand for climate policy – not in a multilateral agreement or a supranational command.

A number of countries have shown a willingness to impose a domestic emission reduction policy, and maintain this for years in the absence of a strong international treaty. Norway, for instance, introduced a carbon tax in the early 1990s. Germany, the Netherlands, and the UK have had domestic policies for two decades now, despite large shifts in the electoral fortunes of political parties. Climate policy apparently meets with approval by the population, as revealed by opinion polls (Brewer, 2004; Kasemir et al., 2000; Kim, 2011; Lofgren and Nordblom, 2010; Nisbet and Myers, 2007; Pidgeon et al., 2008; Reiner et al., 2006) and willingness-to-pay estimates (Berrens et al., 2006; Cai et al., 2010; Cameron, 2005; Lee and Cameron, 2008; Li et al., 2004; Viscusi and Zeckhauser, 2006) even if such action is unilateral.

5.2   International Confidence Building

However, unilateral emission reduction is expensive (Babiker, 2001, 2005; Bernard and Vielle, 2009; Boehringer et al., 1998; Boehringer et al., 2006; Boehringer et al., 2010; Hoel, 1991; Kuik, 2005; Kuik and Gerlagh, 2003). Emission abatement increases the cost of energy, and this puts a country at a competitive disadvantage at product and capital markets. A country would therefore hesitate to ambitiously reduce its emissions if it suspected that other countries would not follow. A rational national planner would need to know about the climate policies of its main trading partners before it could decide on the stringency of its own policy. An effective international climate policy thus requires a forum through which countries can gain confidence in each other's climate policies. The annual meetings of UN Framework Convention on Climate Change provide a platform for such a system of pledge and review.

The UNFCCC also regulates reporting of standardized emissions data, so that countries can monitor each other's progress.

5.3   International Flexibility Mechanisms

The costs of emission reduction vary widely between countries. Roughly, the demand for lowering emissions is greater in richer countries, while it is cheaper to reduce emissions in poorer countries. The third element required for international climate policy, therefore, is a mechanism to arbitrage supply and demand. There are three options (bar international tax harmonization, which is a non-starter even in the EU). The first option is a multilateral fund, financed by voluntary contributions, that would purchase emission reduction wherever it is cheapest (Bradford, 2008). This option is used, *inter alia*, for peacekeeping operations, for health care and for development aid. The World Bank and regional development banks are already engaged in this for greenhouse gas emission reduction.

Another option is to link two or more domestic markets for emission permits (Anger, 2008; Flachsland et al., 2009; Haites and Mehling, 2009; Jaffe et al., 2009; Rehdanz and Tol, 2005; Tuerk et al., 2009). For a brief period before the European Trading System was created, there was bilateral trade in emission permits between Denmark and the UK. Permits are government licences rather than commodities, so governments would need to recognize each other's permits. Permit trade is as advantageous as trade in commodities. Indeed, it is profitable for countries with low emission abatement costs to create a permit market just for the export opportunities (Tol and Rehdanz, 2008). Exchange rates are needed to account for differences in the legal definition of a 'tonne of carbon dioxide', and perhaps for differences in monitoring, enforcement and stringency (Rehdanz and Tol, 2005). Rating agencies could play a useful role in that.

The Kyoto Protocol also has 'international flexibility mechanisms', through which private and public actors create and trade emission reduction credits (Ellis et al., 2007; Greiner and Michaelowa, 2003; Holm Olsen, 2007; Michaelowa and Dutschke, 1998; Michaelowa and Jotzo, 2005), also in countries without emissions targets. This has created a burgeoning industry with a lively trade in emission credits, particularly under the Clean Development Mechanism (CDM). The CDM is overly bureaucratic, which brings high transaction costs and excludes least developed countries. The large bureaucracy has been unable to stop fraud and abuse (Wara, 2007, 2008), which may be teething problems.

All three flexibility options are needed to cater for the vastly different demands and capabilities of countries. The multilateral route is probably the only feasible one in least developed countries, as multilateral agencies handle most of the project management. The CDM has proved to be reasonably successful in emerging economies. Joined-up domestic permit markets are an option for countries with an emissions cap and the wherewithal to regulate a market in government licences.

6.   CONCLUSION

In sum, there is domestic support for greenhouse gas emission reduction, a global public good. The UN Framework Convention on Climate Change provides a platform through which countries can gain confidence in each other's climate policies. The Kyoto Protocol created mechanisms through which countries could invest in emission abatement abroad. We have all the tools needed for international climate policy: electoral demand for action; international monitoring of emissions data; a forum to pledge domestic action and review progress; and international flexibility in emission reduction.

Future international climate negotiations should therefore focus on refining existing agreements, particularly on the instruments of international climate policy. This would quickly bore the media circus, so that negotiators could focus on the issue at hand instead of trying to impress the voter at home. The most fragile element is the appetite for climate policy among the electorates, even if that means higher taxes and dearer energy.

This is a radical conclusion. Much policy effort is spent on an international climate agreement, with the tens of thousands of people attending the annual COP and almost continuous preparatory meetings. I argue here that the international climate policy regime is complete. Much of that effort is wasted. Instead, the international negotiations should focus on refining the CDM while countries should focus on domestic climate policy.

REFERENCES

Anger, N. (2008), 'Emissions trading beyond Europe: linking schemes in a post-Kyoto world', *Energy Economics*, **30**(4), 2028–49.

Arrow, K.J., W.R. Cline, K.-G. Maeler, M. Munasinghe, R. Squitieri and J.E. Stiglitz (1996), 'Intertemporal equity, discounting, and economic efficiency', in J.P. Bruce, H. Lee and E.F. Haites (eds), *Climate Change 1995: Economic and Social Dimensions – Contribution of Working Group III to the Second Assessment Report of the Intergovernmental Panel on Climate Change*, Cambridge: Cambridge University Press, pp. 125–44.

Asheim, G.B., C. Bretteville Froyn, J. Hovi and F.C. Menz (2006), 'Regional versus global cooperation for climate control', *Journal of Environmental Economics and Management*, **51**, 93–109.

Azar, C. and T. Sterner (1996), 'Discounting and distributional considerations in the context of global warming', *Ecological Economics*, **19**, 169–84.

Babiker, M.H. (2001), 'Subglobal climate-change actions and carbon leakage: the implication of international capital flows', *Energy Economics*, **23** (2), 121–39.

Babiker, M.H. (2005), 'Climate change policy, market structure, and carbon leakage', *Journal of International Economics*, **65**(2), 421–45.

Barrett, S. (1994), 'Self-enforcing international environmental agreements', *Oxford Economic Papers*, **46**, 878–94.

Beckerman, W. and C.J. Hepburn (2007), 'Ethics of the discount rate in the Stern Review on the Economics of Climate Change', *World Economics*, **8**, 187–208.

Bernard, A. and M. Vielle (2009), 'Assessment of European Union transition scenarios with a special focus on the issue of carbon leakage', *Energy Economics*, **31**(Suppl. 2), S274–S284.

Berrens, R.P., A.K. Bohara, H.C. Jenkins-Smith, C.L. Silva and D.L. Weimer (2006), 'Information and effort in contingent valuation surveys: application to global climate change using national Internet samples', *Journal of Environmental Economics and Management*, **47**, 331–63.

Birdsall, N. and A. Steer (1993), 'Act now on global warming – but don't cook the books', *Finance & Development*, **30**(1), 6–8.

Boehringer, C., C. Fischer and K.E. Rosendahl (2010), 'The global effects of subglobal climate policies', *B.E. Journal of Economic Analysis and Policy*, **10**(2), 13.

Boehringer, C., A. Loeschel and T.F. Rutherford (2006), 'Efficiency gains from "what"-flexibility in climate policy: an integrated CGE assessment', *Energy Journal* (Multi-Greenhouse Gas Mitigation and Climate Policy Special Issue), 405–24.

Boehringer, C., A. Voss and T.F. Rutherford (1998), 'Global CO_2 emissions and unilateral action: policy implications of induced trade effects', *International Journal of Global Energy Issues*, **11**(1–4), 18–22.

Bradford, D.F. (2001), 'Time, money and tradeoffs', *Nature*, **410**, 649–50.

Bradford, D.F. (2008), 'Improving on Kyoto: greenhouse gas control as the purchase of a global public good', in R. Guesnerie and H. Tulkens (eds), *The Design of Climate Policy*, Cambridge: MIT Press, pp. 13–36.

Brewer, T.L. (2004), 'US public opinion on climate change issues: implications for consensus-building and policymaking', *Climate Policy*, **4**(4), 359–76.

Broome, J. (1992), *Counting the Cost of Global Warming*, Cambridge: White Horse Press.

Cai, B., T. Cameron and G. Gerdes (2010), 'Distributional preferences and the incidence of costs and benefits in climate change policy', *Environmental and Resource Economics*, **46**(4), 429–58.

Cameron, T.A. (2005), 'Individual option prices for climate change mitigation', *Journal of Public Economics*, **89**(2–3), pp. 283–301.

Carraro, C. and D. Siniscalco (1992), 'The international dimension of environmental policy', *European Economic Review*, **36**, 379–87.

Carraro, C. and D. Siniscalco (1993), 'Strategies for the international protection of the environment', *Journal of Public Economics*, **52**, 309–28.

Chen, Z. (1997), 'Negotiating an agreement on global warming: a theoretical analysis', *Journal of Environmental Economics and Management*, **32**, 170–88.

Cline, W.R. (1993), 'Give greenhouse abatement a fair chance', *Finance & Development*, **30**(1), 3–5.

Cole, D.H. (2008), 'The Stern Review and its critics: implications for the theory and practice of benefit-cost analysis', *Natural Resources Journal*, **48**(1), 53–90.

Davidson, M.D. (2006), 'A social discount rate for climate damage to future generations based on regulatory law', *Climatic Change*, **76**(1–2), 55–72.

Davidson, M.D. (2008), 'Wrongful harm to future generations: the case of climate change', *Environmental Values*, **17**, 471–88.

Dellink, R.B., M. Finus, and N. Olieman (2008), 'The stability likelihood of an international climate agreement', *Environmental and Resource Economics*, **39**, 357–77.

Ellis, J., H. Winkler, J. Corfee-Morlot and F. Gagnon-Lebrun (2007), 'CDM: taking stock and looking forward', *Energy Policy*, **35**(1), 15–28.

Escapa, M. and M.J. Gutierrez (1997), 'Distribution of potential gains from international environmental agreements: the case of the greenhouse effect', *Journal of Environmental Economics and Management*, **33**, 1–16.

Evans, D.J. and H. Sezer (2005), 'Social discount rates for member countries of the European Union', *Journal of Economic Studies*, **32**(1), 47–59.

Eyckmans, J. and H. Tulkens (2003), 'Simulating coalitionally stable burden sharing agreements for the climate change problem', *Resource and Energy Economics*, **25**, 299–327.

Fearnside, P.M. (2002), 'Time preference in global warming calculations: a proposal for a unified index', *Ecological Economics*, **41**, 21–31.

Finus, M. and B. Rundshagen (1998), 'Renegotiation-proof equilibria in a global emission game when players are impatient', *Environmental and Resource Economics*, **12**, 275–306.

Flachsland, C., R. Marschinski and O. Edenhofer (2009), 'To link or not to link: benefits and disadvantages of linking cap-and-trade systems', *Climate Policy*, **9**(4), 358–72.

Frederick, S., G. Loewenstein and T. O'Donoghue (2002), 'Time discounting and time preference: a critical review', *Journal of Economic Literature*, **40**(2), 351–401.

Greiner, S. and A. Michaelowa (2003), 'Defining investment additionality for CDM projects – Practical approaches', *Energy Policy*, **31**, 1007–15.

Haites, E. and M. Mehling (2009), 'Linking existing and proposod GHG emissions trading schemes in North America', *Climate Policy*, **9**(4), 373–88.

Hoel, M. (1991), 'Global environmental problems: the effects of unilateral actions taken by one country', *Journal of Environmental Economics and Management*, **20**, 55–70.

Holm Olsen, K. (2007), 'The Clean Development Mechanism's contribution to sustainable development: a review of the literature', *Climatic Change*, **84**, 59–73.

Jaffe, J., M. Ranson and R.N. Stavins (2009), 'Linking tradable permit systems: a key element of emerging international climate policy architecture', *Ecology Law Quarterly*, **36**(4), 789–808.

Kasemir, B., U. Dahinden, A.G. Swartling, R. Schale, D. Tabara and C.C. Jaeger (2000), 'Citizens' perspectives on climate change and energy use', *Global Environmental Change*, **10**(3), 169–84.

Kemfert, C., W. Lise and R.S.J. Tol (2004), 'Games of climate change with international trade', *Environmental and Resource Economics*, **28**, 209–32.

Kim, S.Y. (2011), 'Public perceptions of climate change and support for climate policies in Asia: evidence from recent polls', *Journal of Asian Studies*, **70**(2), 319–31.

Koopmans, T.C. (1967), 'Objectives, constraints, and outcomes in optimal growth models', *Econometrica*, **35**, 1–15.

Kuik, O.J. (2005), 'Climate change policies, international trade and carbon leakage: an applied general equilibrium analysis', Vrije Universiteit.

Kuik, O.J. and R. Gerlagh (2003), 'Trade liberalization and carbon leakage', *Energy Journal*, **24**(3), 97–120.

Lee, J. and T.A. Cameron (2008), 'Popular support for climate change mitigation: evidence from a general population mail survey', *Environmental and Resource Economics*, **41**(2), 223–48.

Li, H., R.P. Berrens, A.K. Bohara, H.C. Jenkins-Smith, C.L. Silva and D.L. Weimer (2004), 'Would developing country commitments affect US households' support for a modified Kyoto Protocol?', *Ecological Economics*, **48**, 329–43.

Lind, R.C. (1995), 'Intergenerational equity, discounting, and the role of cost–benefit analysis in evaluating global climate policy', *Energy Policy*, **23**(4/5), 379–89.

Lind, R.C. and R.E. Schuler (1998), 'Equity and discounting in climate change decisions', in W.D. Nordhaus (ed.), *Economics and Policy Issues in Climate Change*, Washington, DC: Resources for the Future, pp. 59–96.

Lise, W. and R.S.J. Tol (2004), 'Attainability of international environmental agreements as a social situation', *International Environmental Agreements: Politics, Law and Economics*, **4**, 253–77.

Lofgren, A. and K. Nordblom (2010), 'Attitudes towards CO_2 taxation – is there an Al Gore effect?', *Applied Economics Letters*, **17**(9), 845–48.

Michaelowa, A. and M. Dutschke (1998), 'Interest groups and efficient design of the Clean Development Mechanism under the Kyoto Protocol', *International Journal of Sustainable Development*, **1**(1), 24–42.

Michaelowa, A. and F. Jotzo (2005), 'Transaction costs, institutional rigidities and the size of the clean development mechanism', *Energy Policy*, **33**, 511–23.

Nisbet, M.C. and T. Myers (2007), 'The polls – trends: twenty years of public opinion about global warming', *Public Opinion Quarterly*, **71**(3), 444–70.

Nordhaus, W.D. and Z. Yang (1996), 'RICE: a regional dynamic general equilibrium model of optimal climate-change policy', *American Economic Review*, **86**(4), 741–65.

Osmani, D. and R.S.J. Tol (2009), 'Toward farsightedly stable international environmental agreements', *Journal of Public Economic Theory*, **11**(3), 455–92.

Osmani, D. and R.S.J. Tol (2010), 'The case of two self-enforcing international agreements for environmental protection with asymmetric countries', *Computational Economics*, **36**(2), 93–119.

Ott, K. (2003), 'Reflections on discounting: some philosophical remarks', *International Journal of Sustainable Development*, **6**(1), 7–24.

Pearce, D.W., B. Groom, C.J. Hepburn and P. Koundouri (2003), 'Valuing the future – recent advances in social discounting', *World Economics*, **4**(2), 121–41.

Pidgeon, N.F., I. Lorenzoni and W. Poortinga (2008), 'Climate change or nuclear power – no thanks! A quantitative study of public perceptions and risk framing in Britain', *Global Environmental Change*, **18**(1), 69–85.

Pielke, R.A., Jr (2010), *The Climate Fix*, New York: Basic Books.

Quiggin, J. (2008), 'Stern and his critics on discounting and climate change: an editorial essay', *Climatic Change*, **89**(3–4), 195–205.

Ramsey, F. (1928), 'A mathematical theory of saving', *Economic Journal*, **38**, 543–9.

Rehdanz, K. and R.S.J. Tol (2005), 'Unilateral regulation of bilateral trade in greenhouse gas emission permits', *Ecological Economics*, **54**, 397–416.

Reiner, D.M., T.E. Curry, M.A. de Figueiredo, H.J. Herzog, S.D. Ansolabehere, K. Itaoka, F. Johnsson and M. Odenberger (2006), 'American exceptionalism? Similarities and differences in national attitudes toward energy policy and global warming', *Environmental Science and Technology*, **40**(7), 2093–8.

Schelling, T.C. (1995), 'Intergenerational discounting', *Energy Policy*, **23**(4/5), 395–401.

Schelling, T.C. (2000), 'Intergenerational and international discounting', *Risk Analysis*, **20**(6), 833–7.

Stern, N. (2010), 'Presidential address: imperfections in the economics of public policy, imperfections in markets, and climate change', *Journal of the European Economic Association*, **8**(2–3), 253–88.

Stern, N.H. (2008), 'The Economics of Climate Change', *American Economic Review*, **98**(2), 1–37.

Stern, N.H. and C. Taylor (2007), 'Climate change: risks, ethics and the Stern Review', *Science*, **317**(5835), 203–4.

Stern, N.H., S. Peters, V. Bakhski, A. Bowen, C. Cameron, S. Catovsky, D. Crane, S. Cruickshank, S. Dietz, N. Edmondson, S.-L. Garbett, L. Hamid, G. Hoffman, D. Ingram, B. Jones, N. Patmore, H. Radcliffe, R. Sathiyarajah, M. Stock, C. Taylor, T. Vernon, H. Wanjie and D. Zenghelis (2006), *Stern Review: The Economics of Climate Change*, Cambridge: Cambridge University Press.

Tol, R.S.J. and K. Rehdanz (2008), 'A no cap but trade proposal for emission targets', *Climate Policy*, **8**(3), 293–304.

Tuerk, A., M. Mehling, C. Flachsland and W. Sterk (2009), 'Linking carbon markets: concepts, case studies and pathways', *Climate Policy*, **9**(4), 341–57.

Viscusi, W.K. and R.J. Zeckhauser (2006), 'The perception and valuation of the risks of climate change: a rational and behavioral blend', *Climatic Change*, **77**, 151–77.

Wara, M. (2007), 'Is the global carbon market working?', *Nature*, **445**(7128), 595–6.

Wara, M. (2008), 'Measuring the clean development mechanism's performance and potential', *UCLA Law Review*, **55**(6), 1759–803.

Yang, Z. (2003), 'Reevaluation and renegotiation of climate change coalitions – a sequential closed-loop game approach', *Journal of Economic Dynamics & Control*, **27**, 1563–94.

15 Designing a Bretton Woods institution to address global climate change
Joseph E. Aldy

INTRODUCTION

Mitigating climate change risks will require some form of global effort to limit emissions, adapt to a changing climate, and geo-engineer the global climate system. These three policy approaches – prevention, adaptation and remediation – involve, to varying degrees, little incentive to account for the external impacts of decision-making by individuals, firms and nation-states. Thus a 'successful' international climate policy architecture, at a minimum, will need to promote collaboration among nation-states that results in sovereign policies and actions to reduce climate change risks by modifying individuals' and firms' incentives.

Successful collaboration requires more than a simple, initial agreement. The dynamics of learning about climate change and learning about the effectiveness of various risk mitigation measures will necessitate a number of rounds of collaboration among nations. Given the strong incentives for free-riding, what are the characteristics of an international climate policy architecture that could enable repeated collaboration in climate change risk reduction efforts?

A climate policy framework that represents an effective effort in combating climate change, is perceived as fair by all participating nations, and is considered legitimate and thus encourages broad participation is more likely to be successful than a regime that lacks these characteristics. An effective, fair, legitimate policy architecture promotes trust and establishes credibility of the agreement. As one nation takes actions to mitigate climate change risks, it can trust that other nations are also taking actions that will effectively reduce the risks of climate change. As one nation takes actions, it can assess the actions of its peers and determine if their actions are comparable and fair. A framework is legitimated by a positive cycle of nations taking fair, effective actions that elicits broader and more complete participation.

The failure to deliver an effective, fair, legitimate climate change policy through the Kyoto framework is evident in the inability of the international community to negotiate a successor to the Kyoto Protocol's first commitment period and the evolution toward a pledge and review system established under the 2009 Copenhagen Accord. Indeed, the excessive focus on normative goals instead of institutional design in international climate negotiations has undermined the legitimacy of the current multilateral regime, and this absence of legitimacy then facilitates questioning about fairness and effectiveness of commitments. It is not that there are disagreements over what constitutes an effective or fair approach (there are); it's that the current regime does not provide the institutional means to even assess outcomes before subjecting these outcomes to a normative assessment of effectiveness and fairness.

The information structure of the climate change policy collaboration problem necessitates the design of institutions to enhance public knowledge about nations' commitments, policies and outcomes. The international community has addressed this kind of problem in a wide array of other contexts from which lessons can be drawn and applied to international climate policy. Based on these experiences and the characteristics of a successful international climate policy architecture, this chapter proposes the design of a 'Bretton Woods climate institution' (BWCI).

Such an institution should have three primary functions to enhance the legitimacy of multilateral climate policy efforts. First, this BWCI should implement a serious system of national and global policy surveillance. This surveillance would include an evaluation by independent experts of the various policy commitments nations make in international negotiations to assess whether nations delivered on their commitments and to examine the impacts of these actions on various climate change risk reduction margins, such as emission abatement and adaptation. It would also include an aggregate assessment of the net effect of nations' efforts to judge whether these actions were sufficient to combat climate change. As in other multilateral contexts, such a surveillance scheme should be consultative in nature, to allow give and take among experts and among nations engaged in the international climate policy effort.

While some may perceive the primary role of the Bretton Woods institutions, the IMF and the World Bank, as one of financing, policy surveillance is an important part of their missions. In the international climate context, policy surveillance is critical because of the need to promote trust and build credibility of an agreement among nations about each of their respective commitments. In this sense, the envisioned BWCI may better reflect the system of international trade policy surveillance, which evolved from a failed attempt to design an international trade organization at Bretton Woods into the current regime under the World Trade Organization (WTO).

Second, the institution should promote best policy practices. This would reflect its surveillance role, from which an assessment and publication of the most effective policies could inform nations' deliberations over the kinds of policies and actions they may consider for domestic implementation. In particular, such assessments should account for the implications of legal, economic and cultural contexts that can impact the relative effectiveness of various policy instruments and actions across nations.

Third, the institution should provide a means to channel some financing for investments in climate change risk mitigation activities in developing countries. By making funds conditional on agreeing to policy surveillance, such an approach would create an incentive for transparent evaluations of policies and actions. Since the majority of international finance associated with addressing climate change will likely run through the private sector, this financing mechanism would serve as a complement to these private sector efforts. Leveraging finance to enhance transparency could also be extended by the setting of eligibility rules for existing developed country export and development finance programs. For example, the US Ex-Im Bank and the US Overseas Private Investment Corporation could agree to provide loan guarantees and political insurance for use only in developing countries that have agreed to policy surveillance under this new institution. Moreover, access to market-based climate policy schemes, such as the Clean Development Mechanism (CDM) and emission trading, could be predicated on countries agreeing to participate in policy surveillance.

The next section of this chapter draws on game theory, international relations and legal scholarship to argue why promoting policy transparency through surveillance can enhance the legitimacy of and foster more effective international climate change policy agreements. The third section details the shortcomings in the current international climate policy framework for reporting, review and verification that undermine its legitimacy. The fourth section surveys various international systems of policy surveillance on economic, trade, energy and environmental issues, and presents lessons to inform a climate policy surveillance regime. The fifth section presents the detailed proposal for a 'Bretton Woods climate institution' and discusses how such an institution could complement bottom–up and top–down approaches to international climate policy architectures. The sixth section concludes.

SURVEILLANCE AND AGREEMENTS

Public Information and Repeated Games

More than 50 years ago, Professor Schelling wrote about the economic attributes of negotiations. Signaling the seriousness of commitment is often a precondition for securing an agreement among multiple parties. This can be challenging in some contexts, such as international negotiations, given the significant deference to sovereigns and limited tools for coercing a state to take actions beyond what it intends to undertake voluntarily. Schelling suggests that transparency of a party's *ex ante* pledge and *ex post* outcome can enhance the credibility of commitments:

> A potent means of commitment, and sometimes the only means, is the pledge of one's reputation . . . But to commit in this fashion publicity is required. Both the initial offer and the final outcome would have to be known; and if secrecy surrounds either point, or if the outcome is inherently not observable, the device is unavailable. (Schelling, 1956, p. 288)

Schelling stresses that agreements may need to be structured on what is observable, even if that is only correlated with the intended objective of the negotiation, in order to ensure that one can observe compliance with the agreement. In addition, Schelling's take on the role of trust in repeated negotiations can inform the structure of climate negotiations given the dynamic nature of the climate change policy problem:

> What makes many agreements enforceable is only the recognition of future opportunities for agreement that will be eliminated if mutual trust is not created and maintained, and whose value outweighs the momentary gain from cheating in the present instance. (Ibid., pp. 301–2)

Indeed, these insights have since been formalized in a variety of papers on the so-called folk theorems in game theory, which show that repeated games with public information can yield socially efficient, stable agreements if parties to these agreements are sufficiently patient (Fudenberg and Maskin, 1986). In the context of international climate coalition games, Barrett's (1994) dismal one-period finding that the country-specific benefits and costs of climate mitigation will likely undermine the stability of any climate coalition in excess of three parties is softened in a repeated-game environment and, with sufficient patience, larger coalitions are possible.

The key to these findings is that actions by all parties are easily and perfectly observed. In the current climate policy regime, this is certainly not the case: most nations' climate policies and greenhouse gas emissions are neither reported, nor observed, nor reviewed formally under the UN Framework Convention on Climate Change (more on this in the next section).

Under weaker assumptions, agreements among parties could be sustained even with imperfect public information (e.g. Green and Porter, 1984; Fudenberg et al., 1994) and 'almost-public' private information (e.g. Malaith and Morris, 2002). Even in these cases, observation of signals of action (whether commonly shared, i.e. public or individually acquired by each party, i.e. private) is still necessary, and as these signals convey less information on a party's actions, it is more difficult to maintain a stable agreement. As Barrett (2003) notes in his discussion of monitoring in international environmental agreements, 'transparency is of fundamental importance in a repeated game' (p. 284).

The Use of Public Information in International Agreements

The information structure of repeated negotiations is a critical element determining the stability of negotiated coalitions. Keohane (1994) emphasizes the importance of information-producing institutions to facilitate collaboration among nations. In particular, he notes that 'more extensive arrangements for monitoring others' behavior' is required in collaboration games and can promote the reciprocity necessary to secure agreement in such games (Keohane, 1994, p. 20). Wettestad (2007) also notes the relationship between public information about nations' performance under an agreement and the trust it builds: 'good monitoring and verification of practices in international institutions are important in building trust between and among cooperating parties, and in strengthening wider societal confidence' (p. 975). The legitimacy of the international agreement may rest on the caliber, credibility and independence of the implementing institutions (Bodansky, 2007).

Public information about a nation's actions can empower its leaders and stakeholders to call on and pressure that nation to deliver on its commitments. Political leaders who push for their nations to take on more ambitious climate change risk reduction policies could benefit from an institution collecting and publicizing information on their actions. By providing an independent assessment of a country's effort and a comparison with the effort of its peers, regular surveillance can legitimize domestic policies (Francois, 2001).

Improving information on commitments and outcomes supports informal and formal mechanisms of peer review and peer pressure (Pagani, 2002). Nations may initiate an informal bilateral dialogue with those nations lagging far behind (to pressure them to do more) or demonstrating progressive leadership and accomplishment (to learn how to follow their lead effectively). Such information may also enable comparisons among nations that facilitate peer pressure in the next round of negotiations.

International institutions of information collection and dissemination can lower the costs of an international agreement. Some international non-governmental organizations have developed the technical capacity to assist governments with monitoring (Hempel, 1996). Technical and financial assistance for key elements of capacity building could enable improved monitoring, reporting and evaluation in developing countries (Keohane, 1994). International institutions can formally undertake monitoring that lowers the transaction costs of an agreement (Haas et al., 1993).

Some nations may send mixed or misleading signals about their outcomes through various public reporting mechanisms. A variety of means has evolved to mitigate this problem that could inform the design of a surveillance institution. Some non-governmental organizations can serve as external checks on national reporting and evaluation programs by providing their own monitoring information to the public and the media. External monitoring of a country's emissions can occur through remote-sensing technologies (Litfin, 1999) as well as through the use of correlated data sources, such as on fossil fuel consumption, economic activity and so on.

The information compiled on parties' actions and outcomes does not need to meet the standard of a legal compliance mechanism. Indeed, few nations would subject themselves to such a legal mechanism; nor do the parameters of international environmental treaties impose such a requirement. Such information may 'contain deviance within acceptable levels' (Klabbers, 2007, p. 1004). The challenge lies in designing legitimate institutions that provide the necessary public information to mitigate deviations from agreements on climate policy.

THE NEED FOR BETTER SURVEILLANCE TO INFORM GLOBAL CLIMATE CHANGE POLICY

The current international climate change policy architecture suffers from a dearth of information on countries' contributions to climate change and the effectiveness of their efforts to combat the problem. Under the UN Framework Convention on Climate Change and the Kyoto Protocol, the responsibility for mitigating emissions and reporting on emissions has fallen almost exclusively on developed countries. Indeed, the policy rationale in UNFCCC agreements for developed countries to take the first mitigation efforts steps has reflected negotiating rhetoric as opposed to the evidence about the relative current and historic contributions to climate change. The failure of what is primarily an emission mitigation treaty to track emissions of all large countries reinforces the rhetorical bias in the negotiations. Moreover, the UNFCCC provides an insufficient institutional process for evaluating countries' efforts and a single, deeply flawed metric for assessing the comparability of effort among nations.

The Disconnect Between the Rhetorical Basis for Policy and Statistical Evidence

The absence of a credible monitoring regime has created an information vacuum in which north–south rhetoric, as opposed to experience and evidence, has informed the design of policy. For example, at the first Conference of the Parties (COP-1) of the Framework Convention in 1995, the global community agreed to the Berlin Mandate (United Nations, 1995). This decision established a negotiating mandate for emission targets for developed (Annex I) countries, while explicitly noting that Non-Annex I countries would not face any new commitments. The mandate notes that this negotiating process 'shall be guided' in part by 'the fact that the *largest share of historical and current global emissions of greenhouse gases has originated in developed countries*' (United Nations, 1995; emphasis added). This contradicts independent estimates that show that annual Non-Annex I countries' greenhouse gas emissions exceeded Annex I countries' emissions by more than 12 percent in 1995 (WRI, 2012).

In the 2010 Cancun Agreements, this developed countries' 'largest share of historical' global emissions reference returned. Yet Annex I countries contributed only to 54 percent of global greenhouse gas emissions over 1900–2005, and Annex I and Non-Annex I contributions to cumulative emissions since 1900 should reach parity by 2020 (Hohne et al., 2007). When the expressly stated rationale for policy design runs contrary to scientific evidence, the legitimacy of the international agreement is undermined. It also reveals a lack of interest in scientific evidence to motivate policy and a lack of interest in collecting and analyzing scientific evidence about the climate change problem.

The Lack of Adequate Information Collection and Analysis

The UNFCCC monitoring regime is grossly inadequate. Under the treaty, nations communicate to the UN reports on their vulnerability to climate change, policies to address climate change, greenhouse gas emissions and so on. Annex I countries have submitted five national communications in the 18 years the UNFCCC has been in effect, while a majority of developing countries – including China and India – have submitted only one such report (through 1 March 2012). While developed countries submit annual emissions reports, developing countries are exempt from reporting annual greenhouse gas emissions to the UNFCCC despite the dramatic growth in such emissions.

For example, China's most recent greenhouse gas emissions report submitted to the UN Framework Convention on Climate Change is for the 1994 calendar year (UNFCCC, 2012, as of 1 March 2012). Between 1995 and 2010, China is estimated to have emitted about 75 billion metric tons of carbon dioxide from fossil fuel combustion and cement production, which exceeds the cumulative emissions of any developed country over the entire twentieth century, except for the USA (calculated by the author using data from Boden and Blasing, 2012 and Boden et al., 2011). This failure to report does not reflect some burdensome or technically challenging emissions record-keeping obstacle. Under the Montreal Protocol, China has reported annual detailed ozone-depleting substances consumption inventory data over the 1990–2010 period (UNEP, 2012).

This reflects the differentiation in the reporting and review regime under the UNFCCC. Developed countries submit annual emission reports, pursuant to established guidelines and subject to expert review. In contrast, developing countries emissions reports are made as a part of their infrequent national communications, and are neither subject to the same data standards as developed countries nor undergo expert review (Breidenich and Bodansky, 2009). Even the regular reporting of developed countries' emissions is insufficient to characterize the effectiveness of emission mitigation actions in these countries (Ellis and Larsen, 2008). The UNFCCC has not published guidelines for the review of developed countries' national communications, and these expert reviews do not provide the means to verify the impact of emission mitigation measures (Breidenich and Bodansky, 2009).

The Absence of Credible Metrics to Assess Comparability of Effort

In order for a set of countries to perceive their contribution to mitigating climate change as fair, they need a means for acquiring information on each other's actions and emissions and a basis for comparing the effort. The absence of credible, legitimate means for comparing effort plays out both in the determination of the set of countries that should take on emission commitments and in the *ex post* evaluation of countries' performance in delivering on those commitments.

Many have argued that the focus of UNFCCC mitigation goals on industrialized countries reflects their greater wealth and higher emissions (current, historic and per capita). Yet about 50 Non-Annex I countries have higher per capita incomes than the poorest Annex I country. In addition, about 40 Non-Annex I countries rank higher on the Human Development Index than the lowest-ranked Annex I country (Aldy and Stavins, 2010a). Forty-four Non-Annex I countries had higher per capita carbon dioxide emissions than the lowest-ranked Annex I country (US EIA, 2012b).

Comparing the effort to mitigate emissions remains a daunting question. The most 'successful' countries, in terms of achieving emissions below 1990 levels, are the economies in transition. The shutting down of much of the old Soviet industrial infrastructure and the transformation from planned to market economies have resulted in dramatic emission reductions. Russia's 2000 greenhouse gas emissions were 32 percent below 1990 levels. In 2007, the year before the start of the Kyoto commitment period, the old EU-15 had aggregate emissions 5 percent below 1990 levels, while the expanded EU-27 had emissions 11 percent below 1990 levels, illustrating how far below 1990 levels Central and Eastern European countries' emissions had fallen. The world cannot learn from this 'success' in designing future climate policy; these countries did not achieve the emission reductions through innovation emission abatement policies, but rather through painful economic restructuring. Thus, comparing current emissions with those of 1990 provides a very noisy signal of a nation's effort.

Is the USA a climate laggard because it has not ratified the Kyoto Protocol? By 2009, US greenhouse gas emissions were 7 percent below 1997 levels. Since the 1997 Kyoto Conference, the growth rate of US greenhouse gas emissions ranks 17th out of the 36 Annex I nations with commitments under Kyoto, and this growth rate is lower than for ten EU member states (Figure 15.1). This is not simply a function of the recent economic recession. In 2010, US fossil fuel carbon dioxide emissions were 11 percent lower (nearly 700 million metric tons of carbon dioxide lower), but GDP was about 9 percent higher than was forecast by the US Energy Information Administration in 1997 in the lead-up to the Kyoto negotiations (US EIA, 1997, 2012a; Figure 15.2).

Adequacy of Effort

The surveillance program under the UNFCCC does not report a global aggregate greenhouse gas emissions measure. Indeed, there are neither reporting nor monitoring mechanisms that would enable the UNFCCC to publish a global emission estimate. For an international treaty focused on stabilizing atmospheric greenhouse gas concentrations, it is striking that it does not track the flow of global greenhouse gas emissions.

Independent estimates of carbon dioxide emissions from fossil fuel combustion provide

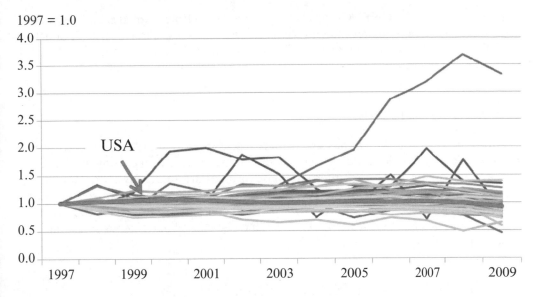

Source: UNFCCC (2012).

Figure 15.1   Annex I greenhouse gas emissions, 1997–2009

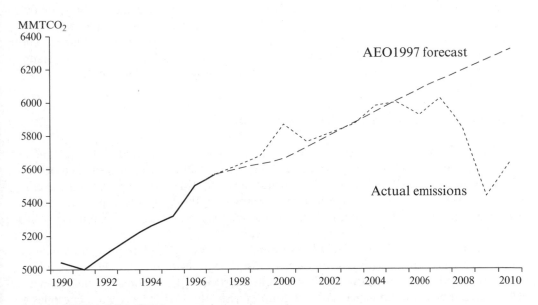

Source: US EIA (1997, 2012a).

Figure 15.2   US carbon dioxide emissions, actual 1990–2010 and forecast 1997–2010

little evidence that the UNFCCC has delivered adequate effort. Global emissions in 2010 were about 48 percent greater than they were in 1992 (Boden and Blasing, 2012; Boden et al., 2011). Global emissions grew about 2.8 percent annually in the decade after the 1997 Kyoto negotiations, dramatically faster than the 1.2 percent annual rate in the lead-up to the Kyoto Conference.

Some Recent Signs of Progress

The international community has taken some recent steps to address this reporting and analysis vacuum. World leaders agreed to design measurement, reporting and verification systems and a system of international consultations and analysis for developing countries mitigation efforts in the 2009 Copenhagen Accord. The initial steps towards implementing the leaders' vision at the 2010 Cancun and 2011 Durban climate talks highlight the potential for a more extensive and rigorous climate policy reporting and review mechanism.

LESSONS ON SURVEILLANCE FROM OTHER INTERNATIONAL POLICY REGIMES

Policy surveillance serves as a key element of a broad array of international policy regimes. International organizations undertake regular assessments of domestic policy design, implementation and outcomes in, *inter alia*, economic policy, trade policy, energy subsidies and trade in endangered species. These experiences can provide important lessons to inform the design of an effective system of climate policy surveillance. The following subsections provide brief summaries of and identify lessons from policy surveillance regimes in a variety of contexts, and a closing synthesis of international policy surveillance.

International Monetary Fund Article IV Consultations

The International Monetary Fund undertakes country-, regional- and global-level economic surveillance (IMF, 2001; Schafer, 2006). Individual country surveillance occurs annually under so-called Article IV consultations. The IMF conducts regular surveillance of the global economy – in effect, an assessment of the aggregate impact of various economic, monetary and fiscal policies of the member countries – and publishes the *World Economic Outlook* typically twice per year.

An Article IV consultation includes an annual visit by IMF economists and experts to the member country, with interim discussions when necessary. Countries are required to produce information to the IMF to enable the review of its economic environment and relevant economic policies. After a country visit, the IMF expert team compiles a report that serves as the basis for a peer review by the executive board, which includes 24 country directors representing member countries or groups of countries. A summary of the board discussion and the report are typically published, and it is the emerging norm that countries agree to the publication of their full Article IV staff report. Making public these reports enables stakeholders to push for better economic policies in their respective countries and improves the quality of the IMF review product by effectively subjecting the reviewers to external assessment (Fischer, 1999).

The IMF also supports standards for data dissemination and codes for good policy practice that can facilitate annual surveillance and also benefit member countries in their design and implementation of economic policy. Such standards provide transparent, timely and measurable metrics for evaluating policy performance and identifying potential economic vulnerabilities. The IMF emphasizes the value in implementing such standards and codes to communicate clearly to the markets and other countries on a country's economic situation. Conditioning financing from the IMF (or even more broadly from other international financial institutions) on the adoption and adherence of standards and codes of practice could improve the quality of the surveillance regime (Fischer, 1999).

OECD Economic Surveys

The OECD facilitates peer reviews of member states' economic policies every one to two years (OECD, 2003; Schafer, 2006). As a part of this effort, a team of experts from the OECD Secretariat compiles a draft report of the relevant policies for the country under review. The expert team typically visits the country under review, draws data from a variety of public and private sources, and employs the latest research to evaluate the country's economic policy program. A delegation from the country under review responds to the draft report in a meeting of all OECD member states. At this meeting, two so-called lead examiners are drawn from the membership of the OECD to initiate the discussion of the draft report's findings, the response by the country under review, and the report's recommendations for policy reforms. After these two peer reviewers have questioned the country under review, the entire membership has the opportunity to discuss various elements of economic policy with that country's delegation. The final report reflects this discussion and must secure agreement among all OECD members before it is completed. The policy reviews are then made available to the public.

A distinctive element of the OECD review mechanism is the focus on peer review. The OECD employs peer review in a number of policy contexts beyond this example of economic policy review, and it has identified several structural elements common to its peer review processes, including an agreed set of principles, standards and criteria for evaluating performance, and an assessment of a country's performance in implementing policy recommendations. Providing a forum for member states to engage one another through peer review can facilitate learning about effective policy practice and promote understanding about countries' individual policy design and implementation. As a review among peers, it clearly serves as a facilitative process, not a tribunal or compliance mechanism, and thus could enable more candid dialogue among participants.

WTO Trade Policy Review Mechanism

WTO members are subject to a regular review of their trade policies (Mavroidis, 1992). The four countries with the largest share of world trade undergo policy review every two years, while the next 16 participate in reviews every four years, and all other countries take part in reviews every six years. Thus differentiation in the frequency of review reflects a country's relative contribution to global trade. The objectives of the Trade Policy Review Mechanism (TPRM) are to promote adherence to commitments in multilateral and plurilateral trade agreements and increase transparency to facilitate a smoother functioning trade regime (Annex 3, Marrakech Accord).

The TPRM involves the preparation of a report by a given member country on its trade policies and a report drafted by a team of experts at the WTO Secretariat. The WTO Trade Policy Review Body provides guidance for country reporting, including a template to structure the regular reports to the WTO. In addition, developing countries may solicit technical assistance from the Secretariat in preparation of their trade policy reports. The WTO Secretariat expert team typically visits the country under review and draws from multiple sources – government data, third-party data sources, data in the public domain – in compiling its assessment of the country's trade policy.

An active give-and-take occurs with the country under review through two procedures. First, the country under review must respond in writing to questions it receives from the WTO expert review team. Second, the Trade Policy Review Body, comprising all WTO members, receives the Secretariat's report on the country under review and then hosts a delegation from that country to participate in a meeting to discuss and explain the findings in the report. The final versions of the reports submitted by the country under review and the WTO expert review team are published.

Montreal Protocol Reporting on Ozone-depleting Substances

Under the Montreal Protocol and related agreements focused on reducing the production and consumption of ozone-depleting substances (ODS), countries report regularly on such ODS outcome data to the United Nations Environment Programme. Up to March 2012, more than 190 countries have reported their ODS consumption data to UNEP for the year 2010, and more than 180 countries have reported annual ODS consumption for at least the past 15 years (UNEP, 2012). This enables credible estimation of global ODS consumption, informed assessment of consumption over time both globally and by nation to illustrate performance relative to policy goals, and a public, transparent record to facilitate identification of policy leaders and laggards. This stands in sharp contrast to the UNFCCC process, in which only three developing countries (two of which are current OECD members, but classified as Non-Annex I under the UNFCCC) had submitted more than one national communication, the document that includes developing-country emissions data and hence only one submission of a national greenhouse gas emission inventory, by the time of the 2009 Copenhagen climate talks (Breidenich and Bodansky, 2009).

The Montreal Protocol surveillance program has a well-defined process and data standards for reporting, as well as a system of reviews of national policies that has strengthened the monitoring and transparency of the agreement (Hampson, 1995). The Montreal Protocol Secretariat analyzes the annual national reports and publishes summary compliance reports before the annual international negotiations under this agreement. In some cases, experts employed data compiled by the World Meteorological Organization's Global Ozone Observing System, established in 1957, to verify the ODS data submitted by individual nations (Wettestad, 2007). This enhanced system of reporting and review can facilitate 'naming and shaming' in the context of an agreement that lacks a strong compliance mechanism, which is true of most international agreements (Hampson, 1995).

The Secretariat also serves to enhance the capacity, especially in developing countries, to monitor and report their ODS consumption data. By elaborating procedures for tracking and reporting ODS data, providing data-reporting templates, and explaining ways to improve domestic monitoring, the Secretariat facilitated developing-country reporting. While these efforts lowered the costs and eliminated potential barriers to timely reporting, the Montreal Protocol surveillance process also increased the costs for failing to submit adequate ODS data reports by linking access to financing for projects to reduce ODS consumption to satisfying the reporting requirements (Wettestad, 2007).

G-20 Fossil Fuel Subsidies Agreement Implementation and Review

In contrast to the previous policy surveillance programs that focus on a reporting and review infrastructure built on treaty organizations' own cadre of experts, several international agreements rely on external, third-party experts to review countries' actions. First, consider the 2009 Pittsburgh G-20 commitment to eliminate fossil fuel subsidies (IEA et al., 2010). The leaders of the 20 largest developed and developing nations agreed to 'phase out and rationalize over the medium term inefficient fossil fuel subsidies while providing targeted support to the poorest' (G-20 Leaders, 2009). The G-20 leaders also called on all nations to eliminate their fossil fuel subsidies, and later that fall at the 2009 Singapore APEC meeting, leaders also echoed the call to phase out such subsidies.

Leaders have agreed to many things in various 'G-#' declarations and communiqués over the years, but the 'G-#' processes do not provide any system for compliance and enforcement. Instead, leaders can increase the cost of failing to deliver on their commitments by making such failure more transparent. In the context of the fossil fuel subsidies agreement, leaders established processes of implementation and third-party expert review to promote such transparency. As a part of the G-20 agreement, leaders tasked energy and finance ministers to identify their nation's fossil fuel subsidies, develop a plan for eliminating these subsidies, and report back to leaders by the following year's summit. The G-20 published a summary report of each member's identified fossil fuel subsidies and the plan for eliminating them at the 2010 G-20 meeting in Toronto. Since then, leaders have continued to task energy and finance ministers to continue their efforts and report back regularly.

To complement this self-reporting, the G-20 leaders also tasked four international organizations – the International Energy Agency (IEA), the Organization for Economic Co-operation and Development (OECD), the Organization of Petroleum Exporting Countries (OPEC), and the World Bank – to undertake their own, joint assessment of fossil fuel subsidies. This includes an examination of individual countries' subsidies as well as the aggregate economic, energy and environmental impacts of the sum of these nations' subsidies. These international organizations published their joint report to G-20 leaders at the 2010 Toronto meeting and have continued to provide analysis and reviews of countries' implementation strategies pursuant to tasking by the G-20.

This third-party reporting provides an independent reference for countries' identification of subsidies and an independent review of their progress in rationalizing fossil fuel pricing. This process enhances transparency on implementation, and can empower domestic stakeholders as well as peer nations within the G-20 to apply pressure and invoke moral suasion to push a country to deliver on its commitment. This process can also highlight the successful efforts to reduce such subsidies and to illustrate possible strategies that other countries could emulate as they attempt to move forward in implementing their commitments. Relying on external experts at established and recognized international organizations also mitigates concerns about politicization of the evaluation and verification mechanism and allows for a rapid ramping up of the review process that would not be possible if a new bureaucracy had to be constructed from scratch. A potential limitation of relying on existing international organizations, however, may be the legitimacy of those with incomplete memberships. For example, some developing countries may question analysis and reviews by the International Energy Agency, whose membership comprises developed nations.

Monitoring Trade in Endangered Species

The implementation of the monitoring system under the Convention on International Trade in Endangered Species (CITES) represents a second example of independent review and verification of an international agreement (Wettestad, 2007). Under CITES, nations submit regular reports on trade in covered species (annually) and on relevant policies impacting trade in endangered species (biennially). In the 1970s, CITES effectively turned to international non-governmental organizations – the World Conservation Union (IUCN) and the World Wildlife Fund (now the Worldwide Fund for Nature) – to provide independent reviews of these annual national reports. This independent verification system has evolved into what is now referred to as the Trade Records Analysis of Flora and Fauna in Commerce (TRAFFIC) monitoring network. In this case, the international policy community agreed to defer to existing expertise outside of an international bureaucracy to assess and verify compliance with international agreements.

Synthesis on International Policy Surveillance Experience

Extensive international policy surveillance occurs in a number of economic, energy and environmental policy contexts. Table 15.1 summarizes the key elements of a sample of these surveillance systems, and compares them with the status quo system of surveillance under the UN Framework Convention on Climate Change.

The table reveals a few stark differences between the climate policy surveillance and these other regimes. First, the IMF, OECD, WTO and G-20 regimes rely on both expert review and peer review, whereas the UNFCCC regime is only an expert review. Second, the expert reviews of economic, trade and subsidy policies are undertaken by career staff of their respective institutions, while the UNFCCC review typically draws in an *ad hoc* nature from government-sponsored experts (from academia, business and government sectors) to conduct a review. Third, the UNFCCC review applies only to the industrialized countries, while the various economic policy surveillance systems apply to all member states of that institution. Fourth, the provision of standards and reporting templates can improve the transparency of the reporting and review, and enhance surveillance effectiveness. While this is common among the economic policy surveillance

Table 15.1   International systems of policy surveillance

Policy	Economic policy	Economic policy	Trade policy	Fossil fuel subsidies	Greenhouse gas emissions	Climate policy
Institution	IMF	OECD	WTO	G-20	UNFCCC	UNFCCC
Review mechanism	Article IV Consultation	Economic and Development Review Committee	Trade Review Policy Board	IEA, OECD, OPEC, World Bank joint report	Subsidiary Body for Implementation	Subsidiary Body for Implementation
Countries reviewed	All IMF members	All OECD members	All WTO members	All G-20 members	Only Annex I countries	Only Annex I countries
Peer review	Yes	Yes	Yes	Yes	No	No
Standards for reporting	Yes	No	Yes	No	Yes	Yes
Technical assistance to least developed countries	Yes	N/A	Yes	N/A	N/A	N/A
Frequency	1–2 years	1–2 years	2–6 years	Annual	Annual	~4 years
Country visit	Yes	Yes	Typically	No	Yes (every 5 years)	No
Published review	Yes	Yes	Yes	Yes	Yes	Yes

and the UNFCCC process for industrialized nations, it does not apply to the occasional reporting of emissions by developed countries under the UNFCCC.

PROPOSAL FOR A BRETTON WOODS CLIMATE INSTITUTION

A Bretton Woods climate institution would provide for surveillance of climate change policies – on emission mitigation, adaptation and, if necessary, geoengineering – and climate-policy-related outcomes. It would draw the lessons of the most effective climate policies from this surveillance and publish recommended best policy practices. International climate finance would be conditioned on a country regularly undergoing surveillance to provide an incentive for participation. After elaborating the details of the proposal, this section closes with a discussion of how this institution could complement an array of bottom–up and top–down international climate policy architectures.

Surveillance Mechanism

The objective of the surveillance mechanism would be to provide transparent, rigorous, credible assessments of individual and global performance in the effort to combat climate change.

The surveillance mechanism would require active engagement by the country under review, an expert review team, and representatives of other countries participating in the peer review. The BWCI would establish standards for monitoring and reporting data on greenhouse gas emissions and climate policy performance. These could be *de novo* standards or the adoption of existing, rigorous standards (e.g. IPCC standards on emission inventories). The BWCI would provide guidance to facilitate monitoring and reporting of the relevant data. In addition, the country under review would submit a report summarizing its climate policy performance that would serve as a key input to the work of the expert review team.

The expert review team would make an in-country visit to meet with government officials and relevant stakeholders, and to collect data. In addition, the expert review team would consult other sources of data to inform its assessment. The BWCI would develop and publish metrics for performance evaluation to frame the country reviews. This team would then draft a report for consideration by the peers of the nation under review.

An executive board that reflects the contributions to the BWCI and geographic diversity would meet regularly to discuss the draft reports by the expert review teams. One developed-country and one developing-country representative would be randomly chosen from the executive board to serve as 'peer examiners' in the discussion of the country's review. A delegation from that country would participate in this discussion and respond to questions raised by the examiners, and then in an open forum with the entire executive board. Based on this discussion, the expert review team would finalize the report, which must be approved by a consensus of the executive board (except for the country under review if it happens to be a board member at that time), and make it publicly available.

Global reviews would occur annually. These would draw from annual reporting of greenhouse gas emissions from every country, the policy surveillance undertaken that year, as well as thematic analyses. These reviews would also provide near-term forecast for greenhouse gas emissions growth and identify policy reforms that could reduce emissions.

The frequency of individual country policy reviews would vary with the level of current greenhouse gas emissions. The largest ten sources of greenhouse gas emissions in 2010 would undergo annual surveillance. The next 20 sources by size would undergo review every two years. All other countries would participate in reviews every three years.

An individual country could solicit a review of another country. If this second country did not consent to review, then the BWCI would undertake the review without an in-country visit. In this case, the BWCI would compile information from available sources and draft a report for review by the executive board. This would be akin to the G-20 tasked reports by the international organizations.

A country would be able to request resources to build capacity to participate more effectively in the surveillance regime. The BWCI would encourage rotations of in-country staff to the institution to refine analytic skills and develop a better understanding for the operation of the surveillance mechanism.

The executive board would identify an external advisory board of experts who would periodically review the surveillance mechanism. This advisory board, by reviewing the reviewers, could identify ways to improve the quality of analysis, enhance the transparency of compiled information, and develop more effective ways to promote best policy practice.

A professional bureaucracy would be recruited to undertake the analysis of the various forms of policy commitments. This would extend beyond simply measuring greenhouse gas emissions. The professional staff would undertake analysis of all countries' climate change policies, including abatement policies, financial contributions to and receipts from the clean technology fund, support for and effectiveness of adaptation programs, research funds for geoengineering and so on. In the near term, such an institution could be staffed by current experts at other international organizations.

This institution would require significant resources to scale up and operate. Countries participating in the BWCI would make regular contributions, akin to existing international organizations. New revenue streams could also be tapped, perhaps based on one of the ideas identified by the Advisory Group on Finance (2010). Countries requesting the review of another country would need to provide the resources for the review.

Best Policy Practice

Developed and emerging economies have substantial experience in implementing policies that can affect the investment in more energy-efficient or low-carbon technologies. Drawing lessons from these successes and failures can inform the design of new policies in all countries, but especially those in developing countries that may lack the institutional capacity to fully evaluate and compare all policy options. Establishing a set of 'best practice' policies can draw from past efforts to promote the deployment of energy-efficient technologies, and tailor guidance for countries' specific economic and cultural circumstances.

The initial set of 'best practice' policies should focus on promoting no regrets and low-cost measures. This would draw attention to policies that facilitate investment in more energy-efficient technologies, since they deliver a stream of returns over time in terms of lower expenditures on fuel and power. It could also highlight opportunities for countries to address multiple policy objectives, such as climate change mitigation and local air quality improvement or climate change mitigation and energy security. These efforts should also consider means to promote the diffusion of technologies to facilitate adaptation to climate change.

Consider the example of reforming energy subsidies. Removing subsidies on the consumption of energy and the production of fossil fuels could significantly reduce carbon dioxide emissions; full implementation of the G-20 fossil fuel subsidies agreement could lower global emissions by 7 percent by 2020 (IEA et al., 2010). Such efforts would free up fiscal resources that could finance other socially desirable activities. Phasing out subsidies could yield comparable price increases on energy as imposing carbon pricing programs in developed countries, and thus deliver a comparable emission abatement impact. Subsidy removal could also deliver economic benefits by stimulating a more efficient allocation of investment that could accelerate economic growth. Rationalizing fossil fuel pricing could also improve air quality.

The BWCI could promote policy learning by supporting randomized field experiments with new policies. For example, the BWCI could provide technical guidance on the ways to design, implement and evaluate such experiments. In addition, the BWCI could provide financing to cover the incremental costs of undertaking such experiments and, to leverage an initial set of experimental efforts, additional resources could be provided.

The BWCI could publicize best policy practices through a variety of outlets. First, the regular reviews of individual countries' climate policy programs could include recommendations of policies appropriate to the geographic, cultural and economic context of that country. Second, best policy practices could be highlighted in annual aggregate surveillance reports and in occasional special thematic reports.

International Climate Finance

The BWCI could establish basic eligibility standards for international climate finance. In particular, a country must participate in the surveillance regime in order to receive climate finance resources. This could reflect a formal integration of the BWCI and the Climate Investment Funds at the World Bank, the Green Climate Fund under the UNFCCC, the climate activities of the Global Environment Facility and so on. A more practical outcome is a set of agreements between the BWCI and these existing financial institutions to ensure that recipient countries participate in policy surveillance. In addition, donor countries could agree to employ the BWCI as the scorekeeper for various forms of climate finance, and thus subject their climate finance programs to surveillance.

Since the international climate finance space is already much further developed – in terms of institutions, resources and projects in progress – than international climate surveillance, I will only briefly describe two ideas for climate finance as a means to leverage participation in the surveillance regime.

First, a clean technology fund could support investment in emission mitigation projects and policies in developing countries. This fund should complement, not substitute, for efforts that would occur otherwise through existing private and carbon markets.

The centralized nature of a fund highlights an opportunity to make investments that can maximize the mitigation potential for a given amount of resources. Specifically, the clean technology fund could address criticisms of the CDM that it has created windfall profits for some firms (Wara, 2007), by financing projects at a level sufficient to provide a reasonable, but not extraordinary, profit for the developing-country firm. A reverse auction could serve as the vehicle for allocating the fund's resources (Keeler and Thompson, 2010). Under such a mechanism, the fund would solicit bids from firms in developing countries. The bids would represent the amount the firm would require in a subsidy in order to invest in and use a new, climate-friendly technology. The fund could start with the lowest-cost bids and work up the bid profile until it has exhausted all the resources dedicated for the specific technology auction. The challenge in effectively implementing a reverse auction lies in identifying technologies and projects that are low-cost but would not have happened in the absence of the fund.

Given fiscal constraints in many OECD countries, the clean technology fund could be financed through an auction of supplemental emission credits to firms regulated by cap-and-trade programs in the developed world (or some other revenue raiser, e.g. those identified by the Advisory Group on Finance, 2010). These credits would not be backed by a ton-for-ton abatement through specific projects. Instead, the expectation is that the broad portfolio of investment from the clean technology fund will deliver emission abatement in developing countries that will more than offset the higher emissions allowed in OECD cap-and-trade programs through the auctioned credits. Prudent management of donors' resources could attract developed-country engagement and secure their participation in surveillance.

Second, financing for adaptation could elicit extensive participation by the least developed countries in the surveillance regime. The *de facto* international 'insurance' for major natural disasters – such as 1998 Hurricane Mitch in Central America and 2008 Cyclone Nargis in Myanmar – is *ad hoc* in terms of the speed, scale and source of support. An alternative approach could be an explicit insurance fund for least developed countries financed through catastrophic risk bonds. The bonds supporting this fund would be issued by developed countries, but payouts of the bonds would require three kinds of actions by developing countries in addition to a qualifying natural disaster. First, to mitigate the incentive for moral hazard, these least developed countries must make some investments to reduce their exposure to the relevant climate-related natural disasters (hurricanes, droughts, floods etc.). Second, participating countries must also implement some 'no regrets' emission mitigation policies identified through the best policy practices assessment of the BWCI. The policies could be fairly modest, and their direct effect on the global climate could be negligible, but they would establish the norm for abatement action by the very poorest countries. This norm then creates the floor for the minimum acceptable abatement effort by wealthier developing countries, so it could encourage more ambitious actions by those with greater resources. Third, the participating countries would agree to undergo surveillance of these risk reduction and emission mitigation efforts.

The least developed countries participating in this insurance fund could make claims for a standard class of disasters; there would be no requirement to attribute an event to climate change. The insurance fund would need to enlist broad, geographic participation in order to diversify risk. Given the lack of actuarial tables for climate change, management of the fund would depend on the improved knowledge base on climate risks and impacts.

Integration with Various Climate Policy Architectures

Such a Bretton Woods-like climate institution could complement a variety of international climate policy architectures (see e.g. Aldy et al., 2003; Aldy and Stavins, 2007; Aldy and Stavins, 2010b). First, the BWCI could serve as a significant scaled-up version, in rigor and scope, of the existing emissions and policy surveillance regime under the UNFCCC. Thus a top–down treaty organization could create such an institution for receiving emissions and policy reports, undertaking evaluations of parties' climate policy programs, and publishing reviews of individuals parties' actions and assessments of the aggregate effort to address climate change.

Second, it could inform the design of new, top–down international agreements on emission mitigation. For example, Frankel (2010) proposes a formulaic approach to determining emission targets as a function of incomes and per capita emissions. Such an approach requires rigorously evaluated data from participating countries in order to set targets, and then to assess compliance with them. The BWCI could play the role of compiling necessary data for the *ex ante* target-setting exercise and the *ex post* performance evaluation. Cooper (2010) advocates for an international climate policy regime premised on harmonization of domestic carbon taxes. The BWCI could serve the role that Cooper would assign to the IMF of reviewing countries' carbon taxes (and related energy tax programs) to ensure that economic activity around the world faces a common carbon price.

Third, a BWCI could facilitate the evolution of a bottom–up climate policy architecture along two dimensions. For example, the pledge and review system reflected in the Copenhagen Accord and the Cancun Agreements would certainly benefit from a more rigorous review program than exists under the *status quo*. Indeed, this is evident in the agreement for 'international consultations and analysis' of developing country mitigation actions, policies and goals, as well as in the assessment of the scaling up of international climate finance. A credible, independent assessment of countries' efforts to deliver on mitigation and financing goals can increase trust among parties and build support for this approach to climate change policy.

In addition, an assessment of domestic cap-and-trade programs by the BWCI could serve as a necessary condition for linking of any two countries' (or regions') cap-and-trade programs. The linking of any two programs would likely require some kind of mutual assessment to ensure comparable integrity. The BWCI could perform this evaluation, as an independent, expert third party, for any two cap-and-trade programs considering linking. Indeed, linking provisions of authorizing legislation or regulations could specifically call on the BWCI to perform this evaluation.

Fourth, various governments could rely on the BWCI for analyses to inform the implementation of their respective domestic climate change policies. A donor country may use BWCI assessments of the policies and actions of those countries receiving international climate finance to determine how to direct a next round of finance. Indeed, a donor country could require any recipient country to undergo periodic evaluation by the BWCI to remain eligible for that donor country's climate financing. Donor countries could reach an agreement as a group on such an approach to create a strong incentive for developing countries to participate in the review process of the BWCI.

In addition, some domestic climate policies may include some kind of border measure, such as a border carbon tax, on the emission intensity of imported goods from countries without comparable domestic climate change programs. For example, the 2009 American Clean Energy and Security Act (H.R. 2454) in the US House of Representatives included such a border measure and required an assessment of the adequacy of other countries' domestic climate programs to determine the applicability of the border measure. A given country could require nations exporting goods to that country to undergo periodic evaluation by the BWCI as a condition for a waiver of the border tax. The BWCI could agree to undertake such a review of trade partners if the country soliciting review undergoes a comparable assessment. This creates an explicit opportunity for comparison of domestic programs consistent with the WTO's non-discriminatory application standard for evaluating a border measure. Thus countries have the incentive to solicit reviews for legitimate (as opposed to political grandstanding) purposes, since an adverse comparison by the BWCI for the soliciting country would likely weaken dramatically that country's position before a WTO tribunal.

CONCLUSION

Designing and implementing a new Bretton Woods climate institution would signal a new seriousness by the international community in its effort to combat climate change. Such an institution, on par with existing institutions to address international financial, trade and development challenges, could serve as an important foundation for the next steps to address climate change.

Countries are likely to participate in an international effort to address climate change that they believe requires them to make a fair contribution to a global effort that is up to the challenge. This will necessitate efforts to assess the comparability of effort among countries and the adequacy of the aggregate effort. To build trust, peers must undertake comparable policies. Such an evaluation requires meaningful metrics of policy implementation and outcomes. The envisioned policy surveillance under the BWCI would provide both an *ex ante* analysis of proposed policy commitments, to promote credibility at international negotiations when countries propose their next steps, and an *ex post* assessment of whether a country complied with its policy commitments.

Assessments of effort can also provide a rigorous basis for identifying best policy practice and to report on the results of various policy experimentation. This can accelerate the learning process, especially for developing and emerging economies, as they gain knowledge from the leaders' initial efforts to abate emissions, promote adaptation and conduct research on geoengineering.

Finally, this institution should be designed and implemented in a way that creates incentives for countries to undergo surveillance willingly. Independent evaluation by experts can provide the legitimacy necessary to elicit support by participating nations. Conditioning international climate finance on participation in policy surveillance could induce broader participation. Providing an open access option for individual countries to request reviews as a component of their domestic policy-making – such as in donor resource allocation and border tax waiver decisions – could induce even the more reluctant nations to submit to BWCI surveillance. Further consideration of the design of the institution could promote even stronger incentives for countries to cooperate and actively support a more rigorous and extensive system of surveillance.

ACKNOWLEDGMENTS

Participants at the 2008 Third Atlantic Workshop on Energy and Environmental Economics provided comments on an earlier draft. Sarah Szambelan and Sarah Cannon provided valuable research assistance.

REFERENCES

Advisory Group on Finance (2010), *Report of the Secretary-General's High-Level Advisory Group on Climate Change Financing*, New York: United Nations.

Aldy, Joseph E. and Robert N. Stavins (eds) (2007), *Architectures for Agreement: Addressing Global Climate Change in the Post-Kyoto World*, Cambridge: Cambridge University Press.

Aldy, Joseph E. and Robert N. Stavins (2010a), 'Introduction', in J.E. Aldy and R.N. Stavins (eds), *Post-Kyoto International Climate Policy: Implementing Architectures for Agreement*, Cambridge: Cambridge University Press, pp. 1–30.

Aldy, Joseph E. and Robert N. Stavins (eds) (2010b), *Post-Kyoto International Climate Policy: Implementing Architectures for Agreement*, Cambridge: Cambridge University Press.

Aldy, Joseph E., Scott Barrett and Robert N. Stavins (2003), 'Thirteen plus one: a comparison of global climate policy architectures', *Climate Policy*, **3**: 373–97.

Barrett, Scott (1994), 'Self-enforcing international environmental agreements', *Oxford Economic Papers*, **46**: 878–94.

Barrett, Scott (2003), *Environment and Statecraft: The Strategy of Environmental Treaty-Making*, Oxford: Oxford University Press.

Bodansky, Daniel (2007), 'Legitimacy', in Daniel Bodansky, Jutta Brunee and Ellen Hey (eds), *The Oxford Handbook of International Environmental Law*, Oxford: Oxford University Press, pp. 704–22.

Boden, Tom and T.J. Blasing (2012), 'Preliminary 2009 and 2010 global and national estimates of carbon emissions from fossil fuel combustion and cement manufacture', Carbon Dioxide Information Analysis Center, Oak Ridge National Laboratory, accessed 19 February 2012, http://cdiac.ornl.gov/trends/emis/perlim_2009_2010_estimates.html.

Boden, T.A., G. Marland and R.J. Andres (2011), 'Global, regional, and national fossil fuel CO_2 emissions', Carbon Dioxide Information Analysis Center, Oak Ridge National Laboratory, accessed 19 February 2012, http://cdiac.ornl.gov/trends/emis/overview_2008.html.

Breidenich, Clare and Daniel Bodansky (2009), 'Measurement, reporting and verification in a post-2012 climate agreement', Pew Center on Global Climate Change Report, April. Arlington, VA: Pew Center.

Cooper, Richard N. (2010), 'The case for charges on greenhouse gas emissions', in J.E. Aldy and R.N. Stavins (eds), *Post-Kyoto International Climate Policy: Implementing Architectures for Agreement*, Cambridge: Cambridge University Press, pp. 151–78.

Ellis, Jane and Kate Larsen (2008), *Measurement, Reporting, and Verification of Mitigation Actions and Commitments*, OECD and IEA Publication COM/ENV/EPOC/IEA/SLT(2008)1, Paris: OECD and IEA.

Fischer, Stanley (1999), 'Reforming the international financial system', *The Economic Journal*, **109**: F557–F576.

Francois, J.F. (2001), 'Trade policy transparency and investor confidence: some implications for an effective trade policy review mechanism', *Review of International Economics*, **9**(2): 303–16.

Frankel, Jeffrey (2010), 'An elaborated proposal for a global climate policy architecture: specific formulas and emission targets for all countries in all decades', in J.E. Aldy and R.N. Stavins (eds), *Post-Kyoto International Climate Policy: Implementing Architectures for Agreement*, Cambridge: Cambridge University Press, pp. 31–87.

Fudenberg, Drew and Eric Maskin (1986), 'The folk theorem in repeated games with discounting or with incomplete information', *Econometrica*, **54**(3): 533–54.

Fudenberg, Drew, David Levine and Eric Maskin (1994), 'The folk theorem with imperfect public information', *Econometrica*, **62**(5): 997–1039.

G-20 Leaders' Declaration (2009), Pittsburgh, PA, 25 September.

Green, Edward J. and Robert H. Porter (1984), 'Noncooperative collusion under imperfect information', *Econometrica*, **52**(1): 87–100.

Haas, P.R., R.O. Keohane and M.A. Levy (eds) (1993), *Institutions for the Earth: Sources of Effective International Environmental Protection*, Cambridge, MA: MIT Press.

Hampson, F.O. (1995), *Multilateral Negotiations: Lessons from Arms Control, Trade, and the Environment*, Baltimore, MD: The Johns Hopkins University Press.

Hempel, L.C. (1996), *Environmental Governance: The Global Challenge*, Washington, DC: Island Press.

Hohne, Niklas, Joyce Penner, Michael Prather, Jan Fuglestvedt, Jason Lowe and Guoquan Hu (2007), 'Summary report of the ad hoc group for Modelling and Assessment of Contributions to Climate Change (MATCH)', 7 November, accessed 17 February 2012, http://www.match-info.net/data/MATCH%20summary%20report.pdf.

International Energy Agency, Organization of Petroleum Exporting Countries, Organisation of Economic Co-operation and Development and the World Bank (2010), 'Analysis of the scope of energy subsidies and suggestions of the G-20 Initiative', Joint report prepared for submission to the G-20 Summit Meeting, Toronto, Canada, 26–27 June 2010.

International Monetary Fund (2001), *Annual Report 2001*, Washington, DC: IMF.

Keeler, Andrew G. and Alexander Thompson (2010), 'Mitigation through resource transfers to developing countries: expanding greenhouse gas offsets', in J.E. Aldy and R.N. Stavins (eds), *Post-Kyoto International Climate Policy: Implementing Architectures for Agreement*, Cambridge: Cambridge University Press, pp. 439–68.

Keohane, R.O. (1994), 'Against hierarchy: an institutional approach to international environmental protection', in Helge Hveem (ed.), *Complex Cooperation: Institutions and Processes in International Resource Management*, Oslo: Scandinavian University Press, pp. 13–34.

Klabbers, Jan (2007), 'Compliance procedures', in Daniel Bodansky, Jutta Brunee and Ellen Hey (eds), *The Oxford Handbook of International Environmental Law*, Oxford: Oxford University Press, pp. 995–1009.

Litfin, K.T. (1999), 'Environmental remote sensing, global governance, and the territorial state', in M. Hewson and T.J. Sinclair (eds), *Approaches to Global Governance Theory*, Albany, NY: State University of New York Press, pp. 73–96.

Malaith, G.J. and S. Morris (2002), 'Repeated games with almost-public monitoring', *Journal of Economic Theory*, **102**(1): 189–228.

Mavroidis, P.C. (1992), 'Surveillance schemes: the GATT's new trade policy review mechanism', *Michigan Journal of International Law*, **13**: 374–414.

OECD (2003), 'Peer review: a tool for co-operation and change', OECD Policy Brief, December.

Pagani, F. (2002), 'Peer review: a tool for co-operation and change – an analysis of an OECD working method', Report SG/LEG(2002)1, Paris: OECD.

Schafer, A. (2006), 'A new form of governance? Comparing the open method of co-ordination to multilateral surveillance by the IMF and the OECD', *Journal of European Public Policy*, **13**(1): 70–88.

Schelling, Thomas C. (1956), 'An essay on bargaining', *American Economic Review*, **46**(3): 281–306.

United Nations (1995), Berlin Mandate. Decision 1/CP.1, UN Framework Convention on Climate Change, http://unfccc.int/resource/docs/cop1/07a01.pdf.

UNEP (2012), 'ODS consumption in ODP tons', accessed 1 March 2012, http://ozone.unep.org/new_site/en/Information/generate_report.php?calculated_field=ODS+Consumption&grouping_option=Cntry&incl_baseline=1&cntry=CN&all_anxgrp=on&anxgrp=AI&anxgrp=AII&anxgrp=BI&anxgrp=BII&anxgrp=BIII&anxgrp=CI&anxgrp=CII&anxgrp=CIII&anxgrp=EI&summary=0&Yr1=1990&Yr2=2011.

UNFCCC (2012), 'Greenhouse gas inventory data – detailed data by party', accessed 1 March 2012, http://unfccc.int/di/DetailedByParty/Event.do;jsessionid=CD2F85D1574D60F06B8B092E62090304.diprod01?event=go.

US Energy Information Administration (1997), *Annual Energy Outlook 1997*, Washington, DC: Department of Energy.

US Energy Information Administration (2012a), *Annual Energy Outlook 2012*, Washington, DC: Department of Energy.

US Energy Information Administration (2012b), 'International energy statistics', accessed 1 March 2012, http://www.eia.gov/cfapps/ipdbproject/IEDIndex3.cfm.

Wara, Michael (2007), 'Is the global carbon market working?', *Nature*, **445**: 595–6.

Wettestad, Jorgen (2007), 'Monitoring and verification', in Daniel Bodansky, Jutta Brunee and Ellen Hey (eds), *The Oxford Handbook of International Environmental Law*, Oxford: Oxford University Press, pp. 974–94.

World Resources Institute (2012), 'Climate analysis indicators tool', accessed 1 March 2012, http://cait.wri.org.

PART V

CARBON MITIGATION POLICIES

16 Fiscal instruments for climate finance
Ian Parry

I. INTRODUCTION

At the 2010 global climate talks in Cancun, Mexico, developed countries reaffirmed a pledge to raise $100 billion per year by 2020, and $30 billion in total between 2010 and 2012, for financing climate mitigation and adaptation projects in developing countries. While efficient use of such revenues is important for environmental effectiveness (and policy credibility), the focus here is entirely on how the funding sources might be mobilized. Furthermore, the discussion is confined to public sources of revenue,[1] as (mostly) applicable to developed countries.[2]

The advisory group on climate finance established by the UN Secretary-General (AGF, 2010a) evaluated a wide range of carbon-related and other fiscal instruments for climate finance, on multiple criteria, especially revenue potential. The instruments examined included domestic, carbon-related taxes (e.g. carbon taxes, taxes on electricity use), broader domestic instruments (e.g. taxes on the financial sector), and charges on international transportation (aviation and maritime) activities. This chapter goes into a bit more detail on some aspects, particularly environmental effectiveness, welfare costs and incidence of these policies, as well as considering some additional instruments (e.g. 'feebates', motor fuel taxes, vehicle taxes and petroleum levies). Possibilities for improving the acceptability of carbon pricing instruments are also discussed. Table 16.1 provides a summary of the main points from the narrative.[3]

The chapter is organized as follows. Section II briefly provides the conceptual rationale for environmentally motivated taxes relative to alternative public sources of finance. Section III discusses specific scenarios for carbon pricing and other domestic, climate-related instruments. Section IV discusses taxes on aviation and maritime fuels. Section V offers brief concluding remarks.

II. CONCEPTUAL RATIONALE FOR CARBON PRICING

Here we summarize, very briefly, the case for (appropriately scaled) carbon taxes relative to broader fiscal instruments, other energy taxes and regulatory approaches, and the (more nuanced) case for taxes over cap-and-trade systems. We also comment briefly on some design issues. Our focus is limited to energy-related CO_2 emissions, as they are easily the dominant greenhouse gas and are (much) easier to monitor than forestry emissions or other greenhouse gases (e.g. from agricultural sources).

Table 16.1 Summary comparison of carbon-related fiscal instruments

Fiscal instrument	Effectiveness (at reducing CO_2 and promoting carbon markets)	Revenue (approximate annual revenue potential for climate finance in 2020)	Cost-effectiveness (relative to broad labor tax and accounting for environmental or other benefits)	Competitiveness effects and cross-country incidence	Overall assessment
$25 per ton CO_2 price (carbon tax or auctioned cap and trade)	Most effective policy	$25 billion (if 10% of revenue allocated)	More cost-effective because charges for environmental damages (assumes revenues used productively)	Potentially significant. May require innovative ways to compensate for higher energy prices	Highly promising technically, but liable to resistance
Per kWh tax on power generation	Limited effectiveness	Fairly large (even if minor fraction allocated)	About the same as labor tax or worse	Same issues as above	Inferior to emissions pricing
$25 per ton tax on CO_2 emissions from power generation	Fairly effective	$10 billion (if 10% of revenue allocated)	Between feebate and carbon tax (if revenues used productively)	Same issues as above	Promising alternative to carbon tax, but high electricity prices may reduce acceptability
Feebates	Fairly effective	$10 billion (but only if all revenue allocated)	More cost-effective than labor tax though less so than carbon tax	Modest impacts (though increase as more revenue is raised)	Promising alternative if carbon pricing resisted
6 cents per liter surcharge on motor fuels	Limited effectiveness	$5 billion (if 10% of revenue allocated)	Can be very cost-effective, but only in limited cases where fuel is currently undertaxed	Opposition to higher fuel prices might be offset by lowering vehicle taxes	Inappropriate in many countries
2 cents per liter levy on all petroleum consumption	Limited effectiveness	$5 billion (if 10% of revenue allocated)	Can be very cost-effective if energy security important	Relatively modest	Promising complement to carbon pricing
$25 per ton charge on international aviation emissions	Modest impact in a sector with a small share of emissions	$8 billion (after developing-country compensation)	More efficient because of charges for environmental damages	Requires international coordination and developing-country compensation	Promising if fuels not covered by comprehensive carbon pricing
$25 per ton charge on international maritime emissions	Modest impact in a sector with a small share of emissions	$15 billion (after developing-country compensation)	More efficient because of charges for environmental damages	Requires international coordination and developing-country compensation	Promising if fuels not covered by comprehensive carbon pricing

The Balance between Carbon Taxes and Broader Fiscal Instruments

Generally speaking, the Ramsey tax literature implies that – environmental objectives aside – government revenues should be raised from broad-based taxes, ideally on final consumption or labor income, and not on intermediate inputs such as fuels (e.g. Sandmo, 1976). Although taxes on labor earnings or general consumption, for example, reduce the returns to work effort, and thereby discourage labor force participation, effort and overtime on the job and so on, they tend to cause less distortion to the economy than narrow-based taxes. For comparable revenues, taxes on dirty energy sources generally cause similar employment effects to those from broader-based taxes, through the effect of higher energy prices on raising the general price level, reducing real household wages, and hence reducing labor supply – this is termed the 'tax-interaction effect' in the environmental tax literature (e.g. Goulder, 1995; Parry, 1995). However, carbon taxes also distort energy markets by causing a shift towards more costly (albeit) cleaner fuels, as well as inducing households and firms to consume less energy (e.g. drive less) than they would otherwise prefer.

Up to a point, however, carbon taxes can be a far more efficient way to raise revenue than broader fiscal instruments when environmental benefits are taken into account. Loosely speaking, and leaving some caveats aside, on efficiency grounds carbon taxes should be set to reflect the marginal environmental damages per ton of CO_2, and further revenue needs should be met through broader fiscal instruments.[4]

The case for using carbon pricing to address externalities is quite distinct from that for using it to generate climate finance. There is no relation between the socially optimal scale of climate finance and the efficient level of carbon taxes. In fact, much of the (large) revenue potential from carbon taxes could be used for domestic deficit reduction (increasing its appeal despite policy makers' current concern with debt and macroeconomic stability), while still generating significant sources for climate finance. Conversely, there is no reason why climate finance has to come from instruments targeted at emissions – some funds might be raised through broader fiscal instruments that have no (direct) mitigation benefits. The appropriate way to raise general revenues from these broader instruments has been discussed extensively elsewhere (e.g. IMF, 2010a, 2010b) – Box 16.1 provides a brief summary of the main points.

Carbon Pricing versus Regulatory Approaches to Emission Control

Comprehensive carbon taxes or cap-and-trade systems are potentially the most effective policies for cutting CO_2 emissions across the economy. Pricing the carbon content of fossil fuels upstream in the fossil fuel supply chain (e.g. at the wellhead, minemouth or point of importation) will cause the price of fuels and electricity to rise, thereby exploiting all (energy-related) emission reduction opportunities across the economy. These opportunities include (i) reducing the emissions intensity of power generation (through switching from high-carbon-intensity coal to intermediate-carbon-intensity natural gas and fuel oil, from these fuels to zero-carbon nuclear and renewables, and by improving plant efficiency); (ii) reducing residential and industrial electricity demand; (iii) reducing the use of, and fuel consumption rate of, transportation vehicles; and (iv) reducing direct fuel use in homes, offices, factories and shops (e.g. for heating). Fuel switching in power generation usually provides the biggest source of low-cost options for reducing CO_2, at least in countries with significant shares of coal generation.

Taxes on electricity use, in contrast, are much less environmentally effective. They exploit only one of the above four CO_2 emissions reductions channels. Taxes on vehicle ownership are especially blunt – not only do they miss out on three out of the four emissions reduction channels, but they also, within the transport sector, fail to reduce miles driven per vehicle and (depending on their design) may have little to no effect on vehicle fuel economy. Pricing major industrial and power plant sources of CO_2 emissions at the point when they are released into the atmosphere, as in the EU's Emissions Trading System (ETS), is far more environmentally effective than a pure tax on electricity. However, this approach still fails to cover emissions from households, transport and small-scale industry, which account for about 50 percent of CO_2 emissions in the EU ETS.[5]

Regulatory policies in isolation are generally poor substitutes for carbon taxes – and do not raise revenues. For example, mandates for minimum generation shares for renewable fuels will not reduce emissions beyond the power sector and will not reduce the emissions intensity of the remaining generation mix (e.g. they will not promote shifting from coal to natural gas or nuclear power). To take another example, even within the transport sector, automobile fuel economy standards will not encourage people to drive less[6] nor reduce emissions from broader transportation vehicles. Combining complementary regulations can substantially improve environmental effectiveness – for example, combining a CO_2 per kWh standard (which reduces the emissions intensity of power generation) with miles per gallon standards for vehicles, building codes and energy efficiency standards for appliances will exploit most of the emissions reductions exploited under a carbon tax. However, not all mitigation opportunities (such as reduced use of vehicles and air conditioners) are encouraged, and these regulatory combinations could involve a considerable loss of cost-effectiveness compared with a carbon tax – unless there are comprehensive credit trading provisions to equate the marginal cost of abatement across emissions sources within a sector and across different sectors.[7]

Carbon Taxes versus Cap-and-trade Systems

There is a large literature comparing these two instruments (e.g. Goulder and Parry, 2008; Hepburn, 2006; Nordhaus, 2007). The general finding is that cap-and-trade systems are reasonable instruments if they are designed to mimic the potential advantages of carbon taxes – mainly this means auctioning allowances to provide a source of government revenue, and including provisions to limit emissions price volatility. See Box 16.2 for further discussion.

Use of Domestic Revenues

Besides the appropriate tax level (see below) and base (see above), using carbon tax revenues productively is also important. From a cost-effectiveness perspective, there is a strong case for using domestically retained revenues from carbon taxes to reduce broader tax distortions (e.g. labor tax distortions), as this can produce large economic benefits. With this recycling, the overall costs of carbon pricing (excluding environmental

BOX 16.1   EFFICIENT WAYS TO RAISE REVENUE FROM
BROADER DOMESTIC SOURCES

All developed countries, aside from Saudi Arabia and the USA, have a value-added tax (VAT). VAT is a relatively efficient source of revenue, generating fewer distortions than many other taxes (e.g. Norregaard and Kahn, 2007). On average VAT raises over 5 percent of GDP in revenue, although in many countries there is a proliferation of exemptions and rate differentiation. Usually, therefore, there is substantial scope for raising revenue through base broadening rather than raising tax rates. Even though the efficiency costs of base broadening may be less than those for raising rates, nonetheless costs are most likely positive, implying this option is less efficient than pricing instruments to internalize the carbon externality.

The corporate income tax (CIT) is not a very promising revenue source. Intensifying international tax competition has led to significant reductions in statutory CIT rates and many countries have also broadened their tax bases by adjusting tax depreciation rules and restricting deductions (e.g. for interest). International coordination over tax rates could perhaps strengthen the revenue potential of the CIT, although the prospects for this appear remote.

The personal income tax (PIT) is considered important for achieving equity objectives, but raising marginal tax rates may have relatively high efficiency costs. Higher tax rates can adversely affect labor supply, for example by discouraging participation among secondary household workers or discouraging effort and overtime on the job (e.g. Blundell and MaCurdy, 1999). Some countries (e.g. the USA) may still have scope for raising revenue at lower efficiency costs, however, through scaling back tax deductions and exemptions (e.g. Saez et al., 2009).

Taxes on the financial sector have received increased attention during recent years. There are several forms, as described in IMF (2010b) and European Commission (2010). 'Bank taxes' have been introduced by several countries over the last two years or so, but their revenue yield is relatively low. Broad-based financial transactions taxes have attracted considerable popular support, and their adoption has recently been proposed by the European Commission. Advocates argue that such taxes can raise substantial revenue at a low rate and may improve financial market performance by discouraging high-frequency trading. Others see drawbacks and risks. For instance, a financial transactions tax would increase the cost of capital with possibly adverse impacts on economic growth; their real burden will likely fall on final consumers rather than on actors in the financial sector; and these taxes are particularly vulnerable to avoidance and evasion. IMF (2010b) concludes that a tax on financial activities is more attractive. This is a tax that would be levied on the sum of profits and remuneration of financial institutions – ideally, it would tax value added in the financial sector and so serve to offset the tendency of the financial sector to become too large because of exemptions for financial services under the VAT.

BOX 16.2   EMISSIONS TAXES VERSUS CAP-AND-TRADE

Emissions taxes and cap-and-trade systems, applied to the same base, are potentially equally effective at reducing emissions. As the price of emissions allowances is (at least in part) passed forward into higher prices for fuels, electricity and so on, a cap-and-trade system would exploit the same behavioral responses across the economy for reducing emissions (reductions in energy demand, shifts towards clean power generation fuels etc.) as under an emissions tax.

Carbon taxes directly raise government revenues. Using this revenue for efficiency-enhancing purposes (e.g. to lower other distortionary taxes) is important for keeping down the overall costs of the policy to the economy. However, cap-and-trade systems can also raise comparable revenues if all allowances are auctioned and revenues accrue to a finance ministry.

Emissions price volatility can be problematic under cap-and-trade systems. Under cap-and-trade, allowance prices are determined in the market and will vary with energy demand, changes in the relative price of clean and dirty fuels, technological advances and so on. This price volatility raises program costs over time and can deter clean technology investments (which often have high upfront costs and provide emissions savings over many years). Usually, price stability provisions (e.g. allowance banking and borrowing, price ceilings and floors) are recommended to make cap-and-trade systems behave more like a carbon tax (which fixes the emissions price).

Carbon taxes are additive to other emissions reduction efforts, while cap-and-trade systems may not be. If emissions are rigidly fixed by a cap, other measures (e.g. energy taxes, efficiency standards) that cover the same emissions sources are environmentally ineffective (they only change the allowance price). For other measures to be effective, the cap must be tightened in response to their implementation. In contrast, under a carbon tax other mitigation efforts are automatically environmentally effective.

---

benefits) are likely modest. But even if domestic revenues are used in this way, the overall impact of carbon pricing is unlikely to have a big impact on overall employment, either way, given the offsetting, adverse employment effects of higher energy prices (Goulder, 2002, part III). On the other hand, if retained revenues are not used productively (i.e. they are not used to offset other tax distortions or fund public spending with comparable efficiency benefits) the revenue-recycling benefit is squandered, and the overall costs of the policy to the economy can be considerable higher (see below).

Promoting Technology Transfer

Comprehensive carbon pricing in developed countries can, via international carbon markets, promote emissions mitigation in, and technology transfer to, developing countries, although monitoring and verifying emissions reductions in those countries

is challenging. In the absence of comprehensive emissions pricing in developing countries, mitigation projects (e.g. adoption of clean technologies) can still be encouraged, if
sources in developed countries subject to pricing are credited for funding such projects.
The demand for offsets (and hence the flow of carbon market funds) is maximized when
all, rather than just a portion, of developed-country emissions are priced. However,
establishing whether mitigation projects in developing countries would have occurred
anyway (in the absence of offset programs) can be difficult. Measuring carbon reductions can also be challenging (e.g. accurate estimates of the carbon contained in many
forested areas are not available at present). And mitigation projects may also lead to offsetting emissions (as, e.g. when forest preserved in one area is offset by forest clearance
elsewhere). Carbon markets should therefore be expanded progressively, as institutional
capacity for monitoring and verifying emissions reductions improves.

III. SPECIFIC CARBON-PRICING PROPOSALS AND OTHER DOMESTIC, CARBON-RELATED ALTERNATIVES

This section discusses carbon-pricing policies based on their environmental effectiveness, revenue potential, cost-effectiveness, incidence and feasibility. Other domestic,
carbon-related instruments (e.g. taxes on electricity and motor fuels) are also considered.
Finally, options for improving the feasibility of carbon pricing – the most promising
option – are discussed.

Carbon-pricing Scenarios

A carbon price of \$25 per (metric) ton (expressed in current dollars) is considered, as
applied to all sources of energy-related CO_2 emissions in 2020 for developed countries.
At least according to the central case from a recent US government study, this is a
reasonable price to put on CO_2 emissions in 2020 from the perspective of charging for
environmental damages.[8] But even if widely implemented in major emitting countries,
and ramped up steadily over time, most likely these emissions prices would not be high
enough to limit eventual mean projected global warming to 2 °C above pre-industrial
levels (the official target agreed among policy makers at international climate meetings).[9]

Carbon taxes imposed on the CO_2 sources already covered in the EU ETS (about
15 percent of developed country emissions – see below) would have no effect on those
emissions (unless the cap were tightened in response). The only impact would be to
lower emissions allowance prices consistent with the cap. In this case the carbon tax is
essentially equivalent to auctioning allowances.

Effectiveness

Gauging the impact of carbon pricing on future emissions cannot be done with accuracy
given that future emissions, the availability of future technologies to lower emissions,
and other policy initiatives that might affect future emissions are all uncertain. These
uncertainties are reflected in wide disparities among energy-climate models in the future
emissions prices associated with emissions control targets (e.g. Clarke et al., 2009).
But based loosely on these models, we assume the \$25 per ton CO_2 price will reduce

developed-country emissions by around 10 percent relative to baseline emissions in 2020 (assuming the tax is phased in prior to that date and factored into long-run investment decisions).

Revenue

A carbon price of $25 per ton applied to developed-country CO_2 emissions would raise annual revenue of around $250 billion in 2020.[10] It is difficult to predict the fraction of these revenues that national governments would be willing to allocate for climate finance in 2020, especially during times of fiscal consolidation. But if only 10 percent of revenues from carbon pricing (or auctioning ETS allowances) were allocated for this purpose, this would provide projected climate finance revenues of $25 billion.

Cost-effectiveness

A CO_2 tax scaled to reflect marginal environmental damages is a more cost-effective way to raise revenue than broad fiscal instruments. As already mentioned, this is because the carbon tax produces environmental benefits which, up to roughly the point where the emissions externality is fully priced, more than compensate for the extra costs of this policy compared with broader fiscal instruments.

Leaving aside these environmental benefits, the costs of a $25 per ton carbon price would be modest – a back-of-the-envelope, partial equilibrium calculation suggests welfare costs in the order of around 0.03 percent of GDP for the average developed country.[11] However, this assumes revenues (for climate finance and domestically retained) are used productively: policy costs are substantially higher if revenues are not used to increase economic efficiency (or revenues are forgone by giving away free allowance allocations). According to Parry and Williams (2011) for the USA (and roughly the same level of CO_2 reductions as assumed here), costs could easily be two to four times as high as just suggested, depending on how much of the potential revenue-recycling benefit is squandered.

Cross-country distribution of revenues

Revenue collected as a percentage of GDP would vary across countries depending on their emissions intensity. Table 16.2 shows climate finance contributions (taken to be 10 percent of revenues) as a percentage of GDP for the $25 per ton CO_2 tax if the tax had been applied for year 2008 to developed countries (we take a retrospective look because detailed emissions projections at the country level are not available for 2020). The revenue allocated to climate finance is around 0.06 to 0.10 percent of GDP for most countries (see column 4). Contributions for Iceland, Norway, Sweden and Switzerland, for whom the bulk of power generation comes from zero-carbon technologies, are less than 0.05 percent of GDP. On the other hand, contributions exceed 0.10 percent of GDP for countries whose power sectors are heavily dependent on (carbon-intensive) coal, such as Australia, the USA, the Czech Republic and Poland.

Impact on energy prices

The energy price impacts are of particular concern to policy makers. There can be strong political opposition to higher energy prices from households (in their role as electricity consumers and motorists) and also industry, including downstream energy users,

Table 16.2   *Energy-related CO₂ emissions, climate finance contributions and GDP per capita by country (year 2008)*

Country	Economy-wide CO_2 emissions, million tons	Emissions intensity of GDP tons CO_2/US$1000	Climate finance contribution $25 tax on CO		GDP per capita (PPP, year 2008$)
			$ billion	% of GDP	
	(1)	(2)	(3)	(4)	(5)
Australia	398	0.59	0.90	0.13	31 581
Austria	69	0.25	0.16	0.06	32 892
Belgium	111	0.34	0.25	0.08	30 561
Bulgaria	48	0.63	0.11	0.14	10 000
Canada	551	0.53	1.24	0.12	31 502
Cyprus	8	0.44	0.02	0.10	22 500
Czech Republic	117	0.54	0.26	0.12	20 673
Denmark	48	0.28	0.11	0.06	31 091
Estonia	18	0.78	0.04	0.18	17 692
Finland	57	0.34	0.13	0.08	31 698
France	368	0.21	0.83	0.05	27 317
Germany	804	0.34	1.81	0.08	28 636
Greece	93	0.34	0.21	0.08	24 464
Hungary	53	0.33	0.12	0.07	16 100
Iceland	2	0.18	0.00	0.04	36 667
Ireland	44	0.29	0.10	0.06	35 000
Italy	430	0.28	0.97	0.06	26 077
Japan	1 151	0.32	2.59	0.07	28 175
Latvia	8	0.24	0.02	0.05	14 348
Lithuania	14	0.26	0.03	0.06	15 882
Luxembourg	10	0.32	0.02	0.07	62 000
Malta	3	0.38	0.01	0.08	20 000
Netherlands	178	0.33	0.40	0.07	33 293
New Zealand	33	0.33	0.07	0.07	23 488
Norway	38	0.20	0.09	0.04	40 417
Poland	299	0.53	0.67	0.12	14 724
Portugal	52	0.28	0.12	0.06	17 736
Romania	90	0.41	0.20	0.09	10 140
Slovak Republic	36	0.38	0.08	0.08	17 778
Slovenia	17	0.35	0.04	0.08	24 500
Spain	318	0.29	0.72	0.07	24 013
Sweden	46	0.15	0.10	0.03	32 043
Switzerland	44	0.17	0.10	0.04	34 545
UK	510	0.28	1.15	0.06	30 000
USA	5 596	0.48	12.59	0.11	38 562
Total/average	11 662	0.39	26.24	0.09	30 105

Note: For illustration, it is assumed that 10 percent of revenues are allocated to climate finance. Revenue projections assume that the tax base (emissions) will fall by 10 percent in response to the emissions price.

Source: OECD (2011b).

intermediate producers of energy and upstream (carbon-intensive) fuel suppliers. Higher energy prices may also undermine distributional objectives to the extent that they impose a disproportionately large burden on poorer households, who tend to spend a relatively large portion of their income on fuels and electricity.[12] Higher energy prices may also cause competitiveness concerns for energy-intensive firms competing in global markets (e.g. aluminum, steel, cement, plastics). Such firms may relocate activities to other countries where energy is cheaper (but perhaps more emissions intensive), causing some moderate offsetting increase in emissions. Leakage is less of a problem if carbon pricing is implemented across developed countries, although this would still exempt some large competitors.

If carbon prices are fully passed forward, a $25 per ton carbon tax would increase electricity prices by around 0.5 to 1.5 cents per kWh across OECD countries, based on year 2008 data (Table 16.3, column 2). On average this would increase residential electricity prices by about 4 percent and industrial electricity prices by about 7 percent, with price impacts smaller for countries like France, Norway and Sweden that use predominantly carbon-free generation technologies (Table 16.3, columns 5 and 8). Gasoline prices would rise by 6 cents per liter, or 3 to 9 percent of (year 2009) pump prices, with the proportionate price increase being higher in countries with low pre-existing retail prices such as the USA (Table 16.4, columns 1 and 2). Natural gas prices would rise by about $1.50 per thousand cubic feet, which amounts to about 15 percent of the average OECD price projected for 2020. The price of coal would increase by about $45 per (short) ton, which would more than double coal prices in countries such as the USA (IEA, 2010b).

Other (Domestic) Carbon-related Instruments

Electricity taxes
These taxes can raise significant revenues. A relatively small tax of US 0.04 cents per kWh on power generation in developed countries, or a tax of $1 per ton on CO_2 emissions from electricity generation (which would average about US 0.04 cents per kWh), would each raise projected annual revenue in 2020 of about $5 billion. Alternatively, a more substantial tax of $25 per ton on CO_2 emissions from the power sector would raise revenues of around $100 billion, or $10 billion for climate finance if 10 percent of revenues are earmarked.[13] However, as already noted, pure electricity taxes (levied per kWh) perform poorly on environmental grounds compared with carbon pricing as, for example, they do nothing to lower the emissions intensity of power generation. If the electricity tax is scaled to CO_2 emissions, environmental effectiveness is greatly improved, although the policy still fails to reduce emissions outside of the power sector.

Electricity taxes are not a complement to carbon pricing. With emissions priced, raising further revenue from electricity taxes could involve higher economic costs than raising that revenue from broader fiscal instruments. Both policies have a (slight) contractionary effect on overall economic activity, but in addition electricity taxes distort energy markets (once emissions externalities have been corrected). Nonetheless, if comprehensive carbon pricing is infeasible, taxing carbon emissions from the power sector would be a promising alternative option (combined with policies for other sectors), given that for many countries the huge bulk of the low-cost opportunities for CO_2 reduction are in the power sector.

Table 16.3 Impacts of carbon pricing on electricity prices (for year 2008)

Country	Emissions rate, gram CO_2/kWh	Increase in generation cost from $25 CO_2 tax, US cents/kWh	Residential				Industrial		
			Current elect. price, US cents/kWh	CO_2 tax	Current excise tax	Current total tax	Electricity price, US cents/kWh	CO_2 tax	Current excise tax
				% of electricity price				% of electricity price	
	(1)	(2)	(3)	(4)	(5)	(6)	(7)	(8)	(9)
Austria	183	0.5	26.2	1.7	13.2	29.9	15.4	3.0	17.2
Belgium	248	0.6	23.5	2.6	8.9	26.2	13.9	4.5	8.1
Czech Republic	544	1.4	19.2	7.1	0.8	16.7	14.8	9.2	1
Denmark	308	0.8	36.5	2.1	35.0	55.0	11.1	6.9	13
Finland	187	0.5	0.5	2.7	7.0	25.1	9.7	4.8	3.7
France	83	0.2	15.9	1.3	10.4	25.0	10.7	1.9	10.3
Germany	441	1.1	32.3	3.4	0	16.0	10.9	10.1	0
Greece	731	1.8	15.2	12.0	0.3	8.7	11.4	16.0	0
Hungary	331	0.8	20.6	4.0	0	18.4	16	5.2	0.8
Ireland	486	1.2	25.5	4.8	0	11.9	16.9	7.2	0
Italy	398	1.0	28.4	3.5	16.0	24.7	27.6	3.6	21.4
Japan	436	1.1	22.8	4.8	1.8	6.5	15.8	6.9	7.3
Luxembourg	315	0.8	23.7	3.3	7.1	12.8	13.6	5.8	9.9
Netherlands	392	1.0	25.8	3.8	12.5	17.2	14.1	7.0	13.3
New Zealand	214	0.5	15.2	3.5	0.0	11.1	7.1	7.5	0

Table 16.3   (continued)

Country	Emissions rate, gram CO_2/kWh	Increase in generation cost from $25 CO_2 tax, US cents/kWh	Residential				Industrial		
			Current elect. price, US cents/kWh	CO_2 tax	Current excise tax	Current total tax	Electricity price, US cents/kWh	CO_2 tax	Current excise tax
				% of electricity price				% of electricity price	
	(1)	(2)	(3)	(4)	(5)	(6)	(7)	(8)	(9)
Norway	5	0.0	13.3	0.1	13.0	33.0	5.9	0.2	20
Poland	653	1.6	16.7	9.8	3.9	21.9	12	13.6	5.4
Portugal	384	1.0	21.5	4.5	0	4.8	12.7	7.6	0
Slovak Republic	217	0.5	23.1	2.3	0	15.9	19.5	2.8	0
Spain	326	0.8	21.2	3.8	4.2	18.0	10.3	7.9	4.9
Sweden	40	0.1	19.4	0.5	17.9	38.0	8.3	1.2	0.8
Switzerland	27	0.1	16.4	0.4	2.5	9.6	9.4	0.7	4.4
UK	487	1.2	20.6	5.9	0	4.8	13.5	9.0	3.7
USA	535	1.3	6.84	19.6	0	0.0	11.6	11.6	0
Average	307	0.8	19.5	4.1	5.9	17.4	12.0	5.9	5.6

Notes: CO_2 tax is calculated by multiplying the CO_2 per kWh by an assumed tax of $25 per ton. This is an approximation, given that emissions rates would fall (moderately) in response to the tax, although on the other hand costs of switching to cleaner-generation fuels would be reflected in higher prices. The total tax for residential electricity includes excise taxes and general sales/value added taxes. Some countries in Table 16.2 for which recent data are not available are not included in the above table. The USA does not impose taxes on electricity at the national level.

Source: IEA (2010b), pp. 73–307, 314, 343; OECD (2011b).

Table 16.4   Impacts of carbon pricing on gasoline prices (for year 2009)

Country	Retail fuel price (unleaded gasoline) US \$/litre (1)	CO$_2$ tax % of retail fuel price (2)	Vehicle excise tax revenue relative to fuel expenditures (3)
Australia	1.08	5.4	8.6
Austria	1.45	4.0	13.4
Belgium	1.83	3.2	14.5
Canada	0.91	6.4	4.4
Czech Republic	1.43	4.0	4.0
Denmark	1.78	3.3	28.0
Finland	1.78	3.2	17.7
France	1.68	3.4	5.7
Germany	1.80	3.2	7.0
Greece	1.40	4.1	26.5
Hungary	1.41	4.1	8.5
Ireland	1.53	3.8	34.0
Italy	1.71	3.4	10.1
Japan	1.29	4.5	22.2
Luxembourg	1.44	4.0	2.3
Netherlands	1.87	3.1	37.8
New Zealand	1.05	5.5	24.5
Norway	1.89	3.1	29.3
Poland	1.32	4.4	0.9
Portugal	1.72	3.4	12.5
Slovak Republic	1.55	3.7	3.3
Spain	1.39	4.2	7.8
Sweden	1.59	3.6	5.3
Switzerland	1.39	4.2	35.2
UK	1.55	3.7	11.8
USA	0.65	8.9	6.3

Source: IEA (2010b).

The distributional incidence of electricity taxes across countries is more even than for taxes scaled to CO$_2$ per kWh. This is because there is greater variation in electricity CO$_2$ emissions than in electricity consumption across countries. For example, if applied to 2008 emissions, the burden of climate finance contributions under the pure electricity tax (assuming 100 percent of revenue is earmarked) is 0.029 percent of GDP for Norway, 0.015 percent for the USA, and 0.011 percent for Poland (Table 16.5, column 3) But if the tax is applied to electricity CO$_2$ emissions, Norway's burden falls to zero, while that for the USA and Poland rises to 0.020 and 0.018, respectively (Table 16.5, column 5).

Taxes on gasoline and petroleum consumption
A \$25 per ton tax on CO$_2$ would add about 22 cents per gallon to the price of gasoline.[14] Suppose a surcharge of this amount, applied only to gasoline, were introduced in developed countries. This would raise annual revenues of around \$5 billion, but reduce emissions only moderately. As indicated in Table 16.6, if applied in 2008, the surcharge

Table 16.5   Cross-country incidence of electricity taxes, selected countries (for year 2008)

Country	Gross electricity production, billion kWh	Burden of 0.04 cents/kWh tax		Burden of $1 per ton tax on electricity CO_2	
		$ billion	% of GDP	$ billion	% of GDP
	(1)	(2)	(3)	(4)	(5)
Austria	67	0.03	0.010	0.01	0.004
Belgium	85	0.03	0.010	0.02	0.006
Czech Republic	84	0.03	0.016	0.05	0.021
Denmark	37	0.01	0.009	0.01	0.007
Finland	77	0.03	0.018	0.01	0.009
France	574	0.23	0.013	0.05	0.003
Germany	637	0.25	0.011	0.28	0.012
Greece	64	0.03	0.009	0.05	0.017
Hungary	40	0.02	0.010	0.01	0.008
Ireland	30	0.01	0.008	0.01	0.009
Italy	319	0.13	0.008	0.13	0.008
Japan	1083	0.43	0.012	0.47	0.013
Luxembourg	4	0.00	0.005	0.00	0.004
Netherlands	108	0.04	0.008	0.04	0.008
New Zealand	44	0.02	0.017	0.01	0.009
Norway	142	0.06	0.029	0.00	0.000
Poland	155	0.06	0.011	0.10	0.018
Portugal	46	0.02	0.010	0.02	0.009
Slovak Republic	29	0.01	0.012	0.01	0.007
Spain	314	0.13	0.011	0.10	0.009
Sweden	150	0.06	0.020	0.01	0.002
Switzerland	69	0.03	0.010	0.00	0.001
UK	389	0.16	0.008	0.19	0.010
USA	4368	1.75	0.015	2.34	0.020

Source: OECD (2011b).

would have raised revenue of about $30 billion in the USA, but less than $20 billion in total from all other developed countries.[15] If 10 percent of revenues were allocated to climate finance, this would provide $5 billion. The economy-wide emission reductions from the change in fuel consumption would be modest relative to those under the carbon tax. Gasoline accounted for about 15 percent of developed-country energy-related CO_2 emissions in 2008 and transportation emissions are less sensitive to pricing than power sector emissions.

However, whether gasoline taxes make sense on economic grounds depends on whether currently existing taxes undercharge for a range of vehicle externalities. Prevailing (federal and state) fuel taxes in the USA appear to be well below levels needed to charge motorists for congestion, accident risk, local pollution and CO_2 emissions (Parry and Small, 2005). Up to a point, this means that higher fuel taxes in the USA would be a more efficient way to raise additional revenue than broader fiscal instru-

Table 16.6   Cross-country incidence of surcharges on gasoline and petroleum consumption (for year 2008)

Country	Gasoline consumption, billion liters	Burden of 6 cents/liter tax		Petroleum consumption, billion liters	Burden of 2 cents/liter tax	
		$ billion	% of GDP		$ billion	% of GDP
	(1)	(2)	(3)	(4)	(5)	(6)
Australia	19.1	1.1	0.17	56.3	1.1	0.17
Austria	2.4	0.1	0.05	16.7	0.3	0.12
Belgium	2.0	0.1	0.04	42.3	0.8	0.26
Canada	42.2	2.5	0.24	132.2	2.6	0.25
Czech Republic	2.7	0.2	0.08	12.5	0.2	0.12
Denmark	2.4	0.1	0.08	10.7	0.2	0.12
Finland	2.3	0.1	0.08	12.6	0.3	0.15
France	12.2	0.7	0.04	114.7	2.3	0.13
Germany	28.3	1.7	0.07	150.4	3.0	0.13
Greece	5.5	0.3	0.12	25.3	0.5	0.18
Hungary	2.1	0.1	0.08	9.4	0.2	0.12
Iceland	0.2	0.0	0.11	1.2	0.0	0.23
Ireland	2.4	0.1	0.09	11.3	0.2	0.15
Italy	15.5	0.9	0.06	96.3	1.9	0.12
Japan	57.9	3.5	0.10	282.3	5.6	0.16
Luxembourg	0.6	0.0	0.11	3.6	0.1	0.23
Netherlands	5.9	0.4	0.06	61.1	1.2	0.22
New Zealand	3.2	0.2	0.19	9.2	0.2	0.18
Norway	1.8	0.1	0.06	13.5	0.3	0.14
Poland	5.7	0.3	0.06	31.5	0.6	0.11
Portugal	2.0	0.1	0.06	17.0	0.3	0.18
Slovak Republic	0.9	0.1	0.06	5.1	0.1	0.11
Spain	8.6	0.5	0.05	91.2	1.8	0.17
Sweden	5.0	0.3	0.10	20.4	0.4	0.14
Switzerland	4.6	0.3	0.10	15.8	0.3	0.12
UK	22.9	1.4	0.07	102.0	2.0	0.11
USA	530.0	31.8	0.27	1149.6	23.0	0.20
Total/average	788.4	47.3	0.16	2493.9	49.9	0.17

Source: EIA (2011).

ments. However, this does not necessarily carry over to countries like Germany and the UK, where high gasoline taxes may overcharge motorists for externalities (Parry and Small, 2005). The problem is that fuel taxes are poorly suited to addressing major problems like traffic congestion, which instead warrant per mile tolls that rise and fall during the course of the rush hour on busy roads.

Gasoline consumption represents only about one-third of petroleum consumption for the average developed country, so even a modest levy on all petroleum consumption would still raise significant revenues. A levy equivalent to 8 cents per gallon on all petroleum products in developed countries (about $3 per barrel of oil or $9 per ton of

CO_2) would raise revenues of around \$50 billion (Table 16.6), again providing about \$5 billion for climate finance if 10 percent of receipts are allocated to climate finance. Such a levy would also (albeit very slightly) address concerns about macroeconomic risks from oil price volatility. Nonetheless, petroleum levies still miss out on the majority of emissions reduction opportunities – for example, they do not raise the price of coal, the most carbon intensive fuel.

In short, there is no substitute for broad-based carbon pricing.

Overcoming Resistance to Carbon Pricing

Given the central importance of carbon pricing in mitigating global climate change, novel strategies need to be developed to enhance acceptability. Appropriate strategies may differ across countries depending on the nature of political obstacles.

One possibility is to preserve the competitiveness of trade vulnerable industries through border tax adjustments, but these are contentious (see Box 16.3).

Alternatively, some of the domestically retained revenues from carbon pricing might be used to adjust the broader tax/benefit system to offset the potentially regressive effects of higher energy prices. For example, revenues might fund both a general reduction in personal income taxes and an earned income tax credit (which helps low-income households). In Australia's prospective carbon-pricing program, much of the revenues will be used to fund a general increase in the threshold for personal income taxes.[16]

Another promising approach is to scale back pre-existing, environmentally ineffective energy taxes. In many OECD countries, the impacts of carbon taxes on electricity prices could be offset by lowering pre-existing excise taxes on electricity use.[17] As indicated in Table 16.3, 12 out of the 24 countries could completely neutralize the impacts of carbon pricing on residential electricity prices by scaling back excise taxes on electricity use (i.e. these taxes exceed, often by a large amount, the potential pass-through of carbon pricing in residential prices). Similarly, ten countries (mostly the same ones) could fully offset the effect of carbon pricing on industrial electricity prices by cutting excise taxes.[18] Furthermore, replacing environmentally ineffective policies with policies targeted more directly at dirty fuels – especially coal – may significantly alleviate local environmental problems.[19]

The burden on motorists from higher fuel prices could also be offset in many countries by lowering vehicle ownership taxes. Annual revenues from vehicle ownership taxes (primarily sales taxes, registration fees and road taxes) are equivalent to more than 5 percent of annual expenditures on motor fuels in 21 out of 26 developed countries and exceed 10 percent of fuel expenditures in 14 countries (Table 16.4, column 3). In fact, in 22 out of 26 developed countries ownership taxes could be scaled back to prevent an increase in the total tax burden (fuel plus vehicle ownership) on the average motorist (compare columns 2 and 3 in Table 16.4).

Feebates: An Alternative to Carbon Taxes

If broad carbon pricing is infeasible, feebates are a possible alternative for the nearer term. Feebates are tax/subsidy combinations that can be used to lower the emissions intensity of power generation and improve the energy efficiency of vehicles, household

BOX 16.3   CARBON PRICING AND BORDER TAX ADJUSTMENTS

In principle, imposing taxes on carbon 'embodied' in imports can be an efficient way to address concerns about emissions leakage. This will ensure that foreign-produced output is treated, in climate terms, the same as domestic production. Such border tax adjustments (BTAs) can therefore help mitigation efforts without impacting the international competitiveness of domestic industry, while potentially also encouraging participation by other countries. And from a wider perspective of global efficiency, such adjustments can have a role to play (Keen and Kotsogiannis, 2011).

Because of the product or sector-specificity required, BTAs could be administratively complex. Determining the appropriate (equivalent) carbon tax for any given product is not straightforward. BTAs should accurately reflect the production process used by the exporting firm, including the use of inputs and their carbon content. Information on this will often not be available and generating it can be costly and easily give rise to arbitrary decisions and inappropriate outcomes. One major operational challenge is that climate-change-related policies often will target plants or enterprises in certain industries, whereas the application of BTAs will center on the products that are cleared through Customs at the border. The precedents in those areas of trade policy where detailed information is needed to determine whether and how much to tax imports – most notably antidumping and countervailing duty mechanisms – are not encouraging (Moore, 2010). Aside from the formidable technical challenges of accurately assessing the carbon emissions that have been generated by producing imported products, the history of the use of contingent trade policies suggests that there is a high probability that domestic import-competing industries will lobby and capture the process to their advantage.

Important legal and equity issues also need to be considered. There has been much discussion and argument regarding the WTO compatibility of trade measures that are related to domestic measures to reduce GHG emissions. In the absence of concrete disputes there is significant uncertainty whether BTAs would be deemed to be consistent with WTO obligations and much will depend on the specifics of any BTA regime.*

Many developing countries oppose a globally uniform carbon tax on the grounds that it is appropriate for those countries that are primarily responsible for global warming to take most action to reduce GHG emissions. The unilateral application of BTAs as part of the implementation of carbon tax regimes could give rise to retaliation. Indeed, some major developing countries have indicated that they will consider such a response.

Moreover, studies suggest that using trade policy to take action against countries that have decided not to apply a carbon tax may do little to further the goal of reducing global GHG emissions (e.g. McKibbin and Wilcoxen, 2009; Dong and Whalley, 2009; Rutherford, 2010). These suggest that BTAs could have adverse consequences on developing countries' output and trade. The magnitude of the effects will depend on how the BTAs are applied. Mattoo et al. (2009), for example, note that BTAs imposed by OECD nations will impact developing-country manufacturers much more detrimentally if tariffs are based on the carbon content embodied in imports than if they are based on the carbon content of competing domestic production in the jurisdiction imposing a carbon tax.

Note: *Many analysts have concluded that BTAs motivated on the basis of a national carbon tax are likely to violate the WTO's national treatment requirement. However, in practice a government has significant scope to justify the use of BTAs under a 'general exceptions' provision that permits countries to use trade policy measures to protect the environment, so long as they apply to all trading partners equally and are least trade-restrictive (see Low et al., 2011).

appliances and so on. A combination of feebates across different sectors can exploit many of the economy-wide CO_2 reduction opportunities (some opportunities, reduced use of vehicles or air conditioners, would be difficult to exploit under this approach). Feebates can be designed to raise revenue – the more revenue they raise, the greater their impact on the price of energy and energy-using products.

For the power generation sector, a feebate would impose a tax (or fee) per kWh on relatively dirty generators, equal to an emissions price times the difference between their CO_2 per kWh (averaged across their portfolio of plants) and a 'pivot point' CO_2 per kWh. And relatively clean generators would receive a subsidy (rebate) per kWh equal to the same emissions price times the difference between the pivot point and their average CO_2 per kWh. The policy is cost-effective because all generators receive the same reward (lowering their tax rate or raising their subsidy) for reducing CO_2 intensity through switching to lower- or zero-carbon fuels and/or improving plant efficiency.

A feebate price of $25 per ton of CO_2 per kWh, and a pivot point set at 90 percent of the projected industry-average CO_2 per kWh, would yield revenues of approximately $6 billion in 2020 for the USA.[20] Given that the USA accounts for about 60 percent of power sector CO_2 emissions in developed countries, total revenues from similar feebate policies applied more broadly could amount to $10 billion. To maintain revenues over time the pivot point would need to be reduced in line with the progressive decline in emissions intensity of power generation (in fact revenues might increase over time with growth in the emissions price).[21]

Feebates can also promote energy efficiency, most obviously for new passenger vehicles. In this case there would be a tax on relatively fuel inefficient vehicles – equal to an emissions price times the difference between their CO_2 per mile and a pivot point CO_2 per mile – and a subsidy for relatively fuel efficient vehicles – equal to the emissions price times the difference between the pivot point and their CO_2 per mile (e.g. Small, 2010). Again, the policy is cost-effective as it provides the same reward across all vehicle classes and across all possibilities for reducing CO_2 per mile (e.g. improvements in engine efficiency, reductions in vehicle size, use of lighter materials). Feebates could also be applied to electricity-using durables (e.g. refrigerators, air conditioners). Across sectors, a set of feebates is cost-effective if a uniform price is imposed on emissions from different sectors.

Feebates may be more acceptable politically than carbon taxes, because they avoid a large increase in energy prices. For example, under the feebate discussed above for the power sector, the increase in residential and industrial electricity prices is only about one-tenth as large as under the carbon tax.[22] However, feebates raise much less revenue – and the more they are designed to raise revenue, the greater their impact on energy prices, as net tax payments are passed forward into prices.

IV. CHARGES ON INTERNATIONAL AVIATION AND MARITIME FUELS

Background

International aviation and maritime fuels are a significant and growing source of emissions. In 2007, international aviation accounted for about 1.5 percent of global (energy-related) CO_2 emissions and international shipping for around 2–3 percent (AGF, 2010b; IMO, 2009).

Both sectors are undertaxed from an environmental perspective. Unlike for motor fuels, these fuels are not currently subject to excise taxes to potentially charge for emissions. Fuels are also undercharged from a broader fiscal perspective. In most countries, international ticket sales are zero-rated under the VAT or general sales taxes, causing excessive consumption of (leisure-related) air travel relative to other consumption goods. And shipping is subject to tax regimes that are in practice seen as more favorable than the normal corporate tax regimes. Therefore, up to a point charges for international aviation and maritime fuels are likely a more cost-effective way to raise finance for climate or other purposes than are broader fiscal instruments.

Underlying the current tax-exempt status of international transportation fuels is a fear that unilateral taxation would harm local tourism, commerce and the competitiveness of national carriers, and would raise import prices and reduce the demand for exports, as well as causing fuelling to take place in countries without similar policy measures. In aviation, multilateral agreements (the 1944 Chicago Convention) and a large number of bilateral service agreements imply legal obstacles to fuel taxation.

Under the auspices of the International Civil Aviation Organization (ICAO) and the International Maritime Organization (IMO), both industries are taking steps to improve the fuel efficiency of new planes and vessels, and economize on fuel use during operations. Nonetheless, raising fuel prices through fuel taxes (or emissions trading systems) would reinforce these efforts while also reducing the demand for transportation and promoting retirement of older, more polluting, vehicles.

International coordination over fuel charges is important, especially for ships, given the mobility of the tax base. However, this may harm developing countries, raising the question of how they might be compensated. This is potentially important for encouraging their participation in pricing schemes, besides meeting the UNFCCC principle of 'common but differentiated responsibilities and respective capabilities'.

Policy Scenarios

Again we consider a carbon charge of $25 per ton of CO_2. This would add about 24 cents per gallon, or about 8 percent, to the price of jet fuel and about 30 cents per gallon, or about 11 percent, to the price of heavy fuel oil used in ships.[23]

In principle, higher fuel prices could reduce aviation emissions from a variety of channels (e.g. Morris et al., 2009) – reducing the demand for passenger trips (as the charge is passed through to ticket prices), technological modifications to improve the fuel economy of new planes, retirement of older (fuel inefficient) planes, and more efficient operations (e.g. reducing time spent idling on runways or circulating airports prior to landing slots becoming available). Nonetheless, the responsiveness might be very modest, given already strong incentives to economize on fuel (which is expensive to carry) and ongoing efforts by ICAO to promote better fuel economy and more efficient operations.

Similarly, we might expect a relatively minor impact on maritime emissions. AGF (2010b, p. 38) projects that the effect of a $25 per ton charge on import prices will be less than 1 percent for most commodities, suggesting little impact on the demand for traded goods. Again, there are various ways to reduce the emissions intensity of maritime activities (analogous to those for aviation), but large responses are not projected for the scale of emissions prices considered here (e.g. IMO, 2011).

ICAO (2009) projected CO_2 emissions from international aviation would be 625 million tons in 2020, although this does not account for recently adopted goals to lower future emissions. If, for example, the combined effect of these efforts and the fuel charge would be to limit emissions to 500 million tons in 2020, the policy would raise projected revenue of $12.5 billion in that year. Suppose that developing countries impose the charge, but they get to keep the revenues they collect as compensation. This would reduce revenues available for climate finance and other purposes to about $8 billion, given that developing countries account for about 35 percent of global fuel use (ICAO, 2009).

IMO (2009) projects CO_2 from international maritime activities of about 1100 million tons in 2020, not accounting for the (modest) effect of prospective measures to reduce emissions intensity and the impact of fuel taxes. A $25 per ton carbon charge might therefore be expected to raise potential revenues in the order of $25 billion for shipping. Again, if developing countries were rebated fuel tax revenues, this would lower projected revenue to about $15 billion, given that they account for about 40 percent of fuel consumption (IMO, 2009).

Compensation and Implementation

The appropriate compensation for developing countries should, ideally, be linked to the effective incidence of the fuel charge. The first step here is to assess how much of the incidence might be 'passed backwards' to oil producers in the form of lower oil prices. Most likely, pass-back is modest, however, as refineries can be retrofitted over the medium term to produce more of other products and less of aviation fuel from a given barrel of crude oil, which effectively flattens the supply curve of jet fuel (even the supply curve for crude oil is upward sloping). The same applies even more strongly for the heavy fuel oil used in ships (see below) because, even with no retrofits to refinery capacity, heavy fuel oil may be further refined into higher-quality and higher-price products, or sold on global markets for use in industry or power stations.

Most of the incidence of international aviation fuel charges is likely borne by passengers in the form of higher ticket prices. And most of the passengers flying in and out of developing countries tend to be residents from developed countries (along with business travelers from developing countries for whom compensation is, presumably, less of an issue). While the issue needs careful study, allowing developing countries to keep revenues from fuel taxes collected at their airports may well overcompensate them from any loss of real income (e.g. due to reduced demand for local tourism) as the tax effectively enacts a transfer from wealthy, foreign passengers to the domestic government.

Maritime is more complicated. Planes typically refuel where they land so there is a close geographical connection between flight activity and fuel disbursements. In maritime, this connection is weaker. For example, a ship may fuel up in one port and drop off cargo at several countries before refuelling. Landlocked countries may not disburse any maritime fuel but may suffer from higher import prices in response to fuel charges if they import goods (by truck or rail) previously shipped to a neighboring country, while hub ports (e.g. Singapore) disperse a disproportionately large amount of maritime fuel relative to their imports. Some have therefore proposed compensation schemes for poor countries linked to their shares in global trade rather than their disbursement of maritime fuels (e.g. Stochniol, 2011). While the pros and cons of alternative compensation approaches (or hybrid approaches) need further study, it should be feasible to design an acceptable scheme.

Other implementation details would also need to be decided. In particular, these schemes could be implemented upstream on refineries (although for aviation a complication is that jet fuel is used for both domestic and international flights); on companies disbursing fuels at airports or ports; or on aircraft and ship operators. There are also subtleties in the choice of instrument – a fuel tax versus an emissions trading scheme (with allowance auctions), although issues are broadly similar to those already noted in Box 16.2. Furthermore, the appropriate balance between monitoring and verifying these policies by national governments, as opposed to international bodies, is not clear. Finally, there are potential complications posed by the prospective extension of the EU ETS to cover flights into and out of the EU, although at the moment this process is subject to litigation.

V. CONCLUSION

Broad-based carbon-pricing policies (carbon taxes and emissions trading systems with allowance auctions) are the most promising source of climate finance. If set appropriately on environmental grounds, and applied comprehensively across developing countries, these policies can raise substantial revenues for climate finance – about $25 billion if 10 percent of revenues are earmarked. These policies exploit all opportunities for reducing CO_2 emissions across all sectors – other carbon-related instruments (including other energy taxes and regulatory approaches) are less effective in exploiting these reduction opportunities. Carbon-pricing instruments also provide across-the-board incentives for clean technology development and (with appropriate crediting provisions) maximize the breadth of international carbon markets for promoting clean technology transfer to low-income countries. Because of all these benefits, carbon-pricing policies of the scale considered here are also a more attractive source of climate finance than broader (non-carbon-related) fiscal instruments.

There are some options for improving the acceptability of carbon pricing. These range from scaling back other (environmentally weaker) taxes in the power and transport sectors to help neutralize the overall burden on households and firms, to broader adjustments to the fiscal system targeted, for example, at low-income households. Another promising option is feebates to promote energy efficiency and de-carbonization of the power sector – under this approach, policy makers can choose the point at which firms pay taxes or receive rebates to trade off raising net revenue against impacts on energy prices. The drawback of these schemes is that they result in lower overall revenues for climate finance, fiscal consolidation or other purposes.

In the absence of broad carbon-pricing policies, charges on international aviation and maritime fuels are another promising option for raising revenue given that these activities are undertaxed from an environmental and fiscal perspective (although their effect on global emissions is obviously tiny relative to that from broad carbon charges). Coordinating charges internationally is important, given the mobility of the tax base (especially for shipping). And enticing developing-country participation would, presumably, require providing them adequate compensation. Nonetheless, these charges should be practically feasible, at least subject to further study of the pros and cons of different compensation schemes and implementation details.

ACKNOWLEDGMENTS

The views expressed here are those of the author and should not be attributed to the International Monetary Fund, its executive board, or its management. I am grateful to Michael Keen, Ruud de Mooij and Jon Strand for very helpful input to the chapter.

NOTES

1. For some discussion of leveraging private sources of finance see AGF (2010a), Ambrosi et al. (2011), Patel (2011), Basu et al. (2011) and World Bank et al. (2011).
2. Developed countries are taken to be the members of the EU and other Annex II countries, all of which have pledged to provide near-term climate finance. This includes the 27 EU member states as well as Australia, Canada, Iceland, Japan, New Zealand, Norway, Switzerland and the USA.
3. The present chapter was written at the same time as a couple of other (more in-depth) reports on fiscal instruments for climate finance prepared by the IMF in response to a request from the G-20 finance ministers (see IMF, 2011a, 2011b). There are a number of further options that are beyond our scope. These include, for example, scaling back agricultural protection programs, taxing satellite launches to reduce the accumulation of space debris, and taxing other sources of externalities such as road congestion, excessive fishing and household waste. Mineral rights in offshore areas that are outside any countries' territory, or in the Antarctic, may offer opportunities for climate finance. For further discussion see Atkinson (2004) and World Bank–IMF (2005).
4. See, for example, papers in Goulder (2002), part III. One of the seminal papers on this topic, Bovenberg and de Mooij (1994), showed that under plausible assumptions the optimal environmental tax equals the

Pigouvian tax divided by the marginal cost of public funds (i.e. one plus the efficiency cost associated with raising an extra dollar of revenue from distortionary labor taxes). This would imply a downward tax adjustment in the order of around 15 percent, under typical values for labor supply elasticities and labor tax wedges for the USA (Bovenberg and Goulder, 1996).

5. It should be noted, however, that because the ETS exploits the low-cost options for fuel switching in the power sector it exploits a lot more than 50 percent of the reductions that would occur if the same price covered all energy-related CO_2 emissions. The EU ETS type system also encourages adoption of carbon capture and storage technologies at coal plants and other facilities, if and when those technologies become economically viable. Nonetheless, an upstream system can easily be designed to reward adoption of these technologies through appropriate crediting provisions.

6. In fact, by lowering fuel costs per mile, they encourage increased use of vehicles, although recent evidence for the USA suggests that this 'rebound effect' is relatively modest (e.g. Small and van Dender, 2006).

7. For an extensive comparison of the effectiveness of carbon taxes over a wide range of other market and regulatory options for reducing future CO_2 emissions (in the USA) see Krupnick et al. (2010).

8. US IAWG (2010) provides a thorough assessment of the social cost of carbon (SCC) under various scenarios and using three leading integrated assessment models. The SCC reflects the discounted value of worldwide damages from the additional climate change over the next 100 years or so caused by an extra ton of emissions. Damage estimates reflect (rudimentary) attempts to measure, for example, future impacts on world agriculture, costs of protecting against rising sea levels, and non-market effects such as human mortality and species loss. Damage estimates are much disputed however (e.g. Stern, 2007), due to different perspectives on the appropriate rate at which to discount projected impacts on future generations and how to incorporate the risks of extreme warming scenarios.

9. See Clarke et al. (2009). Limiting projected long-run warming to 2 °C would require stabilizing global atmospheric concentrations of CO_2 and other greenhouse gases (where the warming potential of the latter gases is expressed in CO_2 equivalents) at about 450 parts per million (ppm). Current CO_2 equivalent concentrations, already around 440 ppm, will overshoot this long-run target, requiring development and widespread use of negative emission technologies (e.g. co-firing biomass in power generation and capturing and storing its CO_2 emissions) to more than offset all other sources of emissions to stand any chance of bringing concentrations back down to 450 ppm.

10. From Table 16.2 below, total CO_2 emissions across developed countries in 2008 amounted to 11.7 billion tons. Based on the 'current policies' scenario for 2020 in IEA (2010a), OECD emissions (which are a reasonable proxy for our developed-country emissions) are projected to decline by about 6 percent between 2008 and 2020 in the absence of policy. Accounting for the 10 percent emissions reduction in response to pricing assumed above, this leaves projected emissions of 9.9 billion tons for 2020 under the $25 per ton scenario. Multiplying by the emissions price gives the above revenue figure. These revenue estimates are approximately consistent with those in AGF (2010a), which projects that each extra $1 tax per ton on OECD CO_2 emissions will raise approximately $10 billion in revenue. See also OECD (2011a). One caveat is that the above figures do not account for offsetting revenue losses as carbon pricing reduces the base of pre-existing taxes (e.g. on electricity and motor fuels).

11. This is based on the standard approximation for the welfare costs of the emission pricing policy, namely one half times the emissions reduction, times the emissions price (e.g. Harberger, 1964). Using this formula and the above assumptions gives a total cost for 2020 of $13.7 billion. GDP for OECD countries in 2008 (which is very similar to that for developing countries) was $41 400 billion (OECD, 2011b). Assuming GDP growth of 1.5 percent a year gives GDP in 2020 of $49 498 billion. Dividing the cost number by this latter figure gives the above percentage.

12. For example, Metcalf (2007) estimated that a $15 per ton CO_2 tax in the USA in 2015 would impose a burden of 3.5 percent of income for the lowest income decile, but only 0.8 percent of income for the highest income decile. See also Burtraw et al. (2009).

13. This estimate assumes the policy would reduce power sector emissions by 20 percent.

14. Combusting a gallon of gasoline produces 0.0088 tons of CO_2 (see http://bioenergy.ornl.gov/papers/misc/energy_conv.html).

15. Revenue is understated slightly to the extent that some (small) developed countries (included in Table 16.2) are excluded from these calculations due to data being unavailable. On the other hand, the revenue is overstated somewhat as account is not taken of the (moderate) reduction in gasoline demand in response to higher fuel prices. The above revenue estimates assume the base of the gasoline tax will be the same in 2020 as in 2008, which seems reasonable given that the dampening effect of higher future oil prices and tighter fuel economy policies may roughly offset higher fuel demand from greater vehicle miles travelled (e.g. Krupnick et al., 2010).

16. On the other hand, tax reliefs (e.g. in the form of lower corporate income tax rates) for energy-intensive firms in trade-sensitive sectors may be problematic as they complicate policy design, may compromise cost-effectiveness, and there is a risk that reliefs will become permanent.

17. Under Energy Directive 2003/96/EC, minimum excise taxes on electricity are imposed in EU countries, although there are current discussions to revise this Directive to target carbon emissions more directly.
18. Value added (or sales) taxes are also applied at the household level. However, from an economic perspective it is generally not desirable to cut value added taxes as (with removal of excise taxes) this would effectively subsidize electricity relative to other goods subject to such taxes.
19. For example, coal combustion is a major cause of local air pollution, which poses mortality risks for vulnerable populations (NRC, 2009).
20. In (spreadsheets underlying) Krupnick et al. (2010), projected power generation for the USA in 2020 is 4125 billion kWh and the projected emissions intensity is 0.000597 tons per kWh. Therefore, on average (after the 90 percent rebate), there will be a net tax payment of 0.15 cents per kWh. This assumes the feebate covers all (new and existing) generation sources, rather than just new sources.
21. The feebate for power generation is equivalent to a tax on generators' CO_2 emissions, plus a subsidy per kWh of generation. The lower the pivot point, the greater the amount of revenue raised, the greater the impact on electricity prices, and the more the feebate resembles a simple tax on power sector CO_2 emissions.
22. Loosely speaking, the price increase is proportional to the amount of revenues that are raised, to the extent that revenues are eventually passed forward into higher prices for end users.
23. Combusting a gallon of jet fuel and maritime fuel produces about 0.0096 tons and 0.0117 tons of CO_2 emissions respectively (EIA, 2011). Fuel prices are for the USA from www.eia.gov/dnav/pet/hist/LeafHandler.ashx?n=PET&s=EMA_EPJK_PTG_NUS_DPG&f=M.

REFERENCES

AGF (2010a), *Report of the Secretary-General's High-level Advisory Group on Climate Change Financing*, New York: United Nations.
AGF (2010b), 'Work stream 2: paper on potential revenues from international maritime and aviation sector policy measures', Addendum no. 2 to *Report of the Secretary-General's High-level Advisory Group on Climate Change Financing*, New York: United Nations.
Ambrosi, Philippe, Klaus Opperman, Philippe Benoit, Chandra Shekhar Sinha, Lasse Ringius, Maria Netto, Lu Xuedu and OECD Secretariat (2011), 'How to keep up momentum in carbon markets?', Washington, DC: World Bank.
Atkinson, Anthony B. (ed.) (2004), *New Sources of Development Finance*, Oxford: Oxford University Press.
Basu, Priya, Lisa Finneran, Veronique Bishop and Trichur Sundararaman (2011), 'The scope for MDB leverage and innovation in climate finance', Washington, DC: World Bank.
Blundell, Richard and Thomas MaCurdy (1999), 'Labor supply: a review of alternative approaches', in O. Ashenfelter and D. Card (eds), *Handbook of Labor Economics*, New York: Elsevier, pp. 1559–695.
Bovenberg, A. Lans and Ruud A. de Mooij (1994), 'Environmental levies and distortionary taxation', *American Economic Review*, **84**(4): 1085–9.
Bovenberg, A. Lans and Lawrence H. Goulder (1996), 'Optimal environmental taxation in the presence of other taxes: general equilibrium analyses', *American Economic Review*, **86**(4): 985–1000.
Burtraw, Dallas, Richard Sweeney and Margaret A. Walls (2009), 'The incidence of U.S. climate policy: alternative uses of revenues from a cap-and-trade auction', *National Tax Journal*, **62**: 497–518.
Clarke, Leon, Jae Edmonds, Volker Krey, Richard Richels, Steven Rose and Massimo Tavoni (2009), 'International climate policy architectures: overview of the EMF 22 international scenarios', *Energy Economics*, **31**: S64–S81.
Dong, Y. and J. Whalley (2009), 'How large are the impacts of carbon motivated border tax adjustments?', EPRI Working Paper (http://economics.uwo.ca/centres/epri/).
EIA (2011), *Voluntary Reporting of Greenhouse Gases Program*, Energy Information Administration, US Department of Energy, Washington, DC, available at www.eia.gov/oiaf/1605/coefficients.html#tbl1.
European Commission (2010), 'Financial sector taxation', Taxation Paper No. 25, DG Taxation and Customs Union.
Goulder, Lawrence H. (1995), 'Environmental taxation and the "double dividend": a reader's guide', *International Tax and Public Finance*, **2**: 157–83.
Goulder, Lawrence H. (2002), *Environmental Policy Making in Economies with Prior Tax Distortions*, Cheltenham, UK and Northampton, MA, USA: Edward Elgar.
Goulder, Lawrence H. and Ian W.H. Parry (2008), 'Instrument choice in environmental policy', *Review of Environmental Economics and Policy*, **2**: 152–74.
Harberger, Arnold, C. (1964), 'The measurement of waste', *American Economic Review*, **54**: 58–76.

Hepburn, Cameron (2006), 'Regulating by prices, quantities or both: an update and an overview', *Oxford Review of Economic Policy*, **22**: 226–47.

IEA (2010a), *World Energy Outlook 2010*, Paris: OCED International Energy Agency.

IEA (2010b), *Energy Prices and Taxes, Fourth Quarter*, Paris: OCED International Energy Agency.

International Civil Aviation Organization (2009), *Update on US Aviation Environmental Research and Development*, Committee on Aviation Environment Protection, Eighth Meeting, International Civil Aviation Organization.

IMF (2010a), 'From stimulus to consolidation: revenue and expenditure policies in advanced and emerging economies', Washington, DC: IMF.

IMF (2010b), 'A fair and substantial contribution by the financial sector', Final Report for the G-20: www.imf.org/external/np/g20/pdf/062710b.pdf.

IMF (2011a), 'Market-based instruments for international aviation and shipping as a source of climate finance', Washington, DC: IMF.

IMF (2011b), 'Promising domestic fiscal instruments for climate finance', Washington, DC: IMF.

IMO (2009), *Second IMO GHG Study 2009*, London: International Maritime Organization.

IMO (2011), 'Reduction of GHG emissions from ships', Report of the Expert Group on Feasibility Study and Impact Assessment of Possible Market-Based Measures, Marine Environment Protection Committee, MEPC 62/5/14, London: International Maritime Organization.

Keen, Michael and Christos Kotsogiannis (2011), 'Coordinating climate and trade policies: Pareto efficiency and the role of border tax adjustments', mimeo, IMF.

Krupnick, Alan J., Ian W.H. Parry, Margaret Walls, Tony Knowles and Kristin Hayes (2010), *Towards a New National Energy Policy: Assessing the Options*, Washington, DC: Resources for the Future and National Energy Policy Institute.

Low, P., G. Marceau and J. Reinaud (2011), 'The interface between the trade and climate change regimes: scoping the issue', WTO Staff Working Paper ERSD-2011-1.

Mattoo, A., A. Subramanian, D. van der Mensbrugghe and J. He (2009), 'Reconciling climate change and trade policy', World Bank Policy Research Working Paper 5123.

McKibbin, W. and P. Wilcoxen (2009), 'The economic and environmental effects of border tax adjustments for climate change policy', Lowy Institute Working Paper (February).

Metcalf, Gilbert (2007), 'A proposal for a U.S. carbon tax swap: an equitable tax reform to address global climate change', The Hamilton Project, Washington, DC: Brookings Institution.

Moore, M. (2010), 'Implementing carbon tariffs: a fool's errand', World Bank Policy Research Paper.

Morris, J., A. Rowbotham et al. (2009), 'A framework for estimating the marginal costs of environmental abatement for the aviation sector', Omega and Cranfield University.

Nordhaus, William D. (2007), 'To tax or not to tax: alternative approaches to slowing global warming', *Review of Economics and Policy*, **1**: 26–44.

Norregaard, John and Tehmina S. Khan (2007), 'Tax policy: recent trends and coming challenges', Working Paper 07/274, Washington, DC: IMF.

NRC (2009), *The Hidden Costs of Energy: Unpriced Costs of Energy Production and Use*, Washington, DC: National Academy Press.

OECD (2011a), *Fossil-Fuel Support*, Paris: OECD.

OECD (2011b), *Development Perspectives for a Post-Copenhagen Climate Financing Architecture*, Paris: OECD.

Parry, Ian W.H. (1995), 'Pollution taxes and revenue recycling', *Journal of Environmental Economics and Management*, **29**: S64–77.

Parry, Ian W.H. and Kenneth A. Small (2005), 'Does Britain or the United States have the right gasoline tax?', *American Economic Review*, **95**: 1276–89.

Parry, Ian W.H. and Roberton C. Williams (2011), 'Moving U.S. climate policy forward: are carbon tax shifts the only good alternative?', in Robert Hahn and Alistair Ulph (eds), *Climate Change and Common Sense: Essays in Honor of Tom Schelling*, Oxford: Oxford University Press.

Patel, Shilpa (2011), 'Climate finance: instruments to engage the private sector', International Finance Corporation.

Rutherford, T. (2010), 'Climate-linked tariffs: practical issues', ETH Zurich Working Paper.

Saez, Emmanuel, Joel B. Slemrod and Seth H. Giertz (2009), 'The elasticity of taxable income with respect to marginal tax rates: a critical review', Working Paper 15012, National Bureau of Economic Research, Cambridge, MA.

Sandmo, Agnar (1976), 'Optimal taxation: an introduction to the literature', *Journal of Public Economics*, **6**: 37–54.

Small, K.A. (2010), 'Energy policies for passenger transportation: a comparison of costs and effectiveness', Discussion Paper, University of California, Irvine.

Small, Kenneth A. and Kurt van Dender (2006), 'Fuel efficiency and motor vehicle travel: the declining rebound effect', *Energy Journal*, **28**: 25–52.

Stern, Nicholas (2007), *The Economics of Climate Change: The Stern Review*, Cambridge: Cambridge University Press.

Stochniol, A. (2011), 'Optimal rebate key for an equitable maritime emissions reduction scheme', Briefing Paper, IMERS, London.

US IAWG (2010), 'Technical support document: social cost of carbon for regulatory impact analysis under Executive Order 12866', Interagency Working Group on Social Cost of Carbon, Washington, DC: US Government.

World Bank and International Monetary Fund (2005), 'Moving forward: financing modalities toward the MDGs', paper prepared for the Development Committee, DC2005-008.

World Bank, IMF, OECD, African Development Bank, Asian Development Bank, EBRD, EIB and Inter-American Development Bank (2011), 'Mobilizing climate finance', paper prepared at the request of G-20 finance ministers, 6 October.

17 How high should climate change taxes be?
Chris Hope

INTRODUCTION

The polluter pays principle tells us that whoever is responsible for producing pollution is also responsible for paying for the damage caused by the pollution (OECD, 1992). So anyone whose activities lead to the emission of a tonne of carbon dioxide (CO_2) should be taxed for the extra damage that is caused by the climate change due to that tonne of emissions.

It can be difficult to determine what the relevant damage includes, particularly in the case of climate change. Some fraction of any greenhouse gas, in particular CO_2, that is emitted today will remain in the atmosphere for many years, mixing thoroughly with all other emissions, and causing impacts across the globe (IPCC, 2007a).

INTEGRATED ASSESSMENT MODELS

The only way in which the sum of the damage caused by a tonne of CO_2 across all regions, all impact sectors and all time periods can be estimated is by the use of an integrated assessment model.

> Integrated Assessment Models of climate change are the formal, computerised, representations that have been created to understand and cope with this complex, global problem . . . Several attempts to define IAMs have been made. They are not totally in agreement, but do provide enough information to come up with a reasonable working definition . . . The most sensible course seems to be to accept the Weyant et al (1996) definition of IAMs of climate change as models that incorporate knowledge from more than one field of study, with the purpose of informing climate change policy. (Hope, 2005, p. 77)

Figure 17.1 shows the general form of an integrated assessment model (IAM) that can be used to estimate the impacts of climate change.

> Some models (eg PAGE, FUND and DICE) omit [at least part of] the first box, and start with exogenously specified emissions (if they are simulation models), or end with optimal tax levels (if they are optimization models). (Hope, 2005, p. 83)

Of course, the whole issue of climate change is surrounded by uncertainty. The climate sensitivity is uncertain, as are the impacts for a given temperature rise (IPCC, 2007a; Stern, 2007). Finding a way to take this uncertainty into account is one of the main tasks facing the designer of an IAM.

> Uncertainty is a common problem for policy making, particularly in the environmental area. But it is almost the main defining characteristic here. Parson and Fisher-Vanden, 1997, state that 'Uncertainty is central to climate change' (p.609). (Hope, 2005, p. 81)

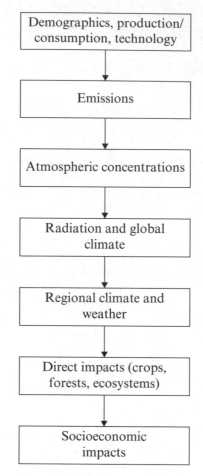

Demographics, production/
consumption, technology

Emissions

Atmospheric concentrations

Radiation and global
climate

Regional climate and
weather

Direct impacts (crops,
forests, ecosystems)

Socioeconomic
impacts

Source: Parson and Fisher-Vanden (1997), p. 596.

*Figure 17.1   The form of an integrated assessment model used to calculate climate
change impacts*

The normal use of an IAM is to calculate the total climate change impacts if the demographics, GDP and emissions follow a specified scenario, such as one of the business-as-usual scenarios defined by the IPCC's *Special Report on Emission Scenarios* (Nakicenovic and Swart, 2000). But it can also be used to find the marginal impact if one extra tonne of CO_2 is emitted, which is what is required by the polluter pays principle.

CALCULATING THE SOCIAL COST OF CARBON DIOXIDE

The calculation is done in the following way. The model is used as normal to find the net present value (NPV) of the total climate change impacts under a particular scenario, such as the A1B business-as-usual scenario. It is then run again with an extra spike of

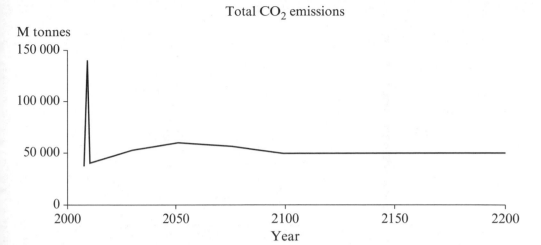

Figure 17.2 *Emissions over time used to find the social cost of CO_2 under the A1B scenario*

emissions superimposed upon the emissions from the scenario, as shown in Figure 17.2, where the spike is 100 gigatonnes of CO_2, and the NPV of the climate change impacts is again found. Subtracting the NPV of impacts from the original scenario gives the extra impacts caused by the spike of emissions. Dividing this by the number of tonnes in the spike, in this case 100 billion, gives the extra impacts from one extra tonne of CO_2 emissions, commonly called the social cost of CO_2 (SCCO2).

RESULTS FOR THE SCCO2

Figure 17.3 shows the full probability distribution of the SCCO2 from the current default version of one leading IAM, PAGE09 version 1.7 (Hope, 2012). PAGE09 uses simple equations to simulate the results from more complex specialized scientific and economic models. It does this while expressing the most important inputs as probability distributions, to reflect the full range of scientific and economic opinion (Hope, 2011b), and calculating probability distributions for all outputs, usually by running the model 100 000 times, and so correctly accounting for the profound uncertainty that exists around climate change. Calculations are made for eight world regions, ten time periods out to the year 2200, for three impact sectors and catastrophes such as the melting of the Greenland or West Antarctic ice sheets, to which the model gives the more neutral name of discontinuities.

New features in PAGE09 include the representation of sea-level rise, the explicit dependence of impacts on GDP per capita, and a new version of equity weighting that gives an appropriate extra weight to losses in poor regions or from discontinuities (Hope, 2011a).

Because the theoretical basis of the PAGE09 model is the calculation of expected utility, the appropriate summary statistic to use in policy making is the expected, or mean, value (Schoemaker, 1982). As can be seen from Figure 17.3, the mean value is

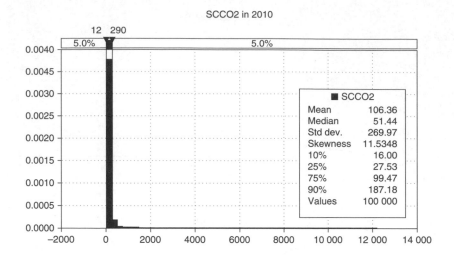

Source: Hope (2012).

Figure 17.3   The SCCO2 from the PAGE09 default model

$106 per tonne of CO_2, with a 5–95 per cent range of $12–290. This assumes GDP, population and emissions follow the A1B business-as-usual scenario. For comparison, the mean SCCO2 is given as $85 per tonne of CO_2 in the *Stern Review* (for emissions in 2001, in year 2000 dollars) (Stern, 2007), and as $21 per tonne of CO_2 (in year 2007 dollars) in a recent EPA report using a single discount rate of 3 per cent per year (US DOE, 2010).

The most obvious feature of the distribution is its large positive skewness, with a few values as high as about $10000 per tonne of CO_2. This is because the PAGE09 model keeps track of whether a discontinuity, such as the melting of the Greenland or West Antarctic ice sheets, has been triggered. These high values result when the small increase in emissions brings forward the date at which a discontinuity occurs.

The skewness is so extreme that it is difficult to see the shape of the distribution when all the values are included. So Figure 17.4 shows the same result but with the top 1 per cent of values omitted.

Comparing the mean value of $84 per tonne for this truncated distribution with the mean value of $106 per tonne when all runs are included shows that the top 1 per cent of runs contribute $22, or about 20 per cent, to the mean SCCO2 value in the default PAGE09 model. This confirms the importance of properly representing uncertainty and appropriately weighting catastrophic outcomes when making estimates of the SCCO2.

The standard deviation of the result in Figure 17.3 is $270, so the standard error of the mean is $270/sq.rt(100000), which is about $0.75. So another 100000 runs would be 95 per cent sure to produce a mean value for the SCCO2 within about $1.5 per tonne of the mean value shown in Figure 17.3. Given this level of precision, it is probably best to describe the result in round numbers as showing that the mean SCCO2, and therefore the recommended level for a climate change tax, is about $100 per tonne of CO_2.

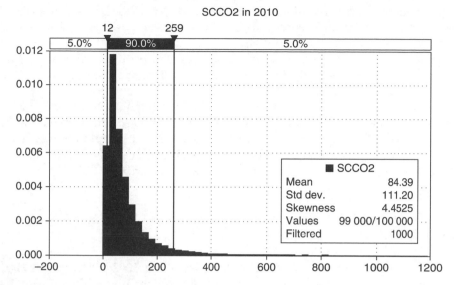

Source: Hope (2012).

Figure 17.4 The SCCO2 from the PAGE09 default model, top 1 per cent of values omitted

EXISTING CO₂ PRICES

Actual prices for CO_2 are lower than this. Sweden has had a tax on CO_2 since 1991; in 2010 its level is $40 per tonne of CO_2. British Columbia in Canada began to tax CO_2 in 2008; the tax is now $30 per tonne of CO_2 (CTC, 2011). Australia has passed a law to institute a tax in 2012, at $25 per tonne of CO_2 (Paton, 2011).

Allowances in the European Union Emission Trading Scheme (EU ETS), although not technically a climate change tax, are trading at about $20 per tonne of CO_2 in 2010. Their price has been as high as about $40 per tonne of CO_2 in 2005 and 2008, and as low as $12 per tonne in 2009 (Ares, 2011).

WHAT DOES A TAX OF $100 PER TONNE OF CO₂ IMPLY?

What changes to the prices of energy and transport does a tax of $100 per tonne of CO_2 imply? Using the UK as an example, with an assumed exchange rate of $1.6 to the pound (Defra, 2007, annex 1), it would add 12 pence to the price of a litre of petrol, which has a typical retail price in 2010 of about 120 pence per litre (AA, 2010). The price rise of about 10 per cent is relatively modest as about 65 per cent of the price of petrol in the UK is already made up of duties and tax, rather than the cost of raw materials (HMRC, 2011).

The tax would add about 1 p/kWh to the price of gas for domestic customers, which has a typical retail price of about 3 p/kWh (DECC, 2011a). It would add about 4 p/kWh to the price of electricity from coal, and just under 2 p/kWh to the price of electricity from gas, both of which have a typical retail price of 11 p/kWh (DECC, 2011a) in 2010. Energy efficiency, renewable energy sources and nuclear power would not attract a climate change tax and so would not rise in price, and may well become cheaper with economies of scale and the reductions of other taxes. The price rises of 20 to 40 per cent for gas and electricity generated from fossil fuels would make these other options much more economically attractive.

Taking into account the extra damage from emissions high up in the sky, which multiply their effect by about a factor of 3 (RCEP, 2002, para 3.35), the climate change tax would add about £60 to the price of a return air ticket from London to Barcelona. Since the advent of low-cost airlines, it is not really possible to give a typical retail price for flights such as this, but £60 is a noticeable increase in the fare.

CLIMATE CHANGE TAX REVENUE AND OTHER TAX REDUCTIONS

The UK's domestic emissions of CO_2 in 2010 are about 500 million tonnes (DECC, 2011b). So a uniform and comprehensive tax of $100 per tonne of CO_2 would raise about £32 billion of revenue in the first year. The UK's share of international air travel emissions of about 35 million tonnes of CO_2 should also be included (DfT, 2010). Doing so, again with a multiple of 3 to take into account its higher impacts, would raise about another £7 billion, giving a total revenue of about £40 billion.

For comparison, other taxes in the UK raise approximately the following amounts: council tax £25 billion, corporation tax £40 billion, value-added tax £75 billion, National Insurance contributions £100 billion, income tax £145 billion. The UK's total tax take is about £500 billion (Chote et al., 2010). So a climate change tax of $100 per tonne of CO_2 would raise about 8 per cent of the UK's tax revenues.

There are good reasons for making any tax on CO_2 fiscally neutral, mainly because the resulting reductions in other taxes would bring down the costs of employment, reduce structural unemployment and increase GDP (Green Fiscal Commission, 2009, ch. 7).

> If the revenues from the taxes are spent on reducing burdensome taxes, then the costs of tackling climate change are reduced by about 2% of GDP, and may even be converted to benefits. (Barker, 2005)

With £40 billion per year of revenue from a climate change tax, the UK government would have many options for reducing other taxes. For instance, it could choose to reduce the standard rate of income tax by about 5 pence in the pound (from 20 pence to 15 pence), losing about £20 billion per year in revenue, and reduce the standard rate of VAT by 4 per cent (from 20 per cent to 16 per cent), losing about £16 billion per year in revenue (Chote et al., 2010). The remaining £4 billion could be divided between welfare payments to the poorest, such as the present £2 billion winter fuel payments to pensioners (Kirkup, 2010), and support for basic research into new, more climate-friendly technologies.

So a tax on CO_2 of $100 per tonne would have significant effects on relative prices and the balance of tax revenues. But there are six reasons for thinking that the appropriate tax on CO_2 could be higher than the $100 per tonne mean result from the PAGE09 model.

WHY THE TAX SHOULD BE HIGHER THAN $100 PER TONNE OF CO_2

Time Horizon

The PAGE09 model has a time horizon of 2200, but some of the CO_2 emitted in 2010 will still be in the atmosphere for many years beyond this. Around 25 per cent of emissions are expected to stay in the atmosphere for several centuries (IPCC, 2001, p. 187). The effect of discounting means that any impacts caused after 2200 will be greatly reduced, but not eliminated. Crudely extending the time horizon of the PAGE09 model to 2300 increases the mean SCCO2 from the A1B scenario by about 20 per cent.

Omitted Impacts

Some impacts are omitted entirely from PAGE09, and all other integrated assessment models. The two most obvious omissions are the effects of ocean acidification (ESF, 2009), and the possible increases in war and large-scale migration from climatic causes, known as socially contingent impacts (Watkiss et al., 2005). By definition, the scale of these omitted impacts is unknown, as the relevant economic studies have yet to be attempted, but they cannot safely be assumed to be negligible, particularly if the climate sensitivity turns out to be towards the high end of what presently appears possible, 5 °C or above (IPCC, 2007a).

Richer Regions

The mean SCCO2 of $100 per tonne is the marginal impact in the year 2010, as measured by someone with the mean GDP per capita in the EU. Some regions are much richer, particularly the USA, with average annual GDP per capita of $42 000 in year 2005 dollars (World Bank, 2011), compared to about $29 000 in the EU, in purchasing power parity terms in year 2005 dollars (Indexmundi, 2011). Once these marginal impacts are corrected, using a mean marginal utility of consumption of just over 1, the appropriate values for the USA will be about 50 per cent higher than for the EU.

Other Taxes are Distortionary

The climate change tax will replace other taxes which are distortionary. For instance, income taxes reduce the incentive to work by reducing the net reward for labour below its marginal product, as assumed by models of perfect competition. There are very few reasons in a rational world why we would want to discourage work. Layard (2005) gives perhaps the only minor reason – that we expect extra income to make us happier than it actually does. So, at the margin, when we are trying to decide between a marginally higher CO_2 tax and a marginally higher income tax rate, it would initially be more economically efficient to raise the CO_2 tax rate slightly above \$100 per tonne, since this brings us best estimate damage reduction benefits of \$100, rather than raise income taxes marginally above 20 per cent, which brings us no compensating benefits.

It is unclear how high the climate change tax needs to be before the balance tilts in the other direction. Would a climate change tax of \$150 be too high? With such a tax, we may have chosen to bring standard rates of income tax down to 15 per cent. So the question that would need to be asked is whether increasing the difference between the climate change tax and the best estimate of impacts to slightly more than \$50 would cause more or less distortion than taxing income at a fraction more than 15 per cent, when we would ideally like it to be taxed at zero.

> All else being equal, we should consider the marginal impact of the final dollar of taxation for each of the taxes in our economy. What's the damage done to the economy from collecting an extra dollar of sales revenue? Of corporate income tax? Of sales tax? (Moffatt, 2011)

(And, of course, an extra dollar of revenue from a climate change tax.) To answer this question requires the use of comprehensive modes of the macroeconomy, such as E3MG (energy–environment–economy) (Barker et al., 2005) and has not yet been attempted.

Present-day Dollars

In PAGE09, the social cost of CO_2 is calculated in year 2005 US dollars, for a tonne of CO_2 emitted in 2010. Converting to year 2010 dollars would increase the values by just over 10 per cent (US Department of Commerce, 2011). If inflation continues at about 2 per cent per year, taxes valued in future years' dollars would need to be increased proportionately.

Increasing over Time

The mean social cost of CO_2 is increasing over time in real terms, as we get closer to the time when the most serious impacts from climate change might occur. The rate of increase is about 2 per cent per year (IPCC, 2007b, ch. 20).

A CLIMATE CHANGE TAX OF \$250 IN THE USA

Pulling together all the adjustments that can be quantified would suggest that the best estimate of an appropriate CO_2 tax in the USA in 2012 would be about $106 \times 1.2 \times 1.5 \times 1.14 \times 1.04 = \$225$ per tonne of CO_2, in year 2012 dollars. This does not include adjustments for the omission of some impacts from the SCCO2 calculation, nor for the taxes being replaced being themselves distortionary. Increasing by a very conservative 10 per cent to adjust for this, gives a tax of about \$250 per tonne of CO_2.

What effect would such a climate change tax have on the balance of taxes in the USA? The USA emits about ten times as much CO_2 as the UK, 5500 million tonnes per year (EPA, 2011). So a uniform and comprehensive tax of \$250 per tonne of CO_2 would raise about \$1400 billion of revenue in the first year. The USA's share of international air travel emissions is harder to estimate but appears to be about 50 million tonnes of CO_2 (OECD, 2010). Including these, again with a multiple of 3 to take into account its higher impacts, would raise about another $35 billion.

For comparison, other taxes in the USA raise approximately the following amounts: business taxes $600 billion, social insurance taxes $900 billion, *ad valorem* taxes $1100 billion, income tax $1500 billion. The USA's total tax take, federal state and local, is about $4500 billion per year (Chantrill, 2011). So a climate change tax of $250 per tonne of CO_2 would raise about one-third of the USA's tax revenues. The reductions in other taxes that would be needed to achieve fiscal neutrality would exceed even the wildest dreams of the deficit hawks in the Republican Party (CBO, 2005).

WHEN THE TAX SHOULD BE LOWER THAN $100 PER TONNE OF CO_2

There are also three reasons for thinking that the appropriate tax on CO_2 could in some circumstances be less than the $100 per tonne mean result from the PAGE09 model.

Lower Emissions

The mean value of $100 applies to one more tonne of CO_2 emitted on top of the business-as-usual emissions from IPCC scenario A1B. But if climate change is recognized as a serious problem, it would be perverse to continue to allow emissions to rise like this, even though that is what seems to be happening at present (Le Quéré et al., 2009). Suppose instead that global CO_2 emissions decline by about 50 per cent by 2050, and 80 per cent by 2100, giving peak CO_2 concentrations of slightly under 500 ppm, and a 50 per cent chance of limiting the global mean temperature rise since pre-industrial times to 3 °C.

An extra tonne of CO_2 emitted in 2010 on top of this lower emissions scenario would have a mean SCCO2 of about $75. The reduced chance of a discontinuity in this scenario, and all that that implies, means that the extra impact from one more tonne of emissions is lower than if emissions are allowed to grow unchecked.

So the choice between a climate change tax of $100 per tonne and one of $75 per tonne depends on one's view of the likelihood of emissions in the rest of the world continuing to rise in a business-as-usual fashion, or being brought under control. Economic theory would suggest that the appropriate climate change tax should assume that CO_2 emissions will follow an optimal path in the future – one that minimizes the mean sum of impacts, abatement costs and adaptation costs (Tietenberg and Lewis, 2009). Exactly what this path looks like is still to be determined, but it could well resemble the lower emissions scenario sketched out here.

Higher Discount Rates

The mean SCCO2 of $100 comes from runs of the PAGE09 model using a range of pure time preference rates and equity weights, whose mean values imply consumption discount rates ranging from the order of 2.3 per cent per year in the USA to 5.5 per cent per year in China. There is great controversy about whether such discount rates are consistent with the rates of return seen on actual investments in the economy.

Nordhaus (2007) argued that real rates of return are closer to 5.5 or 6 per cent per year in the USA, and that consumption discount rates in integrated assessment models should reflect this. As others have pointed out,

> the historic average return . . . includes the last 40 years or so, a period during which total credit market debt in the U.S. has doubled five times, and now stands at over 350% of GDP . . . The reason these returns existed in the past was because we were experiencing debt fuelled, unsustainable rates of rapid economic growth. (CIVFI, 2011)

Real rates of return on the Standard and Poor 500 with dividends reinvested have been about −3.4 per cent per year from 2000 to 2010 (SSI, 2011).

Again it is not possible to settle these controversies here. But we can see the effect of using higher values for the pure time preference rate and the equity weights, giving mean consumption discount rates of about 4 per cent per year in the USA, by running the PAGE09 model with those input value distributions. They reduce the mean SCCO2 by about $50, to around $55 per tonne of SCCO2.

Poorer Regions

The mean SCCO2 of $100 per tonne is the marginal impact in the year 2010, as measured by someone with the mean GDP per capita in the EU. Some regions are much poorer, particularly the developing world, with average global annual GDP per capita of about $10000, compared to about $29000 in the EU, in purchasing power parity terms (Indexmundi, 2011). Once these marginal impacts are weighted, using a mean marginal utility of consumption of just over 1, the appropriate climate change tax values for regions with the global average income per capita will be about 70 per cent lower than for the EU, at about $30 per tonne of CO_2. In the poorest regions of the world, such as India, with per capita income of about $4000 per year (Indexmundi, 2011), the appropriate initial climate change tax would be about $10 per tonne of CO_2.

CONCLUSION

If the best current scientific and economic evidence is to be believed, and climate change could be a real and serious problem, the appropriate response is to institute today a climate change tax equal to the mean estimate of the damage caused by a tonne of CO_2 emissions. The raw calculations from the default PAGE09 model suggest the tax should be about $100 per tonne of CO_2 in the EU. But correcting for the limited time horizon of the model, and bringing the calculations forward to 2102, in year 2012 dollars, brings the suggested tax up to about $150 per tonne of CO_2.

There are good arguments for setting the initial tax at about $250 per tonne of CO_2 in the USA, while starting off at a much lower level, maybe $15 per tonne of CO_2, in the poorest regions of the world, all in the year 2012, in year 2012 dollars.

That such policy advice would not pass the laugh test, particularly in the USA, shows that the rhetoric about getting to grips with climate change has not been seriously thought through to its logical conclusion. As a result, rather than falling, greenhouse gas emissions are continuing to rise (Le Quéré et al., 2009). A fiscally neutral significant climate change tax is the best chance we have of bringing the climate change problem under control.

REFERENCES

(All websites accessed July 2011.)

AA (2010), 'Petrol and diesel price archive 2000 to 2010', http://www.theaa.com/motoring_advice/fuel/fuel-price-archive.html.

Ares, E. (2011), 'Carbon price support, standard note: SN/SC/5927', UK House of Commons, http://www.parliament.uk/briefing-papers/SN05927.pdf.

Barker, T. (2005), *Evidence to the House of Lords Select Committee on the Economics of Climate Change*, London: The Stationery Office Ltd, http://www.publications.parliament.uk/pa/ld200506/ldselect/ldeconaf/12/5022202.htm.

Barker, T., Pan, H., Köhler, J., Warren, R. and Winne, S. (2005), 'Avoiding dangerous climate change by inducing technological progress: scenarios using a large-scale econometric model', in H.J. Schellnhuber, W. Cramer, N. Nakicenovic, T. Wigley and G. Yohe (eds), *Avoiding Dangerous Climate Change*, Cambridge: Cambridge University Press, ch. 38.

CBO (2005), *Analyzing the Economic and Budgetary Effects of a 10 Percent Cut in Income Tax Rates*, Congressional Budget Office, http://www.cbo.gov/ftpdocs/69xx/doc6908/12-01-10PercentTaxCut.pdf.

Chantrill, C. (2011), 'US Government revenues', http://www.usgovernmentrevenue.com/.

Chote, R., Emmerson, C. and Shaw, J. (eds) (2010), *The IFS Green Budget: February 2010*, Institute for Fiscal Studies, London, http://www.ifs.org.uk/publications/4732.

CIVFI (2011), 'Why real rates of return must fall', http://civfi.com/2011/05/16/why-real-rates-of-return-must-fall/.

CTC (2011), 'Where carbon is taxed', http://www.carbontax.org/progress/where-carbon-is-taxed/.

DECC (2011a), 'Energy price statistics', http://www.decc.gov.uk/en/content/cms/statistics/energy_stats/prices/prices.aspx.

DECC (2011b), 'UK greenhouse gas emissions', http://www.decc.gov.uk/assets/decc/Statistics/climate_change/1515-statrelease-ghg-emissions-31032011.pdf.

Defra (2007), 'The social cost of carbon and the shadow price of carbon: what they are, and how to use them in economic appraisal in the UK', http://www.decc.gov.uk/assets/decc/what%20we%20do/a%20low%20carbon%20uk/carbon%20valuation/shadow_price/background.pdf.

DfT (2010), 'UK transport and climate change data', UK Department for Transport, http://www2.dft.gov.uk/pgr/statistics/datatablespublications/energyenvironment/latest/climatechangefactsheets.pdf/.

EPA (2011), 'Inventory of U.S. greenhouse gas emissions and sinks: 1990–2009, executive summary', US Environmental Protection Agency, http://epa.gov/climatechange/emissions/downloads11/US-GHG-Inventory-2011-Executive-Summary.pdf.

ESF (2009), 'Impacts of ocean acidification', European Science Foundation, http://www.solas-int.org/resources/ESF__Impacts-OA.pdf.

Green Fiscal Commission (2009), *The Case for Green Fiscal Reform*, Final Report of the UK Green Fiscal Commission, http://www.greenfiscalcommission.org.uk/images/uploads/GFC_FinalReport.pdf.

HMRC (2011), 'Fuel duty rates', http://www.hmrc.gov.uk/budget2011/tiin6330.pdf.

Hope, C. (2005), 'Integrated assessment models', in D. Helm (ed.), *Climate Change Policy*, Oxford: Oxford University Press, ch. 4.

Hope, C. (2011a), 'The PAGE09 integrated assessment model: a technical description', Judge Business School Working Paper 04/2011, submitted to *Climatic Change*, http://www.jbs.cam.ac.uk/research/working_papers/2011/wp1104.pdf.

Hope, C. (2011b), 'The social cost of CO2 from the PAGE09 model', Judge Business School Working Paper 05/2011, submitted to *Economics*, http://www.jbs.cam.ac.uk/research/working_papers/2011/wp1105.pdf.

Hope, C. (2012), 'Critical issues for the calculation of the social cost of CO_2: why the estimates from PAGE09 are higher than those from PAGE2002', *Climatic Change*, doi:10.1007/s10584-012-0633-2.

Indexmundi (2011), 'GDP - per capita (PPP)', http://www.indexmundi.com/european_union/gdp_per_capita_(ppp).html.

414 *Handbook on energy and climate change*

IPCC (2001), *Climate Change 2001: Working Group I: The Scientific Basis*, Intergovernmental panel on Climate Change, http://www.grida.no/publications/other/ipcc_tar/?src=/climate/ipcc_tar/wg1/index.htm.

IPCC (2007a), *Climate Change 2007. The Physical Science Basis. Summary for Policymakers, Contribution of Working Group I to the Fourth Assessment Report of the Intergovernmental Panel on Climate Change*, Geneva, Switzerland: IPCC Secretariat.

IPCC (2007b), *Climate Change 2007. Impacts, Adaptation and Vulnerability, Contribution of Working Group II to the Fourth Assessment Report of the Intergovernmental Panel on Climate Change*, Geneva, Switzerland: IPCC Secretariat.

Kirkup, J. (2010), 'Winter fuel payment cuts to hit millions of pensioners', *The Telegraph*, 17 August 2010, http://www.telegraph.co.uk/news/politics/7951203/Winter-fuel-payment-cuts-to-hit-millions-of-pensioners.html.

Layard, R. (2005), *Happiness: Lessons From a New Science*, London: Penguin.

Le Quéré, C., Raupach, M.R., Canadell, J.G., Marland, G. et al. (2009), 'Trends in the sources and sinks of carbon dioxide', *Nature Geoscience*, **2**, 831–6.

Moffatt, M. (2011), 'What is the optimal carbon tax rate?', http://economics.about.com/od/incometaxestaxcuts/a/carbon_tax.htm.

Nakicenovic, N. and Swart, R. (2000), *Special Report on Emissions Scenarios (SRES)*, Cambridge: Cambridge University Press.

Nordhaus, W.D. (2007), 'A review of the Stern Review on the Economics of Climate Change', *Journal of Economic Literature*, **45**(3), 686–702.

OECD (1992), *The Polluter-pays-principle*, OECD Monograph, Paris: OECD.

OECD (2010), 'Reducing transport greenhouse gas emissions trends & data 2010', International Transport Forum, http://www.internationaltransportforum.org/Pub/pdf/10GHGTrends.pdf.

Parson, E.A. and Fisher-Vanden, K. (1997), 'Integrated assessment models of global climate change', *Annual Review of Energy and the Environment*, **22**, 589–628.

Paton, J. (2011), 'Gillard sets a $23 carbon tax to reduce Australia's fossil fuel dependence', http://www.bloomberg.com/news/2011-07-10/gillard-unveils-australia-carbon-tax-to-cut-coal.html.

RCEP (2002), *The Environmental Effects of Civil Aircraft in Flight*, Royal Commission on Environmental Pollution, London.

Schoemaker, P. (1982), 'The expected utility model: its variants, purposes, evidence and limitations', *Journal of Economic Literature*, **20**(2), 529–63.

SSI (2011), 'S&P 500: total and inflation-adjusted historical returns, simple stock investing', http://www.simplestockinvesting.com/SP500-historical-real-total-returns.htm.

Stern, N. (2007), *The Economics of Climate Change: The Stern Review*, Cambridge and New York: Cambridge University Press.

Tietenberg, T. and Lewis, L. (2009), *Environmental & Natural Resource Economics*, 8th edn, New York: Addison Wesley.

US Department of Commerce (2011), 'Gross domestic product', US Department of Commerce, Bureau of Economic Analysis, Washington, DC, http://www.bea.gov/national/index.htm#gdp.

US Department of Energy (2010), 'Final rule technical support document (TSD): energy efficiency program for commercial and industrial equipment: small electric motors, Appendix 15A (by the Interagency Working Group on Social Cost of Carbon): social cost of carbon for regulatory impact analysis under Executive Order 12866', available online at http://www1.eere.energy.gov/buildings/appliance_standards/commercial/sem_finalrule_tsd.html.

Watkiss, P., Downing, T., Handley, C. and Butterfield, R. (2005), 'The impacts and costs of climate change', European Commission DG Environment, http://ec.europa.eu/clima/studies/package/docs/final_report2.pdf.

Weyant, J. et al. (1996), 'Integrated assessment of climate change: an overview and comparison of approaches and results', in IPCC Working Group III report, Climate Change 1995, *Economic and Social Dimensions of Climate Change*, Cambridge: Cambridge University Press, ch. 10.

World Bank (2011), 'GDP per capita (current US$)', http://data.worldbank.org/indicator/NY.GDP.PCAP.CD.

18 State-contingent pricing as a response to uncertainty in climate policy
Ross McKitrick

1. INTRODUCTION

Suppose we have a time machine that allows us to visit the year 2040 just long enough to collect some climate data. Figure 18.1 shows the post-1979 globally averaged lower tropospheric air temperature anomaly averaged over the two satellite series developed by, respectively, Spencer and Christy (1990) and Mears and Wentz (2005). This is only one of many data series people use to try to represent the global climate as a univariate time series, but it will do for the current illustration. Figure 18.1 shows the observed data from 1979 up to the end of 2010 (shown by the vertical line), and then runs the series forward using assumed trends and random numbers to conjecture two quite different futures. In the circles the next three decades exhibit continued variability but no upward trend, and even a slight downward trend. The squares show variability and a strong upward trend. Now suppose that, given an identical future greenhouse gas emissions trajectory, the data we collect in 2040 will look like one of those two paths. If we could find out which one would be observed, would it affect today's policy choices?

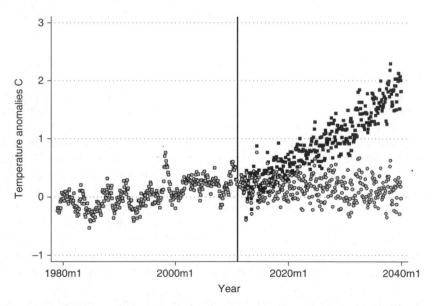

Note: Left of the vertical line are observations from weather satellites of the global average lower troposphere temperature anomaly. Right are conjectures using trends and random numbers.

Figure 18.1   Two conjectured atmospheric warming paths, 1979–2040

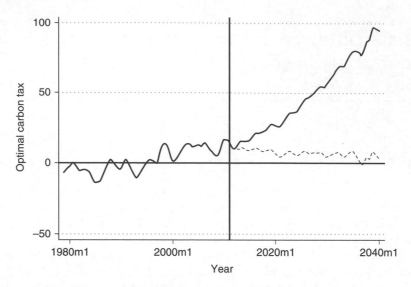

*Figure 18.2   Possible optimal emission price paths corresponding to future warming
scenarios in Figure 18.1*

Obviously the answer is yes. The fact that we do not know what the graph will look like
has led to longstanding and well-known political difficulties in devising policy strategies. In this chapter I will critically review the main current approaches to dealing with
the uncertainty, and then propose an alternative that I believe is more likely to lead to
the right policy outcome than any others currently being examined. Briefly stated, my
argument is as follows:

- Forecast-based proposals (such as from integrated assessment models, or IAMs)
  for making optimal climate policy decisions effectively assume that we can agree
  what the data to the right of the line will probably look like, and we only need to
  resolve the time-sequence of emission pricing. While the optimal time-sequence is a
  significant puzzle to be solved, framing the issue in this way assumes away all the real
  uncertainty that makes the problem difficult in the first place. If we make a commitment to a long-term policy based on IAM analysis, when we get to, say, 2040, there is
  a high probability we will realize that we followed the wrong emission pricing path.
- Bayesian updating and other learning strategies involve placing bets on the
  unknown future, then observing the effects of the policy decisions and revising our
  strategy when we have learned enough to figure out if the bet was right or wrong.
  The main lesson of these approaches is that, in the climate case, this kind of learning will be too slow to be of any use in guiding policy now or in the foreseeable
  future. Consequently, when we get to 2040, we will probably not know if we were
  on the correct path or not.
- Each of the futures in Figure 18.1 implies a corresponding optimal emissions price
  path, which, for instance, might look something like those in Figure 18.2. If we
  knew the future temperatures with certainty, we would, in principle, be able to
  work out the optimal emission tax path.

- The state-contingent approach involves starting an emissions tax at the current best guess as to its optimal level, then specifying a rule that updates it each year based on the observed climate state. At present we do not know what the path will look like, but if we choose the rule correctly, we can know today that as of 2040 we will have followed the closest possible approximation to the optimal price path. Furthermore, the greatest economic gains will accrue to agents that make the most accurate forecasts about the climate state and, hence, the emissions price.

- Under a state-contingent pricing rule, the need for accurate forecasts of the future tax for the purpose of guiding investment decisions will create market incentives to make the maximum use of available information and the most objective climate forecasts to guide optimal behavioural responses to the policy. Consequently, while the state-contingent price path will only track the actual optimum within error bounds, there is no information currently available that could identify a better price path than the one generated by information markets induced by a state-contingent pricing rule.

The rest of this chapter explains these ideas in more detail.

2.   UNCERTAINTY AND INERTIA PROBLEMS IN CARBON DIOXIDE EMISSION PRICING

Sources of Uncertainty

In one respect, analysis of carbon dioxide (CO_2) emission pricing is simple compared to other air emissions such as sulphur dioxide (SO_2) or particulate matter (PM). Since there is no CO_2 scrubber technology, knowing the amount of fuel consumed yields a close estimate of the total CO_2 released, whereas fuel consumption can be quite uncorrelated with other emissions, depending on the pollution controls and combustion technology in place. For that reason, CO_2 emissions are easily represented in empirical and computational economic models, as long as the consumption levels of coal, oil and natural gas are resolved. However, the time element that connects CO_2 to its external costs is considerably more complex than for other emissions. SO_2 and PM do not stay aloft very long after release (days or weeks), and investment in a scrubber today will yield potentially large emission reductions within a year, so from a planning point of view the time path of control policies for these pollutants can be considered as a sequence of short-run decisions.

In the CO_2 case, however, time complicates the planning problem in several ways.

(a)   The atmospheric residency of CO_2 is measured in decades, so emissions today could have effects many years into the future, and each year's emissions have marginal effects that accumulate with those of other years.

(b)   The response of the climate system to changes in the atmospheric stock of CO_2 may be slow, especially if the ocean acts as a flywheel, delaying effects for decades or centuries.

(c)   Since there are no scrubbers, emission reductions must take the form either of changes in combustion efficiency, fuel-switching or reductions in the scale of output, all of which take time to plan and implement.

(d)   Economically viable technology for generating electricity and converting fossil fuels into energy is subject to innovation over time, and while it is reasonable to assume that some innovations and efficiency gains will be realized, the effects and timing of changes can only be conjectured.

(e)   Since the stock of CO_2 mixes globally, actions of individual emitters are negligible. The only policies that would affect the atmospheric stock must be coordinated among all major emitting nations, and such processes are slow and subject to uncertain success.

Parson and Karwat (2011) examine these issues under the headings of 'inertia' and 'uncertainty'. If we faced only inertia, we could sit down today and devise an optimal intertemporal policy plan that would yield the right sequence of interventions at the right time through the future. This is something similar to what IAMs do: assuming that we know the important parameters of the system, we can solve for the optimal intertemporal emission pricing path. On the other hand, if we faced only uncertainty, we could make short-term decisions on the expectation that new decisions would be made at each point in the future as circumstances changed. It might be argued that this is more like what climate policy has been in practice for the past 20 years: a series of short-term decisions that resolve momentary political pressures, but which do not seem rooted in an overall intertemporal plan. Faced with both uncertainty and inertia, Parson and Karwat conclude that sequential decision making is necessary, although they do not spell out how such a process would work in practice. The state-contingent approach, it will be shown, attempts to create a formal structure for sequential decision making in light of both uncertainty and inertia.

With regard to climate change there are two very large sources of anxiety that have fuelled decades of intense controversy. On one side are those who believe that the threat from CO_2 and other greenhouse gas emissions is substantial, and who fear that inadequate policy actions are being taken, so that future generations will experience serious welfare losses due to global warming. On the other side are those who believe that the threat from CO_2 and other greenhouse gas emissions is small, and who fear that implementation of policies sufficiently stringent to achieve large emission reductions will impose costs on current and future generations far larger than any benefits they yield. For the first group, the fear is that by the time enough information is obtained to resolve uncertainty about the environmental effects of CO_2 it will be too late to avert intolerable environmental damages. For the second group, the fear is that if we act now to try to prevent such damages we will have incurred intolerable economic costs by the time they are shown to have been unnecessary.

So-called 'no regrets' policies are sometimes invoked to try to make this wrenching dilemma disappear, but they are irrelevant to the discussion. There is a strain of argument that says, in light of the threat of catastrophic (or even somewhat harmful) global warming, we must act, and the actions we propose would actually make us better off by saving energy and reducing air pollution anyway, so on balance it is better to implement them. This argument fails once the details are examined. The scale of emission reductions necessary to substantially affect the future stock of global atmospheric CO_2 is quite large, namely worldwide reductions of some 50 per cent or more, and marginal local changes in energy efficiency would not begin to be sufficient. Improvements in energy efficiency that actually make consumers and firms better off are automatically adopted by rational economic decision makers anyway, yet CO_2 emissions continue to rise globally as population and income rise. And air pollution is already subject to regulation throughout the developed (and much of the underdeveloped) world. If we assume that households, firms and governments have already made reasonably efficient decisions as to energy efficiency and pollution reduction, further large-scale reductions in CO_2 emissions must be, on net, costly. In other words, policies that might have a trivial cost will only have trivial climatic effects. The policies that actually have an effect on the climate must entail a large economic cost. The dilemma is real.

Integrated Assessment Models and Pseudo-optimal Solutions

The IAM approach of Nordhaus (2007) and co-authors yields a solution that can be described as 'pseudo-optimal'. It assumes that the modeller knows the key parameters that govern the economy and the climate, and solution of the model yields a smooth policy 'ramp' in the form of an escalating tax on CO_2 emissions over time. This solution can only be considered optimal if we assume that the model parameters are correct. But strong assumptions about key functional forms and parameter values are not put to the test by implementation of the policy. If decision makers were to commit to a policy path based on the IAM analysis, it would amount to acknowledging the inertia but not the uncertainty in the policy problem. The lack of recognition of the extent of uncertainty in the IAM approach is one of the bases of the criticism of Weitzman (2009).

Bayesian Learning Models

Kelly and Kolstad (1999) and Leach (2007) introduced learning into the IAM framework by supposing that we observe the response of the climate to policy innovations, and then we use such information in a Bayesian updating routine. The goal is to accumulate enough information that the policy maker can decide, at 5 per cent statistical significance, whether or not to reject the hypothesis that the correct policy is being implemented. Uncertainty and inertia interact in an interesting way: uncertainty about even one or two key inertia (lag) parameters is sufficient to delay for hundreds of years the identification of an expected-optimal policy rule. With only two model parameters subject to uncertainty, Leach (2007) showed that the learning time ranges from several hundred to several thousand years, depending on the base-case emissions growth rate. An expanded version of the model, incorporating simple production and an intertemporal capital investment structure, not only yields a time-to-learn measured in centuries, even when most model parameters are assumed known, but depending on which of several climate data sets is used to form the priors, the policy path may never converge on the correct target.

It is an illusion to suppose that the IAM, or the pseudo-optimal approach, is better, because we apparently follow an optimal path from the outset. The difference between them is that in the Bayesian approach we eventually learn if we are on the wrong path and in the IAM approach we never do.

Insurance and Fat Tails

Weitzman (2009) looked at the global warming problem as one of trying to price an insurance contract when there is a non-trivial probability of extreme damages. Geweke (2001) had shown that a basic insurance problem can become degenerate if a few features of the set-up are chosen in a particular way. If the risk is distributed normally and utility is of the constant relative risk aversion form, and the change in consumption over the insured interval is expressed as e^C, where C is future consumption relative to current consumption, then the expected cost of insuring future consumption under some general conditions can be shown to take the form of a moment-generating function for the t distribution, which does not exist, or is infinitely large, making it impossible to place a finite value on a full insurance contract. Weitzman's adaptation of this model to the climate case depends on some specific assumptions, some of which are conventional and some of which are not. One unusual assumption is that there is a possibility of infinite (+ or −) climate sensitivity, or in other words, that while the possibility of an extreme change in the climate (20 degrees or more) may be small, it cannot be ruled out, no matter how large. To perform conventional cost–benefit analysis it is necessary either to truncate the range of climate sensitivities or assume that the distribution has 'thin tails'. But, as Weitzman points out, this implies that the optimal insurance policy depends on assumptions about the distribution of possible climatic changes in regions where there are too few observations to know for sure. Hence cost–benefit analysis using IAMs assumes away extreme risks, and cannot therefore provide an economic case for ignoring them. Nordhaus (2009), Pindyck (2011) and others have critiqued the Weitzman model, especially for its assumption of infinite marginal utility as consumption gets very low.

The State-contingent Approach

McKitrick (2010) proposed an alternative approach to the pricing of complex intertemporal externalities which focuses on developing an adaptive pricing rule, rather than a long-term emissions path. In the standard economic model of pollution pricing, current damages are a direct function of current emissions (see Figure 18.3).

In simple problems of this sort, the solution is to impose an emissions price

$$\tau = D'(e_t) \tag{18.1}$$

where D is the damage function and e_t is total current emissions, which are assumed to be observable. In the presence of inertia, emissions may have lagged effects, and the length of the delay may itself be unknown. If so, then we are currently experiencing the effects not only of present-day emissions, but also of emissions that have occurred at

Figure 18.3   Relationship of current emissions and current damages

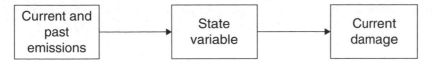

Figure 18.4   Revised relationship of current emissions and current damages

some point in the past. A lag process may arise from stock effects in which the decay rate is less than 100 per cent each period. But here we are interested in the case in which, instead of directly causing damages, emissions affect some aspect of the environment (such as the average air temperature, possibly with lags due to geophysical processes), and those changes cause damages. In that case Figure 18.3 would be redrawn as Figure 18.4.

Emissions affect an observable state variable $s(t)$ and current-period damages are a function of the current state variable $D(s(t))$. The state variable, in turn, is determined as

$$s(t) = s(e_t, e_{t-1}, \ldots, e_{t-k}) \qquad (18.2)$$

where the lag length k is unknown. Equation (18.2) is sufficiently general to handle stock-flow problems. If the stock of emissions is denoted M_t, then the state may be described as the function $s(M_t)$, where $M_t = \delta M_{t-1} + e_t$ and δ is the stock decay rate. But since the emissions flow is the policy target it will be necessary to include e_t in the specification of (18.2). A form such as $s(M_t)$ would not lead to a policy-relevant conclusion since the policy maker cannot control M_t (or past values of e_t).

Note also that $s(t)$ is not, in most cases, the same as the stock of the pollution (i.e. the atmospheric concentration). It is therefore necessary to identify the correct state variable. Hsu (2011) discusses the state-contingent pricing option for CO_2 emissions and proposes a list of possible candidates for the state variable. The list will be critically evaluated below, but for the moment our point is simply that for a complex issue like global warming it can be difficult to agree on what to measure, and not all proposed values of the state variable make sense for the purpose of configuring an optimal policy mechanism.

Setting that aside for now, given a definition of s, the discounted present value (DPV) of damages is

$$V(t) = \sum_{j=0}^{\infty} \beta^j D(s(t + j)) \qquad (18.3)$$

where β^j is the discount factor j periods ahead. The optimal emissions price is

$$\tau(t) = \frac{\partial V(t)}{\partial e_t} \qquad (18.4)$$

The influence of current and past emissions on the state variable is complex and uncertain. While this adds to the difficulty of determining how current emissions ought to be priced, it also implies that the state variable contains information about the effect of emissions over time, and this information can be used to reduce uncertainty. The next section explains how to use observations on s to approximate $\tau(t)$.

3.   DERIVATION OF THE STATE-CONTINGENT PRICING RULE

Required Assumptions

Following McKitrick (2010), the derivation requires a number of assumptions.

(A1)   The state function s is homogeneous of degree c.

This implies

$$s(\lambda e_t, \ldots, \lambda e_{t-k}) = \lambda^c s(e_t, \ldots, e_{t-k})$$ (18.5)

We do not necessarily assume linear homogeneity.

(A2)   Over the interval $(t - k, \ldots, t + k)$, ∇s is locally autonomous, that is, $\frac{\partial s(t + i)}{\partial e_t} = \frac{\partial s(t)}{\partial e_{t-i}}$ for all $i = 0, \ldots, k$

This imposes slightly more structure on s, as it implies the marginal effects over a lag of length i are independent of t. For example, suppose s depends on emissions out to three lags:

$$s(t) = s(e_t, e_{t-1}, e_{t-2}, e_{t-3})$$

At time t, the partial derivative with respect to the second argument is $\partial s(t)/\partial e_{t-1}$. At time $t + 1$, the partial derivative with respect to the second argument is $\partial s(t + 1)/\partial e_t$. (A2) requires that these partial derivatives be equal. This will be true, for instance, if s is a function of a weighted sum or moving average of the es. It will not be true if s is non-linear in the individual es, in which case it will be true only approximately, where the approximation will depend on how much 'curl' s has over time.

(A3)   The function $s(t)$ is locally linear in e_t, and current-period emissions must have a non-zero effect on $s(t)$.

This assumption states that in the direction e_t, $\partial s/\partial e_t = v$ in the neighbourhood of e_t, where v is a positive constant and may be arbitrarily small.

(A4)   At each time t, the damage function $D(t)$ can be approximated by a step-wise quadratic (in emissions) function, such that at time t, $\partial D(t + j)/\partial e_{t+j} \approx \delta e_t$ for $j = 0, \ldots, k$.

This is different from, and slightly more restrictive than, assuming that D is quadratic. It states that the slope can be extrapolated forward at time t over $k-1$ subsequent

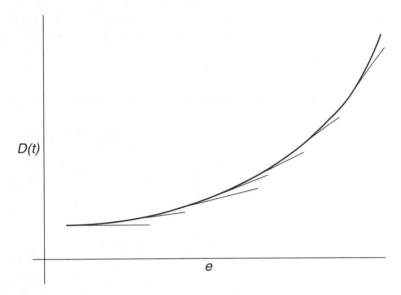

Figure 18.5 Approximation of damage function embodied in (A4)

periods, and that the extrapolation will be re-set each period. (A4) can be pictured as an approximation for D by a porcupine-like pattern of tangent lines, as shown in Figure 18.5. The approximation will get worse the larger is k. However, in cases where k is large, effects mix slowly across long time spans and we would not expect total damages to be strongly non-linear (convex) in current emissions, so the extrapolation is automatically used across a shorter interval in circumstances where it is less accurate over long intervals. In other words, the larger is k (implying greater potential inaccuracy in the slope extrapolation), the less is the likely curvature in D (implying less inaccuracy in the slope extrapolation). If D has a strong curvature, then current emissions must have a rather strong immediate effect, so we expect the tangent lines in Figure 18.5 to be shorter, mitigating the extrapolation error. (A4) also implies that marginal damages are zero when emissions are zero, which is a common and reasonable assumption.

(A4) combined with (A3) implies

$$\frac{\partial D(t + j)}{\partial s_{t+j}} = \theta e_t \text{ for } j = 0, \ldots, k \tag{18.6}$$

where $\theta = \delta/\nu$, a positive constant. Neither c nor θ is typically encountered in ordinary environmental policy models. θ measures the change in damages due to a change in the state variable, per unit of emissions. c is the degree of homogeneity of the state variable, and thus equals the sum of the partials of $s(t)$ with respect to k emission lags, with each term multiplied by $e_{t-k}/s(t)$. If we assume that emissions are constant for k periods, the product $c\theta$ can be shown to equal the sum of the partials of the damage function with respect to k lags of e, all divided by $s(t)$. In other words, $c\theta$ can be thought of as, approximately, the marginal damage rate, or marginal damages proportional to the value of the state variable.

The next assumption eliminates the role of discounting when summing marginal damages over the lag interval k. While it introduces a form of imprecision in the tax instrument, it is conservative in the sense that it will tend to overstate rather than understate the value of the sequence of damages.

(A5)   In the evaluation of τ using equation (18.3), β^j is set equal to unity for $j = 1, \ldots, k$.

The discounted present value of damages is defined by equation (18.3). Since emissions only have an effect for up to k lags we can write this as

$$\frac{\partial V(t)}{\partial e(t)} = \sum_{j=0}^{k} \beta^j \frac{\partial D(s(t + j))}{\partial s(t + j)} \frac{\partial s(t + j)}{\partial e_t} \tag{18.7}$$

$$= \theta e_t \sum_{j=0}^{k} \beta^j \frac{\partial s(t + j)}{\partial e_t} \quad \text{by (18.6)}$$

$$= \theta e_t \sum_{j=0}^{k} \frac{\partial s(t)}{\partial e_{t-j}} \quad \text{by (A2) and (A5)}$$

Now denote, respectively, moving sums, moving averages and weights of e_t as $E_t \equiv \sum_{i=0}^{k} e_{t-i}$, $\bar{e}_t \equiv E_t/(k + 1)$ and $\varepsilon_i(t) \equiv e_{t-i}/E_t$. Note that $\Sigma \varepsilon_i = 1$.
   (A1) implies (by Euler's theorem)

$$c \cdot s(t) = \sum_{j=0}^{k} e_{t-j} \frac{\partial s(t)}{\partial e_{t-j}}$$

Hence

$$c \cdot s(t) = E_t \sum_{j=0}^{k} \varepsilon_{t-j} \frac{\partial s(t)}{\partial e_{t-j}} \tag{18.8}.$$

Equation (18.8) states that $cs(t)/E_t$ equals a weighted average of the partial derivatives of s. In situations where the emissions do not vary too much in percentage terms, over the k-period interval an unweighted mean will be a reasonable approximation to the weighted mean; hence:

(A6) $c \cdot s(t)/E_t \approx \dfrac{1}{k + 1} \displaystyle\sum_{j=0}^{k} \frac{\partial s(t)}{\partial e_{t-j}}$

Combining (A6) and equation (18.7) yields

$$\frac{\partial V(t)}{\partial e_t} \approx \theta e_t (k + 1) c \cdot s(t)/E_t$$

Hence

$$\tilde{\tau}_t = \gamma \frac{e_t}{\bar{e}_t} s(t) \qquad (18.9)$$

where $\gamma = c\theta$ and $\sim$ denotes the approximation to the optimum.

Equation (18.9) is the state-contingent pricing rule. It is an easily calculated approximation to the marginal damages of the complex intertemporal externality shown in equation (18.3). Comparing it to equation (18.4), the trick to its usefulness has to do with how knowledge of the future is represented. In a system with inertia, emissions today will have an effect on the state variable over an uncertain span into the future. But this means that the current value of the state variable must also reflect the influence of past emissions. If the set of lag relationships extending over the past is structurally similar to the set of lag relationships that will extend into the future, then the current observation of s contains information about how past emissions affected today's climate, and hence how today's emissions will affect the future climate. Equation (18.9) uses that information to guide the emission price path.

Note that although (18.9) only uses information dated at time t, the underlying form of the tax is determined by equation (18.4). That means that, in principle, equation (18.9) charges firms today for the future value of their emission damages as well as the current value.

There remains one assumption to invoke: $\bar{e}_t$ is not known exactly since it depends on the unknown lag length k, so a lag length must be selected. However, trailing averages are smoothing devices, so unless the emissions series is extremely volatile, $\bar{e}_t$ will be relatively stable across a range of choices of lag length.

Figure 18.6, taken from McKitrick (2010), shows the implied value of the state-

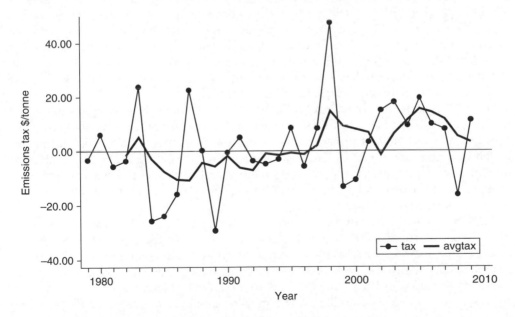

Note: 'avgtax' denotes three-year moving average.

Figure 18.6   Value of state-contingent tax on greenhouse gas emissions since 1979

contingent emissions tax over the 1979–2010 interval, using the mean temperature of the tropical troposphere (see next section) and calibration of the free parameter to yield a value of $15 per tonne in 2002. The thick line is a three-year moving average.

McKitrick (2010) presented synthetic examples calibrated to the stylized facts of global warming, in which a series of complex and varied simulations into the future is shown, and in each case the state-contingent mechanism $\tilde{\tau}_t$ closely followed the unobservable optimum, even when known errors in approximation were introduced, with correlations typically in the >95 per cent range.

Choice of State Variable

(A1) requires that the state variable be chosen carefully, so that while it represents the factor that influences damages, as much as possible it represents the influence of emissions rather than exogenous factors. To give an example, while forest fires are costly, and potentially influenced by global warming, it would not make sense to use the number and extent of forest fires as the state variable for global warming since they are mainly influenced by causes unrelated to greenhouse gas emissions. Denote these other causes as $x(t)$. If the state variable for forest fires is $s(t) = \alpha x(t) + (1 - \alpha)e_t$ and α is close to 1, then (A1) can easily be seen to fail.

On the other hand, even though (A1) would certainly hold if we simply used total emissions (or the atmospheric stock) as the state variable, this would not make sense unless it were known that the sensitivity term $ds(t)/de(t) = 1$ for all t, which is not the case, or at least it is not known to be the case. Imposing it by assumption means that we are assuming away all the uncertainties over the magnitude of $ds(t)/de(t)$. In the extreme, it would imply that e_t should still be priced the same regardless of whether $ds(t)/de(t) = 0$ or $ds(t)/de(t) \to \infty$, which is clearly not credible. CO_2 has not historically been regulated as an air pollutant because it is naturally occurring and harmless for humans except at extremely high concentrations. It is only a candidate for policy intervention if it turns out to have a strong effect on the climate by changing atmospheric temperatures. So the state variable should be chosen to capture that contingency.

Hsu (2011) correctly points out that the state variable must be non-manipulable and reliably, regularly and uncontroversially measurable. He then proposes a basket of measures arising from the tort principle of focusing on the harms done by climate change, which would include (a) the global mean temperature, (b) counts of days of 'unusually' high or low temperatures, (c) counts of extreme rainfall or drought events, (d) rises in sea level, (e) ocean acidity and (f) numbers of hurricanes above a certain intensity level. The weighting scheme is not specified.

McKitrick (2010), by contrast, proposes just one measure, the mean temperature of the tropical troposphere. This is based on the argument that, as well as being non-manipulable and reliably measurable, the state variable must represent an index of the unique effect of emissions with as rapid a response time as possible. The climate modelling work surveyed in IPCC (2007) and Karl et al. (2006) clearly points to the tropical troposphere as a place in the climate system where an unusually strong and rapid signal of the effects of greenhouse gases should be observable. Thorne et al. (2011) and Fu et al. (2011) provide updated analyses emphasizing the importance of the response to greenhouse forcing in the tropical troposphere as a metric of climate sensitivity, since this is where climate model behaviour is strongly constrained. The Arctic surface temperature is also believed to be strongly influenced by greenhouse gases, but is also strongly influenced by natural phenomena and, being an oceanic region, is subject to poor spatial sampling.

The Hsu basket approach includes measures that can be criticized on the grounds mentioned above. The tort concept is not the appropriate criterion for picking a state variable; instead the connection to greenhouse gases is. The tort concept matters when it comes to estimating parameters of the damage function, but that is a different step.

Items (b), (c) and (f) are sub-grid-scale weather phenomena with poorly understood and controversial connections to greenhouse gases. The history of global weather shows that extended intervals of elevated drought conditions or excessive moisture can and do arise purely due to natural variability, so these measures do not satisfy (A1). In other words, such index terms could go up (or down) for extended periods even if greenhouse gases turn out to have no effect on the climate. Item (e) is not an effect of global warming; it is potentially a measure of the atmospheric CO_2 level since acidification (or reduced alkalinity) would occur if large enough quantities of CO_2 were to dissolve in the world's oceans. While use of this measure would make sense if the state variable were ocean alkalinity (and the damage function were defined accordingly), if the damages are connected to atmospheric warming the use of ocean alkalinity would, at best, be equivalent to making emissions the state variable, which is a flawed concept for the reason noted above. Item (d) has a somewhat controversial connection to global warming, as Hsu notes, with sharply varying projections as to the rate of rise over the coming century. The main deficiency as a state variable is that sea levels have apparently followed a steady upward trend for centuries, and what is at issue is not whether this will continue but whether global warming will cause it to accelerate. So the more relevant concept would be the acceleration of sea-level rise. But the oceanic system is so slow to respond that any acceleration may not be apparent for a long time, due to the inertia involved. Finally, item (a) suffers from controversy and data quality problems (see, e.g., de Laat and Maurellis, 2006; McKitrick and Nierenberg, 2010; Fall et al., 2011), especially over the ocean (Thompson et al., 2008; Christy et al., 2001), as well as being subject to many influences other than greenhouse gases, which is why the satellite-based mean temperature of the tropical troposphere would be a more accurate state variable in this context (see discussion in McKitrick, 2010).

These are all very large issues and the above summary paragraph is not sufficient to dispose of them all. Indeed, one of the benefits of the state-contingent approach is that it forces a discussion on the issue of how climate change ought to be measured. The fact that it is hard to come up with a simple answer provides some needed context to the policy discussion, namely that there is work to be done simply to clarify what we are talking about before supposing we are in a position to measure the costs and benefits of policy. If we do not agree on how to measure global warming, how would we know if a policy, once enacted, was making a difference?

4.   A PERMITS-BASED PREDICTION MARKET TO GENERATE FUTURE PRICES

Use of a state-contingent, myopic pricing rule does not mean that we put a price on emissions only after the damage is done, since the approximation is to the discounted present value of current and future emissions. Also, businesses are forward-looking and investment plans are based not only on today's prices, but on expectations of future prices. While those forecasts may be wrong, there is no incentive for firms to make systematically biased forecasts, since in the future the actual prices will be revealed and mistakes will be costly. However, the basic proposal in McKitrick (2010) does not reveal the expectations in the market, except as they are implicit in firms' investment decisions.

In order to reveal the future price expectations, Hsu (2011) proposed extending the tax rule of McKitrick (2010) to include a tradable exemption permits market for future periods. The permits market would allow firms in year t to purchase a per-tonne exemption from paying the tax in years $t + 1$, $t + 2$ and so on, out to the limit of the forecast horizon. Hence it would establish, in effect, a prediction market for the state-contingent tax and, by implication, the state variable. As with any prediction market, the maximum financial rewards would accrue to those who make the best forecasts, so the published list of price futures would then imply an optimal and objective set of climate forecasts. If agents believe that the permit price for year $t + 10$ (say) is below the likely value of the tax rate, or in other words if agents believe that people are underestimating the amount of likely warming between now and then, they will have an incentive to buy permits, and will thereby drive up the price. And likewise, agents who believe that warming is being overestimated will have an incentive to sell permits, or short the market.

Hsu (2011, pp. 212–13) captures well the incentives this would create for basing forward investment behaviour on the most accurate possible scientific information about climate:

> The price bid by emitters for say, permits to emit in 2020, would speak volumes about private expectations of the consequences of climate change, free from, as climate skeptics claim, conspiracies by climate scientists to shore up their research grant fiefdoms, or desires by radical environmentalists who really wish to use climate change as an excuse for imposing environmental restrictions . . . Without an obvious ideological horse in the race, emitters like [American Electric Power, AEP] will brutally and honestly evaluate the credibility of climate science, and spend its climate investigation money carefully. It is the participation of large emitters in a cap-and-trade program for emissions futures that is likely to make or break the credibility of climate science. In essence, this proposal uses markets to turn the evaluation of climate science over to those emitters that will potentially rely on those permits for their emitting operations. And such a liability could be very significant: in 2005 AEP emitted approximately 161 million tons of CO_2; if one assumed a very modest carbon tax that was set to five dollars per ton at current climate outcomes, AEP's annual carbon tax liability would be about $805 million. If climate outcomes increased by say, twenty-five percent, its annual carbon tax liability would top one billion dollars. All 101 electricity generators in the EPA's Egrid database would have a combined current carbon tax liability (assuming a rate of five dollars per ton of CO_2) of $8.75 billion. Environmental advocates may chafe at the notion that the greatest greenhouse gas emitters will have such a large say in evaluating the quality of climate science, but $8.75 billion is a lot of impetus for honestly evaluating climate science.

Some of the concern expressed by authors like Stern (2006) and Weitzman (2009) is that the climate may be subject to non-linear effects of greenhouse gases (Hsu, 2011 also raises this concern). Future greenhouse warming may be subject to sudden, rapid acceleration into catastrophic levels of environmental damage. But when we consider how to integrate this possibility into current decision making, we confront the same dilemma as before, namely between the fears of highly costly potential environmental damages and highly costly potential policy mistakes. The prospect of sudden catastrophic change only amplifies the magnitude of each of these two apparently awful options. But now imagine a state-contingent tax is implemented along with a 30-year sequence of prediction markets. Suppose initially the prediction markets show a slow, smooth ramp in CO_2 emission prices. This would indicate that the market discounts the possibility of a serious bifurcation or (so-called) 'tipping point' into catastrophic change. But to the extent that anyone can construct a credible scientific argument that a bifurcation is approaching, they would know that those emission futures are underpriced and they would be able to invest profitably in them, knowing that once their scientific arguments were digested by the prediction market, there would be a predictable ramp in prices. The prediction market would be the most reliable instrument available for generating a rational signal of a coming bifurcation. Again, the reason is that firms would have a strong financial incentive to get the climate science right, to whatever extent possible, and not to make a systematic forecast error or adopt a disingenuous view of the underlying problem. If an objective information-processing system like a prediction market existed, research warning of a climate non-linearity would be 'brutally and honestly' evaluated, and could be the basis for a spike in emission price futures, thereby providing an instantaneous signal of the gravity of the threat. For this reason, while the state-contingent price mechanism is not guaranteed to resolve the uncertainty at the core of Weitzman's analysis, it provides incentives for the maximum possible resolution; in other words, no other policy path could provide a more objective basis for forming expectations about possible future catastrophes.

The tradable futures market would not allow firms to evade paying the tax on future emissions, since they would have to buy the exemption permits. Instead it would allow firms to trade on changes in expectations about the future path of the state variable and the emissions price. This illustrates another distinction between the state-contingent approach and the IAM approach. In the latter, it is assumed that we possess correct, unbiased scientific information regarding climate, and all we require is a mechanism to implement the optimal price. No allowance is made for the possibility that key parameter estimates (such as climate sensitivity) might be biased due to distorted incentives for scientists who prepare such forecasts. In the IAM approach, implementation of the policy does not induce an improvement in the scientific basis of policy. But the approach in this chapter does not assume that we have correct scientific information. It works even if the information we have today is incorrect, either through technical inadequacy or researcher bias. The policy mechanism rewards those agents that eliminate bias and inaccuracy in their forecasting work by allowing them to trade in futures markets on the difference between the current market price and their expectations of how it will evolve.

Some implementation issues that need to be addressed for other types of policy would also need to be addressed in the present analysis. For instance, ideal implementation would be at the global level, with each country imposing the tax based on damage valuations over the whole world, rather than at the national level. But incentives favour free-riding, making it difficult to ensure global participation. On the other hand, the tax approach has the advantage that revenues can be retained domestically to reduce the excess burden of the tax system, reducing the macroeconomic cost of implementation, and the state-contingent nature of the policy provides reassurance that the stringency will be increased only if the underlying problem is shown objectively to merit such tightening, and both these features may increase incentives for multilateral cooperation compared to alternative policies.

Another potential issue is that as new information becomes available, policy makers may decide that the initial calibrated value of the tax was incorrect and must be revised. This too would be a problem for any policy mechanism. In the state-contingent case an argument can be made that incentives favour efficient acquisition of the information required to optimally calibrate the tax. In this regard the feedback between policy and science would be particularly fruitful, since emitters subject to the tax will have an incentive to pay for the best, most objective information they can get, and there is no assumption that the forecasts made prior to implementing the policy are correct, or will be validated in the future.

5.   FURTHER EXTENSIONS TO THE BASIC CONCEPT

Endogenous Emissions Response

Since no information about abatement costs is used in deriving the tax t, it may seem that it cannot be a complete policy prescription. The tax paths derived in integrated assessment models are solutions to a two-sided optimization problem, with intertemporal damages netted against intertemporal abatement costs. However, it is important to bear in mind that the formula above does not prescribe a policy *path*; it yields a *rule* that ties the tax rate to the environmental state. The actual path of taxes over time will be determined by the evolution of the state variable, and the ensuing level of abatement will be determined by emitters who respond to the current and expected future tax rates according to their current and future marginal abatement costs. If the capital stock is highly variable, then firms will respond to current emission tax rates as they would to any variable input costs. If capital is fixed and time-to-build lags are long, firms will need to form forecasts of the future values of the tax rate, which in turn will depend on future values of the temperature variable, and the usual structure of optimal investment under uncertainty will ensue.

The simulations in McKitrick (2010) assume that the emissions tax $\tau(t)$ is a function of the future state $s(t)$, but not vice versa. The coherence between the approximate tax, given by equation (18.9), and the actual optimum is demonstrated with this assumption in force. However, it is likely to be the case that $s(t)$ is a function of $\tau(t)$ as well. This will certainly be true if emission tax increases reduce future temperatures, but it will also be true as long as they reduce future emissions since $e(t)$ enters equation (18.9) directly, as well as through $s(t)$, and the emissions path in McKitrick's simulations are exogenous. The rationale is that emissions are added up globally whereas the tax is imposed by single governments, and no one country can do much to reduce global total emissions through unilateral action. However, if all countries (or a substantial majority) were to enact the tax, total emissions would be affected by the path of $\tau(t)$. In that situation it has not been shown that the state-contingent pricing rule would yield a stable approximation to the true optimum. This is something that needs to be addressed in subsequent research.

Coalition-formation

An interesting feature of the state-contingent tax is its potential ability to appeal to a broad coalition of interests.[1] People with conflicting expectations about the future evolution of the state variable will nevertheless each expect to observe his or her preferred policy path. Those who think emissions have no effect on climate will expect low emission taxes to prevail in the future, and those who think they have strong effects will expect the tax to increase rapidly. Since each agent expects to get his or her preferred outcome, it may be easier to get agreement for implementation. One of the challenges of climate policy is the need to get agreement at the global level. Different regions have different views on the urgency of the problem and how it compares with their domestic economic priorities, which makes it all but impossible to get agreement on emission targets, or to ensure compliance with earlier agreements. Asking policy makers around the world to agree on a state-contingent tax might be easier. The tax revenue would stay within each country, reducing the burden of inequality across different nations. And during the negotiations, there would be no reason for countries that took opposing views on the likely future path of temperatures to take opposing views on whether the tax is desirable, since each party will expect to get what they consider to be the 'correct' outcome.

Suppose the stringency, and hence costliness, of a policy can be summarized as a parameter z, where a higher value of z corresponds to a more stringent policy. A potential voter (person i) has a private view of the optimal value of z given his or her beliefs about the marginal effect of emissions on $s(t)$, which we denote by s^i. Their preferred policy is thus $z_i(s^i)$. If $z > z_i(s^i)$, then the proposed policy is deemed too strict, and vice versa.

Typical median voter models only require $z \geq z_i(s^i)$ to ensure voter i's support, namely people are satisfied as long as z equals or exceeds their preferred policy. But suppose the voter's support for a policy z declines based on the distance $(z > z_i(s^i))$; that is, z can be too strict even for someone who prefers a relatively high value. In this case, obtaining majority support can be difficult since it faces two-sided opposition. For example, a moderate emissions price might be opposed by those who prefer it to be much higher, as well as much lower. But proposals to adjust z up or down may alienate as many supporters as they would attract, making it impossible to get a majority.

Suppose a potential voter uses a quadratic loss function such as $L_i = [z_i(s^i) - z]^2$ to determine his or her degree of opposition to the policy. Then the greater the variance of beliefs about s^i, the smaller the coalition of support for any policy. The state-contingent approach can potentially alleviate this problem, however. The policy maker no longer proposes a fixed value of z, but instead proposes a function of the observed state $z(s(t))$ over time. Each agent will then expect future values of s to be correlated with s^i; hence the sum of the expected loss terms L_i will be smaller than before, even if the variance of beliefs about s^i remains large. Intuitively, by proposing a policy target that is dependent on the actual future state, each agent 'expects' to get his or her preferred outcome. The one who expects the emissions to have a large effect expects the policy to end up being stringent, while the one who expects emissions to have little effect expects the policy to end up being lax. Consequently both types of agents expect small losses from the policy, and have an equally strong incentive to support it, even though they have conflicting views about the form it will actually take.

6.   CONCLUSIONS

Uncertainties over the future path of global warming and the underlying severity of the problem make derivation of an intertemporally optimal emissions price on CO_2 both theoretically and politically very difficult. IAM-based approaches to the problem assume knowledge of a great many key parameters, while learning models suggest that it will take too long to resolve those uncertainties to be of much use in the current debate. If all conceptually possible climate risks are considered, it may be impossible to place a finite value on full insurance against those risks without arbitrarily truncating the range of extreme outcomes being considered.

A fundamental problem with the existing analyses of CO_2 pricing is that agents and policy makers cannot commit to a long-term emissions price. The issue is polarized such that fear of two very large potential mistakes seems to have paralysed the decision-making process, namely fear of climate catastrophe due to failure to act, and economic catastrophe from inept action. The nature of the climate issue makes these fears justified.

This chapter explores an alternative approach based the concept of state-contingent pricing, in which agents commit to a pricing *rule* rather than a *path*. The rule connects current values of the emissions price to observed temperatures at each point in time. In essence, if the climate warms, the tax goes up, and vice versa. A derivation is provided showing how such a rule yields an approximation to the unknown optimal dynamic externality tax, yet can be computed using currently observable data. A recently proposed extension coupling the state-contingent tax with a tradable futures market in emission allowances would then yield not only a feasible mechanism for guiding long-term investment, but an objective prediction market for climate change.

There are many potential advantages of the state-contingent approach. For one thing, people with divergent views on the nature of the climate issue can still commit to the same instrument, since each one would expect to get his preferred outcome. The rule is structured such that, however the future unfolds, in retrospect we will know that we followed a reasonably good approximation to the optimum, and the incentives along the way favour the use of unbiased forecasts of the pricing path to guide investment decisions. Consequently there are informational, as well as theoretical and practical, advantages to the state-contingent approach, which make it worth exploring in more depth as a potentially viable tool for implementing sound climate policy.

NOTE

1. As an anecdotal illustration, Hsu (2011) and McKitrick (2010) hold very different views on the underlying threat of global warming, yet both advocate the same policy mechanism, albeit with different recommendations as to the appropriate state variable.

REFERENCES

Christy, John R., David E. Parker, Simon J. Brown et al. (2001), 'Differential trends in tropical sea surface and atmospheric temperatures since 1979', *Geophysical Research Letters*, **28**(1): 183–6.

de Laat, A.T.J. and A.N. Maurellis (2006), 'Evidence for influence of anthropogenic surface processes on lower tropospheric and surface temperature trends', *International Journal of Climatology*, **26** (June): 897–913.

Fall, S., A. Watts, J. Nielsen-Gammon, E. Jones, D. Niyogi, J. Christy and R.A. Pielke Sr (2011), 'Analysis of the impacts of station exposure on the U.S. Historical Climatology Network temperatures and temperature trends', *Journal of Geophysical Research*, 116.

Fu, Qiang, Syukuro Manabe and Celeste M. Johanson (2011), 'On the warming in the tropical upper troposphere: models versus observations', *Geophysical Research Letters*, **38**, L15704, doi:10.1029/2011GL048101.

Geweke, John (2001), 'A note on some limitations of CRRA utility', *Economic Letters*, **71**: 341–5.

Hsu, Shi-Ling (2011), 'A prediction market for climate outcomes', *University of Colorado Law Review*, **83**(1): 179–256.

IPCC (2007), *Climate Change 2007: The Physical Science Basis, Contribution of Working Group I to the Fourth Assessment Report of the Intergovernmental Panel on Climate Change*, S. Solomon, D. Qin, M. Manning, Z. Chen, M. Marquis, K.B. Averyt, M. Tignor and H.L. Miller (eds). Cambridge, UK and New York: Cambridge University Press.

Karl, T.R., Susan J. Hassol, Christopher D. Miller and William L. Murray (2006), *Temperature Trends in the Lower Atmosphere: Steps for Understanding and Reconciling Differences*, Synthesis and Assessment Product. Climate Change Science Program and the Subcommittee on Global Change Research, http://www.climate science.gov/Library/sap/sap1-1/finalreport/sap1-1-final-all.pdf, accessed 3 August 2010.

Kelly, D.L. and C.D. Kolstad (1999), 'Bayesian learning, growth, and pollution', *Journal of Economic Dynamics and Control*, **23**(4): 491–518.

Leach, A.J. (2007), 'The climate change learning curve', *Journal of Economic Dynamics and Control*, **31**(5): 1728–52.

McKitrick, Ross R. (2010), 'A simple state-contingent pricing rule for complex intertemporal externalities', *Energy Economics*, doi:10.1016/j.eneco.2010.06.013.

McKitrick, Ross R. and Nicolas Nierenberg (2010), 'Socioeconomic patterns in climate data', *Journal of Economic and Social Measurement*, **35**(3,4): 149–75, doi: 10.3233/JEM-2010-0336.

Mears, C.A. and F.J. Wentz (2005), 'The effect of diurnal correction on satellite-derived lower tropospheric temperature', *Science*, **309**: 1548–51.

Nordhaus, William D. (2007), 'To tax or not to tax: alternative approaches to slowing global warming', *Review of Environmental Economics Policy*, **1**(1): 26–44. doi:10.1093/reep/rem008.

Nordhaus, William D. (2009), 'An analysis of the dismal theorem', Cowles Foundation Discussion Paper No. 1686.

Parson, Edward A. and Darshan Karwat (2011), 'Sequential climate change policy', *WIREs Climate Change*, doi: 10.1002/wcc.128.

Pindyck, Robert (2011), 'Fat tails, thin tails and climate change policy', *Review of Environmental Economics and Policy* **5**(2): 258–74.

Spencer, R.W. and J.R. Christy (1990), 'Precise monitoring of global temperature trends from satellites', *Science*, **247**(4950): 1558–562.

Stern, Nicholas (2006), *Stern Review on the Economics of Climate Change*, London: Her Majesty's Treasury.

Thompson, David W.J., John J. Kennedy, John M. Wallace and Phil D. Jones (2008), 'A large discontinuity in the mid-twentieth century in observed global-mean surface temperature', *Nature*, **453**(7195): 646–649. doi:10.1038/nature06982.

Thorne, Peter W., Philip Brohan, Holly A. Titchner, Mark P. McCarthy, Steve C. Sherwood, Thomas C. Peterson, Leopold Haimberger, David E. Parker, Simon F. B. Tett, Benjamin D. Santer, David R. Fereday and John J. Kennedy (2011), 'A quantification of uncertainties in historical tropical tropospheric temperature trends from radiosondes', *Journal of Geophysical Research – Atmospheres*, **116**, D12116, doi:10.1029/2010JD015487.

Weitzman, Martin L. (2009), 'On modeling and interpreting the economics of catastrophic climate change', *Review of Economics and Statistics*, **91**(1): 1–19.

19 Climate change, buildings and energy prices
Alberto Gago, Michael Hanemann, Xavier Labandeira and Ana Ramos*

1. INTRODUCTION

Buildings are a crucial sector for controlling energy demand and, therefore, greenhouse gas (GHG) emissions. Buildings currently account for around 40 percent of the final energy use in the world (IEA, 2008; IPCC, 2007) and, as indicated by Figure 19.1, in developed countries such as the USA they are responsible for 30 percent of total energy consumption and for around 20 percent of carbon dioxide (CO_2) emissions. The importance of buildings for energy and environmental policies also arises from the fact that they constitute a 'stock' of future energy consumption and emissions. For example, around 60 percent of the existing buildings in the UK, the USA, or Spain were built before 1980 (Sweatman and Managan, 2010) and therefore are likely to have lower energy efficiency and higher GHG emissions than modern buildings. Thus, failing to retrofit old buildings to improve their energy and environmental performances, or an inadequate construction of new buildings, may endanger GHG mitigation.

In this sense, there is a particular concern with emerging economies where increasing population and economic growth may lead to a renewed activity in building construction that could considerably expand future energy consumption and its associated GHG emissions. Higher consumption and emissions would be brought about by a combination of more energy-inefficient buildings (stock) and the increasing flow demanded by households with rising incomes.[1] If this is the case, future energy systems may be clearly unable to comply with the current objectives to reduce GHG emissions because

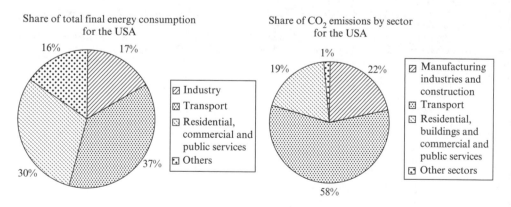

Source: IEA (2012) and World Bank (2012).

Figure 19.1   Main sources of US energy consumption and CO_2 emissions (2009)

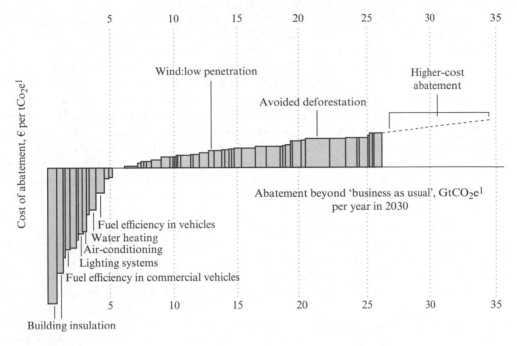

Source: McKinsey (2009).

Figure 19.2   Global GHG abatement cost curve beyond business-as-usual, 2030

they actually rest on significant energy savings from this sector: up to 30 percent of the baseline emissions by 2030 (IPCC, 2007). In fact, the last *World Energy Outlook* of the IEA (2011a) emphasizes the critical importance of energy efficiency to overcome the rising GHG emissions trend and indicates that the stock component of such emissions, together with a lack of significant action, is already closing the window of opportunity to avoid large increases in GHG atmospheric concentrations.

The building sector presents not only a challenge but also an opportunity, because of the apparent cost-effectiveness of energy efficiency measures in buildings. Expert engineers indicate the existence of a large potential to reduce energy consumption in new facilities, through proper design and construction, and in existing buildings through retrofitting. Some studies suggest energy abatement possibilities of up to 75 percent in new buildings and of 20–50 percent through retrofitting old facilities (IPCC, 2007; European Commission, 2011). Not only there is large potential for energy savings in the building domain: these measures, especially those applied on new buildings, are usually ranked among the most cost-effective alternatives to reduce energy consumption. This is depicted in Figure 19.2, where some energy-saving measures in the building sector show even negative costs ('win–win' options) in the global abatement cost curve for GHGs in 2030. While this information should be treated with care, given the several methodological and empirical shortcomings of GHG abatement cost curves such as this (Linares et al., 2012), it identifies promising possibilities in this area.

Indeed, from a technical point of view, most of the energy efficiency measures in buildings are already mature as they have been widely studied and applied since the energy crisis of the 1970s. For instance, Laquatra (1986) and Gilmer (1989) analyze the market effects of energy efficiency investments for units constructed through the Energy Efficiency Housing Demonstration Program carried out by a Minnesota agency in the 1970s. Nowadays, many possibilities are already available in the market, mostly related to the general lighting and heating/cooling facilities, insulation techniques (envelope and windows) and decentralized renewables. For instance, improvements in space heating, which are very important because this end use is responsible for around 30 percent of energy use in residential buildings (IEA, 2008), may be achieved through a combination of insulation, improved heating/cooling methods and the use of different types of renewables. This chapter is interested in this type of combination of alternatives because, although we recognize the importance of energy consumption and emissions from other household appliances, their characteristics and regulation raise different issues from those associated with buildings and discussed in this chapter.

Given the focus of this book, we have emphasized the close relationship between energy consumption by buildings and GHG emissions. However, energy efficiency in buildings would bring about a richer array of benefits. It would definitely reduce other types of pollutants in urban areas, an increasing problem in many emerging economies (those, as indicated above, that may see a larger expansion of their building stock), thus improving health conditions for the population and reducing social damages. Reducing energy use in buildings may also contribute to reducing dependence on foreign supplies of energy, having a positive effect on the so-called energy security of countries. Moreover, energy conservation may result in net savings for households and firms, thus increasing their disposable income and profits. Energy efficiency measures in buildings may not only reduce energy expenditures but also may improve living conditions and indoor air quality (e.g. through substitution of inefficient wood or coal usage) can also be achieved through energy efficiency measures in homes and buildings. In the commercial sector there is evidence of the positive influence of more energy-efficient buildings on indoor air quality and eventually on worker productivity (Leaman and Bordass, 1999). There is abundant evidence that firms can achieve positive reputational effects through eco-friendly behavior. Furthermore, energy efficiency in buildings offers large possibilities for new businesses such as ESCOs[2] and for owners and real-estate investors due to its positive effect on property prices (Eichholtz et al., 2010; Fuerst and McAllister, 2011). Lastly, the economy as a whole may benefit from the creation of jobs that are related to the implementation of energy efficiency measures in buildings.[3]

Last but not least, a widespread introduction of energy efficiency improvements in buildings can generate significant distributional benefits. They may be related to those households with a limited access to basic energy sources and services, quite common in emerging or developing countries. But they may be also linked to overcoming energy poverty in developed countries by allowing households to live in comfortable temperature conditions at affordable costs (UKDECC, 2011a).

Despite the socioeconomic and environmental importance of these matters, in practice energy efficiency measures and techniques have barely been introduced in commercial or residential buildings. A number of factors explain the real-world barriers to energy efficiency improvements, in buildings and elsewhere, which are actually the reason for persistent public intervention in this area. However, buildings and dwellings have particular characteristics that differentiate them from other products and sectors: they are long-lived and costly assets, and many agents are usually involved in this market. These particularities must be taken into account for a proper understanding of this sector, a necessary condition for the design of effective energy efficiency policies. This is the setting for the chapter, which will first provide a general description and discussion on the barriers to energy efficiency in buildings (Section 2.1) and then present the main policy tools available to tackle this problem (Section 2.2). With that information, in Section 3 we will propose a policy package that can simultaneously overcome most of the preceding barriers and provide incentives for a cost-effective implementation of energy efficiency measures in this important sector. Section 4 concludes with a summary of the main implications of our findings and policy proposal.

But before further discussing strategies and policies, it is necessary to emphasize the significant heterogeneity within this sector. First, as indicated above, there is a crucial distinction between new buildings and existing buildings: this difference is very relevant for the design and implementation of energy efficiency measures. For instance, measures are likely to be cheaper and with a higher savings potential if introduced in new buildings. Moreover, some measures are only applicable to new buildings, such as construction or planning codes. Finally, the agents involved in decisions vary in new and existing facilities: in new buildings decisions on energy efficiency are largely in the hands of builders and investors, whereas in existing buildings energy efficiency measures may involve owners and tenants. A second characteristic of buildings is the heterogeneity of users, including both residential and commercial (which may also involve public and governmental, and sometimes industrial); this heterogeneity has important implications for the design and application of energy efficiency policies. Lastly, even residential or commercial buildings (whether new or old) may have plenty of internal heterogeneity: single-unit residences, multi-property houses with common areas or services, different uses in the so-called commercial sector and so on. But complexity does not end there because the heterogeneity also interacts with geographic and climatic variation. Essential factors for energy efficiency such as the ownership turnover period, occupancy rates or the stock of existing buildings are clearly dependent on climate and country/region.

2. ECONOMICS OF ENERGY IN BUILDINGS

2.1 Market Failures and Barriers

If there is a limited implementation of energy efficiency measures in buildings, as suggested in the previous section, the negative cost figures in Figure 19.2 would clearly reflect the so-called 'energy efficiency paradox' (Jaffe and Stavins, 1994). Why are agents reluctant to invest in energy efficiency measures in buildings despite their negative costs? One possibility is that the net costs are not negative: either the outlays are understated or the energy savings benefits are overstated, or perhaps some of the benefits do not accrue to those who must make the outlays (Figure 19.2 fails to consider the *distribution* of benefits and costs). But it is also possible that the net costs are negative but that barriers exist in the real-estate sector that strongly affect investment decisions on energy efficiency. These barriers have been generally identified as absence of information, the conflict of interest between principal and agent, the difficulty to financing a high up-front cost, and uncertainty about the reliability of the energy efficiency device. We now elaborate on these explanations for the suboptimal allocation of resources to energy efficiency before proceeding to identify some possible corrective measures.

Generally, agents have substantial lack of information on future energy consumption at the moment of purchasing or renting a commercial or residential building. This is not something that is easily measured. Future energy use depends partly on the behavior of future occupants and on investments they might make in energy efficiency; it also depends on the aging of the building and the energy-using equipment within it. Another point is that building owners and occupants may not have the same economic interests, and thus problems of asymmetric information and lack of trust could arise. As a consequence, agents may give less consideration to expenditures for energy efficiency, and be less inclined to value the implementation of such measures.

The conflict in interests among the agents that operate in the building sector pertains not only to information. In addition, there is the well-known principal–agent problem, where the party that takes the investment decisions on energy efficiency would not benefit from their returns. In new buildings, for example, the builder may be interested in achieving the highest revenues at the lowest costs without paying attention to the effects on the stream of future energy expenditures and associated pollution emissions. In existing buildings the principal–agent problem is also present in the owner–tenant relationship: tenants may have incentives to promote energy efficiency to reduce their bills but they are unlikely to invest in items that become permanent fixtures in buildings they do not own and may occupy for only a limited period of time, leaving these fixtures for the owners once they quit the building. If the owners do not pay their tenants' energy bills, they themselves have little incentive to invest in energy savings fixtures, except to the extent this would enable them to charge sufficiently higher rents to recoup their investment. This problem also arises in commercial buildings and it becomes even more important in multi-unit residences because the number of actors increases.

There are several other barriers that, while not necessarily market failures, could be mitigated by public actions. First is the large up-front capital cost often required for energy efficiency investments in existing residential and commercial buildings to pay for things like replacing heating systems, installing new windows, better insulation or renewable energy that discourages these investments, since it may be difficult to find a way of financing the capital outlay. Second is the so-called bounded rationality that potentially affects all agents participating in this sector, in new and existing buildings or in commercial and residential units. Bounded rationality is partially due to the above-mentioned lack of information but it is also affected by cultural or idiosyncratic habits that make consumers unaware of their energy use and associated emissions (Brounen et al., 2012; Palmer et al., 2011). Moreover, uncertainty also prevents agents from investing either because of changing legislation or because first movers cannot benefit from lower prices resulting from learning-by-doing effects or economies of scale (NRTE, 2009). On top of this, there might be significant hidden costs that affect the adoption of energy efficiency measures in existing buildings: transaction costs and the inconvenience and nuisance associated with retrofitting (and the possible need for another temporary dwelling). Finally, the literature has also mentioned long payback periods and high discount rates as other barriers to the adoption of energy efficiency in buildings.

It should be emphasized that whether or not these factors are considered market failures is immaterial. What is important is that they are potentially amenable to policy action. Market failure signifies a violation of economic efficiency (typically based on the Kaldor–Hicks criterion of a potential Pareto improvement). But economic efficiency is defined by reference to an existing set of actors, with existing choice sets, existing preferences, existing production technologies, existing constraints, existing information sets and existing behavioral rules for the economic actors. If one can change some of these components of the existing economic system, there will be a different outcome that may be judged superior *ex post*, but this does not make the prior outcome inefficient relative to the economic system that applied then. Thus, creating a new financing mechanism (e.g. local government financing of energy efficiency retrofits of buildings repayable through a supplement to property taxes), providing new and more transparent measurement of energy use, providing information/advertising that makes energy use more salient to economic actors and similar interventions change the economy in such a way as to generate a different economic outcome, one not attainable prior to the intervention.

2.2 Policy Instruments

The preceding market barriers and market failures just described, together with the growing importance of energy efficiency in environmental and energy policies, provide a clear justification for public intervention. The catalogue of policy instruments to promote energy efficiency is well known in the economic literature (Linares and Labandeira, 2010), but the applications of these instruments to the buildings sector are surprisingly scarce. This may be related to the fact that, for some economists, proper energy prices should do the job. In this sense, suboptimal energy efficiency efforts in buildings may be related to artificially low energy prices (either because of subsidies or due to partial coverage of external costs) and, as a corollary, getting the prices 'right' would solve the problem.

Although we agree that a proper level of energy prices is a necessary condition for successful implementation of energy efficiency policies throughout the economy, the complexities associated with buildings (heterogeneity, market failures and barriers, stock consumption and emissions and so on) make it hardly a sufficient condition in this sector. This is a key argument of the chapter, which is also related to the usually low elasticities of household energy demand reported by the literature and to the possible exacerbation of energy poverty through higher prices (Gillingham et al., 2009; Ürge-Vorsatz and Herrero, 2012).

Given this general setting, we next provide a brief review of the three main regulatory alternatives to promote energy efficiency in buildings, with an evaluation of their effects as available in the literature. This will serve as a basis for the definition of a policy proposal that, unlike these partial approaches, can tackle the multiple challenges and problems that exist in this complex sector.

2.2.1   Command and control

As in other energy and environmental areas, command-and-control approaches have been widely used by many governments, with a varying level of stringency, to promote energy efficiency in the building sector. These usually take the form of building and/or planning codes that, therefore, are usually restricted to new buildings. Sometimes there are requirements that have to be satisfied when buildings are resold. However, some standards can be introduced on the heating/air-conditioning systems of buildings and thus can be effective in fostering improvements in both new and existing units.

The effectiveness of these building codes depends on how they are designed and also on the relative importance of new buildings in the overall stock. For instance, building codes with minimum energy performance standards have a bigger impact in emerging countries, like China and India, where there is rapid construction of new structures in parallel with economic and population growth. This phenomenon was confirmed by Chan and Yeung (2005), who report a significant decrease in commercial electricity consumption after the introduction of building codes in Hong Kong. Yet building codes in developing countries, although increasing in number, are considerably more lenient than in the developed world (Iwaro and Mwasha, 2010). In particular, these instruments have a more limited effectiveness in industrialized countries where old and inefficient buildings are in the majority. For instance, Aroonruengsawat et al. (2012) analyze the effects of building codes for 48 US states on per capita electricity consumption in residential units between 1997–2006 and find only modest reductions in energy use in the range of 3–5 percent in 2006.

Command-and-control approaches have usually been considered inefficient or cost-ineffective in energy and environmental policies, due to the tendency to make the regulation a uniform requirement (as a consequence of asymmetric information between the regulator and the regulated). In the case of buildings, the remarkable heterogeneity would intensify this problem and demands flexible or differentiated codes and standards (Galvin, 2010). This is particularly important regarding climatic variations and their interactions with the different nature and use of buildings.

2.2.2 Taxes, permits, white certificates and subsidies

Energy and environmental taxes (or equivalent tradable emissions permit systems) would contribute to higher energy prices and thereby stimulate the adoption of energy efficiency devices in a cost-effective (flexible) way. In practice, however, fiscal instruments designed to enhance energy efficiency in buildings have tended to take the form of tax deductions for investments. This may be related to the fact that, although energy taxes and tradable permits have been widely applied in a number of countries for many years, they have usually been targeted at the transport sector and energy-intensive users. Sometimes this has led to the application of complementary instruments that attempt to reach sectors that are not covered (or just partially covered) by energy and environmental taxes or tradable permits, such as the UK climate change levy (a tax on energy usage for industrial, commercial and public sectors) or the carbon reduction commitment (tradable permits on non-energy intensive sectors such as hotels, supermarkets etc.). Even though these complementary approaches may foster the adoption of energy efficiency measures in buildings, as they usually take place in commercial or residential domains that are hardly affected by general pricing instruments (such as the EU Emissions Trading Scheme), there is an obvious risk of negative interactions with other policy tools being applied at the same time (Labandeira and Linares, 2011).

Some examples of significant tax deductions to promote energy efficiency in buildings can be found in the US 2009 Recovery Act. On the one hand, the Non-Business Energy

Property Tax Credit applies when space-heating and air-conditioning devices, biomass heating systems or other insulation measures with efficient properties are purchased. On the other hand, the Energy Efficiency Property Tax Credit offers deductions on the installation costs of solar panels, wind or geothermal systems of renewable energy. Gillingham et al. (2006) survey the literature on tax deductions to promote energy efficiency, finding mixed empirical evidence. More recently, McKibbin et al. (2011) compare household tax credits aimed at promoting energy efficiency purposes with carbon taxation, and demonstrate that the latter has greater environmental effectiveness, although higher costs for the industrial sector.

Another market-based instrument that can be applied in this area is the so-called white certificate scheme, already implemented in several developed countries such as France, Italy, Denmark or the UK during the last few years (Bertoldi et al., 2010). This consists of obligations on energy producers or sellers to foster cost-effective energy efficiency improvements in commercial and residential consumers that are tradable. Mundaca and Neij (2009) provide an assessment of the effectiveness of the British Energy Efficiency Commitment, a white certificate scheme that focuses on buildings; they find a significant increase in certified energy savings during its life span (although they are unable to identify what share of the energy savings was due to a business-as-usual trend for improvements).

Tax deductions are not dissimilar to direct subsidies for energy efficiency measures in buildings, or to preferential interest rates (i.e. partly funded by the public sector) for energy efficiency investments. In all these cases, the effects on long-run investment will depend on the perceived continuity of the subsidy over the usually long payback period involved. Moreover, as with building codes, subsidy programs policies should be adjusted to fit the climate variability within the country to achieve cost-effective outcomes. For example, using uniform grants or tax deductions across the country to promote more efficient air conditioning would not allow for different levels and intensities of use in different climatic regions, and would therefore probably be cost-ineffective. Finally, fiscal deductions may bring about problems of free-riding when they are applied indiscriminately because some agents would have taken the energy-efficient alternative without the tax incentive (NRTE, 2009). Linking an indicator of economic capacity, or in some cases of geographical/climatic variation, to the definition of the fiscal deduction could overcome this problem, although probably at higher administration and compliance costs. The empirical evidence on the effectiveness of subsidies for energy efficiency finds mixed results: Nair et al. (2010) found that the initial investment cost was quite important for the decisions of Swedish consumers and thus energy efficiency subsidies could play an important role (although necessarily complemented by campaigns that provide information on energy savings and available financial facilities), whereas Kemp (1997) found limited effects of subsidies for energy efficiency in the Netherlands during the 1970s and 1980s.

2.2.3   Energy performance certificates
In the last few years the use of certificates or labeling systems for many products (such as renewable energy, ecological food, household appliances or buildings) has seen a remarkable expansion with the objective of providing consumers with environmental and energy information that is not easily available in the market. In the case of buildings,

where such informational problems certainly exist and constitute one of the main barriers for energy efficiency (Section 2.1), energy performance certificates offer detailed information about the future demand of energy that would be necessary to maintain a standard comfortable level of temperature within the unit. Consumers would then have direct access to reliable energy information of the building that can be added to their preferences, and thus are likely be more energy conscious when taking decisions.

Although energy certification systems for buildings may be voluntary or mandatory, depending on the country and/or type of building, they always follow the same basic methodology. First, some private or public experts compute the energy consumption of the building, taking into account factors such as the level of insulation, air conditioning, heating and lighting systems, and the presence of any source of renewable energy. Later, an energy index (usually ranking facilities from A/platinum – most efficient – to G/certified – less efficient[4]) is used for rating the building while controlling for geographic and other structural factors that allow for a consistent comparison with other units (CEN, 2005).

So far, energy performance certificates have been widely used in the case of US commercial buildings through two voluntary rating systems: the Leadership in Environmental and Energy Design (LEED), created in 1998 by the US Green Building Council, and the EnergyStat Program, developed in 1995 by the US Environmental Protection Agency and the Department of Energy. But other countries, such as Australia, Canada or the EU, have also introduced labeling systems for buildings (Laustsen, 2008). The European Commission moved a step further by promoting the 2002 Energy Performance of Buildings Directive (EPBD) that demands member states to require a certificate of energy performance for all commercial and residential buildings when sold or rented out.

In addition to the information function, energy performance certificates also generate incentives for investing in energy efficiency because it is reasonable to expect price increases in those buildings with better certifications (linked to a lower flow of future energy expenses). If this is the case, energy efficiency investments in buildings may become more attractive both to real-estate investors and to property owners who would get higher sale or rental prices. Although there are important data limitations to evaluate the effects of certificates on selling and rental prices, there is a growing literature computing the price premium associated with more efficient buildings. Using hedonic prices, Eichholtz et al. (2010), Kok et al. (2011) and Fuerst and McAllister (2011) reported the existence of a price increase of around 2–6 percent in effective rents and of 13–16 percent in selling prices for US commercial buildings. After the introduction of the EPBD in Europe, Brounen and Kok (2011) found that the residential Dutch housing market has capitalized the information of the certificates into the prices of houses. Other authors such as Banfi et al. (2008), Kwak et al. (2010) and Leung et al. (2005) used stated preferences methods to analyze the willingness to pay for energy efficiency retrofits revealing positive figures for the replacement of old air-conditioning or heating systems by residential and commercial consumers.

The power of this instrument to overcome several of the problems advanced in Section 2.1 is thus evident. It would, first of all, mitigate the informational problem and, by doing so, would reduce the asymmetric information that is present in the sector. Moreover, this instrument seems to generate incentives to investment by tackling some of the principal–agent problems. It is therefore an essential policy instrument to promote energy efficiency in buildings, although it is still unable to solve some of the problems identified in Section 2.1, such as access to capital or fragmented property, and their effectiveness may be affected by voluntary schemes or restricted to singular purchases/ rentals. A coordinated combination of this instrument with other policy alternatives could produce a successful policy package, as we will show in the next section.

3.   A NEW POLICY PACKAGE

As explained above, the design and implementation of energy efficiency policies in buildings is a rather difficult but important task, which probably explains the limited and often ineffective experience seen so far. Market failures and barriers, huge heterogeneity within this sector that interacts with climatic and geographical variations, efficiency and distributional concerns are all factors that undermine conventional and piecemeal regulation in this area. The application of isolated policy instruments, such as those described in Section 2.2, to foster energy efficiency in such a complex and difficult context is likely to fail or to produce suboptimal results.

In this section we propose a policy package that aims to tackle the main obstacles to the adoption of energy efficiency measures in buildings, namely problems of information, incentives, reliability and access to capital. The proposed policy package offers the simultaneous and coordinated application of several instruments: a mandatory energy performance certificate for all buildings (EPC), a new energy inefficiency tax (EIT), a fund to provide capital for energy efficiency measures (EEF), and other complementary tools (smart metering, building codes etc.). The package is not just a combination of instruments to avoid negative interactions and to promote positive synergies towards a cost-effective promotion of energy efficiency, but also an institutional device, built around an energy efficiency organization (EEO) and firms providing energy services, that attempts to respond both to efficiency and distributional concerns. Before these are described in detail, Figure 19.3 depicts the main components of the scheme, with their linkages and expected outcomes.

3.1   Energy Performance Certificate System

The EPC system is a central piece of the policy package as, besides its important specific roles, it works as a kind of linking mechanism to other instruments (see below). As explained in Section 2.2.3, EPC systems provide the essential information for consumers on energy efficiency characteristics of buildings and, indirectly, they create incentives for agents to invest in them as long as real-estate markets capitalize their information. EPC systems are also flexible tools, able to adapt to heterogeneous geographical and climatic conditions, because ratings explicitly take account of unit-specific factors to allow for comparability.

In this proposed policy package the EPC system should be mandatory to promote behavioral change among consumers due to improved information and better incentives. Voluntary EPC systems do not work properly when a significant proportion of agents do not expect to be selling or renting their properties in the short or medium run. Voluntary EPC systems would also prevent the general application of other instruments of the

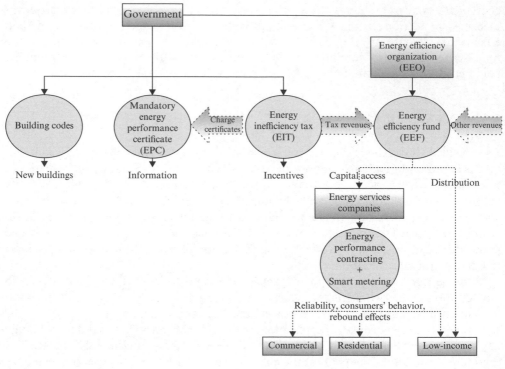

Source: The authors.

Figure 19.3   The policy package to promote energy efficiency in buildings

policy package that depend on their existence, as seen below. Moreover, voluntary EPCs could bring about undesirable distributional effects if only high-income individuals can take advantage of the system because only they are able to invest in high-efficiency buildings and thereby obtain better ratings with higher associated prices/rentals. The system should be, moreover, applied to all types of buildings for the same reasons that merit a compulsory application of the scheme. Finally, it would be desirable to have a periodic review and update of the building EPCs in order to provide updated information and promote continuous efficiency improvement.

3.2   Energy Inefficiency Tax

Even mandatory EPCs do not provide meaningful incentives for energy efficiency improvements to agents who have only a limited involvement in transactions related to buildings (purchases or rentals). To solve this problem we propose a new recurrent tax on energy inefficiency in buildings (EIT), also able to generate revenues for a partial or full funding of the package through the energy efficiency fund (EEF). This tax is closely related to the mandatory EPC, which is used for the definition of a unit-specific tax rate, thus contributing to the central importance of that instrument within the package.

  As any tax, the EIT applies a tax rate to a tax base. To avoid legal challenges and

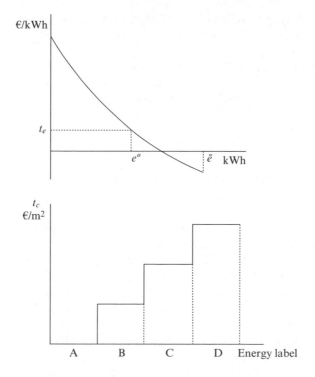

Source: The authors.

Figure 19.4   Energy inefficiency tax rate and base

to send the proper economic incentives, we propose a scheme with a progressive rate that depends on the grade of energy inefficiency of the building and is also related to a general energy tax that is supposed to attain an exogenously determined level of energy demand. Figure 19.4 depicts the tax rate of the general energy tax (t_e) and of the energy inefficiency tax (t_c) on buildings. The general energy tax is designed to move energy consumption from an unrestricted level, where the existence of some win–win options is reflected in the chosen shape of the marginal abatement cost curve, to the exogenously defined level (e^o). As indicated in Section 2, if energy prices alone provided the incentives for a successful adoption of energy efficiency, no other instruments would be necessary. However, the already mentioned presence of several barriers to the adoption of energy efficiency measures justifies the EIT within a broader package. Yet by linking this new tax to a general energy tax, we use a non-discretionary tax rate that can be related to energy/environmental objectives elsewhere and reduces the possibility of promoting too-high or too-low energy efficiency efforts in the building sector.[5] Equation (19.1) shows the linkages between the tax rates of the general energy tax and the EIT.

$$t_c = (C_X - C_A)t_e \tag{19.1}$$

where C_X and C_A are respectively the estimated energy consumption provided by the EPC (used as the basis for grading), per square meter, of a building unit awarded with certificate X and A (the most efficient). This means that, as shown in Figure 19.4, t_c would increase with the level of inefficiency of labels, with a zero value when the unit has the highest energy efficiency rating.

Once the tax rate of the EIT is defined, in euros per square meter, its application is straightforward as the tax base should be the built area (S, in square meters) to provide the adequate level of incentives. This would lead to tax revenues of T_c for a unit with a label X, as shown in equation (19.2).

$$T_c = t_c S \qquad (19.2)$$

It should be stressed that the actual application of the EIT would not be burdensome from administrative or compliance perspectives. As long as the EPC is mandatory, the information on the ratings and associated energy consumption would be available for every building unit. Moreover, information on built areas should be easily retrieved from the property taxes on buildings that are common elsewhere and generally used as indicators of property value. Indeed, although EIT could be implemented by almost any jurisdiction given its reduced administration costs, it would probably be best to allocate them to the government level that manages taxes on building properties (in many cases local authorities). Of course, the preceding does not preclude the likely existence of political and social opposition to the introduction of a new tax.

Although this tax would provide revenues that are earmarked for the EEF, another essential instrument within the policy package, its main objective should be the continuous promotion of energy efficiency. As long as building codes are properly implemented and lead to good energy efficiency ratings, the EIT will be especially useful to promote retrofitting of old units. Given that old buildings are prevalent in many advanced countries, EIT may have a particularly important role here. However, building ratings are not static because they usually get more stringent due to technological advances and stricter regulations or because the energy properties of buildings are subject to obsolescence, which is why we proposed a periodic reassessment of EPCs in the previous section. In this setting, the EIT works even better because it provides continuous incentives to improvement and may also promote voluntary re-evaluations of energy grades to avoid tax payments. Indeed, the EIT has the advantages and flexibility of pricing mechanisms as agents would only pay the tax when its costs are lower than the difference between benefits (reduced consumption) and costs (investments associated to a higher grade) of energy efficiency improvements: hence the economic importance of a non-discretionary approach to tax rate setting.

3.3   Complementary Tools: Building Codes, Smart Metering and Energy Performance Contracting

Although the preceding instruments constitute the core of our proposal, other policy alternatives are also necessary to guarantee the effectiveness and good performance of the package. As indicated by Figure 19.3, a coordinated use of three other tools is particularly necessary: building codes with minimum energy performance standards, smart metering for residential and commercial buildings and energy performance contracting.

As shown in Section 2.2.1, building codes should play an important role in fostering more energy-efficient new buildings as energy efficiency measures introduced at this stage are usually cheaper and they also prevent the increase of the stock of inefficient buildings. Their main problem is related to the command-and-control nature, which may lead them to obtain the desired objectives without cost minimization due to uniform approaches in a very heterogeneous setting. In our policy package, however, building codes should only provide some minimum level of energy efficiency because the existence of the other policy instruments would provide incentives, even at the moment of construction, to improve the standard if deemed necessary. Therefore flexible building codes may increase the level of energy efficiency in new buildings without compelling investments that are over or under the socially desirable outcome.

Smart metering is a useful complement for two reasons. First, it improves consumers' awareness and may make energy use more salient to them.[6] Once consumers know their energy consumption at each point in time, they can make adjustments if they wish to lower their energy use. Behavioral adjustment would be reinforced if there were some form of time-differentiated pricing, which is hardly worthwhile in the absence of real-time metering. The combination of smart metering and time-varying pricing has the potential to overcome bounded rationality barriers (see Section 2.1). Second, smart metering makes it easier for building managers or ESCOs to manage energy use in retrofitted buildings.[7] A possible way to introduce this technology in buildings is by adding an obligatory clause in the retrofitting contract that requires the installation of a smart metering system.

Finally, many governments and institutions are promoting the diffusion of energy performance contracting as a way to facilitate efficient investment (IPCC, 2007; European Commission, 2011). These contracts determine that payments for the services provided by energy services companies (see below) are subject to the performance of the new installation and to the expected energy savings. Using this performance-based form of acquisition, improvements in energy efficiency are paid through the energy savings derived from the investment. Energy performance contracting is also included in this policy package to provide confidence for those doubtful consumers who do not believe in the potential to reduce energy consumption. Indeed, the incorporation of such contracts contributes to overcome another major barrier to energy efficiency: lack of reliability.

3.4 Energy Efficiency Fund and Institutional Arrangements

Another important component of the policy package, and certainly related to some of the preceding instruments, is the energy efficiency fund (EEF). Given its partial or total funding through the energy inefficiency tax,[8] the EEF should have clear links with the government, ideally through a public or semi-public non-profit energy efficiency organization (EEO). The fund must provide access to capital, another of the main barriers to energy efficiency indicated by Section 2.1, to low-income households and to the previously mentioned energy services companies (including ESCOs) whose role in the policy package will be described below.

The EEF is thus the tool that should take care of the distributional concerns and objectives associated to energy efficiency policies. As stated before, energy efficiency is not just concerned with economic efficiency because it can considerably improve the economic conditions and quality of life of the so-called energy-poor (see Section 1). Moreover, a policy package without the EEF would lead to undesirable distributional and efficiency effects from the EIT because, given the particularly limited access of poor households to capital for energy efficiency investments in their dwellings, they would keep paying the burdens of the tax without simultaneous energy efficiency gains.

In addition, the EEF can provide financial resources to companies that manage energy efficiency retrofits in buildings. Energy services companies would need a sizable initial capital to start the operation of such a system. This is the particular case of ESCOs, with an increasing importance in countries such as the USA and Germany (IPCC, 2007), which guarantees energy savings (reliability) through the above-mentioned energy performance contracts. The links of energy services companies with the EEF explains the compulsory introduction of smart metering and use of energy performance contracts for those households and businesses that benefit from the EEF.

Although, given their regulatory characteristics, the EPC system and the EIT should be logically implemented by the government (at national, state or local level), an EEO of a public or semi-public nature should be in charge of the EEF. There might be reasons for private involvement in the EEO, such as the possibility to attract external contributions to the resources of the EEF, although its central role in the policy package (right-hand side of Figure 19.3) certainly demands public involvement.[9] Actually, the EEO should define the characteristics that make households and companies eligible for funds of the EEF and should play the role of information provider on energy efficiency for all sectors of the economy. Although the EEO does not itself conduct retrofitting operations, it should maintain an official list of energy services companies that are judged to have done successful retrofits. In this way, the EEO is a matchmaker between energy services companies and consumers. Furthermore, the EEO should oversee and guarantee a proper functioning of the energy performance contracts and, in general, of the liaisons between energy services companies and final energy consumers.

In closing this section we should note that, although we have presented a novel and comprehensive policy proposal that has not been tested in reality, some of the components are related to a successful package applied in the US state of Vermont with the same objectives. The scheme is based on a volumetric energy efficiency charge paid by all retail electric consumers and the fund Efficiency Vermont, an entrepreneurial NGO selected through an auction by the state administration with a performance-based contract. The main objective of Efficiency Vermont is to influence energy-related decisions through the most cost-effective strategies that can be devised and implemented. All utility customers should have the opportunity to participate and benefit from the energy efficiency programs, especially those with high barriers, as low-income, seniors or small businesses (Hamilton, 2010). Moreover, in mid-2011 the European Commission launched the European Energy Efficiency Fund, which also has similarities with our proposed EEF.[10] Finally, the so-called Green Deal, a funding scheme for residential and commercial energy consumers that should be implemented during 2012 in the UK, attempts to promote cost-effective improvements in energy consumption in buildings through expert assessment and compulsory funding (UKDECC, 2011b). However, all the preceding schemes lack the comprehensive and coordinated use of several policy instruments of our policy proposal, which tackles all the market barriers and market failures associated with energy efficiency in buildings.

4. CONCLUSIONS

Buildings are crucial to control present and future energy demand, and therefore GHG concentrations in the atmosphere, as they are responsible for a sizable share of global energy consumption and GHG emissions and are associated with an important 'stock' of future energy consumption and emissions. This phenomenon is likely to worsen due to the irruption of emerging economies whose increasing population and economic growth will lead to a renewed activity in building construction and use. Enhanced energy efficiency in buildings will thus bring about environmental (less emissions and damages) and economic (savings to households and firms) benefits, but also improvements in energy security and positive distributional effects (a mitigation of energy poverty).

However, despite the socioeconomic and environmental importance of these matters, in practice energy efficiency measures and techniques have hardly been introduced in commercial or residential buildings. This is somehow surprising because, from a technical point of view, most of the energy efficiency measures in buildings are already technologically and economically mature. Indeed, many of the measures applicable to buildings are apparently among the most cost-effective within the energy efficiency domain.

A number of issues explain the real-world barriers to energy efficiency improvements, which are not restricted to this sector and are actually the reason for persistent public intervention in this area. However, buildings and dwellings have particular characteristics that differentiate them from other products and sectors: they are long-lived and costly assets, and many agents are usually involved in this market. Indeed, absence of information, high financial requirements and lack of reliability have been identified as the major barriers to the adoption of energy efficiency measures in buildings.

In this chapter we suggest that, due to a number of general and specific barriers to energy efficiency policies in this area, energy prices and conventional energy and environmental policy instruments may not achieve the desired outcomes unless they are introduced in a comprehensive and coordinated package of instruments that can simultaneously tackle the existing problems of information, split incentives among agents, uncertainty or access to capital. The proposed policy package is defined around energy certification of buildings, uses flexible building codes and smart metering, and employs a new tax on energy inefficiency to foster continuous incentives towards energy efficiency improvements and to provide revenues for an energy efficiency fund that provides capital to firms and poor households.

We feel that the proposed policy package is, first of all, of easy application: most of the instruments already exist and the novel tax on energy inefficiency would have low administration and compliance costs due to its easy integration within existing tax systems. Second, and more importantly, the package can be defined as long, loud and legal because it provides a set of legally feasible instruments that generate reliability and strong incentives on agents for the adoption of energy efficiency measures. This is especially needed, given the size and likely evolution of the problems associated with energy consumption and emissions from buildings.

One can think of the policy package that we have proposed as a means of commodifying energy efficiency. While energy is clearly a recognized commodity that is sold in existing markets, as are the individual pieces of equipment that contribute to energy efficiency, for the typical residential consumer, or the typical small business owner, energy efficiency is not something he sees as amenable to his control. It is not an item that he can simply go out and purchase, the way he can purchase a light bulb or a gallon of gasoline. He would not know how to transform his home, or his office, to make it use less energy, how much it would cost him to do this, how much money it would save him. He does not know how to make the transformation happen, or even where to acquire that information. The portfolio of measures that we propose makes the energy efficiency transformation accessible, visible and affordable to these consumers. By certifying the energy efficiency of buildings, and taxing the energy efficiency gap relative to a high-efficiency building, our package makes the lack of energy efficiency very visible and highly salient. This is the precondition for energy efficiency to become a marketable commodity. The other measures in our package lower the cost and reduce other barriers to the purchase of energy efficiency as a commodity. In the successful Vermont experience, for instance, homeowners do not have to leave their home: representatives of Efficiency Vermont come to their door with the offer of analyzing their energy use, identifying solutions that conserve energy, bringing in architects, engineers and contractors to implement these solutions, and offering a financing package that eliminates any large up-front cost and makes it affordable. This goes beyond lowering the cost of energy efficiency to the individual energy user: it makes energy efficiency a commodity that is accessible as well as affordable.

NOTES

*   We thank Roger Fouquet for his comments and suggestions on an earlier version of the chapter. We are also grateful to ERDF and the Spanish Ministry of Finance and Competitiveness (project ECO2009-14586-C02-01), and to Alcoa Foundation program on Advancing Sustainability Research. The usual disclaimer applies.
1.   The current breakdown of household and commercial energy demand in emerging economies reveals a heterogeneous pattern. For instance, the Chinese distribution of energy use in residential and commercial buildings is close to that observed in developed countries; however, other emerging societies, such as Mexico, show remarkable differences (IPCC, 2007; Rosas-Flores et al., 2011). This variation is probably due to specific climatic and cultural conditions that will obviously affect the potential evolution (and savings) of energy uses in buildings within the developing world.
2.   An ESCO is a company that offers energy services, such as analysis, audits, management, implementation or maintenance. In contrast to other firms or institutions, ESCOs also offer guaranteed energy savings (see Section 3.4).
3.   The European Commission (2005) estimates that a 20 percent reduction in EU energy consumption may directly and indirectly create one million new jobs in the EU.
4.   Following the grades awarded by the EU EPBD and the US LEED programs (see below).
5.   It can be argued that a simultaneous presence of both taxes may cause negative interactions and efficiency losses through a sort of double taxation. However, the EIT is just a tax to promote the highest level of energy efficiency whereas the general energy tax is levied on consumption. Actually, the presence of the general energy tax could contribute to the limitation of the rebound effect that could be generated by higher levels of energy efficiency (see note 7).
6.   This is confirmed by an OECD (2011) survey that shows that consumers who are charged differentially for their electricity consumption based on the time of use (peak versus off-peak) exhibit more responsible behavior with regard to energy use.
7.   This is also likely to mitigate any rebound effect, whereby higher efficiency energy-using appliances lower the real cost of energy services and stimulate an increase in their demand.
8.   The EEF may obtain further funds from the government or from energy utilities or companies that may be compelled to use part of their revenues to foster energy efficiency.
9.   This fund is partially funded by the EU and several financial institutions to back projects in the field of energy efficiency and renewable energy.

10. This is also likely to mitigate any rebound effect, whereby higher efficiency energy-using appliances lower the real cost of energy services and stimulate an increase in their demand.

REFERENCES

Aroonruengsawat, A., Auffhammer, M. and Sanstad, A. (2012), 'The impacts of state level building codes on residential electricity consumption', *Energy Journal*, **33**: 31–52.

Banfi, S., Farsi, M., Filippini, M. and Jakob, M. (2008), 'Willingness to pay for energy-savings measures in residential buildings', *Energy Economics*, **30**: 503–16.

Bertoldi, P., Rezessy, S., Lees, E., Baudry, P., Jeandel, A. and Labanca, N. (2010), 'Energy supplier obligations and white certificate schemes: comparative analysis of experiences in the European Union', *Energy Policy*, **38**: 1455–69.

Brounen, D. and Kok, N. (2011), 'On the economics of energy labels in the housing market', *Journal of Environmental Economics and Management*, **62**: 166–79.

Brounen, D., Kok, N. and Quigley, J. (2012), 'Residential energy use and conservation: economics and demographics', *European Economic Review*, **5**: 931–45.

CEN (2005), 'Energy performance of buildings. Methods for expressing energy performance and for energy certification of buildings', European Committee for Standardization, Brussels.

Chan, A. and Yeung, V. (2005), 'Implementing building energy codes in Hong Kong: energy savings, environmental impacts and cost', *Energy and Buildings*, **37**: 631–42.

Eichholtz, P., Kok, N. and Quigley, J. (2010), 'Doing well by doing good? Green office buildings', *American Economic Review*, **100**(5): 2494–511.

European Commission (2005), *Green Paper on Energy Efficiency or Doing More with Less*, COM (2005) 265 final, Brussels.

European Commission (2011), Energy Efficiency Plan 2011, COM (2011) 109 final, Brussels.

Fuerst, F. and McAllister, P. (2011), 'Green noise or green value? Measuring the effects of environmental certification on office values', *Real Estate Economics*, **39**(1): 45–69.

Galvin, R. (2010), 'Thermal upgrades of existing homes in Germany: the building code, subsidies and economic efficiency', *Energy and Buildings*, **42**: 834–44.

Gillingham, K., Newell, R. and Palmer, K. (2006), 'Energy efficiency policies: a retrospective examination', *Annual Review of Environment and Resources*, **31**: 161–92.

Gillingham, K., Newell, R. and Palmer, K. (2009), 'Energy efficiency economics and policy', *Annual Review of Resource Economics*, **1**: 597–619.

Gilmer, R. (1989), 'Energy labels and economic search: an example from the residential real estate market', *Energy Economics*, **11**: 213–18.

Hamilton, B. (2010), 'A Vermont case study and roadmap to 2050', paper presented at the Transatlantic Energy Efficiency Workshop, Centre for European Policy Studies, Brussels.

IEA (2008), 'Energy efficiency requirements in building codes. Energy efficiency policies for new buildings', Information Paper, International Energy Agency, OECD, Paris.

IEA (2011a), *World Energy Outlook*, Paris: IEA/OECD.

IEA (2011b), 'Energy efficiency policy and carbon pricing', Information Paper, Paris: IEA/OECD.

IEA (2012), '2009 energy balance for the US', Database, Paris: IEA/OECD.

IPCC (2007), *Fourth Assessment Report, Intergovernmental Panel on Climate Change*, Geneva: IPCC.

Iwaro, J. and Mwasha, A. (2010), 'A review of building energy regulation and policy for energy conservation in developing countries', *Energy Policy*, **38**: 7744–55.

Jaffe, A. and Stavins, R. (1994), 'The energy efficiency paradox. What does it mean?', *Energy Policy*, **22**(10): 804–10.

Kemp, R. (1997), *Environmental Policy and Technical Change: A Comparison of the Technological Impact of Policy Instruments*, Cheltenham, UK and Northampton, MA, USA: Edward Elgar.

Kok, N., McGraw, M. and Quigley, J. (2011), 'The diffusion of energy efficiency in buildings', *American Economic Review*, **101**: 77–82.

Kwak, S., Yoo, S. and Kwak, S. (2010), 'Valuing energy-saving measures in residential buildings: a choice experiment study', *Energy Policy*, **38**: 673–7.

Labandeira, X. and Linares, P. (2011), 'Second-best instruments for energy and climate policies', in I. Galarraga, M. González-Eguino and A. Markandya (eds), *Handbook of Sustainable Energy*, Cheltenham, UK and Northampton, MA, USA: Edward Elgar, pp. 441–51.

Laquatra, J. (1986), 'Housing market capitalization of thermal integrity', *Energy Economics*, **3**: 134–8.

Laustsen, J. (2008), 'Energy efficiency requirements in building codes, energy efficiency policies for new buildings', Information Paper, International Energy Agency, OECD, Paris.

Leaman, A. and Bordass, B. (1999), 'Productivity in buildings: the "killer" variable', *Building Research Information*, **27**(1): 4–19.

Leung, T., Chau, C., Lee, W. and Yik, F. (2005), 'Willingness to pay for improved environmental performance of the building envelope of office buildings in Hong Kong', *Indoor Built Environment*, **14**: 147–56.

Linares, P. and Labandeira, X. (2010), 'Energy efficiency: economics and policy', *Journal of Economic Surveys*, **24**: 573–92.

Linares, P., Labandeira, X., Pintos, P. and Würzburg, K. (2012), 'Costs and potential of energy efficiency measures: an application to Spain', Working Paper 09-2012, Economics for Energy.

McKibbin, W., Morris, A. and Wilcoxen, P. (2011), 'Subsidizing household capital: how does energy efficiency policy compare to a carbon tax?', *Energy Journal*, **32**: 111–27.

McKinsey (2009), *Pathways to a Low-Carbon Economy*, New York: McKinsey & Co.

Mundaca, L. and Neij, L. (2009), 'A multi-criteria evaluation framework for tradable white certificate schemes', *Energy Policy*, **37**: 4557–73.

Nair, G., Gustavsson, L. and Mahapatra, K. (2010), 'Owners' perception on the adoption of building envelope energy efficiency measures in Swedish detached houses', *Applied Energy*, **87**: 2411–19.

NRTE (2009), 'Geared for change: energy efficiency in Canada's commercial building sector', National Round Table on the Environment and the Economy and Sustainable Development Technology Canada. Ottawa.

OECD (2011), *Greening Household Behavior: The Role of Public Policy*, Paris: OECD.

Palmer, K., Walls, M., Gordon, H. and Gerarden, T. (2011), 'Assessing the energy efficiency information gap: results from a survey of home energy auditors', Discussion Paper 11-42, Resources for the Future.

Rosas-Flores, J., Rosas-Flores, D. and Gálvez, D. (2011), 'Saturation, energy consumption, CO_2 emissions and energy efficiency from urban and rural households appliances in Mexico', *Energy and Buildings*, **43**: 10–18.

Sweatman, P. and Managan, K. (2010), 'Financing energy efficiency building retrofits', Climate Strategy and Partners, Madrid.

UKDECC (2011a), *Annual Report on Fuel Poverty Statistics*, London: Department of Energy and Climate Change.

UKDECC (2011b), *The Green Deal: A Summary of the Government's Proposals*, London: Department of Energy and Climate Change.

Ürge-Vorsatz, D. and Herrero, S. (2012), 'Building synergies between climate change mitigation and energy poverty alleviation', *Energy Policy*, **49**: 83–90.

World Bank (2012), 'World dataBank: world development indicators & global development finance', Database, Washington, DC.

20 Using micro data to examine causal effects of climate policy

Caterina Gennaioli, Ralf Martin and Mirabelle Muûls

1 INTRODUCTION

Much of economic research on climate change is concerned with assessing the effects of climate change policies using simulation or general equilibrium models.[1] It is in the nature of such models to rely on strong assumptions[2] and parameters that are only loosely linked to the underlying economy. These models nevertheless make it possible to quantitatively analyze a wide range of scenarios and policy options even if none of them has ever been implemented (e.g. a world carbon tax). For a long time that was the only way forward, as governments were not too keen to put in place any concrete policies to mitigate the risk of climate change. However, in the wake of the first UN conference on climate change in Rio (1992), some governments started implementing various pieces of climate policies. Policy implementation further increased in the wake of the Kyoto conference in 1997, which resulted in the Kyoto Protocol. Worldwide emissions have since been increasing nevertheless. From an efficiency point of view, it would be desirable to have market-based policies imposing a globally uniform price on greenhouse gas (GHG) emissions. The present situation on the ground is a far cry from this ideal. Policies vary greatly in stringency and design between countries, as well as within countries for different emitters. Even in European countries which have gone furthest in their effort to curb emissions, policies are typically believed to fall short of what is necessary. Worse, there is concern that emissions might 'leak', both across space and time, implying that existing climate policies could make things worse. This presents both a need and an opportunity for economic research. First, there is an urgent need to better understand the effects existing policies have had, both in terms of their desired outcomes, such as emissions, but also in terms of undesired outcomes such as negative effects on competitiveness or carbon leakage. Second, the existence of a wide variety of different policy instruments presents a unique possibility to study different policy options, as well as interactions of different policies.

Much of the existing empirical – that is, non-simulation-based – literature examines the impact of policies in terms of time-series variation only. For instance, in 2001 the UK government introduced the Climate Change Levy – a tax on energy consumption for UK businesses. Early studies on the subject looked at energy consumption for the UK industry as a whole or at the sectoral level and examined if, relative to pre-2001 trends, there was a change in post-2001 (see Ekins and Etheridge, 2006). The issue with such an approach is that it confounds any actual effects of the policy with all other factors that might have changed in a non-linear way after 2001. This could include other policies – for instance 2001 also saw the introduction of the UK Carbon Trust, a government-funded initiative to consult businesses about energy savings – or other time-varying shocks to

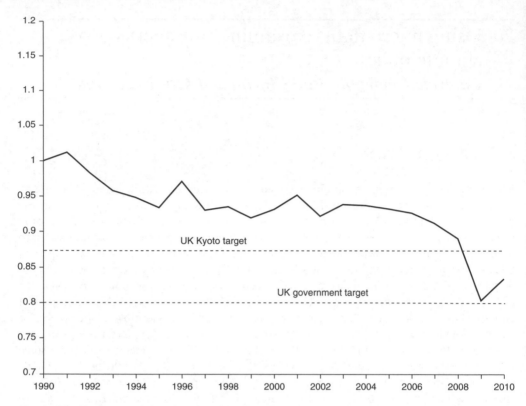

Notes: The Kyoto target is in terms of total GHG emissions. The government target is in terms of CO_2.

Source: Department of Energy and Climate Change.

Figure 20.1   UK CO_2 emissions relative to 1990

the economy. The dramatic effect the recent recession has had on emissions underlines this most clearly, as can be noted in Figure 20.1. Emissions dramatically dropped after 2008 so that the UK was comfortably within the requirements of the Kyoto Protocol and almost met its domestic target of a 12 percent reduction relative to 1990, which had seemed completely out of reach before 2008. However, it would be very misleading to celebrate this as a success of climate policy.

The purpose of this chapter is to highlight that, for many climate change policies, we can do better in assessing their causal effect than by using time-series variation only. This is made possible by the fact that these policies often apply to a large number of units (firms or individuals) with the presence of comparable units that are not subject to the policy. Sometimes, it is the policy design itself that varies across units. For instance, in the case of the Climate Change Levy, the policy applies to all firms; however, they are not all equally affected as some receive a partial exemption from the tax. A comparison can therefore be made between what happened to the energy consumption of firms that were not exempt compared to those that were. In order to interpret this as a causal effect of a higher tax rate, the assumption would be that the only factor that has affected the difference in energy consumption between non-exempt and exempt firms is the Climate Change Levy. Clearly, this is an assumption that might be controversial and it is shown below how it can be further relaxed. However, the point here is that it is clearly a considerably weaker assumption than saying that the only factor affecting energy consumption that has changed from 2001 is the Climate Change Levy. Similar arguments can be made for climate policies affecting households or individuals. In addition, such policies also vary across space. The applied econometrics literature has made a great deal of progress in recent years in dealing with data and questions of this kind, inspired mostly by applications from labor economics and more recently development economics (see, e.g., Angrist and Lavy, 2002; Angrist and Pischke, 2008; Duflo and Hanna, 2005; Duflo, 2004, 2006).

In this chapter we will give an overview of the main approaches that have emerged for policy evaluation in recent years. We will discuss them by way of example, drawing on the small, but growing, literature of applications of these methods to the area of climate change. We close with a discussion of further and ongoing implementations of these approaches. Our chapter is intended as a first introduction for researchers in the social sciences related to climate change unfamiliar with econometric program evaluation.[3] We also hope to raise the profile of econometric evaluation among policy makers implementing various climate change policies. Most policy evaluations rely on the collaboration of policy makers to provide data and information about a policy. This will be facilitated if policy makers have a basic understanding and appreciation of this field of inquiry.

2   ECONOMETRIC POLICY EVALUATION

The objective of econometric policy evaluation is to establish the causal effect of a policy – e.g. a carbon tax – on an outcome y_i – e.g. firm-level energy consumption. The ideal way to conduct policy – evaluation is by accessing different universes. In one universe the policy is applied to firm i, in the other one not. Comparing the values of y_i in the two cases gives us the causal impact of the policy. Formally:

$$\beta_i = y_i(D_i = 1) - y_i(D_i = 0)$$

D_i is an indicator variable equal to 1 if i is subject to the policy. Typically, we would be interested in the average effect across treated firms:

$$\beta = \frac{1}{N_D}\sum_i \beta_i$$

In practice and in most contexts, we do not have several universes at our disposal. A simpler alternative is to compare the average outcomes in firms that were subject to the policy – also referred to as the treated group[4] – with those that were not – also known as the control group. In other words, we estimate $y_i(D_i = 0)$ for firms with $D_i = 1$ by looking at firms with $D_i = 0$.

It is often suggested that such a comparison can only be made if firms in the control group are very similar to those in the treatment group in all respects but the policy treatment. While this would improve the precision of the estimate, a less strong assumption is sufficient to get an estimate of the treatment effect that is not systematically wrong: all factors that determine y_i apart from the policy need to be systematically non-correlated with the policy. For instance, we must be sure that the policy was not applied to some firms and not to others because they differed in terms of outcome y_i. For example, many climate policies for the business sector are applied differently to firms with high energy or emissions intensity.

A more formal but instructive way to express this is to note that estimating β by comparing treated and non-treated average outcomes is algebraically equivalent to running a regression of y_i on the policy variable D_i. Hence the regression equation becomes

$$y_i = \beta D_i + \varepsilon_i \qquad (20.1)$$

where the error ε_i captures all aspects other than D_i that determine y_i. We thus require that ε_i is not correlated with D_i.

Randomization

A simple way to ensure this non-correlatedness is by applying the policy treatment randomly, that is by conducting a randomized controlled trial as it would be done for instance to test the effectiveness of a medical drug. This is likely to be more difficult in the context of policy evaluation, as governments might find it hard to justify giving out certain favors at random. However, it is not impossible, as demonstrated by a flourishing industry of field experimentation in development (see Banerjee and Duflo, 2008) and labor economics (see Angrist and Pischke, 2010). More recently, researchers have begun to use randomized trials in areas related to climate change. For instance, Allcott (2011) analyzes data from one of the largest field experiments in history run by a company called OPOWER, with 600 000 households in treatment and control groups across the USA. The experiment consisted in sending home energy report letters to residential utility customers comparing their electricity use with that of their neighbors. He finds that the average program reduces energy consumption by 2 percent. Despite similar non-price interventions' effects usually being short-lived, Allcott's results also last through time. The effect of the home energy reports is constant or increasing for the first two years of treatment.

A similar contribution from Costa and Kahn (2010) analyzes the effect of the OPOWER program in a district of California where it started in 2008. The size of the sample is obviously much smaller than in Allcott's paper, consisting in 35 000 households treated and 49 000 households in the control group, but the empirical evidence is compelling. Matching the electricity billing data of the households with voter registration and marketing data, the authors study whether treatment effects of the program vary according to ideology and other individual characteristics. Interestingly, they find that a registered liberal who pays for electricity from renewable sources, who donates to environmental groups, and who lives in a liberal neighborhood reduces consumption by 3.1 percent after receiving the electricity report. On the contrary, a registered conservative who does not pay for electricity from renewable sources, who does not donate to environmental groups, and who lives in a non-liberal neighborhood increases consumption by 0.7 percent following the treatment. This study can be considered as complementary to the previous one: by providing evidence on the presence of heterogenous treatment effects, it suggests that in order to be effective, non-price interventions need to be targeted to particular population subgroups.

Controlling on Observables

If randomization is not an option, there are several ways to ensure that the estimate of β represents a causal effect of treatment. The simplest way – which is almost always feasible – is to condition on further observed variables that might explain why the policy treatment was not random. For instance, in the example of firms facing an energy tax, governments might have exemption rules varying on a sectoral basis, with energy-intensive sectors facing less stringent rules. This could be accounted for by including sectoral dummies; thus one would run regressions of the form

$$y_i = \beta D_i + X_i \beta_x + \varepsilon_i \tag{20.2}$$

where X_i is a vector of sector dummies. We can include a range of further variables in X_i which we suspect of determining firm i's selection into either treatment or control group. It is important, however, not to include variables that might themselves be influenced by the policy treatment. For example, it might be the case that governments are more likely to exempt smaller firms from the tax. Hence one might be inclined to include firm size in terms of employment as an additional control. The difficulty here is that, in response to the energy tax, firms might adjust employment as well as their energy consumption. If we include this adjusted employment variable in our regression in 20.2, we would consequently get a biased estimate of the selection effect induced by employment, which, in turn, would bias our estimate of β. However, if we have panel data,[5] we can include a firm's employment from a period before the policy treatment so that it cannot have been affected by the treatment as well.

An alternative way to use observables to get round non-random treatments is matching approaches. This typically involves a two-stage procedure: in a first stage, a propensity score is estimated for each unit in the sample. That represents the probability of treatment given the observable variables X_i; for example we run a probit or logit on the event $D_i = 1$.

One can then predict for each firm the probability of being treated as a function of X_i: $\hat{P}(X_i)$. In the second stage, for every treated firm, one (or several) non-treated firm $j(i)$ is matched as being similar in terms of $\hat{P}(X_i)$. The mean treatment effect is then estimated by

$$\hat{\beta}_M = \frac{1}{N_D} \sum_i (y_i - y_{j(i)}) w_i$$

where w_i is a set of weights and N_D is the number of treated firms. The effect of the policy is calculated as the average difference in outcomes between the treated and non-treated.

If the variables in the vector X_i are all discrete and fully control for all finite states of a firm in the underlying population, then the regression estimator derived from equation 20.2 boils down to the matching estimator $\hat{\beta}_M$ with a specific choice of weights.[6] Thus both estimators are valid under the same conditions, and determining the most preferred one depends on what weights are more relevant. Moreover, if we can assume that the β_is do not vary much across firms, then the two estimators should lead to very similar values.

Often X_i may include continuous variables so that functional form restrictions – such as linearity – need to be imposed in order to estimate either an outcome equation, or a selection equation. The choice between matching estimator or outcome regression approach then depends on the equation for which we feel more comfortable making these additional assumptions. If this is the selection equation, then the matching approach is preferable.

As an example of using observables, Jacobsen and Kotchen (2010) provide evidence on the effect of a change in the residential building energy code introduced in Florida in 2002, on energy consumption.[7] Focusing on the city of Gainesville, they combine monthly residential billing data on electricity and natural gas with data on a set of observable characteristics for each residence. The analysis first compares consumption levels of residences constructed within three years before and three years after the energy code change was implemented, controlling for differences in observable characteristics (square footage, roof type, number of rooms etc.). The energy code stringency increase resulted in a 4 percent decrease in residential electricity consumption. The authors also derive the private payback time of the additional investment required by the home energy code change. Under the best-case scenario, this is 6.4 years, while the social payback time, which includes the benefits of less pollution, is estimated between 3.5 and 5.3 years.

Anderson and Newell (2004) examine the response of firms to energy audits offered by the US government. Using data on the recommendations and the projects implemented within two years of the audit for each firm, they observe that only 53 percent of recommended projects were adopted. Besides, they find higher adoption rates for projects with lower costs, shorter paybacks, greater annual savings and greater energy conservation. The authors also estimate that the firms evaluate energy audit recommendations by typically using an investment threshold of about one- to two-year payback, equivalent to a hurdle rate of 50–100 percent for projects lasting ten years or more. These are consistent with other results in the literature, but are higher than usually assumed in climate policy analyses. What remains unclear in this study is if adoption actually had an effect on energy efficiency or consumption.

Zachmann and Abrell (2011) use matching to evaluate the impact of the European Union Emissions Trading System (EU ETS) on firms' value added, profits and employment. Using a financial database of firms accross Europe, they identify 2101 firms included in the EU ETS. To each of these 'treated' firms they match another non-treated but similar firm, based on a set of pre-treatment characteristics (working capital, number of employees, fixed capital, intermediate consumptions, remuneration of employees). Comparing both types of firms before and after the introduction of the EU ETS, they show that the policy had no effect on value added, profits or employment. This goes against the arguments that such a policy induces carbon leakage. However, the particularities of their data set do not allow the authors to do matching on the basis of the sector of activity of the firm, meaning that their results might only be reflecting sectoral dynamics.

Average Treatment Effect and Treatment Effect on the Treated

Consider once more equation 20.1. We suggested earlier that if policy treatment is uncorrelated with the shock ε_i, then a regression of equation 20.1 leads to an estimate of the causal average treatment effect. It makes sense to qualify a bit further the kind of average we are estimating here. It could be the case that, while ε_i is uncorrelated with D_i, β_i is not. For instance, units that are more likely to respond favorably to a policy treatment or benefit more from it are more likely to participate in a scheme, even if participants and non-participants are not systematically different to begin with. If this is the case, we still would identify a causal effect. However, it would not be the average treatment effect for the whole population, but only the sub-population that actually participates in the policy treatment. We refer to this as the average treatment effect on the treated (ATT), as opposed to the average treatment effect (ATE). Of course in many practical applications D_i might be correlated with both ε_i and β_i. We will come back to this issue below.

Fixed Unobserved Factors

Ideally we conduct policy evaluation using time-series (panel) data on individual units (e.g. firms or households) with information from before and after a policy change. In such a case we can control not only for *observed* but also *unobserved* differences between treated and non-treated units. The unobserved differences must be fixed through time, however. In terms of equation 20.1, we are dealing with a model of the form

$$y_{it} = \beta D_{it} + \alpha_i + \varepsilon_{it} \tag{20.3}$$

where t indexes time and it is assumed that policy treatment is only correlated with α_i but not ε_{it}. An unbiased estimate of β is obtained by estimating 20.3 in terms of deviations from means (fixed effect estimator) or in terms of differences (first differences estimator). Either approach would remove the fixed effect α_i correlated with D_{it}. If the treatment is binary,[8] an alternative way of implementing this involves defining a non-time-varying dummy $D_i^{MAX} = I\{D_{i\tau} = 1 \text{ for any } \tau\}$; that is, D_i^{MAX} is equal to 1 if a unit is treated even in periods when it is not. Including D_i^{MAX} as additional regressor – that is, in a regression of the form

$$y_{it} = \beta D_{it} + \beta_{MAX} D_i^{MAX} + \widetilde{\varepsilon}_{it} \tag{20.4}$$

will capture all unobserved heterogeneity from α_i that could bias our estimate of β. Hence β_{MAX} captures the fixed difference in the level of y_{it} between the group of treated units compared to the group of non-treated units. Regression of equation 20.4 is implicitly comparing how the mean difference between treated units ($D_i^{MAX} = 1$) and non-treated units ($D_i^{MAX} = 0$) changes when treatments starts. Hence we look at the difference in the mean difference, which is why this approach is often referred to as differences-in-differences (diff-in-diff) estimation. Note that we can implement equation 20.4 even in some cases where we have a repeated cross-section but not genuine panel data. It is also feasible if treatment is based on a fixed characteristic of individual units – for instance, in the case of firm treatment being confined to specific industrial sectors.

Kotchen (2010) uses diff-in-diff techniques in a climate-change-related context. He uses panel data to evaluate the effects of the Connecticut Clean Energy Communities program (CCEC) on household decisions to voluntarily purchase electricity generated by renewable sources. According to the so-called 'Option Program', a municipality is entitled to receive free photovoltaic panels or other clean technology in proportion to the number of households that decide to purchase green energy. A municipality can participate in the CCEC if at least 10 percent of households are committed to purchase 20 percent of their energy from renewable sources. Municipality-level data on the total number of households participating in the Options Program from 2005 to 2009 are matched with data on municipality characteristics such as the education level and income distribution. Using the resulting panel data set, the author estimates the effect of CCEC on participation rates in the Option Program using fixed effect estimators. Controlling for fixed unobserved factors substantially reduces endogeneity concerns. The main result of the paper is that within municipalities enrolled in the CCEC program, the household participation rates in the Option Program increased by 35 percent. With this simple analysis, Kotchen (2010) provides evidence that relatively low-cost government programs can promote voluntary 'green' behavior by households.

Besides, as described above, Jacobsen and Kotchen (2010) seek to evaluate the effect of the policy change of building codes. In a second step of their analysis, the authors derive diff-in-diff estimates measuring how weather variability affects pre- and post-code change residences: is the effect of the code strongest when demand for heating and cooling is greatest? This is confirmed in the results, also confirming that the policy change reduced consumption of electricity for air-conditioning and reduced consumption of natural gas for heating.

In a paper by Bjorner and Jensen (2002), firm-level data on energy consumption and value-added for Denmark are used to estimate demand for energy. Besides, the paper investigates the impact of various policy instruments on energy consumption. There appears to be no significant effect of subsidies given to firms for investments in energy-saving projects. They also assess the effect of a policy package including a CO_2 tax, with a reduced rate for companies entering a voluntary agreement with government to carry out certain energy-saving activities. The authors find a positive effect of negotiated agreements on energy efficiency. However, the number of 'treated' firms is low, and their empirical approach does not control for selection into negotiated agreements based on time-varying unobservables.

Instrumental Variables (IV) and Natural Experiments

How can we deal with reverse causality if we do not have panel data? Similarly, how can we deal with reverse causality that is due to time-varying unobserved factors even if panel data are available? Consider for example a government scheme where firms can enter and make voluntary reductions in emissions over a target period.[9] It is plausible that such an option is more attractive for firms, which for completely unrelated reasons expect that their emissions are decreasing over the target period. For instance they might just have installed a new heating system, not because of any policy intervention but because their old system has broken down. Because of this self-selection we could consequently find a completely spurious negative link between participation in the voluntary scheme and emissions.

The IV approach deals with such issues by trying to mimic the random assignment idea of randomized controlled experiments, even if no formal experiment has been undertaken. This can be achieved by finding factors that had a random impact on policy treatment. While in a randomized trial every treatment is fully determined by randomization, the IV approach suggests that it is really enough if the assignment is somewhat random as long as we can observe this random factor. The variable capturing this random factor we then call an instrument. For instance, in a recent paper, Gennaioli and Tavoni (2011) examine if government subsidies for wind power in Italy – which were phased in via a green certificate system in 1999 – had the negative side effect of increasing corruption, measured as the number of charges made by the police force to the judiciary for the crime 'criminal association'. One might investigate this by verifying whether Italian regions that received higher levels of subsidies for wind power also had a higher increase in corruption. However, the concern is that regions with higher levels of corruption might also be the ones that managed to divert larger amounts of government subsidies, thereby leading to a reverse causality. The study accounts for this by building an instrumental variable on the basis of the windiness of different regions. It might not be immediately apparent why windiness is similar to randomly selecting regions. However, what is important is that it is random or rather uncorrelated with respect to the outcome variable – in this case corruption – while having a strong effect on the policy variable – that is, wind farm subsidies. In other words: the central assumption is that there is no systematic relation between windiness and corruption – besides the mechanism implied by the wind farm subsidies.[10] The windiness example makes it very apparent why the instrumental variable approaches are also referred to as a natural experiments – that is, experiments that were not arranged by deliberate *ex ante* design of a researcher. Such cases would rather occur unintentionally by historical accident or as a side effect of bureaucratic necessities.[11] In the case of windiness, this would indeed happen through the idiosyncratic nature of the weather gods.

Gennaioli and Tavoni (2011) find that windier regions experienced a significantly higher increase in recorded criminal association offences over the 1999 introduction of wind subsidies (0.8 offenses more per 100 000 inhabitants).

To understand the IV estimators a bit more formally, it is useful to express them as a two-stage regression. That is, suppose we have an instrument Z_{it} for our policy variable D_{it}. Rather than regressing the outcome y_{it} directly on D_{it}

$$y_{it} = \beta D_{it} + \varepsilon_{it} \qquad (20.5)$$

we first regress D_{it} on Z_{it}, that is,

$$D_{it} = \gamma Z_{it} + v_{it} \qquad (20.6)$$

In a second stage we then regress the outcome on the predicted value of the policy variable $\hat{D}_{it} = \hat{\gamma} Z_{it}$:

$$y_{it} = \beta \hat{\gamma} Z_{it} + \beta v_{it} + \varepsilon_{it}$$

Hence, for this to lead to an unbiased estimator of β, we need that first, Z_{it} is not correlated with ε_{it} and second that, Z_{it} is not part of the equation explaining y_{it} (20.5) to begin with. It is easy to show that this implies that the estimate for β is computed as the correlation between the outcome and the instrument over the correlation between the treatment and the instrument:

$$\hat{\beta} = \frac{Cov\ (y_{it}, Z_{it})}{Cov\ (D_{it}, Z_{it})} \tag{20.7}$$

Another nice idea for an instrument is used in Aichele and Felbermayr (2011). They examine the impact of the Kyoto Protocol using country-level panel data; that is, they compare the performance of countries that signed up to the Kyoto Protocol with those that did not. Clearly, there is potentially a substantial selection problem: countries expecting lower emission growth rates irrespective of Kyoto are more likely to sign up to Kyoto. To address this, Aichele and Felbermayr construct an instrument on the basis of a country's membership with the International Criminal Court (ICC). Thus they identify causal effects on the basis of the assumption that ICC participation is not correlated with changes in emissions between pre- and post-1997. ICC participation is however a strong predictor in explaining participation in Kyoto, perhaps reflecting a country's general preference for multilateralism. Using this strategy leads significant negative causal effects of the Kyoto Protocol on emissions.

An instrumental variable approach is also pursued in the previously mentioned evaluation of the UK Climate Change Levy (CCL) by Martin et al. (2011). Similar to the work by Bjorner and Jensen (2002) for Denmark, they look at the change in firm-level outcomes in firms that were subject to an energy tax (the Climate Change Levy, introduced in 2001), compared to firms that received a partial exemption. The exemption, referred to as Climate Change Agreements or CCAs, came with firm-specific targets on energy consumption. Again, it is a voluntary decision by firms to pay the full tax or go for the exemption with targets. Clearly, the worry is that this leads to a self-selection into CCAs of firms which for unrelated reasons would have reduced their energy consumption anyway. To address this, Martin et al. use an instrument that is based on a rather idiosyncratic eligibility criterion for participation in the CCAs. Only firms that were subject to earlier pollution, prevention and control (PPC) regulation – that is, pollution other than greenhouse gas emissions – could apply for CCAs. The government department implementing the CCL and CCAs came up with this rule because it provided an institutional and legal shortcut.[12] Identification of the treatment effect is consequently coming from the difference in energy consumption – as well as other outcome variables – between eligible and non-eligible firms rather than between firms that were and were not exempt from the tax. Martin et al. find that compared to a naive diff-in-diff regression, the negative impact of the tax on energy consumption increases. This is in line with the mechanism leading to self-selection suggested earlier: firms that would have reduced energy consumption irrespective of the policy are more likely to sign up to CCA. This might explain the contrasting result of the Bjorner and Jensen (2002) study; they find – without correcting for self-selection – that firms receiving a tax exemption reduce emissions by more. A central concern with unilateral climate policies such as the CCL is the potentially negative effect on competitiveness. With that in mind, Martin et al.

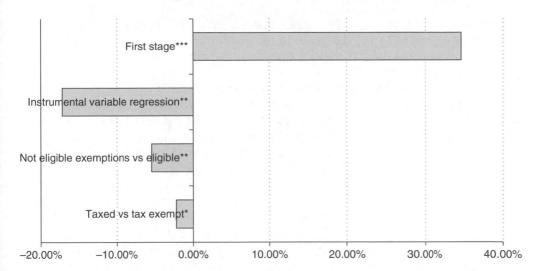

Notes: Based on results in Martin et al. (2011). * = significant at 10%, ** significant at 5%, *** significant at 1%.

Figure 20.2 Estimates of the impact of the Climate Change Levy on energy intensity

also examine the effect of the CCL on economic performance variables such as output, employment or TFP (total factor productivity). They cannot detect any impact on those variables, however, suggesting that a risk to competitiveness is not of primary concern.

As an illustration, Figure 20.2 reports the estimates for energy intensity (energy expenditure over revenue) from Martin et al. (2011). The interpretation of the figures is the differences in growth rates. Hence we see that when comparing taxed with tax exempt firms, we find that the taxed firms reduced their energy intensity by 2 percentage points more, a figure that is significant at the 10 percent level. When comparing firms not eligible for tax exemption with eligible ones, we find a difference that is larger (about 5 percentage points) and more significant. This regression estimate of the outcome on the instrument is sometimes referred to as the reduced form. We also see that eligible firms are more likely to actually enter the discount scheme; that is, the first-stage regression suggests that the likelihood that an eligible firm is in the discount scheme (i.e. an CCA agreement) is more than 30 percent. The IV estimate uses this first stage to take into account what fraction of the eligible population are actually expected to be treated to scale the reduced-form coefficient. Hence we find that the IV estimate is about 17 percent.

In the CCL study both the treatment variable and the instrument are binary. In such a case we can rewrite the IV formula 20.7 as

$$\hat{\beta} = \frac{E\{y_{it}|Z_{it} = 1\} - E\{y_{it}|Z_{it} = 0\}}{E\{D_{it}|Z_{it} = 1\} - E\{D_{it}|Z_{it} = 0\}}$$

This provides an intuitive interpretation of the IV approach. The numerator is the average difference in outcomes between the two instrument categories observations can fall into (e.g. in the CCL case a firm is or is not subject to PPC regulation). This reduced-form effect is divided by the difference in the probability of being treated for the same two instrument categories. In other words, it is the difference in outcomes per extra amount of treated units.

One aspect we have to be careful about when adopting an IV approach is the case of heterogeneous treatment effects across units. We suggested earlier that this implies that we might only identify the average treatment effect on the treated (ATT) as opposed to an average treatment effect (ATE). In the IV case it gets a bit more confusing as we might only identify what is known as the local average treatment effect (LATE).[13] Its intuitive interpretation in the binary context is as follows: it is the average treatment effect for those treated units that change their treatment status if the instrument changes status. Put differently, it's the average treatment effect for units that are *not* always takers or never takers (of treatment). An interesting result[14] emerges when instruments are based on eligibility for a treatment as in the CCL study: when firms are not eligible and $Z_{it} = 0$, we must also have that $D_{it} = 0$. Hence we have no 'always takers' and all treated would change their treatment status if eligibility were changed. In other words, in this case, LATE is equal to ATT.

Regression Discontinuity

Another strategy to identify causal effects of policy treatment, even if treatment is non-random, are discontinuities in relationships between the outcome and other variables that can be attributed to treatment; i.e. suppose there is a (running) variable X_{it} and we think that in absence of any policy intervention there is a linear relationship between X_{it} and outcome variable y_{it}:

$$y_{it} = \beta_X X_{it} + \eta_{it} \tag{20.8}$$

Many policy treatments depend on thresholds. For instance, many social policies are means tested so that only individuals or households below a certain income threshold – say X^T – become eligible for a benefit program. If the program has any effect on outcome y_{it}, we would get a discontinuity in the linear relationship of equation 20.8 which we can model as

$$y_{it} = \beta_X X_{it} + \beta D_{it} + \eta_{it} \tag{20.9}$$

This looks suspiciously like equation 20.2. However, there is an important difference. Unlike earlier, we assume that treatment is a deterministic function of X_{it}; that is, $D_{it} = I\{\text{if } X_{it} < X^T\}$.

To get regression discontinuity right we need to get the relationship between X_{it} and y_{it} (equation 20.8). It might not be linear, in which case we can add more non-linear terms to equation 20.8. We are also bound to make smaller errors in approximating the true relationship between y_{it} and D_{it} if we focus on a subset of the range of X_{it} close to the threshold X^T. Finally, note that, in many applications the thresholds define eligibility for a policy rather than the actual policy treatment. In order to compute the ATT, we can then simply use the threshold indicator variable as instrument in an IV design; that is, $Z_{it} = I\{\text{if } X_{it} < X^T\}$.

Martin et al. (2012) use a regression discontinuity design in a climate policy context. They examine the impact of providing free allocations to some firms within the third phase of the EU ETS. According to theory, initial allocations should only have distributional implications but not affect decision making in the firm. For the third trading phase (starting in 2013) the allocation of permits is based on thresholds that are defined in terms of sectoral carbon and trade intensity. Martin et al. (2012) examine if there are any discontinuities at these thresholds in the response to climate-change-related R&D. They find a large and significant negative effect; that is, firms that receive permits for free undertake less R&D. This is consistent with the idea that making firms pay for permits – rather than giving them away for free – has an important signaling function. Figure 20.3 provides a graphical representation of their study. First, Figure 20.3(a) illustrates the criteria the EU Commission will be using after 2012 to allocate free permits. They are defined in terms of the (non-within-EU) trade intensity and carbon intensity of a (4-digit) sector. Firms that fall above the dashed line will continue to receive free permits after 2012. Firms that fall below will have to buy permits covering their carbon emissions. Martin et al. (2012) examine if there are discontinuities along the two dimensional threshold that consequently arises (i.e. the dashed line). In Figure 20.3(b) we examine only one sub-segment of this threshold, so that the problem reduces to a simpler one-dimensional problem; that is, we examine what happens to firm-level climate R&D scores[15] as firms move over the threshold of 30 percent trade intensity. We focus on an equally spaced band around the 30 percent threshold ranging from 0 to 60 percent. It is visually apparent that there is a lower fraction of firms with higher R&D scores to the right of the threshold. Fitting a regression line with a discontinuity at 30 percent confirms that this reduction is statistically significant.

Allcott's (2011) study discussed earlier also uses a regression discontinuity design for parts of his study. The OPOWER reports for households which he analyzes include in addition to quantitative performance data so-called injunctive norms. These are smiley – or frowny – faces that are added on the letters received by customers, depending on a household's performance. The cut-off points in terms of energy consumption provide arbitrary thresholds which he examines. There do not appear to be any discontinuities at these thresholds, meaning that the smiley faces categorization did not change the response to the ranking information and energy conservation tips also contained in the letter.

3   CONCLUSION AND THE ROAD AHEAD

With an increasing number of policies being implemented in the name of climate change, it is important to understand the causal effects of these policies. This will help to improve these policies and make them more effective and efficient. Rigorous evaluation of pioneering programs will also provide an important knowledge spillover from early to late movers.

For a successful evaluation, a number of things have to come together. First, we need a great many data. At best we need panel data on performance for both participants and sufficiently similar non-participants from before and after an intervention. Second, we need a convincing identification strategy. In most cases randomized controlled trials

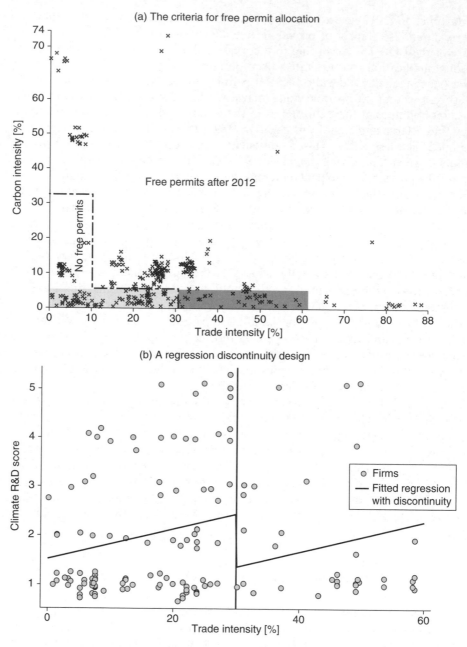

Notes: Based on results and data from Martin et al. (2012). In (a) the dashed line outlines the criteria the EU Commission will be using after 2012 to allocate free permits, defined in terms of the (non within-EU) trade intensity and carbon intensity of a 4-digit sector. Each cross represents one firm of the sample. In (b) only the trade intensity sub-segment of the EU threshold (30 percent) is examined. Each point represents one firm. The y-axis measures a firm-level climate R&D score ranging from 1 to 5 based on interview data. Fitted regression lines are represented with a discontinuity at 30 percent.

Figure 20.3   The impact of free permit allocation

would be ideal.[16] Using additional control variables can mitigate reverse causality problems when randomized trials are not an option. Results will be more reliable if we can use a fixed effect or diff-in-diff design to remove fixed unobserved differences between treated and non-treated units. We can hope to identify even better estimates of causal effects if we can uncover a good natural experiment. In any case it is good idea to be in close contact with the policy makers or agencies executing a policy. It is typically they who maintain at least part of the data that are crucial to undertake evaluations (e.g. the data on who participated in a scheme). They are also most likely aware of institutional details of a scheme that might be exploited as natural experiments. If approached ahead of the actual implementation of a policy, the most adventurous and forward-looking of such policy makers might even be interested in introducing some form of randomization in a scheme. This could be done in the context of a pilot phase with only a small sample of the eventually treated population. Or treatment could be randomly delayed for some participants, although this option often meets strong opposition.[17] Of course, closeness to policy makers can also be a vice. They more often than not have vested interests in the outcome of a study, which might lead them to try to influence results, block a study entirely by not releasing data or hold a veto over outputs. However, in most situations there is an area of common ground between policy implementers and researchers. Particularly in times of austerity there is huge pressure on policies to show that they deliver value for money. Clearly, the most robust way to do that is an independent econometric evaluation study.

It is perhaps too early to draw overall conclusions or distill stylized facts on the basis of the evaluation studies which have been undertaken to date and which we have been discussing in this chapter. The different studies examined a wide range of policies and interventions. However, one theme which might be emerging is a horse race between 'hard' incentives such as forms of carbon pricing and more 'soft' interventions such as energy reports, free energy advice or voluntary agreements. We have discussed results showing that either type of intervention can lead to successful reductions in energy consumption and thereby emissions. However, an important question for policy makers is if the soft interventions can be a substitute for the typically less popular hard interventions. The most direct evidence on this so far comes from the Bjorner and Jensen (2002) and Martin et al. (2011) studies. The former concludes that a voluntary reduction scheme which has a not-so-voluntary backstop can be more effective, whereas the latter comes to the opposite conclusion. However, the Bjorner and Jensen study does not address the potential self-selection of firms into the 'agreement' group. This on balance would give support to the hypothesis that exemption rules and individually negotiated soft agreements will be exploited by the regulated to continue with business as usual and are no match for 'hard' incentives. Further evidence that exemptions and indiscriminate subsidies are less effective than hard financial sticks comes from the Martin et al. (2012) study exploiting the ETS exemption rules. Besides inquiries into the effect of policies on emission mitigation, a central theme for the future will be undesirable side effects of climate policies such as deterioration of competitiveness. At present, there is no evidence that existing policies have had any such effects. The prevalence of this topic in the public debate is therefore more likely the consequence of affected industries' efforts to lobby for a transfer of rents. There is however evidence of some more surprising negative side effects, such as the increase in corruption due to wind farm subsidies (Gennaioli and Tavoni, 2011). This result would also reinforce the conclusion that a transition to a carbon-free economy is best achieved by carbon pricing, rather than direct subsidies. Finally, an important theme in going forward must be the effect of policies on 'clean' innovation. In the current situation, where only a small fraction of global sources of emission are subject to any form of meaningful regulation, it is hard to justify such partial interventions if they do not lead to the discovery of new ways to mitigate emissions more cheaply. If this happens, there is a chance that non-polluting alternatives become attractive enough to be adopted widely, even without any supporting regulation. This is essentially a mechanism for countries and regions, such as the EU, that are pressing ahead faster with stringent regulation to punch above their carbon weight.

At any rate, there is no shortage of future work when it comes to econometric evaluations of climate change policies. There is a variety of existing policies and schemes that would be very suitable for econometric *ex post* evaluation which have not been examined yet. Examples in the UK alone include the Carbon Trust Energy Audits,[18] the Carbon Reduction Commitment[19] or the Warm Front Scheme.[20] Our own efforts in the near future will be focused on conducting a comprehensive evaluation of the EU ETS using firm-level panel data from a variety of sources.

NOTES

1. There exists a wide range of so-called integrated assessment models (IAMs) that combine equations characterizing the climate and the economy as well as their interaction. They aim to predict how greenhouse gas emissions impact the economy and vice versa. This in turn allows them to analyze potential policies and advise on the optimal emissions abatement choices. Stanton et al. (2009) present a survey of 30 such models.
2. For instance, many early models assumed non- or exogenous technological change. Another limitation is the lack of modeling trade in many models. For detailed discussions of these limitations and discussion of controversial assumptions see Kohler et al. (2006), Stern (2006), Nordhaus (2007) or Weitzman (2009).
3. Similar reviews focusing on environmental policies more widely include Greenstone and Gayer (2007), Ferraro (2009) and Frondel and Schmidt (2001).
4. Econometric program evaluation borrows many ideas, as well as language, from medical trials.
5. Panel data contain observations for the same firms or individuals over multiple time periods.
6. Namely weights that correspond to the variance of D_i in each treatment group.
7. The energy code sets a minimum energy efficiency standard for space heating, space cooling and water heating of a 'model' home, as well as 'points' for different features based on how important they are considered for energy efficiency. In order to obtain a building permit, a newly constructed residence must exceed the number of points of a 'model' home.
8. This can easily be extended to the case where D_{it} can take on more than two but a finite number of values.
9. For example, as in the Bjorner and Jensen (2002) study we discussed above.
10. Because this implies that in a model equation describing the outcome – i.e. corruption – there is no need to include the instrument; this is also sometimes referred to as an exclusion restriction.
11. A nice example here is the studies on the effect of military service exploiting draft lotteries based on social security numbers (Angrist, 1990; Angrist et al., 2011).
12. Creating a new legally sound definition would have delayed the legislation further, so they opted for adopting an already approved categorization of firms.
13. Imbens and Angrist (1994).
14. Due to Bloom et al. (1997).
15. Derived from in-depth interviews with managers in those companies. For details see Martin et al. (2012).
16. This must not always be the case as the fact of a controlled setting might induce changes in behavior that would otherwise occur.
17. The implementation of the PROGRESA program in Mexico represents a suggestive example; the Mexican government took advantage of the budgetary constraints to first launch a pilot program benefiting only some randomly selected rural communities. The pilot was then evaluated by several independent

studies. The program, proved to be effective, was subsequently extended, reaching 2.6 million families in the rural area. However, it was impossible to start with a pilot for the urban version of PROGRESA, due to the strong opposition to delay some people's access to the program (see Duflo, 2004 for a discussion).
18. Free energy-saving advice for companies provided by the Carbon Trust. The Carbon Trust is a good example of a bad example; that is, an agency hostile to making data available and not facilitating sound evaluation of its programs.
19. A benchmarking scheme for industry where companies are compared on their emission reduction performance.
20. A range of subsidies for poorer households for energy-saving investments.

REFERENCES

Aichele, R. and Felbermayr, G. (2011), 'What a difference Kyoto made: evidence from instrumental variables estimation', Technical Report Ifo Working Paper No. 102.
Allcott, H. (2011), 'Social norms and energy conservation', *Journal of Public Economics*, **95**(9–10), 1082–95.
Anderson, S.T. and Newell, R.G. (2004), 'Information programs for technology adoption: the case of energy-efficiency audits', *Resource and Energy Economics*, **26**(1), 27–50.
Angrist, J.D. (1990), 'Lifetime earnings and the Vietnam era draft lottery: evidence from social security administrative records', *American Economic Review*, **80**(3), 313–36.
Angrist, J.D. and Lavy, V. (2002), 'The effect of high school matriculation awards: evidence from randomized trials', Technical Report 9389, National Bureau of Economic Research.
Angrist, J.D. and Pischke, J.-S. (2008), *Mostly Harmless Econometrics: An Empiricist's Companion*, Princeton, NJ: Princeton University Press.
Angrist, J. and Pischke, J.-S. (2010), 'The credibility revolution in empirical economics: how better research design is taking the con out of econometrics', Technical Report 15794, National Bureau of Economic Research.
Angrist, J.D., Chen, S.H. and Song, J. (2011), 'Long-term consequences of Vietnam-era conscription: new estimates using social security data', *American Economic Review*, **101**(3), 334–8.
Banerjee, A.V. and Duflo, E. (2008), 'The experimental approach to development economics', Technical Report 14467, National Bureau of Economic Research.
Bjorner, T.B. and Jensen, H.H. (2002), 'Energy taxes, voluntary agreements and investment subsidies – a micro-panel analysis of the effect on Danish industrial companies' energy demand', *Resource and Energy Economics*, **24**(3), 229–49.
Bloom, H.S., Orr, L.L., Bell, S.H., Cave, G., Doolittle, F., Lin, W. and Bos, J.M. (1997), 'The benefits and costs of JTPA Title II-A programs: key findings from the National Job Training Partnership Act study', *The Journal of Human Resources*, **32**(3), 549–76.
Costa, D.L. and Kahn, M.E. (2010), 'Energy conservation "nudges" and environmentalist ideology: evidence from a randomized residential electricity field experiment', National Bureau of Economic Research Working Paper Series, No. 15939.
Duflo, E. (2004), 'Scaling up and evaluation', in F. Bourgignon and B. Plesovič (eds), *Accelerating Development*, New York: World Bank and Oxford University Press, pp. 342–67.
Duflo, E. (2006), 'Field experiments in development economics', prepared for the World Congress of the Econometric Society.
Duflo, E. and Hanna, R. (2005), 'Monitoring works: getting teachers to come to school', Technical Report 11880, National Bureau of Economic Research.
Ekins, P. and Etheridge, B. (2006), 'The environmental and economic impacts of the UK climate change agreements', *Energy Policy*, **34**(15), 2071–86.
Ferraro, P.J. (2009), 'Counterfactual thinking and impact evaluation in environmental policy', *New Directions for Evaluation*, **122**, 75–84.
Frondel, M. and Schmidt, C.M. (2001), 'Evaluating environmental programs: the perspective of modern evaluation research', Technical Report 397, Institute for the Study of Labor (IZA).
Gennaioli, C. and Tavoni, M. (2011), 'Clean or "dirty" energy: evidence on a renewable energy resource curse', Technical Report 2011.63, Fondazione Eni Enrico Mattei.
Greenstone, M. and Gayer, T. (2007), 'Quasi-experimental and experimental approaches to environmental economics', Technical Report 0713, Massachusetts Institute of Technology, Center for Energy and Environmental Policy Research.
Imbens, G.W. and Angrist, J.D. (1994), 'Identification and estimation of local average treatment effects', *Econometrica*, **62**(2), 467–75.

Jacobsen, G.D. and Kotchen, M.J. (2010), 'Are building codes effective at saving energy? Evidence from residential billing data in Florida', Technical Report 16194, National Bureau of Economic Research.

Kohler, J., Grubb, M., Popp, D. and Edenhofer, O. (2006), 'The transition to endogenous technical change in climate-economy models: a technical overview to the innovation modeling comparison project', *The Energy Journal* (Special Issue on Endogenous Technological Change and the Economics of Atmospheric Stabilization), 17–56.

Kotchen, M.J. (2010), 'Climate policy and voluntary initiatives: an evaluation of the Connecticut clean energy communities program', Technical Report 16117, National Bureau of Economic Research.

Martin, R., de Preux, L.B. and Wagner, U.J. (2011), 'The impacts of the Climate Change Levy on manufacturing: evidence from microdata', Working Paper 17446, National Bureau of Economic Research.

Martin, R., Muûls, M. and Wagner, U. (2012), 'Carbon markets, carbon prices and innovation: evidence from interviews with managers', mimeo, London School of Economics.

Nordhaus, W.D. (2007), 'A review of the "Stern Review on the Economics of Climate Change"', *Journal of Economic Literature*, **45**(3), 686–702.

Stanton, E.A., Ackerman, F. and Kartha, S. (2009), 'Inside the integrated assessment models: four issues in climate economics', *Climate and Development*, **1**(2), 166.

Stern, N. (2006), *The Economics of Climate Change. The Stern Review*. Cambridge: Cambridge University Press.

Weitzman, M.L. (2009), 'On modeling and interpreting the economics of catastrophic climate change', *The Review of Economics and Statistics*, **91**(1), 1–19.

Zachmann, G. and Abrell, J. (2011), 'Assessing the impact of the EU ETS using firm level data', Working Paper 579, Bruegel.

21 Carbon trading: past, present and future
Julien Chevallier

1 INTRODUCTION

The European Union Emissions Trading Scheme (EU ETS) was created on 1 January 2005 to reduce by 8 per cent CO_2 emissions by 2012, relative to 1990 emissions levels. This aggregated emissions reduction target in the EU has been achieved following differentiated agreements, sharing efforts between member states based on their potential of CO_2 emissions reduction.

The introduction of a tradable permits market has been decided to help member states in achieving their targets in the Kyoto Protocol. Among the members of Annex B, these agreements include CO_2 emissions reductions for 38 industrialized countries, with a global reduction of CO_2 emissions by 5.2 per cent. The commitment period of the Kyoto Protocol goes from 1 January 2008 to 31 December 2012.

Industrial operators may cut the costs of reducing their emissions by using credits issued from the Kyoto Protocol Clean Development Mechanism (CDM), called certified emissions reductions (CERs). These CERs correspond to one tonne of CO_2 emissions avoided in the atmosphere, and may be obtained through projects developed in non-Annex-B countries of the Kyoto Protocol that allow to reduce emissions compared to a baseline scenario. Once credits have been issued by the CDM executive board (EB) of the United Nations, they may be sold by project developers on the market, and thus become secondary CERs (sCERs).

As the CDM has revealed the strong potential of CO_2 emissions abatement in countries such as Brazil, China or India, the main issue of post-Kyoto negotiations is linked to achieving the largest possible level of cooperation in order to avoid the well-known free-rider behaviour, and to preserve the global public good that constitutes the climate.

When dealing with climate policies, the EU has clearly adopted a leadership position, which contrasts with its early reluctance during the first steps of the negotiation of the Kyoto Protocol. In January 2008, the European Commission extended the scope of its action against global warming by 2020 with the 'Energy and Climate Change' package. It aims at reducing GHG emissions by 20 per cent, at increasing the use of renewable energy in energy consumption to 20 per cent, and at saving 20 per cent of energy by increasing energy efficiency.

The EU ETS, which is currently in Phase II (2008–12), has been confirmed until 2020. Besides the shift to auctioning, its scope has been extended to major sectors in terms of CO_2 emissions growth, such as aviation and petrochemical industries during 2013–20.

These repeated public policies in favour of climate protection aim at correcting the negative externality attached to the release of uncontrolled greenhouse gases emissions in the atmosphere and thus, according to the well-known principle in economics, at internalizing the social cost of carbon.

These initiatives reveal the difficulty to create a scarcity condition regarding CO_2

emissions. These emissions indeed were not limited in the pre-existing institutional environment, and thus could not be considered as a scarce resource.

As often in economic studies, we need to study the roots of the current emissions trading schemes in order to potentially indicate the future developments at stake, as well as the main barriers to the establishment of a truly unified 'world' carbon market.

In 1975, the US Environmental Protection Agency (EPA) used permits trading in order to fight air pollution. In 1995, the US Acid Rain Program was established (under Title IV of the Clean Air Act Amendments of 1990) to drastically reduce SO_2 emissions by electric utilities. This ambitious programme sets a permanent cap to SO_2 emissions by utilities at about half of their annual emissions in 1980.

Following the US Acid Rain Program, considered as the first 'grand policy experiment' in terms of emissions trading (Stavins, 1998), various schemes have emerged at the regional and world levels, more especially the EU ETS and the Kyoto Protocol's CDM.

This chapter discusses the extent to which these markets are overlapping and intertwined, and to what extent they might as well yield to the extension of regional initiatives instead of a globally unified carbon market (in absence of a successor to the Kyoto Protocol).

The remainder of the chapter is organized as follows. Section 2 recalls the main principles behind carbon trading. Sections 3 and 4 contain, respectively, the latest developments on the EU ETS and the CDM. Section 5 concludes.

2   REVIEW OF THE MAIN CHARACTERISTICS OF CAP-AND-TRADE SYSTEMS

Broadly speaking, a cap-and-trade system consists for the regulator in defining the 'size of the pie' (i.e. the maximum global amount of pollution), and then to distribute the rights to pollute to market participants, so that they can engage in trading and determine the price of the commodity based on the balance between supply and demand.

In what follows, we deal with various issues arising when creating such a tradable permits market.

2.1   How is a Tradable Permit Different?

Inevitably, the introduction of a tradable permits market raises the debate about the notions of 'right to pollute' and of 'marketability of the environment'. As Sandel (1997) puts it, what difference does it make which places on the planet send less greenhouse gases to the sky? It may make no difference from the standpoint of the heavens, but it does make a political difference.

Turning pollution into a commodity to be bought and sold removes the moral stigma that is properly associated with it. A fee, on the other hand, makes pollution just another cost of doing business, like wages, benefits and rents.

Virtually any manufacturing activity entails the creation of some pollution. If there is to be pollution, the regulator will try to provide economic incentives in order to fight against the negative consequences on the environment. Such a trade-off is facilitated by tradable rights.

Thus, maintaining a moral stigma on pollution makes sense for hazardous substances where polluters have choices for reducing the pollution. But global warming is not such a situation. Do we need to feel ashamed when we cook dinner, switch on a light or turn on a computer to write an article? These daily habits may not be associated with immoral behaviours!

Since the 1980s, we have witnessed a trend towards increased reliance on markets: in Europe, many fields have been deregulated (telecoms, electricity etc.). It could be concluded that the same reasoning applies to environmental goods. Besides, high-income societies tend to value environmental goods more, according to the environmental Kuznets curve.

On top of these issues, the usual command-and-control regulation policies have been unsuccessful in regulating many aspects of the environment (Rees, 1990). This situation may contribute to explaining why we have witnessed an increasing influence of the market as a regulatory tool for environmental externalities over the last decades.

Therefore we may broadly distinguish between organized markets (i.e. the government mainly enforces regulation) versus constructed markets (i.e. the government's role consists in creating the market for individual needs, and then not coming back to regulate). A tradable permits market would correspond to the second category. The theoretical properties of tradable permits have been derived by Dales (1968) and Montgomery (1972).

Hahn (1984) shows that, at the equilibrium permits price, price-taking agents adjust emissions until the aggregated marginal abatement cost is equal to the permit price. Therefore the role of the regulator should be restricted to create the allowance trading mechanism, and then to let the market operate on its own.

2.2   Circumstances Favourable to the Adoption of Tradable Permits

The literature on tradable permits usually requires the following conditions to introduce such a policy instrument (Tietenberg, 2003):

1.   a large number of agents to regulate,
2.   an asymmetrical and strategic access to information,
3.   a high level of heterogeneity of costs and opportunities across decentralized agents,
4.   uncertainty about the shape of cost and damage curves.

Concerning the Kyoto Protocol and the EU emissions trading mechanism, these conditions are considered by most academics to be fulfilled (Newell et al., 2000).

Regarding the debate whether taxes would be more appropriate than permits to regulate the emissions of pollutants, Weitzman (1974) demonstrates that taxes are superior to permits unless the marginal benefit of abatement is steeper than the marginal cost.[1]

Therefore, in a situation where the marginal cost of abatement rises rapidly and future abatement costs are very uncertain, a carbon tax would be superior to tradable permits in theory. But permits are easier to introduce in practice for political reasons.

In the next section, we discuss various design issues linked to the creation of a tradable permits market.[2]

2.3   Spatial and Temporal Market Design Issues

Let us deal first with spatial limits: there are scaling issues. Increasing the scale of the cap-and-trade system not only increases economic efficiency, but also reduces trade security. The regulator also needs to take into account deposition constraints, by avoiding exceedance of critical loads in specific geographical zones (as in the US Acid Rain Program; see Ellerman et al., 2000). Another concern lies in the proper design of national emission ceilings.

Second, the literature highlights the importance of designing the temporal limits of the scheme. Banking (saving permits for future use) and borrowing (borrowing from future allocation) may be used to equalize marginal costs in present value, when the Hotelling conditions apply. It forms another dimension of efficiency by adding the time dimension to cost savings.

The authorization of these provisions appears desirable in a tradable permits market, since they allow firms to achieve their depolluting objective at least cost by smoothing emissions over time. However, they may also change the temporal profile as well as the magnitude of environmental damages (see Chevallier, 2012a for a survey).

From the regulator's viewpoint, the best configuration of the intertemporal flexibility mechanism therefore consists in authorizing banking without restrictions, and in penalizing borrowing by using a non-unitary intertemporal exchange ratio (Rubin, 1996; Kling and Rubin, 1997).

Next, we deal with an equally important topic: the definition of the initial allocation methodology.

2.4   Initial Allocation

In a brilliant text, Raymond (1996) recalls that initial allocation reveals social norms and what society considers acceptable regarding how to distribute the newly created permits. There is a need for specific allocation strategies according to sociopolitical settings:

- grandfathering is advocated by the possessory viewpoint (i.e. Hume);
- auctioning is advocated by the instrumental viewpoint (i.e. Cohen);
- baseline emissions are advocated by the intrinsic viewpoint (i.e. Locke);
- per capita allocation is advocated by the egalitarian viewpoint (i.e. Proudhon).

From this discussion, we observe that depending on the property theory, different allocation methodologies may be selected. The nexus of the problem is that any free allocation system will be deficient in both efficiency and equity terms. Yet this principle is strongly favoured by a key stakeholder, namely business. As reported by the US Congressional Budget Office (2000), the sale of permits, with revenue recycling, has the potential to satisfy both efficiency and equity objectives but faces deep opposition from business. Therefore the 'grandfathering' approach decided in Kyoto is understandable to achieve political acceptability.

More generally, it appears problematic to distribute permits following the appropriate criteria, be it emissions (total or per capita), GDP (total or per capita), population size, or the historical relative responsibility in the growth of greenhouse gases emissions (for a detailed study, see Chevallier et al., 2009a).

As Parry (2002) puts it, grandfathered permits create windfall gains for shareholders, who are concentrated in high-income groups, because such policies hand out a valuable asset to firms for free. There is no windfall gain to wealthy households under auctioned permits or emissions taxes; instead, the government obtains revenues that can be recycled in tax reductions that benefit everyone or disproportionately favour the poor.

On this issue, Newell et al. (2000) further comment that tradable permits create rents, and grandfathering distributes those rents to existing firms while also erecting barriers to entry. Direct allocation of grandfathered permits offers a degree of political control over the distributional effects of regulation, enabling the formation of majority coalitions.

In a similar vein, Chevallier et al. (2009a) argue that, according to production optimization decisions, the capital return is determined not only by the capital marginal productivity, but also by the market value of the free endowment of pollution permits given to the firm, valued at the market price.

The latter represents what is called windfall profits in the literature. According to Sijm et al. (2006), at €20/tonne of CO_2, the windfall profits vary between €5.3 and 7.7 billion. In another study, Sijm et al. (2008) evaluate that the annual impact on profits of European power producers in Phase I comprise between €3.5 and 5.4 billion (at the average price of €12/tonne of CO_2).

With free permits, the rent is given to the firms, which distorts the capital market. This adverse effect would not appear with price or command-and-control regulations. If all the permits were auctioned, then the regulation with tradable emissions permits would be strictly equivalent to a regulation with an emissions fee (Goulder, 1995; Oates, 1995).

The main interest of auctioning permits consists in the income transfer, which may take the form of a lump sum or a tax rebate. This recycling revenue effect (also called double dividend) takes place when auctioning permits allows decreasing taxes and reducing pre-existing distortions.

This should have a positive effect on the supply of capital and labour, but the net effect on high- or low-income quintiles depends on the tax structure. If the tax cut benefits richer people more, then there is a clear trade-off between efficiency and equity (Dinan and Rogers, 2002).

According to Chevallier et al. (2009a), the most efficient free allocation methodology (maximizing the world's production for a given emissions level) consists in distributing permits based on the quantities of efficient labour.

A more equitable solution consists in distributing permits to each production factor proportionally to its share in production.

Grandfathered permits do not necessarily create windfall gains for shareholders if their allocation is judiciously made to all production factors.

In this section, we reach the global conclusion that tradable permits markets need to be carefully designed by the regulator if the implementation is to be successful (Chevallier, 2009a). We point to the critical discussion about the assignment of property rights on the pollution.

In what follows, we detail the creation of the EU emissions trading system in practice.

3   THE EU ETS: THE EU FLAGSHIP ENVIRONMENTAL POLICY

This section provides an overview of the main features of the EU ETS, the price development and main economic mechanisms at stake behind carbon pricing. It builds on the previous work by Chevallier (2011a) and Ellerman et al. (2010).

3.1   Main Features

A new class of carbon assets has been created by the European Commission since 1 January 2005 during three distinct periods: Phase I (2005–07), Phase II (2008–12) and Phase III (2013–20). One EU allowance (EUA) is equal to one tonne of CO_2 emitted in the atmosphere. EUAs therefore constitute a new commodity market, attracting increasing academic interest, in a moving institutional context (Chevallier, 2009b).

Allowances are tradable all around Europe on exchanges and by over-the-counter (OTC). Various European-based marketplaces exist for carbon trading (BlueNext in Paris, European Climate Exchange in London, European Energy Exchange in Leipzig, Nordpool etc.). Every year, industrial installations have the obligation to restitute as many allowances as actual CO_2 emissions.

Under the current provisions, there is unlimited and free banking and borrowing between Phases II and III. This was not the case between Phases I and II, which caused the spot price to gear asymptotically towards zero near the end of 2007 (for more details, see Section 3.2).

Market participants include industrials, investment banks, traditional brokers and small specialized brokers (Chevallier, 2012b). In terms of financial contracts, spot, futures and options prices are exchangeable on organized markets.

Another interesting characteristic of EUAs is revealed by Chevallier et al. (2009b): the carbon market is a commodity market. However, the cash-and-carry rationale linking futures and spot prices does not hold. The main reason behind this non-validity of the theory of storage (and the non-existence of a convenience yield) is that carbon allowances only exist in the balance sheet of regulated companies. Hence there are no 'physical' storage costs associated with carbon trading, which makes it a singular commodity among energy markets (in sharp difference with the oil market for instance; see Chevallier, 2012b).

3.1.1   Scope

The EU ETS was established by Directive 2003/87/EC, and launched for a trial period from 2005 to 2007.

Across its 27 member states, the EU ETS covers large plants from CO_2-intensive emitting industrial sectors: power generation, mineral oil refineries, coke ovens, iron and steel and factories producing cement, glass, lime, brick, ceramics, pulp and paper, and all combustion activities with a rated thermal input exceeding above 20 MW (Alberola et al., 2008a, 2009).

Overall, this scheme includes approximately 13 000 industrial installations. During the first two years, it covered 25 member states. Bulgaria and Romania have integrated the trading scheme in 2007.

3.1.2   Allocation

A total of 2.2 billion of allowances per year were allocated during 2005–07, and 2.08 billion quotas per year will be distributed during 2008–12. Most permits will be auctioned as of 2013.

During 2005–07, allowances distributed have more than covered verified emissions, with a net cumulated surplus of 156 million tonnes.

Verified emissions in the EU ETS sectors have been rising by 2.5 per cent in 2010, from 1869 million tonnes of CO_2 in 2009 to 1915 million tonnes of CO_2 in 2010. The amount of surplus allowances was equal to 55 million tonnes. The quantity of Kyoto credits used for compliance was equal to 140 million tonnes, that is, approximately 7 per cent of the allowances surrendered.

3.1.3   Calendar

The delivery of allowances is made on a yearly basis:

- on 28 February of year N, European operators receive their allocation for the commitment year N;
- 31 March of year N is the deadline for the submission of the verified emissions report during year $N - 1$, from each installation to the European Commission;
- 30 April of year N is the deadline for the restitution of quotas utilized by operators during year $N - 1$;
- 15 May of year N corresponds to the deadline of the official publication by the European Commission of verified emissions for all installations covered by the EU ETS during year $N - 1$.

To sum up, according to this calendar, installations need to report by the end of March their verified emissions that occurred during the preceding year. This information becomes publicly available when the EC officially publishes its report by mid-May, which is typically called a 'yearly compliance event' in the literature (Chevallier et al., 2009b). These events can be used as the cornerstone of major changes in agents' expectations.[3] That is why we can expect a decrease in the level of the EUA price volatility after the diffusion of information by the EC.

As on other energy and financial markets, we obtain the classical property that the diffusion of reliable information is crucial to the formation of anticipations regarding current and future price levels. Installations have indeed a fair amount of information between the publication of their own report and the compilation of verified emissions by the EC to approximate the global level of emissions relative to allowances allocated and to adjust their anticipations (Ellerman and Buchner, 2008).

Moreover, there is a period in between the two yearly allocation periods where utilities benefit from two vintages of allowances (N and $N - 1$) during which they can effectively smooth their emissions over time, thanks to the banking and borrowing provisions.

3.1.4   Transactions

According to Point Carbon (2011), the volume of transaction has been increasing steadily from 262 million tonnes in 2005 to 809 million tonnes in 2006, 1455 million tonnes in 2007, 2713 million tonnes in 2008, and 5016 million tonnes in 2009. These estimates account for exchange-based trading as well as over-the-counter trading of emissions rights. In 2011, the volume of transactions amounted to 6053 million EUAs (up 20 per cent compared with 2010).

It is worth noting that an options market was created by ECX in October 2006. The variability of call and put prices exchanged by market participants reflects various levels of risk aversion which are embedded within the trading of options prices (Chevallier et al., 2009b). As for any financial market, the emergence of liquid options markets constitutes another derivative asset that may be used by energy companies, brokers and investment banks in order to insure themselves against unwanted price movements. The introduction of an options market for carbon assets has improved the liquidity of the market, without negative effects on the volatility of the carbon price (Chevallier et al., 2011a).

3.1.5   Penalties

During 2005–07, if an installation does not meet its emissions target during the compliance year under consideration, the penalty is equal to €40/tonne of CO_2 in excess, plus the restitution of one allowance during the next compliance period.

During 2008–12, this amount corresponds to €100/tonne of CO_2, following the same principle.

Having detailed the functioning of the EU cap-and-trade system, we investigate next the behaviour of the carbon price.

3.2   Price Development

The price development of EUAs is shown in Figure 21.1. EUAs were characterized by a high level of spot price volatility during 2005–07, due to heterogeneous anticipations from market participants concerning the 'true level' of the carbon price. In April–May 2006, a severe market correction followed the initial compliance event by the European Commission, revealing that the market was long by approximately 4 per cent. In June–September 2006, the carbon market stabilized at around €15/tonne of CO_2.

During October 2006–March 2007, Alberola and Chevallier (2009) explain the reasons behind the divorce between Phases I and II prices.[4] The price of EUAs has been declining at far lower levels than expected during Phase I (2005–07). Previous literature identifies among its main explanations over-allocation concerns, early abatement efforts in 2005, and possibly decreasing abatement costs in 2006. Low allowance prices may also be explained by banking restrictions between 2007 and 2008, which undermined the ability of the EU ETS to provide an efficient price signal for emissions abatement.

Consequently, Phase I price fell to €0.5/tonne of CO_2, while Phase II price remained in the range €15–20/tonne of CO_2. Stricter allocation caps led to Phase II prices rising to €25/tonne of CO_2 in May 2007.

During March 2007–August 2008, Phase II prices have stabilized over €20/tonne of CO_2, following the European Council decision to maintain the EU ETS at least until 2020.

In absence of post-Kyoto climate agreement, the current medium-term price signal fluctuates around €10/tonne of CO_2.

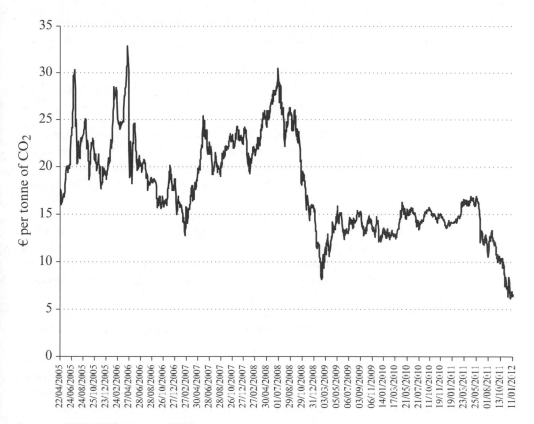

Sources: ECX (2007); Ecotrade (2012).

Figure 21.1 Prices of EUA (€ per tonne of CO₂), 2005–12

3.3 Drivers of EUA Prices: Main Mechanisms

The EU carbon price aims at helping industrials to reduce their CO_2 emissions. In theory, the carbon price is equal to the marginal cost of emissions abatement (Ellerman et al., 2000; Ellerman and Montero, 2007). On the supply side, the amount of CO_2 allowances distributed each year according to National Allocation Plans (NAPs) is known in advance. On the demand side, carbon prices are affected by energy prices, because they relate to the producing process of the utilities regulated by the EU ETS.

In the previous section, we highlighted that the European carbon price has experienced strong volatility. Researchers therefore need to understand the drivers of carbon prices to assess abatement decisions, as well as the impact of institutional decisions disclosed by the European Commission (as decisions regarding EUA allocations have consequences for the demand and supply of allowances).[5] The literature also identifies the relationship between carbon prices changes, energy price changes[6] and temperatures.[7]

Indeed, the theoretical literature review by Springer (2003) and Christiansen et al. (2005) reveals broadly two kinds of carbon price drivers:

1. Policy and regulatory issues of the EU ETS: NAPs, auctioning share of allowances, banking and borrowing allowances possibilities, new entrant reserve, new covered sectors and so on.
2. Emissions levels: economic activity (industrial production by covered installations, electricity power demand by others sectors), energy prices (Brent, natural gas, coal), weather conditions (temperature and rainfall).

The main results from this rapidly growing literature (see, among others, Mansanet-Bataller et al., 2007; Alberola et al., 2008b; Hintermann, 2010), may be summarized as follows. First, structural breaks in EU carbon prices may be explained by official communications from the European Commission, such as the publication of 2005 verified emissions in April 2006 or the announcement of stricter NAPII allocation for 2008–12 in October 2006. Second, energy prices forecast errors have statistically significant impacts on EU carbon prices: Brent (positive sign), natural gas (positive sign) and coal (negative sign) forecast errors and various spreads for power production[8] are found to play a statistically significant role. Third, weather conditions are expected to affect the price path of carbon in so far as they influence energy demand. If temperatures conform to seasonal averages, then market participants (brokers, analysts, investors etc.) can properly anticipate them. Unanticipated temperature changes have an impact on EU carbon prices: these effects channel through the non-linearity of the relationship between temperatures and carbon price changes.[9]

Another strand of literature focuses on the link with the macroeconomic environment. As industrial production increases, associated CO_2 emissions increase and therefore more CO_2 allowances are needed by operators to cover their emissions. This economic logic results in carbon price increases, due to tighter constraints on the demand side of the market. Alberola et al. (2008a, 2009) suggest linking the expected impacts of production peaks and compliance positions on carbon price changes: if a sector combines a net short (long) position and/or an increasing (declining) variation of activity, then this sector is a net buyer (seller) of allowances and the impact on the allowance price will be positive (negative). Industrial sectors which record a higher (lower) production growth than their baseline projections over 2005–07 are expected, due to their deficit (surplus) of allowances, to be net buyers (sellers) of allowances and should have a positive (negative) impact on EUA price changes. By using monthly industrial production indices from Eurostat in the various sectors[10] covered by the EU ETS, the authors are able to explain that production indices influence CO_2 price changes in three sectors (combustion, iron and paper), which represent roughly 80 per cent of allowances allocated. The authors also advance a potential justification for the non-significant sectors: allowance pooling may occur between or within firms, so that the considered sectors are globally in compliance.[11]

In the light of the recent periods of economic expansion (2005–07) and recession (since 2008), Chevallier (2011c, 2011d) confirms that the carbon market adjusts to the macroeconomic environment with a delay due to the specific institutional constraints of the EU ETS. It is interesting to relate these states to the underlying business cycle. Indeed, EU industrial production had been falling since July 2007. However, the carbon market seems to have adjusted to this situation only in October 2008, when most operators were looking to sell allowances in exchange of cash. Switches from 'high-growth' regime to 'low-growth' regime are especially perceptible during January–April 2005, April–June 2006, October 2008, April 2009 and May 2010, related with the annual compliance event. Moreover, the author indicates that the carbon–macroeconomy relationship may fade for some periods. One possible cause is changes unique to the carbon market that diminish its ability to react to macroeconomic factors. That is, sometimes changes in demand and supply fundamentals of allowances initiated by regulatory changes will disturb this relationship. To improve the understanding of the link between growth and carbon pricing, researchers need to delve further into the timing of business cycles and the reasons that may explain why the carbon price reacts immediately (or not) to these fluctuations.

In addition, the reaction of the carbon price to changes in macroeconomic fundamentals can be understood from the perspective of financial markets, such as equity and bond markets. Chevallier (2009b) emphasizes how the volatility of the carbon price is affected when financial markets enter 'bull' or 'bear' periods. By estimating various GARCH-type volatility models, carbon futures prices may be weakly forecast on the basis of two variables from the stock and bond markets, that is equity dividend yields (returns on stocks) and the 'junk bond' premium (spread between BAA- and AAA-rated bonds). Moreover, by assessing the transmission of international shocks to the carbon market,[12] Chevallier (2011e) shows that carbon prices tend to respond negatively to an exogenous recessionary shock on global economic indicators. In consequence, for investment managers, carbon assets such as EUA appear to be well suited to portofolio diversification since they do not match the business cycle exactly (Chevallier, 2009c).

Having detailed the functioning of the EU ETS, which will enter the auctioning phase during 2013–20, we now detail the main empirical characteristics of the CDM.

4 THE CDM: THE UN PROXY FOR A 'WORLD' CARBON PRICE

This section deals with the Clean Development Mechanism, which has been introduced under the framework of the Kyoto Protocol. The CDM has become the second-largest emissions trading system in the world to date. By analysing critically its current state of development, we aim at understanding its role in shaping a 'world' carbon market in the near future.

We present first the main characteristics of CER contracts (along with their price development), second the CER price drivers, third their relationship with EU allowances (including a discussion on arbitrage strategies that may arise from trading the CER–EUA spread), and fourth an assessment of the main barriers towards the establishment of the CER as the 'world' carbon price.

4.1 CER Contracts and Price Development

CER means a unit issued pursuant to Article 12 of the Kyoto Protocol. These are tradable units generated by projects in non-Annex-1 parties under the Clean Development Mechanism. They may be counted by Annex 1 parties towards compliance with their UN and EU emissions target and are equal to one tonne of CO_2 equivalent gases. CERs are valid during 2008–12 under the Kyoto Protocol. Secondary CERs (exchanged on the secondary market by firms) are valid up to 13.4 per cent on average for import within the EU ETS. CERs can be used during 2008–12 for compliance in the EU ETS. While the EU ETS has been confirmed until 2020, the existence of CERs is tied to that of a post-2012 international agreement (Chevallier, 2011b).

In 2011, the volume of transactions amounted to 1418 million CERs (up 53 per cent compared with 2010). Similarly to the EUA market, it is worth noting that the use of options prices has been steadily growing from 68 million tonnes in 2008 to 92 million tonnes in 2009 for CER futures. Strike prices range from €10 to €50 per tonne of CO_2 for CER call options, and from €6 to €16 per tonne of CO_2 for CER put options. The level of uncertainty attached to CER options prices also appears quite high in a context of post-Kyoto negotiations (and the fact that CERs are not confirmed to be operating after 2013 to date).

Throughout the historical data available for the scheme, CER prices have been varying in the range of €15 per tonne of CO_2 (before the financial crisis); that is, they trade at a discount compared to EUAs. The main reasons behind the price differences between EUAs and CERs have been studied by Mansanet-Bataller et al. (2011). We recall them briefly in Section 4.3.

In the next section, we detail the main CER price drivers.

4.2 CER Price Drivers

On the supply side, CER prices are determined by the decisions of the CDM Executive Board (EB) of the United Nations, which needs to validate the delivery of credits once the project has been completed. The main task of the CDM EB consists in assessing the 'additionality'[13] of the CO_2 emissions reduction achieved through the project versus a counterfactual situation without investments in depolluting activities. The global information concerning CER supply is difficult to obtain, despite the existence of the 'CDM pipeline' which tracks the status of each project from origination to validation by the United Nations.[14]

To sum up, the supply of sCERs is unknown due to the following reasons (Chevallier, 2011b):

1. the supply of primary CERs is unknown and difficult to estimate (as it depends on several risks related to the issuance of primary CERs);
2. the amount of primary CERs that will be converted into sCERs is also difficult to assess,
3. there exists an upper limit on imports within the EU ETS of 1.4 Gt by 2012 (1.9 Gt by 2020) which has been fixed by the European Commission;
4. there are quite often methodological changes on delivery rules from the CDM EB.

On the demand side, CER prices are determined by various factors such as the decisions of the European Commission (fungibility within the European system), and the CERs demand from governments to meet their compliance within the Kyoto Protocol. Compared to EUAs, the demand for CERs comes from a larger number of participants (investors, industrials and Annex-B countries). In practice, most of the CERs' demand to date comes from European industrials, which are limited to 13.4 per cent (on average) of surrendered allowances for compliance during Phase II of the EU ETS. Nevertheless, it is important to underline that Annex-B countries of the Kyoto Protocol may also use CERs for compliance. For instance, countries with a potential deficit of assigned amount units (AAUs valid under the Kyoto Protocol[15]) in 2012 (such as Japan) are involved in CER purchases.

Mansanet-Bataller et al. (2011) have studied econometrically the main CER price drivers. Their conclusion reveals that the price fundamentals of CERs are roughly the same factors as those affecting EUA prices (since both assets may be used for compliance in the EU ETS):

- energy prices (Brent, gas and coal);
- the switching price variable (arising from power producers' fuel-switching behaviour);
- temperature variables;
- variables related to production levels, market volatility; and
- dummy variables related to the announcements concerning the status of the CDM in Phase III of the EU ETS and the ITL–CITL connection (see more details in Section 4.3).

Finally, the authors uncover the statistical influence of a new specific variable, called momentum$_{sCER}$, for the indication of bullish and bearish periods on the CER market.[16]

To finalize our presentation of the CDM as a potential 'world' carbon market, we detail its current linkages with the EU ETS.

4.3 Relationship with EU Emissions Allowances

In their econometric analysis, Mansanet-Bataller et al. (2011) find that past values of sCER prices statistically affect EUA prices in a vector autoregressive framework (while sCER prices are best explained by an autoregressive process). In addition, a positive Granger causality runs from CERs to EUAs, but not conversely. Finally, by conducting an impulse–response analysis, they identify that EUAs react rapidly and positively to a shock on CERs (while CERs are negatively affected by a shock on EUAs).[17] These results show the importance of CER prices in the current sphere of carbon markets: not only do they grow in importance in terms of geographic scope and volume of transactions, but they can also be found to have an impact on the European market. These statistical results suggest that there are strong linkages between the EUA and CER markets.

In fact, the registries from these two markets – the European Commission's 'Community Independent Transaction Log' (CITL) on the one hand, and the United Nations' 'International Transaction Log' (ITL) on the other hand – have been connected since 19 October 2008. These operations allowed the delivery of sellers' international credits from the Kyoto Protocol (such as CERs) into buyers' EU national registries.[18] It has offered to EU investors in CDM projects the possibility to monetize their assets, by using CERs towards their own compliance or by exchanging them on the European market as secondary CERs.[19]

Since the ITL–CITL connection, Kyoto credits may be imported into the EU ETS, and are valid for compliance up to 13.4 per cent on average. During Phase II of the EU ETS (2008–12), the EU Commission has announced that the use of CERs by industrial installations will be capped at 270 million tonnes per year. For Phase III (2013–20), the limit for existing installations will be 40 million tonnes per year. Altogether, Phase III CER imports will add up to 300 million tonnes. Thus the post-2008 EU ETS CER import limit has been fixed at 1.7 billion tonnes.

Simultaneously, the rise of CER and EUA trading has fostered the development of a new arbitrage strategy consisting in taking advantage of the CER–EUA spread. It is defined as the premium between CERs and EUAs. Traders watch the spread closely since it represents the cost of swapping EUAs for CERs for use in meeting compliance targets under the EU ETS. Utilities trading desks in the carbon finance industry have long been interested in this spread, because they may benefit from 'free lunch' arbitrage opportunities simply by using discounted CDM credits for their compliance under the EU ETS. Mansanet-Bataller et al. (2011) consider the CER–EUA spread as a speculative product, which may be explained by financial microstructure and traders' activity variables. Indeed, when the CER–EUA spread is high, the volumes of EUAs and CERs exchanged are high as well. The authors also point out that the ability to benefit from the CER–EUA spread is limited to market agents which have their own registry and their own trading desks.

To complete this discussion, we need to emphasize that, due to the high level of uncertainties concerning international climate policy agreements, the European Commission is waiting for the outcome of the post-Kyoto negotiations before taking final decisions concerning CERs use in Phase III of the EU ETS (2013–20). Therefore the relationship between EUAs and CERs is subject to various regulatory changes.

4.4   Beyond 2012, an 'Age of Wars' on Carbon Markets?

The purpose of this section is to evaluate the hypothesis of potential unification of regional carbon market initiatives into a 'world' carbon market.

First, concerning the development of CERs as the 'money' for carbon trading worldwide, we may notice a few caveats. CERs are certainly affected by a delivery risk, and the performance for operating projects, to which we may add a high volume of registered projects, as well as registration and methodological risks for proposed projects. Besides, as briefly discussed in Section 4.2, one particularly striking source of uncertainty concerning the validation of CDM projects lies in the so-called 'additionality' of greenhouse gases emissions reductions claimed.

How can we overcome these specific CER risk factors? On the primary market, there needs to be an increased predictability of issuance of CERs by the CDM EB (which implies a stable methodology and schedule), that is, to enact clear guidelines of project validation. On the secondary market, market agents need to account for limitations on the use of certain 'quality' of CERs for compliance under the EU ETS.

Second, the development of new emissions trading schemes should increase CER demand, and thus push CER prices to converge with other carbon prices under compliance schemes. For instance, in the New Zealand emissions trading scheme, emitters need to buy New Zealand units from domestic suppliers or Kyoto units (CERs and ERUs) from abroad to cover their emissions every year. Since the CERs are transferable into this scheme (in addition to the EU ETS), it becomes more and more plausible that they will play a central role in the extension of the carbon market sphere.

Other regional carbon market segments which exist independently of the UN-based markets can be found in North America (Regional Greenhouse Gas Initiative (RGGI), Western Regional Carbon Action Initiative (WCI), Midwestern Greenhouse Gas Reduction Accord (MGGRA) and AB32 Bill entitled 'California Global Warming Solutions Act'). In Australia, the negotiations on the implementation of the Carbon Pollution Reduction Scheme (CPRS) have stalled. In Japan, the 'Law on Global Warming' is unlikely to pass given the recent earthquakes. In South Korea, the 'Green Growth Law' which passed in 2009 includes the creation of an ETS. Last but not least, in China, some pilot emissions trading schemes will be launched in six cities and provinces, with a national ETS to be introduced by 2016. All of these regional ETSs are planning to include compatibility features with CERs.

On the other hand, the existence of other mechanisms can also limit the influence of the CER as the 'world' carbon price. Beyond the CDM, there are indeed a number of proposals for how to use market mechanisms to reduce greenhouse gases emissions in developing countries. The COP/MOP Meetings[20] in Copenhagen and Cancun have paved the way for an instrument to Reduce Emissions from Deforestation (REDD), whereas Japan in particular has been pushing for bilateral offsets. Sector crediting and credits from Nationally Appropriate Mitigation Actions (NAMAs) are also in preparation.

Third, CERs are obviously affected by the uncertainties concerning the post-2012 climate regime. Typically, the transfer of CERs from a non-Annex-1 country requires a Letter of Approval (LoA) from that country's government, permitting project participants to receive issued CERs. In most cases, LoAs are issued for the duration of a project's crediting period, be that for seven or ten years. China, however, usually grants LoAs only for credits expected to be generated during the 2008–12 period. What will happen at the end of the Kyoto Protocol's trading period? According to Point Carbon (2011), China will nevertheless continue to supply CERs to the market after 2012.

Unfortunately, the fear of being confronted with segmented carbon markets from 2012 onwards has not been eliminated by the COP/MOP Meeting in Durban in December 2011. Under the Durban agreements,[21] the Kyoto Protocol should be extended to 2017 for those that will ratify it, and all parties have agreed 'to develop a protocol, another legal instrument or an agreed outcome with legal force under the United Nations Framework Convention on Climate Change (UNFCCC) applicable to all governments' that will enter into force by 2020.

Therefore, based on the outcome of Durban,[22] the CDM should continue post-2012, along with a new market mechanism operating under the guidance and authority of the COP. Under this veil of uncertainty, there is a set of provisions that will need to be defined in the coming year and beyond. How these issues are resolved will determine the future direction of the carbon market.

According to Marcu (2011), Durban must be seen as a step forward. It has answered questions regarding the continuity of the CDM, but it has little to say about the extent of the demand for units emanating from it, and consequently their future use. It has finally launched work for a new mechanism under the UNFCCC but has not provided direction on how the new mechanisms that are emerging from the USA and Japan, for example, would interact with the emerging international framework. This will prove critical for market dynamics, for demand for units from domestic programmes, such as the one in California or Japanese bilateral programmes, as well as for general market liquidity.

5   CONCLUDING REMARKS

This chapter aims at reviewing the main characteristics of tradable permits systems, with an eye on current and future developments. Since the EU ETS and the CDM stand out as the two largest emissions trading mechanisms in the world to date, we focus on the main characteristics, dynamics and linkages between these two schemes.

Our analysis suggests that some lessons can be learnt from a policy viewpoint. First, from the debate on initial allocation, we get the insight that the free permits distributed to existing firms on a 'first-come', 'first-served' basis under the EU ETS represent a market value of €40 billion (evaluated at €20/tonne of CO_2) that was created at the same time as CO_2 emissions were capped. This creates a major distortion between market agents (existing companies and new entrants), which can only be solved by a shift to auctioning as the preferred allocation methodology (from a theoretical viewpoint).

Second, during Phase I of the EU ETS (2005–07), EUAs experienced a high level of volatility around each compliance event. This fact was especially true during April 2006 (Alberola et al., 2008b) and April 2007 (Chevallier et al., 2009b). It highlights the central role played by the European Commission, which echoes the voice of the 'central bank', in order to reveal reliable information about the overall tightness of this emissions market.

Third, we have detailed the functioning of the CDM market under the Kyoto Protocol. Most interestingly, we have established various linkages between EUAs (tradable under the EU ETS) and CERs (tradable under the CDM). These linkages suggest that CERs may be considered as a transferable unit between various regional carbon market initiatives (e.g. Europe, USA, Asia and Australia).

Finally, we have highlighted the main barriers towards the establishment of the CER as the 'world' carbon price. These barriers can be technical, but they are especially of a political nature following the COP/MOP meetings in Durban. Our conclusion may sound pessimistic: in absence of a unified framework for allowance trading post-2012, the spectre of segmented carbon markets – building through a bottom–up approach – looms on the horizon.

NOTES

1. Since CO_2 is a global persistent stock pollutant, the damage from CO_2 today is effectively the same as tomorrow. Hence the marginal benefit of abatement is essentially flat.
2. Note that, in this chapter, we deliberately leave out the discussion regarding price caps – that is, the debate concerning the 'safety valve' (the regulator offers to sell permits in whatever quantity at a predetermined price) – and price floors.
3. Besides, note that, to produce, installations do not need to physically hold allowances during the year, but must only match the required number of allowances with verified emissions for their yearly compliance report to the European Commission. Consequently, the probability of a potential illiquidity trap exists if market participants face a market squeeze during the compliance event.
4. Based on a Hotelling-CAPM (capital asset pricing model) type analysis, EUA spot prices do not meet

equilibrium conditions in the intertemporal permits market. There is statistical evidence that, during the negotiation of National Allocation Plans for Phase II, the French and Polish decisions to ban banking contributed to the explanation of low EUA Phase I prices. Finally, the cost-of-carry relationship between EUA spot and futures prices for delivery during Phase II does not hold after the enforcement of the inter-period banking restrictions.

5. The EU Commission can intervene to amend the functioning of the EU ETS. According to modern finance theory on market efficiency, participants are expected to fully react to news. Any policy intervention from the regulator has an immediate impact on the carbon price.

6. Energy prices are the most important drivers of carbon prices due to the ability of power generators to switch between their fuel inputs. An increase in the Brent and natural gas prices should play a positive influence on the carbon price, whereas an increase of coal prices (as the most CO_2-intensive fuel) should influence negatively the carbon price. Carbon prices are costs for electricity production, and should pass through power prices. For a thorough definition of the clean dark spread, clean spark spread, and switching price behind power producers' fuel-switching mechanism, see Chevallier (2011a, 2012b).

7. Extreme higher and lower temperatures should increase CO_2 emissions.

8. Power producers are key players on the carbon market, as they received globally around 50 per cent of allocation each year. The fuel-switching behaviour (producing one unit of electricity based on coal-fired or gas-fired power plants, and switching between inputs as their relative price varies) is a central factor in explaining the variation of the carbon price. With carbon costs, the marginal cost for each plant contains the emission factor which depends on the fuel burnt.

9. Therefore unanticipated temperature changes matter more than temperatures themselves during extreme weather conditions (extremely hot and/or cold winters/summers).

10. That is, paper and board; iron and steel; coke ovens; refineries; ceramics; glass; cement; metal; production and distribution of electricity, gas and heating.

11. On this issue, see also the theoretical analysis by Chevallier et al. (2011b).

12. Factor models use principal components analysis (PCA) in order to extract and summarize the information from large datasets (Stock and Watson, 2002a, 2002b). Then the econometrician can look at their explanatory power for the returns of carbon prices.

13. That is, project developers need to demonstrate the greenhouse gases emissions reductions claimed would not have been achieved in the absence of the project.

14. The CDM pipeline is a spreadsheet where CDM projects are registered by the CDM EB, and displayed in the CDM pipeline along the several steps from registration to validation.

15. AAUs are units derived from an Annex 1 party's assigned amount. They are tradable units that Annex 1 parties may count towards compliance with their emissions target. Each AAU is equal to one tonne of CO_2 equivalent gases. AAUs are issued each year during 2008–12. No post-2012 international agreement has been reached to date.

16. It is obtained as the difference between the sCER variable at time t and at time $t - 5$.

17. The authors also accept the presence of at least one cointegrating relationship between EUAs and CERs, with a strong error-correction mechanism to the long-term equilibrium.

18. Before that date, issued CERs remained in the UN CDM registry, waiting for the connection with the EU registry to be completed.

19. Note that some types of CERs credits are restricted from use in the EU ETS, such as land use, land use change and forestry (LULUCF) projects. Note also that, based on technological grounds, CERs generated through HFC destruction or large hydro projects have been banned from import within the EU ETS.

20. That is, Conferences of the Parties of the Kyoto Protocol (COP), and Meeting of the Parties of the Kyoto Protocol (MOP).

21. Draft conclusions by the Chair on the outcome of the work of the Ad Hoc Working Group on Long-Term Cooperative Action (AWGLCA) to be presented to the COP for adoption at its seventeenth session (FCCC/AWGLCA/2011/L.4) and Outcomes of the Work of the Ad Hoc Working Group on Further Commitments for Annex 1 Parties under the Kyoto Protocol at its sixteenth session FCCC/KP/AWG/2011/L.3.

22. At this point, it may be interesting to relate the state of climate negotiations to the business cycle. The world economies have entered a severe financial crisis since 2008. The recession in the EU and Japan, which are the main demand centres of greenhouse gases compliance instruments, may explain why carbon prices have fallen to record low levels since that date (without price or demand support from other areas of the world).

REFERENCES

Alberola, E. and Chevallier, J. (2009), 'European carbon prices and banking restrictions: evidence from Phase I (2005–2007)', *The Energy Journal*, **30**(3), 51–80.

Alberola, E., Chevallier, J. and Cheze, B. (2008a), 'The EU Emissions Trading Scheme: the effects of industrial production and CO_2 emissions on European carbon prices', *International Economics*, **116**, 93–126.

Alberola, E., Chevallier, J. and Cheze, B. (2008b), 'Price drivers and structural breaks in European carbon prices 2005–07', *Energy Policy*, **36**(2), 787–97.

Alberola, E., Chevallier, J. and Cheze, B. (2009), 'Emissions compliances and carbon prices under the EU ETS: a country specific analysis of industrial sectors', *Journal of Policy Modeling*, **31**(3), 446–62.

Chevallier, J. (2009a), 'Emissions trading: what makes it work?', *International Journal of Climate Change Strategies and Management*, **1**(4), 400–406.

Chevallier, J. (2009b), 'Carbon futures and macroeconomic risk factors: a view from the EU ETS', *Energy Economics*, **31**(4), 614–25.

Chevallier, J. (2009c), 'Energy risk management with carbon assets', *International Journal of Global Energy Issues*, **32**(4), 328–49.

Chevallier, J. (2011a), 'The European carbon market (2005–2007): banking, pricing and risk-hedging strategies', in I. Galarraga, M. Gonzalez-Eguino and A. Markandya (eds), *Handbook of Sustainable Energy*, Cheltenham, UK and Northampton, MA, USA: Edward Elgar, pp. 395–414.

Chevallier, J. (2011b), 'The Clean Development Mechanism: a stepping stone towards world carbon markets?', in I. Galarraga, M. Gonzalez-Eguino and A. Markandya (eds), *Handbook of Sustainable Energy*, Cheltenham, UK and Northampton, MA, USA: Edward Elgar, pp. 415–40.

Chevallier, J. (2011c), 'Evaluating the carbon–macroeconomy relationship: evidence from threshold vector error-correction and Markov-switching VAR models', *Economic Modelling*, **28**(6), 2634–56.

Chevallier, J. (2011d), 'A model of carbon price interactions with macroeconomic and energy dynamics', *Energy Economics*, **33**(6), 1295–312.

Chevallier, J. (2011e), 'Macroeconomics, finance, commodities: interactions with carbon markets in a data-rich model', *Economic Modelling*, **28**(1–2), 557–67.

Chevallier, J. (2012a), 'Banking and borrowing in the EU ETS: a review of economic modelling, current provisions and prospects for future design', *Journal of Economic Surveys*, **26**(1), 157–76.

Chevallier, J. (2012b), *Econometric Analysis of Carbon Markets: The European Union Emissions Trading Scheme and the Clean Development Mechanism*, Berlin: Springer.

Chevallier, J., Jouvet, P.A., Michel, P. and Rotillon, G. (2009a), 'Economic consequences of permits allocation rules', *International Economics*, **120**, 77–90.

Chevallier, J., Ielpo, F. and Mercier, L. (2009b), 'Risk aversion and institutional information disclosure on the European carbon market: a case-study of the 2006 compliance event', *Energy Policy*, **37**(1), 15–28.

Chevallier, J., Le Pen, Y. and Sevi, B. (2011a), 'Options introduction and volatility in the EU ETS', *Resource and Energy Economics*, **33**(4), 855–80.

Chevallier, J., Jouvet, PA. and Etner, J. (2011b), 'Bankable pollution permits under uncertainty and optimal risk management rules', *Research in Economics*, **65**(4), 332–9.

Christiansen, A., Arvanitakis, A., Tangen, K. and Hasselknippe, H. (2005), 'Price determinants in the EU emissions trading scheme', *Climate Policy*, **5**, 15–30.

Congressional Budget Office (2000), 'Who gains and who pays under carbon-allowance trading? The distributional effects of alternative policy designs', *Report*, Washington, DC.

Dales, J. (1968), *Pollution, Property and Prices*, Toronto, Canada: Toronto University Press.

Dinan, T. and Rogers, D.L. (2002), 'Distributional effects of carbon allowance trading: how government decisions determine winners and losers', *National Tax Journal*, **5**(2), 199–221.

Ecotrade (2012), 'Prices of EUA', http://www.ecotrade.pt/?m=201201&cat=10&lang=en.

ECX (2007), 'Prices of EUA', http://www.ecx.eu/ECX-Historical-Data.

Ellerman, A.D. and Buchner, B.K. (2008), 'Over-allocation or abatement? A preliminary analysis of the EU ETS based on the 2005–06 emissions data', *Environmental and Resource Economics*, **41**, 267–87.

Ellerman, A.D. and Montero, J.P. (2007), 'The efficiency and robustness of allowance banking in the U.S. Acid Rain Program', *The Energy Journal*, **28**(4), 66–87.

Ellerman, A.D., Joskow, P.L., Schmalensee, R., Montero, J.P. and Bailey, E. (2000), *Markets for Clean Air: The US Acid Rain Program*, New York: Cambridge University Press.

Ellerman, A.D., Convery, F. and De Perthuis, C. (2010), *Pricing Carbon: The European Emissions Trading Scheme*, Cambridge: Cambridge University Press.

Goulder, L.H. (1995), 'Environmental taxation and the double dividend: a reader's guide', *International Tax and Public Finance*, **2**, 157–83.

Hahn, R.W. (1984), 'Market power and transferable property rights', *Quarterly Journal of Economics*, **99**, 753–65.

Hintermann, B. (2010), 'Allowance price drivers in the first phase of the EU ETS', *Journal of Environmental Economics and Management*, **59**, 43–56.

Kling, C. and Rubin, J. (1997), 'Bankable permits for the control of environmental pollution', *Journal of Public Economics*, **64**, 101–15.

Mansanet-Bataller, M., Pardo, A. and Valor, E. (2007), 'CO$_2$ prices, energy and weather', *The Energy Journal*, **28**(3), 67–86.

Mansanet-Bataller, M., Chevallier, J., Herve-Mignucci, M. and Alberola, E. (2011), 'EUA and sCER Phase II price drivers: unveiling the reasons for the existence of the EUA–sCER spread', *Energy Policy*, **39**(3), 1056–69.

Marcu, A. (2011), 'Post Durban: moving to a fragmented carbon market world?', CEPS Commentary, Center for European Policy Studies, Brussels, Belgium.

Montgomery, W.D. (1972), 'Markets in licenses and efficient pollution control programs', *Journal of Economic Theory*, **5**, 395–418.

Newell, R., Pizer, W. and Zhang, J. (2000), 'Managing permit markets to stabilize prices', *Environmental and Resource Economics*, **31**, 133–57.

Oates, WE. (1995), 'Green taxes: can we protect the environment and improve the tax system at the same time?', *Southern Economic Journal*, **61**, 915–22.

Parry, I.W.H. (2002), 'Are tradable emissions permits a good idea?', Resources for the Future, Issues Brief 02-33, Washington, DC.

Point Carbon (2011), 'Carbon 2011: Annual Survey', PointCarbon Report, Thomson Reuters.

Raymond, L. (1996), 'Private rights in public resources: equity and property allocation in market-based environmental policy', *Resources For The Future*, Washington, DC.

Rees, J. (1990), *Natural Resources, Allocation, Economics and Policy*, London: Routledge.

Rubin, J. (1996), 'A model of intertemporal emission trading, banking, and borrowing', *Journal of Environmental Economics and Management*, **31**, 269–86.

Sandel, M.J. (1997), 'It's immoral to buy the right to pollute', *New York Times*, 15 December, p. A29.

Sijm, J.P.M., Bakker, S.J.A., Chen, Y., Harmsen, H.W. and Lise, W. (2006), 'CO$_2$ price dynamics: a follow-up analysis of the implications of EU emissions trading for the price of electricity', Working Paper 06-015, Energy Research Centre for the Netherlands.

Sijm, J.P.M., Hers, S.J., Lise, W. and Wetzelaer, B.J.H.W. (2008), 'The impact of the EU ETS on electricity prices', Working Paper 08-007, Energy Research Centre for the Netherlands.

Springer, U. (2003), 'The market for tradable GHG permits under the Kyoto Protocol: a survey of model studies', *Energy Economics*, **25**(5), 527–51.

Stavins, R.N. (1998), 'What can we learn from the grand policy experiment? Lessons from SO$_2$ allowance trading', *Journal of Economic Perspectives*, **12**(3), 69–88.

Stock, J. and Watson, M. (2002a), 'Forecasting using principal components from a large number of predictors', *Journal of the American Statistical Association*, **97**, 1167–79.

Stock, J. and Watson, M. (2002b), 'Macroeconomic forecasting using diffusion indexes', *Journal of Business and Economic Statistics*, **20**, 147–62.

Tietenberg, T. (2003), *Environmental and Natural Resource Economics*, New York: Pearson Education.

Weitzman, M.L. (1974), 'Prices vs. quantities', *Review of Economic Studies*, **41**, 477–91.

22 Moral positions on tradable permit markets[1]
Snorre Kverndokk

1. INTRODUCTION

Permit trading is a preferred environmental policy instrument among economists, and has become a popular tool in environmental treaties in recent decades. The reason is that both in theory (Montgomery, 1972) and in practice (Schmalensee et al., 1998), market-based policy instruments such as permit trading have been shown to foster cost-effectiveness. However, many non-economists have not embraced permit trading as the right way to attack environmental problems, and even among economists there are many arguments for other market-based instruments such as taxation.[2] Environmental organizations, political parties and individuals have expressed concerns about permit trading. Some of these concerns are over specific implementations of permit trading systems or more general practical obstacles to a successful permit trading.[3] Others, however, see permit trading as morally wrong or problematic in principle. For instance, some consider it a way of avoiding one's obligations, to pay others to clean up, or to reward indulgence; see, for example, Goodin (1994). However, concerns may vary among countries, cultures and religions. While we report results from Norway below, where people express strong concerns about permit trading, most polling shows large support for emissions permit trading in the USA.[4]

The concerns may have had impacts on politics. Even if permit trading has advantages when it comes to the costs of meeting a certain emissions target, governments as well as existing multinational tradable permit schemes have put restrictions on permit trading. Norway's broad-based political agreement on climate policy from 2008 specifies that two-thirds of emissions reductions up to 2020 should be taken nationally when reforestation is included. In the Kyoto Protocol, trade in pollution permits is allowed, but only as a supplement to national mitigation.[5] Also in the European Emissions Trading Scheme (ETS), access to buying emission reductions in third-party countries (JI – Joint Implementation for economies in transition – and CDM – Clean Development Mechanism for developing countries) is limited.[6] These restrictions could mean that the cost-effective volume of trade may not be within reach, and the emissions reductions will be achieved at a higher cost than necessary.

There may be several reasons why such restrictions have been introduced. One reason may be that signatories have been reluctant to allow full trading due to concerns about permit trading among those they represent. In a democracy, signatories or governments represent the people, and if their voters express concerns about permit trading, they may want to restrict this option. Restrictions on permit trading can in this case be seen as a trade-off between the benefits from trading (in terms of reduced costs) and the costs of becoming involved in something that is not preferred. However, there may be other explanations. Eyckmans and Kverndokk (2010) present a model where such restrictions are effective in reducing negative environmental impacts of moral concerns about permit trading. Thus restrictions may not be set because of moral concerns, but to reduce the impacts of such concerns. The reason is that moral arguments against trading may lead to higher instead of lower global emissions in an international climate agreement with permit trading, where the total number of pollution permits is determined endogenously when the different signatories decide the amount of permits to be allocated to their own domestic industries. Restrictions on permit trading may reduce the incentives of the signatories to allocate the high amounts of permits to their industries that may follow from moral concerns. Finally, arguments have also been put forward that do not deal with moral concerns, such as means to reduce market power in the permit market; see Ellerman and Wing (2000).

We will present some empirical evidence on people's concerns about permit trading and discuss possible reasons for this. Could the concerns be based on sound reasoning, or do they just follow from lack of economic competence?

The chapter is organized as follows. In the next section we show some results from two recent experiments where the participants are asked about their attitude to emissions permit trading. These examples may help explain some of the attitudes shown in national politics and international environmental negotiations. Section 3 organizes the arguments according to consequential and non-consequential ethics, and Section 4 concludes.

2. SOME RESULTS ON RELUCTANCE TO TRADE EMISSION PERMITS

To illustrate some of the reluctance towards permit trading, we start by referring to some views on permit trading that have been expressed in popular media.

A cartoon by Ruben Bolling (Bolling, 1992) starts by referring to the idea of pollution permit trading under the Clean Air Act, and asks the question 'Where will this lead . . . ?'. It continues with the following story: A person is woken up by the sound of a burglar. He surprises him, points the gun at him and claims that 'I have every right to kill you – I'm in my own dwelling, and I fear for my own life'. However, he has a better idea and decides to contact a 'crime broker' to sell his right to kill. The broker finds a buyer, and 'The same number of deaths result, but with a more efficient allocation'.

In an ironic radio monolog on Norwegian Public Radio (NRK) in 2010,[7] the journalist suggests, in a similar way as Ruben Bolling, several extensions of permit trading to include areas such as sickness leave, infidelity and traffic offenses.

It is not hard to come up with reasons why emissions permit trading is very different from the entertaining examples above. Institutions such as criminal law or marriage will be undermined, while emissions permit trading on the other hand is a way of introducing institutions to an area where they are lacking. Examples like this may however indicate that people find trading a 'bad', and immoral.

To test whether this is the case, two experiments in economics have recently been conducted at the University of Oslo with students as subjects. Both experiments are public-good games, but are framed differently. In the first (Bråten et al., 2011), the subject can choose to use stickers that will inflict harm on the other subjects in their group, but will be beneficial to themselves, while in the other experiments (Hauge et al., 2011), subjects choose how much to give to a public good, where giving is costly

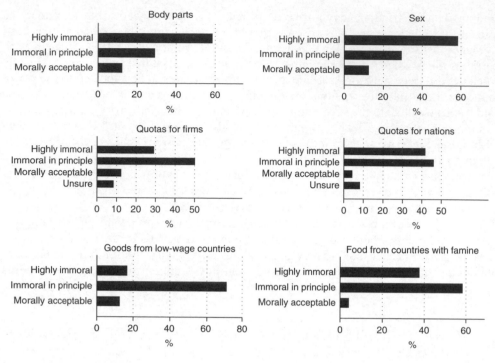

Source: Bråten et al. (2011).

Figure 22.1   Reluctance to trade in different types of goods

(quadratic cost function), but benefit all in their group. Both experiments were first run without any regulation, while in the second round regulations were put on the public good (how much you could harm others, or how much you had to contribute), but with a possibility to trade the rights within the group. In Bråten et al. (2011), the idea was to test whether subjects would vote against the possibility to trade a good that would harm others, while Hauge et al. (2011) wanted to test whether the distribution of commitments was important for the choice to trade. However, both experiments found that there was no significant reluctance to trade the right or the commitment.

The rights or the commitments in the experiment were abstract public goods where no connections to environmental problems were explicitly made. But after both experiments were finished, subjects had to fill in an exit survey where they were asked about their attitudes to trade certain goods. In spite of the behavior in the experiments, few subjects found trade in permits between poor and rich countries morally acceptable. Figure 22.1 shows the results from Bråten et al. (2011) with 44 subjects. In their survey, less than 5 percent of the subjects found permit trading among nations morally acceptable, a lower number than for trade in body parts and sex. Permit trading among firms was found more acceptable, but still a significant number of people found it immoral. Hauge et al. (2011) found a similar pattern with 87 subjects. However, in their survey permit trading between countries was found morally acceptable by a larger number of subjects, close to 30 percent. These questions were followed up by reasons to oppose permit trading between rich and poor countries, and reasons such as 'It is morally wrong', or 'It increases inequality and exploits global poverty' got widespread support.

The results from both experiments may indicate that people are not too concerned about the consequences of trading a public bad in general. However, when the public bad is associated with certain goods such as emissions permits, people tend to find it immoral. We will comment more on this in the next section.

3. EMISSIONS PERMIT TRADING AND ETHICAL REASONING

In ethical reasoning, there are two ways to determine if an action is good or bad. The first is to refer to the consequences (teleology or substantive fairness). Based on this, an action is good if it is the best way to attain the aim we strive for (e.g. maximize welfare, reduce greenhouse gas emissions). However, another way of moral thinking argues that consequences alone do not guide us whether something is right or wrong (deontology or procedural fairness). It is not enough to know that the action is the most effective way to attain the aim. Concerns about permit trading can be organized according to these two strands of ethical reasoning.[8]

Consequences Matter

Standard economic analysis is basically about consequences and if the consequences of a particular policy are positive (i.e. increased welfare), economists will recommend it. This is the case with emission permits. The basic argument in favor of permit trade is cost-effectiveness (Montgomery, 1972). Parties involved in permit trade would get lower abatement costs than if they had to mitigate the emissions within their geographical boundaries. Thus cost-savings will be welfare improving (or at least give potential Pareto improvements), everything else equal. This approach is also referred to as a utilitarian approach, where the aim is to seek people's greatest happiness or utility. If permit trading maximizes the resources available for the individuals in the economy (for a given environmental target), total utility may be maximized.

A contrast to the utilitarian approach is an approach that also considers fairness.[9] A utilitarian approach focuses on efficiency by maximizing the total amount available of resources or utility. However, there may be a trade-off between efficiency and justice. Would a permit trading system be unjust in the way that it increases the pollution gap between rich and poor countries? Would it lead to a more unequal distribution of income in the world? One way of thinking about this is that countries may be opposed to permit trade because of inequality aversion. If for instance they are concerned about income inequality and if inequality increases in a trade regime, they may prefer not to trade. This means that if countries or individuals are not only concerned about their own income, but also about income distribution, they may oppose trade even if their own income may increase because of the trade regime. This hypothesis was tested in the experiment by Hauge et al. (2011) referred to above, but they could not find support for it. However, in the exit survey, more than 50 percent of the subjects agreed or strongly agreed that permit trading between rich and poor countries increases inequality and that this is a reason to oppose permit trading.

Explanations based on allegedly negative side effects of a permit market may also be plausible. Buying CDM quotas, that is, greenhouse gas pollution permits in countries that did not subscribe to binding emission limits in the Kyoto Protocol, may have adverse effects based on lack of an emission baseline, lack of incentives to undertake emissions reductions by the developing countries, transaction costs and carbon leakages. As a result, CDM projects may not fully offset emissions; see Rosendahl and Strand (2009, 2011), and total emissions will be higher with permit trading than without. Related to this is cheating or non-compliance in the permit market, which also has negative environmental impacts. This has been analyzed in van Egteren and Weber (1996), and over the last few years the media have reported several examples of cheating and fraud in the permit markets that may reduce confidence in markets (see, e.g., Harvey, 2006; Davies, 2007). It has also been reported that CDMs offer an incentive to produce greenhouse gases (Pearce, 2010). Further, 'hot air', meaning that some countries receive an initial emission quota allocation that exceeds their actual emissions, has also been mentioned as a reason to avoid emission trading as trading hot air will not reduce emissions. In the Kyoto Protocol hot air exists as several Eastern European countries have quotas that are higher than their 1990 emission levels, and these countries have also not been successful in selling their permits.

Some papers argue that if the permit allocation is set in a non-cooperative equilibrium, permit trading may actually lead to higher emissions; see Helm (2003) and Holtsmark and Sommervoll (2008). This can be the case particularly for sellers of permits that may allocate more permits to their industries than in the non-cooperative case without permit trade. One example of this may be 'hot air', as mentioned above, when several countries got allowances higher than their business-as-usual emissions. The idea is that if the permit price is not very sensitive to increases in permits and if the marginal damage of the country is relatively low, the benefits from overallocation are higher than the costs.

Abating at home instead of buying emission permits may also be perceived as a better policy based on consequences, and arguments against permit trading may, therefore, be based on the benefits of abating at home and not necessarily against permit trading *per se*. Some arguments that have been raised in this debate are the positive local spillover effects of technology development by national abatement as well as the ancillary benefits (reduction in local emissions, traffic accidents, congestion etc.) of abating at home. Related to this is the environmental justice argument, that minority groups suffer from permit trading because they live close to polluting facilities and will, therefore, not benefit from potential ancillary benefits of abating at home if the government decides to buy permits abroad (see, e.g., Kverndokk and Rose, 2008, for a survey). Another argument is that to meet the goals for greenhouse gas emissions in the long run, which have been raised over the last few years,[10] new infrastructure investments have to be made to replace a large part of the existing infrastructure such as roads, railways, buildings and so on. By buying permits abroad instead of abating at home, the incentives for making such investments may be smaller, and, therefore, also the possibility of meeting future goals. It may be a way to postpone a necessary restructuring of the economy.

It is finally argued that unilateral abatement has positive effects for the global environment as it may lead to similar behavior by other countries, it may affect positively the negotiation climate in the international policy arena, and it may reduce the conflict of interest within a country as it actually shows the true costs of abatement, a cost that economic agents have an incentive to exaggerate; see Hoel (1991) and Golombek and Hoel (2004).

Consequences may not be that Important

To look at the consequences may not be the only way of evaluating certain actions. Based on a procedural justice approach, consequences alone do not guide us whether something is right or wrong. One way of approaching this may be virtue ethics, associated with Aristotle. Virtue ethics cares about the moral standing of those engaged in the activity. Will the moral stigma attached to pollution be different due to pollution permits? One possibility is that if one can pay for the right to pollute, the moral stigma may be reduced; pollution is not 'wrong' anymore. We no longer share the responsibility for pollution reduction as we can pay somebody else to take over the responsibility. One illustration of this is given in Brekke et al. (2003), where people can pay an organization instead of doing voluntary work. If they think that the payment is enough to pay professionals to do the job, they do not feel responsible anymore, and they may choose the market solution. But if they think that the payment is not enough, they feel that they still are responsible for having the task done.[11]

Related to this is crowding out of moral motivation as permit trading may reduce incentives to behave in a 'green' way. As Hansen (2009) points out, individual actions to reduce the carbon footprint will not have any impacts, as all you do is to free up emission permits for someone else as the total amount of emissions is set by government. Thus the motivation to behave in a green way may be reduced.

Duty is also often referred to when discussing climate change. Some argue that industrialized countries have created the global warming problem, and that it is their duty to reduce the consequences of it, even if this does not minimize overall costs of taking action. This can be used as an argument against developing countries selling permits to industrialized countries because the permit trade would not lead to abatement in the countries responsible for the problem.[12] Another argument is based on unfair background conditions (see Kverndokk, 1995; Eyckmans and Schokkaert, 2004). Even if two parties agree to trade permits, the trade may not be justified on ethical grounds. A voluntary agreement between two parties is not necessarily fair if it is entered into under conditions that are not fair (Pogge, 1989). Background justice is not preserved when some participant's basic rights, opportunities or economic positions are grossly inferior. Some examples can be kidney trade or an agreement between a prostitute and her/his customer. Even if these trades may be beneficial for both sellers and buyers, they may not be right due to the unfair background conditions. Under the Kyoto Protocol, for instance, some may argue that this is the case for some CDM contracts, as it is a trade between poor and rich countries.[13] The view of unfair background conditions may also get some support from the surveys referred to in Section 2 above, where a large number of the subjects supported the claim that permit trade between rich and poor countries exploits global poverty which may, therefore, be a reason to oppose permit trading.

Markets for pollution permits have been recognized by several authors as a case in which there may exist some reluctance or even repugnance against transactions; see, for example, Goodin (1994), Bénabou and Tirole (2007) and Roth (2007). Some activities are considered repugnant because they violate traditional values or religious and moral prohibitions. What is considered repugnant may vary from place to place and may change over time. Slavery is an example of a market that used to exist in large parts of the world, but is now repugnant and illegal in most places. On the other hand, there have been more positive attitudes over time towards life insurance (Zeliner, 1999) and legalized prostitution. It may be difficult to predict what is considered repugnant, but introducing money into the exchange may often be what people do not like and find immoral. It may just be 'unnatural' in such settings; everything may not be subject to market pricing. Examples of 'priceless' or 'sacred' goods may be life, freedom, love, friendship, children, religion and democracy. Other examples are connected to the human body; you may donate an organ, but payment to living kidney donors is highly debated and often prohibited; giving blood is considered a deed, but not to sell your blood in an open market; having a one-night stand is accepted in most societies, but selling sexual services is often met by legal restrictions. In a similar way, introducing the environment into the marketplace is also repugnant to many people. Such a view may, for instance, be consistent with ecophilosophy or the 'deep ecology' movement (see, e.g., Næss, 1973), that is, respect for nature and the inherent worth of other beings.[14] In the experiments discussed in Section 2 above, the repugnance to trade goods connected to both the human body and the environment gets support.

4.   CONCLUSIONS

This chapter has studied several reasons for the observed concern about international permit trading. All policy judgments are based on ethical considerations, so even if there are advantages of permit trading in terms of lower costs of reaching an environmental target, there may be good reasons for reluctance based on both consequentialistic and non-consequentialistic ethics. Several ethical arguments can be used in the debate, but based on newly conducted experiments, it may seem that one main argument is that international permit trading is simply immoral, even if the argument was rationalized when the subjects were presented with several reasons to oppose it. Introducing the market to international environmental problems may just be wrong to many people; it may be considered a taboo.

One important question is, however, if the best may be the enemy of the good. Would, for instance, abandoning or restricting permit trading mean that it would be harder to reach international climate agreements due to the higher costs? As mentioned in the introduction, abandoning permit trading or introducing restrictions could mean that the cost-effective volume of trade may not be within reach, and the emissions reductions will be achieved at a higher cost than necessary. Thus reaching an agreement will be more expensive, and may therefore get less support. On the other hand, it may be easier to reach an agreement that has ethical support. Eyckmans and Kverndokk (2010) found that moral concerns about permit trading most likely will increase global emissions, as countries may set their caps at levels that reduce trade. In this case, however, the negative environmental impacts may be offset by introducing restrictions on permit trading.

One parallel may be kidney sale, which is prohibited by federal law in the USA.

According to Cohen (2008), about 4400 people died while waiting for a kidney in the USA in 2006, and Becker and Elias (2007) have calculated the price of a kidney to eliminate the waiting list if a free market existed. Is following the ethical rule of not involving money in the transactions worth the deaths? When it comes to climate change, such trade-offs will probably not play the same role. Market-based instruments such as taxes and permits have been introduced, and will probably play an important role also in the future. But there may be a tendency that the optimism of using market-based instruments has been reduced, and other options are now also discussed among economists; see, for example, Aldy and Stavins (2007) for a number of alternative approaches, and Barrett (2008) for sectoral agreements. Whether or not this is due to an ethical debate is still an open question.

NOTES

1. This chapter is partly based on Eyckmans and Kverndokk (2010) and Kverndokk (2010), and is funded by the NORKLIMA program at the Research Council of Norway. I am indebted to Roger Fouquet and Ole Røgeberg for comments. While carrying out this research I have been associated with CREE – Oslo Center for Research on Environmentally friendly Energy. CREE is supported by the Research Council of Norway.
2. See, e.g., http://www.openforum.com.au/content/top-climate-scientists-opt-carbon-taxes-slam-ets for examples of scientists and economists who favor taxation over cap and trade.
3. For a report on practical problems with a permit market see, e.g., Friends of the Earth (2009).
4. See, e.g., references in http://en.wikipedia.org/wiki/Emissions_trading#Public_opinion.
5. Article 6.1 of the original Kyoto Protocol text states 'The acquisition of emission reduction units shall be supplemental to domestic actions for the purposes of meeting commitments under Article 3'. However, later meetings of the Conference of the Parties (CoP) have not been able to find a consensus on a more precise or quantitative meaning of this supplementarity requirement.
6. Under Phase II of the ETS (2008–12), some EU member states have limited access to CDM credits for the installations on their territory. For Phase III (2013–20), a stricter limitation is in place requiring that no more than 50 percent of the total EU reduction effort over the period 2008–20 can be covered by credits generated by project-based mechanisms in third countries.
7. 'Morgenkåseriet', 22 November 2010, by Ulf-Arvid Mejlænder.
8. Some, however, argue that we do not necessarily know why something is right or wrong, but that moral reasoning is based on intuitions or emotions, and that we rationalize these intuitions later. See, e.g., Haidt (2001). For an example on incest, see also http://www.polipsych.com/tag/disgust/.
9. The arguments below share similarities with the Kantian approach. However, the Kantian approach is usually not connected to consequentialism.
10. The EU aims, for instance, to reduce domestic emissions by 80 to 95 percent by the mid-century (see, e.g., European Commission, 2011).
11. See also Gneezy and Rustichini (2000) for a similar argument when a fine is introduced for late-coming parents in day-care centers.
12. The problem may, however, be more complicated as other countries than those directly involved may also have benefited from the early industrialization via spillover effects; see Kverndokk (1995).
13. Some argue that it is not fair that the developed countries take all the 'low-hanging fruits' and the developing countries are left with the more expensive mitigation options in a possible future agreement. An economic treatment of this low-hanging-fruit argument can be found in Narain and van 't Veld (2008).
14. Some of the goods mentioned above are called taboo goods (Fiske and Tetlock, 1997). Taboos are meant to protect individuals and societies 'from behaviour defined or perceived to be dangerous' (Tannenwald, 1999), and breaking a taboo is usually met by social sanctions or repercussions. Incommensurability may also be a problem, meaning that there may not be a common measure to compare the goods (O'Neill, 1993, ch. 7). Some examples may be friendship or love. A market may destroy these goods, as setting a price on them may reduce their value.

REFERENCES

Aldy, J.E. and R.N. Stavins (eds) (2007), *Architectures for Agreement. Addressing Global Climate Change in the Post-Kyoto World*, Cambridge: Cambridge University Press.

Barrett, S. (2008), 'Climate treaties and the imperative of enforcement', *Oxford Review of Economic Policy*, 24(2): 239–58.

Becker, G.S. and J.J. Elias (2007), 'Introducing incentives in the market for live and cadaveric organ donations', *The Journal of Economic Perspectives*, 21(3): 3–24.

Bénabou, R. and J. Tirole (2007), 'Identity, dignity and taboos: beliefs as assets', Discussion Paper No. 2583, January, IZA Bonn.

Bolling, R. (1992), 'Tom the dancing bug: tales of market-driven crimes'.

Bråten, R.H., K.A. Brekke and O.J. Røgeberg (2011), 'Can we buy a right to do wrong?', Ragnar Frisch Centre for Economic Research, Oslo, draft.

Brekke, K.A., S. Kverndokk and K. Nyborg (2003), 'An economic model of moral motivation', *Journal of Public Economics*, 87, 1967–83.

Cohen, P. (2008), 'Economists dissect the "yuck" factor', *The New York Times*, 31 January.

Davies, N. (2007), 'Abuse and incompetence in fight against global warming', *The Guardian*, 2 June, http://business.guardian.co.uk/print/0,,329965233-108725,00.html.

Ellerman, D. and S. Wing (2000), 'Supplementarity: an invitation for monopsony', *The Energy Journal*, 21(4): 29–59.

European Commission (2011), *A Roadmap for Moving to a Competitive Low Carbon Economy in 2050*, Brussels, 8 March.

Eyckmans, J. and E. Schokkaert (2004), 'An "ideal" normative theory for greenhouse negotiations?', *Ethical Perspectives*, 11(1): 5–22.

Eyckmans, J. and S. Kverndokk (2010), 'Moral concerns on tradable pollution permits in international environmental agreements', *Ecological Economics*, 69(9): 1814–23.

Fiske, A.P. and P.E. Tetlock (1997), 'Taboo trade-offs: reactions to transactions that transgress the spheres of justice', *Political Psychology*, 18(2): 255–97.

Friends of the Earth (2009), 'Subprime carbon? Re-thinking the world's largest new derivatives market', Report, http://libcloud.s3.amazonaws.com/93/77/4/452/SubprimeCarbonReport.pdf.

Gneezy, U. and A. Rustichini (2000), 'A fine is a price', *The Journal of Legal Studies*, 29(31): part 1, 1–17.

Golombek, R. and M. Hoel (2004), 'Unilateral emission reductions and cross-country technology spillovers', *Advances in Economic Analysis & Policy*, 4(2): article 3.

Goodin, R.E. (1994), 'Selling environmental indulgences', *Kyklos*, 4: 573–96.

Haidt, J. (2001), 'The emotional dog and its rational tail: a social intuitionist approach to moral judgment', *Psychological Review*, 108: 814–34.

Hansen, J. (2009), 'Cap and fade', opinion article, *New York Times*, 9 December.

Harvey, F. (2006), 'Beware the carbon offsetting cowboys', *Financial Times*, 26 April, http://www.ft.com.

Hauge, K.E., S. Kverndokk and A. Lange (2011), 'Punishing the good or bad guys?', Ragnar Frisch Centre for Economic Research, Oslo, draft.

Helm, C. (2003), 'International emissions trading with endogenous allowance choices', *Journal of Public Economics*, 87: 2737–47.

Hoel, M. (1991), 'Global environmental problems: the effects of unilateral actions taken by one country', *Journal of Environmental Economics and Management*, 20: 55–70.

Holtsmark, B.J. and D.E. Sommervoll (2008), 'International emissions trading in a non-cooperative equilibrium', Discussion Papers 542, Statistics Norway.

Kverndokk, S. (1995), 'Tradeable CO_2 emission permits: initial distribution as a justice problem', *Environmental Values*, 4(2): 129–48.

Kverndokk, S. (2010), 'One economist's thoughts about moral concerns over cap-and-trade', *Weekly Policy Commentary*, Resources For the Future, Washington, DC, 20 September.

Kverndokk, S. and A. Rose (2008), 'Equity and justice in global warming policy', *International Review of Environmental and Resource Economics*, 2(2): 135–76.

Montgomery, D.W. (1972), 'Markets in licenses and efficient pollution control programs', *Journal of Economic Theory*, 5: 395–418.

Næss, A. (1973), 'The shallow and the deep, long-range ecology movement', *Inquiry*, 16: 95–100.

Narain, U. and K. van 't Veld (2008), 'The Clean Development Mechanism's low-hanging fruit problem: when might it arise, and how might it be solved?', *Environmental and Resource Economics*, 40(3): 445–65.

O'Neill, J. (1993), *Ecology, Policy and Politics – Human Well-Being and the Natural World*, London: Routledge.

Pearce, F. (2010), 'Carbon trading tempts firms to make greenhouse gas', *NewScientist*, 16 December, http://www.newscientist.com/article/dn19878-carbon-trading-tempts-firms-to-make-greenhouse-gas.html.

Pogge, T.W. (1989), *Realizing Rawls*, London: Cornell University Press.
Rosendahl, K.E. and J. Strand (2009), 'Simple model frameworks for explaining inefficiency of the Clean Development Mechanism', Policy Research Working Paper 4931, The World Bank, Washington, DC.
Rosendahl, K.E. and J. Strand (2011), 'Carbon leakage from the Clean Development Mechanism', *The Energy Journal*, 32(4): 27–50.
Roth, A.E. (2007), 'Repugnance as a constraint on markets', *Journal of Economic Perspectives*, 21(3): 37–58.
Schmalensee, R., P.L. Joskow, A.D. Ellerman, J.P. Montero and E.M. Bailey (1998), 'An interim evaluation of sulfur dioxide emissions trading', *The Journal of Economic Perspectives*, 12: 53–68.
Tannenwald, N. (1999), 'The nuclear taboo: the United States and the normative basis of nuclear non-use', *International Organization*, 53(3): 433–68.
van Egteren, H. and M. Weber (1996), 'Marketable permits, market power, and cheating', *Journal of Environmental Economics and Management*, 30: 161–73.
Zeliner, V. (1999), *Morals and Markets: The Development of Life Insurance in the United States*, New York: Columbia University Press.

23 The European CO_2 allowances market: issues in the transition to Phase III

Christian de Perthuis and Raphaël Trotignon

INTRODUCTION: THREE LEVELS OF REGULATION

The European Union Emissions Trading Scheme (EU ETS) was launched in 2005, organized in such a way that considerable autonomy was given to member states through the principle of subsidiarity. This decentralized architecture was a political condition for its launch, and the result was dramatic: the market grew rapidly, becoming a key standard in terms of global carbon price (see Ellerman et al., 2010). The downside was the weakness of its regulation by the public authority, the full extent of which became apparent with the large-scale frauds and embezzlement of 2009 and 2011.[1] Today no one disputes the need to strengthen this regulation, but many unknowns arise as soon as one attempts to put it into practice, since the term 'regulation' can have three different meanings.

Regulation first refers to the set of rules ensuring the security of the market infrastructure. In the case of an allowances market, it mainly concerns the registries system, the registration procedures for recording transactions, and the conditions for market access. These rules have been greatly reinforced following the misappropriation of funds occurring in the market, and will be further strengthened with the transition to Phase III (2013–20), which will include setting up a single registry at a European level as from 2013.

A second aspect of regulation concerns what is traditionally called market oversight by a regulator who has to guarantee the fluidity and transparency of transactions by preventing the risk of manipulation. In the EU ETS, this oversight raises complex questions of harmonization and legal definition of the compliance instruments that were created within the framework of climate policy (see Jickling and Parker, 2010; Pew Center, 2010). In France it gave rise to an innovative sharing of responsibilities between the financial regulator and the energy regulator following the recommendations of the Prada report published in 2010. For convenience, ways of strengthening this oversight in Europe should be cast more along the lines of the regulation of financial markets, although some points would need to be adapted for industrial actors.

A third possible level of intervention by the public authority concerns actions designed to influence price formation in order to change its equilibrium level or reduce its volatility. In a system of regulation by quantities, the role of the public authority is to set the emissions cap, leaving it to the market to set the price. Other than a revision of the cap, we see no *a priori* reasons that would lead to an intervention to change the market price if the first two levels of market regulation are implemented effectively. The principle of an intervention of this type has, however, been under discussion for several months, both in the European Parliament and in the various committees dealing with the climate issue at the European level.

The present chapter focuses on this third level of regulation. It first outlines the contextual elements explaining the rationale for an intervention by European public authorities intended to raise the price of CO_2 allowances in the market. Using the ZEPHYR-Flex model,[2] it then simulates the various options available to the public authority by showing how improbable it is that an intervention would settle once and for all a problem whose recurrence is attributable to intrinsic weakness of governance. In the third section, the chapter explores the options for renewed market governance, based on the establishment of an independent regulatory authority (IRA), whose mandate, drawing on lessons from monetary policy, would be to proactively manage the supply of allowances by ensuring that the different time horizons of climate policy are properly articulated. In the absence of such a mandate, it is a safe bet that improvised interventions to restore the 'right carbon price' would lead first to the weakening and then to the marginalization of an instrument that until now has been a key tool in the European climate strategy.

1. IN SEARCH OF THE 'RIGHT' CARBON PRICE SIGNAL

One of the major difficulties in implementing climate policies is the treatment of uncertainty, which concerns both the evaluation of potential damage from climate change and the cost of the actions taken to reduce emissions. This uncertainty makes it impossible to know *ex ante* the 'right' price for carbon, that is, one whose implementation would over time lead to the equalization of the marginal costs of action taken to reduce climate risk and the marginal benefits that would accrue to society. The role of the EU ETS is precisely to reveal this price *ex post* from the level of constraint imposed *ex ante* in the form of an overall cap limiting the emissions of industries subject to the scheme. The experience of the first eight years of the market leads to a result often seen in allowances markets tried out in other contexts: an *ex ante* debate dominated by fears of an overly high constraint level that makes allowance prices skyrocket at the risk of damaging the competitiveness of the economic entities subject to the system; and the *ex post* problem of managing the consequences of the trade-offs and compromises that led to limiting the credibility of the constraint and consequently the blurring of the predictability of the price signal.

1.1 What Eight Years of Market Operation Reveals

When Europe launched its CO_2 emissions trading scheme in 2005, the information available on historical emissions was of very varied quality, depending on the countries and sectors concerned. Choosing a constraint level in the form of an overall emissions cap was thus carried out through trial and error, within the framework of the three phases of the market.

The constraint during the first phase (2005–07) was defined as part of a highly decentralized process, leading to rather similar trade-offs from one country to another: a more restrictive cap for installations in the electricity sector than for industries open to competition from outside Europe, which were generally well provided with allowances distributed free of charge; and a higher constraint level in countries such as Germany and the UK, which were subject to more ambitious Kyoto commitments than France

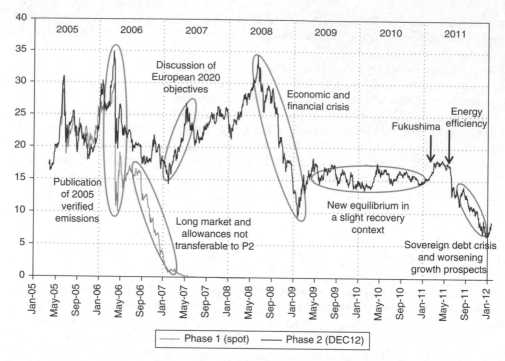

Source: Climate Economics Chair, based on BlueNext and ICE ECX Futures data.

Figure 23.1   CO₂ emissions prices since the launch of the ETS in 2005 (€/tCO₂)

or the countries of Eastern Europe. This second group therefore generated large surpluses of allowances, the price of which tended towards zero because it was impossible to transfer them from the first to the second phase (Trotignon and Solier, 2011; Figure 23.1).

The allocation process for the second phase (2008–12), characterized by the use of rules that were much more harmonized at the European level, led to a tightening of the cap by about 10 per cent. Simultaneously, industries subject to the cap were allowed to ensure their compliance through credits from the Kyoto projects within the framework of an average limit of 13.5 per cent at EU level. When that decision was taken, no one knew very clearly how many credits were likely to be placed on the market up to 2012: the pace of development of projects around the world and the level of demand for these credits outside the European system were both very uncertain. As in the first period, the start of this phase was marked by expectations of high allowance prices, driven up by the climate and energy package negotiations in Europe and the revival of federal carbon market projects in the US Congress. These expectations did survive the deterioration of economic conditions, which led to a sharp market correction in the second half of 2008, at the same time as discussions on the rules of the third phase of the market were getting under way.

In the initial project, the third phase (2013–20) should see a tightening of the cap and a change in the allocation method, whereby the auctioning of allowances is likely to become the rule and free allocation the exception. Strengthening the constraint will involve applying a 1.74 per cent annual cap decrease factor from 2013, leading to a 21 per cent reduction in emissions by 2020 compared to 2005. In addition, the Commission is aiming to reduce the amount of Kyoto credits used by industry by imposing qualitative restrictions on certain industrial projects as from 2013. The measure will tend to speed up their entry in 2011 and 2012. This tightening of the constraint was in line with the Commission's intention to 'shore up' CO_2 allowance prices in the second phase (2008–12) by encouraging industry to make precautionary savings of allowances for the third phase (2013–20), as quotas are transferable from one phase to the next. But the market reacted to fundamentals and did not conform to the wishes of the people running the scheme: the sovereign debt crisis and the prospect of strengthening policy instruments for energy efficiency again caused allowance prices to fall to 'crisis' levels in the second half of 2011 (Figure 23.1).

Two fairly basic lessons emerge from this brief retrospective. The first is that the market reacts very quickly to any changes, even minor ones, in the balance between supply and demand for allowances: changing short-term meteorological conditions affecting heating and/or air-conditioning needs; changes in relative energy prices making primary sources with varying carbon content less or more attractive; changes in the pace of economic activity that can drastically change the workload of large high-emitting, intermediary industries; and regulatory decisions that tend to become more complex and overlap at the expense of efficiency and clarity. This responsiveness is all the stronger because in a compliance market the supply and demand for allowances, in the absence of so-called banking or borrowing regulatory mechanisms (to be discussed later), may become very inelastic.

The second lesson draws on a number of observations of other emissions allowances cap-and-trade systems. As shown in the first two stylized images of Figure 23.2, actors and governments at the beginning of the first (2005) and second (2008) periods anticipated a market starting from an initial position close to instantaneous equilibrium between the supply of and demand for allowances, and in which the hypothetical growth of future emissions, by increasing the allowances deficit and the need for action to reduce emissions, would pull the price up. In both cases, the correction of these

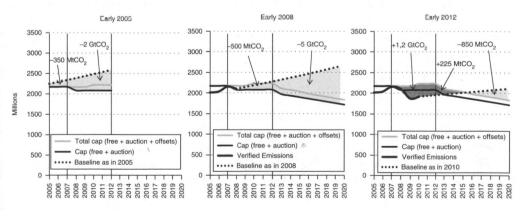

Source: Climate Economics Chair.

Figure 23.2   *Perception of the market in 2005, 2008 and 2012*

expectations during the period concerned caused a major downward revision of the market equilibrium price.

The graph's third stylized image shows the different expectations on the part of market actors and public authorities in the third phase. Due to the accumulation of past surpluses and a strongly revised downward projection of anticipated emissions in the scenario resulting from the worsening economic climate, there is a general expectation of a surplus of allowances on the market until 2020. Hence current plans for an exceptional intervention by public authority in the third period with the intention, one way or another, of raising the equilibrium price. We will now look more in detail at the conditions for the transition to the third phase.

1.2   The Transition to Phase III: Acrobatic Conditions

The preceding analysis of the history of the EU ETS, though brief, revealed the many factors influencing the supply/demand equilibrium of allowances. These factors make it very difficult to anticipate the future trajectory of allowance prices. As a result we need to approach interpretations of prospective assessments of the development of the third phase with a degree of caution. Three key parameters should in our view be taken into account: the management of surplus allowances inherited from the past, in a context where prior banking and borrowing behaviour will be altered by the transition to auctions in the electricity sector; uncertainties as to economic growth; and possible overlapping with other European policy instruments that are themselves affected by the macroeconomic and budgetary context (see Chevallier et al., 2011).

One of the few points that is no longer in dispute concerns the assessment of surplus allowances or credits not used in the second phase that installations will transfer to the third phase. The estimates obtained from the ZEPHYR-Flex model result in a magnitude of about 1.2 Gt of CO_2, slightly more than half the 2012 emissions cap. This amount can be viewed as a safety margin that companies want to keep unchanged from one period to the next. In this case, it might be argued that the amount of surpluses would not alter the market equilibrium conditions in the third period. This view, which is difficult to maintain even in a situation where the allocation rules are not modified in the third phase, becomes particularly unrealistic with the extension of the allowances auction system, which is likely to significantly change the banking and borrowing behaviour of installations subject to the constraint.

As Figure 23.3 shows, the great majority of allowances until now have been retained from one year to the next in a passive way: if an installation receives free of charge more allowances than it needs for compliance and does nothing, it will bank them without any active decision on its part. In an auction system where companies must buy all their allowances, this routine banking system disappears, thereby, one, reducing the surplus of allowances that until now have been almost automatically transferred from one year to the next and, two, weakening one of the traditional levers for supporting prices in a market with a surplus of allowances. In the basic ZEPHYR-Flex scenario for the third phase, it was assumed that most of this passive banking would be replaced by actors' active decisions to buy allowances in excess of their compliance needs (Figure 23.3). If such a process were not to occur, we could see a more rapid reduction in surpluses accompanied by a new period of allowance price weakness. At the same time, the application of

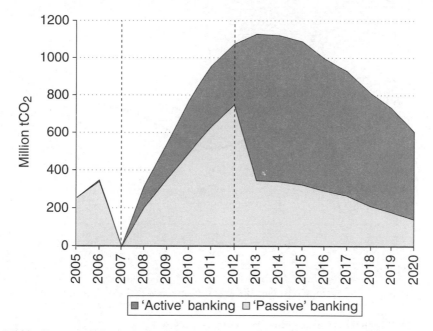

Note: Within the stock of allowances carried over from one year to the next, the ZEPHYR-Flex model makes a distinction between, on the one hand, allowances allocated free of charge to installations and retained (so-called passive banking) and, on the other, allowances or credits bought by an installation as a precautionary measure (so-called active banking).

Source: Climate Economics Chair, ZEPHYR-Flex model.

Figure 23.3   Estimation of the amount of allowances and credits carried over from one year to another

payment by auction should limit the previously existing possibility of borrowing quotas free of charge for up to a year, by superimposing the period for that year's allocation of allowances with the period for the surrender of allowances from the previous year. By restricting the previous possibilities of freely carrying out banking and borrowing operations, the transition to auction, unless it is accompanied by new flexibility vectors,[3] may therefore increase the instability of allowance prices. While economic reasoning allows the direction of expected changes to be detected, our tools may have difficulty predicting their magnitude and timing.

The second element to include concerns hypotheses about economic growth. As the simulations carried out with the ZEPHYR-Flex model reveal, the demand for allowances by the industrial and energy production sectors is very sensitive to changes in the growth rate of GDP, with an elasticity greater than one reflecting the pro-cyclical nature of the major intermediate industries. In the third phase, an increase of two points in the growth assumption leads, *ceteris paribus*, to a rise of nearly €10 in the equilibrium price of allowances over the whole period (Figure 23.4). This positive relationship between economic growth and allowance prices is also highly desirable if it does not interfere with the long-term expectations of the actors: it reduces compliance costs for businesses in lean times and asks them to increase their tribute when business is good. All the same,

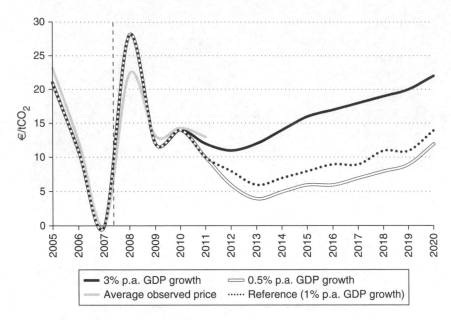

Source: Climate Economics Chair, ZEPHYR-Flex model.

Figure 23.4   Sensitivity of allowance prices to assumptions of economic growth

uncertainty about economic growth up to 2020 and the probably bearish bias of forecasts made by actors in a negative economic climate may result in a number of surprises during the third phase, especially if the economic and budgetary context disrupts the implementation of public policy instruments outside the cap-and-trade system.

The third challenge faced by market participants in anticipating the balance between the supply of and demand for allowances arises from the risk of superimposing public policy instruments. This issue was latent from the formulation of the energy climate package with two binding targets – the first regarding the reduction of greenhouse gas emissions, the second in terms of market penetration by renewables – without there being a clear specification as to the type of instrument applied to each of them. The orientations consistent with making the objectives of the Directive on energy efficiency binding may make the overlapping of instruments acting on the real level of emissions more likely. In this context, any progress in terms of energy efficiency or renewable energy obtained by measures outside the allowances system may result in decreased demand for allowances and lower prices. If we want to avoid the cap-and-trade system becoming a residual tool, allowance caps need to be adjusted according to the observed or expected effects of other instruments.[4] But we must be extremely careful regarding the forecasts, because it is not the intentions announced by the legislator that count, but the ability of governments to make use of the instruments available given the economic and budgetary context.

Implemented in a particularly uncertain economic and financial context, with the rules regarding policies for energy efficiency and renewables still to be settled and with the modified basic allocation rules liable to increase market instability, the transition to the third phase is in fact taking place under acrobatic conditions. How consequently do we calibrate the public authority's intervention to spontaneously alter the expected market equilibrium? With the help of the ZEPHYR-Flex model, the second part of this study examines the various options that have been put on the table for such an intervention.

2. THE DIFFERENT PROPOSALS FOR MARKET INTERVENTION

Three types of intervention proposal have been advanced to change the equilibrium price of allowances during the third phase. The first would take advantage of the transition to auction in the electricity sector to establish a reserve price that would *de facto* play the role of a price floor. The second would be to operate a set-aside, that is to say, to distribute differently over time the amount of allowances available, without explicitly changing the reduction target associated with the system. The third would lengthen the market's time horizon, quickly setting a binding cap for 2030 in order to change the long-term expectations of industry. To our knowledge these three options have not yet been subject to quantified evaluations. We propose here to use the ZEPHYR-Flex model (Box 23.1) to link each option with figures as to their main impacts on the functioning of the market up to 2020 (see Trotignon, 2012). To do this, we start from the representation of market equilibrium simulated in the central scenario of our forecasting exercise.[5] Each proposal is then treated as a variant on this central scenario.

2.1 Setting a Reserve Auction Price

The transition to the third phase will be accompanied by a sharp increase in the amount of allowances auctioned. The way auctions are organized was designed by a regulation whose main objective is to ensure that the auction system, which will introduce a genuine primary market into the scheme, does not disturb the equilibrium of the secondary market. In other words, auctions must be 'neutral' in relation to market prices, as economic theory also makes clear. The proposal to establish a reserve price at a level well above the current price has been advanced by some economists[6] and endorsed by CDC Climat. In this alternative approach, the aim would be to use the auction system in order to act directly on the equilibrium price. The Zephyr-Flex model allows us to analyse its possible implications.

Establishing a reserve price at an auction means that a sale is not concluded unless the price reaches a value determined *ex ante* by the public authority. In our simulation, the reserve price is set at €20/tCO₂. If this option is implemented from 2013, the first consequence is that half the allowances do not enter the market in 2013, which raises the allowance price to €20/t. For this price to become a market price responsive to changes in supply and demand would require that emissions be immediately reduced by 50 per cent, which is not possible at less than €20/t given the marginal abatement cost curves used in the model. A gap is thus created – which cannot be absorbed over the period as a whole – between the reserve price that will prevail due to the removal of allowances by the allocating authority and the price that theoretically would appear on the market in the absence of this intervention (Figure 23.6, top left).

The difference between the new price and the theoretical market price spontaneously

BOX 23.1　DESCRIPTION OF THE ZEPHYR-FLEX MODEL

The Zephyr-Flex model developed by the Climate Economics Chair is a tool that simulates the price formation mechanism and allowances trading on the European carbon market. Its specificity is to operate within an economic framework that takes into account the particularities of the European cap-and-trade system (operating rules and characteristics of the area covered), as well as the expectations and attitudes to compliance of the entities covered (Figure 23.5). It allows us to test different configurations, for example by varying the growth scenario or the dividing up of emissions reduction targets over time.

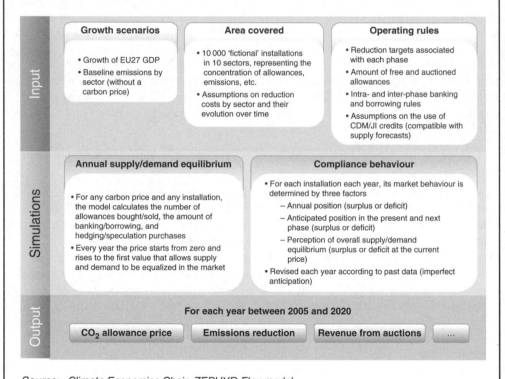

Input

Growth scenarios
- Growth of EU27 GDP
- Baseline emissions by sector (without a carbon price)

Area covered
- 10 000 'fictional' installations in 10 sectors, representing the concentration of allowances, emissions, etc.
- Assumptions on reduction costs by sector and their evolution over time

Operating rules
- Reduction targets associated with each phase
- Amount of free and auctioned allowances
- Intra- and inter-phase banking and borrowing rules
- Assumptions on the use of CDM/JI credits (compatible with supply forecasts)

Simulations

Annual supply/demand equilibrium
- For any carbon price and any installation, the model calculates the number of allowances bought/sold, the amount of banking/borrowing, and hedging/speculation purchases
- Every year the price starts from zero and rises to the first value that allows supply and demand to be equalized in the market

Compliance behaviour
- For each installation each year, its market behaviour is determined by three factors
  - Annual position (surplus or deficit)
  - Anticipated position in the present and next phase (surplus or deficit)
  - Perception of overall supply/demand equilibrium (surplus or deficit at the current price)
- Revised each year according to past data (imperfect anticipation)

Output

For each year between 2005 and 2020

| CO_2 allowance price | Emissions reduction | Revenue from auctions | ... |

Source:　Climate Economics Chair, ZEPHYR-Flex model.

Figure 23.5　The ZEPHYR-Flex model

Each proposal for intervention by the public authorities in the allowances market can be simulated in Zephyr-Flex and compared with a reference scenario, corresponding to the situation that can currently be represented in the absence of intervention in the market.

- The objective of reducing emissions by 21 per cent in 2020 compared to 2005, and the implicit continuation of the cap-and-trade system after 2020 (continuity of the linear reduction factor of the cap and transferable allowances into the next phase).
- Auctioning of all allowances in the electricity sector from 2013 and a decreasing share of free allowances in other sectors, except those identified as being at risk from carbon leakage.
- Nearly total use of the amount of carbon credits authorized during the 2008–20 period, or 1.5 billion credits, such use being compatible over time with the qualitative restrictions agreed in 2011 and the likely supply of credit over the period.
- A baseline emissions growth scenario for each sector (i.e. as they would have been without a carbon price) on the basis of a low-growth scenario of European GDP from 2012 (+1 per cent per year until 2020).

tends to grow: the floor price of €20/tCO_2 'forces' emissions reductions that were not necessary to attain the environmental objective, which remains unchanged because the cap has not been altered. This pseudo-tax lowers CO_2 emissions below the limit represented by the total number of allowances and credits available (Figure 23.6, top right).

The decrease in emissions resulting from a price that is twice the level dictated by supply/demand equilibrium automatically leads to additional emissions reductions and hence a reduced demand for allowances. But as the auctioning authority cannot sell allowances at a price below the reserve price, the allowance price remains tied to the reserve price throughout Phase III. The authority auctioning allowances is thus forced to accumulate an ever-increasing amount of allowances that it is unable to sell without lowering the price below €20. Our simulation puts this quantity at about 500 Mt in 2013, increasing to a total of 1.1 Gt unsold over Phase III (Figure 23.6, bottom).

The direct intervention in the price of allowances by establishing a reserve price at auctions thus leads to the 'freezing' of 1.1 Gt, or the equivalent of more than a year's auctions. The great unknown in so acting is obviously the future of those allowances that have not been put onto the market. Here the Zephyr-Flex model recalls a key mechanism: as soon as the auctioning authority places these quantities on the market without reserve clause, the price immediately tends to zero. A radical option would then be to simply cancel these allowances – in other words, to modify the cap. But in that case why not directly adopt this option, thereby sending a clear signal to industry?

In short, setting a reserve price of €20/t means removing allowances from the market, but without deciding in advance the quantity withdrawn. If maintained throughout the period, this reserve price gives actors a temporary visibility on the price thus 'forced', but totally clouds their medium-term perspective in that it has not been clearly decided what would happen to these allowances in the succeeding period. If the fixed reserve price is abandoned during the period, it causes a collapse of market prices because the imbalance between supply and demand for allowances has been widened by the price floor that generated additional reductions. Many such lessons may be found in the following analysis of different variants of set-aside.

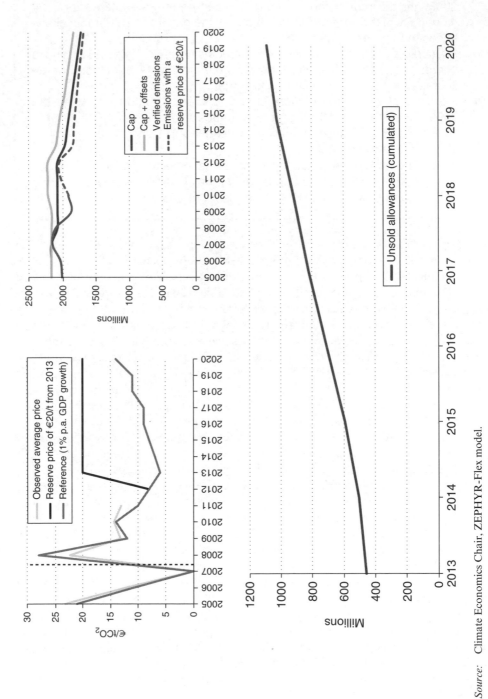

Source: Climate Economics Chair, ZEPHYR-Flex model.

Figure 23.6 The implications of a reserve auction price

2.2   Did you Say Set-aside?

For institutional reasons, if an intervention by the public authorities is agreed, it will probably involve a set-aside decision. Specifically, the draft Directive strengthening energy efficiency objectives in the EU envisages going ahead with such a withdrawal,[7] made all the more necessary in the European legislator's thinking in that the implementation of said Directive could have a further depressive effect on the carbon market.

Such a measure is actually not as distant as it first seems from the introduction of a reserve price. It consists simply in setting *ex ante* the quantity of allowances that will be withdrawn from the market and then observing the resulting price. The simulations carried out using Zephyr-Flex lead to the same kind of conclusion: in both cases, the impact of intervention by the public authorities in the medium term depends primarily on what happens to the allowances that were set aside. To clarify this point, we simulated three possible options for a withdrawal of 1.1 Gt over the whole period: in variant 1, the allowances are removed from the market at the start of Phase III and then reintroduced before 2020; in variant 2, they are removed between 2013 and 2020 and reintroduced in the next phase; and in variant 3, they are removed permanently from the market.[8]

- Variant 1 involves withdrawing 275 Mt of allowances a year from 2013 to 2016 and then putting them back on the market between 2017 and 2020. The intervention causes a fairly rapid rise in the allowance price, which reaches €20 in 2016, as shown in Figure 23.7. The rise generates additional emissions reductions from

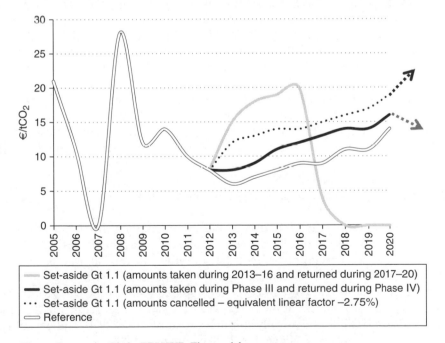

Source: Climate Economics Chair, ZEPHYR-Flex model.

Figure 23.7   The implications of different types of set-aside

companies, which proportionally reduce their demand for allowances, thereby accelerating the fall in the price of allowances following their return to the market from 2017. The effect is quite similar to what would occur in the case in which the auction authority introduces a reserve price of €20 at the start of the early period and then drops it in 2017. In both cases, the removal of allowances at the start of the period drives the price up, which leads to more emissions reduction. The price then falls dramatically when the allowances are put back on the market.

- In variant 2, 138 Mt of allowances are withdrawn each year and banked in the following period. In its current state of development the ZEPHYR-Flex model does not describe market equilibrium after 2020 and therefore cannot simulate the detailed impacts of different choices adopted for the return of allowances to the market after 2020. But it is evident that the problematic is the same as in the previous case. However, we have incorporated this uncertainty into agents' expectations regarding the market in the second phase. The result is an increase in the equilibrium price of CO_2 allowances of around €5/t over the entire third period compared to our baseline scenario (Figure 23.7). Note that the baseline scenario already includes an implicit assumption that the scheme is extended after 2020 with continuity of the linear allocations reduction factor, as well as the possibility of retaining Phase III allowances for use after 2020.
- In variant 3, it is assumed that the allowances removed will not be returned to the market. This corresponds to a scenario in which the linear annual reduction factor of the emissions cap increases from 1.74 per cent to about 2.75 per cent. In this variant (Figure 23.7), the quantity of allowances supplied is the same as in the previous case, but companies are no longer uncertain as to the possible return of allowances to the market from 2021. This reduction in uncertainty leads them to increase their demand for allowances in order to build up precautionary savings (banking) in view of the expected tightening of the constraint in Phase IV.

The main lesson from these simulations is that any set-aside action, if it is to avoid blurring the signals given to industry, must be very explicit on the future allowances withdrawn from the market. Rules that are unclear or inappropriate in this area would be likely to disrupt industry's medium-term outlook and trigger undesirable shocks in the market.

2.3 Setting a 2030 Target Compatible with Roadmap 2050

The third possible type of intervention would be to lengthen the time horizon of the market, and quickly decide on the amount of allowances available until 2030 so as to change the long-term expectations of industry. For now, the total allowance cap is determined only up until 2020. The planned reduction of the cap is set by the Directive and represents an annual decrease of 1.74 per cent compared to the average cap for the period 2008–12. According to the text of the Directive, if the cap's linear reduction factor is not changed in Phase III, it automatically continues to apply after 2020, and constitutes an implicit reduction target of 38 per cent in 2030 compared to 2005.

Since the adoption by the European Council of a long-term objective for the EU leading to a reduction in emissions of at least 80 per cent in 2050 compared to 1990 (see

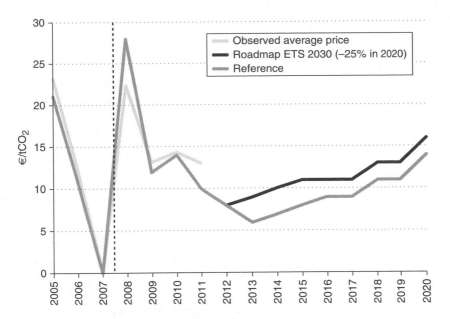

Source: Climate Economics Chair, ZEPHYR-Flex model.

Figure 23.8   The implications of an EU ETS 2030 target compatible with Roadmap 2050

European Commission, 2011), the public authorities have been discussing the inclusion of this new target in official documentation. The trajectory aimed for is detailed in the European Commission's Roadmap 2050 published in March 2011. On the basis of impact studies carried out for the Commission, the implementation of the roadmap would require emissions reductions of 43–48 per cent in 2030 compared to 2005 in the sectors covered by CO_2 allowances.

One of the solutions proposed to raise the allowance price in Phase II is to immediately ensure that the cap's linear reduction factor is consistent with the long-term European objective by establishing the amounts of allowances that will be available until at least 2030. This intervention would entail revising the cap's linear reduction factor from 2013 by raising it from 1.74 per cent to about 2.15 per cent. This change would have an impact on the 2020 reduction target, increasing it to 25 per cent below the 2005 figure as against 21 per cent currently.

The ZEPHYR-Flex simulation (Figure 23.8) shows that this reassessment of reduction targets would in fact raise the allowance price from 2013 and throughout Phase III by about €3–4/tCO_2. Taking into account the expectations of market participants ensures that the price increase is smoothed out over time, leading to a price increase compared to the baseline scenario that is higher at the beginning of the period than at the end. Nevertheless, everything depends on the nature of the actors' expectations, particularly the element of surprise that this measure may have in a context of imperfect forecasts.

Table 23.1 summarizes the main results obtained from the ZEPHYR-Flex model for each type of intervention envisaged. It is apparent that all the measures cause the price to rise within the 2020 time horizon if maintained throughout the period. But they

Table 23.1   Summary of the implications for Phase III of a public intervention

Scenario	Allowance price in €/tCO$_2$		Banking in MtCO$_2$		Verified emissions in MtCO$_2$	
	2013	2020	2013	2020	2013	2020
Reference	6	14	1075	610	2080	1900
Reserve price of €20/t	20	20	900	1550	1860	1680
Set-aside – version 1	15	0	990	945	1950	2130
Set-aside – version 2	8	16	1200	305	2040	1825
Set-aside – version 3	13	19	1365	880	1980	1740
2030 target compatible with Roadmap	9	16	1210	700	2030	1835

Source: Climate Economics Chair, ZEPHYR-Flex model.

significantly alter the quantities by increasing the amount of allowances carried over to the future, which potentially can generate market disruption. Although market actors acquire a better perception of prices in the short term, their long-term outlook becomes considerably less clear. Moreover, past experience suggests that forecasts made at the beginning of the period may be disappointed later, thus casting doubt on the idea that a single intervention by the public authorities can permanently put allowance prices on the 'right' path. For this reason it seems to us that such an intervention could be wisely carried out only through establishing the dynamic management of the supply of allowances conducted within a new market governance framework.

3.   THE NEED FOR RENEWED GOVERNANCE

After eight years of operation, the EU ETS has not yet fully established its credibility. Interventions by the public authority carried out in a hurry to raise allowance prices risk further weakening the scheme unless they are accompanied by changes in its governance. This has been noted by several analysts; see for example Climate Strategies (2012) or Curien and Lewis (2011). This section explores ways of enhancing its credibility by establishing an independent regulatory authority (IRA), whose mandate would draw on various lessons from monetary policy.

3.1   A Parallel with Monetary Policies

In some ways, a CO$_2$ allowance amounts to a new currency, characterized by the curious feature that it can be used to purchase only one good: the right to emit one tonne of CO$_2$ into the atmosphere. The result is a certain similarity between the operation of the carbon market and that of the money market.

Every year, the issuing of carbon currency (the primary money supply) is implemented through the process of allocating CO$_2$ allowances to industry. It can be done on a free-of-charge basis or by auction. The total amount of the currency created is determined by the emissions cap, which must not be exceeded. Once issued, this currency can circulate freely. It is withdrawn from circulation at the time of compliance, when installations must surrender to the public authority as many allowances as the number of tonnes of CO$_2$ they have emitted. In the case of 'over-allocation', the value of the carbon currency is eroded – a kind of 'carbon inflation'. And just as inflation weakens the economy, so over-allocation impairs the capacity of the carbon price to trigger the specified emissions reductions. In a symmetrical way, if there is a liquidity crisis, the drying up of the money in circulation may cause a systemic crisis: in such a situation, the central bank typically acts as lender of last resort to prevent the collapse of the financial system and of economic activity. Similarly, in the absence of flexibility mechanisms, a shortage of carbon currency could generate a surge in the price of carbon, leading to economic breakdown (see Sustainable Prosperity, 2011).

Since 2008, industries subject to allowances quotas can use Kyoto credits, which they import from outside the system for part of their compliance. These credits are equivalent to a currency, the use of which may affect the value or stability of the domestic currency. This raises the traditional question of the degree of convertibility of the domestic currency and exchange rate management. Again, the parallel between the carbon market and money market still holds.

Given these similarities, we can draw a parallel between the functioning of a central bank in the money market and the role that an IRA could have in the emissions trading system. For decades, the main central banks have relied on quantitative management to achieve their targets for interest rates, mainly through the supply of base money and open market operations in the money market. The instruments available to an IRA would be fairly similar, as shown in Table 23.2, especially in the case where the majority

Table 23.2   Comparison of the money market and the carbon market

	Money market	Carbon market
Final objective	Monetary stability in the long term	Emissions reduction at the lowest costs in the short and long term
Market oversight	Integrity and liquidity of transactions	Integrity and liquidity of transactions
Price instrument	Interest rate	Carbon price
Quantitative regulation		
Primary market	Central money supply	Auction of allowances
Secondary market	• 'Open market' (sale and purchase of monetary assets) • Exchange rate	• Sale and purchase of carbon assets • Links with other markets (international credits)
Role in a liquidity crisis	Lender of last resort	Additional supply borrowed on the future (borrowing)
Periodic communications to the public authorities (European Council, European Parliament, European Commission)	• Annual and quarterly reports on the financial and economic situation • Public hearings in the European Parliament + Council	• Reports on the carbon price and the long-term trajectory • Public hearings

Source: Authors.

of allowances are auctioned. Dynamic management of the supply of allowances by auction should prevent the risk of 'carbon inflation' in the short term, probably limiting the need to intervene in the secondary market. In the event of a 'liquidity crisis', the IRA could play a similar role to that of lender of last resort, by a farsighted use of borrowing.[9] Of course, strict requirements for public information and reporting would be imposed on the IRA, as on any central bank (Table 23.2).

As Whitesell (2011) has shown with great acumen, there are not only technical similarities between currency and climate, but also a shared problematic of articulating very different time horizons. The implementation of monetary policy and climate policy involves trade-offs in relation to the long-term view that fall outside the usual concerns of government: non-inflationary growth for the first, and transition to a low-carbon economy for the second. Experience shows that governments' time horizons are too short to effectively combat the risk of inflation. This short horizon can create an instability in political decision making that runs counter to the confidence needed for the proper functioning of markets. This is the reason for the creation of 'independent' central banks to which the political authority mandates the task of making painful decisions in the short term to protect society against inflation in the long run. Similarly, the customary political horizons are too short to act effectively against the risks of climate change. The implementation of climate policies therefore requires innovative forms of governance in order to take account of the long view in decision making. The EU ETS is no exception to this rule: its credibility would be greatly enhanced by setting up an IRA (see De Perthuis, 2011a, 2011b), the mandate of which we attempt to outline below.

3.2   Outline of the Mandate of an Independent Regulatory Authority (IRA)

As with the definition of a central bank's mandate, it is first necessary to separate the ultimate objective, which must remain the prerogative of the political authority, from the intermediate objectives, which are better delegated to the independent authority.

The ultimate objective of monetary policy – achieving the best non-inflationary growth path – is not the responsibility of the central bank. Its role is to decide how much money to put into circulation, ensuring that the amount is large enough to stimulate growth, but not so large that it leads to traditional inflationary processes or speculative bubbles.

Similarly, the ultimate goal of climate policy – bringing emissions levels onto a path that can mitigate climate change – is not the responsibility of the IRA. That is a sovereign prerogative falling within the remit of the political authority. Concretely, this target was set in Europe in the form of caps defined with great precision through to the 2020 horizon and of a decarbonization trajectory by 2050, whose interim targets (2030 and 2040) are in the preparatory stage. The role of the IRA is not to intervene in defining these objectives, but to ensure that the carbon market puts the economy on the right track to reduce emissions – one in which industries subject to the ETS achieve the required reductions in each period and make the necessary investments to prepare for the reductions of subsequent periods.

The IRA's mandate is therefore to reconcile the different time horizons, ensuring that the market generates a carbon price that reflects changing short-term market conditions and simultaneously sends a signal triggering long-term investments. In case of an unanticipated shock, such as the 2009 recession, its role would not be to prevent a decline in allowance prices – altogether desirable in regard to short-term conditions – but to ensure that the change does not alter industrialists' expectations and their low-carbon investment programmes. Where there are threats of soaring prices due to insufficient carbon currency, the IRA could smooth out the bumps by anticipatory releasing onto the market allowances earmarked for a future period (borrowing).

One of the key conditions for the success of this mandate is that the IRA assembles the necessary expertise to understand and anticipate both the dynamics of the functioning of the carbon market and the dynamics of emission reduction trajectories. The independence of its mandate is in fact conditional upon this technical competence, which alone should guide its choices and allow it to build up its credibility *vis-à-vis* market actors. Of course, this technical credibility and the discretionary powers resulting from it must be balanced by strict obligations in regard to reporting, to both the public and the political authorities.

One of the IRA's priority mandates will be to increase the temporal depth of the European carbon market. In the industrial and energy sectors concerned, there is an enormous stock of fixed capital. Investment decisions determine the volume of future emissions for several decades. We cannot therefore be satisfied with a goal for the allowances market on a 2020 time horizon, as is the case now. Following the European Council's commitment to an emissions reduction target of at least 80 per cent in 2050 (compared to 1990), the Commission has thought long and hard as to the best way to achieve this target. It comes within the mandate of the IRA to convert this long-term goal and its intermediate targets for 2030 and 2040 into credible changes in the emissions cap in the market. To be credible with industrialists and energy producers, who would then be cognizant of the emissions cap for 40 years, a number of possible clause revisions must be foreseen, depending on the highly unpredictable future evolution of technologies, energy prices, the international trading environment and economic conditions.

An important component of the IRA mandate should be to incorporate developments outside the system into its dynamic management of the supply of allowances. Progress in international climate negotiations is likely to expand the range of market tools encouraging emissions reduction. Connecting the EU ETS to these international instruments is highly desirable, but doing so calls for specific rules to be established under the responsibility of the IRA, if we are to avoid a repetition of the unfortunate experience of the introduction of Kyoto credits, followed by their restriction. The same applies to coordination between the operating rules of the allowances market and the introduction of other climate policy instruments. The dynamic adaptation of the ETS to the effects of these various instruments is a priority mission for the IRA, which must avoid the inefficiencies resulting from the unwanted overlapping of public policy instruments.

The final aspect of the division of roles between the public authority and the IRA concerns the allocation of the proceeds from auctions. It goes without saying that the entire amount must be returned to the public authority, whose prerogative is to set the climate policy goal and to collect taxes or quasi-taxes. Any allocation of a portion of auction proceeds to the IRA would generate conflicts of interest and be totally unjustified. Without the emergence of an IRA, it is moreover likely that the transition to auction will considerably complicate decision making by the 27 governments of the EU, which are increasingly experiencing a conflict of interest in dividing up the financial windfall of the tens of billions of euros expected to be generated annually from the auction of CO$_2$ allowances as of 2013.

CONCLUSION

If we are to maintain and strengthen the role of the carbon market in the European policy for combating climate change, three conditions appear to be necessary: (1) ensuring that climate policy instruments are complementary; (2) taking a long-term view of the market; and (3) making transactions and compliance procedures secure in the short term.

All three of these conditions need to be fulfilled if confidence in the market is to be restored. To achieve this, the lessons of monetary policy teach us that it is risky to rely on current interventions by the public authorities, whose limited time horizon leads to decisions favouring the short term. It is for this reason that in this chapter we attempt to outline the mandate of a prospective independent regulatory authority (IRA) responsible for the dynamic management of the supply of allowances in the market.

In the absence of progress in market governance, although the courses of action currently under discussion to boost carbon prices may be able temporarily to increase price visibility, they risk further clouding the medium- and long-term visibility required by market actors. As the simulations using the model ZEPHYR-Flex show, establishing a reserve price in allowances auctions or creating a set-aside system simply defers the problem to the future. On the other hand, setting an ambitious goal now for 2030 would not suffer from this disadvantage – although its impact on the market would depend on participants' expectations and on the credibility of the public authority, which in the current market governance context has been weakened.

Although institutional constraints make setting up an IRA difficult for the next few years, would it not make sense to undertake action, however imperfect, in the form of an exceptional withdrawal of allowances within the framework of the adoption of the Directive on energy efficiency? We know that in the flawed world in which we live, the best can be the enemy of the good. Completely excluding an intervention until such time as the IRA is operational would no doubt be somewhat dogmatic. On the other hand, it would advisable to insist that such a measure be accompanied by an impact study clearly analysing the various long-term effects of the different options, including those resulting from the introduction of the dynamic management of the supply of allowances under the mandate of the IRA.

NOTES

1. On this topic, see Climate Economics Chair (2011).
2. An economic model developed by the Climate Economics Chair that simulates the price formation and trading of CO_2 allowances in the European cap-and-trade system up until 2020. It is described in more detail in Box 23.1.
3. Note that in a generalized auction system, an alternative (or a complement) to banking carried out at installation level arises if the authority retains a proportion of the allowances that it could put up for sale in the market.
4. On this point, see Baron (2012).
5. See 'EUA price forecast', January 2012, downloadable from the Climate Economics Chair website: http://www.chaireeconomieduclimat.org/?page_id=2259&lang=en.
6. See, for example, Grubb and Laing (2009).
7. The text voted through by European Parliament ENVI Committee in late 2011 explicitly refers to set-aside; and the MEPs of the ITRE Committee proposed in early 2012 that the Commission present a report on the impact of energy efficiency measures. The MEPs are also asking the Commission to consider whether

or not to take measures that 'may include withholding of the necessary amount of allowances' before the beginning of Phase III.

8. In variant 2, in the absence of the possibility of taking allowances from Phase IV into Phase III, the reduction target associated with Phase III is increased, while this is compensated by easing the Phase IV target. This is not set-aside in the strictest sense of the term. In variant 3, the quantity of allowances available is explicitly altered since the allowances withdrawn will never be reintroduced, leading *de facto* to a change in the cap.

9. For example, by anticipatory auctioning of allowances initially planned for subsequent years, without changing the total quantity of allowances allocated over the period. Care needs to be taken, however, as this increases the short-term supply at the cost of increasing demand in the longer term. One might also imagine a 'buffer' reserve system, of a size determined in advance or which is fuelled by the market, where allowances would be released in the event of price hikes. For example, the cost control reserve due to be introduced in the California market is one such mechanism, as was the projected reserve in the Waxman–Markey bill in Congress.

REFERENCES

Baron, R. (2012), 'Politiques d'efficacité énergétique et prix du carbone: Quelle logique?', presentation to the workshop 'Marché du carbone: nécessité et forme d'une intervention des pouvoirs publics et coordination des politiques énergie-climat', February.

Chevallier, J., Etner, J. and Jouvet, P.A. (2011), 'Bankable pollution permits under uncertainty and optimal risk sharing management rules', *Research in Economics*, **65**(4), 332–9.

Climate Economics Chair (2011), 'Failings of the European CO$_2$ emissions trading market', March, available at http://www.chaireeconomieduclimat.org/?page_id=1111&lang=en.

Climate Strategies (2012), 'Strengthening the EU ETS', Climate Strategies, March, http://www.climatestrategies.org/research/our-reports/category/60/343.html.

Curien, I. and Lewis, M. (2011), 'EU emissions: what is the value of a political option?', Deutsche Bank Global Markets Research, November.

De Perthuis, C. (2011a), 'Carbon market regulation: the case for a CO$_2$ central bank', Climate Economics Chair 'Cahiers', Information & Debates series, No 10, August.

De Perthuis, C., (2011b), 'Pourquoi l'Europe a besoin d'une banque centrale du carbone', *Revue de l'OFCE / Débats et politiques*, 120.

Ellerman, D., Convery, F. and de Perthuis, C. (2010), *Carbon Pricing: The European Union Emission Trading Scheme*, Cambridge: Cambridge University Press (published in French by Pearson under the title *Le prix du carbone*).

European Commission (2011), 'A roadmap for moving to a competitive low carbon economy in 2050', COM(2011) 112 final, March.

Grubb, M. and Laing, T. (2009), Price floors – getting some perspective', presented at the CEPS/IETA Workshop: 'Pushing and Pulling: What has the EU learned about the efficient combination of carbon market and low-carbon investment support', 3 September.

ITRE Committee (2012), EU Parliament Press Release, 'Energy savings: committee backs binding national targets and CO$_2$ set aside plan', February.

Jickling, M. and Parker, L. (2010), 'Regulating a carbon market: issues raised by the European carbon and US sulfur dioxide allowance markets', Congressional Research Service, February.

Pew Center (2010), 'Carbon market design & oversight: a short overview', February, http://www.pewclimate.org/.

Prada, M. (2010), 'CO$_2$ markets regulation', www.minefe.gouv.fr/services/rap10/101004prada-report.pdf.

Sustainable Prosperity (2011), 'A carbon bank: managing volatility in a cap-and-trade system', University of Ottawa, Policy Brief, August.

Trotignon, R. (2012), 'In search of the carbon price: The European Union Emissions Trading Scheme: from ex ante and ex post analysis to the projection in 2020', PhD thesis, Paris Dauphine University, October.

Trotignon, R. and Solier, B. (2011), 'The European market on the road to Phase 3', in *Climate Economics in Progress*, Economica.

Whitesell, William C. (2011), *Climate Policy Foundations, Science and Economics with Lessons from Monetary Regulation*, Cambridge: Cambridge University Press.

PART VI

LOW-CARBON BEHAVIOUR
AND GOVERNANCE

24 The role of behavioural economics in energy and climate policy

*Michael G. Pollitt and Irina Shaorshadze**

1. INTRODUCTION

While energy-efficiency and conservation have been important tenets of energy policy for decades, concerns about climate change have put these issues at the forefront of policy dialogue. The International Energy Association (IEA, 2010) estimates that, by 2020, about 34 per cent of the global decrease in carbon emissions in a '450 scenario' (limiting the long-term concentration of greenhouse gases in the atmosphere to 450 ppm CO_2 equivalent) compared to the reference scenario should stem from direct end-use energy-efficiency measures. This goal calls for a step change in how individuals consume energy and make energy-efficiency purchases. Energy consumption, energy-efficient investment and pro-environmental actions involve consumer decision making and behaviour. These aspects have generated increased interest in designing policy interventions that target energy demand, and interest in assessing the responsiveness of consumer behaviour to these interventions. Behavioural economics can provide new perspectives that can inform policy design on how individuals evaluate options, make decisions and change behaviour.

It is important to point out that energy policy is not just about climate change, but also about security of energy supply and affordability. Climate policy significantly interacts with both of these elements of energy policy via the introduction of expensive and intermittent renewable electricity and heat. If consumer behaviour can be changed to reduce energy demand or to make energy demand more responsive in time and space to weather-induced shortages of energy, it could be a significant contribution to facilitating the introduction of climate policy induced renewable energy. By contrast, failure to address public concerns about the security of supply or affordability implications of climate policy may jeopardize the achievement of ambitious carbon emissions reduction targets.

Behavioural economics uses insights from psychology to increase the explanatory power of economics. According to neoclassical economics, agents maximize expected utility using exponential discounting, and they have access to information that they can assess freely and completely. While this is a parsimonious representation of how economic decisions are made, experimental settings and empirical observations indicate that behaviour deviates systematically from what traditional models would predict. Some of the puzzles that traditional economics struggles to explain are the following: why are returns on equity much higher than returns on bonds (equity premium puzzle); why are there untapped opportunities to reduce (energy) expenditure through increased (energy) efficiency (efficiency gap); and why do individuals indulge in immediate gratification, knowingly compromising their long-run well-being (substance abuse)? Behavioural economics challenges one or more of the assumptions of the neoclassical economics, and offers an alternative way to model decision making. These alternative models often better match empirical observations and have higher predictive power than models based purely on neoclassical assumptions.

While traditional economics assumes that individuals always behave rationally, behavioural economists often stress the 'irrational' aspect of decision making, often referred to as 'behavioural failures'. These behavioural failures may make individuals act against their own long-term interest. Thaler and Sunstein (2008) argue that if individuals do not always choose what is best for them in the long run, it is welfare-enhancing for policy makers to ensure that the set of choices that individuals face is such that a long-term, welfare-maximizing outcome becomes more likely. This may be done through proper framing choices, setting appropriate (limited if necessary) choice sets and providing appropriate 'default options'. In essence, individuals are 'nudged' towards a welfare-maximizing outcome, even if their freedom to choose is still respected. Thaler and Sunstein (2003) called this approach 'libertarian paternalism'. Libertarian paternalism would argue for policy interventions in the face of behavioural failures, even if market failures are absent.

Traditionally, economics has focused on how changes in prices affect behaviour. Research in behavioural economics and psychology has demonstrated that non-pecuniary interventions compare favourably to monetary interventions in changing consumer behaviour. It was also shown that judiciously applied pecuniary interventions increase the impact of monetary interventions if used in combination. This has increased interest in research in behavioural economics as a guide for policy making in areas as diverse as public health, finance and law. That behavioural economics can inform decision making in energy policy has increasingly been recognized by policy makers and researchers (Allcott and Mullainathan, 2010; DEFRA, 2010; OFGEM, 2011).

In order to realize energy savings and emissions reductions necessary to address climate change, decision makers have to consider tapping into behavioural transformation strategies. Behavioural economics provides insights that can inform this effort. Behaviours that are relevant to household energy consumption encompass three broad areas: (1) energy consumption, curtailment and habits; (2) energy-efficiency investments; and (3) contribution to public goods (i.e. green energy) and pro-environmental behaviour. These three aspects of energy consumption are interrelated; for example, pro-environmental attitudes may make efficiency investments more likely, and these investments may reduce energy consumption in the long run. However, these topics differ in terms of the decision making and behaviours involved, and warrant separate reviews.

The rest of this chapter is organized as follows. Section 2 presents the major concepts that distinguish behavioural economics from neoclassical economics. Sections 3, 4 and 5 form the heart of the chapter and discuss how behavioural economics relates to energy consumption and policy. Three broad areas related to energy are discussed: energy consumption, curtailment and habits (Section 3); energy-efficiency investments and purchases (Section 4); pro-environmental behaviour and public goods (Section 5). Section 6 provides concluding remarks.

2.   BEHAVIOURAL ECONOMICS VERSUS NEOCLASSICAL ECONOMICS

The main departures from neoclassical economics proposed by behavioural economics can be grouped under four main areas: (1) time-varying discount rates, (2) prospect theory and importance of reference points; (3) bounded rationality, and (4) pro-social behaviour and fairness. Below we briefly discuss each of these areas.

2.1   Time-varying Discount Rates

Experiments show that individuals use a higher discount rate over a longer time horizon than over a shorter time horizon (Thaler, 1981; Benzion et al., 1989; Holcomb and Nelson, 1992). To deal with this apparent anomaly, behavioural economics proposes hyperbolic discounting. Under hyperbolic discounting, individuals have higher discount rates for short horizons, but low discount rates for long horizons (Laibson, 1997). This implies that people will be far-sighted when planning if both costs and benefits occur in the future. However, they will make short-sighted decisions if costs or benefits are immediate (Camerer and Loewenstein, 2004). Some of the manifestations of the time-inconsistent preferences are inability to lose weight, stop smoking and save enough for retirement (Wilkinson, 2007).

If individuals have time-varying discount rates, at some point in the future their preferences change. Preferences between two future rewards can reverse in favour of a more proximate reward, if the time to both rewards diminishes. An individual may prefer $110 in 31 days over $100 in 30 days, but prefer $100 now over $110 tomorrow (Frederick et al., 2004). This is inconsistent with exponential discounting used by neoclassical economics in expected utility models. If agents were to discount future utilities exponentially, time preferences would not reverse, because the delay of 30 days is shared between the two options. Time-varying discount rates could explain the tendency to procrastinate. When given a choice between performing 5 hours of an unpleasant task today and 5.5 hours of an unpleasant task tomorrow, most people choose the second option and delay the task. On the other hand, when given a choice of 5 hours of unpleasant work in a month versus 5.5 hours of unpleasant work in a month and a day, most people take the first option. However, if an individual decides to do the work in a month, when the day comes, he or she would again prefer to delay it until the next day (O'Donoghue and Rabin, 2000).

Some individuals may be aware of their tendency to procrastinate, and may value the opportunity to make a commitment. The fact that people value commitment devices has been demonstrated empirically. Ashraf et al. (2006) show that when given the choice of depositing money in a savings account from which they can draw freely and a bank account that pays the same interest rate but restricts when funds can be withdrawn, some individuals choose the latter option. Individuals who chose commitment saving accounts increased saving rates by 82 per cent. Traditional economics finds it hard to explain why some people would choose an illiquid asset over a liquid asset, even if they pay the same interest rate.

2.2   Prospect Theory and Importance of Reference Points

In standard economic theory, an individual's preferences among different commodity bundles depend on wealth and prices, but are independent from the composition of their current endowment (assets) or their current consumption. Prospect theory, developed by Kahneman and Tversky, states that welfare changes should be evaluated according to certain reference points (Kahneman and Tversky, 1979). The following are some of the manifestations of the important reference points for welfare evaluation.

Loss aversion
Traditional economics assumes that individuals are risk-averse or risk-neutral but place the same value on losses and gains of equal amount. Kahneman and Tversky (1979) argue that valuation of losses is the mirror image of valuation of gains and refer to this phenomenon as the 'reflection effect'. Decision making will exhibit the reflection effect when the individual is risk-averse in the face of potential gains, but risk-seeking in the face of potential loss. It has been demonstrated empirically that individuals tend to value losses more than gains. This is found in contingent valuation studies that show that willingness to accept (WTA) is typically higher than willingness to pay (WTP) (Shogren and Taylor, 2008). Shefrin and Statman (1985) demonstrated that investors hold on too long to the stocks that lost value, but are eager to sell the stocks that gained in value, and argue that this is due to reluctance to sell at a loss.

Endowment effect
This refers to the extra value that individuals attach to goods they already own or services they already receive. In essence, the endowment point is the reference point, and agents have a kink in the valuation around this point (Thaler, 1980). Heberlein and Bishop (1986) found that hunters were willing to pay $31 for a particular hunting permit but were not willing to let go of the same permit for less than $143.

Status-quo bias
Individuals tend to stick to the default option chosen for them. For example, in countries where organ donation is conducted under presumed consent (i.e. consent by default, unless explicit opposition was registered by donor), participation rates are 25–30 per cent higher than in countries where donation is conducted under informed consent (i.e. no consent is presumed unless it was made explicit) (Abadie and Gay, 2006; Johnson and Goldstein, 2003). Samuelson and Zeckhauser (1988) showed that when new healthcare options were offered to Harvard University faculty, new faculty members were more likely to choose them, but older faculty members were unlikely to modify their current plans.

2.3   Bounded Rationality

Bounded rationality refers to the phenomenon that agents are rational but have cognitive constraints in processing information (Simon, 1986). Therefore they deviate from rationality in certain circumstances. Some of the manifestations of bounded rationality are the following: (1) choice overload; (2) heuristic decision making; and (3) failure to assess statistical probabilities.

'Choice overload' refers to the difficulty individuals have in making a choice when presented with too many options. Studies show that more shoppers make a purchase of a jam when they are presented with six choices than when they are presented with 24 choices (Iyngar and Lepper, 2000). Traditional economics struggles to explain this tendency, and assumes that more choices are always preferred to fewer choices.

'Heuristics' are shortcuts to decision making, such as via a rule of thumb. Traditional economics assumes that individuals use concepts of statistical sampling and statistical rules (e.g. Bayes's rule) for updating probabilities of future events in the face of new evidence (Camerer and Loewenstein, 2004). Experiments demonstrate that individuals often make choices in a way that departs from the Bayesian assessment that they are supposed to make under traditional economics. This departure may be systematic (biased) rather than idiosyncratic. Individuals may categorize purchases under different categories and have different discount rates for these categories. For instance, higher one-off expenditures may be categorized under a separate mental account and have different discount rates than multiple smaller expenditures. Thaler (1999) showed that an individual's willingness to spend earned income, windfall income or saved income is not the same, even if money can be used interchangeably. This contradicts the assumption of traditional economics that money is fungible. Heath and Soll (1996) suggest that mental accounting can explain why individuals make apparently suboptimal consumption choices.

There is some indication that consumers do not correctly process statistical information and probabilities. They are swayed by vivid and salient information more than by simply convincing, statistically correct information. Research shows that individuals overstate small probabilities of catastrophic losses or large gains. Julien and Salanie (2000) find that on horseraces, there is bias towards betting on 'longshots', implying that gamblers like to gamble, but they are disproportionally afraid of small chances of losing when they bet on heavy 'favourites'. Cook and Clotfelter (1993) find that lotteries are popular because players are more sensitive to large jackpots than to the probability of winning.

2.4   Pro-social Behaviour and Fairness

Neoclassical economics assumes that an agent makes choices that depend only on his or her own monetary payoff and consumption (Pesendorfer, 2006). However, in experimental games and empirical studies, it has been demonstrated that individuals seem to value fairness and often act pro-socially. Kahneman et al. (1986) show that consumers have strong feelings about the fairness of a firm's short-run price decisions, and suggest that this prevents firms from exploiting their full monopoly power. Individuals often act pro-socially, contribute to charities, and engage in pro-environmental behaviour, even if this imposes costs on them. Behavioural economics challenges the view that economic agents are purely selfish.

If individuals are not as selfish as traditional economics has assumed, this unselfishness has important implications for understanding the private provision of public goods. According to neoclassical economics, individuals care only about their own consumption of public goods but do not directly benefit from their own contribution, nor are they directly affected by other people's consumption or contribution (Bernheim and Rangel, 2007). Traditional economics claims that people will have a tendency to free-ride, and that public goods will be underprovided unless provisions are mandated through taxation. Furthermore, only the very wealthy are predicted to make voluntary contributions, and as population increases contributions should converge to zero. These assumptions provide testable hypotheses that are contradicted by empirical evidence (Andreoni, 2006; Bernheim and Rangel, 2007). Behavioural economics provides an alternative view to help explain why and when individuals make private provision of public goods.

Two behavioural explanations relate to people's attitudes towards providing for public goods. First, individuals are not purely selfish, but place value on social goods. In essence, they value not only their own consumption, but also the consumption of others (they are other-regarding). Public goods may still be underprovided, but this is because individuals think it is not fair that they bear the burden for their provision at the expense of others. Ostrom (1998) finds that in public-goods games (i.e. experiments where individuals are given the opportunity to contribute to the provision of a public good) most individuals are conditional cooperators – they will contribute if they are sure that others will do the same. Second, individuals contribute to the social good because of the 'warm-glow' effect. Warm-glow effect refers to the idea that individuals might participate in public goods (such as a green electricity programme) because it makes them feel good (either because they feel better about themselves, or because they care about what others think of them), but not necessarily because they care about the public benefit *per se* (Andreoni, 1990; Bernheim and Rangel, 2007). In essence, besides valuing their own consumption of public goods, individuals value having contributed to public-good provision.

The motivations for providing for public goods are hard to test empirically. As a result, 'warm glow' is often treated as a reduced form of deeper underlying processes (social norms, social signalling, reciprocity, altruism etc.). Disentangling these processes is challenging and their representation may vary among different policies and contexts (Bernheim and Rangel, 2007). Nevertheless, different underlying motivations may have different policy implications. If an individual would provide for public goods only with the assurance that others do not free-ride (i.e. showing a concern for fairness), then increased contribution by others will increase the likelihood of his or her contribution. If, on the other hand, an individual contributes as a status symbol (e.g. because of a desire for a warm glow or prestige), then provision by others may reduce the likelihood of his or her contribution. In this case, a monetary incentive for contributing may crowd out the altruistic incentive to contribute.

Researchers indeed find that monetary rewards sometimes crowd out intrinsic motivation, especially if the monetary rewards are small. For example, when a small monetary payment was offered for blood donation, blood donations actually decreased (Titmuss, 1987[1971]). It was also found that volunteers perform better when they are not compensated than when they receive small monetary compensations (Gneezy and Rustichini, 2000). These findings are hard to reconcile with traditional economics, but can be explained by behavioural economics, as they take away from the warm-glow satisfaction of giving.

3. ENERGY CONSUMPTION, CURTAILMENT AND HABITS

In this section we discuss topics that involve repetitive or continuous efforts to reduce energy consumption or to change energy-use habits. The policy interventions discussed in this section focus primarily on promoting energy curtailment in the household energy sector. In addition, we discuss how changing the tariff or billing structure affects energy consumption patterns.

3.1 Rate Structure: Flat versus Dynamic Tariffs

The marginal cost of producing electricity varies through the day, and the wholesale electricity price is high during the time of peak electricity usage. Traditionally, residential electricity customers faced flat electricity tariffs, and were insulated from fluctuating wholesale electricity prices through the day. If electricity demand becomes 'flatter', utilities will save on energy costs by minimizing usage of peaking plants, which are usually less efficient and produce more carbon emissions. Further down the line, having a smoother demand will result in reduced investments for building peaking plants. Cost savings can eventually be recycled back to the consumers.

Recently, policy makers and utilities have been looking at the potential of introducing tariffs that vary by the time of usage. Some possible variable tariffs are the following: (1) time-of-use tariffs (TOU), when customers face different tariffs according to the time of day; (2) critical peak pricing (CPP), when customers face higher tariffs during certain critical peak times through the year; (3) peak-time rebate, when customers pay flat electricity tariffs but receive rebates if the electricity usage is reduced compared to a certain benchmark at critical times through the year; (4) real-time pricing, when customers' electricity tariffs fluctuate in real time according to the wholesale electricity prices.

Time-varying tariffs require advanced (smart) meters that can measure consumption in real time. The UK, Italy and the state of California in the USA have legislated large-scale deployment of smart meters, with other jurisdictions planning to follow their lead (Faruqui and Sergici, 2009). Extensive research is under way by policy makers, utilities and academics to gauge the potential for demand-side response (DSR) to time-varying tariffs. The value of DSR hinges on the potential of customer behavioural response and is an area that can draw important insights from behavioural economics. Below are some of the concepts from behavioural economics that are relevant to the decision to implement variable rates.

- *Endowment effect* Bill payers currently enjoy the benefit of being insulated from variable rates during the day. Proper design and marketing of the dynamic tariffs will be critical for overcoming consumers' resistance to changing the cost–benefit structure of the way they consume electricity. Individuals are attached to their routines and daily habits and may be inflexible to modify them, or demand high compensation to do so.
- *Status-quo bias* Research shows that when presented with a utility bill with a default choice, most consumers will not change it (Brennan, 2006). Those who object to having the dynamic tariff either as a default or mandatory option mention distributional considerations. They argue that most households will remain on the

default plan even if it is not optimal for their consumption patterns. Vulnerable households, such as the elderly and disabled, will not be able to vary their load and will be losers under the dynamic tariffs if that is set as a default (Felder, 2010).

● *Time-varying discount rates*   Introducing dynamic tariffs raises concerns about short-term cost versus the 'lag' in long-term gain (Simshauser and Downer, 2011). Dynamic pricing will result in a 'rate shock', as bills of some consumers will sky-rocket in the near term, before behavioural adjustments, or before households acquire enabling technologies or replace old appliances with ones that better accommodate varying tariffs. Even if the long-term costs of smart meter infra-structure proves to be beneficial, the long term may be really long (Hanser, 2010). Since individuals tend to have higher discount rates for the future, they may not think that the costs are worth the benefits, especially if the savings are initially small or nil.

● *Loss aversion*   If individuals value (negatively) losses more than they value gains, rate increases during peak periods may have to be compensated with larger rate decreases during off-peak periods.

● *Concern for fairness*   Opponents of mandatory dynamic tariffs cite fairness considerations towards the vulnerable. It is argued that vulnerable households (elderly, disabled and poor) will not be able to shift consumption to off peak, since they have minimal electricity consumption to begin with and are often homebound. On the other hand, proponents for the dynamic tariffs state that it is not fair that 'peaky' households are being subsidized by 'less peaky' households through flat tariffs (Faruqui, 2010).

High electricity bills are salient and raise vociferous opposition that results in a media outcry. Moving customers *en masse* to dynamic prices may bring adverse consequences and may result in a strong political backlash. Alexander (2010) gives examples of large-scale time-of-use (TOU) tariff rollouts in the USA that did not meet expectations:

● Central Maine Power Company had implemented a mandatory TOU structure in the 1980s. However, in the 1990s, the TOU rate structure was changed to reflect higher peak electricity costs, but the increased bills caused consumer opposition, and TOU rates were made voluntary.

● Puget Sound Energy of Washington State implemented mandatory TOU pricing for residential customers in 2001. However, the programme actually resulted in higher bills under the new rate structure. The programme was halted in 2002.

Alexander (2010) suggests that having a peak-time rebate (PTR) is a more attractive option compared to mandatory CPP or TOU. A PTR scheme would leave the under-lying rate structure unchanged, but provide rebates or credit to those customers who reduce usage during critical peak hours. A rebate option should be viewed as 'carrot only' rather than 'stick only' (Maine Public Utilities Commission, 2007; Alexander, 2010). With CPP, 'peaky' customers will see their rates increase if they do not change behaviour. If, as suggested by prospect theory, consumers focus more on the downside risk of higher bills than the upside potential, they will dislike CPP. More consumers would choose to take advantage of PTR than volunteer for CPP. Letzler (2007) argues that an incentive-compatible rebate addresses the heuristics of consumer decision making better than CPP.

In 2008, Baltimore Gas & Electric (BGE) conducted a dynamic pricing pilot which showed that consumers responded to PTR as well as to CPP. As a result, BGE abandoned CPP and conducted a trial of only PTR in 2009. Faruqui and Sergici (2009) review 15 pilot experiments with dynamic pricing of electricity. Across the range of experiments studied, TOU rates induced a drop in peak demand that ranged between 3 per cent and 6 per cent, and CPP induced a drop in peak demand that ranged between 13 per cent and 20 per cent. When combined, the drop in peak demand was in the 27 per cent to 44 per cent range.

3.2   Billing and Payment Methods

How customers pay their utility bills may have implications for how they consume energy. This was demonstrated by a study of consumption and meter top-up behaviour of the households in Northern Ireland that use prepayment meters (Brutscher, 2011a, 2011b). Brutscher (2011a) shows that consumers with prepayment meters tend to consume more electricity. Households tend to purchase relatively small amounts of top-ups, and adjust to increases in tariffs by increasing their number of top-ups, rather than by increasing the amount. However, exogenous increases in minimum top-up amount result in decreased energy use. This suggests that consumers perceive costs differently according to how large they are. They have different mental accounts for larger purchases, and are more aware of the consumption after they have made a large top up. Increasing minimum top-up amount would therefore probably result in decreased energy consumption.

Brutscher (2011b) finds that low-income households use electric heat rather than oil because they have liquidity constraints – heating oil requires bulk purchase, whereas electricity meters can be prepaid in small amounts. As oil heating is more efficient and cheaper in the long run, this could be explained by time-varying discount rates, which prevent individuals from saving for bulk heating oil purchase. If the bulk purchase necessary for heating oil involves saving money for a future large purchase, the money being saved will be subject to temptation to make alternative purchases with more immediate gratification. If individuals have higher discount rates for the far future than for the near future, they will repeatedly succumb to temptation, even if they are aware that saving money will buy them a more efficient source of fuel in the long run. A heat stamp programme (where consumers buy non-fungible credits towards future purchases of oil) is a potential solution for this behavioural failure. Heat stamp programmes, currently operating in various communities in Northern Ireland, let consumers gradually collect stamps that they can redeem for the bulk oil purchase.

3.3   Non-pecuniary Incentives to Conserve Energy

Non-pecuniary interventions have been attempted to elicit reductions in energy consumption, often in combination with monetary incentives and information provision. Where interventions were combined with monetary incentives, consumption feedback, or energy-saving information, it is hard to disentangle the effects of these interventions.

Competition has been effective in incentivizing individuals to reduce energy consumption. McClelland and Cook (1980) studied the effect of competition between master-metered residential buildings at the University of Colorado, USA. The buildings, where occupants were not individually metered, were competing on which building would save more electricity. Contestants received information on how to save electricity and feedback on savings of their usage, as well as the usage of the other groups. The winning building received a reward of $80. The contest groups used 6.6 per cent less electricity than control groups. However, the savings decreased with time, suggesting that the effect of the reward was short-lived. Four buildings and 228 families participated in the study.

Pallak and Cummings (1976) studied whether they could induce reduction in energy consumption through soliciting public commitment. The study was carried out in Iowa City, USA. People who signed a public commitment showed lower rates of increase in gas and electricity use than those who signed a private commitment or those in the control group.

Energy savings can also be motivated by assisting consumers with goal setting. Becker (1978) gave households a relatively difficult goal (20 per cent) or a relatively easy goal (2 per cent) to reduce electricity use. All households received information on which appliances used more electricity, but only some households received consumption feedback. Only the households that had the difficult goal and received feedback had a significant change in electricity consumption (15 per cent savings). This study involved 100 families who lived in identical townhouses in central New Jersey, USA.

Most of the research conducted on non-pecuniary incentives has involved small samples, and it is not clear if these interventions are scalable. Most of the studies do not monitor interventions for a prolonged period of time, and it is not certain if habits were changed or behaviours eventually returned to pre-intervention norms. Where follow-up studies were conducted, it was typically found that the behavioural changes were not sustained. It is worth noting that in the Former Soviet Union countries, where price signals were ideologically frowned upon, prizes were used extensively to promote energy saving via goal setting, competition and public appeals. However, anecdotal evidence suggests that these incentives, even if initially heeded, became eventually ineffective where price signals were absent.

3.4 Influence of Social Norms on Household Energy Consumption

Social norms affect individual actions through providing guidelines as to what is acceptable or 'normal' behaviour. Some behavioural interventions aim to influence consumer energy consumption through increasing awareness of social norms. A number of studies have attempted to change energy consumption of households through providing them with the consumption information of their peers, as an indicator of social norms.

Nolan et al. (2008) left door hangers at 271 homes in San Marcos, California, USA, with different, randomly assigned energy conservation messages. Door hangers that compared a given household's energy demand to that of their neighbours led to 10 per cent more energy demand reduction than door hangers that gave only energy conservation tips. Schultz et al. (2007) left door hangers in 286 homes in the same city. Residents who had lower energy consumption than average increased consumption (the 'boomerang effect'). However, this effect was eliminated when a smiley face was drawn next to their energy consumption. The author postulates that the smiley face was interpreted as a normative signal and resulted in behavioural change.

While the above studies were based on small sample sizes, their findings were consistent with the results of a programme run by OPOWER, one of the largest randomized field experiments in history. OPOWER mailed home energy report letters to customers, comparing their energy usage to that of their neighbours. These letters also gave customers energy conservation tips. OPOWER ran a programme for 23 utilities, including six of the largest ten utilities in the USA and 600 000 households. The study found that the intervention reduced average energy demand by 1.11 per cent to 2.78 per cent from the baseline usage (Allcott and Mullainathan, 2010).

It is not clear if the behavioural changes resulting from these interventions can be sustained in the long run, or if the novelty of the social comparison would eventually wear off. Costa and Kahn (2010) analysed the OPOWER data and found, in fact, that the programme effects were heterogeneous: while the electricity conservation 'nudge' of providing feedback to households on their own and peers' home electricity usage works with political liberals, it backfires with political conservatives. The large-scale OPOWER experiment gained considerable media publicity and was hailed by leading policy makers as a testament that behavioural economics should motivate viable policy alternatives. However, Loewenstein and Ubel (2010) cautioned against being overly reliant on these types of interventions, pointing out that the energy savings they generated were very small. They stated that traditional mechanisms, such as a carbon tax, would be far more effective even if politically more difficult to implement, as they would increase the price of carbon in line with its true cost.

3.5   Influence of Information Provision

Providing energy-saving information and energy-consumption feedback is successful in eliciting behavioural changes. The residential electricity market has traditionally suffered from asymmetric information. Traditional electricity meters provide cumulative consumption information and individuals do not always know which appliances consume most electricity and when. New technologies, such as smart appliances and smart meters, provide innovative ways to access consumption information. Having access to disaggregated consumption information through a variety of media (i.e. the Internet) makes electricity consumption more tractable and easier to manage. Information asymmetry is typically assumed to be a market failure, and is studied under the paradigm of traditional economics. However, behavioural economics finds that not only is the information important, but so also the way it is presented or framed.

If the communication of information takes into account the behavioural failures and heuristic decision making of consumers, messages can be crafted to solicit a sharper behavioural response. Since individuals are affected more by salient information rather than simply accurate information, then visual cues and vivid descriptions are important. For example, Thaler and Sunstein (2008) found that when the energy company in Southern California gave its customers an 'Ambient Orb' that glowed in red when energy consumption was high (salient signal), orb users reduced peak energy demand by 40 per cent. With large-scale deployment of new technologies that provide innovative ways of communicating information, providing appropriate vivid cues will become increasingly important. If consumers are affected by losses more than by gains, the effective message should stress money wasted by missing the opportunity to save energy rather than emphasizing energy-saving behavioural change. The former formulation provides the same information but is more effective, although less welcomed, especially by elderly and vulnerable customers.

The credibility and trustworthiness of information sources makes a difference. Craig and McCann (1978) showed that when consumers received identical letters giving energy conservation advice but on different letterheads, the letter from the local energy commission had higher impact than the letter from the local utility. In communication, simple, salient and personally relevant information is more effective than detailed, technical and factual information (Wilson and Dowlatabadi, 2007).

3.6   Choice of Electricity Suppliers or Tariffs

The unregulated monopolist does not have the incentive to keep prices down. Competition between suppliers results in lower prices if the buyers are able to shop for the best deal and change suppliers. However, there are indications that when presented with many suppliers, consumers do not switch. If consumers do not shop around for the best suppliers, opening markets that were formerly regulated can reduce welfare if the incumbent supplier increases prices above those that were set by regulators (Brennan, 2006). Behavioural economics explains this phenomenon by status-quo bias, and/or information overload.

Traditional economics often assumes that information is costless and freely available. But when allowing for cost of searching and obtaining information, traditional economics would explain consumers' unwillingness to choose suppliers by the fact that search is costly (electricity bills are notoriously hard to read) and that the product (electricity) is homogeneous. Wilson and Waddams Price (2005) provide the behavioural explanation that consumers are irrational and fail to switch. This has implications for the offering of more choices to energy consumers, which might be facilitated by the deployment of smart meters. Having more choices implies more opportunities for mistakes to be made by customers by selecting inappropriate suppliers for their needs. It may also reduce switching from incumbents due to information overload.

4.   ENERGY-EFFICIENT INVESTMENTS AND PURCHASES

Increasing energy-efficiency can play a significant role in reducing overall energy consumption and associated emissions. Efficiency investments involve one-time, large monetary costs but result in cost savings over the long run through lower energy consumption. It has been shown that efficiency improvements could result in substantial long-run cost savings. However the 'energy-efficiency gap' puzzle remains. The gap normally refers to the difference between the observed level of energy-efficiency and what is considered optimal energy use (Jaffe et al., 2004; Gillingham et al., 2006, 2009). Sometimes the energy-efficiency gap is illustrated by comparing the market discount rate and the 'implicit discount rate' imputed from appliance purchase decisions, taking into account cost of appliances and their energy-efficiency (Hausman, 1979). Studies have shown that the implicit discount rate is between 25 and 100 per cent (Sanstad et al., 2006; Train, 1985). Below we discuss behavioural explanations of the energy-efficiency gap and effectiveness of policy interventions that aim to tackle it.

4.1 Behavioural Explanations for the Energy-efficiency Gap

It is not costless to assess how a new technology fits into one's home, or to find a reliable supplier and installer. Furthermore, future energy prices and future savings are uncertain. The transaction costs of new technology adoption can still be significant, so that the 'purchase price' is only the lower bound of adoption cost (Jaffe and Stavins, 1994). Below we discuss some of the behavioural explanations of the efficiency puzzle.

- *Time inconsistency*   Individuals have a high discount rate for future cost savings, but a small discount rate for large initial investment outlay. Alternatively, individuals may want to invest in energy-efficiency but are procrastinating, or do not have the discipline to save money to pay for the initial investment.
- *Endowment effect*   Households are attached to the appliances they currently own, and are not willing to replace them, even if it is efficient to do so. This can partly be rationalized by the certainty that the new equipment works against the risk of problems in installation or operation.
- *Salience*   Yates and Aronson (1983) suggest that individuals place disproportional weight on vivid and observable factors. This tendency may result in placing too much emphasis on initial investment costs, and underinvestment in energy-efficiency. They recommend giving salient examples of energy savings and state that energy-saving advice should demonstrate the experience of a 'highly efficient household' for the message to be better retained.
- *Heuristics*   Kempton and Montgomery (1982) use a survey and find that consumers use simple heuristics to assess their energy consumption, which leads to systematic underinvestment in energy-efficiency. They conclude that this is the example of bounded rationality, when people adapt known methods to solving new problems, even if the known methods are not optimal for the new situation. This may be the optimal strategy, as it avoids the effort of analysing the new situation (Simon, 1955). For example, families used current energy prices to calculate expected savings from efficiency investment, thus not taking into account future price increases. When comparing consumption over the years, households compared their highest utility bills to estimate their consumption totals. Kempton et al. (1992) show that consumers systematically miscalculate payback periods for air conditioners, and this leads to overconsumption of energy.

4.2 Incentives for Energy-efficient Investments

Since energy-efficiency investments are subject to market and behavioural failures, policy makers and utilities have devised various incentives to overcome them. Behavioural economics can shed light on which efficiency-promoting incentives are more effective. Since high upfront costs appear to discourage efficiency investments, policy makers often attempt to incentivize these investments through tax credits or efficiency programmes administered by utilities.

Income tax credits or deductions have been used in the USA as an instrument to encourage energy-efficient investments. However, evidence of their effectiveness is mixed. The US Energy Tax Act of 1978 (ETA78) provided a federal tax credit for residential energy-efficiency investments and encouraged investment in solar, wind and geothermal energy technologies. Carpenter and Chester (1988) conducted a survey with over 5000 respondents from the USA. Around 89 per cent of the respondents were aware of ETA78's federal tax credit, but only around one-third of them filed this claim. Out of those who did file the claim, 94 per cent were going to make the investment even without tax incentives (free-riding). However, Hassett and Metcalf (1995) found that tax credits do encourage efficiency investments – a change of 10 percentage points in the tax price of the investment increases probability of investment by 24 per cent. Williams and Poyer (1996) found that tax credits have a significant effect in improving energy conservation, even with free-riders. They suggest that this may be due to spillovers, as some households make the investment because of the tax credit, but then fail to file for the credit when it comes time to file their taxes. This explanation is consistent with time-varying discount rates and tendency to procrastinate.

In summer 2011, the UK government ran a pilot trial of the 'Green Deal' – a scheme to encourage homeowners to upgrade their buildings by installing energy-saving improvements at no upfront cost. Repayments for these investments would be made via a charge paid from savings made on a customer's energy bills. The trial took place in the London borough of Sutton, and involved 400 households that responded to the advertisement. Of the 126 households that eventually received home energy audits, only 60 signed up for the scheme, even if the subsidy represented 40 per cent. The households that did sign up for the scheme indicated that financial incentives were not the primary motivation (BioRegional, 2011).

Studies show that even when a utility offered to subsidize 93 per cent of the cost of home insulation, consumer take-up varied from 1 per cent to 20 per cent, depending how the subsidy was communicated to the consumer (Stern et al., 1985). Stern (2000) suggests that incentives and interventions interact, and the joint effect of combining them is often bigger than the sum of each intervention on its own.

4.3   Appliance Standards and Building Codes

If consumers indeed make biased decisions in their efficiency investment, and they do not take operating cost into consideration, then appliance standards may be welfare enhancing. Mandatory appliance standards can define the minimum energy-efficiency standard that will be required by law for a given appliance. Building codes would define certain minimum efficiency characteristics that buildings should have. In essence, consumers will not be able to choose inefficient appliances or homes, as they will not be on the market. Mandatory standards and codes will encourage manufacturers to provide better energy-efficiency in the context where it is not the most salient feature for consumers (DEFRA, 2010).

Koomey et al. (1999) found that each dollar of US federal expenditure on implementing the appliance energy-efficiency standards contributed $165 of net present-valued savings to the US economy over the 1990 to 2010 period. They estimated the average benefit–cost ratios for these efficiency standards to be about 3.5 for the USA as a whole. However, households that are low appliance users may prefer less efficient appliances, and suffer welfare losses when standards are imposed (Morss, 1989).

5.   PUBLIC GOODS AND PRO-ENVIRONMENTAL BEHAVIOUR

Supporting green energy and combating global warming is a public good. Behavioural economics can help understand why and under which circumstances individuals are willing to contribute to these public goods voluntarily (Kotchen and Moore, 2004). These contributions may be monetary (i.e. when individuals pay a premium for green energy) or non-monetary (i.e. when individuals act pro-environmentally but sacrifice their comfort).

5.1   Provision Point Mechanisms and Green Energy

Behavioural economics claims that many individuals are 'conditional cooperators' and value fairness. Individuals would be willing to contribute to public goods if they knew that others did not free-ride and also contributed (this is what a tax system can achieve formally). In essence, the difference between a 'behavioural' explanation and a 'traditional' explanation is that under the former, individuals do not want to free-ride, they are just concerned that others do and think that this is not fair; under the latter, individuals have an innate tendency to free-ride, and that will be the case even if others contribute.

Moskovitz (1992, 1993a, 1993b) argued that customers would voluntarily sign up and pay higher electricity rates if the additional money collected were earmarked to support renewable energy projects and environmental activities. Since then, utilities in many jurisdictions have offered green energy tariffs. These green tariffs represent a contribution to public good. If individuals are indeed willing to make voluntary contributions, public policy may harness this tendency through devising mechanisms that make these contributions more likely.

Public economics has dealt with the problem of underprovision of public goods through such mechanisms as taxes and provision point mechanisms (PPMs). Under PPMs, individuals make voluntary contributions to a project, with the disclaimer that if the necessary benchmark amount is not collected, the contributions will be refunded (Steven et al., 2002). Rondeau et al. (2005) compared the voluntary contribution mechanism (VCM) with the PPM in the laboratory and in a small field experiment, and found that the PPM was able to achieve higher contributions. Traverse City Light and Power in Michigan, USA, successfully built a windmill using this mechanism (Holt, 1996a); the city of Fort Collins in Colorado raised money for three separate wind turbines using it (Holt, 1996b).

Rose et al. (2002) used laboratory and field experiments to test the use of a PPM to finance a renewable energy programme run by Niagara Mohawk Power Corporation in the USA. In the laboratory experiment, the PPM increased the rate of participation in a green energy programme substantially above that of a treatment group. When the PPM was tested in the field, sign-up rates observed were much higher than those from other green-pricing programmes that solicit voluntary contributions.

One of the problems with selling green energy in some jurisdictions (such as the UK) is that of proving additionality. Fixed targets for given quantities of renewables set by the government mean that buying green electricity simply assigns renewable output that would have been forthcoming anyway to a given group of customers. This is because all customers will be made to pay for the renewable energy anyway, regardless of whether any make voluntary or conditional contributions.

5.2   Crowding out Intrinsic Motivation

Behavioural economics suggests that if individuals are motivated by the 'warm-glow' effect, then giving monetary incentives would reduce their motivation for contributing to public goods. On the other hand, if individuals pay a fine for behaviour that diminishes public goods, their intrinsic motivation for avoiding this behaviour may be reduced. This 'crowding-out hypothesis' is intellectually appealing, but so far few studies have demonstrated this effect empirically in the context of energy.

Jacobsen et al. (2010) use billing data of participants and non-participants in a green electricity programme in Memphis, Tennessee. They find that households participating at the minimum threshold level increase electricity consumption by 2.5 per cent after enrolling in the plan. They explain this with the 'buy-in' mentality of these households. A household's guilt of generating high emissions is reduced by buying into green electricity at the minimum threshold, and payment for green electricity crowds out their motivation to reduce energy consumption. However, the effect was not large enough to offset the environmental benefit of paying for green electricity. Therefore the net effect was a reduction in emissions.

5.3   Voluntary Contributions and Public Image

Individuals may be more likely to provide public goods if their contributions are 'publicly' acknowledged. Thus token gifts (i.e. pins, mugs and stickers) given in exchange for the contribution are one way to encourage provisions.

Yoeli (2010) conducted a large-scale experiment in collaboration with PG&E, a regulated investor-owned utility in Northern California. Consumers were sent letters inviting them to volunteer to install a device in their homes that would allow the utility to control their air conditioners when electricity supply was tight. Volunteers had to sign up for the programme on a sheet that was displayed publicly near mailboxes. The treatment group of households was requested to write their names on the sign-up sheet, while the control group of households provided a unique, anonymous numerical identifier. Households that had to provide full identity information had a higher sign-up rate (however, the difference was not statistically significant).

5.4   Public Appeals to Conserve Energy

When increasing prices is considered socially or politically unacceptable, governments have occasionally resorted to public appeals through the mass media to induce energy conservation. In the absence of price signals, traditional economics would not expect public appeals to change behaviour, because individuals would have already optimized their consumption choices for the given price (and would free-ride on others' contributions to the public good of avoiding blackouts). However, behavioural economics suggests that public appeals may result in increased awareness, and may induce altruistically motivated individuals to conserve more energy. In addition, some behavioural economists postulate that public appeals affect social norms.

Reiss and White (2008) used household-level data on energy conservation in California during the energy crisis of 2000–2001. Data consisted of a five-year panel of San Diego Gas & Electric Company households' utility bills. The sample consisted of 70 000 accounts. The prices increased sharply as a response to the crisis – electricity prices more than doubled in a span of three months. As a result, the average household electricity use fell more than 13 per cent in 60 days. Following the initial price increases, prices were rolled back and capped, and consumption rebounded to former levels. Subsequently, to avoid blackouts, the government used televised public appeals to urge households to reduce energy consumption. Public appeals were accompanied by energy-saving advice. Public appeals resulted in a 7 per cent decline in energy use over six months. It should be noted that the households in California had heightened awareness of the consequences of energy shortages. Electricity prices had doubled in their recent memory. Therefore increasing prices was a credible threat for the consumers, and may have made them more receptive to social appeals.

IEA (2005) provides some international case studies where public appeals were successful in reducing energy consumption quickly. Some of the successful solicitations to reduce energy occurred in 2001 in New Zealand, Australia and Brazil, as a result of shortfalls caused by drought. At the beginning of New Zealand's 2001 shortfall, the government calculated that blackouts could be avoided if everybody reduced their consumption 10 per cent for ten weeks. Thus '10 for 10' became the goal. The government distributed advice on how to obtain those savings but gave no incentives towards more efficient equipment or reduced bills. New Zealand employed public appeals through many extremely short reminders on television. The conservation goal was reached within six weeks.

IEA (2005) states that while a price increase is the first-best scenario to deal with shortfalls, it may not be politically feasible in the short run. In designing successful public appeals to reduce electricity quickly, it is important to educate consumers, raising their awareness, and making conservation a matter of civil duty or prestige. This is particularly important in electricity markets where richer, less price-sensitive consumers, by not reducing their consumption in a time of crisis, impose a negative consumption externality on the poor. This is in contrast to most markets where consumption of the rich creates a positive consumption externality for the poor by allowing economies of scale to be exploited in production and facilitating increased competition.

5.5 Public Policy Instruments and Environmental Morals

When economic activity has a negative externality, traditional economics suggests that the following instruments can be used to curb the extent of harmful behaviour: taxes, restrictions (i.e. quotas or bans) and tradable permits. Taxes will discourage the harmful behaviour through increasing its price, while quotas discourage the behaviour through restricting the quantity of the 'bad' allowed or available. Taxes levied on the activity that generates negative externalities are called Pigovian taxes, and have traditionally been applied to goods such as cigarettes and alcohol, or to activities such as pollution. With tradable permits, the overall allowable level of activity is established, and the permits for this activity are allocated to entities (usually firms). These allowances can be traded, so that the entities with the lowest marginal costs of reducing the activity towards the allowable level will be the ones that do so. Tradable allowances have been used to regulate fisheries and air pollution (Frey, 2005; Tietenberg, 2003). Traditional economics assumes that the difference among these three policy instruments is purely in terms of economic and administrative efficiency.

Psychologists argue that another important distinction between these instruments is that they vary to the extent that they send the signals that crowd out intrinsic motivation. Frey (1999) calls intrinsic motivation for pro-environmental activity 'environmental morale'. He argues that both tradable permits and taxes will have two opposing effects on consumers – an increase in price of activity will discourage behaviour, but intrinsic environmental morale will also go down as a result (crowding out). He argues that environmental moral will be reduced by tradable permits more than by taxes: tradable permits may be viewed as being similar to indulgences sold for sins in the Middle Ages (Goodin, 1994). These permits may convey the impression that it is acceptable to sin as long as one pays the price for it. Frey also proposes that both low and high environmental taxes are more effective than medium-level taxes. He argues that, with low taxes, consumers may feel that protecting the environment is something that has to be done from moral obligation. On the other hand, high environmental taxes make harmful behaviour prohibitively costly and dominate the crowding-out effect. Meanwhile medium environmental taxes result in crowding out intrinsic motivation, but are insufficient to cause reduction in behaviour due to extrinsic motivation.

With climate policy the negative activity to be curtailed is carbon emissions. Taxes, restrictions and tradable permits are used as climate policy instruments, often simultaneously. The EU has the largest tradable emissions scheme in the world (EU ETS). So far, tradable permits have been assigned to the businesses, but not to final consumers. On the other hand, consumers in many jurisdictions are subject to carbon prices directly (effectively a tax on emissions) and quotas indirectly (through emissions standards). Unfortunately, the evidence of crowding out environmental morale due to the signalling effect of taxes and tradable permits has not been evaluated empirically, but studies have established the existence of the phenomenon in the laboratory and field experiments (Frey and Jegen, 2001; Deci et al., 1999). However, its relevance to the actual behaviour of consumers, as well as the magnitude of the effect remains, is yet to be determined. Meanwhile, public perception of the effect of these instruments is relevant for the political economy of public policy. To the extent that the public (particularly proenvironmental activists) perceive that taxes and tradable permits are morally inferior because they seem to sanction pollution and emissions, government may find it politically harder to implement these mechanisms.

6.   CONCLUDING REMARKS

Economics studies how agents interact and allocate limited resources. In essence, economics is about behaviour by definition. The traditional economics discipline is based on behavioural assumptions and axioms which allow models to explain the key phenomena of interest parsimoniously but with sufficient clarity and accuracy. What is referred to as 'behavioural economics' is the modification of traditional assumptions by drawing insights from psychology. Normally, behavioural economists start from observations of how individuals actually behave and then show how this behaviour violates traditional assumptions. Researchers then proceed to offer models based on alternative assumptions that better match the observed phenomena. Behavioural economics is a growing and thriving field, but there are some theoretical and empirical gaps, many of which were mentioned in this survey. Below we discuss some of the future directions for behavioural economics in general, and its application to energy policy in particular. Technological innovations, such as smart meters and smart appliances (and, in the future, widespread use of electric vehicles), provide new ways to study how consumer behaviour responds to monetary and non-monetary interventions.

Theoretical and Empirical Work Aimed at Sorting out the Interactions of Various Behavioural Phenomena

Behavioural studies uncover anomalies in behaviour that are inconsistent with neoclassical economics. However, these anomalies are often studied on an *ad hoc*, case-by-case basis. Synthesizing behavioural anomalies within a consistent framework would be welcome. Some of the questions that need to be addressed are the following: How do we disentangle behavioural explanations from conventional information effects? What are the interactions between the alternative behavioural insights and between behavioural failures and market failures? For example, how do loss aversion and hyperbolic discount rates interact? Sorting out these interactions will be important in designing the optimal package of behaviour-change-inducing measures to reduce or shift energy consumption.

Increased Reliance on Empirical Research and Impact Evaluation

Much behavioural work has been based on experimental studies. However, ultimately economics is concerned about actual behaviour in markets, and some of the behavioural failures may no longer hold in a real market setting where behavioural anomalies cancel out. It is important to conduct empirical studies in order to uncover how relevant behavioural anomalies are to the way the markets work in non-ideal conditions. There seems to be good scope for linking experimental work with empirical trials to see the extent to which strong experimental results can be reproduced in fieldwork.

Study of Large-scale Interventions

Much behavioural research, in both experimental and field settings, has been carried out on small sample sizes. This raises two concerns: how scalable are the interventions and, even if behavioural anomalies do exist, do they represent a fundamental departure from the way individuals make choices, or is it only about the tails of the 'behavioural' distribution? If the observation is only regarding the tails of the distribution, how thick are the tails (Shogren and Taylor, 2008)? What is needed is much more attention to whole population interventions (i.e. the general rollout of smart meters) or natural experiments (such as the New Zealand's campaign for electricity demand reduction in 2001). Conducting randomized, controlled experiments involving large-scale interventions would help provide answers to these questions. Widespread use of smart meters will make this sort of research much easier to implement in the future.

Increased Study of the Relationship between Short-run and Long-run Behavioural Changes

While short-run behavioural changes may be extremely useful and important in energy, there are also issues to do with whether certain behavioural changes can be sustained over a prolonged period. With energy consumption this may be a consideration in designing the school curriculum with a view to increasing awareness of energy conservation, or in keeping effective repeated interventions aimed at dealing with successive short-run energy shortages. Most studies of energy consumption behaviour do not resample behaviour over a prolonged period, but clearly there may be opportunities to design studies to monitor the effects of long-term interventions in the future.

Integrating Behavioural Economics within the Framework of Traditional Economics

While behavioural economics provides insights about the way decisions are made, some of the departures from neoclassical economics are hard to reconcile within the traditional framework. For instance, how do we evaluate the impacts of policy if we can't make neoclassical assumptions about discounting? If individuals are really self-conflicted in their evaluation of costs and benefits over time, then what is the rationale for the policy makers to cater to their far-sighted selves rather than their near-sighted selves? While hyperbolic discounting is intellectually appealing, it makes equilibrium models intractable. Further theoretical work to reconcile these differences is needed. Institutional economics (following North, 1990) includes the study of the impact of long-run behavioural differences between economic agents. It may be that behavioural economics is a way of understanding how short-run behaviour can be changed, while institutional economics studies how long behavioural differences can be sustained via the creation of appropriate institutions. Thus understanding the process of institutionalizing behavioural change may allow a reconciliation of more traditional economics with behavioural economics. This may be important in suggesting ways in which differences in energy consumption per capita currently attributed to vaguely defined 'institutional' differences (such as between the UK and Denmark) can be narrowed.

Behavioural economics can provide valuable insights into how individuals make their decisions. These insights can be used to increase effectiveness of traditional interventions in energy policy. However, it is important that behavioural interventions do not crowd out more effective traditional interventions (Loewenstein and Ubel, 2010). Behavioural economics should complement, not substitute for, more substantive economic interventions, such as those based on influencing energy pricing (e.g. via taxation) or energy investment (e.g. via subsidy schemes).

To come back to where we began, behavioural economics seems unlikely to provide the magic bullet to reduce energy consumption by the magnitude required by the IEA (2010) recommended 450 climate policy scenario. However, it does offer exciting new suggestions as to where to start looking for potentially sustainable changes in energy consumption. It may also be that its most useful role within climate policy is in addressing issues of public perception of the affordability of climate policy and in facilitating the creation of a more responsive energy demand, better capable of responding to weather-induced changes in renewable electricity supply.

NOTE

* The authors wish to thank the EPSRC Flexnet project for financial support and Roger Fouquet for his encouragement to write the chapter. We acknowledge the helpful comments of Roger Fouquet and an anonymous referee. The usual disclaimer applies.

REFERENCES

Abadie, A. and Gay, S. (2006), 'The impact of presumed consent legislation on cadaveric organ donation: a cross country study', *Journal of Health Economics*, **25**(4): 559–620.
Alexander, B. (2010), 'Dynamic pricing? Not so fast! A residential consumer perspective', *The Electricity Journal*, **23**(6): 39–49.
Allcott, H. and Mullainathan, S. (2010), *Behavioral Science and Energy Policy*, Cambridge, MA: Ideas42. http://ideas42.iq.harvard.edu/publications/behavioral_science_energy_policy.
Andreoni, J. (1990), 'Impure altruism and donations to public goods: a theory of warm-glow giving', *Economic Journal*, **100**: 464–77.
Andreoni, J. (2006), 'Philantropy', in S.-C. Kolm and J. Ythier (eds), *Handbook of the Economics of Giving, Altruism, and Reciprocity*, vol. 2, Amsterdam: Elsevier, ch. 18.
Ashraf, N., Karlan, D. and Yin, W. (2006), 'Tying Oddyseus to the mast: evidence from a commitment savings product in the Philippines', *The Quarterly Journal of Economics*, **121**(2): 635–72.
Becker, L. (1978), 'Joint effect of feedback and goal setting on performance: a field study of residential energy conservation', *Journal of Applied Psychology*, **63**(4): 428–33.
Benzion U., Rapoport, A. and Yagil, J. (1989), 'Discount rates inferred from decisions: an experimental study', *Management Science*, **35**(3): 270–84.
Bernheim, D. and Rangel, A. (2007), 'Behavioral public economics: welfare policy analysis with nonstandard decision-makers', in P. Diamond and H. Vartiainen (eds), *Behavioral Economics and Its Applications*, Princeton, NJ: Princeton University Press, pp. 7–77.
BioRegional (2011), 'Helping to inform the Green Deal: green shoots from Pay as You Save', London: BioRegional Solutions for Responsibility.
Brennan, T. (2006), 'Consumer preference not to choose: methodological and policy implications', RFF Discussion Paper 05-51. Washington DC: Resources for the Future (RFF).
Brutscher, P. (2011a), 'Liquidity constraints and high electricity use', Working Paper EPRG1106, Cambridge, UK: Electricity Policy Working Group, Cambridge University.
Brutscher, P. (2011b), 'Payment matters? An exploratory study into pre-payment electricity metering', Working Paper EPRG1108, Cambridge, UK: Electricity Policy Working Group, Cambridge University.
Camerer, G. and Loewenstein, G. (2004), 'Behavioral economics: past, present, future', in C. Camerer, G. Loewenstein and M. Rabin (eds), *Advances in Behavioral Economics*, New York/Princeton, NJ: Sage/ Princeton University Press, pp. 3–51.
Carpenter E. and Chester, T. (1988), 'The impact of state tax credits and energy prices on adoption of solar energy systems', *Land Economics*, **64**(4): 347–55.

Cook, P. and Clotfelter, C. (1993), 'The peculiar scale economies of lotto', *American Economic Review*, **83**(3): 634–43.

Costa, D. and Kahn, M. (2010), 'Energy conservation "nudges" and environmentalist ideology: evidence from a randomized residential electricity field experiment', NBER Working Paper No. 15939, Cambridge, MA: National Bureau of Economic Research (NBER).

Craig, S. and McCann, J. (1978), 'Assessing communications effects on energy conservation', *Journal of Consumer Research*, **5**(2): 82–8.

Deci, E., Koestner, R. and Ryan, R.M. (1999), 'A meta-analytic review of experiments examining the effects of extrinsic rewards on intrinsic motivation', *Psychological Bulletin*, **125**: 627–68.

DEFRA (2010), *Behavioural Economics and Energy Using Products: Scoping Research on Discounting Behaviour and Consumer Reference Points*, London: Department for Environment, Food and Rural Affairs.

Faruqui, A. (2010), 'The ethics of dynamic pricing', *Electricity Journal*, **23**(6): 13–27.

Faruqui, A. and Sergici, S. (2009), 'Household response to dynamic pricing of electricity – a survey of the experimental evidence', *Journal of Regulatory Economics*, **38**(2): 193–225.

Felder, A. (2010), 'The practical equity implications of advanced metering infrastructure', *Electricity Journal*, **23**(6): 56–64.

Frederick, S., Loewenstein, G. and O'Donoghue, T. (2004), 'Time discounting and time preference: a critical review', in C. Camerer, G. Loewenstein and M. Rabin (eds), *Advances in Behavioral Economics*, New York/ Princeton, NJ: Sage/Princeton University Press, pp. 162–222.

Frey, B. (1999), 'Morality and rationality in environmental policy', *Journal of Consumer Policy*, **22**(4): 395–417.

Frey, B. and Jegen, R. (2001), 'Motivation crowding theory: a survey of empirical evidence', *Journal of Economic Surveys*, **15**: 589–611.

Frey, B. (2005), 'Excise taxes: economics, politics, and psychology', in S. Cnossen (ed.), *Theory and Practice of Excise Taxation: Smoking, Drinking, Gambling, Polluting, and Driving*, Oxford, UK: Oxford University Press, ch. 8.

Gillingham, K., Newell, G. and Palmer, K. (2006), 'Energy-efficiency policies: a retrospective examination', *Annual Review of Environment and Resources*, **31**: 161–92.

Gillingham, K., Newell, G. and Palmer, K. (2009), 'Energy-efficiency economics and policy', Discussion Paper 09-13, Washington, DC: Resources for the Future.

Gneezy, U. and Rustichini, A. (2000), 'Pay enough or don't pay at all', *Quarterly Journal of Economics*, **115**(3): 791–810.

Goodin, R. (1994), 'Selling environmental indulgences', *Kyklos*, **47**: 573–96.

Hanser, P. (2010), 'On dynamic prices: a clash of beliefs?', *The Electricity Journal*, **23**(6): 36–8.

Hassett, K. and Metcalf, G. (1995), 'Energy tax credits and residential conservation investment: evidence from panel data', *Journal of Public Economics*, **57**(2): 201–17.

Hausman, J. (1979), 'Individual discount rates and the purchase and utilization of energy-using durables', *Bell Journal of Economics*, **10**(1): 33–54.

Heath, C. and Soll, J. (1996), 'Mental budgeting and consumer decision', *Journal of Consumer Research: An Interdisciplinary Quarterly*, **23**(1): 40–52.

Heberlein, T. and Bishop, C. (1986), 'Assessing the validity of contingent valuation: three field experiments', *Science of the Total Environment*, **56**: 99–107.

Holcomb, J. and Nelson, P. (1992), 'Another experimental look at individual time preference', *Rationality and Society*, **4**: 199–220.

Holt, E. (1996a), 'Green pricing and restructuring', vol. 3. Montpelier, VT: The Regulatory Assistance Project. Search from http://www.raponline.org/search.

Holt, E. (1996b), 'Green pricing newsletter', vol. 4. Montpelier, VT: The Regulatory Assistance Project. Search from http://www.raponline.org/search.

IEA (2005), *Saving Electricity in a Hurry: Dealing with Temporary Shortfalls in Electricity Supplies*, Paris: International Energy Agency.

IEA (2010), *World Energy Outlook*, Paris: International Energy Agency.

Iyngar, S. and Lepper, M. (2000), 'When choice is demotivating: can one desire too much of a good thing?', *Journal of Personality and Social Psychology*, **79**(6): 995–1006.

Jacobsen, G., Kotchen, M. and Vanderbergh, M. (2010), 'The behavioural response to voluntary provision of an environmental public good: evidence from residential electricity demand', NBER Working Paper No. 16608, Cambridge, MA: National Bureau of Economic Research (NBER).

Jaffe, A. and Stavins, R. (1994), 'The energy paradox and the diffusion of conservation technology', *Resource and Energy Economics*, **16**(2): 91–122.

Jaffe, A. Newell, R. and Stavins, R. (2004), 'The economics of energy-efficiency', in C. Cleveland (ed.), *Encyclopedia of Energy*, Amsterdam: Elsevier, pp. 79–90.

Johnson, E. and Goldstein, D. (2003), 'Do defaults save lives?', *Science*, **302**: 1338–9.

Julien, B. and Salanie, B. (2000), 'Estimating preference under risk: the case of racetrack bettors', *Journal of Political Economy*, **108**(3): 503–30.

Kahneman, D. and Tversky, A. (1979), 'Prospect theory: an analysis of decision under risk', *Econometrica*, **47**(2): 263–91.

Kahneman, D., Knetsch, J. and Thaler, R. (1986), 'Fairness as a constraint in profit seeking: entitlement in the market', *American Economic Review*, **76**(4): 728–41.

Kempton, W. and Montgomery, L. (1982), 'Folk quantification of energy', *Energy*, **7**(10): 817–27.

Kempton, W., Feuermann, D. and McGarity, A. (1992), '"I always turn it on super": user decisions about when and how to operate room air conditioners', *Energy and Buildings*, **18**: 177–91.

Koomey, J., Mahler, S., Webber, C. and McMahon, J. (1999), 'Projected regional impacts of appliance efficiency standards for the US residential sector', *Energy*, **24**(1): 69–84.

Kotchen, M., and Moore, M., (2004), 'Private provision of environmental public goods: household participation in green-electricity programs', *Journal of Environmental Economics and Management*, **53**(1): 1–16.

Laibson, D. (1997), 'Golden eggs and hyperbolic discounting', *Quarterly Journal of Economics*, **112**(2): 443–77.

Letzler, R. (2007), 'Applying psychology to economic policy design: using incentive preserving rebates to increase acceptance of critical peak electricity pricing', Working Paper 162. Berkeley: Center for the Study of Energy Markets, University of California.

Loewenstein, G. and Ubel, P. (2010), 'Economics behaving badly', *The New York Times*, 14 July.

Maine Public Utilities Commission (2007), 'Rebuttal Testimony of Stephen S. George, Ph. D., Nov. 9, 2007', Maine Public Utilities Commission, Central Maine Power Company, Request for New Alternative Rate Plan, Docket No. 2007- 215.

McClelland, L. and Cook, S. (1980), 'Promoting energy conservation in master-metered apartments through group financial incentives', *Journal of Applied Social Psychology*, **10**(1): 20–31.

Morss, F. (1989), 'The incidence of welfare losses due to appliance efficiency standards', *The Energy Journal*, **10**(1): 111–18.

Moskovitz, D. (1992), 'Renewable energy: barriers and opportunities: walls and bridges', Report for the World Resources Institute.

Moskovitz, D. (1993a), 'Green pricing: customer choice moves beyond IRP', *The Electricity Journal*, **6**(8): 42–50.

Moskovitz, D. (1993b), 'Green pricing: experience and lessons learned', The Regulatory Assistance Project, Gardiner, ME.

Nolan, J., Schultz, P., Cialdini, R., Goldstein, N. and Griskevicius, V. (2008), 'Normative social influence is underdetected', *Personality and Psychology Bulletin*, **34**(7): 914–23.

North, D. (1990), *Institutions, Institutional Change and Economic Performance*, Cambridge, UK: Cambridge University Press.

O'Donoghue T. and Rabin, M. (2000), 'The economics of immediate gratification', *Journal of Behavioral Decision Making*, **13**(2): 233–50.

OFGEM (2011), *What Can Behavioural Economics Say about GB Energy Consumers?* London: Office of Gas and Electricity Markets.

Ostrom, E. (1998), 'A behavioral approach to the rational choice theory of collective action', *American Political Science Review*, **92**(1): 1–22.

Pallak, M. and Cummings, N. (1976), 'Commitment and voluntary energy conservation', *Personality and Psychology Bulletin*, **2**(1): 27–31.

Pesendorfer, W. (2006), 'Behavioral economics comes of age: a review essay on *Advances in Behavioral Economics*', *Journal of Economic Literature*, **44**(3): 712–21.

Reiss, P. and White, M. (2008), 'What changes energy consumption? Prices and public preasures', *RAND Journal of Economics*, **39**(3): 636–63.

Rondeau, D., Poe, G. and Schulze, W. (2005), 'VCM or PPM? A comparison of the performance of two voluntary public goods mechanisms', *Journal of Public Economics*, **89**(8): 1581–92.

Rose, S., Clark, G., Poe, D., Rondeau, D. and Schulze, W. (2002), 'Field and laboratory tests of a provision point mechanism', *Resource and Energy Economics*, **24**:131–55.

Samuelson, W. and Zeckhauser, R. (1988), 'Status quo bias in decision making', *Journal of Risk and Uncertainty*, **1**: 7–59.

Sanstad, A., Hanemann, M. and Auffhammer, M. (2006), 'End-use energy-efficiency in a "post-carbon" California economy: policy issues and research frontiers', The California Climate Change Center at UC-Berkeley, Berkeley, CA, USA.

Schultz, P., Nolan, J., Cialdini, R., Goldstein, N. and Griskevicius, V. (2007), 'The constructive, destructive, and reconstructive power of social norms', *Psychological Science*, **18**(5): 429–34.

Shefrin, H. and Statman, M. (1985), 'The disposition to sell winners too early and ride losers too long: theory and evidence', *The Journal of Finance*, **XL**(3): 777–90.

Shogren, F. and Taylor, L. (2008), 'On behavioural–environmental economics', *Review of Environmental Economics and Policy*, **2**(1): 26–44.

Simon, H. (1955), 'A behavioural model of rational choice', *The Quarterly Journal of Economics*, **69**(1): 99–118.

Simon, H. (1986), 'Rationality in psychology and economics', *The Journal of Business*, **59**(4): S209–S224.

Simshauser, P. and Downer, D. (2011), 'Limited-form dynamic pricing: applying shock therapy to peak demand growth', Working Paper No. 24 – Dynamic Pricing. North Sydney, Australia: Applied Economic and Policy Research, AGL Energy Ltd.

Stern, P. (2000), 'Toward a coherent theory of environmentally significant behavior', *Journal of Social Issues*, **56**(3): 407–24.

Stern, P. Aronson, E., Darley, J., Hill, D., Hirst, E., Kempton, W. and Wilbanks, T. (1985), 'The effectiveness of incentives for residential energy conservation', *Evaluation Review*, **10**(2): 147–76.

Steven, K., Rose, A., Clark, J., Poe, G., Rondeau, D. and Schulze, W. (2002), 'The private provision of public goods: tests of a provision point mechanism for funding green power programs', *Resource and Energy Economics*, **24**: 131–55.

Thaler, R. (1980), 'Toward a positive theory of consumer choice', *Journal of Economic Behavior and Organization*, **1**(1): 39–60.

Thaler, R. (1981), 'Some empirical evidence of dynamic inconsistency', *Economics Letters*, **8**(3): 201–7.

Thaler, R. (1999), 'Mental accounting matters', *Journal of Economic Behavior and Organization*, **12**(3): 183–206.

Thaler, R. and Sunstein, C. (2003), 'Libertarian paternalism is not an oxymoron', *The University of Chicago Law Review*, **70**(4): 1159–202.

Thaler, R. and Sunstein, C. (2008), *Nudge: Improving Decisions about Health, Wealth, and Happiness*, New Haven, CT: Yale University Press.

Tietenberg, T. (2003), 'The tradable-permit approach to protecting the commons: lessons for climate change', *Oxford Review of Economic Policy*, **19**(3): 400–419.

Titmuss, R. (1987[1971]), 'The gift of blood', in B. Abel-Smith and K. Titmuss (eds), *The Philosophy of Welfare: Selected Writings by R. M. Titmuss*, London: Allen and Unwin, pp. 254–68.

Train, K. (1985), 'Discount rates in consumers' energy-related decisions: a review of literature', *Energy*, **10**: 243–53.

Wilkinson, N. (2007), *An Introduction to Behavioral Economics: A Guide for Students*, New York: Palgrave Macmillan.

Williams, M. and Poyer, D. (1996), 'The effect of energy conservation tax credit on minority household housing improvement', *Review of Black Political Economy*, **24**(4): 122–34.

Wilson, C. and Dowlatabadi, H. (2007), 'Models of decision making and residential energy use', *Annual Review of Environmental Resources*, **32**: 169–203.

Wilson, C. and Waddams Price, C. (2005), 'Irrationality in consumers' switching decisions: when more firms may mean less benefit', Working Paper 0509010. East Anglia, UK: ESRC Center for Competition Policy, University of East Anglia.

Yates, S. and Aronson, E. (1983), 'A social-psychological perspective on energy conservation in residential buildings', *American Psychologist*, **38**(4): 435–44.

Yoeli E. (2010), *Does Social Approval Stimulate Prosocial Behavior? Evidence from a Field Experiment in the Residential Electricity Market*, Ann Arbor, MI: ProQuest LLC.

25 Valuing nature for climate change policy: from discounting the future to truly social deliberation[1]

John M. Gowdy

Human activity is set to leave an indelible mark on the geological record. Deforestation, mining and road building have unleashed tides of sediment down rivers and onto the ocean floor. Fossil-fuel use and land clearance have already emitted perhaps a quarter as much carbon into the atmosphere as was released during one of the greatest planetary crises of the past, the Paleocene-Eocene Thermal Maximum 55 million years ago. Now, as then, corals and other organisms are recording a global carbon-isotope shift. The increasing acidification of the oceans as they absorb carbon dioxide will dissolve carbonate from deep sediments, and what is likely to be the sixth great mass extinction in earth's history will gather speed, adding vivid new markers to the record. (*Nature*, **473**(254), editorial, 19 May 2011)

I. INTRODUCTION

The recent human impact on the environment is so unusual in the geological record that the official geological body that defines the division of geological time, the International Commission on Stratigraphy, is considering designating a new geographical epoch called the Anthropocene (Jones, 2011, p.133). This will call attention to the global impacts that humans, and particularly the human economy, are having on the earth's biological, atmospheric and geological systems. The driving force behind previous major geological transitions has been natural processes such as meteor impacts, massive volcanic activity and continental shifts. As the quotation above shows, the current human perturbation of the global environment will have an impact of similar magnitude (*Nature*, 2011). Yet most people are unconcerned about the looming possibility of environmental devastation and the resulting social chaos in the years to come. What is it about our way of living and associated ways of thinking that put so little value on the future of the planet? A major reason is the narrow logic of the global market economy, which values nature solely on its contribution to the discounted present value of economic activity. Following the logic of the market, the dominant economic model views the natural world from a financial investment perspective. Using the example of climate change, it is argued below that the magnitude, suddenness and long-term consequences of the current human abuse of the natural world call for a radical new approach to valuing nature, one based not on individual choice in the immediate present but rather on a socially embedded 'deeper sense of time' (Wing, 2011). Such an approach would move beyond attempts to 'correctly price' nature based on imputed market values and would instead rely on shared social values and a concern for future generations. These shared social values can be made concrete through discursive processes drawing upon our long evolutionary history of collectively solving the problem of intergenerational sustainability.

One of the most dramatic indicators of the Anthropocene is the increase in greenhouse gases over the past few decades. Over the past 800000 years atmospheric concentrations of CO_2 varied between 180 ppm and 280 ppm (Lüthi et al., 2008). CO_2 levels during this period are tightly correlated with temperatures and sea levels. These 50 ppm fluctuations around the average of 230 ppm were enough to push the earth between warm periods comparable to today's climate to extremely cold ice age conditions. In May 2011 atmospheric CO_2 levels at Mauna Loa, Hawaii passed 394 ppm, an increase approaching 100 ppm since the middle of the twentieth century (the Mauna Loa data are available at ftp://ftp.cmdl.noaa.gov/ccg/co2/trends/co2_mm_mlo. txt). When CO_2 levels were this high in the past, the earth's climate was dramatically different from today's. Tripati et al. (2009) report that during the Middle Miocene, some 10–14 million years ago, CO_2 levels were slightly lower than today's (around 350 ppm) but temperatures were 3 °C to 6 °C warmer and sea levels were 25 to 40 meters higher. Further back in time, around 56 million years ago, the earth experienced the Paleocene–Eocene Thermal Maximum (PETM) when temperatures rose by 8 °C. The PETM was probably triggered by volcanic activity which caused a release of CO_2 and frozen methane (Kump, 2011, pp. 58–9). The estimated release of greenhouse gases (5000 petagrams) then was about the same as the total projected release due to human activity in the industrial era. But the PETM event took about 20000 and the rate of heating was estimated to be 0.025 °C per 100 years compared to the projected rate of 1–4 °C per 100 years over the next few centuries. After the PETM episode it took the earth about 200000 years to recover (Kump, 2011). If past climate regimes are approximate indicators of what we can expect in the future, large, abrupt and unpredictable changes will occur for centuries to come.

In spite of international efforts to curb greenhouse gases, CO_2 emissions have grown at an annual rate of 3 percent per year since 2000, compared to 1.1 percent per year in the decade of the 1990s (Raupach et al., 2007) and reached record levels in 2010 (go. nature.com/rtgd7f). In view of the magnitude of emission increases, and the inertia of the world's economic and political systems, the chances of limiting the CO_2 level to one consistent with the Holocene's stable climate regime are bleak. By some estimates CO_2 levels could reach 2000 ppm within a few centuries if the readily available coal, petroleum and natural gas are burned (Kump, 2002). Kasting (1998) believes that the most likely scenario is that atmospheric CO_2 will peak at about 1200 ppm sometime in the next century. A climate-carbon model developed by Bala et al. (2005) has the business-as-usual CO_2 peak occurring around the year 2300 at 1400 ppm. Emissions scenarios by the IPCC include a worst-case, carbon-intensive scenario projecting a level of 1370 ppm by 2100 (Kintisch, 2008). Obviously, if CO_2 levels reach these extremes, abrupt and catastrophic climate events are likely.[2] The scientific consensus is that delaying substantial emission reductions for even a few more years may be disastrous (Jaeger et al., 2008).

The growing seriousness of the climate change threat and the release of the *Stern Review* (Stern, 2007) of the economics of climate change led to a vigorous debate among economists as to the merits and limitations of the standard economic model[3] used to value nature. This ongoing debate has done much to illuminate the assumptions of the standard economic model and their consequences for estimating the costs and benefits of climate change policies. At first the Stern debate centered primarily on the 'proper' discount rate to apply to future costs and benefits of climate change mitigation (Ackerman, 2008; Dasgupta, 2006; Mendelsohn, 2006; Yohe and Tol, 2007). As the debate progressed it became clear that there was more to the economics of climate change than choosing the 'correct' discount rate. Several prominent environmental economists came to the conclusion that the standard model offers an inadequate framework to analyze environmental issues characterized by irreversibilities, large uncertainties and very long time horizons (Gowdy and Juliá, 2010; Quiggin, 2008; Weitzman, 2009). The 'key messages' in the *Stern Review* (Stern, 2007, 7 ch. 2, p.25) make it clear that standard economic analysis embodies assumptions that make its application to climate change problematic: (1) climate change is a global phenomenon with global consequences; (2) its impacts are long-term and irreversible; (3) pure uncertainty is pervasive; (4) changes are non-marginal and non-linear; and (5) questions of inter- and intragenerational equity are central.

The latest views of leading environmental economists suggest that a profound reformulation of the economic valuation of the natural world is needed. Regarding the human impact on nature, we are in uncharted waters where the costs of mitigation may be large but the cost of inaction is potentially infinite, namely the extinction of our species and a catastrophic reorganization of the earth's climate and biosphere. Marginal analysis of near-to-equilibrium changes in market activity is a woefully inadequate approach to address a problem of the magnitude of today's massive reorganization of the earth's climate and biosphere.

One positive development during the last few decades is the recognition of the global impacts of *Homo sapiens* on the earth's biophysical systems and the danger these impacts pose to the viability of our species. Another positive development is the recognition of the extent to which human nature and human institutions, as well our physical characteristics, have been shaped by the forces of natural selection. Our 'social brain' (Frith and Frith, 2010) evolved in part to give humans the ability to change customs and technology to adapt to a quickly changing resource base compared to other animals that depended on more purely genetic adaptation. Richerson and Boyd (2005) argue that culture and complex brains were an evolutionary advantage for humans during the extreme climate volatility of past ice age transitions. The ability to use culture as an adaptive mechanism creates another source of variety – in addition to genes – upon which Darwinian selection can work. The ability of humans to adapt material culture, value systems and behavior associated with these values to changing conditions offers some hope in successfully managing the coming environmental transition. Judging from historical records of hunting and gathering societies (Gowdy, 1998) and behavioral and neurological evidence, for most of our evolutionary history our value systems placed much more emphasis on the social good and much less on individual-based materialism. Understanding the uniqueness (among mammals at least) of the degree of sociality among humans may hold the key to moving toward a sustainable way of living on our finite planet. Successful policies will require an evolutionary perspective on valuing the natural world, one that goes beyond proximate causes of resource use (prices and markets) to examine ultimate causes (institutional responses to resource availability and biophysical constraints and opportunities).[4]

II.   TRULY SOCIAL CHOICE: THE MISSING PERSPECTIVE IN ECONOMIC VALUATION

The climate change debate has forced economists to rethink the market-based approach to valuing the natural world (Ackerman, 2008; DeCanio, 2003; Nelson, 2011; Spash, 2002; Weitzman, 2009). The economic model is a financial investment model designed to show how a perfectly rational individual should allocate resources so as to maximize the discounted flow of income evaluated at a particular point in time (Gowdy, 2010b). The basic inadequacies of the standard economic approach include the failure to address the existence of pure uncertainty, threshold effects, incommensurability of values and non-substitution. These shortcomings have been widely discussed and will not be dealt with here, with one exception, namely, the reliance of standard economic valuation on the assumption of individual, self-referential behavior.[5] The economic model values future states of the environment using a so-called social discount rate. But this discount rate is merely the individual discount rate adjusted for external effects (Krall and Gowdy, 2012). The future is valued from the perspective of an isolated individual at a specific point in time. The 'social good' is simply the sum of the well-being of self-regarding individuals. Valuation decisions are stripped of their social context.

In contrast to the economic model, recent evidence from such diverse fields as anthropology, behavioral science, psychology and neuroscience has established that humans are unique among mammals as to their degree of sociality (Chapais, 2011; Hill et al., 2011; Wexler, 2006). This research may point the way toward more effective environmental policy design. Dealing with climate change will require cooperation on an unprecedented scale. It is encouraging that the evolutionary success of our species is based on our ability to manage scarce resources through cooperation and collective valuation of the future consequences of our actions (Richerson and Boyd, 2005; Sober and Wilson, 1998). Before getting to the details of these new findings about human sociality it is useful to briefly review the standard economic approach to valuing the future.

In the standard economic model future monetary costs and benefits are converted into 'present values' by discounting them at the rate r defined by the so-called Ramsey equation:

$$r = \rho + \eta \bullet g \tag{25.1}$$

The discount rate r is determined by the rate of pure time preference (ρ), the elasticity of consumption η, and the rate of growth of per capita consumption (g). In intuitive terms, individuals discount future economic benefits: (1) because they are impatient, and (2) because they have a declining marginal utility of money – since income and consumption levels are expected to rise, one additional unit of future consumption will provide less satisfaction than one additional unit of consumption today.

The Ramsey equation and the interpretation of its parameters show the unresolved tension between private and social valuation. The discounting debate between Stern and his critics has centered on the term ρ, the rate of pure time preference. Nordhaus (2007), for example, uses a relatively high value for ρ while Stern uses a value near zero. Stern's argument for this is clear and convincing from a social point of view, namely that the well-being of someone born in the future should not count less than that of someone born today. For Stern the choice of ρ is a social and ethical one; for Nordhaus it is a private investment decision. If the perspective is that of a self-referential individual at a point in time, then ρ is positive – I care only about the lower value to me of something I get in the future as compared to having it now. If the perspective is that of the well-being of human society, then there is no reason to count the utility of one generation more than that of another, so ρ should be near zero.[6] Likewise the second term η*g* may also be given a private or social interpretation. The elasticity of substitution η is usually given a value of 1 so that a 10 percent increase in income is given the same weight no matter what the absolute magnitude (see the discussion of the political economy of the use of this term in Stanton, 2011). The value of *g* answers the question 'how well off will future generations be?' In the standard model *g* is a private investment concept, the growth of per capita income, and is usually assumed to be the growth rate of income over the past 100 years or so (as in both the Stern and Nordhaus models). The values of *g* in the Stern report and in the most widely used climate change models range between 1.5 percent and 2.0 percent (see the discussion in Quiggin, 2008). Assuming rapid economic growth will continue, discounting today's negative impact on those in the future is justified by the assumption that those living in the future will be better off than those living today (Pearce et al., 2006). The use of average income as a universal welfare measure allows economists to sidestep questions about substitutability, irreversibility, quality of life, relative income and many other issues plaguing the standard model. By contrast, using well-being or quality-of-life estimates of *g* would highlight the negative consequences of market-induced environmental changes and the social responsibility of those in the present for those in the future.

In spite of heroic attempts to include social values in the discount rate, reducing environmental policy to choosing the 'correct' discount rate reduces an intergenerational and social problem to one of individual choice at a point in time. Economic estimates of environmental values assume that people care only about absolute income, not their income relative to others. By contrast, experimental results show that economic behavior is based on preferences that depend on the relationship of the evaluator to others. This has been demonstrated in thousands of behavioral, game-theoretic and neurological experiments (Gintis, 2000). Other relevant valuation considerations include inequality aversion, pure altruism, spiteful or envious preferences and altruistic punishment (Fehr and Fischbacher, 2002). Recent research results from experimental economics, behavioral psychology and decision theory have the potential to make economic analysis and policy recommendations more reflective of actual human decision making but contemporary environmental economics largely ignores these recent theoretical and empirical advances (Knetch, 2005).

As society considers how to motivate humans to address the challenges of global environmental change, increasing attention is turning to psychological and biological insights into human behavior. Current research in behavioral science and neuroscience is confirming what critics of standard economic theory have long argued: humans are highly social mammals whose behavior deviates significantly from strict 'rationality' because of social norms and evolutionary history. Our species has survived because we have evolved biological features (including brain structure) and institutional arrangements that promote cooperation. Behavioral insights have been slow to penetrate the economic policy world, but these insights may help shape more effective policy approaches to redirecting individual and system behavior.

III.   VALUATION AND THE DEEP SOCIAL STRUCTURE OF HUMANKIND

New findings about human behavior and the deep social structure of the human species have important implications for the valuation of nature and the formulation of environmental policies, particularly policies having very long-run impacts such as climate change and biodiversity loss. Behavioral economics is still a new field but it has already challenged standard theory, and standard valuation practices, as much as any theoretical revolution since the 1930s. For improving valuation techniques the most important contribution of behavioral economics and neuroscience has been to establish that human behavior is social. Humans make decisions not only as individuals addressing immediate individual problems, but also as members of groups with highly evolved institutions to insure cohesion and the long-run stability of the group.

The Deep Social Structure of Human Society

Primates are exceptional in their degree of sociality, but scientists are just beginning to discover the uniqueness of human sociality. For example, to a degree not seen in other primates, humans are able to form long-term cooperative bonds with non-kin. Hill et al. (2011) looked at co-residence patterns in 32 present-day foraging societies and found that humans, compared to other primates, are unique in that (1) either sex may remain with their parental group, (2) adult brothers and sister may co-reside, (3) most members of a residential group are unrelated, and (4) preferential bonds are maintained with spouses' relatives and relatives' spouses. Generally, primary kin make up only 10 percent of a residential band. In other primate groups most members are closely related. Thus human cooperation cannot be explained entirely by kin selection (Trivers, 1971) or inclusive fitness (Hamilton, 1964). Commenting on the Hill et al. study, Chapais (2011, p. 1277) writes:

> Cooperation in other primates is limited to the coordination of individuals belonging to the same group. The advent of the primitive tribe moved cooperation to substantially higher levels of complexity. It paved the way for the coordination of whole social groups, hence creating the nested character of human social structure.

Evidence suggests the existence of a kind of collective intelligence of human groups related to group composition but unrelated to the characteristics of individuals within the groups. In a recent study of group decision making, Woolley et al. (2010) found evidence for what they called a 'collective intelligence factor'. In two different studies, groups of two to five people were assigned a variety of tasks, then the groups were ranked according to their performance of these tasks. The authors found that a collective intelligence factor explained the groups' performance and that

The 'c-factor' is not strongly correlated with the average or maximum individual intelligence of group members but is correlated with the average social sensitivity of group members, the equality in distribution of conversational turn-taking, and the proportion of females in the group. (Woolley et al., 2010, p. 686)

This research is still in its infancy but it seems to corroborate theories that humans evolved deep social, non-kin bonds as a way to adapt to environmental change. It suggests that a case can be made for valuation processes allowing for interaction and deliberation. Such deliberation can capture information and deal with uncertainty in ways that isolated individuals cannot. There may also be an 'ideal' composition of groups for making critical decisions. For example, is there an ideal mix of selfish individuals and altruists in collective decision making? What role does gender play in successful group composition? Does voting based on isolated individual decisions preclude solutions based on group deliberative valuation that might result in better outcomes? Many mammals are highly social animals with a variety of behavioral attributes that evolved to facilitate social interaction, but humans seem to be unique in their degree of sociability. Evidence for the existence of the social brain (Fehr, 2009; Frith and Frith, 2010; Singer, 2009) suggests the need for a theory of valuation and decision making that is deliberate, other regarding, and consistent with human sociality.

The Neuroscience of the Social Brain

The uniqueness and importance of human sociality has been confirmed and enriched by neuroscience. The way the brain is organized and develops provides strong evidence for the evolutionary importance of human sociality. Most of the neurons in the human brain develop after birth and the way they are configured depends critically on how a child is socialized. It is another way that variability can be introduced into evolutionary mix. Wexler (2006, p. 3) writes about the evolutionary advantages of brain plasticity:

[T]he distinctive postnatal shaping of each individual's brain function through interaction with other people, and through his or her own mix of sensory inputs, creates an endless variety of individuals with different functional characteristics. This broadens the range of adaptive and problem-solving capabilities well beyond the variability achieved by sexual reproduction.

Another finding from neuroscience is the presence in the human brain of von Economo neurons. These specialized neurons, also called spindle neurons, apparently evolved to enable people to make rapid decisions in social context (Allman et al., 2005, p. 370; Sherwood et al., 2008, p. 433). Deficiencies in the number of these neurons have been implicated in diseases affecting social interactions such as autism and Alzheimer's. These neurons are also found in a few other species – great apes, elephants, and whales and dolphins – although in much smaller numbers.[7] These other species are highly intelligent with complex social systems (Semendeferi et al., 2010). In humans, about 85 percent of von Economo neurons are formed after birth. This again points to the blurred line between heredity and socialization. Sherwood et al. (2008, p. 433) write:

It is interesting that these specialized projection neuron types have been identified in cortical areas that are positioned at the interface between emotional and cognitive processing. Given their characteristics, it has been speculated that Von Economo neurons are designed for quick signaling of an appropriate response in the context of social ambiguity (Allman et al. 2005). Enhancements of this ability would be particularly important in the context of fission–fusion communities, such as those of panids [chimpanzees and bonobos] and possibly the LCA [last common ancestor], with complex networks of social interactions and potential uncertainties at reunions.

A fission–fusion community is a kind of social group where the members gather together in one locality to sleep, but split into smaller groups to forage. Among human hunter-gatherers – a type of society that characterized most of human existence – fusing and splitting can be seasonal, with small bands being the group type for most of the year but coming together to form a larger group when resource availability permits (Gowdy, 1998). In this kind of social organization groups are continually changing in composition and Allman et al. (2005, p. 370) argue that von Economo neurons help humans to adjust quickly to rapidly changing social situations:

> We hypothesize that the VENs and associated circuitry enable us to reduce complex social and cultural dimensions of decision-making into a single dimension that facilitates the rapid execution of decisions. Other animals are not encumbered by such elaborate social and cultural contingencies to their decision-making and thus do not require such a system for rapid intuitive choice.

Neurological evidence confirms the uniqueness of the degree of sociality in humans. The success of our species (so far) may be largely the result of our ability to cooperate and to harness the advantages of collective decision making. This is in sharp contrast to the economic view of the sanctity of individualistic rational choice and it has important implications for climate change and other environmental policies affecting the future of our species.

Whatever the recent successes of civil society organizations in helping to address such challenges, it seems that current responses are incommensurate with the scale of the problems we confront. It is increasingly evident that resistance to action on these challenges will only be overcome through engagement with the cultural values that underpin this resistance.

A report published by the UK Worldwide Fund for Nature (Crompton, 2010) argues that current approaches to solving global challenges are failing because they do not engage with cultural values. It seems clear that, in trying to meet these challenges, civil society organizations must champion some long-held (but insufficiently esteemed) values, while seeking to diminish the primacy of many values which are now prominent – at least in Western industrialized society (Crompton, 2010, p. 5). These values include the importance of family and social relationships, concern for future generations and empathy toward others. These values are particularly important in addressing 'bigger-than-self' problems – problems important to individuals but whose solution is unlikely to be justified by self-interest alone. 'Immediate-self-interest' problems, by contrast, are those whose solutions are justified in terms of personal gains alone. Related to this is the distinction between intrinsic values and extrinsic values (see Sheldon and McGregor, 2000). Intrinsic values are those that do not depend on competitive comparisons with others – a sense of community, enjoyment of friends and family, and self-actualization. Extrinsic values relate to things that have zero-sum comparisons, like material wealth and power. Advocates of aggressive climate change policies may have missed the boat by focusing exclusively on extrinsic motivations.

IV.   HARNESSING HUMAN SOCIALITY: ENVIRONMENTAL GOVERNANCE THROUGH DELIBERATIVE VALUATION

Behavioral and neuroscience evidence suggests that the degree of human sociality is unique. We apparently evolved to make critical decisions in social settings, not as isolated individuals. Understanding the social nature of decision making may be a key both to formulating successful social and environmental policies and to gaining public acceptance of these policies. The human species has been so successful precisely because we have created elaborate social institutions to decide the common good (Richerson and Boyd, 2005; Sober and Wilson, 1998). But concern for the common good has been eroded by economic theories and public policies increasingly focused on the individual, not the good of the group. The contrast between standard economic theory and behavioral economics is mirrored in the conflict between neoliberal democracy and deliberative democracy (Quiggin, 2010).

Neoliberal Democracy

This governance model is consistent with the 'isolated selfish individual' underlying the standard economic valuation of nature. This valuation framework has been consciously used to discourage public support for any sort of cooperative, collective public policy, the very kinds of policies that are critical for addressing climate change. 'Social choice' is simply the sum of choices made by individuals who are 'free to choose' in private markets. The value of nature is determined by the ability of ecosystem services to contribute to economic output. This is the economic view of 'weak sustainability' – sustainability as non-declining GDP (Gowdy, 2000).

Over the past few decades the dictates of the market have come to dominate the valuation discussion. Bromley (2007, p. 677) describes the takeover of reasoned public discourse and democratically chosen public policies by the let-the-market-decide mentality:

> Suddenly, it seems that public policy is not what we thought it was. Democracy as public participation and reasoned discourse is somehow suspect – not to be trusted. It seems that the public's business cannot be properly conducted unless it adheres to the precepts of individualistic models of 'rational choice' applied to collective action . . . It is a quest for public policy in which applied micro-economics is deployed as the only way to impose 'rationality' on an otherwise incoherent and quite untrustworthy political process. This is not a clash of worldviews. It is a clash of contending truth claims about how to figure out what is to be done in the public sphere – it is a confrontation between prescriptive consequentialism and reasoned public debate over how to get to the future.

The neoliberal public policy prescription is to set markets in motion and then let efficiency in allocation determine the socially optimal outcome. In terms of the valuation of nature, this prescription requires only that prices be 'correct' and that property rights are fully specified. Moreover, it moves 'democracy' from 'one person, one vote' to 'one dollar, one vote'. And as this happens, social stability and environmental sustainability are eroded in the name of efficiency and economic growth.

But human sociality offers a way out of the straitjacket of individual-based valuation. Human society is more than a collection of isolated individuals acting only in their narrowly defined self-interest. Bromley (2007) argues that since future generations cannot express their preferences to us, we have a duty to act as regents on their behalf. We have a responsibility to protect the economic, biophysical and social conditions that will allow them to achieve their aspirations even though we have no way of knowing what these aspirations are. The question then becomes not *how much* to preserve but rather *what* to preserve. An alternative global governance model consistent with the behavioral and neurological evidence about human sociality and cooperation is one based on deliberative discourse.

Global Governance through Deliberative Discourse

Deliberative discourse recognizes that there is more to democracy than individuals voting. Dryzek and Stevenson (2011) sketch out some essential features of a global deliberative governance framework. These include empowered space, transmission, accountability and meta-deliberation. It is hard to envision exactly what a global deliberative discourse environmental authority would look like. But this emerging framework for earth systems governance is consistent with the view of human nature as outlined in the behavioral and neuroscience literature. Deliberative valuation served our species well for 100 000 years or more. A global governing body to regulate our use of nature seems to run counter to the anti-government reaction now prevalent in Western democracies. But it is a mistake to see the widespread distrust of governments as a rejection of collective action based on democratic deliberative discourse.

Behavioral studies can give some clues about governance structures that facilitate cooperation. The most cooperative societies are those that have the most efficient mechanisms for punishing free-riders (Henrich et al., 2006). People are more likely to accept decisions they do not agree with if the deliberative process is perceived to be fair. Group decisions lead to better outcomes if the composition of the group is balanced in terms of gender and personality types (Woolley et al., 2010). Deliberative discourse is consistent with behavioral science in that it recognizes that reasoned judgment is not something undertaken by a lone organism in isolation (Nelson, 2011).

Does deliberative discourse lead to the sustainable use of nature? Dryzek and Stevenson (2011) give several examples of successful consensual systems that might be a model for a global agency. Norway, for example, has integrated social and environmental movements and has had some success in moving toward environment-friendly economic policies. Deliberative polls held in Texas led to greater investment in renewable energy and conservation policies (Fishkin, 2009). Focusing public policy on well-being rather than exclusively on per capita income may also have positive consequences for the environment. Rangel (2003) argues that providing social goods like health care, job security and a minimum income may play a crucial role in sustaining investment in 'forward intergenerational goods' such as environmental preservation. Focusing policies on subjective indicators of happiness, rather than on per capita income, would pay a double dividend. People would be happier and also more willing to support polices promoting environmental sustainability. To fully develop a viable alternative to the neoclassical notion of sustainability, scientific measures of the factors contributing to human well-being are needed as well as indicators of the physical and biological requirements for long-term human survival.

V.   SUMMARY AND CONCLUSIONS

To summarize the above discussion, the current reorganization of the earth's life support systems is almost unprecedented in its magnitude and speed of occurrence. A major threat to the viability of the human species is rising temperatures due to the buildup of greenhouse gases in the atmosphere. Policies to address this problem have focused on correcting market 'externalities' using individual-based incentives and they have been ineffective – CO_2 and other greenhouse gas emissions continue to grow exponentially. It is argued above that environmental policy formulation should embrace our basic nature as social animals concerned with the good of the group. Current research has confirmed that humans are unusual in their degree of sociality and this may be the primary reason humans have been so evolutionarily successful to this point. The question today is whether or not we can harness our cooperative nature on a global scale to meet the unprecedented challenges we face. Our success depends on how we value the natural world and how we construct institutions to articulate our social values.

It is encouraging to note that humans lived sustainably as hunter-gatherers within the confines of local ecosystems for 95 percent of our existence as a species. We survived by creating institutions that served the well-being of the group as well as the individual. For a variety of reasons, including tapping into the stock of the earth's stored carbon energy, we broke out of those local confines. We now find ourselves once again coming up against biophysical limitations, this time imposed by the entire finite planet (Eldredge, 1995). We may be able to escape this dilemma by drawing on the unique social characteristics that define our species, the ability to cooperate and the ability to construct social, technological and economic systems in harmony with the biophysical systems that sustain us. But this will require expanding the notion of the 'group' to include the entire human species. In the words of Georgescu-Roegen (2011, p. 102):

A new ethics is what the world needs most. If our values are right, everything else – prices, production, distribution and even pollution – has to be right. At first man has heeded (at least in a large measure) the commandment 'Thou shalt not kill,' a little later 'Love thy neighbour as thyself.' The commandment of this era is 'Love thy species as thyself.'

NOTES

1.   The author would like to thank Jack Hanich and Lisi Krall for helpful comments on an earlier draft. Parts of this chapter were adopted from Gowdy (2008, 2010a) and Gowdy et al. (2010).
2.   A recent re-examination of 'hothouse earth' climate regimes indicates that CO_2 levels during these periods were around 1000 ppm, not 3000–4000 ppm as previously thought (Newton, 2010).
3.   By standard economics I mean the Walrasian general equilibrium model, also called the dynamic stochastic general equilibrium model (DSGE), the new welfare economics, or simply neoclassical economics.
4.   The distinction between ultimate and proximate causation (Tinbergen, 1963) stresses the need for two

separate and complementary explanations for all products of genetic and cultural evolution. Ultimate causation explains why a given trait exists, compared to many other traits that could exist, based largely on the winnowing action of selection. Proximate causation explains how the trait exists in a mechanistic sense. Excessive amounts of CO_2 are being pumped into the atmosphere because prices are too low (proximate cause) but more importantly because of the way industrial capitalism evolved in terms of production techniques dependent on fossil fuels, the concentration of economic and political power, and the culture of consumption (ultimate causes). It is especially important to recognize the many-to-one relationship between proximate and ultimate causation, whereby many functionally equivalent solutions can evolve in response to a given environmental challenge. Failing to distinguish between design features and specific implementation of design features can result in the inability to detect correlations and why policies work in some situations but not in others (Wilson and Gowdy, 2010).

5.   The term 'self-referential' is critical. The standard model assumes that behavior is not influenced by the behavior of others, one's social position in relation to others, or cultural norms. The standard economic assumption of self-regarding behavior strips away everything that makes *Homo sapiens* unique.

6.   Geogescu-Roegen (1974, p. 32) called for a zero discount rate for using irreplaceable resources or causing irreducible pollution. He argued further: 'The only way to protect future generations, at least from the excessive consumption of resources during the present bonanza, is by reeducating ourselves so as to feel some sympathy for our *future* fellow humans in the same way in which we have come to be interested in the well-being of our *contemporary* "neighbors."'

7.   An intriguing exception was found in an autopsy of the lowland gorilla Michael. Michael lived in a rich social environment interacting with humans and was a companion of Koko, another gorilla who taught him sign language (Patterson and Gordon, 2002). Michael was found to have considerably more VENs than other gorillas autopsied, approaching the lower end of the human range (Allman et al., 2010, p. 501).

REFERENCES

Ackerman, F. (2008), 'The new climate economics: *The Stern Review* versus its critics', in J.M. Harris and N.R. Goodwin (eds), *Twenty-First Century Macroeconomics: Responding to the Climate Challenge*, Cheltenham, UK and Northampton, MA, USA: Edward Elgar, pp. 32–57.

Allman, J., T. McLaughlin and A. Hakeem (2005), 'Intuition and autism: a possible role for Von Economo neurons', *Trends in Cognitive Science*, **9**, 367–73.

Allman, J., N. Tetreault, A. Hakeem, K. Manaye, K. Semendeferi, J. Erwin, S. Park, V. Goubert and P. Hof (2010), 'The von Economo neurons in frontoinsular and anterior cingulated cortex in great apes and humans', *Brain Structure & Function*, **214**, 495–517.

Bala, G., K. Caldeira, A. Mirin, M. Wickett and C. Delire (2005), 'Multicentury changes in global climate and carbon cycle: results from a coupled climate and carbon cycle model', *Journal of Climate*, **18**, 4531–44.

Bromley, D. (2007), 'Environmental regulations and the problem of sustainability: moving beyond "market failure"', *Ecological Economics*, **63**, 676–83.

Chapais, B. (2011), 'The deep social structure of humankind', *Science*, **331**, 1276–7.

Crompton, T. (2010), 'Common cause: the case for working with our cultural values', Worldwide Fund for Nature UK, available at http://assets.wwf.org.uk/downloads/common_cause_report.pdf (accessed 31 August 2011).

Dasgupta, P. (2006), 'Commentary: *The Stern Review*'s economics of climate change', *National Institute Economic Review*, **119**, 4–7.

DeCanio, S. (2003), *Economic Models of Climate Change: A Critique*, New York: Palgrave Macmillan.

Dryzek, J. and H. Stevenson (2011), 'Global democracy and earth system governance', *Ecological Economics*, **70**, 1865–74.

Eldredge, N. (1995), *Dominion*, Berkeley, CA: University of California Press.

Fehr, E. (2009), 'Social preferences and the brain', in P. Glimscher, C. Camerer, E. Fehr and R. Poldrack (eds), *Neuroeconomics: Decision Making and the Brain*, London: Academic Press, pp. 215–32.

Fehr, E. and U. Fischbacher (2002), 'Why social preferences matter – the impact of non-selfish motives on competition, cooperation and incentives', *Economic Journal*, **112**, C1–C33.

Fishkin, J. (2009), *When the People Speak: Deliberative Democracy and Public Consultation*, Oxford: Oxford University Press.

Frith, U. and C. Frith (2010), 'The social brain: allowing humans to boldly go where no other species has been', *Philosophical Transactions of the Royal Society B*, **365**, 165–75.

Georgescu-Roegen, N. (1974), 'Energy and economic myths', Reprinted in *Energy and Economic Myths*, 1976, San Francisco, CA: Pergamon Press, pp. 3–36.

Georgescu-Roegen, N. (2011), 'The steady state and ecological salvation', in Mauro Bonaiuti (ed.), *From Bioeconomics to Degrowth: Georgescu-Roegen's New Economics in Eight Essays*, Oxford and New York: Routledge, ch. 3.
Gintis, H. (2000), 'Beyond *Homo economicus*: evidence from experimental economics', *Ecological Economics*, **35**, 311–22.
Gowdy, J. (1998), *Limited Wants, Unlimited Means: A Reader on Hunter-Gatherer Economics and the Environment*, Washington, DC: Island Press.
Gowdy, J. (2000), 'Terms and concepts in ecological economics', *Wildlife Society Bulletin*, **28**, 26–33.
Gowdy, J. (2008), 'Behavioral economics and climate change policy', *Journal of Economic Behavior and Organization*, **68**, 632–44.
Gowdy, J. (2010a), 'Behavioral economics, neuroeconomics, and climate change policy', Baseline Review for the Garrison Institute Initiative on Climate Change Leadership, Garrison, New York.
Gowdy, J. (2010b), *Microeconomic Theory Old and New: A Student's Guide*, Stanford, CA: Stanford University Press.
Gowdy, J. and R. Juliá (2010), 'Global warming economics in the long run', *Land Economics*, **86**, 117–30.
Gowdy, J., R. Howarth and C. Tisdell (2010), 'Discounting, ethics, and options for maintaining biodiversity and ecosystem services', in Pushpam Kumar (ed.), *The Economics of Ecosystems and Biodiversity: Ecological and Economic Foundations*, An output of TEEB: The Economics of Ecosystems and Biodiversity, London: Earthscan, pp. 257–83.
Hamilton, W.D. (1964), 'The genetical evolution of social behaviour I and II', *Journal of Theoretical Biology*, **7**, 1–16 and 17–52.
Henrich, J., R. McElreath, A. Barr, J. Ensminger, C. Barrett, A. Bolyanatz, J. Cardenas, M. Gurven, E. Gwako, N. Henrich, C. Lesorogol, F. Marlowe, D. Tracer and J. Ziker (2006), 'Costly punishment across human societies', *Science*, **312**, 1767–70.
Hill, K., R. Walker, M. Božičević, J. Elder, T. Headland, B. Hewlett, M. Hurtado, F. Marlowe, P. Wiessner and B. Wood (2011), 'Co-residence patterns in hunter-gatherer societies show unique human social structure', *Science*, **331**, 1286.
Jaeger, C., H. Schellnhuber and V. Brovkin (2008), 'Stern's Review and Adam's fallacy', *Climatic Change*, **89**, 207–18.
Jones, N. (2011), 'Human influence comes of age', *Nature*, **473**, 133
Kasting, J. (1998), 'The carbon cycle, climate, and the long-term effects of fossil fuel burning', *Consequences*, **4**, 15–27.
Kintisch, E. (2008), 'IPCC tunes up for its next report aiming for better, timely results', *Science*, **320**, 300.
Knetch, J. (2005), 'Gains, losses, and the US-EPA *Economic Analyses Guidelines*: a hazardous product?' *Environmental & Resource Economics*, **32**, 91–112.
Krall, L. and J. Gowdy (2012), 'An institutional and evolutionary critique of natural capital', in Rolf Steppacher and Julien-François Gerber (eds), *Toward an Integrated Paradigm in Heterodox Economics: Alternative Approaches to the Current Eco-Social Crises*, Basingstoke: Palgrave-Macmillan, pp. 127–47.
Kump, L. (2002), 'Reducing uncertainty about carbon dioxide as a climate driver', *Nature*, **419**, 188–90.
Kump, L. (2011), 'The last great global warming', *Scientific American*, **305**, 57–61.
Lüthi, D. et al. (2008), 'High resolution carbon dioxide concentration record 650,000–800,000 years before present', *Nature*, **453**, 379–82.
Mendelsohn, R. (2006), 'A critique of the Stern Report', *Regulation*, **29**, 42–6.
Nature (2011), Editorial, 'The human epoch', *Nature*, **473**, 254.
Nelson, J. (2011), 'Ethics and the economist: what climate change demands of us', Global Development and Environment Institute, Working Paper no. 11-02.
Newton, A. (2010), 'Insights from earth', Nature reports Climate Change, 4 (February), www.nature.com/reports/climatechange.
Nordhaus, W. (2007), 'A review of the *Stern Review on the Economics of Climate Change*', *Journal of Economic Literature*, **45**, 686–702.
Patterson, F. and W. Gordon (2002), 'Twenty-seven years of project Koko and Michael', in B. Galdikas, N. Briggs, L. Sheeran, G. Shapiro and J. Goodall (eds), *All Apes Great and Small*, vol. 1, Amsterdam and New York: Kluwer, pp. 165–76.
Pearce, D., G. Atkinson and S. Mourato (2006), *Cost Benefit Analysis and the Environment: Recent Developments*, Paris: OECD.
Quiggin, J. (2008), 'Stern and the critics on discounting and climate change: an editorial essay', *Climatic Change*, **89**, 195–205.
Quiggin, J. (2010), *Zombie Economics: How Dead Ideas Still Walk Among Us*, Princeton, NJ: Princeton University Press.
Rangel, A. (2003), 'Forward and backward generational goods: why is social security good for the environment?', *American Economic Review*, **93**, 813–34.

Raupach, M., G. Marland, P. Ciais, C. Le Quéré, J. Canadell, G. Klepper and C. Field (2007), 'Global and regional drivers of accelerating CO_2 emissions', *Proceedings of the National Academy of Sciences*, **104**, 10288–93.

Richerson, P. and R. Boyd (2005), *Not by Genes Alone*, Chicago, IL: University of Chicago Press.

Semendeferi, K., K. Teffer, D. Buxhoeveden, M. Park, S. Bludau, K. Armunts, K. Travis and J. Buckwalter (2010), 'Spatial organization of neurons in the frontal pole sets humans apart from great apes', *Cerebral Cortex*, **20**, 1485–97.

Sheldon, K. and H. McGregor (2000), 'Extrinsic value orientation and "the tragedy of the commons"', *Journal of Personality*, **68**, 383–411.

Sherwood, C., F. Subiaul and T. Zadiszki (2008), 'A natural history of the human mind: tracing evolutionary changes in brain and cognition', *Journal of Anatomy*, **212**, 426–54.

Singer, T. (2009), 'Understanding others: brain mechanisms of theory of mind and empathy', in P. Glimscher, C. Camerer, E. Fehr and R. Poldrack (eds), *Neuroeconomics: Decision Making and the Brain*, London: Academic Press, pp. 251–68.

Sober, E. and D. Wilson (1998), *Unto Others: The Evolution and Psychology of Unselfish Behavior*, Cambridge, MA: Harvard University Press.

Spash, C. (2002), *Greenhouse Economics: Value and Ethics*, London: Routledge.

Stanton, E. (2011), 'Negishi welfare weights in integrated assessment models: the mathematics of global inequality', *Climatic Change*, DOI: 10.1007/s10584-010-9967-6 (accessed 19 June 2011).

Stern, N. (2007), *The Economics of Climate Change: The Stern Review*, Cambridge: Cambridge University Press.

Tinbergen, N. (1963), 'On aims and methods in ethology', *Zeitschrift für Tierpsychologie*, **20**, 410–33.

Tripati, A., C. Roberts and R. Eagle (2009), 'Coupling of CO_2 and ice sheet stability over major climate transitions of the last 20 million years', *Science*, **326**, 1394–7.

Trivers, R. (1971), 'The evolution of reciprocal altruism', *Quarterly Review of Biology*, **46**, 35–57.

Weitzman, M. (2009), 'On modeling and interpreting the economics of catastrophic climate change', *The Review of Economics and Statistics*, **XCI**, 1–19.

Wexler, B.E. (2006), *Brain and Culture*, Cambridge, MA: MIT Press.

Wilson, D.S. and J. Gowdy (2010), 'The relevance of evolutionary science for economic theory and policy', White Paper for NSF SBE Program Initiative 'Framing Research for 2020 and Beyond', available at http://evolution-institute.org/files/NSF-EvoEco-White-Paper.pdf.

Wing, S. (2011), 'We need a deeper sense of time', *Science*, **333**, 825.

Woolley, A., C. Chabris, A. Pentland, M. Hashmi and T. Malone (2010), 'Evidence for a collective intelligence factor in the performance of human groups', *Science*, **330**, 686–8.

Yohe, G. and R. Tol (2007), 'The *Stern Review*: implications for climate change', *Environment*, March, 49–52.

26 Individual consumers and climate change: searching for a new moral compass
Tanya O'Garra

1. INTRODUCTION

> climate change appears to be a perfect moral storm because it involves the convergence of a number of factors that threaten our ability to act ethically. (Gardiner, 2006, p. 398)

There is an almost worldwide consensus that human activities associated with the burning of fossil fuels are contributing significantly to changing the world's climate. The IPCC report (2007) outlines the predicted and actual impacts from this changing climate, which include increasing severity of floods, melting permafrost and increased heat-related human mortality rates. Mitigation of these impacts will require serious reductions in our greenhouse gas emissions, of which carbon dioxide (CO_2) is the most important source.

So far, climate change mitigation policies worldwide have tended to focus on reductions from industrial and commercial sources, largely ignoring emissions from individuals and households. However, personal transport and domestic heating represent 30–40 per cent of all carbon emissions in the USA (Vandenbergh et al., 2008) and about 32 per cent of carbon emissions in the UK (DECC, 2011; DfT, 2010).

Despite the significant contributions of individuals and households to global emissions, policy makers are reluctant to introduce regulations at this level. This is due to a number of factors, including fear of electoral protest and a short-term approach to policy linked to government's limited period in office (Lorenzoni et al., 2007). Thus emissions reductions at the individual and household level will rely essentially on voluntary approaches.

Voluntary approaches, such as the UK's 'Act on CO_2' campaign and the US Energy Star programme, typically use a combination of information, labelling and pecuniary and/or non-pecuniary incentives to motivate behavioural change. Despite some modest successes reported in the literature (e.g. Murray and Mills, 2011; Kotchen, 2010), in general, voluntary schemes have had minimal impact on behaviour (Lorenzoni et al., 2007). Furthermore, in an empirical study of intentionality and behaviour, Whitmarsh (2009) finds that those who take action to conserve energy generally do so for reasons unrelated to climate change or the environment (e.g. to save money).

The question is: why are individuals failing to engage? In many cases, the lack of behavioural change may be due to structural and/or financial constraints. However, there are many individuals for whom the alternatives are available and accessible, and who still fail to change their behaviour. What is happening?

The considerable literature on the public perception of climate change (mostly Europe and US-based) may provide some clues. A common finding across most survey studies is that, despite widespread concern about climate change, it is perceived as a distant threat – one that affects people in faraway countries, and/or future generations – with little personal relevance (Lorenzoni et al., 2007; Leiserowitz, 2006). There is also wide-spread confusion about the causes, consequences and solutions to climate change, as well as distrust of major sources of information and scepticism about the science behind climate change (Lorenzoni et al., 2007; Lorenzoni and Pidgeon, 2006). Add to this mix the large number of psychological barriers to behavioural change (Gifford, 2011), a general reluctance to change lifestyle and the ever-present collective action problem, and one wonders why anyone would engage in behavioural change at all.

In this chapter, I propose that voluntary engagement by individuals and households in carbon-reducing behaviours might be significantly enhanced if climate change were to be framed clearly, and unequivocally, as a moral issue. This is because, typically, people do not want to be associated with behaviours they find morally wrong. In economic terms, one might say that the greater the moral cost of a behaviour, the less likely that behaviour (Levitt and List, 2007).

There is a wealth of empirical evidence indicating that moral concerns do indeed influence behaviour. Laboratory experiments show that fairness concerns and inequity aversion operate under most conditions, such that players will cooperate even when this goes against self-interest (Fehr and Schmidt, 1999; Fehr et al., 2006). A review of the extensive literature on dictator games, for example, indicates that about 60 per cent of the time, an average 20 per cent endowment is made by the dictator (List, 2007). Furthermore, participants in economic experiments are more likely to cooperate if the game is framed in terms of 'giving' rather than 'taking' (List, 2007), or if they are playing 'Community' rather than 'Wall Street' (Ross and Ward, 1996), suggesting that a moral frame will significantly influence behaviour.

These findings are not just relegated to the laboratory. In real-world settings, people regularly engage in pro-social behaviours, such as donating to charity and buying ethical goods. One might contend that these behaviours are motivated by a desire for social approval rather than any inherent moral concern. Many lab-based studies confirm this to be the case: as player anonymity increases, cooperation and giving decrease (e.g. List et al., 2004; Hoffman et al., 1994). However, outside of the lab, people often engage in pro-social behaviours that are unobservable. For example, many people vote, consume green electricity and clean up after their dogs on quiet streets. It would be hard – if not implausible – to try to explain the wide range of voluntary behaviours solely in terms of social approval.

This chapter argues that voluntary approaches to climate change reductions at the individual and household level may have a greater chance of succeeding if people perceive climate change to be a moral issue. This is not a novel idea. In the UK, and more recently in the USA, discourses about climate change have been framed in an increasingly moralistic way, such that certain behaviours are presented as good or responsible, and other behaviours as bad or irresponsible (Butler, 2010). However, these designations are rarely backed up with a clear moral argument that individuals can find meaningful.

That climate change is a moral issue is, I will argue below, indisputable. However, as we shall see, it has a number of features that make it difficult to apprehend as a typical moral problem. The purpose of this chapter is to identify these problematic features, to consider them in terms of moral philosophy, and to discuss how they might be recast within the structure of the archetypal moral problem, so that individuals may start to perceive climate change as a moral issue. This chapter also serves as a rudimentary review of the ethics literature relevant to climate change.

2.   THE MORAL DIMENSION OF CLIMATE CHANGE

The climate change problem may be summarized thus: human activity affects the climate, and the ensuing climate change harms vulnerable others (the 'others' including poorer countries and future generations). Put in these terms, climate change is indisputably a moral issue. However, climate change has a number of features that make this particular moral problem very difficult to address.

In a standard moral problem, an agent acting intentionally imposes harm on another individual. Both individuals and the harm are usually identifiable, and the causal link between the harming action and the harm itself are also clearly identifiable (Jamieson, 2007).

Now consider the following case: millions of people worldwide, driving their cars, flying on holiday and heating their homes, add to carbon emissions in the atmosphere which lead to increasingly extreme storms, floods and droughts, as a result of which millions of other people will die. In this example, it is undeniable that some people are harming others. However, it is not clear how much one has contributed to what harm, and to whom. Additionally, there appears to be no clear intention to do harm: the harms are imposed as a secondary effect of other behaviours   namely, the consumption of energy services.

These various characteristics of climate change make it difficult for individuals to perceive it as a moral problem, and this in turn hinders personal behavioural change based on moral concerns.

As noted by Jamieson (2010) and Greene (2003), our value system, which evolved in low-population, low-technology societies, is inadequate for dealing with moral problems of the type posed by climate change. Our 'moral intuitions' are just evolutionary mechanisms that reflect the environment in which our social instincts evolved – an environment in which significant and meaningful social interactions took place face to face, and not across vast distances with complete strangers.

The present chapter does not purport to suggest where and how our value system and sense of moral responsibility must evolve. Rather, keeping with the idea that climate change must be framed as a moral issue in order for individuals to engage in behavioural change, we will focus on the various features of climate change that distinguish it from standard moral problems, with a view to identifying how best to recast climate change within a conventional moral framework. To summarize, these are:

1.   The nature of harm, and causal link between harming behaviours and harms, is unclear.
2.   Harms are (perceived as) removed in space and time with respect to harming behaviour.
3.   Harms are produced as a secondary effect of other actions, and not intended.
4.   Harms result from collective behaviour.

Notably, ethics has dealt with each of these 'problematic' features individually. There is a large literature in ethics on: the issue of ignorance with regard to morally charged behaviours (discussed in Section 3); the moral permissibility of unintended secondary effects of one's actions (Section 4); and the influence of temporal and spatial distance on the perception of moral problems (Section 5). There is rather less literature on the ethics of collective action problems, with most of the focus being on the attribution of responsibility (discussed in Section 6). Although some of these complex issues are still under debate, there is at least a considerable intellectual foundation on which we can rely for the current analysis.

This chapter addresses two key questions. First, do these various 'problematic' features outlined above reasonably justify behaviours that are responsible for harmful outcomes? For example, does ignorance about the effects of one's behaviour justify harming behaviours?

Second, and more importantly: how exactly do these 'problematic' features lessen – if at all – the perception of moral permissibility of the harming action? By understanding the mechanisms by which these features reduce or eliminate the perception of moral responsibility in morally impermissible situations, it may be possible to approach these features with scissors and glue, and reframe them so that the climate change problem might be perceived, clearly and unequivocally, by individuals as a moral problem.

3.   NATURE OF HARM UNCLEAR

A defining characteristic of the public response to climate change is a lack of knowledge. In an extensive review of survey studies carried out in Europe, the USA and Japan, Lorenzoni et al. (2007) found that survey respondents have a limited understanding of the causes of climate change, often identifying ozone depletion and air pollution as the main causes. Interestingly, survey respondents rarely attribute climate change to their own personal behaviour, and mostly refer to generic causes such as 'traffic' and 'burning fossil fuels' (e.g. Bulkeley, 2000). Knowledge about the impacts of climate change is also vague and generic; Leiserowitz (2006) found that respondents mostly associate the term 'climate change' with generic descriptors, such as 'rising temperatures' and 'world devastation'. All in all, the empirical evidence suggests that public knowledge about climate change is vague and generic, and often confused.

What does this mean for the perception of climate change as a moral problem?

Standard moral problems involve actions that result in clearly defined harms, and in which the causal link between the acting agent and the victim is clear. But what if the acting agent is unaware of the harms that they cause?

Lack of knowledge or understanding can be used to absolve someone from moral responsibility. For example, it is used in criminal cases involving minors, who are treated with leniency because they are considered to lack understanding. The flip-side of this reasoning is that it incentivizes ignorance. This very interesting issue – that of moral culpability under conditions of ignorance – has been discussed at length by Moody-Adams (1994), Zimmerman (1997), Rosen (2003) and Guerrero (2007), among others. I will not attempt to present the many complex arguments discussed in these papers, but will focus on the main points below.

Ethics of Moral Ignorance

The basic premise, proposed by Zimmerman (1997) and elaborated on by Rosen (2003), is that when a person acts wrongly out of ignorance, he is guilty only by virtue of being guilty of his ignorance. The key question to ask in such a case would be: should he have known? The implication is that when actions potentially have a moral consequence, we are obliged to ask questions, investigate, think and take the necessary steps so as to act in the morally correct manner. Rosen refers to this as 'management of opinions'; if it is our opinions that lead us to act in one way or another, then we are responsible for managing them.

Guerrero (2007) builds on this idea by classifying ignorance about a wrong action into three types: (1) ignorance because the person has never thought about the issue; (2) ignorance because, even though the person has thought about the issue, she has come to false beliefs; and (3) ignorance because, even though the person has thought about the issue, she doesn't know what to believe. He proposes that, in some circumstances, the first two types of ignorance may be excusable; the third type, he argues, is never excusable.

To illustrate, consider the example of Tom, who lives in the 1950s, has bought a car, and drives it around. He is completely ignorant of the fact that, in 50 years' time, climate change resulting from his actions will contribute to catastrophic consequences for future individuals. Given the lack of information on this topic, and the fact that the causal link between his action and the consequence is so complex and removed in time and space, there is very little reason to believe that he should have thought about the issue. We would therefore consider him blameless. He is, in effect, ignorant of the fact that he is ignorant of the impacts of his actions.[1]

Now consider Don, who buys his car in 2012. Assume that there is an extensive public transport service where he lives and works, and that Don is wealthy enough to buy a clean-fuel vehicle. However, he drives his car everywhere, completely ignorant of the consequences of his actions. In the year 2012, there is ample information available in the public domain on the impacts of car driving on climate change and air pollution, and, therefore, Don has a moral obligation to ask questions and become informed. In this sense, Don is guilty of his ignorance, and therefore his actions are morally wrong.

This type of ignorance ('affected ignorance') has been discussed at length by Moody-Adams (1994), who rebuts the view that people are blameless for wrong actions that are considered acceptable within their cultures. Thus, even though it is considered culturally acceptable for Don to drive his car as much as he wants, according to Moody-Adams he is not blameless because he has a moral obligation to question his actions (and as a result of his questioning, modify his actions appropriately), even if his culture condones such actions.

Now, consider that Don becomes thoroughly informed, and as a result has developed incorrect notions about the effect of his driving on climate change. He may be blameless, unless he actively avoided engaging with both sides of the debate, and sabotaged his own attempts at discovery, in which case his ignorance would similarly be considered 'affected'.

Finally, consider that – despite reading up extensively on both sides of the debate – Don is unsure as to what is right and wrong. Guerrero (2007) argues that, in the absence of a belief, and given the potential of real harm, he should not drive his car. To do so would be morally wrong. This is, in effect, a moral precautionary principle.

As I will discuss below, ignorance in all its forms is one of the main justifications used by individuals to continue engaging in carbon-emitting behaviours, and it is unlikely that a precautionary principle will change much.

Public Ignorance about Climate Change

As noted earlier, empirical studies have found that, despite widespread awareness about climate change, there is confusion about the causes, consequences and solutions to the problem. Thus the types of ignorance predominant with respect to climate change fall under one of two categories: (2) (ignorance from false beliefs) and (3) (ignorance from confusion) – particularly the latter.

As discussed, the culpability of individuals who fall under category '2' (i.e. the climate change sceptics) depends on whether they really aimed to become informed, or whether their scepticism is motivated by private interests, such as a reluctance to change their lifestyles. Only the latter group are considered morally guilty of their false beliefs.

As for 'confused' individuals (category '3'), Guerrero argues that, in the absence of a firm belief, they should abstain from acting just in case. To continue driving and air-conditioning their homes would be morally wrong. In reality, however, this judicious argument is weak; it does not have the moral weight that an actual belief in negative consequences might have. For example, if I believe that killing animals is wrong, then I will find it harder to eat meat than if I am merely unsure of my opinion on this matter.

Unfortunately, one of the main reasons that there is widespread public confusion about climate change is, very simply, that information in the public domain is mostly confusing and contradictory. Until quite recently, media coverage of climate change largely focused on scientific uncertainties and political disagreements (Carvalho and Burgess, 2005), ignoring the growing scientific consensus on the anthropogenic contributions to climate change (Boykoff, 2007). These practices had the effect of confusing the public, and undermining public trust in science.

It seems evident, then, that the first step towards reframing climate change as a moral problem is to tackle this general confusion among individuals. In particular, the link between specific individual behaviours and the impacts of these behaviours on vulnerable others, must be – if possible – simplified and made clear.

It might be argued that climate change is so complex that there is no way to simplify the causal link between actions and impacts. However, there are countless examples of complex cause–effect relationships that are presented to the public in simple terms. For example, we accept that drinking milk is good for our bones, although the process by which drinking milk benefits our bones is extremely complex and involves a large amount of biological and chemical detail. Similarly, we accept that passive smoking causes lung cancer. However, the process by which inhaled smoke from other people's cigarettes then turns into a cancer is largely ignored by the public.

It seems that the public information on climate change is bogged down with details about processes, and this renders the issue both confusing and impersonal to individual members of the public. What people want to know is: what are the specific impacts of my actions? Who or what gets harmed by my energy-related behaviour?

The question of impacts has only really been addressed over the last decade or so, prior to which most analyses of climate change focused on one variable – the globally averaged surface temperature (Nordhaus, 1993). Although this is useful as an index of change, it is the impacts that result from these temperature changes – such as increased flooding, mortality rates, migration – that are meaningful to people.

Today, there is an increasing wealth of evidence on specific impacts of anthropogenic climate change, including human health effects (Patz et al., 2005; WHO, 2002), effects on the Arctic ice (IPCC, 2007; ACIA, 2004), effects on migration patterns (IPCC, 2007) and impacts on agricultural output (Reilly et al., 2003). Optimally, this information would be used to formulate clear, coherent messages on the linkages between impacts and specific energy-related behaviour. Such a message might say something along the lines of: every year, by driving into work, you produce (say) one tonne of carbon. One tonne of carbon has approximately X effect on crops/arctic ice/ flooding severity/other. By reframing climate change in these simple terms of cause and effect, we come one step closer to engaging individuals, both cognitively and morally.

4. HARMS PERCEIVED AS DISTANT

A common perception among members of the public across the developed world is that climate change is not personally threatening (Lorenzoni et al., 2007; Lorenzoni and Pidgeon, 2006). Even in areas that are considered potentially vulnerable to the effects of climate change, members of the community show little concern about the problem (Bickerstaff et al., 2004). However, most respondents do consider climate change a serious threat – but only for future generations, or people living far away.

This is interesting, because the issue of harm to others tends to be a key driver of moral thinking, especially in Western society (Haidt and Graham, 2007). Part of the success of the anti-smoking campaign, for example, is attributed to a change in how the issue was framed. Where smoking had previously been portrayed as a personal choice with hazardous health consequences (but a personal choice nonetheless), the publication of the 1986 Surgeon's General Report, which documented the hazards of second-hand smoke, made smoking a strictly moral issue. This reframing of the debate helped shift public opinion in favour of the anti-smoking lobby (Rozin and Singh, 1999).

Why is it, then, that, despite recognizing that climate change poses a significant threat to vulnerable others, individuals fail to respond as they do to other situations involving harm to others?

Ethics and 'Distant' Harms

Consider the classic trolley problem devised by Philippa Foot in 1967 and expanded on by Judith Jarvis Thompson (1978): a trolley (i.e. train) is running out of control down a track. In its path are five people who have been tied to the track by a mad economist. In one scenario, you are given the option of flicking a switch, which will force the trolley down another track. Unfortunately, there is a single person tied to that track. What should you do? Flick the switch or do nothing? In another scenario, instead of a switch, you have the choice of pushing a fat man in front of the trolley; he will die, but his body will stop the trolley from killing the five. Again: what do you do?

Both alternatives involve killing one to save five, and yet they elicit very different responses from people. Empirical evidence indicates that whereas people largely favour flicking the switch, they do not agree with pushing the man onto the tracks (e.g. Greene et al., 2001).[2] This apparent inconsistency has been discussed and debated at length by philosophers, with a view to identifying a principle that would explain the different responses.

However, recent research in neuro-ethics suggests that the solution to the trolley conundrum lies not in our intuitive moral thinking or in any rational moral principle, but in the way our brains are wired. In experiments carried out by Joshua Greene and colleagues, subjects' brain activity was scanned using functional magnetic resonance imaging (fMRI) while responding to questions about the trolley problem. Findings indicated that the fat man scenario generated greater activity in the affective/emotional parts of the brain, while the switch scenario activated parts of the brain associated with cognitive/controlled reasoning (Greene et al., 2001).

Pushing a fat man to his death with our bare hands sets off emotional alarm bells in our brains – and as result, we reject this action as highly immoral. However, flicking a switch is a hands-off affair, and so the emotional alarm bells are not activated. The upshot of this is that we are more likely to reject harming actions that involve a strong personal component compared to those which are more impersonal, even when the outcomes are identical. There have been a number of studies since, confirming the link between emotion and moral judgement (for a review, see Prinz, 2006).

Irrespective of the question of whether it is acceptable or desirable to base moral evaluations on emotional factors (for an impassioned argument against emotion-based moral judgement, in the context of climate change, see Grasso, 2011), the relevant issue in this chapter is: what does this mean for the perception of climate change as a moral problem?

Public Perception of Climate Change as 'Distant'

It is fairly evident that climate change, as framed at present, is unlikely to trigger emotional responses in people, which in turn means that they are unlikely to evaluate climate change in strongly moral terms. Of course, the ideal would be that we, as individuals, could make reasoned, utilitarian moral assessments without the interference of our emotions, such that pushing the fat man onto the tracks would be considered morally akin to flicking the switch. Arguably, such a rational, utilitarian-minded world might be a better place.

However, the simple fact is that utilitarian reasoning does not inspire strong moral convictions, and hence decisive actions based on these convictions. As noted by Rozin and Singh (1999), the most effective way of assigning a moral value to an action is through the recruitment of a strong, negative emotion, such as disgust. By associating a morally questionable action with a strong, negative emotion, it is much more likely that individuals will cease to engage in that behaviour. As Daniel Gilbert humorously puts it, 'global warming is bad, but it doesn't make us feel nauseated or angry or disgraced, and thus we don't feel compelled to rail against it as we do against other momentous threats to our species, such as flag burning'.[3]

Maybe climate change fails to make us feel nauseated or angry because we cannot see its victims, and we have no personal connection with them. It is noted by Walvin (2007), with reference to the abolition of slavery in Britain, that one of the key elements leading to abolition was the appearance in the 1780s of large numbers of freed blacks on the streets of London. Their presence, he argues, helped focus the debate on slavery. More recently, the successful claim in 1992 by the James Bay Cree Indians against Hydro-Québec, which was planning to dam the James Bay in Northern Quebec, was partly due to their 'up close and personal' approach, which took them to the streets and schools and legislature of New York (projected to be one of the main consumers of the electricity) to talk to people and explain their situation. On 28 March 1992, Mayor Cuomo cancelled the $20 billion contract with the Quebec government (Heinzerling, 2007).

Of course, it may not be feasible (not to mention rather carbon-intensive) to move large groups of climate change victims around the world to convince others to curb their carbon emissions. However, maybe they could become more visible via the media.

The 'face of climate change' in the media has undoubtedly been that of the polar bear swimming to a sure death in a melting Arctic. The image has emotional content, so our emotional alarm bells are (at least, somewhat) stimulated – and yet there has been no evidence of major reductions in individual carbon emissions since the appearance of this image in our homes. Why? Some argue that ultimately, no one cares that much about polar bears because they have no direct experience of them (Shanahan, 2007).

It might also be because, ultimately, it is not clear how we – as individuals – can act to save that particular drowning polar bear. The link between our energy-related behaviour and that polar bear is vague and uncertain (as discussed in Section 3), and the effectiveness of our actions as individuals depends on other individuals acting similarly (discussed in Section 6). It might also simply pander to the increasing public cynicism and distrust of climate scientists and journalists in general. Pulling the heartstrings of the public is perhaps no longer effective, and, in fact, may be detrimental to the public willingness to engage in climate change.

It is suggested that the media might more effectively focus on the courtroom, where victims of climate change are increasingly taking their plight. Environmental lawyers and grassroots organizations have started to use litigation and other legal procedures as the means to effect change in this area (Martel, 2007; Aminzadeh, 2007). In the USA alone, 431 cases have been filed related to climate change litigation (Gerrard and Howe, 2012). Climate change litigation is considered by many to be the next big target for the legal profession (Schwarte and Byrne, 2010). Although there have been no successful climate change claims made by victims, Dahl (2007) claims that it is just a matter of time.

There have, however, been some recent victories in the courtroom in which prevention of climate change was used as the defence. For example, in September 2008, six Greenpeace climate change activists were cleared of causing £30 000 of criminal damage at a coal-fired power station in Kent, UK. Their defence rested on the argument that they were trying to prevent climate change from affecting property around the world (the power station, incidentally, produced 20 000 tonnes of CO_2 daily) (Vidal, 2008).

Media attention to these climate-change-related court cases may serve to raise the profile of the victims, which, in turn, may go some way towards increasing our sense of personal connection with them – and perhaps the perception that we are effectively 'pushing them onto the train tracks'. Furthermore, successful claims could signal quite strongly the moral impermissibility of carbon-emitting behaviour (and might directly affect public pockets, if governments or companies have to pay compensation).

5.   HARMS AS EXTERNALITIES

Environmental issues, such as climate change, have been critical in drawing attention to the problem of third-party impacts, or 'externalities'. Negative spillover effects from economic activity are on the increase as populations grow and economies become increasingly interlinked. Furthermore, competitive pressures in a growing global market incentivize firms to externalize costs whenever and wherever possible. As Hahnel (2007) puts it: externalities are pervasive in everything we do.

The question is: is it morally permissible to harm others as a secondary effect of another activity, when the harm is not actually intended?

Ethics and Externalities

Guidance within normative ethics on the acceptability of third-party impacts is provided by the much-disputed doctrine of double effect (DDE), according to which secondary effects of actions are acceptable only if not intended. Thus, for example, it may be permissible for Tom to kill Sue as a secondary effect of his behaviour only if he did not intend it, and subject to the following constraints (Driver, 2007):

1.   The act must be good in itself, or at least be indifferent.
2.   The agent does not intend the third-party impact, and if they could avoid it, they would.
3.   The agent is not seeking the third-party impact as an end, or means to an end.
4.   The good end must be proportional to the third-party impact.

DDE is particularly used in medicine, and warfare, where certain actions have foreseeable negative consequences that are not intended. For example, DDE might be used to explore the moral permissibility of a national vaccination programme, in which it is foreseen that a small number of individuals might die due to negative reactions to the vaccine. DDE would condone the vaccination programme (assuming it saved many people) because the death of the few was strictly not intended.

However, as pointed out by McIntyre (2001), there may not be a morally relevant distinction between intention and foresight. For example, if I foresee that my heavy smoking will affect the baby sleeping next to me, but I do not intend to harm the baby, am I free from blame? Most would agree not. The smoker can avoid harming the baby (stop smoking, move baby), and it is debatable whether the good end (nicotine rush) is proportional to the negative impact on the baby (coughing, increased chances of respiratory conditions).

Now, consider the example of someone driving their car to work each day, thus indirectly contributing to extreme climatic events in vulnerable areas. The act is good in itself

– or at least indifferent (it gets the driver to work so she can provide for her family); she does not intend to harm anyone.[4] However, maybe she could avoid the harms by purchasing a cleaner vehicle, driving less, car-pooling or taking public transport? All of these options involve expenditures of time and/or money, and, admittedly, some people simply do not have the resources to change their behaviour; in these situations, the behaviour might be considered morally acceptable. However, many people do have the resources to change their behaviour and fail to do so. The actions of these people are therefore morally impermissible.

Furthermore, it could be argued that the driver who chooses to keep her personal costs of driving down by externalizing those costs onto the environment, is in effect using the environment as a means to an end. The atmosphere is absolutely essential for the driver's enjoyment of cheaper driving; she is using it to absorb the toxic gases from her car, which allows her to keep driving cheaply. And she knows (or indeed, she has a responsibility to know, as discussed in Section 3) that this is bad for others. She is, in effect, externalizing costs onto others. She could argue that she did not intend to harm others; it was merely foreseen as a result of her using the atmosphere to absorb the toxic gases from her car. And thus we come up against the 'distance' problem inherent in the trolley example: put a switch (or in this case, an atmosphere) between the acting agent and the victim, and morality seems to disappear from the picture.[5]

Finally, there is the proportionality constraint (number 4), which asks: is the value of the externality proportional to the value of the intended outcome? Take, as an example, our car driver, and another ten million car drivers. The value to them of using the air conditioning when they drive, say, is a function of comfort. Now, imagine that a small village in another country will be hit by extreme storms resulting from climate change, and that all the villagers will die (and let us assume, in this case, that it was the actions of those ten million and one car drivers using their air conditioning that contributed to this particular climatic event). Although results of a simple cost–benefit analysis may well indicate that the proportionality constraint holds – that is, the total value to the ten million car drivers of using their air conditioning outweighs the total value to the 150 villagers of losing their lives – this leads to a situation somewhat akin to Derek Parfitt's 'repugnant conclusion' (Parfit, 1984); that is, the aggregate value of a tiny change in utility for a million people outweighs the aggregate value of a huge utility change for 150 people, such that we favour the million.

There has been much discussion in the literature on the validity and fairness of aggregation in welfare analysis (see ch. 6 in Yeager, 2001 for a summary of the debate), so I will not attempt to address this here. Suffice to say that DDE would probably allow the car driver to use her air conditioning in the above example, due to the proportionality constraint holding, despite what appears to be a morally questionable outcome.

But I Didn't Intend it!

DDE is problematic for the various reasons outlined above (i.e. intention and foresight overlapping, proportionality constraint favouring majorities with small utility gains); however, in the absence of any other normative theory on third-party impacts, DDE will remain the primary source of guidance on the morality of third-party impacts. Using DDE as a guide, harms posed to others as secondary effects of carbon-emitting behaviours such as car driving may be considered morally unacceptable – if only in reference to constraint 2 (if they could avoid it they should).

Thus the moral permissibility of externalities resulting from energy-related behaviour hinges on the avoidance issue. Take, for example, the success of smoking bans worldwide. The moral argument – that smoking harms others – was coupled with the fact that smokers did not have to smoke in public places, and this made for a very compelling argument.

In the case of energy-related behaviour, the argument is less poignant: many people need their cars, appliances and central heating to secure basic standards of living. Should the moral message about climate change harming others successfully come across, it is not clear that individuals will be able to avoid engaging in the harmful behaviour, simply due to a lack of options.

6.   HARMS FROM COLLECTIVE BEHAVIOUR

The collective action question is possibly the most complex issue to tackle with regard to individual reductions in carbon emissions. Here we have multiple agents, often acting in ways that are not harmful in and of themselves, all contributing to a greater harm. Any one of these actions individually will not cause harm, and it is only through the combination of all these actions that the harm occurs.

As we have seen, one of the features of archetypal moral problems is a clearly defined acting agent. In the example of Tom who kills Sue, the acting agent is clearly defined, and so the question of attribution of responsibility, or blame, is simple. The moment we have multiple agents acting individually, and contributing negligible amounts to a greater overall harm, we have an agency problem. Who exactly is responsible? And for what fraction of the harm? In order for there to be personal moral responsibility, there must be a clearly defined acting agent.

The question of where to locate responsibility for collective action problems is the subject of intense debate in moral philosophy, and will be reviewed below. In addition, I will briefly review the extensive literature on solutions to collective action problems, mostly generated in economics, social psychology and game theory. This literature may provide some indication as to how the mechanism of the collective action problem reduces the perception of individual moral responsibility, and, consequently, may provide some guidance as to how to overcome this problem.

Ethics and the Collective Action Problem

The climate change collective action problem is often portrayed as a prisoner's dilemma (PD), in which the individual is faced with two choices: to cooperate, that is, restrict one's emissions, or to defect, that is, not restrict one's emissions. To cooperate would require the individual to make personal short-term sacrifices for collectively long-term benefits, and this would only make rational sense if all (or a significant most) other individuals also cooperate.

From an ethical perspective, cooperation is generally considered the morally superior position: a Kantian approach would enjoin the participants to always cooperate, whatever the final outcome, the reasoning being that we must always do what we would want everyone else to do (the categorical imperative); a utilitarian perspective would conclude that it is better to cooperate because it leads to the greater good; a contractarian perspective (e.g. Gauthier, 1986) would consider cooperating not just the moral choice, but the self-interested choice – cooperating with the expectation that the other participant cooperates. In summary, there is a general consensus that, in collective action situations, cooperation is the morally right, or good, choice.

What is less clear is the question: where do we locate responsibility for initiating, and sustaining, cooperation and, hence, positive collective action? Do we locate this responsibility in the collective itself, or in the individuals that make up the collective?

There is an ongoing debate among moral philosophers about this issue (although it is generally approached as an *ex post* question, after the fact of non-cooperation). Proponents of 'collective responsibility' argue that groups and collectives can be considered moral agents in their own right, who cause, and are blameworthy for, morally questionable actions (Smiley, 2011). However, it is generally agreed that not all collectivities are appropriate sites for moral responsibility, and only groups with well-ordered decision-making procedures in place ('conglomerate collectivities') are considered appropriate. This is primarily because these types of groups contain an identifiable moral agent (i.e. the governing or representative body) that makes the decisions that inform the group actions (Smiley, 2011). Random groups of individuals ('aggregate collectivities') with no formal decision-making structure are generally considered inappropriate sites for moral responsibility.

Critics of the idea of collective responsibility highlight the fact that it is individuals who contribute to a collective action, and that there is no such thing as a 'collective mind' that can produce a collective intention. Only individuals, they argue, can have moral agency. Their main criticism, however, is directed at the non-distributional nature of collective responsibility. Specifically: should a whole collectivity be held responsible for the actions of particular group members?

This is a fascinating debate, but the relevant question for this chapter (should moral responsibility about climate change impacts be directed at individuals, or at collectives?) remains unanswered. Given that ethics does not help us resolve this question satisfactorily, we must look elsewhere for some guidance on how best to allocate responsibility.

Solutions to the Collective Action Problem

There is a vast literature in economics and social psychology, mostly involving experimental methods or field research, which explores potential solutions to collective action situations. This literature has been reviewed extensively in Kollock (1998), so this will not be repeated here. The aim of this section is to briefly review the literature so as to identify how individual-level solutions fare in comparison to group-level solutions. This may give us some indication as to where to locate responsibility for the climate change problem.

Individual-level solutions to collective action problems – mostly identified in lab-based studies – typically aim at enhancing reciprocity between individuals. This can be done by: (a) encouraging communication between individuals; (b) ensuring interactions between individuals are frequent or durable; (c) increasing identifiability of individuals and transparency of their actions; and (d) providing mechanisms by which individuals can punish non-cooperators. Generally, the more conditions are met, the greater the likelihood of cooperation.

In a laboratory setting, the desire to cooperate can be assisted by facilitating the above conditions in the game. This may be a simple process when the game situation involves a handful of agents. The challenge in the climate change context is how to facilitate these conditions among the vast, dispersed and non-unified collection of individuals and institutions that contribute in uncertain ways to climate change. As the size of the collective increases, individuals find it harder to communicate with each other, to monitor each other's behaviour, and to sanction the behaviour of others, leading to a decline in cooperation. Furthermore, as group size increases, individuals tend to feel increasingly powerless to effect meaningful change.

This perceived lack of efficacy has been identified by researchers as one of the main reasons that individuals do not cooperate in collective action problems. Kollock (1998) reviews a number of studies that find that, if a collective action problem is structured in such a way that individuals perceive themselves as able to have a significant effect on the outcome, then the chances of cooperation increase (e.g. Bornstein et al., 1990). One way in which the collective action problem can be restructured is to introduce thresholds, by using a step-level production function for the public good. In such functions, actions by up to k individuals make no difference to the outcomes, but actions by k or more individuals shifts the benefits upwards. In this situation, each of k individuals' contributions is crucial to reach the threshold at which provision of the public good becomes positive, and none can free-ride (or else the threshold is not reached). In these situations, individuals are much more likely to cooperate (Ostrom, 2002).

Additionally, the creation of a minimal number of individuals who can affect change (k) confers an important sense of group identity upon those individuals, and this has been shown to significantly increase the chances of cooperation within groups (Brewer and Kramer, 1986; Chen and Li, 2009). In fact, the impact of group identity is so strong that it increases cooperation even if the group is composed of strangers assigned to arbitrary groups, as in the seminal studies by Henri Tajfel and colleagues (Tajfel et al., 1971, Tajfel, 1974). Furthermore, experimental studies by Rapoport et al. (1989) and Bornstein et al. (1990) have found that combining a step-level function and group identity, with groups competing against each other for a prize, stimulates intragroup cooperation even further.

Public Moral Responsibility in the Climate Change Collective Action Problem

Climate change, as it is currently presented to the public, suffers from a serious agency problem. There is no clear indication about *who* is responsible for *what*. As noted at the beginning of this section: for there to be personal moral responsibility, there must be a clearly defined acting agent. So how do we locate our acting agent?

Framing the climate change collective action problem in terms of individual responsibilities is problematic for the very simple reason that individual contributions to climate change are almost negligible. Consequently, individuals are unlikely to have a strong sense of moral responsibility in this context. On the other hand, framing climate change in terms of collective responsibilities can be problematic too. One of the major arguments in ethics against the concept of collective responsibility is that it liberates individuals from personal responsibility. This is particularly important with regard to climate change, which involves a very large and dispersed collective made up of individuals who have little influence on each other's behaviour. Personal responsibility is likely to be highly diluted in this context.

However, individuals are also members of smaller collectives, such as their workplace, local neighbourhood, sports centre, residents' association, parents' group, religious group and so on. The important work by Elinor Ostrom (1990) on communal tenure arrangements for the management of environmental ecosystems, such as fisheries, demonstrated that collective action problems could be overcome by small-scale communities without any need for external support or coercion. In these communities, the aforementioned conditions for cooperation between individuals (communication, transparency, repeated interactions and ability to punish defectors) were always present, and sanctioning costs were modest.

Given the difficulty of convincing millions of individual energy users to cooperate with complete strangers, it might be more effective to encourage cooperation within these small-scale collectives (e.g. workplace, residents' association), such that they engage in carbon-reducing actions. Incentives for cooperation might come in the form of formal public recognition for efforts, such as certification schemes. Other incentives might include infrastructural assistance, subsidies or tax breaks. Furthermore, these approaches would serve to enhance the social capital of these small-scale collectives and of the wider society that they belong to, thus enhancing the ability of individuals within these collectives to deal with long-term collective action problems.

In short, by reframing the climate change problem in terms of attainable thresholds and associated group actions by small-scale collectives, we might help to overcome the agency problem inherent in this collective action problem. How possible this is in the context of climate change is another matter. Will it be possible to translate the energy use of a collective such as a university, for example, to a measurable impact – such as the loss of X acres of land in low-lying islands? This, yet again, is a matter for the climate scientists.

7. RECASTING CLIMATE CHANGE WITHIN A STANDARD MORAL FRAMEWORK

In ethics, climate change is viewed as a 'perfect moral storm' (Gardiner, 2006) because it presents so many challenges to conventional moral thought. Ideally, we would become aware of the pitfalls in our conventional moral appraisal processes, most of which are rooted in our evolutionary past, and upgrade our values and sense of responsibility to adjust to our new highly populated and globalized world. However, this chapter argues that, in the meantime, while we adjust our moral thinking to the new world we live in, much of the moral confusion associated with climate change might be resolved by reframing the problem.

Specific recommendations, aimed at policy makers, activist groups and the media, for how this might be achieved include:

1. *Make the nature of the harm clear* This would require more effective communication efforts, involving a whittling away of the huge amount of detail (which, should the public want it, can be made readily available) and development of a clear and coherent message focusing on final impacts. Climate science is producing an increasing wealth of evidence on impacts, and it is up to experts in the public communication of science to formulate this information in usable and coherent forms that are meaningful to people.

2. *Reduce the perception of temporal and spatial distance of impacts by increasing visibility of affected human communities and groups* Consider the example of the James Bay Cree action against Hydro-Québec, discussed in Section 4. Activist groups and other interested organizations could follow suit and assist affected groups so as to increase their visibility to the general public.

    Another approach, such as that taken up by environmental law firms, is to embark on court cases – these are likely to raise the profile of the victims of climate change and raise public awareness. Furthermore, successful claims could signal the moral impermissibility of carbon-emitting behaviour (and might directly affect public pockets if governments or companies have to fork out large sums of money). The perception of distance between acting agent and victim may therefore be shortened considerably.

3. *Emphasize the moral impermissibility of avoidable behaviours and make clear which options are available* The impacts of climate change on third parties are not intended; they are merely (foreseen) negative spillover effects from our use of energy to secure basic standards of living. The moral fibre of this issue hinges on the question: is the externality avoidable? If it is avoidable (if temporal, financial and other relevant constraints allow), then, morally, to continue such behaviour is impermissible.

    This is an important argument as it places greater responsibility onto those who can avoid certain behaviours – and for the most part, this refers to higher-income individuals and families, who are more likely to own cars and large houses with high energy requirements. However, as noted, avoidability with regard to essential behaviours, such as travelling to work, implies the existence of alternatives, such as adequate public transport, car-pooling options, cycling lanes, clean-fuel refuelling stations and so on. As alternatives become readily available, it becomes easier to avoid harmful behaviours.

4. *Reframe the problem in terms of moral responsibilities of small-scale collectives and groups* The problem should be reframed, not in terms of individuals adding to a global calamity through tiny actions, but in terms of empowered groups of individuals who can effect real and measurable change.

8. CONCLUSION

The aim of this chapter was to review the relevant ethics literature in order to identify how best to recast climate change within a conventional moral framework. It is suggested that much of the moral confusion associated with climate change might be resolved by reframing the problem in terms of specific impacts on highly visible groups of victims that can be avoided by groups of empowered individuals who can – and, morally, should – effect real and measurable change.

There is of course a danger in resorting to moralizing in order to change people's behaviour. The very argument used in this chapter to justify the framing of climate change in moral terms – that it will make people change their ways – is also a dangerous argument, because it implies that the question of good/right versus bad/wrong has a very powerful hold on people – and this can be used to ill effect. Throughout time, vested interests have used morality to justify prejudiced and oppressive policies. The liberation movements of the 1960s and 1970s partially liberated us from institutionalized forms of oppression – racism, homophobia, oppression of women – that had been couched for centuries in moral terms (see Hamilton, 2008 for an interesting discussion). Most people do not want to go back there.

However, this does not mean that we should throw away morality for good. The question of whether something is right or wrong is at least a level above the question of what is in one's self-interest. In a sense, the question is a tool for obtaining an 'ideal' – using Bertrand Russell's definition of ideal as 'something having (at least ostensibly) no special reference to the ego of the man who feels the desire, and therefore capable, theoretically, of being desired by everybody' (Russell, 1961, p. 132). An ideal might be wishing there was no hunger in the world; or wishing there was no crime in your cities; or wishing that there was no war, and so forth. These are worthy desires, and should not be ignored out of hand.

It is considered that the role of morality in the context of climate change will be valuable for several reasons: it may help to reduce the impacts of climate change caused by individuals and families; it may serve to highlight the far-reaching harmful consequences of our energy-consumption behaviour; and, it is hoped, it will help to start shifting our dated moral value system towards one that is more suited to our overpopulated and globalized world.

NOTES

1. Of course, the fact that the car emitted smoke from the exhaust would surely alert 1950s Tom to the fact that his car might be polluting the air. Therefore he might not be considered fully blameless, but for the purpose of this exposition, we will consider him mostly blameless.
2. It is worth noting that philosophers had decided, *a priori*, that pushing the fat man was not acceptable while flicking the switch was indeed acceptable. There is a growing literature questioning the traditional 'intuitive' approach used by many moral philosophers to judge moral dilemmas. This approach starts with a moral intuition, and assuming that one's moral intuition is correct, sets out to construct and identify theories to defend it (Copp, 2010; Nichols and Mallon, 2005).
3. D. Gilbert, 'If only gay sex caused global warming', *The Los Angeles Times*, 2 July 2006. Many thanks to Grasso (2011) for this quote.
4. She might even argue that she could not foresee harm either – but as discussed earlier, she would be guilty of such ignorance, and therefore culpable for actions based on this ignorance.
5. Of course, we could argue that there are two actions in this example that should be considered separately: (1) driving a car and (2) polluting the atmosphere. If we focus on the action of polluting the atmosphere, then the foreseen harm to the third party might be morally inadmissible (assuming the polluting action is avoidable).

REFERENCES

ACIA (2004), *Impacts of a Warming Arctic: Arctic Climate Impact Assessment*, Cambridge: Cambridge University Press.

Aminzadeh, S.C. (2007), 'A moral imperative: the human rights implications of climate change', *Hastings International and Comparative Law Review*, **30**(2), 231–65.

Bickerstaff, K., Simmons, P. and Pidgeon, N.F. (2004), 'Public perceptions of risk, science and governance: main findings of a qualitative study of five risk cases', unpublished Working Paper, Centre for Environmental Risk, University of East Anglia, Norwich, UK.

Bornstein, G., Erev, I. and Rosen, O. (1990), 'Intergroup competition as a structural solution to social dilemmas', *Social Behavior*, **5**, 247–60.

Boykoff, M.T. (2007), 'Flogging a dead norm? Newspaper coverage of anthropogenic climate change in the United States and United Kingdom from 2003 to 2006', *Area*, **39**(2), 1–12.

Brewer, M.B. and Kramer, R.M. (1986), 'Choice behaviour in social dilemmas: effects of social identity, group size, and decision framing', *Journal of Personality and Social Psychology*, **50**, 543–9.

Bulkeley, H. (2000), 'Common knowledge? Public understanding of climate change in Newcastle, Australia', *Public Understanding of Science*, **9**, 313–33.

Butler, C. (2010), 'Morality and climate change: is leaving your TV on a risky behavior?', *Environmental Values*, **19**(2), 169–92.

Carvalho, A. and Burgess, J. (2005), 'Cultural circuits of climate change in UK broadsheet newspapers, 1985–2003', *Risk Analysis*, **25**(6), 1457–69.

Chen, Yan and Li, Sherry Chin (2009), 'Group identity and social preferences', *The American Economic Review*, **99**(1), 431–57.

Copp, David (2010), 'Experiments, intuitions, and methodology in moral and political theory', UC Davis Working Paper.

Dahl, R. (2007) 'A changing climate of litigation', *Environmental Health Perspectives*, **115**(4), 204–7.

DECC (2011), 'UK climate change sustainable development indicator: 2010 greenhouse gas emissions, provisional figures and 2009 greenhouse gas emissions, final figures by fuel type and end-user', Department of Energy and Climate Change, 31 March.

DfT (2010), 'Transport statistics Great Britain: 2010', London: Department for Transport.

Driver, J. (2007), *Ethics: the Fundamentals*, Oxford, UK: Blackwell Publishing.

Fehr, E., Naef, M. and Schmidt, K.M. (2006), 'Inequality aversion, efficiency, and maximin preferences in simple distribution experiments: comment', *The American Economic Review*, **96**(5), 1912–17.

Fehr, E. and Schmidt, K.M. (1999), 'A theory of fairness, competition, and cooperation', *Quarterly Journal of Economics*, **114**, 817–68.

Foot, P. (1967), 'The problem of abortion and the Doctrine of Double Effect in Virtues and Vices', *Oxford Review*, **5**, 5–15.

Gardiner, S.M. (2006), 'A perfect moral storm: climate change, intergenerational ethics and the problem of moral corruption', *Environmental Values*, **15**, 397–413.

Gauthier, D. (1986), *Morals by Agreement*, Oxford: Oxford University Press.

Gerrard, M.B. and Howe, J.C. (2012), 'Climate change litigation in the U.S.: chart of litigation cases', Arnold and Porter LLP, available at www.climatecasechart.com.

Gifford, R. (2011), 'The dragons of inaction: psychological barriers that limit climate change mitigation and adaptation', *American Psychologist*, **66**, 290–302.

Grasso, M. (2011), 'The ethics of climate change: with a little help from moral cognitive neuroscience', CISEPS Research Paper No. 7/2011 Working Paper, University of Milan, Bicocca.

Greene, J. (2003), 'From neural "is" to moral "ought": what are the implications of neuroscientific moral psychology?', *Nature Reviews*, 847–50.

Greene, J., Sommerville, R., Nystrom, L., Darley, J. and Cohen, J. (2001), 'An fMRI investigation of emotional engagement in moral judgment', *Science*, **293**, 2105–8.

Guerrero, A.A. (2007), 'Don't know, don't kill: moral ignorance, culpability and caution', *Philosophical Studies*, **136**, 59–97.

Hahnel, R. (2007), 'The case against markets', *Journal of Economic Issues*, **XLI**(4), 1139–59.

Haidt, J. and Graham, J. (2007), 'When morality opposes justice: conservatives have moral intuitions that liberals may not recognize', *Social Justice Research*, **20**, 98–116.

Hamilton, C. (2008), *The Freedom Paradox: Towards a Post-Secular Ethics*, New South Wales, Australia: Allen & Unwin

Heinzerling, L. (2007), 'Why care about the polar bear? Economic analysis of natural resources law and policy', Working Paper, Georgetown University Law Center, US, accessed from http://scholarship.law.georgetown.edu/fwps_papers/46.

Hoffman, E., McCabe, K., Shachat, K. and Smith, V. (2004), 'Preferences, property rights, and anonymity in bargaining games', *Games and Economic Behaviour*, **7**(3), 346–80.

IPCC (2007), *Climate Change 2007: Impacts, Adaptation and Vulnerability. Working Group II Contribution to the Intergovernmental Panel on Climate Change, Fourth Assessment Report. Summary for Policymakers*, Geneva: IPCC.

Jamieson, D. (2007), 'The moral and political challenges of climate change', in S. Moser and L. Dilling (eds), *Creating a Climate for Change: Communicating Climate Change and Facilitating Social Change*, New York: Cambridge University Press, pp. 475–82.

Jamieson, D. (2010), 'Ethics, public policy, and global warming', in S.M. Gardiner, S. Caney, D. Jamieson and H. Shue (eds), *Climate Ethics Essential Readings*, New York: Oxford University Press, pp. 77–86.

Kollock, P. (1998), 'Social dilemmas: the anatomy of cooperation', *Annual Review of Sociology*, **24**, 183–214.

Kotchen, M.J. (2010), 'Climate policy and voluntary initiatives an evaluation of the Connecticut Clean Energy Communities Program', NBER Working Paper No. 16117, June.

Leiserowitz, A. (2006), 'Climate change risk perception and policy preferences: the role of affect, imagery, and values', *Climatic Change*, **77**, 45–72.

Levitt, S.D. and List, J.A. (2007), 'What do laboratory experiments measuring social preferences reveal about the real world?', *Journal of Economic Perspectives*, **21**(2), 153–74.

List, J.A. (2007), 'On the interpretation of giving in dictator games', *Journal of Political Economy*, **115**(3), 482–93.

List, J.A., Berrens, R., Bohara, A. and Kerkvliet, J. (2004), 'Examining the role of social isolation on stated preferences', *American Economic Review*, **94**(3), 741–52.

Lorenzoni, I., Nicholson-Cole, S. and Whitmarsh, L. (2007), 'Barriers perceived to engaging with climate change among the UK public and their policy implications', *Global Environmental Change*, **17**, 445–59.

Lorenzoni, I. and Pidgeon, N.F. (2006), 'Public views on climate change: European and USA perspectives', *Climatic Change*, **77**, 73–95.

Martel, J.S. (2007), 'Climate change law and litigation in the aftermath of Massachusetts v. EPA', *BNA Daily Environment Report*, **7**(214), 1–11.

McIntyre, A. (2001), 'Doing away with the double effect', *Ethics*, **111**(2), 219–55.

Moody-Adams, M.M. (1994), 'Culture, responsibility, and affected ignorance', *Ethics*, **104**(2), 291–309.

Murray, A.G. and Mills, B.F. (2011), 'Read the label! Energy Star appliance label awareness and uptake among US consumers', *Energy Economics*, **6**, 1103–10.

Nichols, S. and Mallon, R. (2005), 'Moral dilemmas and moral rules', *Cognition*, **100**(3), 530–42.

Nordhaus, W.D. (1993), 'Reflections on the economics of climate change', *Journal of Economic Perspectives*, **7**(4), 11–25.

Ostrom, E. (1990), *Governing the Commons: The Evolution of Institutions for Collective Action*, Cambridge: Cambridge University Press.

Ostrom, E. (2002), 'Property-rights regimes and common goods: a complex link', in A. Windhoff-Héritier (ed.), *Common Goods: Reinventing European and International Governance*, Oxford: Rowman and Littlefield, pp. 29–58.

Parfit, D. (1984), *Reasons and Persons*, New York: Oxford University Press.

Patz, J.A., Campbell-Lendrum, D., Holloway, T. and Foley, J.A. (2005), 'Impact of regional climate change on human health', *Nature*, **438**(17), 310–17.

Prinz, J. (2006), 'The emotional basis of moral judgments', *Philosophical Explorations*, **9**(1), 29–43.

Rapoport, A., Bornstein, G. and Erev, I. (1989), 'Intergroup competition for public goods: effects of unequal resource and relative group size', *Journal of Personality and Social Psychology*, **56**, 748–56.

Reilly, J., Tubiello, F.N., McCarl, B., Abler, D., Darwin, R., Fuglie, K., Hollinger, S., Izaurralde, C., Jagtap, S., Jones, J., Jagtap, S., Jones, J. Mearns, L. Ojima, D., Paul, E., Paustian, K., Riha, S. Rosenberg N. and Rosenzweig C. (2003), 'U.S. agriculture and climate change: new results', *Climatic Change*, **57**:43–69.

Rosen, G. (2003), 'Culpability and ignorance', *Proceedings of the Aristotelian Society*, **103**(1), 61–84.

Ross, L. and Ward, A. (1996), 'Naive realism in everyday life: implications for social conflict and misunderstanding', in E.S. Reed, E. Turiel and T. Brown (eds), *Values and Knowledge*, Mahwah, NJ: Lawrence Embaum, pp. 103–35.

Rozin, P. and Singh, L. (1999), 'The moralization of cigarette smoking in the United States', *Journal of Consumer Psychology*, **8**(3), 321–37.

Russell, B. (1961), *History of Western Philosphy*, London: Routledge.

Schwarte, C. and Byrne, R. (2010), 'International climate change litigation and the negotiation process', Foundation for International Environmental Law and Development Working Paper, October.

Shanahan, M. (2007), 'Talking about a revolution: climate change and the media. An IIED Briefing', Institute for Environment and Development, December.

Smiley, M. (2011), 'Collective responsibility', in Edward N. Zalta (ed.), *The Stanford Encyclopedia of Philosophy (Fall 2011 Edition)*, available at http://plato.stanford.edu/archives/fall2011/entries/collective-responsibility/.

Tajfel, H. (1974), 'Social identity and intergroup behaviour', *Social Science Information*, **13**(2), 65–93.
Tajfel, H., Billig, M., Bundy, R. and Flament, C. (1971), 'Social categorization and intergroup behaviour', *European Journal of Social Psychology*, **1**, 149–78.
Thompson, J.J. (1978), 'The trolley problem', *The Yale Law Journal*, **95**, 1395–415.
Vandenbergh, M.P., Barkenbus, J. and Gilligan, J. (2008), 'Individual carbon emissions: the low-hanging fruit', *UCLA Law Review*, **55**, 1701.
Vidal, John (2008), 'Not guilty: the Greenpeace activists who used climate change as a legal defence', *The Guardian*, 11 September.
Walvin, J. (2007) 'Slavery – the emancipation movement in Britain', Lecture delivered on 5 March 2007 at Gresham College, London, available at http://www.gresham.ac.uk/lectures-and-events/slavery-the-emancipation-movement-in-britain.
Whitmarsh, L. (2009), 'Behavioural responses to climate change: asymmetry of intentions and impacts', *Journal of Environmental Psychology*, **29**, 13–23.
WHO (2002), *The World Health Report 2002*, Geneva: The World Health Organization.
Yeager, L.B. (2001), *Ethics as Social Science: The Moral Philosophy of Social Cooperation*, Cheltenham, UK and Northampton, MA, USA: Edward Elgar.
Zimmerman, M.J. (1997), 'Moral responsibility and ignorance', *Ethics*, **107**(3), 410–26.

27 Decentralization of governance in the low-carbon transition
Nick Eyre

1. INTRODUCTION

Addressing climate change is now widely accepted as a key global challenge for the twenty-first century. It is largely an energy challenge, as the global energy system is the largest emitter of carbon dioxide (CO_2) and therefore a low-carbon economy implies radical changes to the energy system. At the same time energy decision makers face challenges with respect to energy security and, in the developing world, providing much wider access to modern forms of energy.

In delivering on these challenges, very large investments in energy technology will be required and energy cost remains critical, for both industrial competitiveness and social goals. So it is tempting to see the issue as a technical and economic challenge. However, the scale of change is so significant, and the role of energy in modern life so central, that it is a social and political challenge as well. If the system is to change, its governance may also need to change.

This chapter argues that the highly centralized systems of energy governance that have developed over the last century are unlikely to prove fit for purpose in meeting the challenges ahead and that more inclusive and decentralized governance will be required, especially to engage 'the demand side' in the process of change.

The remainder of the chapter begins in Section 2 with a brief assessment of the different roles the demand side might be required to play in the transition to a low-carbon energy system. Section 3 addresses the related and controversial question of the broader consumption trends that drive energy demand and thence energy supply. Section 4 outlines the existing dominant system of governance in the energy sector, Section 5 explores the ways in which this might be inadequate to meet future challenges and Section 6 sets out some of the potential elements of a new governance system. Section 7 identifies some of the new actors who might play a larger role in this revised system. Section 8 draws some preliminary conclusions.

2. THE DEMAND SIDE IN THE LOW-CARBON TRANSITION

Consumers of energy are frequently described, particularly in the electricity sector, as 'the demand side'. Whilst most of the debate about energy and the changes needed in the energy system focuses on energy supply, there is increasing evidence that the demand side will need to play a key role in the transition to low-carbon economies.

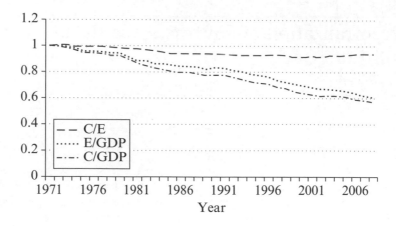

Source: Author calculations based on IEA (2011).

Figure 27.1   Global carbon, energy and GDP ratios, 1970–2008

2.1   Energy Efficiency

Energy efficiency is usually defined, at the level of the individual energy-using process, as the ratio of the energy service derived to the energy used. The importance of energy efficiency is well established: globally the 40 per cent reduction in the carbon/GDP ratio achieved since the first oil crisis is very largely attributable to changes in energy demand, with decarbonization of the energy mix playing a minimal role (see Figure 27.1). There is no prospect of this changing rapidly, indeed in recent years the carbon intensity of energy supply has increased, largely due to the key role of coal in powering the expansion of the major Asian economies. Energy efficiency improvement will therefore remain central to hopes of controlling carbon emissions, at least in the medium term and more probably for even longer.

Despite energy efficiency improvements, we are not in danger of 'running out' of energy efficiency opportunities. Global assessments indicate that the total efficiency of energy conversion from primary energy to useful energy at the point of use is approximately 11 per cent of the theoretical optimum (Cullen and Allwood, 2010) and the efficiency with which services are then provided by that energy is then only 25 per cent of what might reasonably be achieved (Cullen et al., 2011). The implication is that the global energy system is less that 3 per cent efficient compared to what might be thermodynamically possible.

Moreover, it is well established that a significant fraction of this potential is relatively cost-effective. Energy efficiency technologies provide a reservoir of opportunities for some low- (and even negative-) cost carbon abatement. The potential for carbon emissions reduction at costs lower than $100/tCO$_2$ in the key end-use sectors (buildings, industry and transport) is 12 GtCO$_2$/year – more than three times as large as in energy supply (IPCC, 2007). Moreover, if low- and zero-carbon energy sources are to take a much increased share of energy supply, unit energy costs are likely to rise. This will further improve the economics of energy efficiency. These opportunities are sometimes identified as 'low-hanging fruit', although this can be rather misleading as a number of barriers, discussed in Section 3 below, related to current governance structures currently constrain their full use.

2.2   Demand Response

There is rapidly growing interest in the role of distributed energy users in real time involvement in electricity markets ('demand response') to enable effective use of inflexible and intermittent generation options.

Other fuels (solids, liquids and gases) can be stored at reasonable costs, but electricity storage is relatively expensive and only used to a limited degree. Supply and demand therefore need to be in balance even over short periods of time. The traditional approach to achieving this has been to ensure that capacity is sufficient to meet maximum demand to a relatively high level of security. Demand has been treated as difficult to manage, essentially because it is highly disaggregated and diverse, and therefore difficult to engage in the same real-time management processes as power station operators. Constraining maximum demand is helpful, and large industrial users have long had rather sophisticated tariffs that promote such behaviour. But the options for small power users have generally been restricted to the rather crude pricing options allowed by conventional metering.

However, there has been a significant increase in interest in demand response in recent years, driven by two factors. The first is the prospect of very large capacities of wind and solar power. These are inevitably intermittent, and this intermittency is made more problematic if the other generation technologies on the system cannot load-follow easily, which is likely to be the case for both nuclear and coal with carbon capture. Whilst backup generation capacity is a perfectly viable prospect, even for systems with very high penetration of intermittent renewables (Fripp, 2011), it is clearly expensive. The economics of demand response are therefore likely to improve. The second factor is the proposed deployment of smart meters in major electricity systems. Whilst these may have some informational benefits for end users, the key drivers for energy utilities are opportunities to incentivize customers to reschedule loads, either through time-dependent prices or by utility control over non-critical loads, for example refrigeration (Vojdani, 2008). This type of opportunity is now being recognized in some electricity markets (Jenkins et al., 2010).

2.3   Electrification

However effectively energy demand is addressed, it will be unable to deliver all the carbon emissions reductions required in developed countries if the global climate is to be stabilized. Energy users will also need to be able to use low-carbon energy vectors to replace fossil fuels, in particular for heating and transportation. Whilst direct use of solar energy and biomass may be able to make substantial contributions, most studies indicate that electricity will be a critical low-carbon vector (Ekins et al., 2010; IEA, 2010). The implication is that the demand side will have to undertake 'electrification', at the same time as becoming more efficient and more flexible.

Some published scenarios imply very rapid rates of heat pump and electric vehicle market penetration, in some cases up to 100 per cent of relevant markets (Ekins et al., 2010). However, these tend to overlook the multiple challenges to technology deployment that can be deduced from an assessment of the history of energy efficiency.

2.4   Microgeneration

Potentially the most significant change to the demand side is the development of small-scale energy supply options ('microgeneration') within the premises of energy users. Electricity generation from photovoltaics is the most obvious technology, but micro-wind turbines also exist and very small-scale combined heat and power, using either Stirling engines and fuel cells, has been demonstrated. Whilst current deployment levels are modest, growth rates are high in many countries, especially where supported by feed-in tariffs to ensure investments are financially attractive. Some scenarios foresee very high rates of deployment (Boardman, 2007; Jardine and Ault, 2008).

Even with significant deployment, microgeneration is likely to make a smaller contribution to meeting energy service needs than improvements in energy efficiency. However, it has been argued that its importance may be more qualitative than quantitative. With microgeneration, actors whose role in the energy system has previously been that of consumers thereby become generators as well, somewhat blurring the traditional distinction between supply and demand, and therefore between consumption and production. They are neither consumers nor producers, but 'prosumers' (Toffler, 1980). This might result is more proactive attitudes and behaviours towards the role of the individual citizen in the energy system (Sauter and Watson, 2007), but this is likely to require substantial institutional change (Bergman and Eyre, 2011).

3.   CONSUMPTION AND THE LOW-CARBON TRANSITION

The previous section shows the potential for individuals and 'non-energy' businesses to engage differently with the energy system as it develops, moving away from being simply a purchaser of energy into a more active player in fuel choice, market size and balancing, and even small-scale production. However, for the foreseeable future, the dominant relationship with the existing system remains that of a consumer.

Of course, it is well known that what the consumer wants is a set of services (usually known as energy services) that are produced by energy-using equipment from energy bought as a commodity. There have been substantial efforts to develop a market in energy services and in some countries there is now an effective market (Bertoldi et al., 2006). However, it is largely confined to larger energy users who have the capacity and resources to establish and manage the appropriate contracts. Smaller energy users still purchase energy and equipment separately and this market structure seems likely to continue.

The quantity of energy demanded is therefore dependent on both the energy efficiency of the energy-using process and the quantity of energy services required. In any conventional economic analysis, the two are clearly distinct – energy efficiency having the possibility of increasing welfare (depending on the cost-effectiveness of the investment), whilst a reduction in energy service demand is necessarily a welfare loss. Most economic analysis on the scope for reducing energy demand has therefore focused on energy efficiency improvement.

The scope for cost-effective energy efficiency improvement is very large (IEA, 2010; IPCC, 2007). This, of course, runs contrary to the assumptions of a perfect market, in which such investment options would be automatically adopted. The necessary conclusion is that there are a number of market failures that result in real markets differing from the simple textbook model (Brown, 2001; Sanstad and Howarth, 1994). This is now widely accepted and a variety of taxonomies of the barriers exists. However, whilst neoclassical economic analysis can identify such failures, it is less well suited to their explanation. Behavioural and institutional economics is beginning to do this (Gillingham et al., 2009), but essentially by drawing on older insights into the same topic from psychology, sociology and political science (Shove, 1998; Stern, 1986). The conclusion that can be drawn is that the failure of existing markets to deliver optimum energy efficiency is not best thought of as a set of independent market failures, all inconveniently working in the direction of lower energy efficiency, but instead a deeply embedded property of existing energy markets in which centralization, commoditization and consumerism prioritize investment in new supply over that in efficient consumption (Eyre, 1997).

In this sense the barriers to efficiency are not distinct from the nature of the consumption process that generates the demand for energy services. And almost all the difficulties in controlling carbon emissions ultimately stem from the increase in energy service demand driven by economic growth. So the inevitability of ever-rising demand for energy services needs to be questioned. Of course, there have been some fundamental challenges to the role of overall consumption in developed economies: for example, whether it is sustainable (Jackson, 2009) and whether it is less important for social welfare than distributional issues (Wilkinson and Pickett, 2009). However, the fundamental position accorded to economic growth in policy making looks unlikely to change in the near future.

More modest challenges to economic orthodoxy, which may be more politically plausible, relate to the role of energy services in social welfare, and specifically whether lifestyle and cultural change can reduce demand for energy services in a manner that is socially acceptable. A number of thought-experiments in the context of climate change mitigation indicate that such change is indeed possible, either reversing trends in developed countries (Eyre et al., 2010; Fujino et al., 2008) or moderating them in industrializing countries (Jiang et al., 2009; Sukla et al., 2008). Whilst such scenarios raise difficult questions for analysts and policy makers, about if and how such 'environmentally desirable' changes might be stimulated or encouraged by public policy, they have the very considerable attraction that the scale of investment required in low-carbon energy systems can be significantly reduced if demand reductions can be achieved. Indeed, the plausibility of the claim that climate change can be stabilized at a cost of 1 per cent of GDP – that is, the lower end of the range of Stern's much-quoted costs of mitigation (Stern, 2006) – seems to depend on changes of this type.

Whatever the plausibility of specific projections, given the importance of energy service demands for carbon emissions and sustainability more generally, it seems very unwise to assume that their trajectory has a fixed relationship to economic activity that cannot be affected by public policy on the decadal timescales of interest. The driving forces that determine the relationship (preferences, behaviour, lifestyle, culture – depending on the chosen disciplinary approach) vary over both space and time. Consumption patterns vary strongly as a function of culture (Wilhite et al., 1996) and even because of different lifestyles between apparently similar neighbouring households (Socolow, 1978). So it is not surprising that consumption patterns change for reasons other than income and price on long timescales; moreover, given the interactions between technology and society, they are likely to change more quickly in periods of rapid technological change. Energy models based on optimizing welfare are generally inadequate alone to explore this insight. A fixed set of preferences for energy services during a systemic energy change, though a standard assumption in such models, is actually highly improbable. So the very large reductions in carbon emissions now under discussion have unavoidable implications for energy service trends.

The public policy implications have not been fully worked through; indeed doing so will necessarily be a learning, and therefore iterative, process rather than one of 'design' followed by 'implementation'. But it is already clear that, at least in liberal democracies, it is likely to require a deeper and wider engagement of civil society with energy. The prospective changes to the demand side outlined in Section 1 offer some opportunities for energy users to move from being solely passive consumers at the end of a pipe or wire to a more active role. And the implications of this section are that such change will be necessary, but not trivial. There are implications not only for energy markets, but also for the structure of decision making in the energy system. Thinking about public policy as an external force capable of 'managing the change', is probably the wrong framing, where the machinery of the process is embedded in the activities that may have to change (Smith et al., 2005). A wider concept of energy system governance is more appropriate.

4.   ENERGY SYSTEM GOVERNANCE

From the beginning of the 'oil and electricity age' in the late nineteenth century, the priority of governments has been to provide access to modern forms of energy at reasonable cost, both the businesses and households. From this desire emerged the key energy transition of the twentieth century, the creation of large network systems that dominate energy supply, particularly of electricity (Hughes, 1983). Ownership took different forms in different countries, but close collaboration between government and major energy companies was the norm. The key goal was initially network growth to provide access with high system reliability – a focus that remains in the developing world today.

With universal access achieved in developed countries and a reduced support for direct state intervention, market liberalization became a dominant trend in the 1990s. This changed the company–government relationship – elements of competition were introduced and regulation became more formal. The interests of the population were reconceptualized away from universal access (now taken for granted) to supply-cost minimization. But the dominance of large utilities and government in decision making remained.

So energy system governance is typically highly centralized. In this context, 'energy decision making' is considered to cover access to primary energy resources, large-scale investments and energy pricing. The energy system implicitly ends at the customer meter, with the interests of the citizen assumed to be coincident with the interests of the consumer – a highly reliable system with a low and stable price for energy. Energy companies seek to deliver such a system, economic regulators are mandated to ensure it, individuals and business expect it and governments worry about any signs that it may not happen. In this context, the key actors in governance are national governments, who remain effectively politically accountable even for privately owned systems, and the large energy corporations that dominate the energy supply sector. Barriers to entry are high, both to the market and therefore to the key roles in system governance (Mitchell, 2007).

The system of governance reflects and reinforces centralization within the decision-making process. Energy is viewed as a key responsibility of the nation state. Even in countries where powers over energy are constitutionally devolved (e.g. the USA, Canada and Australia), federal governments have a tendency to accrete powers, based on the interstate nature of the physical systems and markets. And powers that are routinely devolved to lower levels of government may be retained for key energy investments, for example land-use planning decisions about large power stations in the UK.

Whilst policy remains formally the preserve of government, the nature of the governance system accords a major role in practice to the energy companies, without whose participation policies can become difficult to implement. Key decisions are made with reference to these actors; in contrast, representatives of energy users (e.g. the cities in which most energy is used) are so routinely marginalized from energy decision making that it is rarely even commented upon. Even policies requiring more decentralized implementation have tended to become the province of the incumbent energy industry, for example through obligations to buy renewable energy (whether through certificate schemes or feed-in tariffs) and to deliver energy efficiency improvements in buildings.

The increasingly transnational character of energy markets has supported the centralization of governance. Oil shocks in the 1970s led to the formation of the International Energy Agency as an international response to oil security. And the rise of climate mitigation as a central concern of energy policy has led to European and international carbon markets as the perceived cornerstone of policy responses in most countries. In neither case was this outcome the only possible response. Measures to reduce energy demand and increase use of indigenous renewable energy are also relevant and have formed part of the policy response in most countries. But the centralized nature of governance means that the principal policy responses tend to be measures at the national and international level, that is, the levels at which the key governance actors operate, rather than more decentralized responses, which tend to be considered 'difficult' and 'unreliable' precisely because of their remoteness from the centres of political power. Even though the majority of energy use and its associated emissions are decentralized, policy solutions that might engage these actors tend to be rejected (Parag and Eyre, 2010). The broad set of actors involved in energy use are generally treated, and consider themselves, as marginal stakeholders, merely energy consumers.

5. THE NEED FOR CHANGE

Patterns of consumption are changing; post-industrial energy consumption is increasingly distributed. In the OECD, the use of energy in industry has not risen since 1970, with the sector's share of total demand falling from 35 per cent to 25 per cent. At the same time, energy use in buildings and transport has increased by more than 50 per cent (IEA, 2011). Energy use is no longer associated principally with economic production, but with consumption, where the scale of individual uses is smaller, more diverse and more distributed. And in power generation, the scale of typical new-generation plant began to fall in the 1980s, as first gas and then renewables (both with characteristic generator sizes smaller than coal and nuclear plant) took a larger share of the generation mix. Most energy futures foresee a continuing relative shift to a more distributed energy future.

The existing key actors in energy markets and governance are not, alone, able to deliver the changes implied by the low-carbon transition. Even in electricity generation, the dominance of large companies arguably inhibits some of the innovation that might be helpful in delivering change (Mitchell, 2007). And transnational companies have a relatively poor record in persuading local communities that wind farms are in their interest, partly at least because the financial benefits do not accrue to the community (Devine-Wright, 2005).

For the 'demand side' (as broadly defined in Section 2), the current governance arrangements are arguably even less appropriate. The dominant actors (national governments and transnational corporations) are physically remote from the sites of demand-side solutions. The types of intervention needed are in urban planning, buildings, transport systems and industrial processes, which are external to the business models of energy companies and generally also outside the purview of the government departments responsible for energy policy. In other words, the dominant governance actors are unable to deliver them very effectively, even if they wish to.

International energy companies are unlikely agents of the deep refurbishment needed to convert the existing building stock to a high level of energy efficiency (Killip, 2011). Although energy companies in some countries have a good track record in delivering energy efficiency obligations, this has been delivered primarily through partnerships with locally based building companies and restricted to relatively simple measures (Eyre et al., 2009).

Similar issues are beginning to emerge in the other potential roles for the demand side identified in Section 2. Microgeneration faces a number of barriers not dissimilar in character from those identified historically for adoption of energy efficiency, in particular the need for active engagement of users (Sauter and Watson, 2007). Electrification of building heating faces a wide range of technical and social challenges, and therefore will be at least as difficult as efficiency improvement (Fawcett, 2011). Whilst rapid deployment is technically feasible, it will depend on concerted action in many areas, and it cannot be assumed that it will happen simply because low-carbon electricity supply is available (Eyre, 2011). Demand response clearly will need energy companies to play a leading role in providing the incentives, but the smart metering technology required to facilitate this communication already faces resistance, notably in the Netherlands, because of lack of trust of householders in energy utilities (Darby, 2010).

Existing governance arrangements are probably most starkly unsuitable for the task of reducing consumption. Energy companies clearly cannot be expected to do this; indeed some consumer resistance to existing energy efficiency programmes results from a suspicion that sellers of energy cannot be expected to want to reduce demand (Eyre, 2008). Much of the existing analysis of the low-carbon transition neglects energy users

(Shove and Walker, 2010). Arguably this is because reconceptualizing the user role as an active shaper of markets presents such a fundamental challenge to consumer societies.

If decentralized actors are needed as active players in future energy policy, there is a strong case for broadening the conception of the energy system and therefore its governance system. This would cover both the scale and scope of energy governance, and it would place more weight upon influencing the decentralized decision-making processes involved in energy use and the practices that generate the demand for energy services in the first place.

6. THE ELEMENTS OF CHANGE

6.1 The Government Process

A key element of any change will be for formal government processes to recognize that energy practices occur across society in almost all aspects of production and consumption. So redefining the energy system to include energy demand and decentralized participation is a very significant step. However, trying to redesign government structures and processes to match such a concept exactly is likely to prove futile: it would simply result in everything being redesignated as energy policy. But what does need to change is to give 'energy' and 'climate' the same status as 'economy', that is, to recognize that a very wide set of activities have an energy impact and to take this into account in policy making.

Critical government functions will include the three sectors that are primarily responsible for energy demand in modern economies – 'transport', 'built environment' and 'industry'. These are typically the responsibilities of other government departments (quite reasonably, because the social goals of these activities do not relate specifically to energy). But any energy policy can be undermined by decisions made in these sectors, so it is critical that policies, programmes, targets and performance are evaluated against energy metrics as well as more conventional sectoral goals. The use of energy/m^2 as a metric of building standards is becoming more common, although still primarily as a design intention rather than an evaluated outcome. Concepts such as food kilometres (or food miles) have entered public discourse, but still hardly feature in formal analysis of transport infrastructure policies. More broadly, trade policy and globalization tend to be treated, depending on political preferences, as either self-evidently beneficial or having some negative social consequences, but rarely with any reference to energy.

The built environment is the largest energy user in modern economies and has the scope for the most low-cost greenhouse gas reduction (Levine et al., 2007). Policies for reducing energy demand will include moving rapidly towards high thermal standards both for new buildings and retrofitting older buildings (Ürge-Vorsatz et al., 2012). In both cases, the key delivery agents will be the buildings sector, for which energy has traditionally been a minor concern. Energy efficiency, microgeneration and electrification imply investment and engineering works at the point of energy use. Much of this will be by the traditional building trades, which not only undertake the work but also frequently act as key advisers to householders and small businesses (Banks, 2001; Killip, 2011). In particular, residential building refurbishment is normally undertaken by small companies or sole traders (or even by householder 'do-it-yourself') with skills sets where low-energy design is not currently prioritized (Nosperger et al., 2011). Government departments and agencies responsible for building standards enforcement and construction sector skills will therefore have a key role.

In the transport sector, energy futures are potentially affected by both climate change concerns and the availability and cost of petroleum-based fuels. Both point to a significant break with the long trend of rising consumption of oil-based fuels for road and air travel. Most sustainable futures point to some combination of reduced travel, modal shift, more efficient vehicles and alternative fuels (Ekins et al., 2010). The policy levers implied range from differential taxation and incentives for low-energy and alternative fuel vehicles to support for low-impact travel modes and urban planning prioritization of dense settlements. Within transport policy as practised, energy and low-carbon policies tend to be reflected primarily in technology policies, rather than broader issues around the need for mobility or modal shift. And other policy drivers of transport energy use fall into other policy domains such as fiscal policy, innovation support and land-use planning.

All of this implies that responsibility for energy transition must inevitably lie partly outside the focus traditional focus of 'energy policy' or even 'energy and climate policy'. Departments with responsibility for construction, land-use policy and transport are also involved. And the departments traditionally central to policy making, through coordination and finance, will inevitably be critical. Some institutionalization of these commitments is likely to be needed. The UK has arguably taken some initial steps in this direction with the allocation of its self-imposed carbon budgets between responsible departments (HMG, 2009). Whether such commitments can endure the inevitable pressures from competing goals remains to be seen.

6.2   Governance Scale and Scope

Whilst broader responses within central government will be necessary, they are not sufficient to constitute effective governance for a decentralized energy system. Changes of scale imply that energy governance becomes more distributed. In some cases, this may be to supranational institutions, for example product policy for standards for vehicles and electrical appliances. However, the main implications are for policy more locally, in local government, most obviously on local land-use and transport policies.

Governance implications also extend far outside the remit of formal government policy. For the finance sector, the shift in investment to decentralized actors will involve more locally based lending, including more informed assessment of sustainable energy investments. New types of infrastructure (e.g. for mass transit, cycling, walking, heat and smart grids) imply an active role for other regulated economic actors, such as operators of power grids and public transport systems. Infrastructure has a major role in constraining and shaping different consumer 'choices' (Unruh, 2000), so any attempt to change end-use technology without reference to its importance is likely to be ineffective.

Greater public engagement implies that many initiatives will need to be delivered, and perhaps even designed, primarily at a local level, for example through local government, third-sector groups or community-based businesses. This represents a major change for energy policy, which has traditionally been highly centralized. A key example lies in energy-saving programmes. It has been known for decades that energy company programmes can achieve greater uptake when local government and community groups are engaged as delivery partners (Hirst, 1989; Stern et al., 1986). But these partnerships were largely lost during decades of reliance on large, market-based solutions and are now having to be reinvented by energy companies facing increasing targets in both Europe and the USA (e.g. Platt, 2011).

6.3 Governance Goals

Changing consumptions patterns towards more sustainable ends also has implications for governance. What is now frequently called 'pro-environmental behaviour' is not readily amenable to direct policy control (at least in liberal democracies). However, it is culturally influenced and therefore affected by social processes. In particular, it may be steered by individuals and organizations that are highly trusted. As neither national governments nor large energy companies expect to fall into this group, other actors are required in the governance system if pro-environmental behaviour is to be encouraged. Engaging a wider set of actors more effectively requires a broad set of policy approaches (Jackson, 2005). These include traditional policy levers (incentives and regulation) to encourage change, but also extend to exemplification (government leadership by example) and more facilitative measures to engage civil society and enable effective action.

Even with the growth of Internet communication, engagement of a wider public in energy issues has proved difficult for government. Significant change is more likely under the influence of friends, family and community exemplars than by remote exhortation. The implication is that government may need to support social as well as technical innovation. Much of innovation theory emphasizes the strong interconnection of technical and social change, but this has had little impact on government-funded innovation programmes, which have largely focused on technology. However, more ambitious community-led investments are now increasingly seen as a way forward. They clearly have the potential to include a wider range of actors and thereby overcome the scepticism that attaches to the initiatives of governments and multinationals. However, 'community projects' are not a panacea. The phrase is used to cover a wide range of initiatives, including many that have little direct engagement with most members of geographical communities, and therefore concerns that the approach is not genuinely sustainable without more fundamental changes to existing institutional structures (Walker et al., 2007).

Enabling change depends to some extent on ensuring that decentralized actors have the financial resources to invest. But non-financial interventions are also required, as action is not purely economically driven. Enabling change therefore involves ensuring that the information, skills and capacity are available to enable positive intentions to be put into practice. General information campaigns are probably the easiest interventions of this type to organize through existing governance systems, but information alone is notoriously ineffective in delivering change. Direct, personalized advice is significantly more effective (Eyre et al., 2011). It requires more local and resource-intensive approaches, including the establishment of local advice agencies that understand not only the physical nature of buildings in their area, but also the dynamics of community and local business engagement. Elderly and disadvantaged energy users can already benefit from higher levels of support than basic telephone advice, and this may become true for more energy users as decentralized energy interventions, for example in the built environment, become more technically complex, as when priorities move from basic insulation measures to low-carbon heating systems.

Traditional energy policy instruments of taxes and regulation are therefore insufficient to secure sustainable change. Enabling and engagement are activities outside the mainstream of the usual policy debates. They involve a less directed and less predictable process than achievable with the traditional tools of government. They therefore sit less comfortably within the usual policy analysis approaches, as they do not assume that costs and benefits are entirely predictable or at the command of central government.

So governments need to think about new 'delivery mechanisms' outside their direct control. This implies not just new actors that will be needed, but new priorities as well. Technological innovation in a broadly defined energy system requires more than a focus on energy supply technology. Technology policy will also need to engage with systems that facilitate lifestyle change. Increased investment will be needed in low-carbon infrastructure and low-carbon end-use technologies, at both community scale (e.g. transit and CHP (combined heat and power) systems) and individual and household scale (vehicles, building fabric, heating systems and appliances). Particularly at these scales, innovation challenges have social as well as technical components – the goal is not just to develop technical systems that work well, but to use them in ways that are consistent with lower-energy and lower-carbon lives.

The concept of social innovation is, in principle, rather straightforward in that it is based on the idea of innovation motivated by social need. It remains relatively unfamiliar in energy analysis, but better known in sectors where more devolved governance has been debated for longer (e.g. Mulgan, 2006). Initial exploration of its application in the energy sector (Bergman et al., 2011) concludes that although its role to date has been limited by the dominance of centralized technology and governance, it could benefit from the type of mechanisms already applied in technology innovation. For example, higher rates of feed-in tariff might be paid to social enterprises, reflecting the logic that organization from outside the mainstream might have greater opportunities to reduce costs over time. However, widespread use of this type of approach is only useful, and therefore likely, if there is general acceptance that existing governance systems are inadequate.

7.   TOWARDS A DECENTRALIZED GOVERNANCE

There are signs of some new actors beginning to rise to the challenges set out above. The critical change has been the widespread recognition (though far from universal acceptance) of climate change as a critical challenge for humanity in the twenty-first century. In addition, concerns about 'peak oil', now with more establishment recognition than a few years ago, have motivated some grass-roots activity.

Local governments in many countries have responded to civil society concerns about climate change by integrating them more clearly into their own policies and programmes. This has received most media attention through the actions of major cities, but there has been a much wider response (although very uneven) of municipalities in many countries (Bulkeley and Betsill, 2005). In most cases the attention to energy issues is primarily through a discourse about climate. This is reflected in real policy action in areas under municipal control – typically planning and regulatory powers in transport, land use and buildings. More direct civic engagement in energy investment remains limited, except where municipal authorities bucked the liberalization trend of earlier decades to retain a stake in energy utilities.

Social and commercial entrepreneurs have made some inroads into energy markets and practices. New energy services models have been tried. However, to date, there has been no significant move away from a commodity model of provision in mass markets, with transaction costs of restructuring contractual arrangements exceeding potential efficiency savings. However, there has been success in specific areas – notably with a facilities management model for heat in industry and large commercial buildings (Bertoldi et al., 2006), and community-based schemes around medium-scale renewable energy projects and heat networks (Walker et al., 2007). Both seem to imply that new forms of governance are likely to be needed to deliver a change away from a commodity market model.

At the individual level, 'green consumerism' is potentially a significant force. However, the undifferentiated nature of most energy commodities has tended to limit progress. The most obvious opportunity is the purchase of renewable energy through green electricity tariffs. These have had some impact, but particularly in countries where limited public policy support has forced renewable energy commerce along these lines. Where public policy initiatives are strong, voluntary approaches tend to have had limited uptake, because both governments and civil society organizations have been concerned that higher voluntary payments by consumers will only increase renewable generation above the level mandated by regulation (Diaz-Rainey and Ashton, 2008; Markard and Truffer, 2006; Rohracher, 2009). Citizen engagement is therefore arguably more important for the discretionary investment and use decisions directly under the control of individuals (e.g. in the built environment) than in reshaping power generation markets.

New products may make individual households a more attractive proposition for mass energy service markets. Microgeneration and electrification, unlike basic energy efficiency measures, both require capital investment on a scale that may make separate financing arrangements a more worthwhile proposition (Hinnells, 2008). Feed-in tariffs provide new value streams around which there might be new commercial offers. Managing intermittent and inflexible supply resources requires the adoption of smart metering, which itself may prove a disruptive technology through the information it enables to both end users and others (Darby, 2010).

Community groups of environmentally concerned citizens have for many decades affected energy systems, but primarily through resisting new supply investments. The increasing availability of smaller-scale investments has opened new opportunities for more proactive engagement. These are beginning to occur (Middlemiss and Parrish, 2010; Mulugetta et al., 2010), although the institutional and governance context is important in determining the level of success (Moloney et al., 2010; Peters et al., 2010).

So there is a growing group of new energy actors. Although currently excluded from the mainstream of energy governance, they potentially form the beginnings of a new more diverse governance system. Engaging new actors in the governance process to address new challenges has a logic that many find attractive. Both government and private sector actors are supportive where decentralized actors provide them with access to activities from which their remoteness would otherwise prove problematic.

But there is a more difficult related agenda – reducing the influence of existing actors. Reconceptualizing energy, as more than a market commodity, and redesigning energy systems, to give a larger role to decentralized actors, both implicitly involve a reduction in control for currently powerful actors. Every kilowatt hour reduced is a kilowatt hour not sold; every decision shared or decision devolved is a decision outside the control of the incumbent actors, who will tend to react to threats to their power (Geels and Schot, 2007).

Despite the signs identified in this section, change to date has been limited. The governance system remains predominantly centralized and the pace of future change is uncertain. Ambitious targets within local government may be increasingly common (e.g. Bulkeley and Betsill, 2005), but there is rarely the effective control over energy decision making in the same institutions that is needed for target delivery. The analysis that governance change is needed to secure sustainable outcomes in the energy system is not to be confused with its inevitability.

There is a strong alternative discourse around the rapid growth of centralized low-carbon power generation – some mix of nuclear, fossil fuels with carbon capture and storage (CCS) and more centralized forms of renewable power generation such as remote solar power. These fit more comfortably within the paradigms of the current dominant actors within the energy system, allowing the maintenance and even growth of a centralized power generation network and requiring more limited governance changes. However, even in this paradigm, decentralized actors have a role. They have to accept the use of new, large-scale power generation technology; and they have to electrify key categories of their own energy demand – heating and personal transport. Neither of these is unproblematic, and therefore disputes can be expected around these issues, reflecting the underlying interests of different actors.

Such disputes about the relative merits of centralized power (particularly nuclear) and demand-side solutions are not new. They were strong in the late 1980s, both about the relative costs (Keepin and Kats, 1988) and underlying more arcane disputes about the rebound effect (see e.g. Brookes, 1990 and Grubb, 1990). The market-based approaches to energy policy in the 1990s marked the decline in such debates. It seems likely that the stronger role of policy intervention required to deliver carbon and energy security goals in this century will see their renaissance.

8.   CONCLUSIONS

The need to transition to a low-carbon economy is increasingly widely recognized. Despite a dominant public discourse about large-scale energy supply technologies, the evidence is clear that a transition will involve fundamental changes to the energy system at the point of energy use. Energy efficiency is likely to be critical for cost-effectiveness; demand response increasingly important for effective utilization of intermittent renewables; substantial electrification of heat and transport demand required to enable use of low-carbon vectors. Microgeneration and greater decentralized production can involve energy users in different ways. The role of consumption more generally remains highly controversial, but the scope for reducing energy service demand through lifestyle change is very significant and potentially critical for cost-effective climate policy.

All these changes imply a more active role for decentralized actors. This has substantial implications, not least for the governance of the energy system. For historical reasons, existing governance systems privilege centralized actors in national government and large energy companies, with energy decision making focusing on energy resources, large investments and energy pricing. If the energy system is to be remade to include decentralized actors, it needs to include them in governance. This will involve reconceptualizing energy as more than a commodity and energy users as more than consumers.

National government energy policy will need to include those departments responsible for buildings, transport and land use. Local governments will need to play a bigger part because of their role in these policy areas. A wide range of other actors will need to be included, from the finance sector and network providers through to social entrepreneurs and community groups.

There are some signs of change in this direction. A range of more locally based energy initiatives is emerging with new actors. They potentially form the beginnings of a new more diverse governance system, and governments and energy companies increasingly recognize the need for partnerships including this wider set of actors to deliver sustainable change. However, change to date is limited, a centralized governance system remains dominant, and the pace and direction of future change is uncertain and contested.

ACKNOWLEDGEMENTS

The author gratefully acknowledges support from the Jackson Foundation. This work was also supported by the Natural Environment Research Council (grant number NE/H013598/1).

REFERENCES

Banks, N. (2001), 'Socio-technical networks and the sad case of the condensing boiler', in P. Bertoldi, A. Ricci and A.T. de Almeida (eds), *Energy Efficiency in Household Appliances and Lighting*, Berlin: Springer pp. 141–55.

Bergman, N. and Eyre, N. (2011), 'What role for microgeneration in a shift to a low carbon domestic energy sector in the UK?', *Energy Efficiency* **4**, 335–53.

Bergman, N., Markusson, N. Connor, P. Middlemiss, L. and Ricci. M. (2011), 'Bottom-up, social innovation for addressing climate change', University of Oxford, ECI Working Paper.

Bertoldi, P., Rezessy, S. and Vine, E. (2006), 'Energy service companies in European countries: current status and a strategy to foster their development', *Energy Policy*, **34**, 1818–32.

Boardman, B. (2007), 'Examining the carbon agenda via the 40% House scenario', *Building Research and Information*, **35**, 363–78.

Brookes, L. (1990), 'The greenhouse effect: the fallacies in the energy efficiency solution', *Energy Policy*, **18**, 199–201.

Brown, M.A. (2001), 'Market failures and barriers as a basis for clean energy policies', *Energy Policy*, **29**, 1197–207.

Bulkeley, H. and Betsill, M. (2005), 'Rethinking sustainable cities: multilevel governance and the "Urban" politics of climate change', *Environmental Politics*, **14**, 42–63.

Cullen, J.M. and Allwood, J.M. (2010), 'Theoretical efficiency limits for energy conversion devices', *Energy*, **35**, 2059–69.

Cullen, J.M., Allwood, J.M. and Borgstein, E.H. (2011), 'Reducing energy demand: what are the practical limits?', *Environmental Science & Technology*, **45**, 1711–18.

Darby, S. (2010), 'Smart metering: what potential for householder engagement?', *Building Research & Information*, **38**, 442–57.

Devine-Wright, P. (2005), 'Beyond NIMBYism: towards an integrated framework for understanding public perceptions of wind energy', *Wind Energy*, **8**, 125–39.

Diaz-Rainey, I. and Ashton, J.K. (2008), 'Stuck between a ROC and a hard place? Barriers to the take up of green energy in the UK', *Energy Policy*, **36**, 3053–61.

Ekins, P., Skea, J. and Winskel, M. (eds) (2010), *Energy 2050: The Transition to a Secure Low Carbon Energy System for the UK*, London: Earthscan.

Eyre, N.J. (1997), 'Barriers to energy efficiency: more than just market failure'. *Energy and Environment*, **8**, 25–43.

Eyre, N. (2008), 'Regulation of energy suppliers to save energy – lessons from the UK debate', British Institute of Energy Economics Conference, Oxford, September.

Eyre, N. (2011), 'Efficiency, demand reduction or electrification?', *Proceedings of the European Council for an Energy Efficient Economy*, Summer Study, Hyeres, France, pp. 1391–400.

Eyre, N., Pavan, M. and Bodineau, L. (2009), 'Energy company obligations to save energy in Italy, the UK and France: what have we learnt?', in *Proceedings of the European Council for an Energy Efficient Economy*, La Colle sur Loupe, France.

Eyre, N., Anable, J., Brand, C., Layberry, R. and Strachan, N. (2010), 'The way we live from now on: lifestyle and energy consumption', in P., Ekins, J. Skea and M. Winskel (eds), *Energy 2050: The Transition to a Secure Low Carbon Energy System for the UK*, London: Earthscan, pp. 258–93.

Eyre, N., Flanagan, B. and Double, K. (2011), 'Engaging people in saving energy on a large scale: lessons from the programmes of the Energy Saving Trust', in L. Whitmarsh, S. O'Neill, and I. Lorenzoni (eds), *Engaging the Public with Climate Change: Behaviour Change and Communication*, London: Earthscan, pp. 141–59.

Fawcett, T. (2011), 'The future role of heat pumps in the domestic sector', *Proceedings of the European Council for an Energy Efficient Economy*, Summer Study, Hyeres, France, pp. 1547–58.

Fripp, M. (2011), 'Greenhouse gas emissions from operating reserves used to backup large-scale wind power', *Environmental Science & Technology*, **45**(21), 9405–12.

Fujino, J., Hibino, G., Ehara, T., Matsuoka, Y., Masui, T. and Kainuma, M. (2008), 'Back-casting analysis for 70% emission reduction in Japan by 2050', *Climate Policy*, **8**, S108–S124.

Geels, F.W. and Schot, J. (2007), 'Typology of sociotechnical transition pathways', *Research Policy*, **36**, 399–417.

Gillingham, K., Newell, R.G. and Palmer, K. (2009), 'Energy efficiency economics and policy', National Bureau of Economic Research Working Paper Series No. 15031.

Grubb, M.J. (1990), 'Energy efficiency and economic fallacies', *Energy Policy*, **18**, 783–85.

Hinnells, M. (2008), 'Technologies to achieve demand reduction and microgeneration in buildings', *Energy Policy*, **36**, 4427–33.

Hirst, E. (1989), 'Reaching for 100% participation in a utility conservation programme', *Energy Policy*, **17**, 159–64.

HMG (2009), *The UK Low Carbon Transition Plan: National Strategy for Climate and Energy*, London: HM Government.

Hughes, T.P. (1983), *Networks of Power*, Baltimore, MD: Johns Hopkins University Press.

IEA (2010), *Energy Technology Perspectives – Scenarios and Strategies to 2050*, Paris: IEA.

IEA (2011), http://www.iea.org/stats/index.asp.

IPCC (2007), *Climate Change 2007: Mitigation of Climate Change, Contribution of Working Group III to the Fourth Assessment Report of the Intergovernmental Panel on Climate Change*, B. Metz, O.R. Davidson, P.R. Bosch, R. Dave, L.A. Meyer (eds), Cambridge: Cambridge University Press.

Jackson, T. (2005), 'Motivating sustainable consumption: a review of evidence on consumer behaviour and behavioural change: a report to the Sustainable Development Research Network', Centre for Environmental Strategy, Guildford, UK.

Jackson, T. (2009), *Prosperity without Growth*, London: Earthscan.

Jardine, C.N. and Ault, G.W. (2008), 'Scenarios for examination of highly distributed power systems', *Proceedings of the Institution of Mechanical Engineers, Part A: Journal of Power and Energy*, **222**, 643–55.

Jenkins, C., Neme, C. and Enterline, S. (2010), 'Energy efficiency as a resource in the ISO New England forward capacity market', *Energy Efficiency*, **4**, 31–42.

Jiang, Y., Tubiana, L., Zhou, W., Mao, Q., Li, Q., Qi, Y., Jiang, Y., Jiang, K., Château, B., Bressand, A., Dhakal, S., Eyre, N., Kross, L., Mukhopadhyay, P. and Major, M. (2009), 'Report of the Task Force on Urban Planning and Energy Efficiency', China Council for International Cooperation on Environment and Development.

Keepin, B. and Kats, G. (1988), 'Greenhouse warming: comparative analysis of nuclear and efficiency abatement strategies', *Energy Policy*, **16**, 538–61.

Killip, G. (2011), 'Can market transformation approaches apply to service markets? An investigation of innovation, learning, risk and reward in the case of low-carbon housing refurbishment in the UK', *Proceedings of the European Council for an Energy Efficient Economy*, Summer Study, Hyeres, France, pp.1185–96.

Levine, M., Ürge-Vorsatz, D., Blok, K., Geng, L., Harvey, D., Lang, S., Levermore, G., Mongameli Mehlwana, A., Mirasgedis, S., Novikova, A., Rilling, J. and Yoshino, H. (2007), 'Residential and commercial buildings', in B. Metz, O.R. Davidson, P.R. Bosch, R. Dave and L.A.Meyer (eds), *Climate Change 2007: Mitigation, Contribution of Working Group III to the Fourth Assessment Report of the Intergovernmental Panel on Climate Change*, Cambridge University Press, Cambridge, UK and New York, USA.

Markard, J. and Truffer, B. (2006), 'The promotional impacts of green power products on renewable energy sources: direct and indirect eco-effects', *Energy Policy*, **34**, 306–21.

Middlemiss, L. and Parrish, B.D. (2010), 'Building capacity for low-carbon communities: the role of grassroots initiatives', *Energy Policy*, **38**, 7559–66.

Mitchell, C. (2007), *The Political Economy of Sustainable Energy*, Basingstoke: Palgrave Macmillan.

Moloney, S., Horne, R.E. and Fien, J. (2010), 'Transitioning to low carbon communities – from behaviour change to systemic change: lessons from Australia', *Energy Policy*, **38**, 7614–23.

Mulgan, G. (2006), 'The process of social innovation', *Innovations: Technology, Governance, Globalization*, **1**(2), 145–62.

Mulugetta, Y., Jackson, T. and van der Horst, D. (2010), 'Carbon reduction at community scale', *Energy Policy*, **38**, 7541–45.

Nosperger, S., Killip, G. and Janda, K. (2011), 'Building expertise: a system of professions approach to low-carbon refurbishment in the UK and France', *Proceedings of the European Council for an Energy Efficient Economy*, Summer Study, Hyeres, France, pp.1365–66.

Parag, Y. and Eyre, N. (2010), 'Barriers to personal carbon trading in the policy arena', *Climate Policy*, **10**, 353–68.

Peters, M., Fudge, S. and Sinclair, P. (2010), 'Mobilising community action towards a low-carbon future: opportunities and challenges for local government in the UK', *Energy Policy* **38**, 7596–603.

Platt, R. (2011), 'Green streets, strong communities', Institute of Public Policy Research.

Rohracher, H. (2009), 'Intermediaries and the governance of choice: the case of green electricity labelling', *Environment and Planning A*, **41**, 2014–28.

Sanstad, A.H. and Howarth, R.B. (1994), 'Normal markets, market imperfections and energy efficiency', *Energy Policy*, **22**, 811–18.

Sauter, R. and Watson, J. (2007), 'Strategies for the deployment of micro-generation: implications for social acceptance', *Energy Policy*, **35**, 2770–79.

Shove, E. (1998), 'Gaps, barriers and conceptual chasms: theories of technology transfer and energy in buildings', *Energy Policy*, **26**, 1105–12.

Shove, E. and Walker, G. (2010), 'Governing transitions in the sustainability of everyday life', *Research Policy*, **39**, 471–6.

Smith, A., Stirling, A. and Berkhout, F. (2005), 'The governance of sustainable socio-technical transitions', *Research Policy*, **34**, 1491–510.

Socolow, R.H. (1978), 'The twin rivers program on energy conservation in housing: highlights and conclusions', *Energy and Buildings*, **1**, 207–42.

Stern, N. (2006), *The Economics of Climate Change*, London: HM Treasury.

Stern, P.C. (1986), 'Blind spots in policy analysis: what economics doesn't say about energy use', *Journal of Policy Analysis and Management*, **5**, 200–227.

Stern, P.C., Aronson, E., Darley, J.M., Hill, D.H., Hirst, E., Kempton, W. and Wilbanks, T.J. (1986), 'The effectiveness of incentives for residential energy conservation', *Evaluation Review*, **10**, 147–76.

Sukla, P.R., Dhar, S. and Mahapatra, D. (2008), 'Low-carbon society scenarios for India', *Climate Policy*, **8**, S156–176.

Toffler, A. (1980), *The Third Wave*, New York: William Morrow.

Unruh, G. (2000), 'Understanding carbon lock-in', *Energy Policy*, **28**(12), 817–30.

Ürge-Vorsatz, D., Eyre, N. et al. (2012), 'Energy end-use: buildings', in T. Johansson, N. Nakicenovic, A. Patwardhan and L. Gomez-Echeverri (eds), *Global Energy Assessment*, Cambridge: Cambridge University Press, pp.649–760.

Vojdani, A. (2008), 'Smart integration', *Power and Energy Magazine*, IEEE **6**, 71–9.

Walker, G., Hunter, S., Devine-Wright, P., Evans, B. and Fay, H. (2007), 'Harnessing community energies: explaining and evaluating community-based localism in renewable energy policy in the UK', *Global Environmental Politics* **7**, 64–82.

Wilhite, H., Nakagami, H., Masuda, T., Yamaga, Y. and Haneda, H. (1996), 'A cross-cultural analysis of household energy use behaviour in Japan and Norway', *Energy Policy*, **24**, 795–803.

Wilkinson, R. and Pickett, K. (2009), *The Spirit Level: Why Equality is Better for Everyone*, London: Penguin.

28 Is a global crisis required to prevent climate change? A historical–institutional perspective
Edward B. Barbier

INTRODUCTION

The threat posed by global warming is symptomatic of the diminishing returns from the way in which the world economy has been exploiting its remaining land and natural resources since the Industrial Revolution (Barbier, 2011a). At the heart of the global warming problem is that the anticipated increasing costs associated with climate change are not routinely reflected in markets. Nor have adequate policies and institutions been developed to handle these costs. All too often, policy distortions and failures compound these problems by encouraging wasteful use of natural resources and environmental degradation. The result is that we have failed to design a comprehensive set of policies and market incentives to ensure an efficient price for carbon, and, as a consequence, the world economy still emits too many greenhouse gases (GHGs).

The global policy stalemate on climate change is epitomized by the failure to achieve a comprehensive successor agreement for the 1997 Kyoto Protocol on curbing global GHG emissions, which expires in 2012. It was widely anticipated that the December 2009 Copenhagen meeting of all the parties to the United Nations Framework Convention on Climate Change would lead to a binding agreement for the post-Kyoto period. Instead, the meeting produced a much weaker Copenhagen Accord, which adopts a target of limiting the increase in global mean temperature to below 2 °C and pledges from high-income countries to reduce their GHG emissions by 2020 to achieve this target. For example, the USA pledged to reduce its baseline 2005 emissions 17 percent by 2020, with a further 30 percent reduction in 2025 and a 42 percent reduction in 2030, in line with the goal to reduce emissions 83 percent by 2050.[1] However, scenario modeling shows that, 'even if countries meet their ambitious objectives under the Copenhagen Accord, global temperatures are unlikely to keep within the objective of 2 °C. This conclusion is reinforced if developing countries delay their full participation beyond the 2030–2050 time frame' (Nordhaus, 2010, p. 11724).

Similarly, progress on achieving a global carbon market has also stagnated. In the case of climate change, the uncompensated damages from GHG emissions are truly global, in that all economies are contributing to emissions without paying fully for the costs, and the economic consequences of the market failure will be felt worldwide. The most efficient solution is to create a global carbon market that establishes a long-term and credible price signal for carbon across world markets as an incentive for the world economy to invest in clean energy technologies and reduce carbon dependency. By creating the first regional carbon market with its Emissions Trading System (ETS), the European Union has demonstrated how international trading can function to provide regional incentives for reducing GHG emissions, but expansion and reform of the ETS is needed if it is to become the basis of a global trading scheme (Convery, 2009; Ellerman and Joskow, 2008; Stankeviciutute et al., 2008). Similarly, the Clean Development Mechanism (CDM) of the Kyoto Protocol has funnelled projects and investments into large emerging market economies, such as Brazil, China, India, South Korea and Mexico, modestly linking them into global GHG emissions trading and financing. As with the ETS, however, reform and expansion of the CDM is essential to cover a broader range of GHG reduction projects and developing economies if it is truly to be the basis for a global carbon market (Barbier, 2010a; Hepburn and Stern, 2008). A number of important economies, such as Australia, Canada, Japan, New Zealand, Norway and Switzerland, have proposed or implemented cap-and-trade systems, which could link into the larger international trading network. On the other hand, cap-and-trade legislation for important global emitters, such as the USA, Russia and China, is still unlikely for political reasons. International trading in carbon is still patchy and limited at best, and is unlikely to evolve into a fully fledged global carbon market any time soon.

There was also hope that, in the wake of the 2008 09 Great Recession, the world economy would move towards a coordinated policy initiative to combat global warming while stimulating economic recovery. For example, in their *communiqué* at the 2 April 2009 London Summit, the leaders of the Group of 20 (G20) largest economies stated: 'We will make the transition towards clean, innovative, resource efficient, low carbon technologies and infrastructure . . . We will identify and work together on further measures to build sustainable economies.'[2] Although by the end of 2009 several G20 economies had incorporated a sizable 'green fiscal' component in their recovery spending, including support for renewable energy, carbon capture and sequestration, energy efficiency, public transport and rail, improved electrical grid transmission, and environmental protection, most G20 governments were cautious in making low-carbon and other environmental investments during the 2008–09 recession, and some did not implement any green stimulus measures (Barbier, 2010a; Robins et al., 2009, 2010). In addition, global fossil fuel subsidies and other market distortions, as well as the lack of effective environmental pricing policies and regulations, diminished the impacts of G20 green stimulus investments not only on long-term investment and job creation in green sectors but also stimulating the transition to low-carbon economies (Barbier, 2010a; Strand and Toman, 2010). Without correcting existing market and policy distortions that underprice the use of natural resources, contribute to environmental degradation and worsen carbon dependency, public investments to stimulate clean energy and other green sectors in the economy will be short-lived. The failure to implement and coordinate green stimulus measures across all G20 economies also limits their effectiveness in 'greening' the global economy. It seems that even the worst economic crisis since the Great Depression of the 1930s has not been enough to instigate the policy changes required to transition to a low-carbon world economy.

Thus it seems that our inability to prevent climate change represents a massive policy failure on a global scale. Given the overwhelming scientific and economic evidence on the anticipated costs of global warming, we can no longer argue that uncertainty over these costs is the main obstacle (IPCC, 2007; Stern, 2007). Instead, this chapter draws on an institutional and historical perspective to explain the continuing policy stalemate on climate change.

Building on the arguments developed in Barbier (2011b), this chapter suggests that the core problem may lie in the intransigence of social institutions – the mechanisms and structures for ordering economic behavior and the means of production within society. To explain this intransigence, or institutional inertia, a transactions cost perspective on the difficulty of implementing the transition to a low-carbon economy is offered. The chapter then explores how high transaction costs and institutional rigidities are reinforced by the political power of vested interests. However, both institutional inertia and vested political interests do not rise up overnight; they have their basis in long-term economic development. Using the example of the USA, this chapter explores how this influence arises through historical patterns of resource-based development. Finally, the chapter ends by discussing whether the shock of a global crisis is the only means of over-coming the transaction costs, institutional rigidities and vested interests that are delaying the transition to a low-carbon world economy.

INSTITUTIONAL INERTIA AND TRANSACTION COSTS

Why has it proven so difficult to overcome the numerous market, policy and institutional failures that drive global warming and climate change? An explanation of this intransigence is the tendency of many important social institutions, broadly defined, to be highly invariant over long periods of time (e.g. see Dixit, 1996, 2003; Hodgson, 1998; McCann et al., 2005; North, 1990, 1991; Williamson, 2000). Barbier (2011b) refers to this rigidity as 'institutional inertia', which is equivalent to what North (1990) first called 'institutional path dependence'. In this and the subsequent section, the methodology developed by Barbier (2011b) is used to show how institutional inertia, transactions costs and vested political interests have interacted to block significant progress in global climate change policy.

Institutions are the mechanisms and structures for ordering the behavior and ensuring the cooperation of individuals within society. They are the formal and informal 'rules' that govern and organize social behavior and relationships, including reinforcing the existing social order, which is a stable system of institutions and structure that characterizes society for a considerable period of time. Consequently, as societies develop, they become more complex, and their institutions are more difficult to change. Institutions help structure the means of production, and how goods and services are produced influences the development of certain institutions. This is a cumulative causative, or mutually reinforcing, process.

One reason for this self-reinforcing process is that institutions and the social order become geared toward reducing the transaction costs – the costs other than the money price that are incurred in exchanging goods or services – of existing production and market relationships. For example, typical transaction costs include search and information costs, bargaining and decision costs, and policing and enforcement costs, or, as summarized by Krutilla (1999, p. 250), 'any costs associated with establishing, administrating, monitoring or enforcing a government policy or regulation'. Several studies have highlighted how such costs routinely hinder the successful implementation of environmental policies (e.g. see Gangadharan, 2000; Krutilla, 1999; Krutilla and Krause, 2011; Mettepenningen et al., 2009; Rousseau and Proost, 2005; Stavins, 1995).

With respect to the global warming problem, it is typically these broad transaction costs that are responsible for the institutional inertia, or path dependence, which is thwarting whole-scale policy change to a low-carbon economy.[3] That is, the existing system of social institutions and structure – the 'social order' – becomes fixed around a stable set of economic institutions, including how production is organized and how certain natural resources are combined with other inputs, such as technology and knowledge, in production. Our institutions and social order therefore evolve to reduce the transaction costs associated with the existing production and exchange relationships, including the same way in which we find, extract and use existing natural resources in combination with other inputs, whereas attempting to change these production and exchange relations incurs relatively high costs. As a consequence, we may become more aware of the economic and ecological consequences of global warming that arise through perpetuating the same pattern of resource-based development, including overreliance on fossil fuels and high energy consumption. But the high relative transaction costs involved in making the necessary corrections to the market, policy and institutional failures necessary for the transition to a low-carbon economy, compared to perpetuating the same pattern of production and energy use, seem prohibitive.

Such transaction costs are additional to the substantial economic costs and barriers associated with switching from fossil fuels to alternative energy sources. As a consequence, the former effectively magnify the latter costs and give the impression that they are insurmountable. Overcoming the market, policy and institutional failures associated with fossil fuel use and high energy consumption seems a formidable task, and the combined economic and transaction costs appear to dwarf the significant environmental externalities associated with conventional energy use. The result is that institutional inertia and the accompanying transaction costs of making the transition to a low-carbon economy are considerable, given the magnitude of the economic transformation involved.

Figure 28.1 illustrates the magnitude of the problem often confronted, with instigating policies to correct market, institutional and policy failures contributing to a major

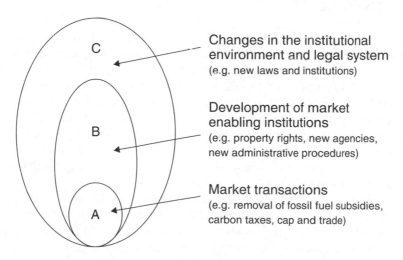

Changes in the institutional environment and legal system (e.g. new laws and institutions)

Development of market enabling institutions (e.g. property rights, new agencies, new administrative procedures)

Market transactions (e.g. removal of fossil fuel subsidies, carbon taxes, cap and trade)

Source: Adapted from McCann et al. (2005).

Figure 28.1   The transaction costs of climate change policy

environmental problem, such as global warming. When a new policy is implemented, such as a tax on carbon, cap and trade in GHG emissions, or removal of fossil fuel subsidies, additional search and information costs, bargaining and decision costs, and policing and enforcement costs are bound to occur (Area A). However, establishing some market-based instruments and trading mechanisms, such as taxes, tradable permit systems and new carbon markets, will also require the establishment or reallocation of property rights to facilitate these instruments, and the setting up of new public agencies and administrative procedures to record, monitor and enforce trades. Thus the full transaction costs of the policies will be areas A and B in Figure 28.1. Finally, if additional changes in the institutional environment and legal system are required, the transaction costs will be larger still, including areas A, B and C.

All three types of transactions costs act as barriers to implementing policies to combat global warming and promote the long-run transition to a low-carbon economy. As several studies have shown, various types of transaction costs are attributed to delaying or inhibiting the implementation of carbon taxes or tradable permits, adding to the costs of technological change promoting clean energy and greenhouse gas (GHG) abatement, and reducing the effectiveness of the Clean Development Mechanism (Grubb et al., 1995; Michaelowa and Jotzo, 2005; Krutilla and Krause, 2011; Schwoon and Tol, 2006). Although it is difficult to determine how significant these transactions costs are compared to the economic costs of, say, switching from fossil fuels to alternative energy sources or abating GHG emissions, the fact that transaction costs are deterring important mechanisms to control global warming is a serious problem. Without the successful implementation of policies to control GHG emissions, spur research and development into clean energy technologies, and disseminate these technologies globally, economies will remain fundamentally dependent on fossil fuel energy for some time to come.

VESTED INTERESTS

As argued by Barbier (2011b), vested interests and political lobbying also reinforce institutional inertia towards implementing climate policy, and will thus help delay the transition to a low-carbon economy. Governments can be influenced by powerful interest groups to block policy reforms that redistribute costs and benefits against their interest. In effect, the role of vested interests, political lobbying, and in some cases outright corruption and bribery, is to 'expand' each of the transaction cost 'bubbles' A, B and C of Figure 28.1. The result is that it becomes even more difficult to implement a new policy to control GHG emissions or reduce fossil fuel use. Figure 28.2 illustrates how this economic incentive against reforms might work in the case of removing fossil fuel subsidies or imposing a carbon tax.

The removal of a fossil fuel production subsidy causes a shift in market supply from S to S' (see Figure 28.2(a)). The quantity sold may decline from Q' to Q, and the market price rises from P' to P. The government may save $(P'' - P') * Q'$, and the present value of future global warming damages and other pollution costs associated with fossil fuels are reduced by area d. Although there is a loss in producer and consumer surplus, as consumers are taxpayers and may also gain from the environmental improvement, they may feel compensated by the policy change. However, as a special interest group

(a) Removal of a fossil fuel subsidy

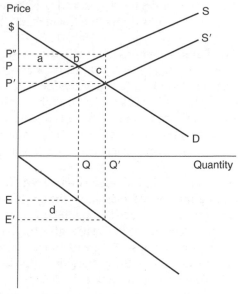

Producer surplus loss = a + b
Efficiency gain = c
Environmental gain = d
Political cost = (a + b)/(c + d)

(b) Carbon tax

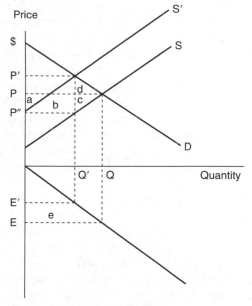

Producer surplus loss = a + b + c
Efficiency loss = c + d
Environmental gain = e
Political cost = (a + b + c)/[e − (c + d)]

Marginal environmental damages

Source: Adapted from Barbier (2011b).

Figure 28.2   The political cost of climate policy

profiting from the subsidy, fossil fuel producers might feel differently. As indicated in Figure 28.2(a), they have a strong economic incentive to block the policy change, as they experience a high relative political cost from the subsidy removal, which amounts to $(a + b)/(c + d)$.

The imposition of a carbon tax, which we assume is imposed initially on fossil fuel producers, also has political cost implications (see Figure 28.2(b)). The tax shifts the supply curve from S to S'. As a result, Q declines to Q', and market price rises from P to P'. The government gains tax revenue equal to $(P' - P'') * Q'$, but there is still an efficiency loss equal to $c + d$. The environmental improvement from reducing fossil fuel pollution and GHG emissions, which is e, may compensate for this loss; however, producers will still be worse off by $a + b + c$. Thus, the political cost of the carbon tax is $(a + b + c)/e - (c + d)$.

The incentive of vested interests to lobby against policy change is therefore strong. In economics, a growing literature is examining the role of such lobbying in influencing environmental policy outcomes (e.g. see Aidt, 1998; Fredriksson, 2003; López and Mitra, 2000; Wilson and Damania, 2005). In all cases, the influence of lobbying by powerful vested interests fosters outcomes that work against the greater social interest and perpetuate environmental damages. The greater the political bargaining power of special interests, the more difficult it is to implement reform. Yet it is clear that such reforms could yield improvements in both global warming outcomes and economic efficiency.

For example, globally, fossil fuel consumption subsidies amounted to $557 billion in 2008 (IEA/OPEC/OECD/World Bank, 2010). Production subsidies accounted for an additional $100 billion. Together, these subsidies account for roughly 1 percent of world GDP. In addition, global fossil fuel consumption in 2008 amounted to 10070 million tons of oil equivalent (BP, 2011), suggesting that the average subsidy amounted to about $66 per ton of oil equivalent. Such fossil fuel consumption and production subsidies are an additional market failure preventing improved energy efficiency in economies. By artificially lowering the cost of using fossil fuels, such subsidies deter consumers and firms from adopting energy efficiency measures that would otherwise be cost-effective in the absence of any subsidies. Removal of such perverse incentives would therefore boost energy savings substantially. Phasing out all fossil fuel consumption and production subsidies by 2020 could result in a 5.8 percent reduction in global primary energy demand and a 6.9 percent fall in greenhouse gas (GHG) emissions.

A global carbon tax on GHG emissions is frequently proposed as an alternative to raising funds through an international cap-and-auction scheme (Hyder, 2008; Nordhaus, 2007, 2010). Under such a tax regime, countries set market penalties on GHG emissions at levels that are equalized across different regions and industries. The tax would be set low initially, and rise steadily over time to reflect the rising damages from global warming. Estimated revenues from such a scheme could range from $318 to $980 billion by 2015 (in 2005 prices) and $527 to $1763 billion by 2030 (Hyder, 2008). Conceivably, some of these revenues could be used to finance clean energy and other low-carbon investments, or simply to offset income and payroll taxes in economies. However, as Nordhaus (2010, pp. 11725–6) has pointed out, international agreements on harmonized taxes on GHG emissions are proving to be more elusive than for IPES or carbon cap and trade: 'Economists often point to harmonized carbon taxes as a more efficient and attractive regime, but these have been generally shunned in negotiations, particularly in the United States, because of the taboo on considering tax-based systems.'

HISTORICAL PATH DEPENDENCE

As indicated previously, today's institutional inertia that reinforces entrenched political interests with respect to climate policy have roots in the historical path dependence of the economy. In the case of global warming and other environmental problems, such path dependence emanates from long-run patterns of natural-resource-based economic development that reinforces carbon-dependent economic growth (Barbier, 2011a). As Table 28.1 shows, evidence of this long-term pattern is reflected in a number of key environmental, economic and demographic trends from the 1890s to the 1990s, including carbon dioxide emissions. Contemporary unease over natural resource scarcity, energy insecurity, global warming and other environmental consequences is to be expected, given the rapid rate of environmental change caused by the global economy and human populations over the twentieth century.

Table 28.1   Magnitudes of global environmental change, 1890s to 1990s

Indicator	Coefficient of increase, 1890s to 1990s
Drivers	
Human population	4
Urban proportion of human population	3
Total urban population	14
World economy	14
Industrial output	40
Energy use	13–14
Coal production	7
Freshwater use	9
Irrigated area	5
Cropland area	2
Pasture area	1.8
Pig population	9
Goat population	5
Cattle population	4
Marine fish catch	35
Impacts	
Forest area	0.8 (20% decrease)
Bird and mammal species	0.99 (1% decrease)
Fin whale population	0.03 (97% decrease)
Air pollution	2–10
Carbon dioxide (CO_2) emissions	17
Sulfur dioxide (SO_2) emissions	13
Lead emissions	8

Source: Adapted from McNeill (2000, pp. 360–61) and McNeill (2005, Tables 1 and 2).

This pattern of environmental change mirrors the emergence of the global 'fossil fuel' era, which began in earnest during the late nineteenth century. The rapid exploitation of these new energy sources by rapidly industrializing economies, starting with coal then followed by oil and natural gas, led to two important global energy trends that first began appearing in the 1870s.

First, world energy consumption began growing exponentially. For Western Europe, which industrialized first, the takeoff in energy consumption started in the mid-nineteenth century but then accelerated after 1870, whereas for the USA, the exponential rise in energy consumption began occurring in the 1870 to 1914 period (see Barbier, 2011a, Figure 7.1). As the Western European and US economies emerged from this era as the dominant global economic powers, their rise and industrialized wealth had led inexorably to finding and successfully exploiting the new vertical frontiers of fossil fuel energy. The growth in world energy consumption has continued ever since, as more countries continue to develop and industrialize (Barbier, 2011a; BP, 2011; Smil, 2000).

Second, the late nineteenth century also saw a dramatic, and so far permanent, change in energy consumption by fuel type. During the nineteenth century industrialization in the Western European and US economies led to the rapid spread of coal consumption and the replacement of charcoal for indoor heating and metallurgy by coal, coke and gas. By the late 1880s, fossil fuels had surpassed biomass in global energy consumption. By 1910, biomass energy had fallen to 35 percent of total energy consumption, and was being rapidly displaced by coal consumption (over 60 percent) and the newest fossil fuel, oil (3–4 percent) (Etemad et al., 1991; Fouquet, 2008; Smil, 1994). Although coal consumption continued to grow in the twentieth century, world consumption of hydrocarbons – oil and natural gas – grew even faster. By 1950, hydrocarbons comprised around 37 percent of global primary energy consumption (Smil, 2000). Today, oil amounts to 32.6 percent of world consumption, natural gas 22.2 percent and coal 29.6 percent (BP, 2011).

Thus the pattern of carbon-dependent economic development has continued largely unchecked over the past 100 years or so. This is hardly surprising given that, since the late nineteenth century, the world economy has embarked on the modern fossil fuel age (Barbier, 2011a). The contemporary era, from 1950 to the present, has corresponded with the peak of that age, and thus ushered in an unprecedented phase of intensive carbon-based development.

No economy better exemplified this pattern of development than the USA, which since 1950 has been the successful model of industrialization and development that all other economies have attempted to emulate. For example, the rapid transition to centrally generated electricity during the first half of the twentieth century was made possible by the abundant and cheap fossil fuel supplies in the USA. The result was an exponential rise in electricity generation's share of fossil fuel use, from less than 2 percent in 1900 to more than 10 percent by 1950 and 34 percent by 2000 (Smil, 2006). Electricity for industrial, commercial and energy use has became omnipresent in the US economy, and centralization of electricity generation and expansion of the grid network has led, in turn, to an exponential growth in energy use by US firms and households (Nelson and Wright, 1992). US fossil fuel abundance, especially in petroleum, also led to the development of the internal combustion engine, the automobile and the use of roads, and like electrification, the automobile industry and a national road network helped transform the entire US economy. The parallel development of the aircraft industry and air transport across the USA spurred further economic integration through increasing the mobility of people, cargo and even the mail (Meinig, 2004; Nelson and Wright, 1992). The development of the petroleum industry in the USA in the 1920s and 1930s led to the rise of the economically important petrochemical industry, and its products, including plastics, oils and resins, chemical fertilizers and synthetic rubber, profoundly impacted the US economy, from the rise of the automobile and aircraft industries to the transformation of US agriculture (Hugill and Bachmann, 2005; Nelson and Wright, 1992).

As the roots of carbon-based development in the USA illustrate the historical path dependence of the world economy, contemporary global economic development has been typified by increasingly high levels of energy use per capita (see Figure 28.3). The USA leads all economies with the highest rates of primary energy consumption per person, although in recent years US consumption has stabilized at 8 metric tons of oil equivalent per capita. Energy use per person has grown in other high-income economies as they have striven to match the pattern of energy use and development typified by the USA. Even developing economies have increased energy consumption per person, which has risen steadily since the 1960s. The result is a steady increase in energy use per capita throughout the world, from just over 1.1 metric tons of oil equivalent per person in 1965 to close to 1.7 tons today.

Because of the world economy's continuing carbon dependency, projections indicate that the growth in GHG emissions for most economies and regions will continue in coming decades (CAIT, 2008). These trends suggest that, despite encouraging signs that the GHG intensity of many large economies is declining, the overall carbon dependency of the global economy is actually increasing. In 2030, a carbon-dependent world economy will produce close to 60 percent more GHG emissions from energy combustion than it does today. Growth in emissions will occur in the high-income OECD economies,[4] but just 17.4 percent higher than today. Japan's emissions might fall, and the EU's emissions may increase by less than 6 percent. Much of the growth in OECD emissions is likely to come from the USA, which may show a 19 percent increase. However, the large increase in global GHG emissions is likely to come from transition and developing economies. Emissions by 2030 will more than double for developing economies, led by large increases in India and China. Emissions from transition economies will rise by nearly 30 percent, led by Russia. By 2030, China's share of GHG emissions could be close to one-third the world total, and all developing economies could account for the majority of emissions.

The prominence of Brazil, Russia, India and China – the so-called 'BRIC' economies – in contributing to global GHG emissions is alarming but not surprising. Through exploiting their available endowments for energy production, and through importing additional fossil fuel supplies on world markets, the BRIC countries have become major global energy consumers to fuel their industrial development and recent economic growth. China accounts for 16.8 percent of global primary energy consumption, which is second only to the USA (21.3 percent). Russia is third with 6.2 percent of world consumption, but in 1989, just before the collapse of the Soviet Union, Russia's energy consumption was almost a third higher than current levels. India (3.6 percent) is already the fifth-largest energy consumer, just behind Japan. Brazil (2.0 percent) now

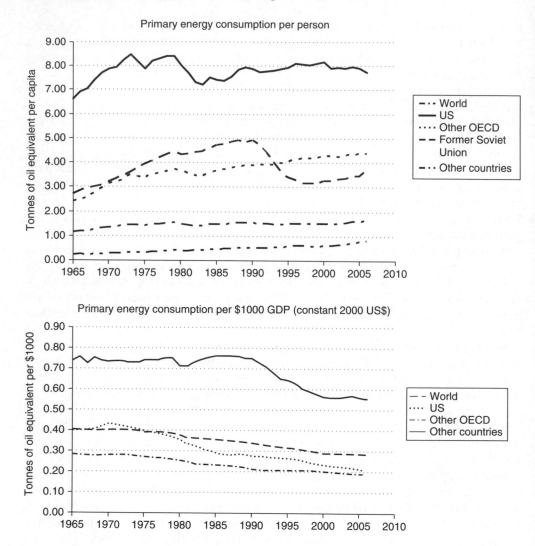

Notes: Other OECD means all OECD countries except the USA. For OECD, see note 4. Former Soviet Union includes Armenia, Azerbaijan, Belarus, Estonia, Georgia, Kazakhstan, Kyrgyzstan, Latvia, Lithuania, Moldova, Russian Federation, Tajikistan, Turkmenistan, Ukraine and Uzbekistan. Primary energy consumption is from BP (2008), and population and real GDP are from World Bank (2008).

Figure 28.3   Global energy use, 1965–2006

exceeds the primary energy consumption of all European economies, except France and Germany.[5]

Thus the BRIC economies have embraced the 'US-style' carbon-dependent development path to industrialization and growth. The BRIC economies, along with two other large emerging market economies, Mexico and Indonesia, are now among the top ten emitters of global GHG emissions (CAIT, 2008). Together, all four BRIC countries account for over 31 percent of global emissions, and the greenhouse gas intensity of these economies (GHG emissions per unit of GDP) exceed those of all major developed countries, including the USA.

Such trends do not bode well for the future. The 2008–09 Great Recession may have slowed down global fossil fuel use and GHG emissions, but the effects have been only temporary. Given the current fossil fuel dependency of the world economy, once global growth returns to long-term trends, energy prices could rise significantly. Projections by the International Energy Agency (IEA, 2008) suggest that, over the long run, fossil fuel demand will rise by 45 percent, and the oil price could reach $180 per barrel. The remaining oil reserves will be concentrated in fewer countries, the risk of oil supply disruptions will rise, and oil supply capacity will fall short of demand growth. Such factors may influence fossil fuel prices even in the short and medium term. For example, events over 2010–11, such as political instability in the Middle East and the nuclear reactor catastrophe in Japan, combined with the nascent world recovery, quickly pushed the price of oil to over $100 per barrel.

Even if demand for energy remains flat until 2030, just to offset the effect of oilfield decline the global economy will still need 45 million barrels per day of additional gross production capacity – an amount approximately equal to four times the current capacity of Saudi Arabia (IEA, 2008). But with long-term world economic growth, fossil fuel demand is unlikely to stay constant, despite the rise in energy prices. Increasing consumption of fossil fuels will worsen global energy security concerns, which will be exacerbated by the increased concentration of the remaining oil reserves in a smaller number of countries, the risk of oil supply disruptions, rising energy use in the transport sector worldwide, and rapid demand growth from emerging market economies (IEA, 2007). From 2007 to 2035, global energy consumption is expected to increase from 495 to 739 quadrillion Btu, with non-OECD countries accounting for the vast majority of this nearly 50 percent projected increase in energy use (US EIA, 2010). And, of course, as growth in the world economy revives fossil fuel consumption, it will also accelerate global climate change.

CONCLUSION: IS A GLOBAL CRISIS REQUIRED TO PREVENT CLIMATE CHANGE?

Given the stranglehold that historical path dependence, institutional inertia and vested interests exert on climate policy, it may well be that a global crisis is required to instigate the transition to a low-carbon world economy. This is an alarming prospect. As noted in the introduction, initially there was considerable hope that the 2008–09 Great Recession would spark global policy cooperation on a 'green recovery' that would be the basis for such a transition. But if the worst global recession since the Great Depression of the 1930s failed to stimulate sufficient clean energy and other low-carbon investments in most G20 economies, then how large a global economic or ecological crisis is needed to spark the transition to a low-carbon world economy?

Of course, speculating on what sort of global crisis is likely to occur is beyond the scope of this chapter. Smil (2008) has made a cogent attempt to summarize the evidence for human-made and natural (and even extra-terrestrial!) disasters that could take place over the next 50 years, and concludes that there is a 10–15 percent chance of a global transformative war occurring. Unfortunately, as Smil also notes, scientific evidence suggests that climate change may soon be producing its own inexorable crises. Without a change in the carbon dependency of the global economy, the atmospheric concentration of GHG could double by the end of this century, and lead to an eventual global average temperature increase of up to 6 °C (IEA, 2008). Such a scenario is likely to cause a sea-level rise between 0.26 and 0.59 meters, and severely disrupt freshwater availability, ecosystems, food production, coastal populations and human health (IPCC, 2007). With 5–6 °C warming, the world economy could sustain losses equivalent to 5–10 percent of global gross domestic product (GDP), with poor countries suffering costs in excess of 10 percent of GDP (Stern, 2007). Across all cities worldwide, about 40 million people are exposed to a 1 in 100-year extreme coastal flooding event, and by the 2070s the population exposed could rise to 150 million (Nicholls et al., 2007).[6]

The world's poor are especially vulnerable to the climate-driven risks posed by rising sea level, coastal erosion and more frequent storms. Around 14 percent of the population and 21 percent of urban dwellers in developing countries live in low-elevation coastal zones that are exposed to these risks (McGranahan et al., 2007). The livelihoods of billions – from poor farmers to urban slum dwellers – are threatened by a wide range of climate-induced risks that affect food security, water availability, natural disasters, ecosystem stability and human health. For example, many of the 150 million urban inhabitants that are likely to be at risk from extreme coastal flooding events and sea-level rise are likely to be the poor living in developing-country cities (Nicholls et al., 2007; OECD, 2008; UNDP, 2008).

Global ecosystems and freshwater sources are already endangered by the widespread environmental degradation that are accompanying current patterns of global economic development. A major concern is that global warming will precipitate these trends and potentially cause an environmental catastrophe (IPCC, 2007; UNDP, 2008). Over the past 50 years, ecosystems have been modified more rapidly and extensively than in any comparable period in human history, largely to meet rapidly growing demands for food, fresh water, timber, fiber and fuel. The result has been a substantial and largely irreversible loss in biological diversity. Approximately 60 percent of the major global ecosystem services have been degraded or used unsustainably, including freshwater, capture fisheries, air and water purification, and the regulation of regional and local climate, natural hazards and pests (MEA, 2005).

Poor people in developing countries will be most affected by the continuing loss of critical ecological services worldwide. The rural poor in developing regions tend to be clustered in areas of ecologically fragile land, which are already prone to degradation, water stress and poor soils (Barbier, 2010b). In addition, by 2019, half of the developing world will be in cities, and by 2050, 5.33 billion people, or 67 percent of the population in developed countries, will inhabit urban areas (PDUN, 2008). This brisk pace of urbanization means that the growing populations in the cities will be confronted with increased congestion and pollution and rising energy, water and raw material demands – and increasingly the threats posed by global warming and climate change. Although such environmental problems are similar to those faced by industrialized countries, the pace and scale of urban population growth in developing countries are likely to lead to more severe and acute health and welfare impacts of any climate-induced environmental crises.

In conclusion, it may take a momentous global crisis, possibly involving a major environmental catastrophe combined with widespread economic disruption and large-scale fatalities, before the world comes to grip with the policy challenges posed by climate change. Perhaps it will even take multiple crises and irrefutable scientific evidence that global warming is a principal cause to instigate concerted change in 'business as usual'. Otherwise, the global policy stalemate on climate change is likely to persist for the foreseeable future, buttressed by institutional inertia, vested interests and historical path dependence.

NOTES

1. See Copenhagen Accord, Appendix I – Quantified economy-wide emissions targets for 2020, available at http://unfccc.int/meetings/cop_15/copenhagen_accord/items/5262.php.
2. From 'London Summit – Leaders' Statement 2 April 2009', available at www.g20.org/pub_communiques. aspx. The members of the G20 include 19 countries (Argentina, Australia, Brazil, Canada, China, France, Germany, India, Indonesia, Italy, Japan, Mexico, Russia, Saudi Arabia, South Africa, South Korea, Turkey, the UK and the USA) plus the EU as a whole.
3. According to Boettke et al. (2008, p. 332), 'path dependence emphasizes the increasing returns to institutions, which "lock in" particular institutional arrangements that have emerged in various places for unique historical reasons' (see also Arthur, 1994; North, 1990; Page, 2006; Pierson, 2000).
4. OECD stands for the Organization for Economic Cooperation and Development, which includes, from Europe, Austria, Belgium, Czech Republic, Denmark, Finland, France, Germany, Greece, Hungary, Iceland, Republic of Ireland, Italy, Luxembourg, Netherlands, Norway, Poland, Portugal, Slovakia, Spain, Sweden, Switzerland, Turkey, and the UK, and from other regions, Australia, Canada, Japan, Mexico, New Zealand, South Korea and the USA. Non-OECD economies are the remaining economies of the world, most of which are low- and middle-income countries.
5. These comparative energy statistics are from BP (2008), *Statistical Review of World Energy 2008*, http:// www.bp.com/statisticalreview.
6. According to the authors, the top ten cities in terms of exposed population are Mumbai, Guangzhou, Shanghai, Miami, Ho Chi Minh City, Kolkata, Greater New York, Osaka-Kobe, Alexandria and New Orleans.

REFERENCES

Aidt, T. (1998), 'Political internalization of economic externalities and environmental policy', *Journal of Public Economics*, **69**: 1–16.
Arthur, B. (1994), *Increasing Returns and Path Dependency in the Economy*, Ann Arbor, MI: University of Michigan Press.
Barbier, E.B. (2010a), *A Global Green New Deal: Rethinking the Economic Recovery*, Cambridge: Cambridge University Press.
Barbier, E.B. (2010b), 'Poverty, development, and environment', *Environment and Development Economics*, **15**: 635–60.
Barbier, E.B. (2011a), *Scarcity and Frontiers: How Economies Have Developed Through Natural Resource Exploitation*, Cambridge: Cambridge University Press.
Barbier, E.B. (2011b), 'Transaction costs and the transition to environmentally sustainable development', *Environmental Innovations and Societal Transitions*, **1**: 58–69.
Boettke, P.B., C.J. Coyne and P.T. Leeson (2008), 'Institutional stickiness and the new development economics', *American Journal of Economics and Sociology*, **67**: 331–58.
BP (2008), *Statistical Review of World Energy 2008*, available at http://www.bp.com/statisticalreview.
BP (2011), *BP Statistical Review of World Energy June 2011*, available at http://www.bp.com/statisticalreview.

Climate Analysis Indicators Tool (CAIT), Version 6.0. 2008, World Resources Institute, Washington, DC.

Convery, F.J. (2009), 'Origins and development of the EU ETS', *Environmental and Resource Economics*, **43**: 391–412.

Dixit, A. (1996), *The Making of Economic Policy: A Transaction-Cost Politics Perspective*, Cambridge, MA: MIT Press.

Dixit, A. (2003), 'Some lessons from transaction-cost politics for less-developed countries', *Economics & Politics*, **15**(2): 107–33.

Ellerman, A.D. and P. L. Joskow (2008), 'The European Union's Emissions Trading System in perspective', prepared for the Pew Center on Global Climate Change, MIT, Cambridge, MA.

Etemad, B., J. Lucini, P. Bairoch and J.-C. Toutain (1991), *World Energy Production 1800–1995*, Centre National de la Recherche Scientifique and Centre d'Histoire Economique Internationale, Geneva, Switzerland.

Fouquet, R. (2008), *Heat, Power and Light: Revolutions in Energy Services*, Cheltenham, UK and Northampton, MA, USA: Edward Elgar.

Fredriksson, P.G. (2003), 'Political instability, corruption and policy formation, the case of environmental policy', *Journal of Public Economics*, **87**: 1383–405.

Gangadharan, L. (2000), 'Transaction costs in pollution markets: an empirical study', *Land Economics*, **76**: 601–14.

Grubb, M., T. Chapuis and M. Ha Duong (1995), 'The economics of changing course: implications of adaptability and inertia for optimal climate policy', *Energy Policy*, **23**: 417–32.

Hepburn, C. and N. Stern (2008), 'A new global deal on climate change', *Oxford Review of Economic Policy*, **24**: 259–79.

Hodgson, G.M. (1998), 'The approach of institutional economics', *Journal of Economic Literature*, **36**(1): 166–92.

Hugill, P.J. and V. Bachmann (2005), 'The route to the techno-industrial world economy and the transfer of German organic chemistry to America before, during, and immediately after World War I', *Comparative Technology Transfer and Society*, **3**(2): 159–86.

Hyder, P. (2008), 'Recycling revenue from an international carbon tax to fund an international investment programme in sustainable energy and poverty reduction', *Global Environmental Change*, **18**: 521–38.

IEA/OPEC/OECD/World Bank (2010), 'Analysis of the scope of energy subsidies and the suggestions for the G-20 Initiative', Joint Report prepared for submission to the G-20 summit Meeting, Toronto (Canada), 26–72 June.

International Energy Agency (IEA) (2007), *Oil Supply Security 2007: Emergency Response of IEA Countries*, Paris: OECD/IEA.

International Energy Agency (2008), *World Energy Outlook 2008*, Paris: OECD/IEA.

Intergovernmental Panel on Climate Change (IPCC) (2007), *Climate Change 2007: Synthesis Report. Contribution of Working Groups I, II and III to the Fourth Assessment* (core writing team R.K Pachauri and A. Reisinger), Geneva: IPCC.

Krutilla, K. (1999), 'Environmental policy and transaction costs', in J.C.J.M. van den Bergh (ed.), *Handbook of Environmental and Resource Economics*, Cheltenham, UK and Northampton, MA, USA: Edward Elgar, ch. 17.

Krutilla, K. and R. Krause (2011), 'Transaction costs and environmental policy: an assessment framework and literature review', *International Review of Environmental and Resource Economics*, **4**: 261–354.

López, R. and S. Mitra (2000), 'Corruption, pollution, and the Kuznets environmental curve', *Journal of Environmental Economics and Management*, **40**: 137–50.

McCann, L., B. Colby, K.W. Easter, A. Kasterine and K.V. Kuperan (2005), 'Transaction cost measurement for evaluation environmental policies', *Ecological Economics*, **52**: 527–42.

McGranahan, G., D. Balk, D. and B. Anderson (2007), 'The rising tide: assessing the risks of climate change and human settlements in low elevation coastal zones', *Environment and Urbanization*, **19**(1): 17–37.

McNeill, J.R. (2000), *Something New Under the Sun: An Environmental History of the 20th-century World*, New York: W.W. Norton.

McNeill, J.R. (2005), 'Modern global environmental history', *IHDP Update*, **2**: 1–3.

Meinig, D.W. (2004), *The Shaping of America: A Geographical Perspective on 500 Years of History. Vol. 4 Global America 1915–2000*, New Haven, CT: Yale University Press.

Mettepenningen, E., A. Verspecht and G. Van Huylenbroeck (2009), 'Measuring private transaction costs of European agri-environmental schemes', *Journal of Environmental Planning and Management*, **52**: 649–67.

Micahaelowa, A. and F. Jotzo (2005), 'Transaction costs, institutional rigidities and the size of the clean development mechanism', *Energy Policy*, **33**: 511–23.

Millennium Ecosystem Assessment (MEA) (2005), *Ecosystems and Human Well-being: Synthesis*, Washington, DC: Island Press, Table 1.

Nelson, R.R. and G. Wright (1992), 'The rise and fall of American technological leadership: the postwar era in historical perspective', *Journal of Economic Literature*, **30**(4): 1931–64.

Nicholls, R.J., S. Hanson, C. Herweijer, N. Patmore, S. Hallegatte, Jan Corfee-Morlot, Jean Chateua and R. Muir-Wood (2007), 'Ranking of the world's cities most exposed to coastal flooding today and in the future: executive summary', OECD Environment Working Paper No. 1, OECD, Paris.

Nordhaus, W.D. (2007), 'To tax or not to tax: alternative approaches to slowing global warming', *Review of Environmental Economics and Policy*, 1: 26–44.

Nordhaus, W.D. (2010), 'Economic aspects of global warming in a post-Copenhagen environment', *Proceedings of the National Academy of Sciences*, **107**(26): 11721–6.

North, D.C. (1990), 'A transaction cost theory of politics', *Journal of Theoretical Politics*, **2**(4): 355–67.

North, D.C. (1991), 'Institutions', *Journal of Economic Perspectives*, **5**(1): 97–112.

Organization for Economic Cooperation and Development (OECD) (2008), *Costs of Inaction on Key Environmental Challenges*, Paris: OECD.

Page, S.E. (2006), 'Path dependence', *Quarterly Journal of Political Science*, 1: 87–115.

Pierson, P. (2000), 'Increasing returns path dependence and the study of politics', *American Political Science Review*, **94**: 251–67.

Population Division of the United Nations Secretariat (PDUN) (2008), *World Urbanization Prospects: The 2007 Revision Executive Summary*, New York: United Nations.

Robins, N., R. Clover and C. Singh (2009), 'Taking stock of the green stimulus', 23 November, HSBC Global Research, New York.

Robins, N., R. Clover and D. Saravanan (2010), *Delivering the green stimulus*, 9 March, HSBC Global Research, New York.

Rousseau, S. and S. Proost (2005), 'Comparing environmental policy instruments in the presence of imperfect compliance – a case study', *Environmental and Resource Economics*, **32**: 337–65.

Schwoon, M. and R.S.J. Tol. (2006), 'Optimal CO_2-abatement with socio-economic inertia and induced technological change', *The Energy Journal*, **27**(4): 25–59.

Smil, V. (1994), *Energy in World History*, Boulder, CO: Westview Press.

Smil, V. (2000), 'Energy in the twentieth century: resources, conversions, costs, uses, and consequences', *Annual Review of Energy and the Environment*, **25**: 21–51.

Smil, V. (2006), *Transforming the Twentieth Century: Technical Innovations and Their Consequences*, Oxford: Oxford University Press.

Smil, V. (2008), *Global Catastrophes and Trends: The Next 50 Years*, Cambridge, MA: MIT Press.

Stankeviciutute, L., A. Kitous and P. Criqui (2008), 'The fundamentals of the future international emissions trading system', *Energy Policy*, **36**: 4272–86.

Stavins, R.N. (1995), 'Transaction costs and tradable permits', *Journal of Environmental Economics and Management*, **29**: 133–48.

Stern, N. (2007), *The Economics of Climate Change: The Stern Review*, Cambridge: Cambridge University Press.

Strand, J. and M. Toman (2010), '"Green stimulus", economy recovery, and long-term sustainable development', Policy Research Working Paper 5163, The World Bank, Washington, DC.

United Nations Development Programme (UNDP) (2008), *Human Development Report 2007/2008. Fighting Climate Change: Human Solidarity in a Divided World*, New York: UNDP.

US Energy Information Agency (EIA) (2010), *International Energy Outlook 2010*, Washington, DC: EIA.

Williamson, O.E. (2000), 'The new institutional economics: taking stock, looking ahead', *Journal of Economic Literature*, **38**(3): 595–613.

Wilson, J.K. and R. Damania (2005), 'Corruption, political competition and environmental policy', *Journal of Environmental Economics and Management*, **49**: 516–35.

World Bank (2008), *Word Development Indicators 2008*, Washington, DC: The World Bank.

PART VII

LOW-CARBON GROWTH

29 Prosperity with growth: economic growth, climate change and environmental limits

Cameron Hepburn and Alex Bowen[1]

1 INTRODUCTION

> Saving the environment will certainly check production growth and probably lead to lower levels of national income. This outcome can hardly surprise. Many have known for a long time that population growth and rising production and consumption cannot be sustained forever in a finite world.[2]

Debate has raged among and between economists, environmentalists and others about whether increases in production and consumption can be sustained for ever, or whether we are eventually destined for a 'stationary state' of income, labour and capital. This debate, at times, has been remarkably heated and *ad hominem*.[3]

Perspectives on the extent to which environmental and resource constraints will limit economic growth can be grouped into three categories. The first view is that environmental factors pose no limitation to economic growth. For instance, Lomborg (2001, ch. 12) claimed that resources are becoming more abundant. Simon (1981) famously asserted that there is no real limit to our capacity to keep growing, and made bets to prove it. Simon (1980) argues that the term 'finite' is 'not only inappropriate but is downright misleading in the context of natural resources'.[4] Until a decade ago, there appeared to be empirical support for the view that commodities were becoming more economically abundant (Johnson, 2000), given the long-term trend of declining commodity prices over the twentieth century (Dobbs et al., 2011). Such views are also in accordance with the central result of most standard neoclassical and endogenous growth models with labour, capital and human capital as the factors of production: provided technological progress continues, economic growth can be sustained indefinitely. And economic growth may generate technological progress, a view stretching back to Adam Smith's argument that the division of labour is limited by the extent of the market (Smith, 1776). Economic growth enlarges markets and permits greater specialization and variety; increasing returns to scale stimulate economic growth (Young, 1928).

The second view is that environmental limitations will at least exert a 'drag' on economic growth. This environmental drag is caused by natural resource limitations and the various negative effects of pollution on productivity and human well-being. Nordhaus (1992) and Bruvoll et al. (1999) attempted to estimate the historical extent of the environmental drag and make tentative forecasts for the future. While these estimates are admittedly still relatively crude, the concept of the environmental 'drag' on economic growth has entered mainstream macroeconomics and is presented in standard graduate macroeconomics textbooks, such as Romer (2006). Jacobs (1991) argued that if sustainable environmental limits are observed by policy makers, the level and composition of economic growth are likely to be constrained, and also argued that it is an empirical rather than theoretical question whether GDP growth would have to be brought to a complete halt.

The third view, reflected by the opening quote from Tinbergen and Hueting (1992), is that environmental limitations are significant enough to prevent sustained growth in consumption and production. This perspective has its origins in the writings of classical economists, such as Malthus (1798), who argued that living standards would ultimately be driven to a bare subsistence level, and Mill (1848), who argued that the economy would eventually reach a stationary state.[5] Notable proponents of the latter viewpoint in the twentieth century have included John Maynard Keynes,[6] Sir John Hicks[7] and Nicholas Georgescu-Roegen, although in some cases with the emphasis on the satiation of wants not Malthusian limits.[8] Ironically, this view gradually became fiercely anti-establishment. It has seen a revival in recent years with popular books such as *Prosperity without Growth* (Jackson, 2009). In a provocative recent paper, Gordon (2012) questioned the assumption that economic growth is a continuous process that will persist for ever, pointing out that there was virtually no growth before 1750, and arguing that there is no guarantee that growth will continue indefinitely. He cited constraints imposed by energy systems and the environment as one of several headwinds likely to slow US growth, with future growth in consumption per capita for the bottom 99 per cent of the US income distribution possibly falling below 0.5 per cent per year for an extended period of decades.

Some economists concede that the environmental impacts of growth may harm well-being but assert that environmental limitations need not concern us in the long run. Ultimately, enough economic growth, through the associated technological progress and shifts in preferences, will cure environmental degradation and hence remove any constraints on growth. Beckerman (1992, p. 482) expresses the first leg of this argument with his usual clarity:

> Furthermore there is clear evidence that, although economic growth usually leads to environmental degradation in the early stages of the process, in the end the best – and probably the only – way to attain a decent environment in most countries is to become rich.

This inverted-U-shape relationship between environmental quality and per capita income is usually labelled the environmental Kuznets curve (EKC), drawing on an analogy with the observations of Kuznets (1955) on the relationship of income inequality to changes in per capita income. If studies were to provide convincing support for an EKC curve across a wide range of pollutants, including carbon dioxide, then we would be tempted to draw the conclusion that economic growth can be sustained indefinitely, as long as the vast majority of countries can graduate to high enough income levels.

However, careful thinking is required. The EKC literature seeks to answer the question of whether economic growth will ultimately lead to specific environmental improvements. This is a related but different question to whether environmental constraints in general will limit economic growth. Even if EKCs have wide applicability and economic growth is shown to benefit the environment in some respects, it is possible that environmental limitations will nevertheless slow growth, perhaps even to the point that we reach a stationary state. In the current economic context, with widespread pessimism about the economic outlook and anaemic growth at best in most major OECD countries, an eventual stationary state is now easier for some commentators to contemplate.

This chapter considers the three views and isolates the arguments that underpin each of them. We draw a distinction between material economic activity (where the laws of thermodynamics and the limits to substitutability in production and consumption may eventually impose significant economic constraints) and intellectual activity. We observe that it is theoretically feasible for sustained increases in utility to result from increases in the level of intellectual development, even if the material economy ultimately attains a stationary state.

The chapter proceeds as follows. Section 2 presents the key concepts and examines the three views on the severity of environmental limitations to economic growth. It also reviews some empirical estimates of 'environmental drag' and discusses the relevance of the literature on the EKC. In Section 3, we attempt to unify the competing perspectives, and argue that, even in a materially stationary state, indefinite growth in well-being is possible because of progress in the intellectual economy. We also directly consider the notion of 'prosperity without growth' and the claim that stopping economic growth is necessary for resolving a problem such as climate change. Section 4 concludes.

2 CONCEPTUAL CLARIFICATIONS

Research into the relationship between economic growth and the environment is often focused on one of two distinct but related questions. First, do environmental constraints limit economic growth? Second, does economic growth improve environmental quality? We consider both questions in turn in this section. Before doing that, however, it is important to define what is meant by 'economic growth'.

2.1 Economic Growth

A growth rate is merely the (proportional) rate of change of a variable. References to 'economic growth' tend to refer to the rate of change in output, or, more specifically, in real gross domestic product (GDP). Real GDP is a measure of the market value of all final goods and services produced in the economy for a given year, adjusted to remove the impact of changes in the general price level. In this sense, economic growth reflects increases in the value of output as assessed by participants in the markets for goods and services. GDP does not measure increases in physical mass moving through the economic system. This is important, because, although there are, ultimately, physical limitations on the material throughput of the economy, it is conceptually possible for growth in the value of goods and services to grow without bound.

A positive (negative) GDP growth rate implies that the market economy is expanding (contracting). An expansion in average GDP per head is very likely to be associated with an increase in average consumption per head, increasing that measure of material well-being. It also appears to be very helpful in contributing to reductions in the incidence of poverty (Kanbur, 2001; Collier, 2007). Faster growth after a period of slowdown can reduce unemployment and increase perceived investment opportunities, stimulating 'animal spirits'. Sharp slowdowns, as recently experienced in many OECD countries, usually lead to increases in unemployment, crime, mental illness and severe reductions in welfare.

Friedman (2005) also argues that economic growth provides broader welfare benefits, in that it fosters 'moral societies' characterized by social and political liberalization, manifested in increased opportunity, tolerance, economic and social mobility, fairness and democracy. Furthermore, he asserts that economic growth is partly responsible for some of the great periods of technological and intellectual advance, particularly the Enlightenment. Even irrespective of these wider considerations, it is evident that economic growth is important for welfare, especially in poorer countries, and that recessions and the absence of per capita growth (at least under existing economic and social arrangements) are damaging.

2.2   Environmental Limitations on Economic Growth

Conceptually, perspectives of the environmental limitations on economic growth can be divided into the three categories set out in the introduction. The first perspective is the most optimistic – unlimited economic growth is possible, driven by technological progress and human ingenuity (Section 2.2.1). The second perspective is that growth will continue but environmental limits will exert a 'drag' (Section 2.2.2). Finally, the third position is that environmental limitations will, or at least might, eventually bring growth to a halt (Section 2.2.3). We consider each perspective in turn.

2.2.1   Sustained growth: neoclassical and endogenous growth models

Most neoclassical and new-growth theory does not explicitly model environmental limitations, and hence never-ending growth is often possible. Solow (1956) showed that capital accumulation alone cannot support sustained growth, because of diminishing returns. However, the incorporation of (exogenous) technological progress allows indefinite economic growth. Extensions of the neoclassical model to include natural resources (e.g. Stiglitz, 1974) or pollution (e.g. Stokey, 1998) still conclude that unbounded growth can be supported by exogenous technological progress. The new breeds of endogenous growth models[9] also generally give rise to limitless increases in economic output[10] resulting from constant (or increasing) returns to ideas overcoming diminishing returns to capital.[11] Again, many of these models abstract from environmental limitations and hence conclude that growth without limit is a possible, and indeed likely, outcome. Sustained growth is even found as an outcome in some models that include environmental limitations (Aghion and Howitt (1998, ch. 5) and Grimaud (1999)). Michel and Rotillon (1995) find growth without limit is possible when the marginal utility of consumption rises with pollution. Smulders (1995) demonstrates that increasing the stringency of environmental policy can both boost growth (by improving the provision of environmental services) and ensure sustainability of growth.

Standard neoclassical models prompted fierce criticism by authors such as Georgescu-Roegen (1975) and Daly (1997). Stiglitz (1997) replies that this criticism stems from a failure to understand the role of analytical models. He states, 'We write down models as if they extend out to infinity, but no one takes these limits seriously – for one thing, an exponential increase in population presents almost unimaginable problem of congestion on our limited planet.'[12] The danger is that, without adequate and very obvious caveats, conclusions of analytical models *are* taken seriously. Opschoor (1997) states bluntly of Stiglitz: 'Well, he had me fooled, for one'. Furthermore, Simon (1981) argues seriously (and capably) that 'there is no meaningful physical limit – even the commonly mentioned weight of the earth – to our capacity to keep growing forever'. Others, however, such as Clark (1997), claim never to have been misled: 'The argument . . . has always struck me as an exercise in inanity, at least when carried to the limit of supporting the possibility of perpetual economic growth'.

In any event, the odds of humanity surviving the implosion of our sun in several billion years, let alone thriving and continuing to grow out 'to infinity' would appear to be vanishingly small, irrespective of whether models such as that of Stiglitz (1997) are to be taken literally or not. The notion that we might enjoy unlimited economic expansion for ever, without some kind of 'environmental drag', serves as a helpful conceptual benchmark. In practice, however, ignoring the environmental drag is increasingly implausible as 'planetary boundaries' are exceeded (Rockström et al., 2009). And environmental and resource pressures seem only likely to increase as the human population swells from 7 billion to 9–10 billion and, critically, as the number of middle-class consumers grows from 1 billion to 4 billion people (Kharas, 2010). The question is: will that drag at some point reduce the growth rate to zero? Is it appropriate to invoke an 'environmental Laffer curve' by describing a future state where environmental degradation and resource consumption would clearly prevent future growth and then argue that there must be a prior turning point at which attempts to grow faster become self-defeating?

2.2.2 Environmental drag

The concept of an 'environmental drag' appears to have been introduced by Nordhaus (1992), in the course of engaging with Meadows et al. (1992).[13] Nordhaus (1992) defines the environmental drag as 'true national income'[14] growth when resources are 'superabundant (but not free)' and there is no pollution, minus actual 'true national income' growth, with scarce resources and pollution. The concept of an environmental drag on growth is now firmly within mainstream economics, as attested by its presentation in Romer (2006, ch. 1), a standard graduate macroeconomic text.

Despite the usefulness of the concept of the environmental drag, few theoretical papers to date have explicitly provided analytical expressions for it. Nevertheless, it would be straightforward to derive the environmental drag for the models of Bovenberg and Smulders (1995), Smulders (1995), Byrne (1997), Stokey (1998), Aghion and Howitt (1998) and Grimaud (1999).

Empirically, the environmental drag can be divided into two components. The first component is the constraint on production due to the earth's limited natural resources. This has been of concern to economists for some time. Indeed, in deriving the rate of optimal extraction of a resource, Hotelling (1931, p. 137) observed:

> Contemplation of the world's disappearing supplies of minerals, forests, and other exhaustible assets has led to demands for regulation of their exploitation. The feeling that these products are now too cheap for the good of future generations, that they are being selfishly exploited at too rapid a rate, and that in consequence of their excessive cheapness they are being produced and consumed wastefully has given rise to the conservation movement.

Table 29.1   Estimates of environmental drag by Nordhaus (1992)

Source of drag	Impact on world growth rate 1980–2050 (basis points per year)	Impact on world output in 2050 (percentage reduction)
Non-renewable resources		
Energy fuels	15.5	10.3
Non-fuel energy	˙2.9	2.0
Entropy	0.0	0.0
Pollution		
Greenhouse warming	2.9	2.0
Local pollutants	4.4	3.0
Land drag	5.2	3.6
Total	30.9	19.4

Before Hotelling, Jevons (1865) was concerned with the effect of depletion of natural resources, warning of dire consequences for British industry upon the inevitable exhaustion of coal stocks. However, both Jevons and Hotelling appear to have underestimated human inventiveness and adaptability – new reserves have been discovered and substitutes have been developed for scarce resources. Indeed, there have been plenty of counter-examples to the assumption that no factor substitution is possible, a few of which are noted in Nordhaus (1973).

Nordhaus (1992) estimates the drag on growth from scarce resources, comparing a 'limited' case with a case where resources are counterfactually 'unlimited'. He finds the largest growth drag to be from limited energy supplies, amounting to only 15.5 basis points per year (0.155 per cent per year). The drag on growth from limited copper supplies was found to be 1 basis point per year, and by extension, 2 basis points for all non-fuel minerals. While these estimates are extremely crude, he proffers them because 'it is hardly interesting to say we don't know'. These estimates are shown in Table 29.1, under the category of non-renewable resources.

A component of the non-renewable resources drag is retardation due to the second law of thermodynamics. This states that all physical processes (including transformation of materials or energy) must increase total entropy, a measure of thermodynamic disorder.[15] One of the first expositions of the relationship between entropy and economic processes was an introductory essay by Georgescu-Roegen (1966). Paul Samuelson was impressed, writing in the preface: 'I defy any informed economist to remain complacent after meditating over this essay'. The entropy analysis was extended in Georgescu-Roegen (1971), where the enormous size of the flows of solar 'negentropy income' (free energy) is noted.[16] In contrast, Nordhaus (1992) relies on these negentropy flows to support the estimate shown in Table 29.1, concluding that, 'as long as the sun shines brightly on our fair planet, the appropriate estimate for the drag from increasing entropy is zero'.[17]

The second component is the drag from pollution. This is increasingly of concern to policy makers. Nordhaus (1992) provides a rough estimate of the drag on world income growth from the greenhouse effect. His focus is on growth as traditionally measured by changes in the market value of output. Assuming a doubling of CO_2 concentrations by 2050,[18] he estimates that the cost of the greenhouse effect would be between 0 and 2 per cent of world income, while policies to prevent it would use between 1 and 5 per cent of income. He takes 2 per cent as a compromise figure. Since Stern (2007) and the IPCC (2007), however, Nordhaus has updated these damage estimates to between 1.2 and 1.7 per cent of global output.[19] Stern (2007) estimates the welfare-equivalent, balanced-growth-equivalent reduction to be between 5 and 20 per cent. Finally, based upon data from the US EPA, Nordhaus (1992) estimates the annual cost of pollution control to be 3 per cent of total output.[20] Based on the estimates collected in Table 29.1, Nordhaus (1992) concludes, subject to reservations about the tentative nature of his work, that 'an efficiently managed economy need not fear shipwreck on the reefs of resource exhaustion or environmental collapse'.

Nordhaus's estimates are now 20 years old and, as with the climate-change impact estimates, it would seem more likely that drag estimates have increased rather than declined since 1992. Somewhat more recently, Bruvoll et al. (1999) evaluated the size of the environmental drag from seven air pollutants in a computable general equilibrium model of the Norwegian economy. The drag is calculated by examining three effects of these pollutants. First, the model assesses the impact of pollutants such as SO_2, NO_x, CO and PM10 on labour supply and productivity losses due to traffic-related externalities and respiratory problems. Second, the corrosion impacts of SO_2 on equipment, in addition to the damage to roads from traffic, are evaluated. Third, the direct effect of air pollutants on consumer utility is estimated. The authors calculate that the environmental drag on welfare growth is 23 basis points annually, and the drag on conventional economic growth is 10 basis points. This is more than twice the 4.4 basis point estimate by Nordhaus (1992). Given that Bruvoll et al. (1999) considered only three negative effects of seven air pollutants, the total drag on growth from pollution would be expected to be significantly larger.

These limited investigations suggest that the environmental drag on growth may be significant, and may be increasing. Also, it may be larger if a more comprehensive measure of output than traditional GDP is used and if account is taken of the depreciation of all forms of capital. The World Bank has attempted to estimate countries' net saving rates, adjusted for major environmental damages and resource depletion (World Bank, 2006).[21] It suggests that these factors significantly reduce net saving more broadly measured. Energy depletion, for example, is estimated to make annual true global net saving about 3.3 percentage points of gross national income lower than the traditional measure of saving; particulate emissions have an impact of 0.7 percentage points (World Bank, 2011). However, as a fraction of economic activity, these percentages are still rather small. The question that arises is whether these environmental limitations will eventually become large enough to the extent that we are ultimately destined to wind down to a stationary state.

2.2.3   A 'stationary state'?
Some endogenous growth models that consider environmental externalities generate an eventual stationary state. For instance, in an 'AK model' with pollution as a consumption externality, Michel and Rotillon (1995) show a stationary state to be optimal where the marginal utility of consumption falls or remains constant with pollution. Withagen (1995) similarly proves that the presence of a stock pollutant can change the optimal path from balanced growth to a stationary state. Stokey (1998) finds that sustained growth is not optimal without endogenous technological progress.[22]

Whether a stationary state is likely or desirable obviously relates to the notion of sustainability. On one definition, 'weak sustainability' requires that for economic growth to be considered sustainable, the total aggregate stock of capital, both physical and natural, should not decline over time. In other words, even if there is an environmental drag created by pollution and resource exploitation, or more broadly the reduction of natural capital, economic growth may still be sustainable provided that the level of physical (and other) capital increases at least as quickly as natural capital is depleted. This appears to be one concept motivating the World Bank's estimates of adjusted net saving (or 'genuine savings'), aimed at determining the dollar-valued change in social welfare (Hamilton and Clemens, 1999). Heal (2011) discusses further the importance of measuring and assessing sustainability. Neumayer (2003) emphasizes the implications of weak sustainability, noting that, 'According to weak sustainability, it does not matter whether the current generation uses up non-renewable resources or dumps CO_2 in the atmosphere as long as enough machineries, roads and ports are built in compensation'. In other words, if natural capital is perfectly substitutable with man-made capital, then welfare can be preserved provided aggregate capital stocks are not decreasing. Weak sustainability may thus be formalized through the concept of Hartwick–Solow sustainability, in which the total sum of all changes in capital stocks must be zero.

In contrast, various definitions of strong sustainability have been advanced (Neumayer, 2003). For instance, Pearce et al. (1996) argue that strong sustainability requires preserving some 'critical' minimum amount of natural capital, and avoiding 'negative genuine saving' of the natural capital that can be exploited. This does not necessarily condemn us to a stationary state – provided that critical natural capital stocks are protected, the rest of the economy might be able to continue growing – but strong limitations in the substitutability among capital stocks would seem to make a stationary state more likely. We return to reviewing the potential to preserve natural capital stocks while growing in Section 3, on decoupling, following an analysis of the related question of the impact of economic growth on the environment.

2.3   The Impact of Economic Growth on the Environment

It is possible to argue that the drag of environmental degradation on growth will diminish in importance because economic growth itself will lead to a better environment. Views on the effect of economic growth on the environment, like those on many environmental issues, are unnecessarily polarized. Green organizations stress, correctly, that economic growth increases waste and pollution. The 1972 Club of Rome report into the 'Limits to growth' modelled a business-as-usual global system collapse by the middle of the twenty-first century due to finite resources (Meadows et al., 1972). Similar arguments were advanced by Schumacher (1973) and Mishan (1967). Turner (2008) argues that the past 30 years have provided evidence broadly in line with their base case. Others argue, again correctly, that economic growth increases incomes, which increases the demand for a cleaner, more attractive environment (other things – especially prices – equal). One variant on this argument is that societies with higher incomes tend to have better ways of dealing with market failures in the environmental sphere (either because they are better at governance in general, which has helped them become richer, or because good gov-

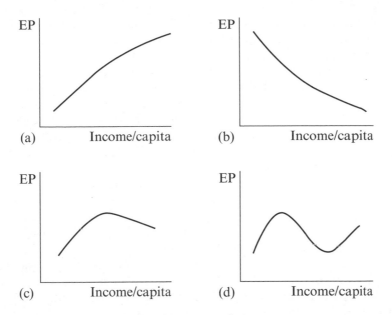

Figure 29.1   Relationships between environmental pressure (EP) and per capita income

ernance is expensive). A corollary is that higher rates of environmental degradation in poorer societies do not necessarily imply that environmental goods are a luxury; apparently high income elasticities of demand for such goods may simply reflect the inability of people in poorer societies to express their preferences with respect to the environment. Direct estimates of income elasticities cast doubt on the assumption that environmental goods are in fact regarded as luxuries (Kriström and Riera, 1996). Another variant invokes a supply-side argument rather than a demand-side argument: richer societies tend to have a comparative advantage in products that are less environmentally harmful, perhaps because they are more intensive in the use of human capital and less in material inputs. Economists have framed the debate by hypothesising that these two opposing tendencies could give rise to a so-called environmental Kuznets curve (EKC) – an inverted-U-shape relationship between per capita income and environmental quality, as illustrated in Figure 29.1(c).

A variety of theoretical justifications for the EKC have been proposed. Stokey (1998) provides perhaps the simplest. She surmises that utility from consumption increases rapidly with diminishing returns, whilst the utility received from environmental quality, measured as the level of pollution, increases slowly with increasing returns, resulting in an inverted-U-shape relationship as income increases over time. The main implication of the model is that the prospects for sustainable growth ultimately depend on whether a constant rate of return on capital is compatible with increasingly strict environmental regulation (Stokey, 1998). Jones and Manuelli (2001) develop a model where societies choose their preferred level of pollution by voting. Assuming that environmental quality is a luxury good, when society votes on the level of effluent charges they predict an inverted U, followed by a pollution increase, as in Figure 29.1(d).[23] Andreoni and Levinson (2001) provide a very simple foundation for the EKC, relying solely on increasing returns to pollution abatement.[24] This has the realistic implication that returns to abatement effort diminish as pollution is abated.

Despite a vast amount of empirical work done on the EKC hypothesis, evidence supporting the hypothesis is at best specific to local pollutants; there is no clear evidence that the EKC holds at the global and general level. Evidence supporting the EKC by Grossman and Krueger (1995), Cole et al. (1997), Shafik (1994) and Selden and Song (1994) shows estimated turning points at per capita incomes of between 1985 US$3280 and $14 700 for a variety of local air and water pollutants. For global pollutants such as CO_2, turning points are modelled to be well above incomes currently achieved by any nation, potentially consistent with Figure 29.1(a) or 29.1(c). Furthermore, work by Shafik (1994) on faecal content and Grossman and Krueger (1995) on urban SO_2 concentrations shows evidence of a cubic relationship, as in Figure 29.1(d), where pollution levels show a second turning point at higher incomes. Grossman and Krueger (1991) also find an 'N-shape' rather than an inverted-U shape for aggregate material inputs per unit output through time. After a thorough review of the literature, Ekins (2000) concludes that the EKC is not unequivocally supported for any individual environmental indicator and is rejected for environmental quality as a whole. Stern (2004) suggests that developing countries are adopting developed-country behaviour with respect to some environmental issues, thus eroding any traditional EKC relationship. This conclusion is supported by Caviglia-Harris et al. (2009), whose analysis of EKC using ecological footprint measures also found no significant EKC relationship between development and growth. Stern (2004) also concludes, 'The evidence . . . shows that the statistical analysis on which the environmental Kuznets curve is based is not robust. There is little evidence for a common inverted U-shaped pathway that countries follow as their income rises'. This echoes Copeland and Taylor (2004, p. 8), who write, 'Our review of both the theoretical and empirical work on the EKC leads us to be sceptical about the existence of a simple and predictable relationship between pollution and per capita income'.

In addition to the lack of support in the empirical literature, there is confusion about the relevance of the EKC hypothesis to the 'limits to growth' debate. In one of the less confused statements, Ekins (2000, ch. 7) states that accepting the EKC hypothesis 'turns the "limits to growth" argument on its head'. Instead of the environment setting limits to growth, these conclusions suggest that growth is a requirement of environmental improvement. While less confused, this is nevertheless misleading in so far as it suggests that proof of the EKC would imply the possibility of unbounded growth. In principle, one can accept that economic growth improves environmental quality while also holding that environmental limits will nevertheless prevent unbounded growth.

Furthermore, there are three other reasons why the nexus between the EKC and the 'limits to growth' debates is not as strong as might be thought. First, as noted above, the EKC appears to be most convincing if it is a microeconomic phenomenon, which is not necessarily applicable in the aggregate. Second, one theoretical basis for the EKC is the fact that pollution reductions are less costly when processes are less efficient to begin with.[25] But, as production processes become more efficient and reach their thermodynamic limits, a lower bound of pollution per unit output will be attained. As such, proof of the EKC hypothesis would not necessarily imply that growth is unconstrained by environmental bounds. Once thermodynamic efficiency is achieved, unless some alternative use for the waste products is discovered, increases in material output will cause corresponding, unambiguous increases in pollution. Third, the independent variable employed in the EKC literature – per capita income – conflates income derived from a range of different sources, some of which are materials intensive and others 'material efficient' (Baptist and Hepburn, 2012). We discuss this further in Section 3. In sum, given these three reasons, the proof or otherwise of the EKC will not decide the question of the environmental limitations on economic growth.

The next section looks in more detail at the feasibility of sustained growth in the value of output. We ask in more detail what would be required to 'decouple' the economy from its material basis so that GDP growth can continue while increases in certain material throughputs (e.g. CO_2 emissions) gradually decline to sustainable levels.

3   DECOUPLING AND THE FEASIBILITY OF SUSTAINED GROWTH

Changes in technology are one of the main factors in determining the development and long-run growth of an economy. For welfare to continue to rise indefinitely with economic growth, eventually the net rate of environmental degradation must approach zero. In other words, increases in economic output eventually must be 'decoupled' from increases in pressure on the environment. If not, perhaps economic growth should stop altogether?

3.1   Is Zero Economic Growth the Answer?

Several scholars and commentators, including Jackson (2009), think that growth in economic activity itself is the fundamental problem, and that we must strive for 'prosperity without growth'. Jackson emphasizes the importance of distinguishing between 'relative' and 'absolute' decoupling of economic output and environmental pressure. Relative decoupling implies a reduction in the environmental pressure per unit of economic output. So-called 'absolute decoupling' is a reduction in environmental pressure. Thus, if economic growth is assumed to continue, relative decoupling is a necessary, but not sufficient, condition for absolute decoupling. Jackson (2009) argues that, in the long run, absolute decoupling is an essential condition for economic activity to remain within ecological limits.

There is clear evidence of relative decoupling in various forms. This is not surprising. Firms have incentives (if imperfect) to stimulate innovation and make improvements in efficiency to reduce input costs, including resource consumption and environmental damage (where this is priced). For example, global energy intensity is now 33 per cent lower than it was in 1970 in the OECD countries (IPCC, 2007). However, for absolute decoupling to occur, resource efficiencies must increase at least as fast as economic output.

Whilst there is frequent evidence of relative decoupling, evidence of absolute decoupling is less common. Absolute decoupling has been observed in some resources. For example, forest cover is increasing, rather than decreasing, in rich countries, flint is no longer needed in axes, saltpetre for gunpowder, or guano for fertilizer. However, in the context of climate change, there has been only relative, not absolute, decoupling of economic growth and greenhouse gas (GHG) emissions. The 'IPAT equation' of Ehrlich and Holden (1971) can be used to explore the relevant relationships. Let I denote total 'impact', measured in tonnes of CO_2e, let P denote global population, let A denote affluence, measured by GDP per capita, and let T denote 'technology' in the form of the CO_2e emissions intensity of GDP.[26] Then it is an accounting identity that

$$I = PAT$$

and hence

$$\ln I = \ln P + \ln A + \ln T$$

Differentiating all terms with respect to time gives the relationship between the growth rates of the variables:

$$\frac{\dot{I}}{I} = \frac{\dot{P}}{P} + \frac{\dot{A}}{A} + \frac{\dot{T}}{T}$$

Empirically, Jackson (2009) notes that since 1990 annual population growth, $\dot{P}/P$, has been 1.3 per cent, annual GDP per capita growth, $\dot{A}/A$, has been 1.4 per cent and emissions per unit of GDP, $\dot{T}/T$, have been falling by 0.75 per cent p.a. The identity indicates that CO_2e emissions have been growing at 2 per cent per annum:

$$\frac{\dot{I}}{I} = 1.3\% + 1.4\% - 0.7\% = 2\%$$

Jackson (2009) employs this identity to argue as follows. First, while there has been relative decoupling ($\dot{T}/T < 0$) over the past 20 years, there is not yet any evidence of absolute decoupling ($\dot{I}/I > 0$) between GHG emissions and economic output. Second, on the basis of the experience over the past 20 years, he concludes that absolute decoupling is not possible, because he dismisses the possibility of a structural shift in technological progress that would lower the emissions intensity of GDP to the degree required. He notes that in order for emissions to fall at the level to achieve the 2 °C global temperature change target, this requires $\dot{I}/I = -4.9\%$. Allowing for an anticipated reduction in reduced annual population growth to $\dot{P}/P = 0.7\%$, this requires annual emissions intensity to fall at the rate $\dot{T}/T = -7\%$, an enormous challenge. Third, given this arithmetic, he (erroneously) concludes from this that the only solution is to target 'prosperity without growth'. That is, his solution is to reduce $\dot{A}/A$ from 1.4 per cent to 0 per cent. Simple addition shows that the implication of the cessation of economic growth, with current rates of population growth and technological change, is that annual CO_2 emissions growth would be

$$\frac{\dot{I}}{I} = 1.3\% + 0\% - 0.7\% = 0.6\%$$

The conclusion is that even halting economic growth does not produce absolute decoupling (as ($\dot{I}/I > 0$), and it certainly does not deliver $\dot{I}/I = -4.9\%$ as required to restrain temperature increases to less than 2 °C. Achieving this, under Jackson's zero growth scenario, would still require a radical, structural shift in technology to $\dot{T}/T = -5.6\%$, implying a dramatic reduction in the emissions intensity of GDP. Indeed, it is precisely this sort of structural shift that Jackson rules out to justify his 'no-growth' world.

It is clear to a very large number of scholars and others that shifting $\dot{T}/T$ from -0.7 per cent to -7 per cent p.a. is an extreme challenge. However, reducing $\dot{T}/T$ to -5.6 per cent while simultaneously $\dot{A}/A$ is reduced to 0 per cent is even more difficult economically (observe the relationship between affluence, R&D investment and the potential for a structural shift), and impossible politically, and is socially undesirable. The consequences of sharply slowing (let alone stopping) growth are observable in the West at present: high unemployment, increased levels of crime and mental illness, large-scale strikes and so on show the social damage wrought by an economic contraction.

Our point is that both paths involve Herculean challenges, and a 'no-growth' world does not solve the problem of climate change or other environmental problems. Rather, for the sake of prosperity and indeed the likelihood of success, it is better to drive increases in technological progress, leading to reductions in intensity, to generate absolute decoupling along with stable growth. Instead of trying to work out how to stop growth at least cost, the significant and important question is how to stimulate a structural shift and a radical change in T. We need 'green growth', not 'no growth'.[27]

3.2   A Conceptual Vision of (Absolute) Decoupling

Consumption, properly defined, is broader than the enjoyments gained merely from the material world. This should not be a controversial statement. When people over the centuries have 'consumed' Rembrandt's *Night Watch*, the novels of Dickens, or Bach's *Saint Matthew's Passion*, many of them have felt awestruck at human capability, and have, in economic jargon, seen their utility increased (even if the impacts might more correctly be seen as being immeasurable). Critically, consumption of such non-material, or 'intellectual', goods – ideas, art, literature, psychological insight, music – is not bounded by the entropy law in the way that material processes must succumb to the laws of nature. As such, there is no physical limit to the progress of the 'intellectual' economy. And as many intellectual outputs are non-rival in consumption, the value of the intellectual economy can expand as access to it expands, without significant extra material input. More prosaically, many activities categorized as services rely only to a relatively small extent on material inputs – for example, the provision of a Facebook page. And some products usually regarded as the outputs of the manufacturing sector, such as CDs or computer memory chips, derive most of their value from their intellectual content.

Quah (1997, 1999) elaborates on some of the implications of what he calls the weightless economy for industrial organization and economic development. He notes 'the increasing importance in national income of knowledge-products – computer software, new media, electronic databases and libraries, and Internet delivery of goods and services' (Quah, 1999). He uses the term knowledge-products to describe such products not because they are knowledge-intensive in production (although they often are), but because their physical properties resemble those of knowledge – infinite expansibility and irrelevance of physical distance. Gordon (2012) is more sceptical about the ability of the weightless economy to deliver rapid growth. He argues that the computer and

Internet revolution ran out of speed quickly and expresses scepticism about the ability to maintain technological progress in key areas such as transport. He thinks that 'invention since 2000 has centered on entertainment and communication devices that are smaller, smarter, and more capable, but do not fundamentally change labor productivity or the standard of living in the way that electric light, motor cars, or indoor plumbing changed it' (p. 2). But he may be underestimating the changes in society and increases in well-being possible as a result of the new information technologies and social media. Also, the very characteristics of knowledge-products that define them tend to loosen the link between prices, marginal cost of production and the marginal utility that they generate, thus making measures of output derived from the costs of inputs possibly an underestimate of the extra utility created.[28] Weitzman (1996) argues that new ideas are generated by bringing different existing ideas together, implying that the set of potential ideas is limitless. John Stuart Mill (1848, p. 129) asserted that while the material economy would attain a stationary state, our intellectual development could increase indefinitely:

> It is scarcely necessary to remark that a stationary condition of capital and population implies no stationary state of human improvement. There would be as much scope as ever for all kinds of mental culture, and moral and social progress; as much room for improving the Art of Living and much more likelihood of its being improved, when minds cease to be engrossed in the art of getting on.

Figure 29.2 presents a simplistic relationship between the economy and the environment, in which the material economy is bounded by the ecosystem (i.e. physical throughput must eventually bump up against limits), while there is the potential for the unrestricted development of ideas in the intellectual economy. Like increases in material consumption, progress in the intellectual economy increases our well-being. In other words, the question to ask is not whether we can sustain limitless growth in material consumption – we cannot – but whether we will be able indefinitely to sustain increases in well-being. Seen through this lens, aspects of all three schools of thought, outlined above, are correct. Proponents of indefinite economic growth would not claim that humans can circumvent the laws of thermodynamics, nor would they contend that unbounded increases to material output are possible with finite matter. Rather, the sensible claim is

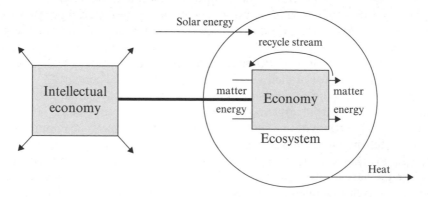

Figure 29.2   The material economy is bounded by the ecosystem; the intellectual economy is not

that technological progress can support 'economic growth' – growth in value to humans – without limit because the intellectual economy is unbounded, even if a stationary material economy is attained in the long run.[29]

The concept of an 'environmental drag' on growth can also be accommodated. In the long run, although the material economy is not growing, it is still producing material output at a constant rate. Constant material production implies non-zero pollution levels, which will exert a drag on utility growth. 'Economic growth' can therefore be sustained with zero material growth and increasing intellectual growth, allowing for a drag exerted by pollution. In short, all three schools of thought can be represented in the conceptual model sketched in Figure 29.2.

In this manner, models with a steady-state optimum, such as Stokey (1998), Michel and Rotillon (1995) and Withagen (1995), can be interpreted as evidence that a stationary *material* state will eventually be attained. The views expressed by Daly (1996) are consistent; he argues that there is a critical distinction between 'growth' (which he defines to be material growth) and 'development' (which essentially constitutes intellectual progress and improvements in well-being). In other words, once the material needs of the society have been satisfied, it is optimal for people to expend their efforts in the intellectual and artistic realms, along the lines envisaged by John Stuart Mill (1848, p. 129) and John Maynard Keynes (1930), the latter of whom suggested in his essay entitled 'Economic possibilities for our grandchildren' that 'a point may soon be reached, much sooner perhaps than we are all of us aware of, when those needs are satisfied in the sense that we prefer to devote our further energies to non-economic purposes'.

3.3   The Evidence on Decoupling

Past evidence of structural shifts in various sectors may provide an insight into the general process of decoupling, and allow progress to be mapped. Understanding the process by which technological change and structural transitions occur can assist by suggesting policies that might be put in place to increase the probability of an accelerated shift, for example (in the case of combating climate change). By definition, past evidence of the absence of absolute decoupling does not and cannot provide proof of the impossibility (or possibility) of any future structural shift.

Historically, periods of technological change have been characterized by substantial reductions in cost and improvements in performance through learning; dynamic competition between technologies and the co-evolution of long-lived infrastructures and technological clusters due to 'network effects' (Grübler et al., 1999). In computing, this process of technological change was so rapid that the number of components that could profitably fit on an integrated circuit had doubled every year from the invention of the integrated circuit in 1958 until 1965 (Moore, 1965). Moore predicted that the trend would continue for at least another decade. In fact, the trend has continued for over half a century now with the number of transistors that can be placed inexpensively on an integrated circuit doubling every 18 months. The question of interest is whether the coming decades might see something similar to Moore's law begin to apply to the energy industry and, if it does, whether we can expect growth to lead a structural shift, resulting in absolute decoupling.

To take one example of several energy technologies where innovation is accelerating,

Moore's law might conceivably apply to the solar photovoltaics (PV) industry in future. The power that can be generated by PV for each dollar of cost has been doubling every seven years. This is nowhere near as rapid as the 18-month doubling time for computing power. The difference may be partly due to lower levels of R&D expenditure on solar PV, the fact that energy is an undifferentiated good, and the stronger competitive pressure on computing firms, which must innovate or cease to exist (in contrast to many regulated energy utilities).

The solar PV industry has, however, had some recent changes that might signal an acceleration in these developments. Costs have fallen 30 per cent over the past 18 months (2009–11) with the emergence of large-scale production facilities in China. The recently increased scale of the industry might signal the start of existential competitive pressure on firms. Also, in recent years, there has been aggressive development of non-silicon-based PV materials (Wadia et al., 2009), driven by fluctuating silicon prices. The likelihood of a Moore's law in scale and cost will depend on various factors, including the judicious selection of photosynthetic materials that are in great abundance, and which are therefore cheap. To take one example, for a given set of assumptions about material density and performance efficiency, it may be that iron sulphide (FeS_2) would produce 10 000 times more electricity than silicon-based PV (Wadia et al., 2009). Nanotechnology solar PV cells have the potential to utilize far less material, bringing about decreasing unit costs and faster production, and therefore a more rapid decline in industry costs.

Ultimately, absolute decoupling of GHG emissions from economic output, while apparently very challenging from the vantage point of 2012, may appear by 2050 to have been a relatively easy challenge. A structural shift that leads to renewable energy becoming significantly cheaper than fossil fuels would trigger such decoupling. Previous structural shifts are evident throughout history. For example, by the seventeenth century, the predominant energy source in England and Belgium was coal. As has been observed, this was due not to a shortage of wood, but to the cost of coal for heating becoming cheaper than that of wood (Fouquet, 2008). A structural shift in renewable technology costs is far from unimaginable if governments accelerate investment in R&D and deployment of such technologies. It is too early to rule out absolute decoupling.

3.4   Rebound Effects and the Impact on Economic Output

A structural shift towards low-cost renewable energy would have various effects on the economy. First, it would reduce GHG emissions and, if big enough, help to solve the challenge of climate change. Second, lower energy prices would increase energy consumption, increasing economic output (Ayres and Warr, 2005) and potentially increasing emissions in other sectors as they expand. The prospect of displacement by renewable energy providers could lead owners of fossil fuel stocks to accelerate their use, pushing down energy prices and encouraging emissions in the near term (Sinn, 2008). These are forms of the 'rebound effect', in which actions taken to increase efficiency reduce the unit cost of use and hence lead to increased demand.[30]

More generally, Baptist and Hepburn (2012) provide evidence that begins to suggest that treading lightly on the planet need not reduce economic growth in value terms. They analyse a panel-data set of 473 manufacturing sectors in the USA over 48 years, and find that sectors with lower material intensity had higher total factor productivity, as did those with higher labour intensity. In other words, using less 'stuff' and more 'human intelligence' increased overall economic productivity. Rebound effects may concern environmentalists, but increases in output imply increases in welfare and should be welcomed if environmental consequences are properly priced.

4   CONCLUSIONS AND POLICY IMPLICATIONS

Intense debate about the environmental limits to growth has been taking place over the past few decades. Three competing schools of thought are identified here: that growth is limitless, that environmental conditions will place a 'drag' on growth, and that economic growth cannot continue indefinitely. We propose an approach that incorporates the essence of the three competing schools of thought. Our conceptual model shows an eventual stationary state in the material economy (which may still be decades or many centuries into the future), with unbounded growth in the intellectual economy, notwithstanding a genuine drag on growth in welfare from environmental constraints.

We engage directly with those who advance a 'zero-growth' world as being necessary to live within environmental limits, and see economic growth as a problem. We demur. We have argued that stopping economic growth (which is measured in terms of value) is neither necessary nor desirable. Indeed, as far as meeting environmental challenges is concerned, it would be counterproductive; recessions have slowed and in some cases derailed efforts to adopt cleaner modes of production. Rather, large leaps in clean technology, triggering a structural shift in the way we produce and consume energy, are required. This is a 'green growth' rather than a 'no-growth' world. The continuation of growth in value to humans is consistent with us living within the material constraints imposed by a finite (if very large) planet, provided that we continue to expand the intellectual economy through innovation, technology development, an increased focus on services and, more fundamentally, the art of living.

The policy implications from this chapter are unsurprising. The large scale of the subsidies spent annually on increasing the size of the material economy should be reduced. As Baptist and Hepburn (2012) note, very approximately US$1 trillion is spent on directly subsidizing the consumption of resources, which includes approximately $400 billion on energy, around $200–300 billion of equivalent support on agriculture, and approximately US$200–300 billion on water. Perhaps another US$1 trillion, very approximately, takes the form of subsidy for the use of the atmosphere as a sink for greenhouse gas emissions. Other subsidies – in the form of incorrect environmental prices – create annual damage in the trillions every year.

In contrast, the scale of the intellectual economy should be increased. Public authorities need to tackle the well-known market failures that tend to lead to insufficient pure research and R&D spending by the private sector, and also consider whether they can help firms reap increasing returns to scale, for example by supporting new networks in their early development. Shifting the tax base towards materials and resources, and away from labour and other intellectual activity, might also contribute to building the intellectual economy and ensuring that its expansion is eventually decoupled from the use of material inputs. It might also increase economic output, given that labour is the factor input that correlates most closely with higher total factor productivity (Baptist and Hepburn, 2012). And it would increase the odds that the rate of reduction in cost of clean technologies would increase, so that solutions are found for our environmental challenges.

NOTES

1. The authors are grateful to Simon Baptist, Sam Fankhauser, Mattia Romani, Hyunjin Kim, Michael Jacobs and Dimitri Zenghelis for comments on this chapter and to Gavin Cameron, Dieter Helm, Ben Irons and Kevin Roberts for helpful suggestions and comments on a much earlier version. Particular thanks are owed to Katerina Kimmorley and especially to Jonathan Colmer for their able and timely research assistance.
2. Tinbergen and Hueting (1992, p. 56).
3. See, in particular, the acerbic remarks of both Daly (1997) and Solow (1997).
4. Within the specific context of energy and fossil fuels, Helm (2011) presents the opinion that the world is awash with fossil fuels. The contrary view, exemplified by Campbell and Laherrère (1998), is the 'peak oil' hypothesis that supply will soon decline as demand increases, leading to higher oil prices. For a more recent expression of this view, see Sorrell et al. (2010) and King and Murray (2012).
5. Mill (1848) further implied that if we did not deliberately guide the economy towards such a stationary state, an environmental collapse would result.
6. See, e.g., 'Economic possibilities for our grandchildren', in Keynes (1930).
7. Hicks (1983, p. 68) thought that once population is controlled, the 'Stationary state is no longer a horror. It becomes an objective at which to aim'.
8. Georgescu-Roegen was a Distinguished Fellow of the American Economic Association and, according to Paul Samuelson, a 'scholar's scholar, an economist's economist'. See the preface to Georgescu-Roegen (1966).
9. Prominent examples include Romer (1986, 1990), Grossman and Helpman (1991) and Aghion and Howitt (1992).
10. Limitless in the sense of output increasing without limit (not in the sense of there being no constraint on annual growth *rates*).
11. General reviews of the early endogenous growth literature can be found in Grossman and Helpman (1994), Romer (1994) and Solow (1994).
12. Solow (1997) similarly felt the need to explain to Daly the purpose of theoretical models.
13. No reference to the concept was found in the literature prior to 1992, and Nordhaus (1992) provides no citation when defining the term. Weitzman (1994) introduces a different 'environmental drag' concept, which represents the cost of environmental regulation, without reference to Nordhaus (1992).
14. This includes appropriately measured consumption, plus the value of net accumulations or decumulations of all 'capital', including physical, human, technical, research and environmental capital.
15. We note that natural scientists would advise us not to employ the term 'disorder' in this context. The precise, statistical definition of 'order' does not equate with the common, intuitive notion, except when applied to ideal gases and dilute solutions: see Sollner (1997).
16. In particular, Georgescu-Roegen (1971) states that, 'as surprising as it may seem, the entire stock of natural resources is not worth more than a few days of sunlight!'
17. This conclusion is supported by the arguments of Young (1991), but disputed by Daly (1992) and Townsend (1992).
18. This is consistent with the high projection by the IPCC (1990).
19. There is a continuing and lively debate about whether these estimates adequately reflect climate-change risks, attitudes to risk, appropriate social discount rates and non-market costs of climate-change impacts.
20. By way of comparison, Adkins et al. (2010) estimate the increase in cost of pollution control following a hypothetical carbon tax of $15t/$CO_2$. They find that for most manufacturing industries, costs are estimated to increase by less than 2 per cent while for other industries, cost increases are larger. For example, refining LPG costs are estimated to rise the most (22.4 per cent), followed by other refining (9.8 per cent) and cement (5.7 per cent).
21. Neumayer (2003) offers a critique of these estimates.
22. In Stokey's model, pollution is an input to production; hence as capital grows the real rate of return can only increase if pollution grows proportionally. Utility is assumed to be concave in consumption and convex in pollution, so such a path is suboptimal.

23. In contrast, when voting is on the dirtiest allowable technology, Jones and Manuelli (2001) find that pollution increases monotonically to a bounded maximum, as in Figure 29.1(a).
24. 'Increasing returns' is used to imply that the more pollution there is before abatement begins, the less costly it is to abate each unit of pollution.
25. See Andreoni and Levinson (2001).
26. An alternative measure is the Kaya identity (Kaya and Yokobori, 1997), which further splits CO_2e intensity into (i) energy use per unit of GDP and (ii) CO_2e emissions per unit of energy consumed.
27. This is the essence of the argument in Jacobs (1991), who considers a broad range of environmental limits.
28. One may wonder what Keynes would have made of Gordon's argument, given his comment in Keynes (1930) that 'We are suffering just now from a bad attack of economic pessimism. It is common to hear people say that the epoch of enormous economic progress which characterised the nineteenth century is over; that the rapid improvement in the standard of life is now going to slow down – at any rate in Great Britain; that a decline in prosperity is more likely than an improvement in the decade which lies ahead of us. I believe that this is a wildly mistaken interpretation of what is happening to us'.
29. Quah (1999) points out that societies may fail to achieve such growth, citing as an example the decline of Chinese inventions and technological progress after the fourteenth century.
30. See Sorrell (2007) and Sorrell and Dimitropoulos (2007) for a review of rebound effects. These effects can be direct (due to income and substitution effects) and indirect (resulting from embodied energy and secondary effects) (Fouquet and Pearson, 2011).

REFERENCES

Adkins, L., R. Garbaccio, M. Ho, R. Moore and R. Morgenstern (2010), 'The impact on U.S. industries of carbon prices with output-based rebates over multiple time frames', RFF Discussion Paper, 10–47.
Aghion, P and P. Howitt (1992), 'A model of growth through creative destruction', *Econometrica*, **60**(2), 323–51.
Aghion, Phillipe and Peter Howitt (1998), *Endogeneous Growth Theory*, Cambridge, MA: MIT Press.
Andreoni, J. and A. Levinson (2001), 'The simple analytics of the environmental Kuznets curve', *Journal of Public Economics*, **80**(2), 269–86.
Ayres, R. and B. Warr (2005), 'Accounting for growth: the role of physical work', *Structural Change and Economic Dynamics*, **16**, 181–209.
Baptist, S. and C. Hepburn (2012), 'Material efficiency, productivity and economic growth', *Philosophical Transactions of the Royal Society A* (forthcoming).
Beckerman, W. (1992), 'Economic growth and the environment: whose growth? Whose environment?', *World Development*, **20**(4), 481–96.
Bovenberg, A. and S. Smulders (1995), 'Environmental quality and pollution – augmenting technological change in a two-sector endogenous growth model', *Journal of Public Economics*, **57**(3), 369–91.
Bruvoll, A., S. Glomsrod and H. Vennemo (1999), 'Environmental drag: evidence from Norway', *Ecological Economics*, **30**(2), 235 49.
Byrne, M. (1997), 'Is growth a dirty word? Pollution, abatement and endogenous growth', *Journal of Development Economics*, **54**(2), 261–84.
Campbell, C.J and J.H. Laherrère (1998), 'The end of cheap oil', *Scientific American*, March, 78–83.
Caviglia-Harris. J., D. Chambers and J. Kahn (2009), 'Taking the "U" out of Kuznets: a comprehensive analysis of the EKC and environmental limits', *Ecological Economics*, **68**(4), 1149–59.
Clark, C. (1997), 'Renewable resources and economic growth', *Ecological Economics*, **22**(3), 275–6.
Cole, M., A. Rayner and J. Bates (1997), 'The environmental Kuznets curve: an empirical analysis', *Environment and Development Economics*, **2**(4), 401–16.
Collier, P. (2007), *The Bottom Billion: Why the Poorest Countries are Failing and What Can Be Done About It*, Oxford: Oxford University Press.
Copeland, B. and S. Taylor (2004), 'Trade, growth, and the environment', *Journal of Economic Literature*, **42**(1), 7–71.
Daly, H. (1992), 'Is the entropy law relevant to the economics of natural resource scarcity? – Yes, of course it is!', *Journal of Environmental Economics and Management*, **23**(1), 91–5.
Daly, Herman (1996), *Beyond Growth: The Economics of Sustainable Development*, Sussex, UK: Beacon Press.
Daly, H. (1997), 'Georgescu-Roegen versus Solow/Stiglitz', *Ecological Economics*, **22**(3), 261–6.
Dobbs, R., J. Oppenheim and F. Thompson (2011), 'A new era for commodities', *McKinsey Quarterly*, November, 1–3.

Ehrlich, P. and J. Holden (1971), 'Impact of population growth', *Science*, **171**(3977), 1212–17.
Ekins, Paul (2000), *Economic Growth and Environmental Sustainability*, London: Routledge.
Fouquet, Roger (2008), *Heat, Power and Light: Revolutions in Energy Services*, Cheltenham, UK and Northampton, MA, USA: Edward Elgar.
Fouquet, R. and P. Pearson (2011), 'The long run demand for lighting: elasticities and rebound effects in different phases of economic development', *Economics of Energy and Environmental Policy*, **1**(1), 83–100.
Friedman, Benjamin (2005), *The Moral Consequences of Economic Growth*, New York: Alfred A. Knopf.
Georgescu-Roegen, Nicholas (1966), *Analytical Economics*, Cambridge, MA: Harvard University Press.
Georgescu-Roegen, Nicholas (1971), *The Entropy Law and the Economic Process*, Cambridge, MA: Harvard University Press.
Georgescu-Roegen, N. (1975), 'Energy and economic myths', *Southern Economic Journal*, **41**(3), 347–81.
Gordon, Robert J. (2012), 'Is US economic growth over? Faltering innovation confronts the six headwinds', NBER Working Paper No. 18315, August.
Grimaud, A. (1999), 'Pollution permits and sustainable growth in a Schumpeterian model', *Journal of Environmental Economics and Management*, **38**(3), 249–66.
Grossman, G. and E. Helpman (1991), 'Quality ladders in the theory of growth', *Review of Economic Studies*, **58**(1), 43–61.
Grossman, G. and E. Helpman (1994), 'Endogenous innovation in the theory of growth', *Journal of Economic Perspectives*, **8**(1), 23–44.
Grossman, G. and A. Krueger (1991), 'Environmental impacts of a North American Free Trade Agreement', National Bureau of Economic Research Working Paper 3914.
Grossman, G. and A. Krueger (1995), 'Economic growth and the environment', *Quarterly Journal of Economics*, **110**(2), 353–77.
Grübler, A., N. Nakićenović and D. Victor (1999), 'Dynamics of energy technologies and global change', *Energy Policy*, **27**(5), 247–80.
Hamilton, K. and M. Clemens (1999), 'Genuine savings rates in developing countries', *World Bank Economic Review*, **13**(2), 333–56.
Heal, G. (2011), 'Sustainability and its measurement', NBER Working Paper No. 17008, May.
Helm, D. (2011), 'Peak-oil and energy policy – a critique,' *Oxford Review of Economic Policy*, **27**(1), 668–91.
Hicks, John (1983), *Classics and Moderns, Collected Essays on Economic Theory*, Volume III, Cambridge MA: Harvard University Press.
Hotelling, H. (1931), 'The economics of exhaustible resources', *Journal of Political Economy*, **39**(2), 137–75.
IPCC (1990), *Climate Change: The IPCC Scientific Assessment*, Cambridge, UK and New York: Cambridge University Press.
IPCC (2007), *Contribution of Working Groups I, II and III to the Fourth Assessment Report of the Intergovernmental Panel on Climate Change*, Cambridge, UK and New York, USA: Cambridge University Press.
Jackson, Tim (2009), *Prosperity without Growth: Economics for a Finite Planet*, London: Earthscan.
Jacobs, Michael (1991), *The Green Economy: Environment, Sustainable Development and the Politics of the Future*, London: Pluto Press.
Jevons, W. (1865), *The Coal Question: an Inquiry Concerning the Progress of the Nation and the Probable Exhaustion of Our Coal Mines*, New York: Augustus M. Kelley.
Johnson, D. (2000), 'Population, food, and knowledge', *American Economic Review*, **90**(1), 1–14.
Jones, L. and R. Manuelli (2001), 'Endogenous policy choice: the case of pollution and growth', *Review of Economic Dynamics*, **4**(2), 369–405.
Kanbur, R. (2001), 'Economic policy, distribution and poverty: the nature of disagreements', *World Development*, **29**(6), 1083–94.
Kaya, Yoichi and Keiichi Yokobori (1997), *Environment, Energy and Economy: Strategies for Sustainability*, New York: United Nations University Press.
Keynes, John Maynard (1930), *Essays in Persuasion*, New York: W. W. Norton & Co.
Kharas, H. (2010), 'The emerging middle class in developing countries', OECD Development Centre Working Paper, 285(January), 1–61.
King, D. and J. Murray (2012), 'A phase shift in oil', *Nature*, **481**, 433–5.
Kriström, B. and P. Riera (1996), 'Is the income elasticity of environmental improvement less than one?', *Environmental and Resource Economics*, **7**, 45–55.
Kuznets, S. (1955), 'Economic growth and income inequality', *American Economic Review*, **45**(1), 1–28.
Lomborg, Bjorn (2001), *The Skeptical Environmentalist: Measuring the Real State of the World*, New York: Cambridge University Press.
Malthus, Thomas (1798), *An Essay on the Principle of Population*, London: W. Pickering (1986).
Meadows, Dennis, Donella Meadows and Jorgen Randers (1992), *Beyond the Limits: Global Collapse or a Sustainable Future*, London: Earthscan Publications.

Meadows, Dennis, Donella Meadows, Jorgen Randers and William Behrens III (1972), *The Limits to Growth: A Report for the Club of Rome's Project on the Predicament of Mankind*, New York: Universal Books.

Michel, P. and G. Rotillon (1995), 'Disutility of pollution and endogenous growth', *Environmental and Resource Economics*, **6**(3), 279–300.

Mill, John Stuart (1848), *Principles of Political Economy with some of their Applications to Social Philosophy*, Oxford: Oxford University Press.

Mishan, Ezra (1967), *The Costs of Economic Growth*, New York: Praeger.

Moore, G. (1965), 'Cramming more components onto integrated circuits', *Electronics*, **38**(8), 114–17.

Neumayer, Eric (2003), *Weak versus Strong Sustainability: Exploring the Limits of Two Opposing Paradigms*, Cheltenham, UK and Northampton, MA, USA: Edward Elgar.

Nordhaus, W. (1973), 'World dynamics: measurement without data', *Economic Journal*, **83**(332), 1156–83.

Nordhaus, W. (1992), 'Lethal Model 2: *The Limits to Growth* revisited', *Brookings Papers on Economic Activity*, **23**(2), 1–60.

Opschoor, J. (1997), 'The hope, faith and love of neoclassical environmental economics', *Ecological Economics*, **22**(3), 281–3.

Pearce, D., K. Hamilton and G. Atkinson (1996), 'Measuring sustainable development: progress on indicators', *Environment and Development Economics*, **1**, 85–101.

Quah, D. (1997), 'Increasingly weightless economies', *Bank of England Quarterly Bulletin*, February, 49–56.

Quah, D. (1999), 'The weightless economy in economic development', Centre for Economic Performance Discussion Paper No. 417, London School of Economics and Political Science, March.

Rockström, J., Steffen, W., Noone, K., Persson, A., Chapin, F. S., Lambin,E. F., Lenton, T. M., Scheffer, M., Folke, C. et al. (2009), 'A safe operating space for humanity', *Nature*, **461**(24), 472–5.

Romer, David (2006), *Advanced Macroeconomics*, New York: McGraw-Hill.

Romer, P. (1986), 'Increasing returns and long-run growth', *Journal of Political Economy*, **94**(5), 1002–37.

Romer, P. (1990) 'Endogenous technological change', *Journal of Political Economy*, **98**(5), 71–102.

Romer, P. (1994), 'The origins of endogenous growth', *Journal of Economic Perspectives*, **8**(1), 3–22.

Schumacher, Ernst (1973), *Small is Beautiful: A Study of Economics as if People Mattered*, London: Blond and Briggs Ltd.

Selden, T. and D. Song (1994), 'Environmental quality and development: is there a kuznets curve for air pollution emissions', *Journal of Environmental Economics and Management*, **27**(?), 147–62.

Shafik, N. (1994), 'Economic development and environmental quality: an econometric analysis', *Oxford Economic Papers*, **46**(0), 757–73.

Simon, J. (1980), 'Resources, population, environment: an oversupply of false bad news', *Science*, **208**(4451), 1431–7.

Simon, Julian (1981), *The Ultimate Resource*, Oxford: Robertson.

Sinn, H. (2008), 'Public policies against global warming: a supply side approach', *International Tax and Public Finance*, **15**, 360–94.

Smith, Adam (1776), *An Inquiry into the Nature and Causes of the Wealth of Nations*, reprinted by Penguin Classics, London, 1986.

Smulders, S. (1995), 'Environmental policy and sustainable growth: an endogenous growth perspective', *De Economist*, **143**(2), 163–95.

Sollner, F. (1997), 'A re-examination of the role of thermodynamics for environmental economics', *Ecological Economics*, **22**(3), 175–201.

Solow, R. (1956), 'A contribution to the theory of economic growth', *Quarterly Journal of Economics*, **70**(1), 65–94.

Solow, R. (1994), 'Perspectives on growth theory', *Journal of Economic Perspectives*, **8**(1), 45–54.

Solow, R. (1997), 'Georgescu-Roegen versus Solow/Stiglitz', *Ecological Economics*, **22**(3), 267–8.

Sorrell, Steven (2007), *The Rebound Effect: An Assessment of the Evidence for Economy-Wide Energy Savings from Improved Energy Efficiency*, London: UK Energy Research Centre.

Sorrell, S. and J. Dimitropoulos (2007), 'The rebound effect: microeconomic definitions, limitations and extensions', *Ecological Economics*, **65**(3), 636–49.

Sorrell, S., R. Miller, R. Bentley and J. Speirs (2010), 'Oil futures: a comparison of global supply forecasts', *Energy Policy*, **38**(9), 4990–5003.

Stern, D. (2004), 'The rise and fall of the environmental Kuznets curve', *World Development*, **32**(8), 1419–39.

Stern, Nicholas (2007), *The Economics of Climate Change: The Stern Review*, Cambridge: Cambridge University Press.

Stiglitz, J. (1974), 'Growth with exhaustible natural resources: efficient and optimal growth paths', *Review of Economic Studies*, **41**(1), 123–37.

Stiglitz, J. (1997), 'Georgescu-Roegen versus Solow/Stiglitz', *Ecological Economics*, **22**(3), 269–70.

Stokey, N. (1998), 'Are there limits to growth?', *International Economic Review*, **39**(1), 1–31.

Tinbergen, Jan and Roefie Hueting (1992), 'GNP and market prices', in Robert Goodland, Herman Daly and

Salah El Serafy (eds), *Population, Technology and Lifestyle: The Transition to Sustainability*, Washington, DC: Island Press, pp. 52–62.

Townsend, K. (1992), 'Is the entropy law relevant to the economics of natural resource scarcity? Comment', *Journal of Environmental Economics and Management*, **23**(1), 96–100.

Turner, G. (2008), 'A comparison to the limits of growth with thirty years or reality', Socio-Economics and Environment in Discussion CSIRO Working Paper Series, 1–49.

Wadia, A., P. Alivisatos and D. Kammen (2009), 'Materials availability expands the opportunity for large-scale photovoltaics deployment', *Environmental Science & Technology*, **43**(6), 2072–7.

Weitzman, M. (1994), 'On the "environmental" discount rate', *Journal of Environmental Economics and Management*, **26**(2), 200–209.

Weitzman, M. (1996), 'Hybridizing growth theory', *American Economic Review*, **86**(2), 207–12.

Withagen, C. (1995), 'Pollution, abatement and balanced growth', *Environmental and Resource Economics*, **5**(1), 1–8.

World Bank (2006), 'Where is the wealth of nations? Measuring capital for the 21st century', Washington, DC: World Bank.

World Bank (2011), *The Little Green Data Book*, Washington, DC: World Bank.

Young, Allyn A. (1928), 'Increasing returns to scale and economic progress', *Economic Journal*, **38**, 527–42.

Young, J. (1991), 'Is the entropy law relevant to the economics of natural resource scarcity?', *Journal of Environmental Economics and Management*, **21**(2), 169–79.

30 Should we sustain? And if so, sustain what? Consumption or the quality of life?*

Humberto Llavador, John E. Roemer and Joaquim Silvestre

1. INTRODUCTION

The rapid growth in greenhouse gas (GHG) emissions and concomitant increase in atmospheric carbon concentration during the past century have raised, in a dramatic way, the spectre of catastrophic effects for the welfare of mankind: in the last century, the only comparable events were the two world wars and worst-case scenarios associated with nuclear proliferation. Unlike these events, the effects of increased atmospheric carbon concentration, mainly due to associated temperature increases, will occur gradually and with a long time lag. Sustainability has gained traction as an ethic and an appropriate goal, as we face the costs of using up our scarce biospheric resource, of a non-carbon-saturated atmosphere.

That ethic is quite pervasive, at least among environmentalists. Perhaps surprisingly, it has influenced economists much less: to wit, the major contributions to the economics of climate change advocate not a 'sustainabilitarian' approach, but a utilitarian one, of maximizing the discounted sum of utilities of the sequence of generations beginning with the present one. Indeed, the two most influential pieces of recent economic analysis, William Nordhaus's (2008) book and Nicholas Stern's (2007) *Review*, both employ versions of discounted utilitarianism, as we will examine below.

There is, however, a literature on sustainability in economics, due to Robert Solow, and built upon by John Hartwick and others. We will review that literature, briefly contrast it with the currently more popular approaches of discounted utilitarianism, and then conclude with a brief description of our own approach, which attempts to rejuvenate the sustainabilitarian research program.

Whether to sustain welfare, or to maximize a discounted sum of welfare levels, is a question concerning the social objective function. The second question we will address is what the arguments of that social welfare function should be: incomes of different generations, their consumptions, their utilities, or something else? In climate change economics – and here the cue is taken from the traditional practice in growth theory – welfare is taken to be some concave function of consumption. We wish to depart from that convention here, as well. Of course, human welfare depends upon more than commodity consumption, but most climate change economists seem to believe that representing the different kinds of inputs to human welfare as a single good is a harmless abstraction. Education and knowledge, in these models, only indirectly impact upon welfare, through their role as inputs in the production of consumption and investment goods. To the contrary, we maintain that ignoring the fact that education and knowledge contribute directly to human welfare is not harmless: for in meeting the challenge of climate change, we should entertain the possibility of shifting our consumption bundle from energy-intensive goods to other goods, such as education, leisure and knowledge, which can be produced in an environmentally more friendly way. This possibility of substitution does not exist in (for example) the models of Nordhaus and Stern, and those analyses may thereby be unnecessarily truncating the set of techniques for adapting to climate change.

We conclude the chapter by demonstrating concretely the consequences of, on the one hand, choosing the optimal path to sustain commodity consumption 'forever', or choosing the optimal path to sustain 'welfare' forever, where welfare includes inputs of knowledge, education and biospheric quality. For both optimization programs, the feasible set of paths is the same – and education, knowledge and GHG emissions enter in production. The two optimal paths turn out to be quite strikingly different, which is to say that a more comprehensive definition of human welfare changes dramatically the policy recommendation.

2.   SOCIAL OBJECTIVES

2.1   GDP, Consumption and the Quality of Life

A large segment of economics, particularly in empirical applications, takes as the index of welfare per capita GDP (gross domestic product) or GNI (gross national income, which includes net income generated abroad) or NNI (net national income, which in addition excludes capital depreciation). Each of these GDP-related measures has two main components: consumption and investment. Consumption is related to current welfare, whereas investment is a contribution to capital, a factor for future welfare. The correlation of any GDP-related measure to consumption is necessarily imperfect: a country, like Singapore, with a high investment-to-GDP ratio has levels of consumption substantially lower than other countries with the same GDP per capita. Hence the two components, consumption and investment, should be separated. One can then ask two distinct questions.

(i)   Is the consumption component of the GDP-related measure an appropriate index of human welfare?
(ii) Is the investment component of the GDP-related measure an appropriate index of the change in capital?

The second question is treated in Section 3.3 below. The present section addresses the first one. What is an appropriate index of human welfare?

The shortcomings of the consumption component of a GDP-related measure as a welfare index have been discussed for decades: see the Sarkozy Report (Stiglitz et al., 2009) for a recent, comprehensive assessment. The criticisms have spurred three different approaches: (a) the inclusion of additional variables besides consumption, in the welfare or utility index, in particular non-marketed goods or services; (b) the measurement of subjective happiness; and (c) the capabilities approach. We refer to Fleurbaey (2009) for a critical review of (b) and (c), and we focus here on (a).[1]

Population size and the level of inequality have potentially important impacts on mankind's well-being. Many of the existing social welfare indices explicitly consider inequality (e.g. the Human Development Index, HDI, by the United Nations Human Development Programme, 2011, Jones and Klenow, 2010, and the recommendations in Stiglitz et al., 2009). Population size is on occasion taken into consideration (see, e.g. Nordhaus, 2008, and Section 3.5.1 below), based on the utilitarian idea that an increase in the total number of people with sufficiently high welfare levels is desirable (Blackorby and Donaldson, 1984). But climate change economics often abstracts from population and distributional ethics: we frequently assume that the number of people in a generation is given, and ignore inequality by representing each generation by a single agent.

The main variables that have been introduced, besides consumption, in various indices of human welfare are leisure, the environment, health and education: observe that these variables may contribute very differently to carbon emissions. It should be noted that a large segment of the economic growth literature, as well as of the economic analysis of climate change literature (e.g. Nordhaus, 2008) does consider improvements in knowledge, education or the environment, but only in so far as they make possible the production of consumption goods with less labor time or produced capital. The present text focuses on their role as direct arguments in the utility function or welfare index.

For instance, Nordhaus and Tobin's (1972) measure of economic welfare (MEW) modifies GDP per capita by excluding depreciation and including leisure and non-market work, in addition to taking congestion and pollution into account. Labor economics has traditionally considered both consumption and leisure as arguments in the utility function, with leisure interpreted as a broad aggregate of unpaid activities, including home production, the care of children and relatives, and home educational activities, in addition to free time and rest and recreation. Jones and Klenow (2010) provide a welfare index where leisure plays an important role in comparing the quality of life across countries.

Environmental economics has emphasized the quality of air and water, as well as the recreational opportunities offered by natural environments. Pollution appears as an argument in Keeler et al. (1972) and, as just noted, in the MEW of Nordhaus and Tobin (1972). The amenity value of the forest was stressed by Hartman (1976), and that of undeveloped ecosystems by Fisher et al. (1972) and Barrett (1992).

Health economics has constructed indices of health and longevity by means of defining QALY (quality-adjusted life years) in order to quantify the benefits of medical intervention. Life expectancy is recognized as a major factor in human welfare. As societies become wealthier, extending life becomes increasingly valuable. Hall and Jones (2007) estimate that, in the USA, 'a sixty-five year old would give up 82 percent of her consumption . . . to have the health status of a 20 year old'.[2] Many recent welfare indices include life expectancy as an argument; see, for example, Jones and Klenow, 2010, and the HDI of the United Nations Human Development Programme (2011).

Education and the accompanying accumulation of human capital have traditionally been considered production factors, increasing the productivity of the time worked and associated wages (see Freeman, 1986). Education increases productivity both in the market and in some areas of home production (Michael, 1972, 1973; Leibowitz, 1975; Gronau, 1986; Ortigueira, 1999; Heckman, 1976). As a result, highly educated women spend more time in child care and education, but less in meal preparation and doing laundry, than their less educated counterparts (Leibowitz, 1975). The literature uses the language of production and output, but, as noted by Michael (1973, p. 307), 'A distinguishing characteristic of human capital is that it is embedded in an individual: it therefore accompanies him whenever he goes, not only in the labor market, but also in the theater, the voting booth, the kitchen and so forth'. Even though the home production literature views education as affecting non-market productivity, he continues, 'one could alternatively argue, for example, that education affects the household utility function . . .'

We have mentioned the Human Development Index (HDI): it was proposed by the first *United Nations Development Report* (United Nations Development Programme, 1990) and has been updated yearly (see the UNDP, 2010, 2011 for the latest issues: the 2010 edition introduced some changes in its composition). It aggregates three dimensions: (a) health, with life expectancy at birth as indicator; (b) education, with mean and expected years of schooling as indicators; and (c) living standards, with GNI per capita as indicator (see, e.g., UNDP, 2010, Figure 1.1). The more recent *Human Development Reports*, as well as Neumayer (2001), among others, emphasize three dimensions of human welfare acknowledged by the authors of the 1990 HDI but neglected in its definition, namely environmental and climate sustainability and human security, rights and freedoms. The 2011 Report introduces an 'adjusted HDI', which reduces the HDI by a factor reflecting income inequality in the country.

2.2   Intergenerational Justice

Sustaining the level of a human welfare forever can be justified by appealing to Rawls's (1971) maximin principle (see Roemer, 1998, 2007). The ethical justification for this type of sustainability is, in our view, that the date at which a person is born is morally arbitrary. Thus every generation has a right to a level of well-being at least as high as that of any other generation. Similarly, Anand and Sen (2000) justify the sustainability of human welfare by an appeal to 'universalism'. Our argument, that the date at which a person is born is morally arbitrary, in turn validates 'universalism'. Maximizing the quality of life of the worst-off generation will often require the maximization of the quality of life of the first generation subject to maintaining that quality of life for all future generations, so that there is no quality-of-life growth after the first generation (see Silvestre, 2002, for possible exceptions to this rule).

2.3   Positive Growth Rates

Alternatively, society may seek a positive rate of growth in the quality of life of future generations at the cost of reducing the quality of life of the present one. It is, however, not obvious how to justify sacrifices of the worst-off present generation for the sake of improving the already higher welfare levels of future ones. Recall that we assume away intragenerational inequality, thereby depriving economic growth of a role in alleviating contemporaneous poverty.

One might argue that parents want their children to have a higher quality of life than they do. Thus welfare growth might be supported by all parents over a constant-utility path. An alternative justification for altruism towards future generations would appeal to growth as a public good: we may feel justifiably proud of mankind's recent gains in, say, extraterrestrial travel, or average life expectancy, and wish them to continue into the far future even at a personal cost.[3] If indeed we do put value on human development, we – the present generation – may choose not to enforce our right to be as well off as all future generations: we may prefer to allow future generations to become better off than we are, at some (perhaps small) cost to ourselves.

Indeed, there is an asymmetry in the way we feel about contemporaneous versus temporally disjoint inequality: a person in a poor country may not wish to sacrifice her quality of life for the sake of improving that of a person in a richer country, while at the same time be willing to make some sacrifices for the welfare of unrelated, yet-to-be born individuals who will as a consequence be richer than she.

3. AN OVERVIEW OF THE LITERATURE

3.1 Steady States with Reproducible Resources

The economic analysis of climate change is rooted in a long tradition that analyzes the tradeoff between present consumption and the enhanced consumption possibilities in the future offered by saving. The pioneering work by Faustmann (1849) sought to determine the best rate of rotation for the harvesting of a forest. Faustmann's work anticipated the literature on the sustainable yields of other renewable resources, such as fisheries, often developed by biologists and involving a large empirical component (see, e.g., Gordon, 1954; Scott, 1955; Clark, 1990). A first objective is the characterization of the dynamic path for the population of a species under various specifications of the natural growth rate and the dependence of the harvest on both the harvesting effort and the size of the population, and the definition of the maximal sustainable yield (MSY). Second, after introducing the disutility of harvesting effort, this literature studies the best dynamic paths, either by a cost–benefit argument or by maximizing the discounted future benefits and costs using the tools of optimal control theory.

The 'sustainability' of the MSY-type solutions must be understood as an attribute of the steady state: consumption, or the harvest, is less than maximal during the transition from the initial conditions to the steady state. If we postulate that a different generation lives at each date, then the earlier generations consume less than the later ones. The same observation applies to the Solow–Swan model of neoclassical growth (Solow, 1956; Swan, 1956), which, contrary to the MSY approach, is often explicitly worded in terms of different generations. The emphasis continues to be on the best steady state, now characterized by the golden rule (which corresponds to the MSY of the renewable resource literature): in Phelps's (1961, p. 642) words: 'each generation invests on behalf of the future generations that share of income that . . . it would have had past generation invest on behalf of it'. Again, consumption is less than maximal during the transition from the initial conditions to the steady state. The golden rule can be viewed as intergenerationally equitable only among the generations that come to be after the steady state is reached.

3.2 Maximin: the Solow–Hartwick Theory

Solow (1974) pioneered the analysis of Rawls's maximin principle in the context of intergenerational equity. As is well known, Rawls himself (1971, Section 44) was reluctant to apply the principle to the intergenerational problem, but Solow (1974, p. 30) decided to be 'plus rawlsien que le Rawls'. Solow (1974) introduced both an exhaustible natural resource and produced capital. Identifying the utility of a generation with its consumption of a single produced good, he worked out the maximin paths of the variables for a Cobb–Douglas technology. In order to maintain equal, maximal utility across all generations, the stock of produced capital has to increase to compensate for the depletion of the non-renewable resource. Solow did not use the term 'sustainability', but of course his objective can be translated as sustaining utility.

Hotelling's (1931) maximization, by variational methods, of the discounted sum of benefits from the depletion of an exhaustible resource yields the well-known 'Hotelling rule', by which the rate of change in the marginal product of the extracted amount must equal the interest, or discount, rate. He considered the exhaustible resource as the only productive input, which of course ruled out any sort of sustainability: in his words (p. 139) 'the indefinite maintenance of a steady state of production is a physical impossibility'. But the Hotelling rule turned out to play a role in Solow's maximin model: Hartwick (1977) showed that Solow's (1974) paths for produced capital and for the depletion of the exhaustible resource could be characterized by a Hotelling-type rule (translated as the equality between the marginal product of the extracted amount and the marginal product of produced capital) together with what is now known as the Hartwick rule:

Investment in produced capital = Marginal product of extracted amount times the extracted amount.

The Hartwick rule can be paraphrased as

Investment in produced capital = Rents obtained by extraction,

that is,

Net investment in produced capital = Value of net depletion of the exhaustible resource,

or

Sum of net investments in all forms of capital = 0.

The precise relation between an appropriately defined index of the sum of net investments and maintaining consumption involves some subtleties: see Dixit et al. (1980) and Asheim et al. (2003), but a formal argument shows that the Hotelling rule and the Hartwick rule together imply stationary consumption in a simple continuous-time model. Let $S^k(t)$ denote the stock of produced capital at t, $m(t)$ be the flow of extraction of the exhaustible resource, $f(S^k,m)$ the production function, and $c(t)$ consumption. Postulate the law of motion of produced capital:

$$\frac{dS^k}{dt} = f(S^k,m) - c,$$

and define

$$\text{Hotelling rule: } \frac{\partial f}{\partial S^k} = \frac{\dfrac{d}{dt}\left(\dfrac{\partial f}{\partial m}\right)}{\dfrac{\partial f}{\partial m}};$$

$$\text{Hartwick rule: } \frac{dS^k}{dt} = \frac{\partial f}{\partial m}m.$$

The differentiation, with respect to time, of the law of motion of capital yields

$$\frac{d^2 S^k}{dt^2} = \frac{\partial f}{\partial S^k}\frac{dS^k}{dt} + \frac{\partial f}{\partial m}\frac{dm}{dt} - \frac{dc}{dt}, \tag{30.1}$$

whereas that of the Hartwick rule yields

$$\frac{d^2 S^k}{dt^2} = \frac{d}{dt}\left(\frac{\partial f}{\partial m}\right)m + \frac{\partial f}{\partial m}\frac{dm}{dt} = \frac{\partial f}{\partial S^k}\frac{\partial f}{\partial m}m + \frac{\partial f}{\partial m}\frac{dm}{dt} \quad \text{(by Hotelling rule)}$$

$$= \frac{\partial f}{\partial S^k}\frac{dS^k}{dt} + \frac{\partial f}{\partial m}\frac{dm}{dt}, \quad \text{(by Hartwick rule)}$$

which together with (30.1) implies that $dc/dt = 0$.

The Hartwick rule, in its interpretation

Sum of net investments in all forms of capital = 0,

plays a central role in the sustainability literature, as we discuss next.

3.3 Weak and Strong Sustainability

The Club of Rome Report (Meadows et al., 1972) and the Report of the United Nations Conference on the Human Environment, Stockholm, 1972 (see the United Nations Environment Programme UNEP, www.unep.org, where the term 'sustainable development' appears) conveyed an increasing concern about the destructive effects of economic activity, which could jeopardize the quality of life of future generations. This motivated the call for sustainability (Brundlandt Report: World Commission on Environment and Development, 1987, and Rio Summit, 1992; see United Nations Department of Economic and Social Affairs, Earth Summit Agenda 21, http://www.un.org/esa/dsd/agenda21/res_agenda21_00.shtml), understood in the oft-quoted terms of the Brundlandt Report: 'sustainable development is development that meets the needs of the present generation without compromising the ability of future generations to meet its own needs'. In academia, this gave rise to a critique of the mainstream, neoclassical economic theories of growth and development, and the flourishing of environmental economics. But environmental economics was often based on discounted utilitarianism (see Section 3.4 below), whereas the new sustainability approach adopts the long-run view.

As we noted, the Solow–Hartwick theory has a legitimate claim to the title of sustainability, because it focuses on maintaining human quality of life for ever, even though it narrowly defines utility as only consumption. The role of undamaged natural environments as a direct factor in human quality of life could easily be modeled by augmenting the list of arguments in the utility function, including, for instance, natural resources, as discussed in Section 2.1 above. The resulting extension of the Solow–Hartwick model to a world where environmental stocks enter the utility function is often identified with weak sustainability (Cabeza Gutés, 1996; Neumayer, 2010, p. 21).

But while the Solow–Hartwick theory focuses on sustaining utility, in accordance with the maximin social welfare criterion, the sustainability approach, even in its weak form, favors focusing upon the 'capacity for utility' rather than 'utility' (Neumayer, 2010, p. 8). The distinction, to some extent semantic, gives the Hartwick rule, appropriately generalized, a central role: sustainability is understood as maintaining capital as an end in itself. Weak and strong sustainability then interpret 'maintaining capital' in different ways. Weak sustainability stays closer to the Hartwick rule, defined as the non-negativity of the sum of investments in all kinds of capital, including both human and natural, renewable and exhaustible. Strong sustainability emphasizes the lack of substitutability among various forms of capital, and advocates the maintaining the physical stocks of some forms of natural capital.

The weak sustainability approach motivates the empirical measure of the sum of net investments in all forms of capital, leading to the Genuine Savings (GS) Index (or 'genuine investment') compiled by the World Bank on a country-by-country basis. The index has received attention: for instance, it is used by Arrow et al. (2004) to conclude that 'we find reason to be concerned that consumption is excessive'.

Any attempt at aggregating the changes in various forms of capital faces the challenge of pricing them. Market prices are patently inadequate, given the severity of the non-internalized negative environmental externalities (see, e.g., Stern, 2008). Hartwick's (1977) appeal to the Hotelling rule amounts to basing the price of the exhaustible resource on Hotelling's (1931) maximization of the sum of discounted sum of future benefits. In a parallel manner, the sustainability literature attempts to derive prices from optimization problems where the objective function is sum of discounted utilities (see, e.g., Neumayer, 2010, Section 5.1.1): the above-mentioned papers by Dixit et al. (1980) and Asheim et al. (2003) show the difficulties involved.

As noted, GDP-type measures combine consumption and investment. Weak sustainability indicators such as Genuine Savings aim at correcting the shortcomings of their investment component. The desire to simultaneously correct for the shortcomings of both their investment and their consumption component leads to measures such as the Index of Sustainable Economic Welfare (ISEW), also known as Genuine Progress Indicator (GPI), the evolution of which has been computed, and compared to that of the GDP per capita, for a variety of countries; see Neumayer (2010, Section 5.2).

As noted, strong sustainability involves a deeper departure from pre-existing economic analysis and denies the substitutability between natural and produced capital (Neumayer, 1999, 2010; Gerlagh and van der Zwaan, 2002).While weak sustainability, or the Solow–Hartwick model, aims at preserving an aggregate of the stocks of produced and natural capital, strong sustainability advocates bequeathing to future generations either an aggregate of stocks of natural capital, or, better yet, physical stocks of certain forms of natural capital. The need for aggregating the various forms of produced capital or even some forms of natural capital remains, but the approach is less precisely defined than weak sustainability, and neither explicitly deals with future paths of the economic variables, nor appeals to optimization. Neumayer (2010, ch. 6) offers detailed arguments as well as a discussion of the construction of the various quantitative indicators of strong sustainability.

3.4 Discounted Utilitarianism

The neoclassical theory of optimal economic growth (Koopmans, 1965; Cass, 1965) adopts the normative criterion, already used by Hotelling (1931) and, reluctantly, by Ramsey (1928), of maximizing the discounted sum of utilities, which in discrete time can be expressed as

$$\sum_{t=1}^{T} u(c_t) \frac{1}{(1 + \delta)^{t-1}}, \tag{30.2}$$

where c_t is consumption per capita at date t, and δ is the discount rate.

The single-date utility function u may be specified as the identity (linear) function, so that utility is synonymous with consumption, or as strictly concave function of consumption. In the latter case, a popular functional form is

$$u(c_t) = \frac{c_t^{1-\eta} - 1}{1 - \eta}, \quad \eta > 0, \eta \neq 1, \tag{30.3}$$

which becomes $u(c_t) = \ln c_t$ for $\eta = 1$.

Population growth is often taken into account, but the analysis is conducted in per capita terms, and, thus, the absolute size of population does not play any role.

The length T of the planning horizon is modeled either as large and finite, or as infinite. Each of these modeling choices presents idiosyncratic analytical challenges: a finite horizon has the problem of the arbitrariness of any terminal conditions, for example, how much capital to leave at the end. On the other hand, some mathematical implications of the infinite-horizon model puzzle our intuitions, anchored in a finite world.

Expression (30.2) can be interpreted in the following ways.

Long-lived consumer
A representative consumer is postulated, who lives for all dates, has a single-date (or instantaneous) utility function with consumption as its argument, and discounts the future. The discount factor $1/(1 + \delta)$ (or the discount rate δ) reflects the consumer's subjective rate of time preference: a more impatient consumer has a larger δ, and attaches little value to a unit of consumption made available far into the future. Expression (30.2) is then the utility function of this consumer. The discount rate δ is sometimes identified with the 'market interest rate', for example by Hotelling (1931), who nevertheless feels

compelled to justify it: he first refers to the productivity of capital (a fallacious argument, since δ should then appear in the technological constraints, and not in the objective function) and to the fact that (p. 145) 'future pleasures are uncertain in a degree increasing with their remoteness in time'.

Dynasty

In this interpretation, a consumer lives for a short number of dates, but derives utility from the utility, or the consumption, of her descendants. For example, the first member of the dynasty may care about her own consumption and her child's utility, discounted; the child in turn cares about her consumption and her child's utility. By expanding this iterated expression, the first member of the dynasty must care about the discounted sum of consumptions of all her descendants, where the discount factor decreases exponentially with time. The discount rate δ reflects each generation's discounting of its child's utility. Note that both in this interpretation and in the previous one, the maximization of (30.2) is equivalent to the maximization of the utility of the present (first) generation, either because it is the only one, or because the utilities of future generations count only to the extent that they affect that of the first generation: under both interpretations, the present generation has the role of a hegemon.[4]

Classical utilitarianism

Utilitarianism, in its original, undiscounted form (Bentham, 1789; Mill, 1848; Sidgwick, 1874), proposed the maximization of the sum of the individual utilities, unweighted because 'each individual must count as one, and none as more than one'.[5] On the other hand, utilitarians postulated that marginal utility is decreasing in consumption. Thus, in the intergenerational context, the classical utilitarian maximandum can be viewed as (30.2) with $\delta = 0$ and a strictly concave utility function u, perhaps of the form (30.3).[6] (Again, each date t corresponds to a different generation.) It should be noted that, when $T = \infty$, undiscounted utilitarianism may face the problem of the divergence of (30.2).

Utilitarianism with uncertainty

Suppose that there is an exogenous probability of p that each generation t of humans will be the last one, should the species have lasted until date t. Thus the probability that the species lasts exactly T generations is $p(1 - p)^{T-1}$. Suppose the social planner is a utilitarian, who desires to maximize the total utility of all generations that exist. A prize, in the von Neumann–Morgenstern terminology, is of the form 'the human species lasts exactly T generations and enjoys the sequence of consumptions $(c_1, c_2, \ldots, c_T)$'. Denote this prize by $(T; c_1, \ldots, c_T)$. As a utilitarian, the von Neumann–Morgenstern utility function of the planner, which is defined on prizes, is given by

$$U(T; c_1, \ldots, c_T) = \sum_{t=1}^{T} u(c_t).$$

Now the planner must choose an infinite sequence of generational utilities to maximize his expected utility, which is

$$\sum_{T=1}^{\infty} p(1 - p)^{T-1} U(T; c_1, \ldots, c_T). \tag{30.4}$$

A straightforward calculation shows that maximizing (30.4) is equivalent to maximizing

$$\sum_{t=1}^{\infty} (1 - p)^{t-1} u(c_t).$$

Again, we have the discounted utilitarian formula where the discount factor is now $1 - p$, or the associated discount rate is $\delta = p/(1 - p)$.

Of these four interpretations, three maximize a discounted sum of generational utilities (all but classical utilitarianism). One sees, however, that the appropriate discount rate may be very different across interpretations. With the 'long-lived consumer' approach, one might choose a discount rate by estimating actual rates of time preference of living humans. But the discount rate that must be used under the last approach has nothing to do with subjective rates of time preference. Moreover, even though we assumed that the probability p in the last interpretation is exogenous, that can be modified: it would be more appropriate to generate the probability of human extinction endogenously, as associated with choice of path.

Axiomatic justifications
Koopmans (1960) provided an axiomatic justification for the discounted-utilitarian social welfare function. Assuming that a preference order on infinite consumption streams obeys certain axioms, he proved that it must be representable by a discounted-utilitarian social welfare function like (30.2) for some choice of u and some choice of δ. But we believe this theorem is not salient for the current debate for two reasons. First, the axioms are not particularly intuitive or compelling, so the theorem does not provide a persuasive argument for discounted utilitarianism. Second, the theorem gives no instruction concerning what the discount rate, or the utility function, should be. Since the choice of the discount rate has been focal in recent discussions, Koopmans's theorem cannot be brought to bear.

3.5 The Nordhaus Model

3.5.1 Discounted utilitarianism with inequality aversion
The work of Nordhaus and his collaborators is central in the economics of climate change (Nordhaus, 2008; Nordhaus and Boyer, 2000). Their analysis is based on maximizing the objective function

$$\sum_{t=1}^{T} L_t \frac{1}{1 - \eta} (c_t)^{1-\eta} \frac{1}{(1 + \delta)^t}, \tag{30.5}$$

where L_t is the number of people in generation t. Nordhaus (2008, pp. 33, 60) calls the δ and η of (30.5) 'central' and 'unobserved normative parameters', reflecting 'the relative importance of the different generations'. The parameter δ is a 'pure social time discount rate': as just observed, a high δ means that the welfare of a generation born far into the future counts very little in the social welfare function. The second one represents 'the aversion to inequality of different generations'. Informally speaking, if the rates of growth turn out to be negative, then δ and η push in opposite directions, a high δ favoring the earlier generations and a high η favoring the later, less well-off, ones. But for positive rates of growth, when the latter generations are better off, high values of either δ or η favor the earlier generations. This is the case for the paths proposed by Nordhaus (2008).

Ignoring the term L_t, (30.5) is formally (30.2) with the utility function of (30.3). But the interpretation of the parameter η is quite different in Nordhaus (2008) than in the utilitarian maximization, discounted or undiscounted. For classical utilitarianism, η embodies the decrease of individual marginal utility, whereas for Nordhaus η is an inequality-aversion parameter of the social welfare function. In fact, as $\eta \to \infty$,

$$\sum_{t=1}^{T} \frac{1}{1-\eta}(c_t)^{1-\eta}$$

tends to the maximin social welfare function $\min_t c_t$.

3.5.2   Nordhaus's parameter calibration

Nordhaus (2008) calibrates η and the annual discount rate $\hat{\delta}$ as follows. He adopts the 'Ramsey equation'

$$\hat{r} = \hat{\delta} + \eta \hat{g}, \tag{30.6}$$

where $\hat{r}$ is the real per year rate of interest on capital and $\hat{g}$ is the per year rate of growth of consumption. Equation (30.6) corresponds to the first-order condition for the solution of the Ramsey program: an infinitely lived consumer under continuous time maximizes $\int_0^\infty u(c(t))e^{-\delta t}dt$ subject to the constraint $dS^k/dt \leq \hat{f}(S^k(t)) - c(t) - \delta^k S^k(t)$, where S^k is produced capital, which depreciates at rate δ^k, u is given by (30.3) and $\hat{f}$ is the production function.

Nordhaus's (2008) calibration method consists in inferring $\hat{r}$ and $\hat{g}$ from 'observed economic outcomes as reflected by interest rates and rates of return on capital' and choosing δ^k and η subject to the Ramsey equation, which gives one degree of freedom. More specifically, Nordhaus (2008, p. 178) takes an 'observed' value of $\hat{r} = 0.055$ (and, implicitly, a predicted growth rate of $\hat{g} = 0.02$).[7] Equation (30.6) is then satisfied by any $(\hat{\delta}, \eta)$ pair satisfying $\hat{\delta} = 0.055 - 0.02\eta$, in particular by the values $(\hat{\delta}, \eta) = (0.015, 2)$ chosen by Nordhaus (2008).

We remark that the constrained optimization of (30.5), with η and $\hat{\delta}$ given, yields endogenous $\hat{r}$ and $\hat{g}$: write them $\hat{r}^N(\hat{\delta}, \eta)$ and $\hat{g}^N(\hat{\delta}, \eta)$. However, there is no reason why (30.6) has to hold, that is, why $\hat{r}^N(\hat{\delta}, \eta) = \hat{\delta} + \eta \hat{g}^N(\hat{\delta}, \eta)$, because the Ramsey program is different: in particular, it does not incorporate any environmental variables. In other words, the Kuhn–Tucker conditions for solving program (30.5) will not include the Ramsey equation, but some possibly much more complicated version of it. Hence, even if $\hat{r} = \hat{r}^N(\hat{\delta}, \eta)$, appealing to the Ramsey equation in order to justify high values for either η or $\hat{\delta}$ (or both) is not justified.

In addition, Nordhaus (2008) takes $\hat{r}$ to coincide with observed historical yearly returns on capital. But there is no reason why the historically observed $\hat{r}$ must equal $\hat{r}^N(\hat{\delta}, \eta)$. It might be argued that the historical $\hat{r}$ is actually the one that is endogenously obtained at the solution of the Ramsey program given $(\hat{\delta}, \eta)$, call it $\hat{r}^R(\hat{\delta}, \eta)$, in which case the Ramsey equation $\hat{r}^R(\hat{\delta}, \eta) = \hat{\delta} + \eta \hat{g}^R(\hat{\delta}, \eta)$ could be used as a restriction on the admissible $(\hat{\delta}, \eta)$ pairs. This could make sense if the historical past were populated by a single infinitely lived consumer, unconcerned with the environment, in which case the parameters $(\hat{\delta}, \eta)$ would be 'positive', rather than 'normative'. But Nordhaus considers a world of many distinct generations, with 'normative' parameters $(\hat{\delta}, \eta)$, and interprets the rates of return observed in the market as depending on these 'normative' parameters, in particular on the aversion, by past and current market participants, to inequality among generations. In sum, Nordhaus's use of the discounted utilitarian formula is not based on any of the four justifications discussed above.

3.6 Cost–Benefit Analysis: *The Stern Review*

Cost–benefit analysis underpins the recommendations of *The Stern Review*, in turn based on the reports of the Intergovernmental Panel on Climate Change (IPCC TAR, 2001) and on Hope (2006). *The Stern Review* does not attempt to solve an optimization program: it is rather a cost–benefit analysis arguing that the 'costs of inaction are larger than costs of action'. Assuming a path of growth for the GDP, and starting from a business-as-usual (*laissez-faire*) hypothesis on the path of GHG emissions, it considers alternative policies that reduce emissions in the present, and eventually stabilize carbon concentration in the atmosphere. *The Stern Review* argues that, properly discounted, the benefits of strong, early action on climate change outweigh the costs.

It should be noted that discount rates have different roles in cost–benefit analysis and discounted-utilitarianism optimization. If the consumer or planner uses the pure time discount rate δ to weight the utilities of the various generations in the utilitarian maximand, cost benefit analysis uses the consumption discount rate $\delta + \eta \tilde{g}$ to evaluate the changes in future consumption streams due to a particular (marginal) investment project, relative to a reference consumption path that exogenously grows at a rate $\tilde{g}$. The project passes the cost–benefit test if the discounted sum of the consumption streams is positive.[8] *The Stern Review* chooses a pure time discount rate of $\hat{\delta} = 0.001$ per annum, together with $\eta = 1$ and $\tilde{g} = \dot{c}/c = 0.013$ (1.3 percent per annum), yielding a consumption discount rate of 0.014 per annum. The key point, however, is that the discount rate used here derives not from the subjective rate of time preference of an agent, but from a postulated probability p that each generation of humans will be the last one. In other words, Stern is using the discounted utilitarian formula associated with the von Neumann–Morgenstern utilitarian social planner who is maximizing the discounted sum of utilities due to the uncertain length of tenure of the human species. Because Stern postulates an annual probability of extinction of $p = 0.001$, he calculates an associated discount rate of $\hat{\delta} = p/(1 - p) = 0.001$ per annum. This implies that the utility of individuals a century from now will be discounted by about 10 percent.

When commentators on *The Stern Review* suggest higher consumption discount rates (Arrow, 2007; Nordhaus, 2007; Weitzman, 2007: see the debate in the *Postscripts to the Stern Review* available at www.sternreview.org.uk, as well as the issue of *World Economics*, 7 (4), October–December 2006, and the subsequent Dietz et al., 2007), they have in mind the long tradition in growth theory that identifies the discounted utilitarian program with the utility of a long-lived consumer, and hence they want a discount rate approximating empirical rates of time preference. But Stern is not using that model, and his choice of a small discount rate is perfectly consistent if we believe that the probability of species extinction is small. Indeed, a choice of annual probability of extinction of $p = 0.001$ is probably too large by at least an order of magnitude. In particular, Stern's choice of p implies that with probability one-half, the human species will survive only another 700 years. This seems too pessimistic. With a choice of $p = 0.0001$, the species will survive 7000 years with probability one-half. We suspect that Stern was cautious, and chose an overly pessimistic value for p because he anticipated criticism over the discount rate.

However, because *The Stern Review* does not solve a full optimization program, its recommendations are in principle open to the criticism, voiced by its critics, that the consumption discount rate should reflect the rates of return of the available investment alternatives: even if, using a consumption discount rate of 0.014, carbon emission reductions pass the cost–benefit test, future generations could conceivably be better off if the current generation avoided incurring the costs of GHG reductions and invested instead in other intergenerational public goods. In defense of the *Review*, Dietz et al. (2007, p. 137) argue that

> it is hard to know why we should be confident that social rates of return would be, say, 3% or 4% into the future. In particular, if there are strong climate change externalities, then social rates of return on investment may be much lower than the observed private returns on capital over the last century, on which suggestions of a benchmark of 3% or 4% appear to be based.

Finally, one must evaluate the debate between Nordhaus and Stern by deciding which model is ethically superior. We think Nordhaus's model is best justified by the model of an infinitely lived consumer, because of his calibration of the discount rate from the subjective rate of time preference of living consumers. Stern is taking the viewpoint of a utilitarian planner facing uncertainty. Given this choice, we prefer Stern, because we see no reason that the fortunes of future generations should be determined by the preferences of the present generation – in particular, by its degree of impatience! That they *will* be so determined is, perhaps, a politically realistic statement, but the Stern approach, of postulating a von Neumann–Morgenstern planner, is ethically superior, in treating each generation equally – 'each as one, and none as more than one'. That future utilities are discounted is not due to the hegemony of the present generation, but to the unalterable fact that time's arrow points in only one direction, that of increasing uncertainty.

Although we prefer Stern's model to Nordhaus's, on ethical grounds, we do not rest with Stern, but substitute a sustainabilitarian planner for Stern's utilitarian one.[9] To this topic we now turn.

4.   A MODEL WITH EDUCATION, KNOWLEDGE AND GREENHOUSE GAS EMISSIONS

We here report on our work on sustainability in the context of GHG emissions and global warming (Llavador et al., 2010, 2011a, 2011b; Roemer, 2011).

4.1   The Quality-of-life Function

As in the literature discussed in Section 3 above, our work abstracts from all conflicts except for the intergenerational one and we assume a representative agent in each generation. Our approach is purely normative, and we do not propose an economic equilibrium model, nor do we attempt to predict what the path would be in the absence of policy.

Contrary to the continuous-time models mentioned in Section 3 above, we work with discrete time, with one date, or period, per generation, and we assume that a generation lives for 25 years. We consider an infinite number of generations, indexed by $t \geq 1$.

More precisely, we formally postulate the following quality-of-life, standard-of-living, welfare or utility function of Generation t, $t \geq 1$:

$$(c_t)^{\alpha_c}(x_t^l)^{\alpha_l}(S_t^n)^{\alpha_n}(\hat{S}^m - S_t^m)^{\alpha_m}, \qquad (30.7)$$

where the exponents are positive and normalized such that $\alpha_c + \alpha_l + \alpha_m + \alpha_n = 1$, and where:

c_t = annual average consumption per capita by Generation t;
x_t^l = annual average leisure per capita, in efficiency units, by Generation t;
S_t^n = stock of knowledge per capita, which enters Generation t's quality-of-life function, as well as the production function (see below), understood as located in the last year of life of Generation t;
S_t^m = total CO_2 in the atmosphere, in GtC, which is understood as located in the last year of life of Generation t; and
$\hat{S}^m$ = 'catastrophic' level of CO_2 in the atmosphere.

Section 2.1 above has referred to the various indices of human welfare and their arguments. We of course keep consumption, the increase of which in advanced countries has significantly improved welfare (Johnson, 2000). Next, we list leisure, which, as indicated in Section 2.1 above, has a long tradition in economics. But we consider leisure in efficiency units, that is, education-enhanced leisure, not only because education improves home productivity (such as in the quality of child-rearing services, a component of the parents' leisure) but also in the sense of 'human capital accompanying the person to the theater' (Michael, 1973, p. 307, quoted in Section 2.1 above). In Wolf's (2007) words:

> The ends people desire are, instead, what makes the means they employ valuable. Ends should always come above the means people use. The question in education is whether it, too, can be an end in itself and not merely a means to some other end – a better job, a more attractive mate or even, that holiest of contemporary grails, a more productive economy. The answer has to be yes. The search for understanding is as much a defining characteristic of humanity as is the search for beauty. It is, indeed, far more of a defining characteristic than the search for food or for a mate. Anybody who denies its intrinsic value also denies what makes us most fully human.

Next, we consider the stock of knowledge, which has the character of a public good. (Education or human capital, on the contrary, is primarily a private good which generates positive externalities.) Knowledge, in the form of society's stock of culture and science, directly increases the value of life, because an understanding of how the world works and an appreciation of culture are intrinsic to human well-being. Moreover, medical discoveries are important factors in the improvements in health status, life expectancy and infant survival (which are not explicit in our specification). Last, the preservation of natural environments is valuable to humans for its direct impact on physical and mental health. Because of our emphasis on climate change, we explicitly list the stock of CO_2 as an argument the quality-of-life function, reflecting our view that environmental deterioration is a public bad in consumption, as well as in production.

Our quality-of-life function follows the spirit of the Human Development Index produced by the UNDP (2011), which considers, as noted in Section 2.1 above, three dimensions; (a) life expectancy; (b) education; and (c) consumption. As just mentioned, health can be thought of as at least partially represented by knowledge. Our quality-of-life function shares with the 1990 *Human Development Report* its abstraction from security, rights and freedoms, but it explicitly includes as an argument the quality of the biosphere.

4.2   Technology and Intertemporal Links: Capital, Knowledge, Education and Emissions

We model commodity production as using as inputs skilled labor, produced capital, accumulated human knowledge, biospheric quality and GHG emissions. For simplicity, we assume that production of knowledge is purely labor-intensive using only skilled labor and past knowledge (think corporate R&D and university research). There are four conduits of intergenerational transmission: capital passes from one generation to the next, after investment and depreciation; knowledge passes in like manner, with depreciation; the stock of biospheric quality diminished by emissions passes to the next generation. The fourth conduit is education: the education effort of one generation increases the efficiency of the labor time of the next one. Even though the total time, in hours, available to a generation for work and leisure is constant, the number of efficiency units of labor–leisure time available to a generation increases with the accumulated investment in education. The education production function plays an important role in the model and is discussed in more detail below.

Formally, feasible paths in our model are constrained by the production function and by the environmental stock–flow relations, modeled as follows:

$$\tilde{f}(x_t^c, S_t^k, S_t^n, e_t, S_t^m) \equiv k_1(x_t^c)^{\theta_c}(S_t^k)^{\theta_k}(S_t^n)^{\theta_n}(e_t)^{\theta_e}(S_t^m)^{\theta_m} \geq c_t + i_t, t \geq 1, \qquad (30.8)$$

with $k_1 > 0$, $\theta_c > 0$, $\theta_k > 0$, $\theta_n > 0$, $\theta_c + \theta_k + \theta_n = 1$, $\theta_e > 0$ and $\theta_m < 0$ (aggregate production function).

$$(1 - \delta^k) S_{t-1}^k + k_2 i_t \geq S_t^k, t \geq 1, \text{ with } k_2 > 0 \text{ and } \delta^k \in [0,1]$$
(law of motion of produced capital),
$$(1 - \delta^n) S_{t-1}^n + k_3 x_t^n \geq S_t^n, t \geq 1, \text{ with } k_3 > 0 \text{ and } \delta^n \in [0,1]$$
(law of motion of the stock of knowledge),
$$x_t^e + x_t^c + x_t^n + x_t^l \equiv x_t, t \geq 1 \text{ (allocation of efficiency units of labor)},$$
$$k_4 x_{t-1}^e \geq x_t, t \geq 1, \text{ with } k_4 > 0 \text{ (education production function)}, \qquad (30.9)$$

with initial conditions (x_0^e, S_0^k, S_0^n), where c_t, x_t^l, S_t^n and S_t^m have been defined in Section 4.1 above, and where:

x_t^c =   average annual efficiency units of labor per capita devoted to the production of output by Generation t,

e_t = average annual emissions of CO_2 from energy (fossil fuels and cement) in GtC by Generation t,

S_t^k = stock of produced capital per capita available to Generation t,

i_t = average annual investment per capita by Generation t,

x_t^n = average annual efficiency units of labor per capita devoted to the production of knowledge by Generation t,

x_t^e = average annual efficiency units of labor per capita devoted to education by Generation t,

x_t = average annual efficiency units of time (labor and leisure) per capita available to Generation t.

We call emissions e_t and concentrations S_t^m environmental variables, whereas the remaining variables are called economic.

The education production function (30.9) plays a significant role in our analysis, and deserves some comparative comments with the literature. Note that the production of education is purely labor-intensive, using only the skilled labor x_t^e of the preceding generation: x_t^e can be interpreted as labor-time multiplied by the amount of human capital embodied in one time-unit of labor, as in Uzawa (1965) and Lucas (1988). Because we assume that $\theta_c + \theta_k + \theta_n = 1$ in (30.8), our production function displays decreasing returns to 'capital' when construed to consist of physical and human capital. But returns would be constant if we broadened the notion of 'capital' to include the stock of knowledge.

As in Uzawa (1965) and Lucas (1988), for simplicity we do not include produced capital as an input in the production of education. This contrasts with Rebelo (1991) and Barro and Sala-i-Martin (1999, p. 179). Adapting the notation of Barro and Sala-i-Martin, their 'human capital production function' is

$$dH/dt = B[(1 - v^k)S^k]^{\widetilde{\eta}} [(1 - v^H)H]^{1-\widetilde{\eta}} - \widetilde{\delta}H, \qquad (30.10)$$

where H is the amount of human capital, $(1 - v^k)S^k$ is the amount of produced capital used in education, $(1 - v^H)$ is the fraction of human capital used in education, and B, $\widetilde{\eta}$ and $\widetilde{\delta}$ are parameters, the last one being the human-capital depreciation factor.

We interpret the labor input in the production of education as that of teachers, rather than students. This departs from the interpretations by Lucas (1988) and Rebelo (1991), but it agrees with the comments in Uzawa (1965) and Barro and Sala-i-Martin (1999). For example, the latter write (p. 179): 'a key aspect of education [is that] it relies heavily on educated people as an input'.

We see the education of a generation as a social investment, in line with Lucas's (1988, p. 19) dictum: 'a general fact that I will emphasize again and again: that human capital accumulation is a *social* activity, involving *groups* of people, in a way that has no counterpart in the accumulation of produced capital'. Also, we adopt a broad view of educational achievement, which in particular bestows the ability to adapt to new technologies, as emphasized by Goldin and Katz (2008).

Our education production function (30.9) can be viewed as a generational version of (30.10) for the parameter values $\widetilde{\eta} = 0$ and $\widetilde{\delta} = 1$ (since, in our model, all adults die at the end of each date), giving

$$H_t - H_{t-1} = B[(1 - v^H)H_{t-1}] - H_{t-1}, \text{ i.e., } H_t = B(1 - v^H)H_{t-1},$$

which is precisely (30.10) under the notational correspondence $H_t \leftrightarrow x_t$, $(1 - v^H) \leftrightarrow \frac{x_{t-1}^r}{x_{t-1}}$ and $B \leftrightarrow k_4$.

One very important intertemporal link is not explicitly modeled, namely the evolution of the stock of GHG from emissions. One might postulate a law of motion for the process by which biospheric quality at date $t + 1$ consists of biospheric quality at date t, partially rejuvenated by natural processes that absorb CO_2, plus the impact of new emissions of GHGs. However, the scientific view on the nature of this law of motion is very much in flux, and so we have elected not to imply a false precision by inserting such a law into our model. Instead, we simply take a path of emissions and concomitant atmospheric concentration of CO_2 computed from the popular Model for the Assessment of Greenhouse-gas Induced Climate Change (MAGICC; a previous version was used by the IPCC AR4, Working Group I; see Meehl et al., 2007), which stabilizes the atmospheric concentration at 450 ppm CO_2, and we constrain our production sector not to emit more than is allowed on this path. That is, we do not optimize over possible paths of future emissions. In a sense, our approach is dual to the cost–benefit method. The latter takes as given a path for the economic variables, and recommends a path for the environmental variables (based on a cost–benefit criterion in the spirit of discounted utilitarianism). We, on the contrary, take as given a path for the environmental variables, and recommend paths for the economic variables based on human sustainability.

We have postulated Cobb–Douglas utility and production functions, which implies an elasticity of substitution of one between 'natural' and 'man-made' variables. As discussed in Section 3.3 above, this type of substitutability is controversial. But because, as just discussed, we exogenously adopt the values for the environmental stocks and flows, the implications of substitutability are less drastic in our analysis than in models that aim at the endogenous determination of both natural and man-made variables.

4.3   Sustaining the Quality of Life

We are concerned with human sustainability, which requires maintaining human welfare, as discussed in Section 2.2 above. Formally, we consider the following Pure Sustainability Optimization Program:

$$\max \Lambda \text{ subject to } (c_t)^{\alpha_c}(x_t^f)^{\alpha_f}(S_t^n)^{\alpha_n}(\hat{S}^m - S_t^m)^{\alpha_m} \geq \Lambda, t \geq 1,$$

and subject to the feasibility conditions given by specific production relations, laws of motion of the stocks and resource constraints, and with the initial conditions given by the relevant stock values in the base year (2000). At a solution of this Pure Sustainability Optimization Program, the path of quality of life will typically be stationary, as noted, and it can be (at least asymptotically) supported by stationary paths in all the arguments of the quality-of-life function.

The main result of our calibrated model shows that human sustainability is achievable: it is possible to sustain for ever human welfare at a level higher than the reference level (year 2000) while keeping emissions at a path that stabilizes CO_2 concentrations at

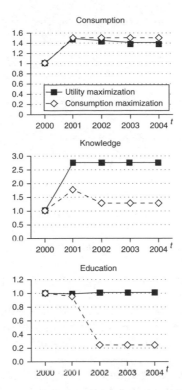

Figure 30.1   Comparison of consumption, stock of knowledge and education paths for quality-of-life maximization and consumption maximization

450 ppm (Llavador et al., 2011b). Figure 30.1 depicts the implied paths for some of the economic variables.

4.4   Sustaining Consumption

Our work departs from a large body of the literature in the definition of the individual quality-of-life function, which includes education, knowledge and CO_2 stocks in addition to consumption. The question arises: what if we conventionally adopted consumption as our index of the individual standard of living? Would the sustainability result still hold? How different would the paths be?

We have repeated our optimization but substituting consumption for quality of life as given by (30.7) above, and assuming that the fraction of the time devoted to leisure is the same as in the solution of the Pure Sustainability Optimization Program.

Our sustainability results carry over to the new specification. Of course, we should expect higher levels of consumption than in (30.9). Figure 30.1 and Table 30.1 compare the values for consumption, the stock of knowledge and education along the paths and for the steady states of the sustainable human welfare solution in the two models. Consumption is indeed higher when the objective is consumption rather than quality of life, but, perhaps surprisingly, not by much (about 7 percent higher).

Table 30.1   Quality-of-life versus consumption maximization

| | Steady-state values (fractions over year 2000) | | | |
	Consumption	Stock of knowledge	Efficiency units in education	Quality of life
Quality-of-life maximization	1.41	2.74	1.01	1.24
Consumption maximization	1.51	1.27	0.25	0.51

Remarkably, when the objective is quality of life instead of consumption, the steady-state stock of knowledge is over twice as large, and steady-state education is over four times as large. Thus the notion of individual welfare has dramatic consequences for the allocation of resources. We believe that our quality-of-life function is a better index of human welfare than just consumption: we therefore recommend substantially deeper investments in knowledge and education than would be optimal in models where these variables only affect productivity.

As just noted, maximizing consumption instead of the quality of life yields a higher value for consumption, but not by much. So if the 'true' social welfare index were consumption, a planner who 'mistakenly' maximized the quality of life would be making a relatively small error. What about the converse? In other words, if (as we believe) our quality-of-life function, as given by (30.7), provides an appropriate welfare index, but the public policy aims at maximizing consumption, what would be the loss in the quality of life? The answer is given in the last column of Table 30.1: the steady-state level of quality of life achieved by its maximization is about two and a half times the level of quality of life reached under consumption maximization. Figure 30.2 compares the paths for the quality of life under the two maximization objectives.

The comparison reinforces the case for sustaining quality of life. Suppose we are uncertain about the appropriate objective we should adopt: quality of life or consumption. If we choose quality of life when we should have chosen consumption, we pay a relatively small cost. On the other hand, if we sustain consumption when quality of life is the appropriate objective, the steady-state level of quality of life falls considerably.

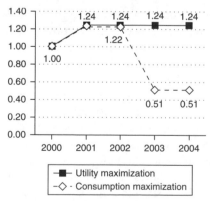

Figure 30.2   Comparison of quality-of-life paths induced by quality-of-life maximization and consumption maximization

Our model is relatively complex, and does not easily offer intuitive reasons for the just-discussed asymmetry between consumption and the quality of life. In order to clarify the result, the following section (more formal than the rest of the text, and which can be skipped by the reader less interested in the mathematical development) provides a stripped-down model which derives a similar result, although with less striking numbers. In any event, the analysis in the following section validates the robustness of our results comparing the maximization of the quality of life with that of consumption.

4.5 A Stripped-down Model Comparing Two Social Objectives: Quality of Life versus Consumption

Consider an economy producing a consumption commodity, using as inputs labor and knowledge (there is no capital good). 'Quality of life' is a function $c^\alpha S^{1-\alpha}$ of consumption (c) and the stock of knowledge (S), and knowledge is produced each period according to a linear production function of labor. Consider the following program:

$$\max \lambda c + (1 - \lambda) c^\alpha S^{1-\alpha}$$

subject to

$$c \le k(x_c)^\theta S^{1-\theta} \text{ (i)}$$

$$1 \ge x_c + x_n \quad \text{(ii)}$$

$$S \le k_3 x_n \quad \text{(iii)}$$

where x_c (resp. x_n) denotes the amount of labor devoted to the production of the consumption good (resp. knowledge) and the weight $\lambda \in [0,1]$ is fixed. Constraint (i) is the commodity production function, constraint (ii) is the labor constraint, and constraint (iii) is the knowledge production function. We may solve the constraints (which will be binding at the solution) and rewrite the program as

$$\max \lambda c + (1 - \lambda) c^\alpha S^{1-\alpha}$$

subject to

$$c \le k \left(1 - \frac{S}{k_3} \right)^\theta S^{1-\theta}. \tag{30.11}$$

There are two Kuhn–Tucker conditions characterizing the solution to (30.11): the first is the constraint in (30.11), which must be binding, and the second one is the dual constraint that, after some algebra, can be written:

$$\left(1 - \frac{S}{k_3} \right) (1 - \lambda)(1 - \alpha) U + (\lambda c + (1 - \lambda) U\alpha) \left(1 - \frac{S}{k_3} - \theta \right) = 0, \tag{30.12}$$

where $U \equiv c^\alpha S^{1-\alpha}$.

Now let $\lambda = 1$, and so Program (30.11) is maximizing consumption only. Condition (30.12) reduces to $c(1 - \frac{S}{k_3} - \theta) = 0$, that is, $\hat{S} = k_3(1 - \theta)$.

On the other hand, let $\lambda = 0$; then Program (30.11) is maximizing quality-of-life, and (30.12) reduces to $S^* = k_3(1 - \alpha\theta)$.

It follows that the ratio of optimal knowledge in the quality-of-life formulation to optimal knowledge in the consumption-only formulation is

$$\frac{S^*}{\hat{S}} = \frac{1 - \alpha\theta}{1 - \theta}. \tag{30.13}$$

If we take (reasonably) $\alpha = 2/3$ and $\theta = 0.9$, then this ratio is 4, a large number.

The expression for consumption is, from the constraint in (30.11)

$$c = k\left(1 - \frac{S}{k_3}\right)^\theta S^{1-\theta}.$$

Hence,

$$\frac{c^*}{\hat{c}} = \frac{\left(1 - \frac{S^*}{k_3}\right)^\theta (S^*)^{1-\theta}}{\left(1 - \frac{\hat{S}}{k_3}\right)^\theta \hat{S}^{1-\theta}} = \frac{(1 - 1 + \alpha\theta)^\theta k_3^{1-\theta}(1 - \alpha\theta)^{1-\theta}}{(1 - 1 + \theta)^\theta k_3^{1-\theta}(1 - \theta)^{1-\theta}} = \alpha^\theta\left(\frac{1 - \alpha\theta}{1 - \theta}\right)^{1-\theta}, \tag{30.14}$$

which for $\alpha = \frac{2}{3}$ and $\theta = 0.9$ gives $c^*/\hat{c} = 0.80$

On the other hand, using (30.14) and (30.13), the quotient for the quality-of-life levels is

$$\frac{\Lambda^*}{\hat{\Lambda}} = \left(\frac{c^*}{\hat{c}}\right)^\alpha\left(\frac{S^*}{\hat{S}}\right)^{1-\alpha} = \alpha^{\alpha\theta}\left(\frac{1 - \alpha\theta}{1 - \theta}\right)^{(1-\theta)\alpha}\left(\frac{1 - \alpha\theta}{1 - \theta}\right)^{1-\alpha} = \alpha^{\alpha\theta}\left(\frac{1 - \alpha\theta}{1 - \theta}\right)^{1-\alpha\theta},$$

which for $\alpha = 2/3$ and $\theta = 0.9$ gives $\Lambda^*/\hat{\Lambda} = 1.37$.

This computation may provide some understanding of why changing the objective function from 'consumption only' to 'quality of life' can have a large effect on the production of knowledge and on the quality of life, but only a rather small reduction in consumption.

4.6   Sustaining Growth

We return to our full-fledged model. As discussed in Section 2.3 above, the present generation may desire that future generations enjoy a steady growth in their quality of life. Instead of maximizing the quality of life of the worst-off generation, it aims at the maximization of the quality of life of the first generation, subject to the condition that quality of life subsequently grows at a given rate g per generation, as it can be simply formalized by the following Sustainable Development Optimization Program (rate g):

$$\max \Lambda \text{ subject to: } (c_t)^{\alpha_c}(x_t^l)^{\alpha_l}(S_t^n)^{\alpha_n}(\hat{S}^m - S_t^m)^{\alpha_m} \geq (1 + g)^{t-1}\Lambda, \quad t \geq 1$$

and subject to the feasibility and initial conditions.

In words, the solution to the Sustainable Development Optimization Program (rate *g*) is the feasible path that maximizes the quality of life of the first generation ($t = 1$) subject to guaranteeing a rate of growth of *g* in quality of life, per generation, for ever. By setting $g = 0$, the Sustainable Development Optimization Program becomes the Pure Sustainability Optimization Program of Section 4.3 above. Our main result on this question is that moderate growth rate *g* can be achieved at the cost of a small reduction in the utility of the first generation, which stays well above the year-2000 reference level (Llavador et al., 2011a).

4.7 Sustainability and Uncertainty

We have discussed in Section 3.4 how uncertainty can be incorporated into the utilitarian ethic. In like manner, we can incorporate it into the sustainabilitarian ethic. Let *p* be the probability (as in Stern) that each generation of humans is the last one. Let the social planner be a sustainabilitarian, with expected-utility preferences. Her von Neumann–Morgenstern utility function on the prize $(T; \Lambda[1], \dots, \Lambda[T])$, where $\Lambda[t]$ is the quality of life for generation *t*, is

$$V^{Sus}(T; \Lambda[1], \dots, \Lambda[T]) = \min\{\Lambda[1], \dots, \Lambda[T]\}.$$

Hence, she maximizes expected utility:

$$\sum_{T=1}^{\infty} p(1 - p)^{T-1} V^{Sus}(T; \Lambda[1], \dots, \Lambda[T]). \qquad (30.15)$$

This is a potentially difficult program to solve. But Llavador et al. (2010) prove that if *p* is sufficiently small, then the solution to the maximization of (30.11) is identical to the solution of the Pure Sustainability Optimization Program. For our parameterization of that program, the Stern value of $p = 0.001$ per annum is sufficiently small.

Therefore the path that we describe in Section 4.4 is also the optimal path of (30.15).

5. CONCLUSIONS

We summarize our main points.

1. Sustainability has a long history in economic analysis. In contemporary climate change economics, however, it has in large part been replaced by discounted utilitarianism. We argue that the two most prominent recent works, those of Nordhaus (2008) and Stern (2007), are based upon different justifications of discounted utilitarianism, and this explains their choice of different discount rates. Of the two approaches, we prefer Stern's, as it is based on a model where each generation is treated equally, whereas in the Nordhaus model the fortunes of future generations depend upon subjective preferences of the first generation concerning how to treat their descendants.

2. But we also object to Stern's approach, as it is based upon the decision of a

utilitarian social planner: we replace this planner with one whose goal is to sustain human welfare at the highest possible level. Our approach rejuvenates earlier work by Solow, Hartwick and others.

3.   Nevertheless, we depart from that earlier work in expanding the conception of human welfare to include not only commodity consumption, but also education, leisure and two public goods – the stock of knowledge and biospheric quality. Our concept of welfare, or quality of life, is inspired by many authors: in particular, by the *Human Development Reports*, and the Human Development Index which they have used, and by many economists, who have included one or all of these arguments in conceptions of human welfare.

4.   We argue that the expansion of the concept of welfare beyond consumption is not a gratuitous generalization, because it renders possible responding to the climate change challenge by changing the consumption bundle of agents, from energy-intensive commodities to less intensive ones, such as knowledge, education and leisure.

5.   We insert into the model uncertainty with regard to the length of existence of the human species, in the manner of Stern (2007), and explain why our optimal sustainable paths for the model with an infinitely lasting human species are, indeed, the optimal paths for that model, as well.

6.   We report on our results, published in several papers, showing that welfare can be sustained for ever at levels higher than present levels, while on a production path that reduces GHG emissions to levels that converge to atmospheric carbon concentrations of 450 ppm. We emphasize the need to keep GHG emissions on track, given the available scientific knowledge.

7.   Our results are encouraging by showing that it is possible to drastically reduce GHG emissions while maintaining the quality of life across generations, but our work shows that only moderate growth rates can be sustained, suggesting slow-growth policies.

NOTES

*   We are indebted to the editor for useful comments, and to Max Tavoni for help in applying the MAGICC model to construct sensible emission-concentration paths. The usual *caveat* applies.

1.   The implications of the capability approach do not differ much from those of ours.

2.   Perhaps surprisingly, life expectancy seems to increase linearly with time. Hall and Jones (2007) estimate that during the last 50 years, life expectancy in the USA has increased by 1.4 months per year. And Oeppen and Vaupel (2002, p. 1029) report that 'female life expectancy in the record-holding country has risen for 160 years at a steady pace of almost 3 months per year'.

3.   See Silvestre (2007).

4.   See Roemer (2011).

5.   See Elster (2008).

6.   The strict concavity of a (differentiable) utility function is equivalent to the property of decreasing marginal utility.

7.   Elsewhere in the book he refers to a $\hat{r}$ of 0.04 (pp. 9–11) and to a $\hat{g}$ of 0.013 (p. 108).

8.   The objective function is (Stern, 2007, p. 51), $\int_0^\infty u(c(t))e^{-\delta t}dt$, where $u(c) = \frac{1}{1-\eta}c^{1-\eta}$ is the utility function of generation t, and δ is the pure time discount rate above. Consider a reference consumption stream $\{c(t), t \in [0,\infty)\}$, that grows at rate $\tilde{g}$, with $c(0) = 1$, and a 'small' project that will modify the consumption stream by $\{\Delta c(t), t \in [0,\infty)\}$. Up to a first-order approximation, the change in the value of the objec-

tive function is $\int_0^\infty u'(c(t))\Delta c(t)e^{-\delta t}dt = \int_0^\infty c(t)^{-\eta}\Delta c(t)e^{-\delta t}dt = \int_0^\infty (e^{\widetilde{g}\,t})^{-\eta}\Delta c(t)e^{-\delta t}dt = \int_0^\infty e^{-(\eta\widetilde{g}+\delta)t}\Delta c(t)dt;$ that is, the discount rate on consumption, defined as the relative change in the discount factor, is $\eta\,\widetilde{g} + \delta$. The project passes the cost–benefit test if $\int_0^\infty e^{-(\eta\widetilde{g}+\delta)t}dt > 0$.

9. Despite their analytical differences, both Stern and Nordhaus have exerted a positive intellectual leadership in the public debate on global warming; see Nordhaus (2012) for a recent contribution.

REFERENCES

Anand, Sudhir and Amartya Sen (2000), 'Human development and economic sustainability', *World Development*, **28**(12): 2029–49.

Arrow, Kenneth J. (2007), 'Global climate change: a challenge to policy', *Economist's Voice, Berkeley Electronic Press*, **4**(3), June (www.bepress.com/ev).

Arrow, Kenneth J., Partha Dasgupta, Lawrence Goulder, Gretchen Daily, Paul Ehrlich, Geoffrey Heal, Simon Levin, Karl-Göran Mäler, Stephen Schneider, David Starrett and Brian Walker (2004), 'Are we consuming too much?', *Journal of Economic Perspectives*, **18**(3): 147–72.

Asheim, Geir B., Wolfgang Buchholz and Cees Withagen (2003), 'The Hartwick rule: myths and facts', *Environmental and Resource Economics*, **25**: 129–50.

Barrett, Scott (1992), 'Economic growth and environmental preservation', *Journal of Environmental Economics and Management*, **23**(3): 289–300.

Barro, Robert J. and Xavier Sala-i-Martin (1999), *Economic Growth*, Cambridge, MA: MIT Press.

Bentham, Jeremy (1789), *An Introduction to the Principles of Morals and Legislation*, London: T. Payne & Son.

Blackorby, Charles and David Donaldson (1984), 'Social criteria for evaluating population change', *Journal of Public Economics*, **25**: 13–33.

Cabeza Gutés, Maite (1996), 'The concept of weak sustainability', *Ecological Economics*, **17**(3): 147–56.

Cass, David (1965), 'Optimum growth in an aggregative model of capital accumulation', *Review of Economic Studies*, **32**: 233–40.

Clark, Colin W. (1990), *Mathematical Bioeconomics* (2nd edn), New York: John Wiley & Sons.

Dietz, Simon, Chris Hope, Nicholas Stern and Dimitri Zenghelis (2007), 'Reflections on the *Stern Review* (1): a robust case for strong action to reduce the risks of climate change', *World Economics*, **8**(1): 121–68.

Dixit, Avinash, Peter Hammond and Michael Hoel (1980), 'On Hartwick's rule for regular maximin paths of capital accumulation and resource depletion', *Review of Economic Studies*, **47**(3): 551–6.

Elster, Jon (2008), *Reason and Rationality*, Princeton, NJ: Princeton University Press.

Faustmann, Martin (1849), 'Berechnung des Werthes, welchen Waldboden, sowie nach nicht haubare Holzbestände für die Weldwirstchaft besitzen', *Allgemeine Forst und Jagd Zeitung*, **15**: 441–55. English translation: 'On the determination of the value which forest land and immature stands pose for forestry', in Michael Gane (ed.), *Martin Faustmann and the Evolution of the Discounted Cash Flow: Two Articles from the Original German of 1849*, Oxford: Commonwealth Forestry Institute, Institute Paper 42, 1968.

Fisher, Anthony C., John V. Krutilla and Charles J. Cicchetti (1972), 'The economics of environmental preservation: a theoretical and empirical analysis', *American Economic Review*, **62**(4): 605–19.

Fleurbaey, Marc (2009), 'Beyond the GDP: the quest for a measure of social welfare', *Journal of Economic Literature*, **47**(4): 1029–75.

Freeman, Richard B. (1986), 'Demand for education', in Orley Ashenfelter and Richard Layard (eds), *Handbook of Labor Economics, Volume I*, Amsterdam: North-Holland, pp. 357–86.

Gerlagh, Reyer and B.C.C. van der Zwaan (2002), 'Long-term substitutability between environmental and man-made goods', *Journal of Environmental Economics and Management*, **44**: 329–45.

Goldin, Claudia and Lawrence F. Katz (2008), *The Race between Education and Technology*, Cambridge, MA: Harvard University Press.

Gordon, H. Scott (1954), 'The economic theory of a common property resource: the fishery', *Journal of Political Economy*, **62**: 124–42.

Gronau, Reuben (1986), 'Home production: a survey', in Orley Ashenfelter and Richard Layard (eds), *Handbook of Labor Economics, Volume I*, Amsterdam: North-Holland, pp. 273–304.

Hall, Robert E. and Charles I. Jones (2007), 'The value of life and the rise in health spending', *Quarterly Journal of Economics*, **102**(1): 39–72.

Hartman, Richard (1976), 'The harvesting decision when a standing forest has value', *Economic Inquiry*, **14**(1): 52–8.

Hartwick, John M. (1977), 'Intergenerational equity and the investment of rents from exhaustible resources', *American Economic Review*, **67**(5): 972–4.

Heckman, J.J. (1976), 'A life-cycle model of economics, learning and consumption', *Journal of Political Economy*, **84**: S11–S44.

Hope, Christopher (2006), 'The marginal impact of CO_2 from PAGE2002: an integrated assessment model incorporating the IPCC's five reasons for concern', *The Integrated Assessment Journal (IAJ)*, **6**(1): 19–56.

Hotelling, Harold (1931), 'The economics of exhaustible resources', *Journal of Political Economy*, **39**(2): 137–75.

Intergovernmental Panel on Climate Change (2001), *Third Assessment Report* (IPCC TAR), http://www.ipcc.ch.

Intergovernmental Panel on Climate Change (2007), *Fourth Assessment Report* (IPCC AR4), http://www.ipcc.ch.

Intergovernmental Panel on Climate Change (2007), *Climate Change 2007: The Physical Science Basis. Contribution of Working Group I to the Fourth Assessment Report of the IPCC*, Cambridge: Cambridge University Press.

Johnson, D.G. (2000), 'Population, food, and knowledge', *American Economic Review*, **90**(1): 1–14.

Jones, Charles I. and Peter J. Klenow (2010), 'Beyond GDP? Welfare across countries and time', NBER Working Paper No. 16352.

Keeler, Emmet, Michael Spence and Richard Zeckhauser (1972), 'The optimal control of pollution', *Journal of Economic Theory*, **4**(1): 19–34.

Koopmans, Tjalling J. (1960), 'Stationary ordinal utility and impatience', *Econometrica*, **28**: 278–309.

Koopmans, Tjalling J. (1965), 'On the concept of optimal economic growth', reprinted in *The Econometric Approach to Development Planning*, Amsterdam: North-Holland, 1966, pp. 225–87.

Leibowitz, Arleen (1975), 'Education and the allocation of women's time', in F. Thomas Juster (ed.), *Education, Income, and Human Behavior*, New York: McGraw-Hill, NBER Books.

Llavador, Humberto, John E. Roemer and Joaquim Silvestre (2010), 'Intergenerational justice when future worlds are uncertain', *Journal of Mathematical Economics*, **46**(5): 728–61.

Llavador, Humberto, John E. Roemer and Joaquim Silvestre (2011a), 'A dynamic analysis of human welfare in a warming planet', *Journal of Public Economics*, **95**(11–12): 1607–20.

Llavador, Humberto, John E. Roemer and Joaquim Silvestre (2011b), 'Sustainability in the presence of global warming: theory and empirics', Human Development Research Paper 5, New York: UNDP–HDRO.

Lucas, Robert E. Jr (1988), 'On the mechanics of economic development', *Journal of Monetary Economics*, **22**: 3–42.

Meadows, Donella H., Dennis L. Meadows, Jørgen Randers and William W. Behrens III (1972), *The Limits to Growth: A Report for the Club of Rome's Project on the Predicament of Mankind*, London: Earth Island Press.

Meehl, Gerard A., T.F. Stocker, W.D. Collins, P. Friedlingstein, A.T. Gaye, J.M. Gregory, A. Kitoh, R. Knutti, J. M. Murphy, A. Nodah, S.C. B. Raper, I.G. Watterson, A.J. Weaver and Z.C. Zhao (2007), 'Global climate projections', in S. Solomon, D. Qin, M. Manning, Z. Chen, M. Marquis, K.B. Averyt, M. Tignor and H.L. Miller (eds), *Climate Change 2007: The Physical Science Basis. Contribution of Working Group I to the Fourth Assessment Report of the Intergovernmental Panel on Climate Change*, Cambridge, UK: Cambridge University Press, ch. 10.

Michael, Robert T. (1972), 'The effect of education on efficiency in consumption', NBER Occasional Paper 116, New York: Columbia University Press.

Michael, Robert T. (1973), 'Education in nonmarket production', *Journal of Political Economy*, **81**(2, Part 1): 306–27.

Mill, John Stuart (1848), *Principles of Political Economy, with Some of Their Applications to Social Philosophy*, London: John W. Parker.

Neumayer, Eric (1999), 'Global warming: discounting is not the issue, but substitutability is', *Energy Policy*, **27**: 33–43.

Neumayer, Eric (2001), 'The human development index and sustainability: a constructive proposal', *Ecological Economics*, **39**: 101–14.

Neumayer, Eric (2010), *Weak versus Strong Sustainability: Exploring the Limits of Two Opposing Paradigms* (3rd edn), Cheltenham, UK, and Northampton, MA, USA: Edward Elgar.

Nordhaus, William D. (2007), 'A review of the *Stern Review* on the Economics of Climate Change', *Journal of Economic Literature*, **45**(3): 686–702.

Nordhaus, William D. (2008), *A Question of Balance: Weighing the Options on Global Warming Policies*, New Haven, CT: Yale University Press.

Nordhaus, William D. (2012), 'Why the global warming skeptics are wrong', *New York Review of Books*, **59**(5): 32–4.

Nordhaus, William D. and Joseph Boyer (2000), *Warming the World*, Cambridge, MA: MIT Press.

Nordhaus, William D. and James Tobin (1972), 'Is growth obsolete?' In National Bureau of Economic Research, *Economic Growth*, New York: Columbia University Press, pp. 1–80.

Oeppen, Jim and James W. Vaupel (2002), 'Broken limits to life expectancy', *Science*, **296**: 1029–31.

Ortigueira, Salvador (1999), 'A dynamic analysis of an endogenous growth model with leisure', *Economic Theory*, **16**: 43–62.

Phelps, Edmund (1961), 'The golden rule of capital accumulation: a fable for growthmen', *American Economic Review*, **51**(4): 638–43.

Ramsey, Frank (1928), 'A mathematical theory of saving', *Economic Journal*, **38**(152): 543–59.

Rawls, John (1971), *A Theory of Justice*, Cambridge, MA: Harvard University Press.

Rebelo, Sergio (1991), 'Long-run policy analysis and long-run growth', *Journal of Political Economy*, **99**: 500–21.

Roemer, John E. (1998), *Theories of Distributive Justice*, Cambridge, MA: Harvard University Press.

Roemer, John E. (2007), 'Intergenerational justice and sustainability under the leximin ethic', in John E. Roemer and Kotaro Suzumura (eds), *Intergenerational Equity and Sustainability*, Basingstoke: Palgrave, pp. 203–27.

Roemer, John E. (2011), 'The ethics of intertemporal distribution in a warming planet', *Environmental and Resource Economics*, **48**(3): 363–90.

Scott, Anthony D. (1955), 'The fishery: The objectives of sole ownership', *Journal of Political Economy*, **63**(2): 116–24.

Sidgwick, Henry (1874), *The Methods of Ethics*, London: Macmillan.

Silvestre, Joaquim (2002), 'Progress and conservation under Rawls's maximin principle', *Social Choice and Welfare*, **19**: 1–27.

Silvestre, Joaquim (2007), 'Intergenerational equity and human development', in John E. Roemer and Kotaro Suzumura (eds), *Intergenerational Equity and Sustainability*, Basingstoke: Palgrave, pp. 252–87.

Solow, Robert M. (1956), 'A contribution to the theory of economic growth', *Quarterly Journal of Economics*, **70**: 65–94.

Solow, Robert M. (1974), 'Intergenerational equity and exhaustible resources', *Review of Economic Studies*, **41**, *Symposium on the Economics of Exhaustible Resources*, 29–45.

Stern, Nicholas (2007), *The Economics of Climate Change: The Stern Review*, Cambridge: Cambridge University Press.

Stern, Nicholas (2008), 'The economics of climate change', *American Economic Review: Papers & Proceedings*, **98**(2): 1–37.

Stiglitz, Joseph E., Amartya Sen and Jean-Paul Fitoussi (2009), *Report by the Commission on the Measurement of Economic Performance and Social Progress* (CMEPSP, 'Sarkozy Report'), http://www.stiglitz-sen-fitoussi.fr/en/documents.htm.

Swan, Trevor W. (1956), 'Economic growth and capital accumulation', *Economic Record*, **32**: 334–61.

United Nations Development Programme (1990), *Human Development Report 1990*, http://hdr.undp.org/en/reports/global/hdr1990/.

United Nations Development Programme (2010), *Human Development Report 2010*, New York: Palgrave Macmillan.

United Nations Development Programme (2011), *Human Development Report 2011*, New York: Palgrave Macmillan.

Uzawa, Hirofumi (1965), 'Optimal technical change in an aggregate model of economic growth', *International Economic Review*, **6**: 18–31.

Weitzman, Martin (2007), '*The Stern Review* of the economics of climate change', Book Review, *Journal of Economic Literature*, **45**(3): 703–24.

Wolf, Martin (2007), 'Education is a worthwhile end in itself', *The Financial Times*, 1 February.

World Commission on Environment and Development (1987), *Our Common Future* (The Brundtland Report), Oxford: Oxford University Press.

31 At the crossroads: can China grow in a low-carbon way?
*Julien Chevallier**

1 INTRODUCTION

Are economic growth and climate change compatible? This chapter addresses this question from the angle of China, the world's fastest-growing economy, where environmental considerations are necessarily becoming a burning issue. Given that it is responsible for 19 per cent of global CO_2 emissions in 2005, and is the largest emitter, future climate change will be affected by the economy's ability to stabilize and reduce emissions. More importantly, because China is becoming a global economic leader and is a centrally planned economy, its path is likely to influence the behaviour of other economies. Against this background, we attempt – through a narrative literature review – to assess whether a transition to a low-carbon economy seems feasible in China, given several aspects of its energy consumption and production.

The concept of a 'low-carbon economy' is defined by Jiang et al. (2010) as an economy that has a minimal output of greenhouse gases (GHG) into the biosphere, that is, an economic model based on small energy consumption, low environmental pollution, and low-carbon emissions. It also entails changes in the country's industrial structure and people's conceptions. Among the key drivers of the low-carbon economy, Carbon Capital (2010) identifies equipments and infrastructures that enable energy efficiency or alternative energy production and use, leading to a reduction of carbon emissions, directly or indirectly (such as smart buildings, smart grids, renewable energy, biofuel vehicles, electric vehicle charging systems etc.).

Given its current state of economic development, China may be seen at a crossroads with respect to the transition to a low-carbon emitting society, which would be in phase with the tenets of sustainable development. China is indeed characterized by its rising global pre-eminence as the leading world economy and, at the same time, by its responsibility to the type of fuel that it will burn. If there is an increased reliance on coal to produce energy (given its relative lower cost with respect to other energy sources), then the earth will become less enjoyable to live in. If, instead, innovative technologies are put forth to reconcile economic growth and climate change (by relying on emissions trading, for instance, as it is burgeoning in China), then there is no need to be overly pessimistic. Jiang et al. (2010) report that, as a symbolic measure, the Chinese prime minister, Wen Jiabao, has announced in 2004 that the Green GDP Index (which deducts the cost of environmental damages and resources consumption from the traditional GDP) will replace the Chinese GDP index itself as a performance measure for government and party officials at the highest levels. Besides, we need to keep in mind the serious problems raised by pollution as an externality of economic growth: as Wang, Y. (2010) puts it, health concerns will become prevalent by 2020 especially due to the emissions of sulphur dioxide (SO_2) and inhalable particulate matter (PM10). Mainly emanating from coal combustion, NO_x emissions have also become very serious in China,[1] and their harmful impacts on health have also been clearly documented (You and Xu, 2010).

It would imply a leap of faith to argue that a centrally planned economy like China has a greater chance to achieving a transition to a low-carbon economy, or is able to make radical decisions. As a matter of fact, our approach remains agnostic with respect to the political system that needs to be put in place in order to achieve most efficiently the transition to a low-carbon economy. Instead, this chapter defends the stance that China has already made significant steps towards achieving a low-carbon economy, and is an example to the world in many respects (energy intensity, renewable energy, forestry). Therefore we will explore in this chapter the extent to which the transition to a low-carbon economy, although challenging, belongs to the 'art of the possible' (Gang et al., 2011).[2]

This chapter will attempt to cover these interdependent issues, while not minimizing the complexity of the transition to a low-carbon society. China's energy challenge during the twenty-first century is unparalleled in the world, and therefore appropriate solutions need to be devised. Among the chief problems, we may cite the increasing scarcity of resources (especially in mining industries), the uncontrolled development of transportations means, and the geographical contamination of pollutants.[3]

Institutions will play a central role in facilitating the transition to a low-carbon society. The Chinese National Energy Administration (NEA) was created in 2006 (as a successor of the National Energy Leading Group Office) to coordinate the country's energy strategy research. At the same time, the Climate Change Division of the National Development and Reform Commission (NDRC) was established to deal with climate change issues. On 29 October 2008, the NDRC published a White Paper on China's Policies and Actions on Climate Change to clarify the country's attitude and actions with regard to climate change. On that matter, China identifies the development of a low-carbon economy as a central priority of its future energy development. It has also explicitly recognized the need to reconcile economic development with environmental concerns (and the control of GHG emissions). On 5 July 2007, the United Nations Secretary General, Ban Ki-moon, called for the world countries to disengage from the dependence on carbon-based fuels and to step on the road of low-carbon economy.[4] Finally, these concerns about the growth–climate change link have been addressed at the highest state level by the Chinese president, Hu Jintao, during the 2008 summit of the Asia-Pacific Economic Cooperation (APEC).[5]

Among the former academic analyses on this topic, we may cite the books by Lin and Swanson (2010), Managi and Kaneko (2010) and Gang et al. (2011). In addition, research articles published in the special issue 'Energy and its Sustainable Development for China' of the academic journal *Energy*, **35**(11) are often used as a reliable source of information in this chapter. The book by Gang et al. (2011) constitutes another immensely helpful source of information on the topic of the economics of climate change in China, along with an overview of the possibility to facilitate the pathway to a low-carbon, high-growth economy. Among other measures, Gang et al. (2011) advocate expenditures with a direct focus on energy use and energy efficiency. Besides, they highlight the need for complementary measures in the financial sector to support China's transition from a model of GHG-intensive industry and exports to an economy based on high value-added exports and services.

Among the institutional communications from the central government on this topic, He et al. (2010a) relate that the main sources of information are the revised Energy Conservation Law in the People's Republic of China (implemented from 1 April 2008), the Renewable Energy Law (implemented from 1 January 2006), and the China Long-term Energy Development Plan (as formulated in the Eleventh Five-Year Plan). In addition, the United Nations Development Programme (UNDP) published its *China National Human Development Report 2009/10* entitled 'China and a Sustainable Future: Towards a Low Carbon Economy and Society'. This report attempts to link economic growth, carbon emissions, technology and human development in China. It highlights that the shift to a low-carbon development pathway is an imperative for China, and that low-carbon policies and choices will make Chinese growth more sustainable.

In the Twelfth Five-Year Plan (2011–16), it appears that the central government will allow the use of emissions trading and/or carbon taxes for flexibility in meeting GHG mitigation targets. A specific section of the chapter is dedicated to the Chinese emissions trading programmes, which may be seen (and encouraged) as promising tools to combat climate change in the near future. We discuss the capacity for China to build a domestic carbon market (both compliance- and voluntary-based), but also talk about the specifics of how five provinces and eight cities will pilot low-carbon economy schemes.

The rest of the chapter is organized as follows. Section 2 contains a summary of the main facts and figures about the Chinese economy and its need for energy. Section 3 provides a systematic review of the available options for energy use in China, based on the latest technological developments and the best scientific evidence to date. Section 4 considers the possibility for China to develop its own Emissions Trading Scheme (ETS), with an objective of capping quantitatively CO_2 emissions. Section 5 concludes with some thoughts on the necessary emergence of a low-carbon society in China if its current economic development is deemed to be sustainable as well.

2   FACTS AND FIGURES

Most of the figures reported in this section are drawn from the *World Energy Outlook* (IEA, 2010) published yearly. We review below the main characteristics of the Chinese situation about (i) population and economic growth, and (ii) the current energy mix and the projected demand for energy.

2.1   Population and Economic Growth

In the coming century, population growth will be one of the biggest factors behind rising energy demand. By 2030, the world's population will reach close to 8 billion, with 85 per cent residing in non-OECD countries. Besides population, the other major influence on energy demand is economic growth. China is a fast-growing economy, with an average annual GDP growth expected to be equal to 6 per cent from today until 2030. By 2030, China and the USA together are expected to contribute about 40 per cent of global GDP growth.

Among non-OECD countries, China will most likely encounter a dramatic climb in energy demand, as the rising prosperity of its large population is reflected in trends such as increased vehicle ownership and higher electricity consumption. Of the 400 million new cars that are forecast to be added to the world's roads between today and 2030, more than one-third of them will be in China. To put it bluntly, the number of personal vehicles in China will nearly quadruple from 2010 to 2030. As personal-vehicle fleets expand, they need to become more fuel-efficient (such as full hybrid, plug-in hybrid and electric vehicles).

Electricity demand tends to rise in conjunction with broader prosperity and rising personal incomes. Cahen and Lubomirsky (2008) reveal that the present need for power is equal to 11 kW/person in the USA, 3.5–5.5 kW/person in Western Europe, and approximately 1 kW/person in India and China (the world average is nowadays 2 kW/person). Of course, this picture does not seem compatible with the future energy needs of China. In fact, about 80 per cent of the growth in global electricity demand by 2030 will likely occur in non-OECD countries. China alone accounts for 35 per cent. According to the Chinese electric power industry development plan (as detailed by Chang et al., 2010 based on the Eleventh Five-Year Plan), the power capacity of the whole country will amount to 852 million kW by 2010, and annual power generation will be 3810 billion kWh. In this context, providing the energy to meet growing electricity demand constitutes a tremendous challenge.[6]

The generating capacity and efficiency of a given power plant each contributes to its ability to provide an economic source to meet 'baseload' and/or 'peak' power requirements as electricity demand shifts throughout the day. Nuclear is a very reliable 'baseload' technology, and up to about 90 per cent of its capacity can be used to generate electricity. Wind- and solar-powered generating facilities, however, are heavily dependent on natural variability in wind and sun conditions, which result in much lower capacity utilization levels. In fact, to meet projected increases in global demand for electricity, all economic fuel sources will need to expand through 2030.

Of course, rising energy needs may be offset by ongoing efficiency gains. These gains may be stimulated in part by government policies that seek to address the risks of climate change by imposing a cost on CO_2 emissions. But rapid economic expansion, especially in China, is very likely to outpace these gains.

2.2 Current Energy Mix and Projected Demand for Energy

What types of fuels will be used to meet the rapidly growing demand for electricity? Answering this question is complex, because power can be generated by a wide range of fuels – traditional sources such as coal and natural gas, and renewable sources such as wind and solar. In 2004, the energy consumption mix of the Chinese steel industry consisted of 69.90 per cent coal, 26.40 per cent electricity, 3.2 per cent fuel oil and 0.5 per cent natural gas (Wang et al., 2007). With so many options, the future mix of fuels depends heavily upon cost. Costs are important because power buyers and utility companies will seek to buy the lowest-cost power first. In most regions of the world, coal and natural gas are the most economical fuels for power generation. Today, about 40 per cent of the world's power comes from coal, while about 20 per cent comes from natural gas. However, many governments are seeking to limit GHG emissions by enacting policies that put a cost on CO_2 emissions. In total, global CO_2 emissions are likely to increase about 25 per cent from 2005 to 2030. By 2030, CO_2 emissions per capita in Europe and China will be nearly equal. In the presence of quantitative limits on CO_2 emissions, coal – which emits far more CO_2 than other fuels – becomes less economically attractive.

In China, coal consumption is expected to rise by nearly 60 per cent from 2005 to 2030. Hence this country crucially needs to diversify the fuels used for power generation. As a relatively 'clean'-burning energy source (compared to coal), natural gas appears as a good candidate to reduce GHG emissions and mitigate environmental impacts. Given its abundance, the expanded use of natural gas – particularly in power generation – can not only help meet growing demand for electricity, but also enable advancement of environmental goals. In many regions, growing demand for natural gas is coming from the power generation sector. By 2030, China's demand for natural gas will be more than six times what it was in 2005. China's growing demand for natural gas is driven more by the residential/commercial and industrial sectors, where distribution lines are being rapidly expanded, and gas is very competitive versus other major fuels. Gas is also used as a raw material for products such as paint, fertilizer and plastics.

In the next section, we move to the analysis of the main technological challenges in energy production and consumption faced by China to enable its transition to a low-carbon society.

3   RECENT DEVELOPMENTS TOWARDS LOW-CARBON ENERGY TECHNOLOGIES

This section investigates the technological developments at stake that would enable in the short and long run the transition to a low-carbon economy in China. We start by exploring the causal link between growth and the consumption of coal.

3.1   'Planet Coal'

China currently generates nearly three-quarters of its electricity from coal-fired power stations. In June 2007, it was reported that construction of an average of two new plants was completed every week (Wang, Z., 2010). China will build 13 large coal bases in the coming years, such as North Shanxi, Middle Shanxi and South Shanxi (Jiang et al., 2010). According to the National Bureau of Statistics of China, in 2005, the total energy production was equal to 2.06 billion tonnes of standard coal, including 2.19 billion tonnes of coal mining, 180 million tonnes of oil, 50 billion m^3 of natural gas, and 2474.7 billion kW of power generation. The Chinese total energy consumption was equal to 2.22 billion tonnes of standard coal, which ranked second in the world after the USA (Xie et al., 2010).[7] One way to resolve the conflict between coal utilization and environmental protection could be to lower energy consumption overall, and to develop advanced pollution control systems,[8] especially for the power sector and for the thousands of small boilers spread throughout China (You and Xu, 2010).

Apergis and Payne (2010a, 2010b) examine the empirical relationship between coal consumption and economic growth for 15 emerging countries (including China) during 1980–2006. In terms of coal production (with 2620 million short tons in 2006) and consumption (with 2584 million short tons during the same year), they document that China ranks the highest based on data from the *Country Energy Profiles* of the Energy Information Administration. With respect to CO_2 emissions from fossil fuel usage, China has again the highest rank (with 6017 million metric tonnes of CO_2 in 2006). Concerning the econometric link between coal consumption and economic growth, Jinke et al. (2008) and Wolde-Rufael (2010) demonstrate the existence of unidirectional causality running from economic growth to coal consumption, whereas Yuan et al. (2009) show bidirectional causality between coal consumption and economic growth. These studies therefore reveal that the reliance on producing energy from burning coal needs to be diminished if one wants to achieve a low-carbon energy society in China.[9]

3.2 Liquefied Natural Gas

Wang et al. (2006b) estimate that China's consumption of gas will reach 100 (200) billion cubic metres in 2010 (2020), but production will only reach 80 (120) billion cubic metres. Therefore about 20 per cent (40 per cent) of the gas consumption will need to be imported. As is standard in the history of energy systems, such a dependence on imports through pipelines from Russia and other Central Asian countries raises serious geopolitical concerns that this chapter does not seek to address (see Victor et al., 2008 for more details).

Lin et al. (2010) argue that the development of liquefied natural gas (LNG) is a necessary part of China's future energy, as it has already profoundly transformed the natural gas market in the USA. Indeed, once well purified and condensed, LNG is easily transported across the sea. The authors report that, in order to meet the increasing demand for natural gas by 2020, China needs to build about ten large LNG receiving terminals, and to import LNG at the level of more than 20 billion cubic metres per year. China's first two LNG receiving terminals are operating in Guangdong and Fujian, while another one is being built in Shanghai. Although the LNG industry is relatively new in China, Lin et al. (2010) consider that overall clean, efficient and safe energy resources can be obtained in that way in abundant quantities as part of China's low-carbon energy future.

3.3 Energy Saving in the Iron and Steel Industry

Guo and Fu (2010) document that the apparent production of crude steel in China is equal to 418.78 million tonnes in 2006 (i.e. about 34 per cent of world steel production). They characterize the iron and steel industry in China as one of the major energy-consuming and -polluting industries. Indeed, it accounts for the consumption of about 15.2 per cent of the national total energy, the generation of 14 per cent of the national total wastewater and waste gas, and 6 per cent of total solid waste materials. Furthermore, the authors reveal that this sector is capable of achieving tremendous gains in energy efficiency. Compared with Japan, for example, energy consumption for China's large and medium firms in 2004 was 705 kgce (coal equivalent) per tonne of steel, 7.5 per cent higher than that in Japan (656 kgce per tonne). For small production units, the energy consumption level was as high as 1045 kgce per tonne of steel. The main energy savings potential occur in the replacement of large-scale, modern blast furnaces.

3.4   Heating Load of Buildings

According to Jiang (2007), buildings represent a large share of energy use in China, especially in Northern China. The heating load of buildings in China is one to three times higher than that of advanced countries. Low energy-efficient heating methods, such as coal stoves and bio-fuel stoves, are still widely used in China, especially in the rural areas. The author points out that this situation has led to significant energy waste. China has one of the world's worst figures for carbon intensity – the CO_2 emitted per unit of economic output. In 2006, the most recent year for which figures are available, the country produced only $435 of GDP for each tonne of CO_2 emitted, compared with $2291 in the USA and $3712 in the EU. The China Academy of Building Research (1995, 2005) estimates that energy intensity could be reduced by 50–65 per cent by using advanced architecture surface materials and improved heating boilers and heating pipe networks. In the urban areas, coal-fired heating boilers need to be replaced by natural gas boilers. In the rural areas, other heating devices, such as coal boilers or electric heaters, need to replace biomass stoves.

3.5   Transportation Sector

As for any developed economy, transportation is a major consumer of oil in China. According to Wang et al. (2006a), road transport takes a major share of gasoline consumption and half of diesel oil consumption. In response to that situation, the Chinese government has been issuing fuel economy standards for vehicles to improve energy efficiency.[10] In addition, He et al. (2005) estimate that, by 2030, 15 per cent of oil could be saved if the government implemented some moderate measures in phases to regulate fuel economy of the targeted fleets, and 23 per cent of oil could be saved if more aggressive standards were implemented.

3.6   The Continued Expansion of Renewables

China has abundant renewable energy resources, and therefore one of the main drivers of the transition towards a low-carbon society is likely to hinge on the continued development of renewable energy. China's Renewable Energy Law was passed by the Congress on 28 February 2005, and took effect on 1 January 2006 (Zhang et al., 2010). It recognizes the strategic role of renewable energies in optimizing China's energy supply mix, mitigating environmental pollution, improving energy supply security, and promoting rural social development. It also directly relates renewable energy development and utilization to China's energy system transition. More importantly, the law largely shapes an integrated renewable energy policy framework by providing a set of directives encouraging renewable energies, including national renewable energy targets, a feed-in tariff, a special fiscal fund, tax relief, and public R&D support as well as education and training.

As a step towards reducing domestic reliance on coal and oil, the National People's Congress is pledging to replace 10 per cent of China's energy consumption with renewable energy resources by 2010 and 15 per cent by 2020 (as seen in the NRDC report 'China's energy conditions and policies').[11] Besides, the Chinese government has undertaken a series of national programmes to promote the development and utilization of renewable energy. These include the Comprehensive Rural Energy Planning and Construction Programme; the Rural Electrification Programme, which focuses on the development of small hydropower plants; the Brightness Programme; the Township Electrification Programme and the Wind Power Concession Programme.

By 2006, the coverage of solar water heaters in China reached 100 million square metres, benefiting about 200 million people. The NRDC 'Medium and Long-Term Development Plan for Renewable Energy in China' expects that the coverage will reach 150 million square metres in 2010.[12] The huge desert and Gobi areas in Northwest China hold enormous potential for large-scale deployment of solar thermal power systems. Therefore solar thermal power can play an important role in meeting the nation's energy demand and reducing GHG emissions (Wang, Z., 2010).[13] In the Eleventh Five-year research project (covering the period 2006–10), the Chinese government aims at encouraging solar energy research for the purpose of developing key technologies involved in the integration of solar thermal systems with buildings. Taken together, over 150 000 000 m[2] solar water collectors will have been put into use by 2010.[14] It is reasonable to expect that solar thermal utilization will play a greater role in building energy systems in the coming years (Wang and Zhai, 2010).[15]

Other renewables are being considered as well. Concerning wind, there are at present many windmill bases in Xinjiang, Inner Mongolia, Shandong, Hebei, Tianjin and Wuxi (Jiang et al., 2010). The utilization of wind energy, a very abundant resource in China, is undergoing rapid development (Xu et al., 2010). According to China Meteorological Administration (2006), the wind energy resource (at 10 m height) is equal to 4350 million kW. China has a coastline length of more than 18 000 km. In the off-shore district, the wind energy resource is abundant (but its amount is still under evaluation). Wind-rich areas are mainly dispersed in the Northwest, North and Northeast regions of China, as well as on the Southeast coast (particularly the sea coast of Zhejiang and Fujian Province) and the islands nearby. In addition, wind energy resources are ample in Western Heilongjiang, Eastern Jilin, Northern Hebei, Eastern Liaoning and some regions in Southwest China. In 2008, the capacity installed by domestic enterprises was equal to 4706 MW (Pengfei, 2009), therefore calling for more fundamental research and policy support in favour of wind energy application.

Concerning hydropower, Chang et al. (2010) state that China's gross amount of hydraulic resource ranks first in the world. Their claim is substantiated by Zhang (2008), who estimates that, by the end of 2007, installed capacity of hydropower in China amounted to 14 823 million kW (which was 20 per cent of the total installed capacity of the whole country), while the annual power generation amounted to 486.7 billion kWh (which was 15 per cent of the gross of the whole country). However, because of the low level of development of this technology (the installed hydropower capacity only takes up 26 per cent of technical exploitable hydropower capacity of China), hydraulic resources have a broad development prospect. According to Chang et al. (2010), the China hydropower project alone aims at an installed hydropower capacity of up to 194 million kW by 2010, accounting for 23.1 per cent of the gross installed power capacity and 35 per cent of hydropower resources. That is why the authors call for the vigorous development of hydropower in this country, in conjunction with worries about environmental pollution and the wish to attain the sustainable development of China's economy.

Among the most controversial projects to develop carbon-free energy, we may cite the construction of the Three Gorges hydroelectric station. It represents the largest hydroelectric project in the world, with an installed capacity of 18.2 million kW, and a mean annual power output of 84.7 billion kWh (Chang et al., 2010). This is roughly equal to the amount of energy produced by burning 40 million tons of coal per year. The construction period of the whole project is 17 years (since 1993), and the total investment as predicted in 1994 was equal to US$25 billion. Among the main drawbacks of such titanic projects, one should mention the relocation of a total of 1.33 million people, while 12 cities and towns have had to be re-established.

Further, solid wastes and waste water discharged from livestock and poultry farms and light industry could be used to produce 31 billion m^3 of biogas (Zhang et al., 2010). The volume of discharge of combustible municipal waste, a renewable source for power generation, has been growing in most Chinese cities. Taking the case of Beijing, for example, it discharged 4.5 million tons of combustible municipal wastes in 2004, and the volume will be growing at 4–5 per cent per year. Approximately 27 million people living in remote and mountainous areas of China did not have access to electricity by the end of 2004, and most of them are among the lowest-income earners. Electricity supply plays a very important role in promoting social and economic development in remote and mountainous areas. Renewable energy power generating options, such as small hydro system, small-sized wind generating systems, and solar photovoltaic systems, are often more cost-effective than the extension of conventional power grids.

Concerning biofuels, China's ethanol production was 1.02 million tons in 2005 (Zhang et al., 2010). As of late 2004, five provinces (Heilongjiang, Jilin, Liaoning, Henan and Anhuid) required ethanol to be mixed with gasoline in a 10 per cent ratio (E10). Four other provinces (Hebei, Hubei, Shandong and Jiangsudwere) were added by late 2005. Ethanol distilleries have already been established in China with a production capacity of 1 million tons. In 2005, China produced 50 000 tons of biodiesel. Most biodiesel production currently comes from waste oil. Sorgo plantations have been set up in Heilongjiang, Inner Mongolia, Xinjiang Uygar A.R., Liangning and Shandong. Sorgo appears to be a promising crop in China for biofuels, as it grows in colder northern climates and is better able to endure drought.

Finally, biomass power generation capacity was about 2000 MW in 2005 (Zhang et al., 2010). Concerning biomass, a great improvement in the rural cooking conditions is to switch from biomass fuels, such as stalks and wood, to cleaner fuels and to promote the use of advanced stoves. China has carried out a National Improved Stove Programme since the 1980s. This programme improved the energy efficiency in cooking and reduced the emissions of particulate matter significantly (Edwards et al., 2007). Meanwhile, cleaner fuels such as biogas have been promoted to replace the traditional stalks and wood (Zhou, 2003). Biogas digesters can not only produce the gas to be used for cooking and lighting in rural households, but also help farmers earn more money from agriculture (Zhang et al., 2010). Geothermal technology can be applied to increase the output and quality of fisheries as well. Solar greenhouses are a common productivity-raising technology in the plantation of vegetables, fruits and flowers.

4 PROSPECTS FOR A CHINESE DOMESTIC CARBON MARKET

This section focuses on the specific development of emissions trading in China, by providing the latest information on regional negotiation and implementation.

4.1 Dramatic Trends in CO_2 Emissions

First and foremost, the importance of developing rapidly emissions trading in China is motivated by its amazing share of world's total CO_2 emissions. According to the International Energy Agency (2010), CO_2 emissions from fossil energy combustion accounted for about 19 per cent of global CO_2 emissions, ranking second after the USA, in 2005. During 2002–04, its increase in CO_2 emissions accounted for about 56 per cent of the increase of the world's total CO_2 emissions. These figures[16] alone make us realize the importance of curbing CO_2 emissions in China in the short run, especially given the potential of economic growth highlighted in Section 2.

According to the National Bureau of Statistics of China (2006), CO_2 emissions per capita in China are quite low (equal to 1.0 t C in 2005, that is, about 91 per cent of the world average, or 35 per cent of the OECD average). The population size comprised in this statistic should not however prevent us from pushing forward drastic action towards the control of CO_2 emissions in this country. The *China Climate Change Review Report* (2007) claims that – based on China's economic structure, stage of economic development, technologies and labour skills – the energy consumption per unit of industrial products is 15–30 per cent higher than that in OECD countries, while the energy intensity of GDP is higher by 3.8 times. Hence there is ample space for energy conservation in China.

Further on this topic, He et al. (2010a) estimate that China's rate of carbon productivity growth is estimated to be equal to 5.4 per cent during 2005–20, while the CO_2 intensity of GDP will reduce by about 50 per cent. However, the authors forecast that CO_2 emissions in 2020 will still be about 40 per cent higher than the emissions levels prevailing in 2005, because of the country's rapid GDP growth.

4.2 Pros and Cons of Environmental Regulation Tools

As is well known, there are several possibilities to exploit the environmental regulation toolbox in order to tackle the problem of CO_2 emissions control. Traditional 'command-and-control' regulation certainly appears feasible in China, given the role played by the central administration of the country. However, that kind of policy measure is often disregarded by academic scholars and policy makers because of a lack of enforcement, a difficult practical implementation and hence a lack of efficiency (see Turner et al., 1994).

Second, there is the possibility to introduce a tax on carbon, which has the economic advantage of sending clear price signals to all participants. Conversely, it suffers from political risks in order to implement such a policy, and from other risks linked to damaging the competitiveness of the country's industries burdened by a tax (while companies in other countries are not necessarily so burdened).

Third, the introduction of an Emissions Trading Scheme (ETS) may be seen as a flexible way to deal with a quantitative limit on CO_2 emissions, while letting the market decide on the price of allowances. As shown by the recent empirical evidence in the USA (Ellerman et al., 2000) and Europe (Ellerman et al., 2010), this latter solution also has some drawbacks, which are concentrated in the allocation mechanism of allowances (for an introduction to the pros and cons of permits markets, see also Chevallier, 2009).

Based on this brief review of the merits of emissions trading relative to other environmental regulation tools, we summarize in the next section the main prospects of the implementation of an ETS in China.

4.3   A Priority: China's Participation in the Clean Development Mechanism

China is one of the major players on the scene of project mechanisms through the Kyoto Protocol's Clean Development Mechanism (CDM; see Chevallier, 2011 for a review). According to the UNEP Risoe CDM Pipeline,[17] with 71 registered projects, another 380 projects are in the pipeline (either requesting registration, or at validation status, or requesting review) and more than 160 million tons of CO_2-equivalent reduction per year, China has become the world's biggest supplier of certified emissions reductions (CERs). Hence the market share of projects hosted by China has been growing over recent years to more than 40 per cent of the total number of projects validated by the CDM executive board. This successful record may be explained by the comprehensive set of policy directives, institutional arrangements, regulations and capacity building to promote CDM activities at the national and the local level (Teng and Zhang, 2010). For instance, in 2005 the NDRC issued measures for operation and management of CDM projects in China in order to improve the transparency and efficiency of CDM project activities.[18] Therefore this first example of participation in a large-scale international agreement to limit CO_2 emissions may be viewed as a very encouraging step for China's transition towards a low-carbon society.

4.4   Low Carbon Pilot Programmes and the Role of Central and Provincial Climate Exchanges

To help deliver the national target of a 40–45 per cent cut in carbon intensity by 2020, China has announced its plan to introduce its carbon trading system in the Twelfth Five-Year Plan (2011–16). This section explores the current landscape for pricing carbon by using tradable permits markets in China, and considers how these markets can be optimally employed around the country, based on the latest available information at the time of writing.

According to Point Carbon (2011), China looks likely to introduce some form of carbon emissions trading scheme in the next five years, although it will be limited in scope to particular regions and industrial sectors. China will launch pilot emissions trading schemes in six provinces before 2013 and set up a nationwide trading platform by 2015. The trading schemes will begin in the cities of Beijing, Chongqing, Shanghai and Tianjin, and in the provinces of Hubei and Guangdong. In August 2010, the National Development and Reform Council asked 13 regional governments to submit proposals for pilot schemes by using tradable permits market mechanisms to control CO_2 emissions. Several governments have submitted their plans covering the period from 2011 to

2015. Guangdong, China's highly industrialized southern province that borders Hong Kong, has proposed a scheme that would regulate emissions by proxy, forcing users to buy credits if they exceed a cap on energy consumption. Meanwhile Hebei province, close to Beijing, is hoping that a cap-and-trade scheme will enable it to recoup a financial benefit from cleaning up its heavy industry. Beijing is expected to impose a gross national energy cap of 4 billion metric tons of standard coal by 2015.

In our view, the successful design and take-off of carbon trading markets in China is likely to play a key role in China's low carbon transition.[19] Gang et al. (2011) underline that it is a politically sensitive issue to introduce a cap-and-trade scheme in China, and that the country has little experience with trading permits and cutting transaction costs. However, we believe that it is a risk worth taking.

5 CONCLUSION

This chapter seeks to explore both the challenges of and the opportunities for achieving a low-carbon society in China, including their implications for environmental sustainability. As He et al. (2010a) put it, China cannot replicate the modernization model adopted by developed countries. Instead, this country needs to coordinate economic development and CO_2 emissions control, while still in the process of industrialization and modernization. Therefore it is central that China evolves to a low-carbon industrialization model. This is the key to the success of sustainable development initiatives in the region, and also by increasingly taking into account climate change considerations during the twenty-first century. Low-carbon transportation policies and technologies (Kahrl et al., 2011), as well as standardized measuring, reporting and verification of GHG emissions, will help tremendously in reducing the carbon intensity of its economy (Li et al., 2011).

Given the scientific evidence reviewed throughout this chapter, it appears reasonable to believe that China can indeed grow in a low-carbon way. The achievement of such a Herculean task will depend on many factors, among which political will and the ability to apply the newest technologies in terms of energy use and production feature prominently. This chapter provides an extensive overview of China's current levels of technological development for the key polluting sectors of its economy. If the dependence on the coal-mining sector and the transport-related oil consumption seems high by today's standards, we believe that we have gathered enough evidence about the development of renewables, heating and energy savings to conclude that a transition towards a low-carbon economy seems feasible. From that perspective, we agree with the findings by Gang et al. (2011).

By implementing an integrated assessment model, He et al. (2010b) estimate that 1469 million tons of reduced emissions of CO_2, 12–32 per cent decline in air pollutant concentrations, and more than US$100 billion of health benefit can be achieved by 2030 if aggressive energy policies are implemented. The challenge must be met if one wants to reconcile economic growth and climate change issues.

Besides the prospect of emissions trading as a way to regulate CO_2 emissions in the near future at the national level, the most promising feature of the transition towards a low-carbon society in China certainly lies in the development of renewable energy. If all renewable energies are taken into account, China's total primary energy supply can be estimated at 2337 Mtce in 2005. Renewable energies contributed approximately 17.8 per cent of primary energy supply, of which traditional biomass (agricultural and forestry residues burned directly for cooking and space heating in rural households) contributed 10.7 per cent, and all new renewable energies (wind, solar energy, biogas, biomass for power generation, biofuels, and hydro) together contributed 7.1 per cent. Electricity dominated the energy supply from renewable energies. In 2005, the total installed capacity of renewable energy power systems was 119.7 GW, accounting for approximately 23.4 per cent of total installed power capacity of China, of which large and small hydro power was 78 GW and 38 GW, respectively (Zhang et al., 2010).

One topic has (voluntarily) not been addressed in this chapter: regional inequalities in Chinese economic development. It is a very sensitive political issue that the economic development of the various regions of China is extremely uneven. According to the *Chinese Statistical Yearbook*, in 2005, the GDP of the 11 Eastern provinces reached US$1512 billion (which accounted for 59.6 per cent of the total GDP), while the GDP of the 12 Western provinces was only US$429.4 billion USD (i.e. about 16.9 per cent of the total GDP). These figures clearly show an uneven concentration of economically developed areas in the two regions. Moreover, the Eastern region has an energy consumption of about 50 per cent of the whole country. The Western region has abundant energy resources, while the Eastern region is noticeably short of them. The Chinese National Bureau of Statistics reports that the coal reserves in the Western region account for about 48 per cent of the nation's total reserves, and its natural gas reserve accounts for about 80 per cent. On the other hand, the Eastern region has an energy reserve of only 8 per cent coal and 15 per cent natural gas. Jin et al. (2010a) warn that this disparity between the energy demand and energy supply of the country's different regions has gradually become a critical problem that embarrasses the otherwise balanced development of China. Also, the prospects for the development of CO_2 capture and storage in China have not been explored, since a preliminary condition for the installation of such technologies lies in the existence of a much more developed energy complex between regions (see Jin et al., 2010b).

To wrap up our discussion, we may suggest (as in Jiang et al., 2010) adopting the following measures as a blueprint for the transition of China to a low-carbon economy: (i) accelerate the modernization of the energy industry system; (ii) strengthen energy resource management skills; (iii) improve the energy standards system; (iv) rationalize the pricing mechanisms (and introduce a price for carbon); and (v) promote innovative systems through the development of science and technology. This, it is hoped, does not read as wishful thinking.

NOTES

*   I wish to thank the editor, Roger Fouquet, as well as anonymous reviewers for their comments on previous drafts which improved the chapter.
1. NO_x emissions from coal-fired plants were equal to 1.3 Mt in 1989, 2.65 Mt in 1995 and 2.85 Mt in 2000. They continued to increase thereafter (Xu et al., 2000). Without major NO_x control measures, the total emissions from coal combustion and vehicle emissions can be evaluated at 12 Mt in 2010, because of increased energy use.
2. Based on our factual approach, it is not possible to conclude whether China will be an example

of the Porter hypothesis (through economies of scale, driving down costs and massively exporting low-carbon technologies such as next generations of solar panels, solar thermals and wind turbines) or if it will become the leader of the low-carbon industrial revolution. We thank a referee for this remark.

3. For example, the contaminants generated from burning coal include waste slag, smoke dust, sulphur and nitric oxides and carbon dioxide. With the rapid increase in coal consumption, SO_2 emissions in China rose from 19.95 million tonnes to 25.49 million tonnes during 2000–05. This results in noxious acid rain (*China Statistical Yearbook 2006*, National Bureau of Statistics of China).
4. See http://www.un.org/apps/news/infocus/sgspeeches/search_full.asp?statID=257.
5. See http://cpc.people.com.cn/GB/64093/64094/6236551.html.
6. Note that, in the Chinese industrial sectors, growth in energy demand comes more particularly from the growth in steel and cement production, and increased manufacturing of goods for local consumption and export.
7. Comparatively, the consumption of oil reached 300 million tonnes, and the net import was equal to 140 million tonnes, which accounted for 44 per cent of total consumption of oil. In 2006, China imported 145 million tonnes of crude oil, accounting for 47 per cent of its total oil consumption (Wang, Z., 2010). In 2020, China's total oil consumption is expected to be nearly 10 per cent of the world's total oil production, and the dependence on import of oil will reach 60–70 per cent of its consumption.
8. For instance, coal washing and sieving regulations could be implemented in all sectors of the coal industry in China to reduce SO_2 emissions and increase combustion efficiency.
9. Of course, coal is not the only pollutant that deserves our attention. For instance, the government also plans to build large crude oil refinery plants in the east coastal areas and oil refinery industry clusters in Bohai Bay, Pan-Yangtze River Delta and Pan Pearl River Delta, as well as to expand the scale of the oil refinery industry in the Northeast and Northwest (Jiang et al., 2010). However, this chapter does not aim to provide an authoritative review on the topic of the Chinese energy mix, but rather to deal with the complex topic of the transition to a low-carbon society as a whole.
10. To control emissions from O_3 pollution in the mega-cities of China, the Chinese government has implemented three stages of emission standards for on-road vehicles in the years 1999, 2003 and 2007.
11. See http://en.ndrc.gov.cn/policyrelease/default.htm.
12. See http://en.chinagate.com.cn/reports/2007-09/13/content_8872839.htm.
13. Photovoltaics are another important area, but they will not be discussed in this chapter.
14. See http://www.gov.cn/jrzg/2007-04/28/content_600212.htm.
15. As mentioned in the introduction, accurate data collection is a challenge to writing any academic paper on China since we are heavily dependent on the availability of official statistics. In this chapter, we have done our best to systematically refer to the most up-to-date scientific sources (from special issues in academic journals or the most recent books) on the topic.
16. As another source for checking these figures, we may cite Jiang et al. (2010, p. 4257): 'In 1990, the GHG emissions related to energy industry was only 2.24 billion t CO_2 in China, 10 per cent of the global GHG emissions; whereas the GHG emissions of US was 4.85 billion t CO_2 in the same year, 23 per cent of the global GHG emissions. However, from 1990 to 2006 the CO_2 emissions of China increased rapidly at the speed of nearly 6 per cent per year and ended in 5.65 billion t CO_2, which was 20.2 per cent of the global GHG emissions and almost equivalent to 5.67 billion t CO_2 of the US, which accounted for 20.3 per cent of the global GHG emissions.'
17. See http://cdmpipeline.org/overview.htm.
18. See http://cdm.ccchina.gov.cn/english/NewsInfo.asp?NewsId=905.
19. Note that we cannot be overly optimistic here, since emissions trading schemes for CO_2 have had little ability (so far) to reduce an economy's emissions. We wish to thank a referee for this comment.

REFERENCES

Apergis, N. and Payne, J.E. (2010a), 'The causal dynamics between coal consumption and growth: evidence from emerging market economies', *Applied Energy*, **87**, 1972–7.
Apergis, N. and Payne, J.E. (2010b), 'Coal consumption and economic growth: evidence from a panel of OECD countries'. *Energy Policy*, **38**, 1353–9.
Cahen, D. and Lubomirsky, I. (2008). 'Energy, the global challenge, and materials', *Materials Today*, **11**(12), 16–20.
Carbon Capital (2010), 'Financing the low carbon economy', Social Intelligence Report, Accenture and Barclays.

Chang, X., Liu, X. and Zhou, W. (2010), 'Hydropower in China at present and its further development', *Energy*, **35**, 4400–406.

Chevallier, J. (2009), 'Emissions trading: what makes it work?', *International Journal of Climate Change Strategies and Management*, **1**(4), 400 406.

Chevallier, J. (2011), 'The Clean Development Mechanism: a stepping stone towards world carbon markets?', in L. Galarraga, M. Gonzalez-Eguino and A. Markandya (eds), *Handbook of Sustainable Energy*, Cheltenham, UK and Northampton, MA, USA: Edward Elgar, pp. 415–40.

China Academy of Building Research (1995), *Energy Conservation Design Standard for New Heating Residential Buildings*, Beijing: China Architecture and Building Press.

China Academy of Building Research (2005), *Design Standard for Energy Efficiency of Public Buildings*, Beijing: China Architecture and Building Press.

China Meteorological Administration (2006), *Report of Chinese Wind Energy Resources*, Beijing: China Meteorological Press.

Compiling Team of the China Climate Change Review Report (2007), *China Climate Change Review Report*, Beijing: China Science Press.

Edwards, R., Liu, Y., He, G., Yin, Z., Sinton, J. and Peabody, J. (2007), 'Household CO and PM measured as part of a review of China's national improved stove program', *Indoor Air*, **17**, 189–203.

Ellerman, A.D., Joskow, P.L., Schmalensee, R., Montero, J.P. and Bailey, E. (2000), *Markets for Clean Air: The US Acid Rain Program*, Cambridge: Cambridge University Press.

Ellerman, A.D., Convery, F. and De Perthuis, C. (2010), *Pricing Carbon: The European Emissions Trading Scheme*, Cambridge: Cambridge University Press.

Gang, F., Stern, N., Edenhofer, O., Shanda, X., Eklund, K., Ackerman, F., Lailai, L., Hallding, K. (2011), *The Economics of Climate Change in China: Towards a Low-Carbon Economy*, London: Earthscan.

Guo, Z.C. and Fu, Z.X. (2010), 'Current situation of energy consumption and measures taken for energy saving in the iron and steel industry in China', *Energy*, **35**, 4356–60.

He, J., Deng, J. and Su, M. (2010a), 'CO$_2$ emission from China's energy sector and strategy for its control', *Energy*, **35**, 4494–8.

He, K.B., Huo, H., Zhang, Q., He, D.Q., An, F. and Wang, M. (2005), 'Oil consumption and CO$_2$ emissions in China's road transport: current status, future trends, and policy implications', *Energy Policy*, **33**(12), 1499–507.

He, K., Lei, Y., Pan, X., Zhang, Y., Zhang, Q. and Chen, D. (2010b), 'Co-benefits from energy policies in China', *Energy*, **35**, 4265–72.

IEA (2010), *World Energy Outlook*, Paris: International Energy Agency.

Jiang, B., Sun, Z. and Liu, M. (2010), 'China's energy development strategy under the low-carbon economy', *Energy*, **35**, 4257–64.

Jiang, Y. (2007), 'Status of building energy consumption and its energy conservation work point in China', *Construction Science and Technology*, **5**, 26–9.

Jin, H., Xu, G., Han, W., Gao, L. and Li, Z. (2010a), 'Sustainable development of energy systems for western China', *Energy*, **35**, 4313–18.

Jin, H., Gao, L., Han, W. and Huong, H. (2010b), 'Prospect options of CO$_2$ capture technology suitable for China', *Energy*, **35**, 4499–506.

Jinke, L., Hualing, S. and Dianming, G. (2008), 'Causality relationship between coal consumption and GDP: difference of major OECD and non-OECD countries', *Applied Energy*, **85**, 421–9.

Kahrl, F., Williams, J., Jianhua, D. and Junfeng, H. (2011), 'Challenges to China's transition to a low carbon electricity system', *Energy Policy*, **39**(7), 4032–41.

Li, F., Dong, S., Li, X., Liang, Q. and Yang, W. (2011), 'Energy consumption–economic growth relationship and carbon dioxide emissions in China', *Energy Policy*, **39**, 568–74.

Lin, T. and Swanson, T. (2010), *Economic Growth and Environmental Regulation: The People's Republic of China's Path to a Brighter Future*, Routledge Explorations in Environmental Economics, Abingdon, UK: Routledge.

Lin, W., Zhang, N. and Gu, A. (2010), 'LNG (liquefied natural gas): a necessary part in China's future energy infrastructure', *Energy*, **35**, 4383–91.

Managi, S. and Kaneko, S. (2010), *Chinese Economic Development and the Environment*, Cheltenham, UK and Northampton, MA, USA: Edward Elgar.

National Bureau of Statistics of China (2006), *China Statistics Yearbook 2006*, Beijing: China Statistics Press.

Pengfei, S. (2009), 'Wind installation in China', *China Wind Energy*, **1**, 16–20.

Point Carbon (2011), 'China planning emissions trading in 6 regions', *Reuters Green Business News*, reported by David Stanway, edited by Ed Lane, 11 April 2011, available at http://www.reuters.com/article/2011/04/11/us-china-carbon-trading-idUSTRE73A1UY20110411.

Teng, F. and Zhang, X. (2010), 'Clean development mechanism practice in China: Current status and possibilities for future regime', *Energy*, **35**, 4328–35.

Turner, R.K., Pearce, D. and Bateman, I. (1994), *Environmental Economics: An Elementary Introduction*, Harlow, UK: Pearson Education.

UNDP (2010), 'China and a sustainable future: towards a low carbon economy and society', *China National Human Development Report 2009/10*, United Nations Development Programme and Renmin University, China.

Victor, D.G., Jaffe, A.M. and Hayes, M.H. (2008), *Natural Gas and Geopolitics: From 1970 to 2040*, Cambridge: Cambridge University Press.

Wang, K., Wang, C., Lu, X.D. and Chen, J.N. (2007), 'Scenario analysis on CO_2 emissions reduction potential in China's iron and steel industry', *Energy Policy*, **35**, 2320–35.

Wang, M., Huo, H., Johnson, L. and He, D. (2006a), 'Projection of Chinese motor vehicle growth, oil demand, and CO_2 emissions through 2050', Argonne National Laboratory Report no. ANL/ESD/06-6, Argonne, IL.

Wang, R.Z. and Zhai, X.Q. (2010), 'Development of solar thermal technologies in China', *Energy*, **35**, 4407–16.

Wang, X.Y., Luo, Y.X., Long, G. and Zhou, Z.B. (2006b), 'Research on China's natural gas supply security strategy', *China Energy Resources*, **28**(2), 23–5.

Wang, Y. (2010), 'The analysis of the impacts of energy consumption on environment and public health in China', *Energy*, **35**, 4473–79.

Wang, Z. (2010), 'Prospectives for China's solar thermal power technology development', *Energy*, **35**, 4417–20.

Wolde-Rufael, Y. (2010), 'Coal consumption and economic growth revisited', *Applied Energy*, **87**, 160–67.

Xie, K., Li, W. and Zhao, W. (2010), 'Coal chemical industry and its sustainable development in China', *Energy*, **35**, 4349–55.

Xu, J., He, D. and Zhao, X. (2010), 'Status and prospects of Chinese wind energy', *Energy*, **35**, 4439–44.

Xu, X.C., Chen, C.H., Qi, H.Y., He, R., You, C.F. and Xiang, G.M. (2000), 'Development of coal combustion pollution control for SO_2 and NO_x in China', *Fuel Processing Technology*, **62**, 153–60.

You, C.F. and Xu, X.C. (2010), 'Coal combustion and its pollution control in China', *Energy*, **35**, 4467–72.

Yuan, J.H., Kang, J.G., Zhao, C.H. and Hu, Z.G. (2009), 'Energy consumption and economic growth: evidence from China at both aggregated and disaggregated levels', *Energy Economics*, **30**, 3077–94.

Zhang, G. (2008), 'On the forum of China hydropower development', *China Power*, 2008–05–09 (in Chinese).

Zhang, X., Ruoshui, W., Molin, H. and Martinot, E. (2010), 'A study of the role played by renewable energies in China's sustainable energy supply', *Energy*, **35**, 4392–9.

Zhou, D.D. (2003), *China Sustainable Energy Scenarios in 2020*, Beijing: China Environmental Science Press.

32 Low-carbon economy: dark age or golden age?
Roger Fouquet

1.   INTRODUCTION

To many, a global economy driven by fossil fuels seems 'unsustainable'. Traditionally, the fact that coal, petroleum and natural gas were non-renewable resources implied that, at least in the very long run, if not sooner, they could not be humanity's dominant energy sources (Jevons, 1865; Hamilton in Chapter 1).

However, humanity's ability to discover, extract and use nearly 500 billion tonnes of oil equivalent of fossil fuels (see Figure 32.1) since the beginning of the Industrial Revolution has led to 1200 billion tonnes of carbon dioxide emissions (see Figure 32.2). The rising global emissions, along with other greenhouse gas emissions, are threatening to accelerate climate change. The threat of climate change has implied that fossil fuels, without worldwide carbon capture and sequestration mechanisms or successful geo-

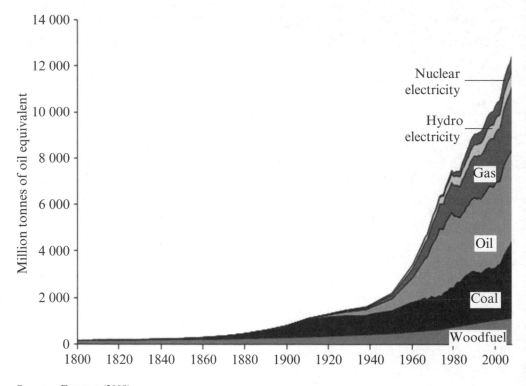

Source: Fouquet (2009).

Figure 32.1   Global primary energy consumption, 1800–2010

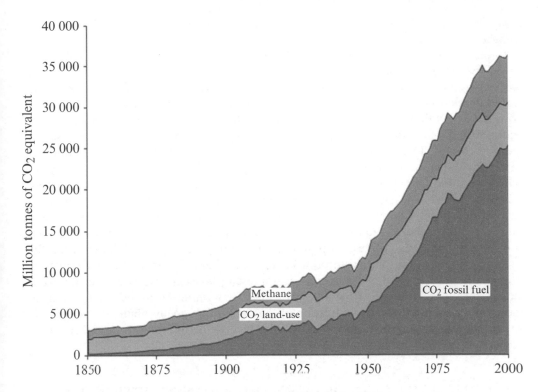

Sources: Marland et al. (2007), Houghton (2008), Stern and Kaufmann (1998).

Figure 32.2   Global carbon dioxide and methane emissions, 1850–2000

engineering projects, will impose a mounting burden on the atmosphere (Hendry and Pretis in Chapter 12). Thus, within the wider context of history, the fossil fuel energy system seems 'unsustainable' and only to be a temporary phase.

So, change is inevitable. A key question is 'what change will occur?' Presumably, what we want is the smoothest transition to the most desirable new state of the world in the long run. There has been a great deal of interest in possible energy transitions[1] (Grubb et al., 2008; Foxon et al., 2008; Smil, 2010; Fouquet, 2010). However, there has been insufficient focus on the outcome of the transition, such as a possible low-carbon economy. Both the transition and the final outcome need to be considered. After all, we want to be sure to avoid leading the global economy into a new 'dark age', a period of long-term economic and social decline, such as in fifth-century Rome or twelfth-century China, or even the collapse of the Mayan empire (Diamond, 2005). Naturally, these were the result of many other forces, although climate change has been linked to at least one of these collapses.

This chapter begins to think about the outcome of the transition, the new economic system that might emerge and its details. Because of the highly uncertain and speculative nature of considering the future states of the world and the detailed implications – whether looking at worlds of intensifying climate change or of possible low-carbon economies – analysts have tended to shy away from such exercises. However, to shy away from them also leads us to limit our awareness and understanding of the potential long-term implications.

In pursuit of detail, this chapter fleshes out some possible future worlds. In Section 2, to set the stage, the chapter considers possible long-run trends in energy consumption associated with power and transport demand. This uses new evidence on how elasticities of demand change with economic development. Then, in Sections 3 to 5, the chapter considers how that demand will be met by outlining three (of many) possible global economies, faced with the atmospheric constraint associated with greenhouse gases and climate change. The ensuing discussion follows a similar line to the one Kahn (2010) takes on adaptation – that is, using reasoned arguments based on economic theory and, where possible, empirical evidence. The first possible outcome is in a global economy where no greenhouse gas mitigation is achieved and adaptation is the only solution. The second possible outcome is one where a low-carbon transition is achieved by meeting all future global energy requirements with nuclear power. The third one is where we do the same but with renewable energy sources. Inevitably, this approach leaves the author open to criticism about the assumptions made. It is, therefore, hoped that the reader will appreciate that the spirit of the exercise is to present three extreme cases in order to consider (speculatively) some of the details of the long-term implications of climate strategies, and possibly to offer an indication of the range of manoeuvrability that might exist to the climate strategists. Section 6 concludes by considering the substantial economic, social and political implications of each outcome.

2.   LONG-RUN TRENDS IN ENERGY CONSUMPTION TO 2100

A starting point for the discussion is considering the long-run trends in population, economic activity, demand for energy services and energy requirements, as well as how these requirements will be met and the associated climatic impact. Given the great uncertainty, such exercises can only be thought-experiments rather than offer predictions about future trends.

Here, the basic model is that (i) energy consumption is a function of the use of energy service and the energy efficiency of the service technology for, say, power or transportation, and (ii) the demand for energy services is a function of the population, personal income and the price of energy services, which is generally determined by the price of energy and the energy efficiency of the service technology (Fouquet, 2011a). This second relationship is mediated by the income and price elasticity of demand for energy services. These projections will focus on the demand for power and transport, and not attempt to cover all energy services.

Figure 32.3 shows the trends in global population and GDP per capita. These are based on predictions of population up to 2100 (UN, 2011) and of GDP (Pwc, 2011) to 2050 with the trend in the growth rate extrapolated to 2100 (see also Table 32.1). Regarding global population, the key point is that it will tend to stabilize at 10 billion during the second half of the century. For GDP, the main assumption is that the global GDP per capita growth rate will tend to decline, from around 4 per cent in the second decade of the twenty-first century to about 2.8 per cent in 2050 and extrapolated down to 1 per cent by 2100. With this growth rate trend, global GDP per capita would have

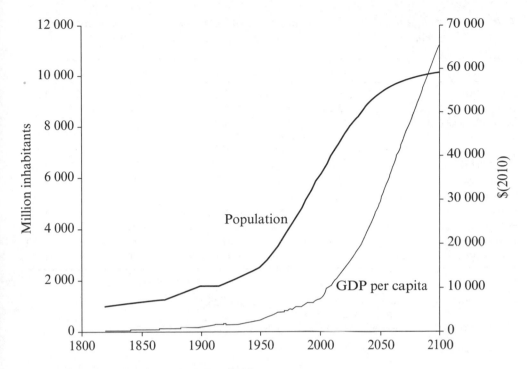

Sources: Population: UN (2011); GDP: Pwc (2011).

Figure 32.3   Long-run trends in global population and GDP per capita, 1820–2100

risen from an average $11 000 in 2010 to $31 000 in 2050 and $69 000 in 2100 (all in 2010 US dollars).

However, to conceive of such trends many factors related to the availability of world resources must be considered to know whether this is possible. Resource availability is not considered in detail in the original studies on population and economic growth. In the present chapter, these projections will be used, but the discussion about the energy consumption that results from those levels of population and economic activity will feed back into the debate in later sections.

As mentioned above, the long-run trend in the price of energy services will be an important driver of energy consumption (Fouquet, 2011a). Here, it is assumed that energy prices will remain constant. Naturally, this is not what will happen. Yet, whether energy service prices will go up as resources become scarcer or fall as new energy reserves are exploited, new sources are used and energy technologies become far more efficient is very unclear.

Nevertheless, an assumption is made that the energy efficiency improvements related to power services (whether computers, appliances, lighting, as well as in industrial activities) doubles between 2000 and 2100 (see Table 32.1). This is a large improvement in energy efficiency, yet over 100 years it is quite small, and assumes a small annual change. Since a linear interpolation between 2010 and 2100 is used to calculate the quantity of energy required per unit of energy service, the annual efficiency improvement falls from 1.6 per

Table 32.1   Assumptions: long-run trends in explanatory variables (% increase per annum) and elasticities of global energy consumption, 2000–2100

	2000	2010	2020	2030	2050	2075	2100
Population	1.28	1.14	0.95	0.74	0.40	0.15	0.05
GDP per capita (%)	2.10	5.02	2.60	2.38	2.42	1.57	0.97
Energy prices (%)	NA[a]	0[b]	0	0	0	0	0
	0.02[c]	0.11[c]					
Efficiency power (%)	−1.19	0.97	0.88	0.81	0.70	0.60	0.52
Efficiency transport (%)	−1.12	1.59	1.37	1.20	0.97	0.78	0.65
Price of power (%)	NA	−0.97	−0.88	−0.81	−0.70	−0.60	−0.52
Price of transport (%)	−1.10	−1.48	−1.37	−1.20	−0.97	−0.78	−0.65
Power income elasticity	NA	1.50	1.25	1.00	0.30	0.30	0.30
Power price elasticity	NA	−0.50	−0.50	−0.50	−0.50	−0.50	−0.50
Transport income elasticity	NA	1.00	0.90	0.80	0.60	0.60	0.60
Transport price elasticity	NA	−1.00	−0.90	−0.80	−0.60	−0.60	−0.60

Notes:
[a] Annual growth rates of global average electricity are not available.
[b] Assumed average growth rate between 2011 and 2020.
[c] Annual growth rate of crude oil prices, 1991–2000 and 2001–10, respectively.

Source: See text.

cent to 0.6 per cent over the period. As a comparison, the energy efficiency in the provision of power services in the UK also doubled between 1900 and 2000 (Fouquet, 2008).

As another comparison, lighting efficiency improved 22-fold between 1900 and 2000 (Fouquet and Pearson, 2012). This improvement was exceptional, involving a number of technological breakthroughs, and will probably not be replicated this century. In addition, the nature of lighting is physically very different from other appliances and technologies using electricity.

During the twentieth century, transport efficiency improved six-fold (Fouquet, 2008). That is, to provide the average passenger-kilometre in 1900 (mostly by railway) required six times more energy than the same service in 2000 (mostly by car). Here, the assumption is that the energy efficiency of transport services will triple between 2000 and 2100.

Given the assumption of constant energy prices and a three-fold improvement in transport efficiency, the price of transport services is assumed to be directly related and to fall three-fold during the twenty-first century, too. Similarly, power service prices are assumed to halve (see Table 32.1).

Income and price elasticity of demand for energy services are assumed to decline. This is based on past evidence. As an example, income elasticities for transport and lighting in the UK were estimated to fall from a peak of over 3 in the second half of the nineteenth century (when per capita GDP was roughly $5000) to about 1.5 in the first half of the twentieth century to unity in 1950 (when per capita GDP was close to $12000, similar to today's global per capita GDP) to well below one at the end of the second half of the twentieth century (Fouquet, 2012; Fouquet and Pearson, 2012). At the beginning of the twenty-first century, British transport and power income elasticity were estimated to be

0.8 and 0.4, respectively (per capita GDP is a little over $30 000). So, today, a 10 per cent increase in income is estimated to lead to a 8 per cent rise in total land transport use and a 4 per cent rise in lighting consumption. Similarly, the (absolute value of) price elasticities of transport and lighting demand fell from around 1.5 in the late 1800s to unity in 1900 and to around 0.6 in the second half of the twentieth century.

If present-day developing countries follow a similar trend, then the elasticities of countries, such as China, are peaking. This implies (in a similar vein as, say, Wolfram et al., 2012) that its energy service (and probably energy) consumption are and will continue (for many years) to increase at a faster rate than its GDP per capita. For the UK (and these specific energy services, lighting and land transport), income elasticities fell below unity around $12 000. Probably, however, present-day developing economies will have smoother trends (that is, a smaller peak) and possibly an earlier peak. The intuition is that access to energy services is greater and prices lower in today's developing economies than in the UK at equivalent levels of GDP per capita. It is proposed that, eventually, all countries experience a decline in elasticities, associated with the changing nature of economic development and saturation effects, implying a slower growth in energy service and energy consumption.

Naturally, each country and sector will be different. Here, crude aggregated assumptions are made to paint a picture. The global average power income elasticities are assumed to fall rapidly between 2010 and 2050 from 1.5 to 0.3. In other words, a 10 per cent increase in per capita GDP raises power demand 15 per cent today but only 3 per cent by 2050. This is intended to reflect that the industrialization process (that, e.g. China and India are undergoing) is very power-intensive (as well as heat-intensive); and that once certain levels of industrialization have been achieved, growth in demand will slow down considerably. The historical evidence suggests that price elasticities in general tend to be more stable, and here they are assumed to stay constant at 0.5 for power demand. This is intended to reflect the proposition that power demand is driven mostly by economic development and less by changes in energy prices or efficiency improvements.

Land transport income elasticities are assumed to fall less than for power demand – from 1 to 0.6 – while, similarly, transport price elasticities are assumed to fall (in absolute terms) from −1 to −0.6. This is intended to reflect the proposition that transport demand is partly driven by economic growth and development, and by price incentives. Also, at current average global levels of per capita GDP (i.e. $12 000), it is assumed that transport income elasticities have fallen from the peaks experienced at lower levels of economic development, and are probably around unity at present.

Based on these assumptions (see Table 32.1), global power and transport demand would increase four-fold and five-fold, respectively, between 2010 and 2050, and then increase 60 per cent and 120 per cent, respectively, between 2050 and 2100 (see Figure 32.4). In other words, both power and transport demands may grow a great deal over the next 30–40 years, and particularly transport demand may continue to grow substantially up to 2100, although the rates of growth may slow considerably in the second half of the twenty-first century (as population growth rates, GDP growth rates, and income and price elasticities are expected to fall).

Although energy efficiency improvements reduce the price of energy services, thus pushing up power and particularly transport use, they will reduce the amount of energy required for each unit of service. As a reminder, it was assumed that energy efficiency

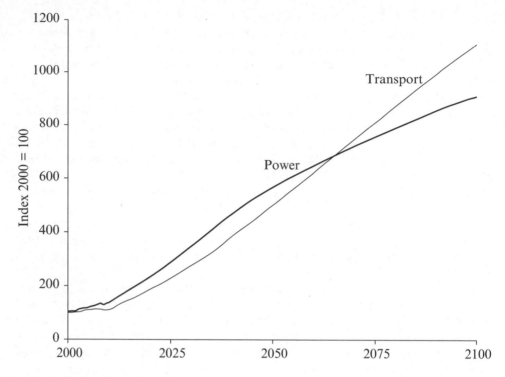

Source: Author's estimates (see text).

Figure 32.4   Long-run trends in global power and transport use, 2000–2100

would double for power services and triple for transport services between 2010 and 2100. This implies that power service consumption will have increased twice as much as the energy consumption between 2010 and 2100. In the same way, transport use will have increased three times as much as the related energy use in that time. Also, because the price elasticities of demand are assumed to be 0.5 for power demand and falling to 0.6 by 2050, energy efficiency improvements are expected to feed through into higher energy service uses, but will still imply that about half of the total improvements generate energy savings.

Indeed, based on the above assumptions (in Table 32.1), energy requirements for power use will triple to 2050, but then only increase by 20 per cent during the second half of the twenty-first century. Meanwhile, energy consumption to meet transport demand will almost triple (i.e. rising 170 per cent) between 2010 and 2050, and then increase a further 50 per cent from 2050 to 2100 (see Figure 32.5).

These projections assume no adoption of electric vehicles. Starting with existing requirements for electric cars (20 kWh per 100 km), and assuming a five-fold improvement (rather than three-fold for the internal combustion engine) in efficiency between 2010 and 2100, a complete switch to electric vehicles was crudely estimated to increase global electricity consumption at the end of the century to around 110000 TWh.

In 2010, global primary energy consumption was a little over 13000 mtoe (million

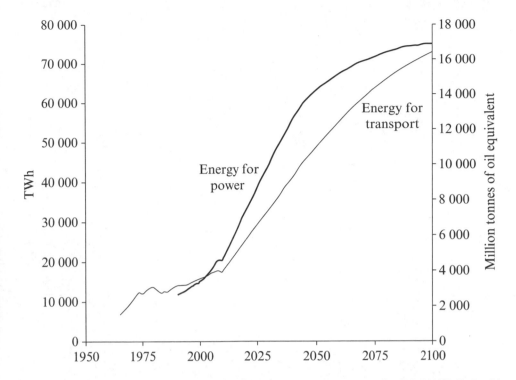

Note: Prior to 2011, the data present global electricity and petroleum consumption (BP, 2011b). It should be noted that, of course, not all the electricity is for power services, nor is all the petroleum for transport purposes. However, for simplicity, it is assumed that the demand for power and transport determine all the post-2010 consumption of electricity and power, respectively.

Sources: Before 2011: BP (2011b); After 2011: author's estimates (see text).

Figure 32.5   *Long-run trends in global energy consumption for power and transport use, 1965–2100*

tonnes of oil equivalent) (BP, 2011b). This is equivalent to 555 exajoules (EJ) or 160 000 TWh. Around 79 per cent (i.e. 10 400 mtoe) originated from fossil fuels – that is, 26.8 per cent from coal, 30.4 per cent from petroleum and 21.6 per cent from natural gas. Nuclear provided 4.7 per cent of the total, hydropower 5.8 per cent and other renewables 1.6 per cent. Roughly 9 per cent, about 1200 mtoe, is still being provided by traditional biomass sources (Fouquet, 2009).

A tripling of global energy consumption by 2050 would imply an annual consumption of nearly 40 000 mtoe. This would be equivalent to 1700 EJ or 480 000 TWh. An increase of, say, one-third in the second half of the twenty-first century would increase consumption in to the 50 000–55 000 mtoe range. Demand for other energy services, such as heating, has been explicitly ignored, because this exercise is not an attempt to produce a usable forecast of energy consumption. The purpose of the exercise was simply to indicate possible long-run trends. Nevertheless, the exercise does suggest that global energy consumption may increase substantially over the next 30 or 40 years, and growth may well slow substantially in the second half of the twenty-first century.

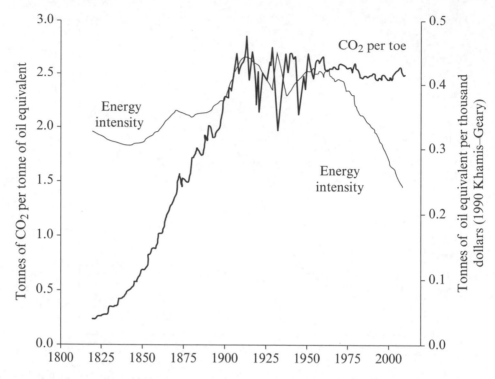

Sources: Fouquet (2009); BP (2011b).

Figure 32.6 *Long-run trends in global energy intensity and carbon dioxide emissions relative to primary energy use, 1820–2010*

A tripling of the global energy system would certainly require huge additional investment in energy production and supply systems. It may put pressure on existing energy resources: the equivalent of 3.4 billion mtoe would be required between 2010 and 2100.

The IEA (2010) baseline scenario suggests that electricity consumption would rise by an annual rate of 2.8 per cent to 37 200 TWh by 2030. BP (2011a) forecasts an annual growth rate of 2.6 per cent. The projection here proposes an annual growth rate of 3.75 per cent to 2030. This implies that electricity consumption would rise to 44 000 TWh in 2030, and would be 20 per cent higher than the IEA outlook. This is explained principally by the assumption of higher income elasticities during the industrialization process of many currently developing countries, including China and India.

This growth in both transportation and power demand would most probably generate a great many more carbon dioxide emissions. Over the last 100 years, since the global economy entered the fossil fuel energy system, each tonne of oil equivalent of energy has emitted 2.5 tonnes of carbon dioxide (see Figure 32.6). At the same rate, another 8500 gigatonnes (Gt) of carbon dioxide would be emitted, putting huge pressures on the earth's carbon cycle and climate. To put this figure in perspective, less than 1150 Gt have been emitted since the beginning of the Industrial Revolution (Marland et al., 2007). Yet those emissions have been associated with climate change. Thus, during the twenty-first century, this estimate suggests that almost six times more emissions might be released into the atmosphere!

3.   A CLIMATE-CHANGING ECONOMY: THE WEALTHY FIRST

Various reasons might be proposed for not taking action on climate change. Some fear that a transition to a low-carbon economy will be costly to the economy, and may even impose sufficient 'drag' to generate a long-term economic decline. Out of fear of taking the wrong road, they propose inaction – perhaps while natural and social scientists gather more information and develop new and more attractive (i.e. less costly) solutions.

Alternatively, the climate sceptic position is to doubt the existence of anthropogenically caused climate change. It is possible that most climate scientists are wrong. But, if they are right, but global greenhouse gas emissions are not reduced, then populations around the world will be faced with substantial and presumably intensifying climate change.

It is possible that society could adapt to life with perpetually changing climates. It would be impossible to consider all the dimensions of reducing climate change impacts and adaptation that would be required or the long-term implications. To tackle the issues, this analysis follows Kahn's (2010) approach (although not necessarily his analysis) on how economies and cities in particular may adapt to climate change.

Historical experience of adaptation to major chronic environmental pollution suggests that wealthier populations have found ways to avoid the damage caused by pollution through markets, although often very slowly – that is, over decades (Kahn, 2010; Fouquet, 2011b). However, the poorer populations were not able to afford the market-based solutions and had to suffer the impacts (Thorsheim, 2002; Fouquet, 2011b).

As an example, during the second half of the nineteenth century, British upper- and upper-middle-class families sought to live away from the crime, sewage and smoke that engulfed Victorian cities by moving to the suburbs (Luckin, 2000). As suburban housing and rail transport supply increased and became cheaper, lower-middle-class households followed, leaving only the poor in the urban centres to face criminality and air and water pollution (Thompson, 1982; Jackson, 2003). Although only one aspect of the 'adaptation' to the social and environmental problems of nineteenth-century Britain, air pollution throughout the second half of nineteenth-century Britain was responsible for annual health damages equivalent to (or greater than) an estimated 10 per cent of GDP at the time, peaking in the 1880s at 20 per cent (Fouquet, 2011b). Based on these estimates of the health costs of air pollution from coal combustion, the total damage between 1850 and 1950 amounted to £3.8 trillion (in 2000 pounds sterling) (or $8.2 trillion). Although lacking other estimates to compare, this probably was one of the most severe large-scale and chronic environmental damages in global history (with air pollution concentrations double those of Delhi in the 1980s). Naturally, over this period, and more recently, many other countries have also experienced and suffered greatly from coal-related air pollution.

For some comparisons of damage, the 1986 Chernobyl nuclear disaster in present Ukraine has been estimated to be in the hundreds of billions of dollars over several decades. Belarus alone estimated its damages at $235 billion (Chernobyl Forum, 2006, p. 33). The economic costs of the Fukushima nuclear accident in 2011 are naturally highly uncertain, but were placed at $250 billion and rising, which effectively bankrupted Tokyo Electric Power Co., the world's fourth-largest utility (Cooper, 2011). While the Chernobyl and Fukushima examples are dramatic, they are associated with acute disasters, with medium- to long-term consequences. The air pollution problem (although obviously different in many ways) was similar to future climate change damage in the sense that it was a chronic and gradually worsening problem. Thus the experience and lessons are pertinent.

The argument here is that the wealthy (and probably wealthy cities, regions or nations) will be the first to protect themselves against environmental disasters. In addition to flooding, droughts and extended heat waves, an increased frequency of storms is one of the likely consequences of climate change that will be of great concern to vulnerable populations, whether rich or poor (IPCC, 2007).

It is difficult to protect property against climate-change-related disasters such as hurricanes. Taking account of property damage only, Hurricane Katrina in 2005 caused $45 billion and was the catastrophe responsible for the largest insurance loss in US history. Apart from the terrorist attacks in 2001 and the California earthquake in 1994, the top ten most damaging catastrophes in US history have been hurricanes (Insurance Information Institute, 2012).

The threats from storms include: damage to crops, damage to property, power outages, disruption of the public water supply, increased risk of deaths and injuries, as well as of water- and food-borne diseases. Industries and communities can be severely disrupted by floods and high winds.

Storms (e.g. hurricanes, cyclones and typhoons) are the biggest financial threat associated with extreme weather events, accounting for 56 per cent of total losses from weather-related catastrophes between 1950 and 2010 (Munich Re, 2012). Other extreme weather events include extreme temperatures, droughts, forest fires and floods. The property damage from weather-related catastrophes between 1950 and 2010 was $1.38 trillion. Contrary to the estimated 4.5 million lives lost from air pollution in the UK between 1700 and 2000 (Fonquet, 2011b), a value has not been placed on the 1.04 million lives lost from these catastrophes – three-quarters of them coming from storms.

Figure 32.7 shows the trend in extreme heat days (as the ratio of observed to expected extreme heat days) and in property damage caused by weather-related catastrophes. Both trends indicate rising extreme temperatures and damage. While before 1985, annual damage was rarely more than $20 billion, after this, annual losses in most years amounted to at least $20 billion, a dozen to more than $50 billion and in one year almost $200 billion was lost (i.e. when Hurricane Katrina hit).

The rising damage may reflect the increase in property destroyed between the mid-twentieth and early twenty-first century (and particularly the property boom in the 1980s and 2000s). However, it also reflects that there has been a rising opportunity cost associated with extreme weather events. Continuing this trend, in the mid-twenty-first century, more property and probably more valuable properties (including prized skyscrapers and other architectural masterpieces) will be at risk than today. Thus the markets, especially property developers and insurers, are becoming increasingly concerned about the threat of storms (Charpentier, 2008; Mills, 2009).

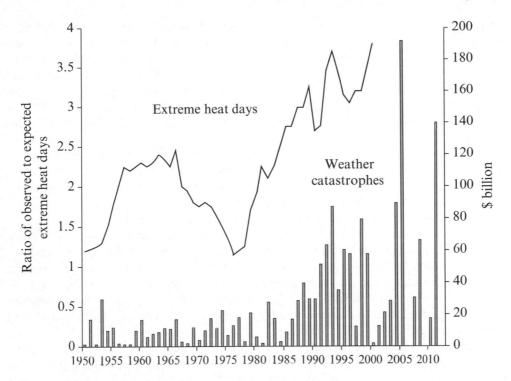

Sources: Extreme heat days: Coumou and Rahmstorf (2012); catastrophes: Munich Re (2012).

Figure 32.7   *Global extreme heat days and weather-related natural catastrophes (1950–2011)*

Furthermore, the case of Hurricane Katrina in New Orleans in 2005 shows that destruction in industrialized cities can heavily skew global financial losses (see Figure 32.7). This is because the economies are more developed and the property value is higher. A number of economies are developing quickly and so will have more property and higher-value property at risk by the mid-twenty-first century than today, and may be more likely to be in the path of tropical storms. Nevertheless, industrialized nations have and will continue to have (for many decades) the most to lose financially from storms and other catastrophes. Thus industrialized nations are likely to dictate the market responses to climate-related threats.

Given the potential threats from hurricanes, cyclones and typhoons around the world, as well as other extreme-weather-related threats, there may be a growing demand for large-scale climate protection (Emanuel, 2005; Webster et al., 2005). To explore possible developments, the potential investment in climate-protected and climate-regulated cities can be used as a highly speculative example.

Climate-protected and -regulated cities could start in vulnerable parts of cities, similar in size to airport terminals or airplane assembly sites. For instance, Terminal 3 at Dubai International Airport is 1.185 million m² (or 1.185 km²), and Boeing's assembly site in Washington State (USA) is the largest building in the world by volume (of 13.3 million m³ and with a surface area of 400 000 m² (or 0.4 km²)). To place this in perspective, Manhattan is 87 km² (34 sq. miles) and central London is a similar size. So, while it seems very ambitious to imagine whole cities to be protected by a single structure, a series of linked structures between parts of cities would be possible, and could act as shelter from extreme weather events and for regulating the inner climate. On the other hand, to consider the construction of linked domes to cover the 1200 km² (500 sq. miles) for Greater London or New York City or even 2200 km² for Tokyo seems beguiling.

Nevertheless, by regulating parts of a city's temperature, humidity and ventilation implies that citizens no longer have to heat or (possibly) air-condition their homes. Thus the provision of these key energy services would be provided and regulated by the city. This would also imply higher city taxes to fund such fixed (e.g. building) and variable (e.g. regulating) costs, but lower domestic and commercial energy bills.

Wealthy neighbourhoods or cities would probably be the first to develop such infrastructure, becoming even more desirable places to live – driving up the value of land and property in such protected cities. For example, cities could regulate their own climates, offering optimal temperatures and weather conditions. This could involve slightly lower temperatures during the week to promote productivity, and higher at weekends. Seasons could also be regulated to reflect expected patterns, ensuring balmy summers and even white Christmases.

As more cities (or parts of cities) adopt this infrastructure, the costs of designing and building such climate-protected and -controlled cities will start to fall. This would accelerate the adoption rate of such protective spheres. As more cities became climate-protected, this would accelerate the current process of urbanization – those living outside the climate-protected bubbles would be at the mercy of increasingly extreme weather conditions. Populations would seek to migrate towards the protective structures, and would become increasingly huddled in urban centres. Given the declining rural populations, the need and demand to protect these environments and ecosystems would decline, making these places increasingly less hospitable. However, because of the huge fixed costs, unprotected poorer cities, especially in developing economies, would be increasingly likely to suffer the impacts of climate change, while developed economies would be increasingly protected from the problem. Meanwhile, cities and countries that are mostly climate-protected will be against heavy investment in low-carbon energy sources and technologies. This political stance towards climate protection rather than stabilization is likely to be reinforced by civil engineering firms (and related industries) that would have become increasingly profitable and sources of employment, and, thus, probably influential in prioritizing policies.

This is just one (highly imaginative) possible path and eventual outcome that might be followed if climate stabilization policies are not adopted by the main emitters of greenhouse gases and climate change intensifies. It indicates that certain paths benefit from increasing returns to scale, but they are also likely to create lock-ins that are increasingly hard to escape, both economically and politically. As Kahn (2010) proposes, climate change can stimulate economic growth. However, it is doubtful whether it is a desirable type of economic growth, since it is likely to magnify extreme inequality between cities and populations around the world.

4.   A NUCLEAR ECONOMY: RESOURCES, WASTE AND RISK

With this in mind, it is worth investigating, instead of a world locked into a fossil fuel energy system intensifying climate change, the nature of possible low-carbon economies. Again, extreme scenarios are presented.

Imagine that global economies invest heavily in nuclear power. Given the existing higher costs of nuclear electricity generation compared with fossil fuels (Davis, 2012), governments in the first half of the twenty-first century would have to heavily subsidize the expansion of nuclear power. Here, although highly unrealistic, the extreme scenario is that during the second quarter of this century there is a transition to a low-carbon economy and that by 2050 almost all electricity in the world is generated by nuclear power.

In this thought-experiment, two related assumption are made. First, the technology used for this expansion is existing pressurized water reactors (PWR). Second, similar to the projections associated with fossil fuel electricity generation in Section 2, power station fuel efficiency doubles during the twenty-first century.

Sovacool (2010) examines a hypothetical transition to a combination of nuclear and renewables by 2030, and highlights that the spectacular costs of such an endeavour and the requirements in terms of financing, land, water, steel, concrete, silicon and other materials would put huge pressures on related markets and production capacity. Here, the concern is not with the transition, which clearly would require the deployment of vast amounts of resources. Instead, the focus is on the new energy system.

So, what would be the implications of a global economy based almost exclusively on nuclear power?

In 2010, global electricity consumption was about 21 000 TWh. The scenario presented in Figure 32.5 suggests that it might rise to about 63 000 TWh in 2050, and grow gradually to 75 000 TWh by 2100. Given that global electricity use doubled between 1990 and 2010, these are probably conservative estimates.

In 2010, there were 375 GW of installed nuclear power capacity in the world, which generated 2767 TWh of electricity (IAEA, 2012; BP, 2011b). If the low-carbon economy was achieved by generating all the electricity from nuclear power by 2050, then clearly a huge amount of new capacity would be needed. The generation to capacity ratio, in the 1990s and 2000s, was between 6.5 and 7.5 TWh of electricity generated per GW of capacity. Although this will certainly change, especially in an almost all nuclear system, using the average ratio of 7.27 as a crude indicator suggests that around 8700 GW of capacity will be needed – or another 8300 GW of nuclear power capacity.

The costs of building nuclear power are very hard to anticipate. Large projects with long construction periods are subject to great uncertainty, including the regulatory regime that will exist (Joskow and Parsons, 2012). Throughout time, and particularly reflecting higher standards of safety, construction costs of nuclear power stations have risen greatly (Koomey and Hultman, 2007; Grubler, 2010; Davis, 2012). Since the beginning of the twenty-first century, the mean value of the capital cost of building nuclear capacity was estimated to be $5800 per kW (Sovacool, 2010). The latest estimates collected for a number of proposed constructions in the USA in 2009 averaged at $7300 per kW (Cooper, 2009). Thus, using the latest estimates would probably be a conservative estimate, given that construction costs are likely to increase due to greater safety requirements following the Fukushima accident (Joskow and Parsons, 2012).

Based on these 2009 average cost estimates, building the 8300 GW of nuclear capacity would cost $60 trillion. Given that at present combined cycle gas turbine (CCGT) power stations can be built for around $1000 per kW (DECC, 2011), the additional cost of building nuclear rather than gas power stations could be around $52 trillion. Because of the long construction times for nuclear power, if this replacement took 20 years between 2030 and 2050, the additional cost of building nuclear power would be $2.5 trillion per year, or 1.5 per cent of global GDP in 2030 and 0.9 per cent in 2050.

Although this is a non-negligible amount, and will be a burden on the global economy, this is a transition cost since it is the expenditure associated with replacing fossil fuel power stations with nuclear power stations. Of course, the replacement of power stations and expansion of global electricity production would also lead to greater construction costs – and these would also impose a continued burden on the economy.

Once built, though, one might expect benefits associated with lower fuel costs. However, recent uranium price trends indicate considerable sensitivity to market forces, as spot prices increased five-fold between 2000 and 2010 – from $20 to $104 per pound equivalent U_3O_8 (Michel-Kerjan and Decker, 2008). More importantly, current estimates of uranium reserves do indicate a major problem with a global economy dominated by nuclear power generated by PWR technology. Uranium reserves are estimated to be 5.5 million tonnes (equivalent to 750 000 TWh of primary energy) for conventional reserves commercially viable up to $130 per kg, 16 million tonnes (equivalent to 2 220 000 TWh of primary energy) for total conventional (including undiscovered but probable) reserves, and 38 million tonnes (equivalent to 5 277 000 TWh of primary energy) for conventional and unconventional reserves (Rogner et al., 2012). With global consumption of electricity likely to be between 60 000 TWh and 100 000 TWh per year, PWR technology would use up all current conventional reserves and most of the current unconventional reserves during the twenty-first century.

In other words, either uranium reserves several times bigger than existing reserves need to be discovered or far more fuel-efficient technologies would need to be adopted. Either is possible. First, exploration only anticipates or responds to the demand, and especially rising prices. The history of coal and petroleum resource discoveries indicates that, at least for the last 500 years (Fouquet, 2011a; Hamilton in Chapter 1), energy producers have managed to find or at least supply more fossil fuels (in the medium run) when prices started to rise. While these past experiences do not preclude long-term shortages in the future, they do warn against concluding that, just because current estimates of uranium reserves are insufficient, they will necessarily be insufficient as the demand for uranium grows, prices rise and further exploration is stimulated.

Second, despite this comment, a cautious commentator might expect that PWR technology (in its current form) could not be the dominant technology in an all-nuclear energy system. Such reactors are very fuel-inefficient, and much of the fuel is unused and turned into waste. So they will need to become much more efficient or nuclear power will need to depend on another technology. For instance, breeder reactors can use the unused fuel, thus potentially increasing the efficiency eight-fold (Rogner et al., 2012). However, breeder reactors remain an unproven technology, with a troubled past. Furthermore, given the history of rising construction costs of nuclear power stations (Koomey and

Hultman, 2007; Grubler, 2010; Davis, 2012), it is probable that the large-scale development of breeder technology would lead to much greater construction costs.

Given the issues related to the availability of uranium, in the same way that oil countries formed a cartel and the potential for international natural gas production or price control was considered by Gabriel et al. in Chapter 3, one might also wonder about the potential for cartel formation in the production of uranium. Naturally, formation is a possibility, but long-term market control and power is another matter. Nevertheless, this is a further threat that could raise the price of electricity.

In addition, generating electricity from nuclear power does have certain additional resource requirements and costs. One important one is water to cool the power stations. While there is variation between types of reactors, an average estimate is that 178 km³ of water (i.e. 47 000 billion US gallons) are used per TWh of electricity (Sovacool, 2010). One might expect that, over the next 50 years, cooling processes for nuclear power will become more efficient. However, this has to be balanced with one impact of climate change, warmer rivers and seas, which will make cooling more difficult. Similarly, droughts may lead to severe water shortages, limiting the ability of nuclear power stations to function – in the summer of 2003, more than 30 nuclear power stations across Europe had to reduce production as a result of water constraints on the cooling process (Rubbelke and Vögele, 2011). Given these uncertainties, this rate of water consumption has been left constant. Based on this rate of water consumption, in 2050, the estimate of 63 000 TWh of electricity would require 11 200 km³ of water. To place this in perspective, the global freshwater resources are estimated at 55 200 km³ (Gleick, 2011). Although seawater is also used for cooling, and the water is effectively recycled during discharge, the possibility that a quantity equivalent to one-fifth of global freshwater will be used for nuclear power is likely to be a severe problem. One of the biggest issues is that the water discharged can be up to 8 °C higher than ambient water, with major effects on aquatic ecosystems (Laws, 1993). This concern is amplified by the water resource availability problems that are likely to be created by climate change over the next century.

Another issue is land use. Nuclear power has relatively low land requirements (0.31 hectares per TWh) relative to the electricity generated. This is a major advantage, especially when compared to many renewable energy resources – 2000 km² per TWh for biomass, 750 km² for hydropower, 137 km² for wind, 28 km² for solar photovoltaic and 0.3 km² per TWh for geothermal (Sovacool, 2010). So, using the estimate of 63 000 TWh of electricity in 2050, only 20 000 km² of land would be required (the size of Israel) to provide all the electricity for the global economy – a modest 0.013 per cent of the planet's land. As a comparison, global urban space is currently around 2 million km² (Menon et al., 2010).

However, because of nuclear power's current huge need for water, these power stations can only be located on rivers or on coasts, which often tends to be more valuable land than the average. So nuclear power will be in competition with other uses of riverside and coastal areas, and will probably have to pay a premium for access to large quantities of water. Nevertheless, the existence of much greater global land along coasts and rivers than the global land required for nuclear power suggests that this is not necessarily a major problem.

A more pertinent land problem is the facility siting issue. Major accidents – such as

Three Mile Island and Chernobyl and, no doubt, also Fukushima – appear to have a long-term effect on public opposition to the siting of nuclear power stations (Rosa and Dunlap, 1994). Thus, planning permission associated with nuclear power will continue to be a problem. This problem may intensify if the number of nuclear power stations increases substantially. Alternatively, with growth, reticence may decline. For example, since siting of nuclear facilities is driven by the potential weakness of opposition (Aldrich, 2008), once the public has accepted the building of one power station nearby, it becomes easier to accept the next one. Thus, despite the additional costs of transmission, it is likely that countries will tend to concentrate their power stations in poorer regions (as they are less likely to oppose siting given the potential economic benefits) with access to the sea (although the recent experience in Fukushima may tend to complicate planning and favour siting away from the sea) or large rivers. One consequence is that the expansion of nuclear power will probably lead to very highly geographically centralized electricity production.

This is likely to intensify the tendency towards industrial concentration. Market concentration is generally expected to depend on the minimum efficient scale of an industry relative to market demand. In the case of nuclear power, the very high capital costs, as well as the great complexity of managing the many aspects of nuclear production, imply that very large levels of production need to be reached before the firm achieves its lowest average costs of production. This implies that the firm with the highest level of production can out-compete other firms and drive them out of the market. Thus, in all but the largest economies (with very high levels of electricity demand), natural monopolies are likely to form; and, in the larger countries, oligopolies are likely.

During the twenty-first century, it is probable that nearby national electricity markets will merge, such as the single European market. This tendency is likely to eventually create a few, very large grids around the world, perhaps forming a pan-European one (possibly linking with the Middle East and Former Soviet Union countries), a North and Central American one, one in South America, one in South Asia, and one possibly linking South East Asia and China. These trends towards large interconnected electricity networks will enhance the potential dominance of a few massive regional nuclear power companies, with important financial, economic and political influence.

These few powerful electricity supply generators are likely to have the ability to influence legislation related to energy markets, reinforcing their position, minimizing the pressure for competition and driving up the costs of electricity. So one of the main characteristics of an all-nuclear low-carbon economy is likely to be a lack of competition. Given the close interrelationship of the nuclear industry and government, both because of their strategic economic importance and their need for subsidies until fossil fuel technology has been abandoned, there may be a tendency in many countries (where a nuclear power firm exists) to nationalize the natural monopoly.

Their influence is also likely to dictate future investment in energy and further research policies related to energy technology – naturally, this will have a tendency to favour further investment and advances in nuclear power, as has happened in the past when the nuclear industry was in a position of influence (Winskel, 2002). Thus technological and institutional lock-ins in an all-nuclear economy are likely to be very strong (even compared with the existing fossil fuel lock-ins).

Having looked at the construction costs, resource availability and market structure, it is important to consider nuclear waste. For the last recorded years (mostly 2008 or 2009), 25 countries' nuclear programmes produced a reported 2.5 million cubic metres (m^3) of waste associated with the generation of 2500 TWh (IAEA, 2012). So, from the average of 25 nuclear power programmes from the latest reported figures, 1 m^3 of waste was produced per GWh of electricity produced. While there was considerable variation in the amount of waste produced (from South Korea and Romania's 0.1 m^3 to the UK 5.9 m^3), France's nuclear programme waste was 0.9 m^3 and the USA's programme produced 1.5 m^3 per GWh. There was also considerable variation in the nature of the waste, but the average amongst these 25 countries was 5 per cent very low-level waste, 67 per cent low-level waste, 19 per cent intermediate-level waste and 9 per cent high-level waste (IAEA, 2012).

A crude estimate suggests that, if nuclear power produced all the global economy's 63 000 TWh of electricity in 2050, it might generate 47 million m^3 of very low-level and low-level waste and 18 million m^3 of intermediate- and high-level waste. That is, more than 60 million m^3 of waste (of which 28 per cent would be intermediate or high level) could be expected and need to be managed each year. If this global nuclear power programme continued for 50 years, it might be expected to generate 3.7 billion m^3 of waste (or 3.7 km^3).

The disposal of low-level waste was reported to cost around \$3150 per m^3 in the UK (Jackson, 2008). The cost or price of dealing with the intermediate and high levels has been estimated to range from \$70 000 to \$370 000 per m^3 (Jackson, 2008). Based on these existing management costs of low-level waste and the 'market' price for high-level waste, this would cost \$148 billion and \$6800 billion, respectively, or a total of 2.4 per cent of global GDP.

The risks associated with intermediate- and high-level waste has led to two problems (related to costs): first, countries find it very difficult to find sites for permanent waste storage or disposal: second, there are often national bans on the market for waste management (Jackson, 2008). The inability to find political consensus in democracies for locations raises the costs of management, as they remain in temporary sites. In the long run, the lack of competitive markets will imply that prices will not fall in the same way that production costs fall for goods and services in competitive markets. As an example, in the first decade of the twenty-first century, the cost of managing 1 m^3 in the UK rose by 9 per cent per year (Jackson, 2008). So, given the tendency for the costs of radioactive waste management to trend upwards, it is probable that the estimates above are in the lower range.

Having explored some of the economic problems associated with a low-carbon nuclear economy, we shall not discuss the risks in any detail. Suffice to say, here, that in energy markets nuclear power is likely to be at a disadvantage, because of the political risks (related to public opposition), the technical risks due to waste disposal issues, the economic risks arising from liabilities for the decommissioning and dismantling of nuclear power plants, and, most recently, the dangers of proliferation (Michel-Kerjan and Decker, 2008). It is doubtful whether these are acceptable risks when escalated to a global economy powered by nuclear fusion based on an energy-inefficient technology.

5.   A RENEWABLE ECONOMY: BALANCING DEMAND AND SUPPLY TOWARDS THE 'WEIGHTLESS ECONOMY'

Having looked at nuclear power, it would be valuable to investigate the implications of a low-carbon economy based almost exclusively on renewable energy sources. However, such an exercise would be slightly different. First, it is unlikely to expect one renewable energy source to dominate the global energy system. There are several reasons for this.

As mentioned earlier, one of the key problems for many renewable sources is the need for land to capture the energy. The land requirements for biomass are estimated to be 2000 km^2 per TWh, for hydropower 750 km^2, for wind 137 km^2, for solar photovoltaic 28 km^2 and for geothermal 0.3 km^2 per TWh (Sovacool, 2010). So, if the economy were to consume the estimated 63 000 TWh of electricity in 2050, 8.6 million km^2 (the size of Brazil) would be required to meet this need with wind power. Although much of this could be placed offshore, it would be greater than the Mediterranean, Caribbean and South China Sea combined. So it seems that wind could only ever provide a share of the total electricity requirements. Solar photovoltaic would need less land; based on the projections in Figure 32.5, around 1.7 million km^2 in 2050 (the size of Libya) to 2.2 million km^2 in 2100 (close to the size of Algeria). Even if much of the land could be in regions where it is relatively cheap, these are still huge requirements. Geothermal is the only renewable energy source that would not require much land – similar to nuclear power.

Second, since different regions have comparative advantages related to a particular source (e.g. North Western Europe has relatively large amounts of wind; North Africa and the Middle East have more potential for solar power; and Iceland and Costa Rica have easy access to geothermal sources). Instead, in certain regions, economies might be dominated by a particular energy resource. Interestingly, features of the economy might be shaped in part by the peculiarities of the energy source dominant in the region.

Third, different countries are backing and investing in different renewable technologies. Given that there are several potential renewable technologies, all of which are less complex than nuclear power technology, for example, the fixed costs of investing in the technologies are smaller. This implies that increasing returns to scale and lock-ins are less great. So renewable technologies are more likely to coexist, without one dominating.

This weakness of individual renewable technologies is also an advantage as a whole because it minimizes the need for policy makers to pick winners on a large scale. They may provide small-scale support for specific renewable technologies, but would not need to invest large quantities of money and wait a long time to observe increasing returns to scale and identify the potential gains from the technology. Instead, the simultaneous pursuit of many renewable energy technologies allows them to develop and mature, first in their own niche (e.g. where the resource is unusually abundant) and perhaps later in more competitive markets (e.g. average resource levels). In time, markets and governments might start to identify the most competitive and valuable technologies.

So, having briefly identified a clear growth constraint related to key technologies (i.e. land), as well as some aspects of the development of the related industries, what are the characteristics of a renewable-driven economy?

First, as mentioned before, for hydro, wind and solar power, land is crucial to the production of energy. Equally, at first, the regions with the most abundant resources will be used first. As renewable energy production increases, less productive sites will be exploited. With the large requirements to meet the electricity demands of a global economy, the marginal product of land will be very low. Ricardo (1817) argued that the declining marginal product of land would tend to drive up the price of the good (potentially rising profits for the producers exploiting the highly productive sites). This will be countered by the economies of scale reducing the costs of production. Thus it will be a race between the exploitation of less productive sites and the development of more effective and cheaper technology.

Second, historical experiences offer some insights into the renewable economy. While Britain made the transition from renewable energy sources to coal, a number of economies, such as Germany and Japan, continued to grow (albeit at a slower pace) well into the nineteenth century, fuelled by a renewable energy system (Fouquet, 2011c). These two economies were able to sustain economic growth because of their management of the demand, supply and trade of woodfuel. Furthermore, in other countries, where governments failed to develop appropriate 'renewable energy' policies, growth and development were particularly limited.

For hydro, wind and solar power, land is a clear constraint (especially if the total area dedicated to global energy provision is equal to the size of Libya, or even of Brazil). Energy needs will be in competition with food requirements, urban spaces and ecosystem services. To expand the production of these energy sources, land needs to be acquired and paid for. Furthermore, at these levels of expansion of renewable energy sources, the marginal product of the land (in terms of its energy generated per km^2) is likely to be low, since the productive lands will be used first. As a result, given the inefficiency with which the technologies will use land and given the marginal costs of increasing capacity, expansion of energy production will be done only in strongly favourable commercial conditions. Also, in the shorter run, the intermittency of renewable energy sources implies that energy production is not constant (Joskow, 2011). So, whether in the short the run or the longer run, it is probable that careful balancing of demand and supply will be an important feature of a renewable economy.

Management already happens in the electricity market, where, on an hourly basis, supply is balanced with the demand according to costs of production (Joskow and Tirole, 2007). While it is possible that storage technologies will help to adjust supply with needs (Johnstone and Haščič in Chapter 6), it is probable that short-run management will necessitate demand flexibility. This balancing from the demand side can be greatly aided by new technologies, such as smart meters (Jacoby et al. in Chapter 5). Similarly, in the longer run, because of constraints on land and, therefore, production and supply, it is likely that demand will also need to be managed.

Traditionally, demand-side management might be undertaken by a firm focused on encouraging the adoption of energy-efficient technologies to raise consumer awareness and address possible market failures related to information provision (Nadel and Geller, 1996; Steinberger et al., 2008). However, the concept of discouraging consumption by setting caps on consumption, is unpopular. Naturally, such policies would be harmful to electricity company profits and would be heavily opposed. More generally, limiting energy consumption is likely to hinder an economy's expansion.

Third, in time, policies might have to be developed to minimize the damage of such constraints. Regulators may encourage the rapid development tools for improving electricity companies' ability to predict renewable energy production. In addition, there might be tradable permit schemes for energy use (note that permits for carbon dioxide emissions may have all but disappeared since, in this example, electricity will be generated from renewable energy sources). Such a scheme would allow firms that have a high willingness to pay to buy permits and those that are more flexible about their time of consumption will sell permits. Such a scheme promotes flexibility either in the short run (e.g. on an hourly basis to reflect intermittency) or in the longer run (e.g. to meet capacity constraints). In other words, an economy driven by intermittent and constrained renewable energy sources is likely to place a high value on flexibility of consumption, as well as on sources of supply that started and stopped to fill the troughs in production.

Fourth, in addition to the declining need for energy to develop, the rising share of renewable energy in the energy mix should reduce an economy's vulnerability to energy price fluctuations.[2] As Hamilton (in Chapter 1) showed, economic growth has traditionally been affected by higher oil prices, and recessions have been triggered by high energy prices. This has been a more general feature of fossil fuel-based economies (Fouquet, 2008).

Finally, the dependence on a renewable energy system might lead to major economic restructuring. Initially, new policies might generate ideas about how to meet service demand with less energy and encourage the adoption of technologies and behavioural choices (see Pollitt and Shaorshadze in Chapter 24). For example, 'inefficiency taxes' provide a powerful way of signalling the value of efficient behaviour and penalizing technological laggards (Gago et al. in Chapter 19). However, to focus on raising energy efficiency is to lower the cost of producing certain energy-dependent services, such as heating, power and transportation, and, thus, increase their consumption.

Also, improving the efficiency of relatively energy-intensive production of goods and services may detract from promoting other activities that require virtually no energy. As Hepburn and Bowen in Chapter 29 and Llavador et al. in Chapter 30 have indicated, the economy might be restructured towards 'weightless' activities. The restructuring might offer an opportunity for new dimensions of economic development.

Inevitably, restructuring will involve bankruptcies and unemployment in other sectors, generating political and social tensions. To ensure that these tensions are turned into positive outcomes, policies might be introduced that promote the accumulation of human capital, which is seen as the key to economic growth (Grimaud and Tournemaine, 2007). Perhaps taxes and tradable permits on energy use can be combined with more popular subsidies on education that turn out to be powerful methods of promoting a major restructuring of the economy and economic development. So a transition to a constrained low-carbon economy might encourage and create some of the conditions to achieve a transition to a 'weightless economy' (Quah, 1997, 1999, 2001).

'Weightless' activities would presumably have very small energy requirements. This implies that the economy would not be subject to the energy resource constraints traditionally associated with, for example, industrialization. Economic growth in the 'weightless' sectors could expand even if energy resources were constrained. Similarly, periods of high economic growth in industrialized economies put pressure on production and supply infrastructure, raising energy prices, which tend to slow down the economy. Thus, both the inflationary and deflationary aspects of the boom–bust cycle resulting from the energy prices–economic growth relationship would be quashed, in a renewable and 'weightless' economy.

6. CONCLUSION

Economic activity has a tendency to expand until limits are reached. Until recently, one of the main constraints on the fossil-fuel-driven economy was the fear of insufficient resources. However, markets have been particularly successful at pushing back the energy resource limits (Fouquet, 2008; Hamilton in Chapter 1).

This success has allowed the global fossil fuel-driven economy to expand spectacularly since the Industrial Revolution (see Figure 32.1). The ensuing prolific use of fossil-fuels has generated 1200 billion tonnes of carbon dioxide emissions (see Figure 32.2), which has intensified the greenhouse effect and is responsible for climate change. Thus, the new limit, constraint or scarcity (as Pearce, 2005 called it), is nature's ability to assimilate the carbon dioxide emissions and other greenhouse gases without disrupting the global climate.

Yet, during the twenty-first century, the anticipated rise in population and economic activity is likely to create an increasing demand for energy services, such as heating, power and transportation. Although the analysis in this chapter is based on relatively simplistic methods, it provides a broad indicator of the likely trends, barring dramatic economic, social or political transformations. With current technologies, or even with substantially more efficient technologies, energy consumption is likely to at least triple or quadruple over the next 50 or 100 years.

Given the great uncertainty about the future, especially one very different from today, it is difficult to investigate the implications of this growth in a scientific manner. Naturally, many authors have managed to develop rigorous models of the future (Cline, 1992; Nordhaus, 1994; Tol, 2002; Hope, 2005; Stern et al., 2006; IPCC, 2007; Nordhaus, 2008; Norgaard, 2011). These models have been critical in improving our ability to anticipate the impacts of climate change and the net benefits of finding solutions.

However, a modelling approach can limit the images of the future that are created. In order to produce a higher-resolution picture of future states of the world, it is generally necessary to perform a less structured analysis. While this approach increases the risk of painting the wrong picture, it may also help us to ask interesting new questions. It might also help us to identify new hypotheses, issues to analyse or possible solutions. Thus this chapter has built on Kahn's (2010) approach to consider three more detailed possible future global economies.

Increasing amounts of carbon dioxide and other greenhouse gas emissions imply that, without worldwide carbon capture and sequestration or successful geo-engineering projects, populations across the world may be faced with more or more damaging natural disasters. These populations will be forced to adapt. One probable outcome is that the increasingly urban populations will seek protection. One possible (if highly imaginative) outcome is that they will develop protected cities that also incorporate climate control. One possible consequence is that cities will become increasingly atomistic and unequal.

It might be expected that because of increasing returns to scale and technological and institutional lock-ins, the further along one path, the harder it will be for the global economy to change onto another path and achieve another outcome. This would pre-scribe care about the path chosen to ensure the most desirable outcome. At present, based on much evidence in this handbook, and elsewhere, many might suspect that 'a climate-changed world' is the most likely outcome. In Section 3, it was indeed proposed that investments in climate protection might crowd out efforts to achieve climate stability. The bold presentation of this hypothesis welcomes support or refutation, even if evidence is hard to produce. Another factor working against this argument is the possible existence of a temperature elasticity of demand for climate stability. That is, as global climate change intensifies, will public concern, national policies and international agreements push harder for a transition to a low-carbon economy? Thus more research is needed to consider the forces working to maintain the fossil fuel economy and ones that seek to move the economy towards a low-carbon economy.

If markets and governments do steer the global economy towards a low-carbon energy system, it would allow us to remove 'the climate constraint' from our energy consumption. It would be naïve to believe, however, that new constraints would not appear. Irrespective of the type of low-carbon economy that develops, the economy would most probably expand until new limits were under pressure (whether of uranium and waste deposit sites in the case of nuclear power, or land in the case of renewable technologies). In addition, each possible low-carbon economy would have specific characteristics, potentially very different from today's economy. Thus a key question is 'What would be the effects of the new constraints and the main characteristics of possible low carbon economies?'

But with regard to the approach used in this chapter to begin to answer this question, a clarification is needed. The objective is not to find faults in specific energy sources, industries or technologies. It is important to appreciate that the images presented in this investigation are extreme cases, and readers should not see them as direct criticisms of particular industries. Presenting extreme cases is simply a tool for thinking about these problems, and it is hoped that readers of this handbook will appreciate the objective of this device and be stimulated to think further about these important issues.

Nevertheless, a nuclear economy (faced with severe resource and waste deposit site constraints and certain unattractive characteristics) is a possible outcome. So the nuclear industry (and more particularly PWR technology) will need to think about how to ensure that it is not an outcome that occurs, and, instead, will perhaps strive for a more desirable nuclear economy. Similarly, the renewable energy industry will need to consider ways to avoid some of the possible constraints and undesirable characteristics associated with renewable economies, and, instead, encourage turning the genuine constraints into positive outcomes.

There are many possible outcomes – some more likely than others; some more desirable than others. In our fear of picking winners, for understandable reasons, there is a risk that economists have become afraid of identifying 'dark ages' that should be avoided. Furthermore, policy makers need to receive advice about what constraints associated with and characteristics of a low-carbon economy are deeply undesirable and should be avoided, what constraints are tolerable and could actually provide positive incentives, and what characteristics are genuinely desirable.

One obvious characteristic that needs to be considered further is risk (although it was not discussed much in this chapter, due to space limitations and the complexity of the related issues). There has been a great deal of debate about external costs, but less about externalized risks. Externalized risks probably need to be internalized too. Otherwise, they are likely to create undesirable incentives, and these market failures are likely to lead to socially suboptimal outcomes. Thus policies may want to consider ways of lowering or internalizing risks within a low-carbon economy.

Another characteristic that is important to consider is flexibility – at present, in a period of great uncertainty, there may be a willingness to pay to not get locked into an undesirable energy system. Thus more research is needed to identify what allows an industry to be flexible. This is particular interesting since there is a market failure associated with flexibility. Firms (especially large ones) and industries have a private incentive to become locked into the economic system. However, while, in the short and medium run, society does benefit from the low average costs, in the longer run, the lock-ins often start to impose excessive external costs – as the dependence on coal did in the second half of the nineteenth century (Fouquet, 2011b).

Finally, as the title of this chapter suggests, crucial to the desirability of the low-carbon economy will be its impact on economic growth and development. In the same way as integrated assessment models do for the impact of climate change on the economy (Nordhaus, 2008), an avenue of future research will be to identify more clearly the effects of these constraints and characteristics on the trend in GDP per capita (as shown in Figure 32.3). Also, as mentioned in the penultimate section, it is possible that a low-carbon economy will be much more than an energy transition. The Industrial Revolution, as well as ushering in the era of fossil fuels, transformed the economy, society, politics and much more. A low-carbon economy could do the same – potentially ushering in a new 'golden age'.

This chapter has sought to identify some of the constraints and characteristics of a low-carbon economy in order to avoid deeply undesirable outcomes and perhaps encourage more favourable ones. Expanding the debate may lead thinkers to imagine new solutions to propel the economy towards a desirable outcome. All this should be done while remembering that an open debate is not an argument for aggressive intervention in energy markets or a further excuse for inaction. Instead, the debate about addressing climate change and a low-carbon future is an opportunity to explore a new world of possibilities.

NOTES

1. See also, for instance, the special issues of the journal *Energy Policy* on transitions to sustainable energy systems in 2008, the role of trust in these transitions in 2010, on the possibility of transitions to hydrogen in 2010 and historical experiences of energy transitions in 2012.
2. Note that, at present, the fuel costs of nuclear power are also very small relative to its total costs, minimizing the impact of uranium price fluctuations. As discussed in the previous section, if nuclear power were to grow substantially, uranium production and supply infrastructure would come under greater pressures, raising fuel prices and making a nuclear economy sensitive to this energy price–macroeconomic cycle.

REFERENCES

Aldrich, D.P. (2008), 'Location, location, location: selecting sites for controversial facilities'. *Singapore Economic Review*, **53**(1), 145–72.
BP (2011a), *Energy Outlook 2030*, London: BP plc, http://www.bp.com/sectiongenericarticle800.do?category Id=9037134&contentId=7068677.

BP (2011b), *Statistical Review of World Energy 2010*, London: BP plc., available at http://www.bp.com/sectionbodycopy.do?categoryId=7500&contentId=7068481.

Charpentier, A. (2008), 'Insurability of climate risks', *The Geneva Papers on Risk and Insurance – Issues and Practice*, **33**(1), 91–109.

Chernobyl Forum (2006) *Chernobyl's Legacy: Health, Environmental and Socio-Economic Impacts*, Vienna: IAEA Publication, http://www.iaea.org/Publications/Booklets/Chernobyl/chernobyl.pdf.

Cline, W.R. (1992), *The Economics of Global Warming*, Washington, DC: Institute for International Economics.

Cooper, M. (2009), 'The economics of nuclear reactors: renaissance or relapse,' Working Paper, Institute for Energy and the Environment, Vermont Law School. Montpellier, VT.

Cooper, M. (2011), 'Nuclear safety and nuclear economics: historically, accidents dim the prospects for nuclear reactor construction'. Institute for Energy and the Environment, Vermont Law School. Montpellier, VT, http://www.nirs.org/neconomics/Nuclear-Safety-and-Nuclear-Economics-Post-Fukushima.pdf.

Coumou, D. and Rahmstorf, S. (2012), 'A decade of weather extremes', *Nature Climate Change*, **2**, 491–6.

Davis, L. (2012), 'Prospects for nuclear power,' *Journal of Economic Perspectives*, **26**(1), 49–66.

DECC (2011), 'Electricity generation cost model – 2011 update', Dept for Energy and Climate Change, http://www.pbworld.com/pdfs/regional/uk_europe/decc_2153-electricity-generation-cost-model-2011.pdf.

Diamond, J.M. (2005), *Collapse: How Societies Choose to Fail or Succeed*, New York: Penguin Books.

Emanuel, K. (2005), 'Increasing destructiveness of tropical cyclones over the past 30 years', *Nature*, **436**, 686–8.

Fouquet, R. (2008), *Heat, Power and Light: Revolutions in Energy Services*, Cheltenham, UK, and Northampton, MA, USA: Edward Elgar.

Fouquet, R. (2009), 'A brief history of energy', in J. Evans and L.C. Hunt (eds), *International Handbook of the Economics of Energy*, Cheltenham, UK, and Northampton, MA, USA: Edward Elgar.

Fouquet, R. (2010) 'The slow search for solutions: lessons from historical energy transitions by sector and service,' *Energy Policy*, **38**(10), 6586–96.

Fouquet, R. (2011a), 'Divergences in long run trends in the prices of energy and energy services', *Review of Environmental Economics and Policy*, **5**(2) 196–218.

Fouquet, R. (2011b), 'Long run trends in energy-related external costs', *Ecological Economics*, **70**(12), 2380–89.

Fouquet, R. (2011c), 'The sustainability of "sustainable" energy use: historical evidence on the relationship between economic growth and renewable energy', in I. Galarraga, M. González-Eguino and A. Markandya (eds), *Handbook of Sustainable Energy*, Cheltenham, UK, and Northampton, MA, USA: Edward Elgar, pp. 9–20.

Fouquet, R. (2012), 'Trends in income and price elasticities of transport demand (1850–2010)', *Energy Policy*, Special Issue on Past and Prospective Energy Transitions.

Fouquet, R. and Pearson, Peter J.G. (2012), 'The long run demand for lighting: elasticities and rebound effects in different phases of economic development', *Economics of Energy and Environmental Policy*, **1**(1), 83–100.

Foxon, T.J., Köhler, J. and Oughton, C. (2008), *Innovation for a Low Carbon Economy: Economic, Institutional and Management Approaches*, Cheltenham, UK and Northampton, MA, USA: Edward Elgar.

Gleick, P.H. (2011), *The World's Water*, Volume 7. The Biennial Report on Freshwater Resources, Washington, DC: Island Press.

Grimaud, A. and Tournemaine, F. (2007), 'Why can an environmental policy tax promote growth through the channel of education?', *Ecological Economics*, **62**(11), 27–36.

Grubb, M., Jamasb, T. and Pollitt, M.G. (2008), *Delivering a Low-Carbon Electricity System: Technologies, Economics and Policy*, Cambridge: Cambridge University Press.

Grubler, A. (2010), 'The costs of the French nuclear scale-up: a case of negative learning by doing', *Energy Policy*, **38**(9), 5174–88.

Hope, C. (2005), 'Integrated assessment models', in D. Helm (ed.), *Climate Change Policy*, Oxford: Oxford University Press, pp. 382–96.

Houghton, R.A. (2008), 'Carbon flux to the atmosphere from land-use changes: 1850–2005', in *TRENDS: A Compendium of Data on Global Change*, Oak Ridge, TN: Carbon Dioxide Information Analysis Center, Oak Ridge National Laboratory, US Dept of Energy.

IAEA (2012), The IAEA Online Information Resource for Radioactive Waste Management, http://newmdb.iaea.org/.

Insurance Information Institute (2012), Media Hot Topics: Catastrophes, http://www.iii.org/media/hottopics/insurance/catastrophes/.

International Energy Agency (2010), *World Energy Outlook 2010*, Paris: OECD.

IPCC (2007), *Climate Change 2007. Impacts, Adaptation and Vulnerability. Contribution of Working Group II to the Fourth Assessment Report of the Intergovernmental Panel on Climate Change, 2007*, IPCC Secretariat, Geneva, Switzerland.

Jackson, A.A. (2003), 'The London railway suburb, 1850–1914', in A.K.B. Evans and J.V. Gough (eds),

The Impact of the Railway on Society in Britain: Essays in Honour of Jack Simmons, Aldershot: Ashgate, pp. 169–80.

Jackson, I. (2008), *Nukenomics: The Commercialisation of Britain's Nuclear Industry*, Sidcup: Progressive Media Markets.

Jevons,W.S (1865), *The Cool Question: Can Britain Survive?*, London: Macmillan.

Joskow, P.L. (2011), 'Comparing the costs of intermittent and dispatchable generating technologies', *American Economic Review, Papers and Proceedings*, **101**(3), 238–41.

Joskow, P.L. and Parsons, J.E. (2012), 'The future of nuclear power after Fukushima', *Economics of Energy and Environmental Policy*, **1**(2), 99–114.

Joskow, P.L. and Tirole, J. (2007), 'Reliability and competitive electricity markets', *RAND Journal of Economics*, **38**, 60–84.

Kahn, M.E. (2010), *Climatopolis: How Our Cities Will Thrive in the Hotter Future*, New York: Basic Books.

Koomey, J. and Hultman, N. (2007), 'A reactor-level analysis of busbar costs for U.S. nuclear plants, 1970–2005', *Energy Policy*, **35**(11), 5360–72.

Laws, E.A. (1993), *Aquatic Pollution*, New York: John Wiley and Sons.

Luckin, B. (2000), 'Pollution in the city', in M.J. Daunton (ed.), *The Cambridge Urban History of Britain: 1840–1950*, Cambridge: Cambridge University Press, pp. 207–28.

Marland, G., Boden, T.A. and Andres, R.J. (2007), 'Global, regional, and national CO_2 emissions', in *Trends: A Compendium of Data on Global Change*, Oak Ridge, TN: Carbon Dioxide Information Analysis Center, Oak Ridge National Laboratory, US Department of Energy.

Menon, S., Akbari, H., Mahanama, S., Sednev, I., and Levinson, R. (2010), 'Radiative forcing and temperature response to changes in urban albedos and associated CO_2 offsets', *Environmental Research Letters*, **5**, 1–11.

Michel-Kerjan, E.O. and Decker, D.J. (2008), *The Future of Nuclear Energy Markets and International Security*, The Wharton School, University of Pennsylvania and Kennedy School of Government, Harvard University.

Mills, E. (2009), 'A global review of insurance industry responses to climate change', *Geneva Papers on Risk and Insurance – Issues and Practice*, **44**(3), 323–59.

Munich Re (2012), *Natural Catastrophes 2011*, http://www.munichre.com/app_pages/www/@res/pdf/media_relations/press_releases/2012.

Nadel, S. and Geller, H. (1996), 'Utility DSM: what have we learned? Where are we going?', *Energy Policy*, **24**(4), 289–302.

Nordhaus, W.D. (1994), *Managing the Global Commons: The Economics of Climate Change*, Cambridge, MA: MIT Press.

Nordhaus, W.D. (2008), *A Question of Balance: Weighing the Options on Global Warming Policies*, New Haven, CT: Yale University Press.

Norgaard, R.B. (2011), 'Weighing climate futures', in J.S., Dryzek, R.B. Norgaard and D. Schlosberg (eds), *The Oxford Handbook of Climate Change and Society*, Oxford: Oxford University Press, pp. 190–204.

Pearce, D.W. (2005), 'Environmental policy as a tool for sustainability', in R.D. Simpson, M.A. Toman and R.U. Ayres (eds), *Scarcity and Growth Revisited: Natural Resources and the Environment in the New Millennium*, Washington, DC: Resources for the Future, pp. 198–224.

Pwc (2011), 'The world in 2050', http://www.pwc.com/en_GX/gx/world-2050/pdf/world-in-2050-jan-2011.pdf.

Quah, D. (1997), 'Increasingly weightless economies', *Bank of England Quarterly Bulletin*, **37**(1),49–56.

Quah, D. (1999), 'The weightless economy in economic development', CEP Discussion Paper CEPDP0417, 417. Centre for Economic Performance, London School of Economics and Political Science, London, UK.

Quah, D. (2001), 'The weightless economy in economic development', in Matti Pohjola (ed.), *Information Technology, Productivity, and Economic Growth*, Oxford: Oxford University Press, pp. 76–89.

Ricardo, D. (1817), *On the Principles of Political Economy and Taxation*, London: John Murray.

Rogner, H.-H. et al. (2012), 'Energy resources and potentials', in N. Nakicenovic, L. Gomez-Echeverri and T.B. Johansson (eds), *Global Energy Assessment*, Cambridge: Cambridge University Press, pp. 423–512.

Rosa, E.A. and Dunlap, R.E. (1994), 'Poll trends: nuclear power: three decades of public opinion', *The Public Opinion Quarterly*, **58**(2), 295–324.

Rubbelke, D. and Vögele, S. (2011), 'Impacts of climate change on European critical infrastructures: the case of the power sector', *Environmental Science & Policy*, **14**(1), 53–63.

Smil, V. (2010), *Energy Transitions: History, Requirements, Prospects*, Santa Barbara, CA: Praeger.

Sovacool, B.K. (2010), 'Exploring the hypothetical limits to a nuclear and renewable electricity future', *International Journal of Energy Research*, **34**(13), 1183–94.

Steinberger, J.K., van Niel, J. and Bourg, D. (2008), 'Profiting from negawatts: reducing absolute consumption and emissions through a performance-based energy economy', *Energy Policy*, **37**(1) 361–70.

Stern, D.I. and Kaufmann, R.K. (1998), 'Annual estimates of global anthropogenic methane emissions:

1860–1994', in *Trends Online: A Compendium of Data on Global Change*, Oak Ridge, TN: Carbon Dioxide Information Analysis Center, Oak Ridge National Laboratory, US Dept of Energy.

Stern, N.H., Peters, S., Bakhski, V., Bowen, A., Cameron, C., Catovsky, S., Crane, D., Cruickshank, S., Dietz, S., Edmondson, N., Garbett, S.-L., Hamid, L., Hoffman, G., Ingram, D., Jones, B., Patmore, N., Radcliffe, H., Sathiyarajah, R., Stock, M., Taylor, C., Vernon, T., Wanjie, H. and Zenghelis, D. (2006), *The Stern Review: The Economics of Climate Change*, Cambridge: Cambridge University Press.

Thompson, F.M.L. (1982), *The Rise of Suburbia*, Leicester: Leicester University Press.

Thorsheim, P. (2002), 'The paradox of smokeless fuels: gas, coke and the environment in Britain, 1813–1949', *Environment and History*, **8**, 381–401.

Tol, R.S.J. (2002), 'Estimates of the damage costs of climate change. Part 1: benchmark estimates', *Environmental & Resource Economics*, **21**(1), 47–73.

UN (2011), *World Population Prospects: The 2010 Revision*, Dept of Economic and Social Affairs, Population Division.

Webster, P.J. et al. (2005), 'Changes in tropical cyclone number, duration and intensity in a warming environment', *Science*, **309**, 1844–6.

Winskel, M. (2002), 'Autonomy's end: nuclear power and the privatization of the British electricity supply', *Social Studies of Science*, **32**(3), 439–67.

Wolfram, C., Shelef, O. and Gertler, P. (2012), 'How will energy demand develop in the developing world?', *Journal of Economic Perspectives*, **26**(1), 49–66.

Index

709